Handbook of

Neurochemistry

SECOND EDITION

Volume 7

STRUCTURAL ELEMENTS OF THE NERVOUS SYSTEM

Handbook of
Neurochemistry
SECOND EDITION

Edited by Abel Lajtha
Center for Neurochemistry, Wards Island, New York

Handbook of Neurochemistry

SECOND EDITION

Volume 7

STRUCTURAL ELEMENTS OF THE NERVOUS SYSTEM

Edited by
Abel Lajtha
Center for Neurochemistry
Wards Island, New York

PLENUM PRESS · NEW YORK AND LONDON

Library of Congress Cataloging in Publication Data

(Revised for volume 7)
Main entry under title:

Handbook of neurochemistry.

Includes bibliographical references and indexes.
Contents: v. 1. Chemical and cellular architecture—v. 2. Experimental neurochemistry—[etc.]—v. 7. Structural elements of the nervous system.
1. Neurochemistry—Handbooks, manuals, etc.—Collected works. I. Lajtha, Abel. [DNLM: 1. Neurochemistry. WL 104 H434]
QP356.3.H36 1982 612'.814 82-493

ISBN-13: 978-1-4684-4588-6 e-ISBN-13: 978-1-4684-4586-2
DOI: 10.1007/978-1-4684-4586-2

Softcover reprint of the hardcover 1st edition 1984
A Division of Plenum Publishing Corporation
233 Spring Street, New York, N.Y. 10013

Contributors

Harish C. Agrawal, Department of Pediatrics, Washington University School of Medicine, St. Louis, Missouri 63178

Gérard Alonso, Laboratoire de Neuroendocrinologie, Université de Montpellier II, 34100 Montpellier, France

Joyce A. Benjamins, Department of Neurology, Wayne State University School of Medicine, Detroit, Michigan 48201

A. Lorris Betz, Departments of Pediatrics and Neurology, University of Michigan, Ann Arbor, Michigan 48109

Elisabeth Bock, The Protein Laboratory, Sigurdsgade 34, Denmark

J. B. Clark, Department of Biochemistry, The Medical College of St. Bartholomews Hospital, University of London, London EC1M 6BQ, England

George H. DeVries, Department of Biochemistry, Medical College of Virginia, Richmond, Virginia 23298

M. Saïd Ghandour, Centre de Neurochimie du CNRS, 67084 Strasbourg, France

Gary W. Goldstein, Departments of Pediatrics and Neurology, University of Michigan, Ann Arbor, Michigan 48109

Giorgio Gombos, Centre de Neurochimie du CNRS, 67084 Strasbourg, France

Boyd K. Hartman, Department of Psychiatry, Washington University School of Medicine, St. Louis, Missouri 63178

Richard L. Klein, Department of Pharmacology and Toxicology, University of Mississippi Medical Center, Jackson, Mississippi 39216

Jack D. Klingman, Department of Biochemistry, School of Medicine, State University of New York at Buffalo, Buffalo, New York 14214

Edward Koenig, Division of Neurobiology/Department of Physiology, State University of New York at Buffalo, Buffalo, New York 14214

Harold Koenig, Neurology Service, Veterans Administration Lakeside Medical Center; and Department of Neurology, Northwestern University Medical School, Chicago, Illinois 60611

O. Keith Langley, Centre de Neurochimie du CNRS, 67084 Strasbourg, France

Kathryn Mack, Laboratory of Cell Biology, Institute for Basic Research in Developmental Disabilities, Staten Island, New York 10314

Henry R. Mahler, deceased. Department of Chemistry and Cellular, Developmental and Molecular Biology Program, Indiana University, Bloomington, Indiana 47405

Charles A. Marotta, Molecular Neurobiology Laboratory of the Laboratories for Psychiatric Research, Mailman Research Center, McLean Hospital, Belmont, Massachusetts 02178; and the Department of Psychiatry, Massachusetts General Hospital and Harvard Medical School, Boston, Massachusetts 02115

Marilyn W. McCaman, Division of Neurosciences, Beckman Research Institute of the City of Hope , Duarte, California 91010

Pierre Morell, Department of Biochemistry, University of North Carolina, Chapel Hill, North Carolina 27514

W. J. Nicklas, Department of Neurology, University of Medicine and Dentistry of New Jersey, Rutgers Medical School, Piscataway, New Jersey 08854

William H. Oldendorf, Veterans Administration Medical Center, Los Angeles, California 90073; and Departments of Neurology and Psychiatry, University of California, Los Angeles, School of Medicine, Los Angeles, California 90024

David Soifer, Laboratory of Cell Biology, Institute for Basic Research in Developmental Disabilities, Staten Island, New York 10314

Maria Spatz, Laboratory of Neuropathology and Neuroanatomical Sciences, National Institute of Neurological and Communicative Disorders and Stroke, National Institutes of Health, Public Health Service, U.S. Department of Health and Human Services, Bethesda, Maryland 20205

Yasuo Takahashi, Department of Neuropharmacology, Brain Research Institute, Niigata University, Niigata 951, Japan

Åsa K. Thureson-Klein, Department of Pharmacology and Toxicology, University of Mississippi Medical Center, Jackson, Mississippi 39216

V. P. Whittaker, Abteilung Neurochemie, Max-Planck-Institut für Biophysikalische Chemie, Göttingen, Federal Republic of Germany

Foreword

Neurochemistry, having the objective of elucidating biochemical processes subserving nervous activity, emerged as an application of chemistry to the investigation of neurobiological problems as a post-World War II phenomenon. However, only in the last 40 years has the chemical community recognized neurochemistry as a distinct, if hybrid, discipline. During this period great strides have been made. However, recently neurochemistry, along with neurophysiology, neuropharmacology, neuroanatomy, and the behavioral sciences, has emerged to form *neuroscience*, a new community of scientists with its own national society, journals, and meetings. Actually, this recently formed hybrid, neuroscience, is in the process of merging with another well-established discipline, molecular genetics (frequently called molecular biology, and itself a hybrid), which appears to have sufficient hybrid vigor to form yet a new community of scientists, which, for want of a more imaginative term, has been called *molecular genetic neuroscience*.

Clearly, advantages resulting from such mergers or hybridizations accrue not only from the merging discipline (neurochemistry in this case) to the new community (molecular genetic neuroscience), but also in the reverse direction. This Foreword will be concerned primarily with examples of this latter process.

Among the first products of the new dispensation was the invention and development of biochemical technologies that have greatly facilitated research in this field. Complex equipment made its appearance academically and commercially early on, such as "gene machines" for nucleotide sequence analysis and synthesis of DNA. Submicrogram quantities of particular proteins (identified by monoclonal antibodies or two-dimensional electrophoresis) sufficed to determine, and even isolate, the gene or genes that synthesize the protein. With such equipment it is possible to pass from phenotype (e.g., a microgram or less of purified protein) to genotype (expressed as the nucleotide sequence of DNA). Similarly, it is possible to pass from genotype to phenotype. A method has recently been published indicating how, from cDNA clones, it is possible to determine nucleotide sequences, hence amino acid sequences, of proteins encoded by brain-specific mRNAs. Antisera to corresponding proteins are used immunocytochemically to localize the protein in the brain. Such proteins may serve as markers for novel neuronal pathways and transmitters used in pathways.

With conventional neurochemical methods it has, over the years, been possible to discover but a small number (circa several dozen) of brain-specific proteins. However, by application of the new technologies, evidence has been obtained for the existence of many thousands of different brain-specific proteins. Many of these proteins may occur very sparsely in the brain, but in the subpopulations of cell types they may occur in high-copy amounts and may play a significant role.

In addition to the approximately ten classical small-molecule transmitters, a large and rapidly growing list of putative peptidergic transmitters or modulators of neurotransmitters has been identified and chemically characterized by application of the new technology, as have other substances of high neurobiological activity (e.g., peptides, receptors, hormones, and factors). Clearly a large flood of new compounds will soon be appearing in the literature, forming a handsome grist for neurochemical study and for function determination.

Neurochemistry will, in the next decade, doubtless pass through a historic transition, which may spin off new communities of scientists with new opportunities for scientific discovery and valuable biomedical application.

Francis O. Schmitt

Contents

Chapter 4

Synaptic Proteins

Henry R. Mahler

Chapter 5

Brain Mitochondria

J. B. Clark and W. J. Nicklas

Chapter 6

The Smooth Endoplasmic Reticulum

Gérard Alonso

Chapter 7

Lysosomes

Harold Koenig

Chapter 8

Brain Nuclei

Yasuo Takahashi

Chapter 11

Neuronal Intermediate Filaments
Charles A. Marotta

Chapter 12

Local Synthesis of Axonal Protein
Edward Koenig

Chapter 13

The Axonal Plasma Membrane
George H. DeVries

Chapter 14

Central Nervous System Myelin
Joyce A. Benjamins, Pierre Morell, Boyd K. Hartman, and Harish C. Agrawal

Chapter 15

Autonomic Ganglia and Their Biochemistry
Jack D. Klingman

Chapter 18

Attenuated Blood–Brain Barrier
Maria Spatz

Chapter 19

Immunohistochemistry of Cell Markers in the Central Nervous System
O. Keith Langley, M. Saïd Ghandour and Giorgio Gombos

Chapter 20

Neurochemistry of Invertebrates
Marilyn W. McCaman

1

*The Synaptosome**

V. P. Whittaker

1. INTRODUCTION

Synaptosomes are detached, sealed presynaptic nerve terminals, often with portions of the postsynaptic membrane still attached. They are formed from the clublike axodendritic and axosomatic nerve terminals of brain tissue in 70–100% yield when this is homogenized under conditions of moderate liquid shear in isosmotic sucrose.[1–3] They retain most of the structural and functional characteristics of the nerve terminals from which they are derived. In peripheral tissues, however, with the partial exception of retina and autonomic ganglia, the mechanical conditions that favor synaptosome formation in brain tissue are not adequately met, and yields of synaptosomes are low or negligible. Even when formed, they may represent only a portion of the nerve terminal and may retain only part of the terminal cytoplasm, resulting in defective function.

Synaptosomes were first isolated in 1956,[4,5] but their nature, provenance, and mode of formation were not properly understood until 1960. In this respect, their history somewhat recapitulates that of liver lysosomes, which were also isolated and characterized biochemically[6] before being identified with a known subcellular structure,[7] in this case the peribiliary dense bodies. It should be realized that electron microscopy at that time was still in its infancy, and very few laboratories had either the equipment or the experience to undertake electron microscopic investigations of subcellular fractions; in addition, most morphologists were skeptical about the significance of any results obtained with "smashed-up tissue."

Synaptosomes were first referred to as nerve-ending particles,[5] abbreviated to the acronym NEP.[8] After an attempt was made in the United States to introduce another acronym, PONE (for pinched-off nerve endings), I felt the need for a name formed along more traditional lines and invented the word

* This chapter is dedicated to Mr. Gordon Dowe, *Cheftechniker, Abteilung Neurochemie*, with gratitude for 25 years of devoted collaboration.

V. P. Whittaker • Abteilung Neurochemie, Max-Planck-Institut für Biophysikalische Chemie, Göttingen, Federal Republic of Germany.

synaptosome (by analogy with lysosome, microsome, etc.). This word first appeared in print (within inverted commas) in 1964[9] and has now entered all languages in which neurochemical results are reported, which include most European languages and Japanese. It has acquired the distinction of being used unreferenced, like mitochondrion.

Although our findings were rapidly confirmed,[10] some time elapsed before the potentialities of the preparation became widely appreciated. However, a survey of the literature made in 1975 indicated that during the previous 5-year period a significant fraction of all neurochemical papers published were concerned with synaptosomes or synaptic vesicles (9% for the *Journal of Neurochemistry*, or about half of all those concerned with the application of tissue fractionation techniques to the nervous system, an approach much stimulated by the discovery of the synaptosome). A key paper[3] has been cited over 1000 times.[11] Thus, the bibliography of a chapter such as this is necessarily highly selective, and many important, even pioneer papers have had to be omitted.

Synaptosomes are primarily of interest as a preparation in which the biochemical basis for chemical transmission at the synapse can be studied without the complications attendant on the use of whole-animal or whole-tissue preparations. When incubated in a suitable balanced salt solution containing an energy source (usually glucose), they behave (Section 5.1) like miniature nonnucleated cells.[12,13] Ionic gradients and ATP levels, reduced by cooling and exposure to a nonionic medium, return to values characteristic of intact neurons; the presence of a normal membrane potential, although not directly measurable by micropipettes because of the small size of the target, has been inferred both from the magnitude of the ion gradients reestablished[14] and by the use of fluorescent probes.[15] Microtubules reassemble,[16] and the synaptosomes strikingly resemble in their morphology the intact terminals from which they were derived.[17] They avidly take up energy-yielding substrates[18] and putative transmitters[19] or their precursors.[20] They release transmitter in a Ca^{2+}-dependent manner when depolarized with K^+ ions or electrically.[21] Evidence for vesicle recycling has been obtained.[22] Thus, most synaptic functions have been conserved.

The intact synaptosome is a valuable tool for the study of the action of drugs and toxins on synaptic processes, as starting material for the attempted identification of new transmitters,[23] for sampling terminal axoplasm in studies of axonal transport,[24] as a source of both pre- and postsynaptic receptors for toxins,[25,26] transmitters,[27] and drugs,[28] and for studying the passive permeability properties of the unmyelinated neuronal plasma membrane.[29]

Controlled disruption and subfractionation of synaptosomes yield preparations of their component parts: synaptic vesicles, presynaptic plasma membranes, intraterminal mitochondria, and terminal cytoplasm are the best characterized of these.[9] The distribution of transmitters and enzymes among these synaptosomal subfractions gives valuable information concerning the molecular organization of the terminal, including intraterminal transmitter pools (Section 4.2), the storage and turnover of transmitters (Section 5.2), and the cellular metabolism (Section 5.1.6) of the terminal region.

Synaptosomes can apparently be formed from any type of central nerve terminal (Section 2.4). Thus, even the mossy fiber endings of the cerebellar cortex are pinched off during homogenization,[8,30–32] although because of their large size the resultant synaptosomes behave anomalously on subcellular fractionation. Synaptosome formation has been observed to occur during the homogenization of the cerebral neocortex, hippocampus, caudate nucleus,[33] spinal cord,[34] hypothalamus,[33,35] and hypophysis[36] and in the mouse,[37] rat, guinea pig,[3] rabbit, coypu,[38] cat,[39] dog,[33] sheep,[40] monkey,[41] man,[42] pigeon, chick,[43] frog,[44], lobster,[45] locust,[46] squid,[47] octopus,[48,49] dogfish,[50] and several species of teleost.[51] However, because of the large number of rather uniform axodendritic terminals in brain cortex, with their easily ruptured postsynaptic spine attachments and fine terminal axons, and the relative ease with which myelin can be removed from the smooth rodent cortex, the latter provides the most homogeneous synaptosome preparation.

The subject of synaptosomes has often been reviewed. One of the first of these reviews has some historic interest for its comprehensive account of the state of the art in 1965,[52] and another provides a laboratory manual with details of subcellular fractionation equipment, bioassay procedures for acetylcholine and marker enzymes, and evaluation of results that has proved successful in a number of advanced practical courses organized by the author.[53] The chapter on the synaptosome in the previous edition of this *Handbook* contains numerous electron micrographs of fractions which have had to be omitted from this edition in order to reserve space for new material.

2. *METHODS FOR PREPARING BRAIN SYNAPTOSOMES*

2.1. *Homogenization of Tissue*

2.1.1. *Conditions of Shear*

Presynaptic nerve terminals must be regarded as the most stable region of nerve cells. Any kind of slicing, chopping, mincing, or dispersion of brain tissue in isosmotic media probably permits the survival of a considerable proportion of them as organized structures.

However, there appears so far to have been no detailed, systematic study of the optimum conditions for the formation of synaptosomes, although some useful estimates of shear stress under different conditions of homogenization have been published.[54] Since bound acetylcholine is a synaptosome marker, the proportion of total brain acetylcholine remaining in the bound state and recovered in the synaptosome fraction may be used as an index of the yield of synaptosomes under varying conditions of shear force.

In a brief study,[55] the homogenization conditions originally selected were found to be superior to those using higher rates of shear. These were the use of a smooth-walled Perspex and glass homogenizer pestle and mortar, 0.32 M sucrose as a suspension medium, a clearance of 0.8%, and a speed of rotation (at 30-mm pestle diameter) of 840 rpm. Hand homogenization gave the same yield of bound acetylcholine, but a higher proportion sedimented in the nuclear-plus-debris fraction, suggesting less complete disintegration of tissue. The yield of synaptosomes determined by a morphological method (Section 2.3.2c) was quite close, using our standard conditions, to the estimated number of nerve terminals in intact tissue,[56] but no comparisons have been made by this method of synaptosome yields under a variety of conditions.

2.1.2. Suspension Media

Synaptosomes are osmotically sensitive structures[9]; thus, in work with lower organisms (elasmobranchs,[50,57] cephalopods,[47–49] and arthropods[45,46]), note must be taken of the osmolarity of the cell environment. It was noted, for example,[48] that the yield of synaptosomes from *Octopus* brain was very low when the tissue was homogenized in 0.32 M sucrose but very high in 0.8–1.0 M sucrose, 1.1 M glucose, or 0.7 M sucrose plus 0.33 M urea.

Like other subcellular particles, synaptosomes are caused to coacervate by low concentrations of divalent ions and somewhat higher concentrations of univalent ions.[3] The threshold concentration for coacervation of synaptosomes is 0.5 mM for calcium and 20 mM for sodium. Since both sucrose and water may contain calcium, it is necessary to use pure reagents and to avoid any appreciable concentration of buffers. Sucrose solutions consisting of analytical grade sucrose and doubly distilled water contain about 12 μM calcium, which is well below the threshold concentration; the concentration can be reduced to less than 2 μM by passage through an ion-exchange resin, but no improvement results from this or from the addition of EDTA. Some authors have advocated the addition of anticoagulants to the sucrose.[36]

2.1.3. Tissue Concentration

Homogenates containing more than 10% wt./vol. of brain tissue do not fractionate satisfactorily, presumably because of the relatively high ionic strength of such homogenates. However, for purposes of analysis, homogenates have sometimes been prepared at a tissue concentration of 20% wt./vol., sampled, and then diluted to 10% wt./vol. with more sucrose before fractionation with satisfactory results.[58]

The presence of large amounts of myelin reduces the efficiency of the separation process, and synaptosomes and mitochondria tend to be held up in the myelin layer. This is particularly evident with spinal cord. It is advisable to remove as much white matter as possible before homogenization: for this reason, the smooth cortices of rodents (rat, guinea pig, and coypu) make excellent starting material; the underlying layer of white matter is readily scraped off with a blunt scalpel before homogenization.

2.2. Separation of Synaptosomes from Other Subcellular Particles

2.2.1. By Moving-Boundary (Differential) Centrifugation

Synaptosomes have broadly similar sedimentation properties to mitochondria and to the smaller fragments of myelinated axons, dendrites, and glial processes formed during homogenization. Thus, the so-called mitochondrial (or P_2) fraction obtained from sucrose homogenates of brain tissue by moving-boundary centrifugation is a mixture of all of these components together with some unavoidable microsomal contamination; this fact largely explains the many anomalous properties that were formerly attributed to brain mitochondria. However, for many purposes this fraction is quite adequate, especially when it can be verified that the results obtained cannot be attributed to other particles present.

Synaptosomes are more labile than mitochondria, and fairly pure functional mitochondria can be prepared from the synaptosome–mitochondrial fraction by repeated washing and sedimentation,[59] apparently as a result of selective breakdown of synaptosomes into smaller fragments, which are washed out along with smaller membrane fragments derived from endoplasmic reticulum, plasma membranes, etc.

2.2.2. *By Density-Gradient Centrifugation*

Synaptosomes may be separated from myelin and free mitochondria in sucrose density gradients by making use of differences in their equilibrium densities. Myelin fragments and many fragments of glia and dendrites are light, unsealed structures that float on 0.8 M sucrose. Mitochondria have a protein-rich matrix and, on exposure to dense sucrose, undergo osmotic dehydration, which allows them to penetrate 1.2 M sucrose. Synaptosomes are sealed structures and also osmotically sensitive, but they are rich in cytoplasm and have a higher water content than mitochondria; thus, they equilibrate between 0.8 and 1.2 M sucrose. A simple step gradient consisting of layers of 0.8 and 1.2 M sucrose provides a reasonably effective separation among myelin (fraction A, floating on 0.8 M sucrose), synaptosomes (fraction B, floating at the 0.8–1.2 M interface), and somatic mitochondria (fraction C, pelletted material). This has become a widely used standard method (Fig. 1).[3] The method may be adapted readily to fixed-angle[60] or zonal[61] rotors.

Synaptosomes prepared in this way, if fixed without further treatment, are shrunken, with a very dense cytoplasm; dilution with an equal volume of water allows rehydration and a more "natural" appearance; still better is a brief incubation in a balanced saline medium fortified with an energy-yielding substrate (Section 2.2.3).[17]

In the days before large-capacity swing-bucket rotors were available, it was convenient to perform this density separation in two steps, centrifuging in an angle rotor through 0.8 M sucrose to remove myelin and then through 1.2 M sucrose to remove mitochondria. A flotation variant of this method has recently been advocated as a "rapid method" of preparing synaptosomes,[62] but the very thorough characterization of the procedure with marker enzymes failed to disclose much difference between synaptosomes prepared in this way and those prepared on a conventional step gradient.

It should be pointed out that step gradients should always be allowed to diffuse for at least 1 h before use; otherwise, the sharp changes of density at the interfaces cause poor separation. The synaptosomes take up some sucrose

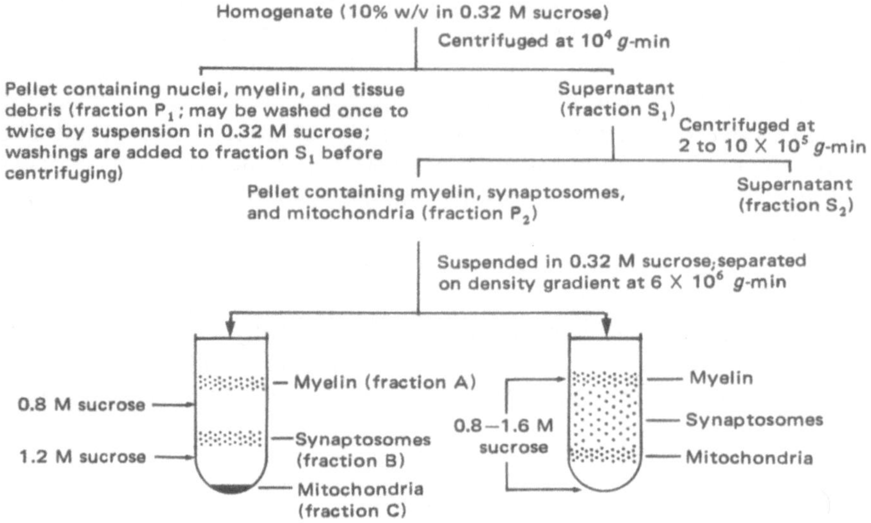

Fig. 1. Flow sheet showing the "classical" step-gradient method of preparing a synaptosome-enriched fraction[3] and a modification of it using a continuous density gradient.[53]

during separation so that they become hyperosmotic to 0.32 M sucrose; rapid dilution from 0.8–1.0 M sucrose to 0.32 M sucrose (*ca.* 1:3) causes some lysis. This is avoided if dilution is only 1:2 or is very slow. As an alternative to sucrose, the X-ray contrast medium sodium diatrizoate has been used.[63]

Some investigators prefer a procedure using an isosmotic density gradient in which varying concentrations of the polysaccharide Ficoll provide the density changes.[64] They claim that synaptosomes prepared in this way perform better metabolically. Unfortunately, Ficoll differs from batch to batch, and since differential osmotic dehydration, an important factor in the efficient separation that occurs in a sucrose density gradient, cannot take place, the synaptosome preparations obtained in Ficoll gradients show heavier contamination. In the author's experience, there is little difference in the metabolic performance of the two types of synaptosome preparation, and this has also been documented by others.[65]

Synaptosomes are intrinsically heterogeneous in size and density, and the now classical B fraction obtained from a 0.8–1.2 M sucrose step gradient is undoubtedly contaminated with other structures, principally dendritic, axonal, and glial fragments (Section 2.3.2d). In a preparation from guinea pig cortex well scraped to remove as much white matter as possible, glial contamination may be as high as 10%; in a preparation from whole rat brain, it is possibly higher. With carbonic anhydrase used as a glial marker, 14% contamination has been estimated.[66] A more pessimistic assessment (over 40% of glial material recovered in the B fraction as sealed "gliosomes")[67] depends on the assumption that labeled cultured C-6 glioma cells will break down on homogenization exactly as normal glial tissue, which may be unwarranted.

Purer synaptosome preparations can be obtained with continuous gradients[68] provided that the much smaller capacity (to avoid "turnover") of

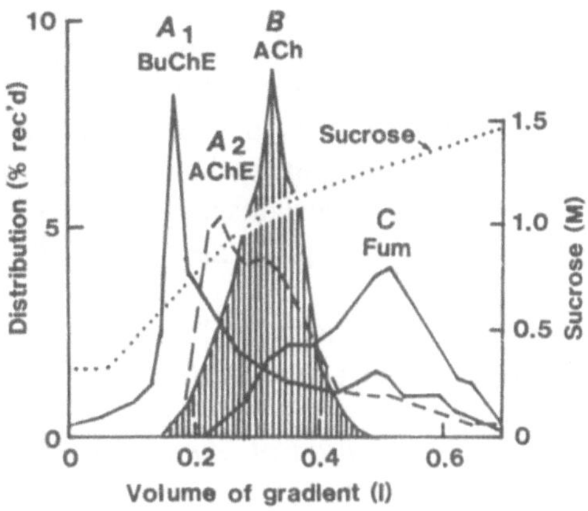

Fig. 2. Separation of synaptosomes from 20 g of scraped guinea pig cortex by density-gradient centrifugation in a 660-ml zonal rotor.[13] Fractions: A_1, glial fragments; A_2, microsomes; B, C as in Fig. 1. Abbreviations: BuChE, butyrylcholinesterase; AChE, acetylcholinesterase; ACh, acetylcholine; Fum, fumarase.

such gradients is respected (Fig. 1). The most efficient way of performing such separations is, however, in a zonal rotor (Fig. 2), obtainable in several sizes. The 330-ml-capacity rotor (Beckman TC) is particularly convenient; the volume of gradient can be adjusted, within limits, to the amount of tissue available by the use of variably sized cushions and overlays. Because of the large volume of gradient, the yield, which is limited in swing-bucket separations by the low capacity of the continuous gradient, is no longer a problem, and the dynamic loading and unloading improves resolution. The separations have been thoroughly characterized.[13,69] The suggestion, made in a recent lengthy review article,[70] that the N-M-L-P-S scheme of de Duve using moving-boundary centrifugation, historically important but long since abandoned by its originator, should replace isopycnic separation techniques for brain fractionation can only be dismissed as quaint.

In spite of some early claims to the contrary,[10] it has proved impossible to separate synaptosomes derived from neurons of different transmitter types from each other; there is too much heterogeneity within each subpopulation. On continuous gradients, the lighter fractions tend to be slightly enriched in cholinergic markers,[68,69] but overlap with noncholinergic markers is extensive. A similar situation applies to adrenergic synaptosomes.[71]

Percoll® gradients have been used to separate synaptosomes.[72] Percoll® is a proprietary preparation of silica gel; the gel particles separate to generate a density gradient under the influence of a centrifugal field. Since the gel must be isosmotic with the parent fraction, differential osmotic dehydration does not occur. However, separations are not truly isopycnic in such gradients. The gel particles are adsorbed by cell membranes. Subcellular structures with a large surface-area-to-volume ratio will adsorb more gel in proportion to their mass than larger structures with a smaller surface-area-to-volume ratio. There is also a sieving effect as a result of the structure of the silica gel. Separations thus depend on a number of physical factors. The most inconvenient aspect of such separations is that the fractions are difficult to prepare for electron microscopy because of the presence of the hard silica particles.

Some but probably not much further improvement in the separation of synaptosomes might be achieved by as yet untried gradient techniques, especially those using dense penetrating solvents such as D_2O, which have proved so successful in lysosome research.[73] A proper study of the water spaces and other physical properties in synaptosomes analogous to those already carried out with synaptic vesicles and chromaffin granules has not been made but might contribute to an improvement in the design of density gradient separations.

Ultrafiltration[74] and electrophoretic[75,76] methods of separating synaptosomes have been described but at present do not seem to have any conspicuous advantages over centrifugation methods.

2.2.3. *Rehabilitation of Synaptosomes after Separation*

As mentioned above, synaptosomes separated from other constituents of the crude synaptosomal–mitochondrial (P_2) fraction in a discontinuous (step) or continuous hyperosmotic sucrose gradient have lost water by partial osmotic dehydration. Morphological examination shows many synaptosomal profiles with varying degrees of condensation of the cytosol. These changes are readily reversible by dilution to approximately 0.45 M sucrose, which gives a more "natural" appearance to the synaptosomes, but such preparations still have K^+ concentrations that are much below normal values for neurons and Na^+ concentrations are much above. This is because of the inhibition of normal metabolism that results from the low temperature and ionic strength to which the synaptosomes have been exposed during isolation. The cold-labile elements of the cytosol microtubules and neurofilaments—have also largely disintegrated. Such changes occur with all techniques of isolation, whether or not osmotic dehydration is superimposed. Fortunately, they are reversible if the synaptosomes are incubated for a short period in a balanced saline medium containing an energy source—usually glucose. Synaptosomes rehabilitated in this way show almost normal ionic gradients and a morphological appearance almost identical with terminals *in situ*.[17] Microtubular elements are reassembled,[16,17] and intraterminal mitochondria return to normal.

2.3. *Evaluation of Fractions*

2.3.1. *Biochemical Evaluation*

A valuable tool for the characterization of subcellular fractions is a "marker" substance, often an enzyme, that is specific for a particular subcellular particle whose isolation and purification it is desired to monitor. Some examples are given in Table I. Many cell components have specific locations in the cell; thus, the enzymes of the tricarboxylic acid cycle and the electron transport chain are known to be localized, in eukaryotic cells, exclusively in mitochondria. Consequently, an enzyme such as fumarase (L-malate hydrolase, E.C. 4.2.1.2) is a marker for mitochondria.[68] Similarly, lactate dehydrogenase (L-lactate:NAD oxidoreductase, E.C. 1.1.1.27) and Na^+,K^+-activated ATPase (ATP phosphohydrolase, E.C. 3.6.1.3) are markers for cytoplasm[3,88–90] and plasma membranes,[66,91,92] respectively.

Table I
Brain Marker Substances[53]

Substances	Location	References and remarks
Acetylcholine (bound)[a]	Synaptosome (labile bound, cytoplasm; stable bound, synaptic vesicles)	1–5,9
Acetylcholinesterase[b]	SPM[c], microsomes derived from cholinergic neuronal cell bodies, cholinergic axonal fragments	52,77–79
Acid hydrolases	Lysosomes	80
Butyrylcholinesterase	Glial fragments	81–83
Cardiolipin	Mitochondria	84
Cerebrosides	Myelin	84,85
Choline acetyltransferase (occluded)[a]	Synaptosome cytoplasm	9; adsorbed onto SPM[a] at low ionic strength 86,87
Glycolytic enzymes (except hexokinases)	Soluble cytoplasm, synaptosome cytoplasm	3,88–90
Malate hydrolase and other enzymes of tricarboxylic acid cycle and electron transport chain	Mitochondria (including intraterminal mitochondria)	5,68,88
Na^+,K^+-ATPase	SPM[c], microsomes derived from plasma membrane	77,91,92
Potassium	Soluble cytoplasm, synaptosome cytoplasm	9,93,94

[a] These components are confined to cholinergic synaptosomes, but because these greatly overlap in sedimentation properties with other synaptosomes, they can also be used as general synaptosomal markers.
[b] Usually associated with cholinergic neurons but not always (e.g., cerebellum).
[c] Synaptosomal plasma membrane.

Since synaptosomes are organized portions of cells containing cytoplasm and one or more small mitochondria enclosed within a plasma membrane, all three of these enzymes are found in synaptosomes, and in any fractionation scheme that separates synaptosomes from other cell components, they will be bimodally distributed. Thus, lactate dehydrogenase will be found mainly in the high-speed supernatant (cell sap fraction) of a homogenate but also in synaptosomes; fumarase is found both in the mitochondrial and synaptosome fractions; Na^+,K^+-activated ATPase is found in the microsome and synaptosome fractions. Even transmitters and the enzymes that synthesize them will not be found exclusively in the synaptosome fraction since in addition to the high concentrations of these components in nerve terminals, they exist in lower concentrations throughout the neuron, and significant amounts may be recovered in the high-speed supernatant.

These remarks will serve to stress the need to evaluate the yield and purity of the synaptosome fraction in any fractionation scheme by determining as many marker substances as possible in all of the fractions.[53] As a test of the validity of the analytical procedure, we must first estimate the recovery of the component in the fractionation scheme. To do this, the content of marker in each fraction must be put on a basis that enables the various fractions to be summed. This is best done by expressing the results as amount of marker per amount of fraction derived from 1 g of tissue, making allowance for any losses resulting from sampling of parent fractions for analysis. Recoveries should be close to 100%; if they are not, the cause must be investigated. Recoveries much below 100% usually imply destruction of the marker during fractionation. This is particularly likely with labile markers such as acetylcholine or ATP; however, inhibition by sucrose or another constituent of the original homogenate is also a possibility with biological assay methods. Recoveries much above 100% may be caused by the synthesis of the component during fractionation, the progressive removal or dilution of an inhibitor of the assay procedure, or inadequate release of occluded marker.

The manipulation of cell fragments during fractionation may have made a portion of the component more fully accessible to the analytical procedure, although it was inaccessible in the original homogenate; thus, an enzyme sequestered within a lipoprotein membrane in the original homogenate, which was not released under the conditions of assay, might have been made more accessible as a result of the mechanical stresses imposed on the membrane during fractionation with its attendant pelletting, resuspensions, and washings. Since many enzymes are partly or wholly occluded in homogenates and subcellular fractions, including those sequestered within synaptosomes, the addition of a noninhibitory detergent to release the enzyme before assay is routine practice.

Even when recovery is 100%, redistribution of a component may occur during fractionation. This is very likely with positively charged soluble enzymes that tend to become adsorbed onto negatively charged membranes under the conditions of low ionic strength that prevail in sucrose homogenates and subcellular fractions; examples are lactate dehydrogenase and choline acetyltransferase (acetyl-CoA:choline acetyltransferase, E.C. 2.3.1.6) in some species. If such adsorption is suspected, washing the particles with isosmotic sodium chloride will often release adsorbed enzyme.

When satisfactory recoveries have been obtained, we may proceed to work out the distribution of the marker among the fractions. This is best expressed as the percentage ratio of the amount of the component recovered in a fraction to the total amount of component recovered. This method of expression makes it easier to compare the results of different experiments, since the inevitable losses of material, errors in the determination of fraction volumes, and analytical errors are averaged over all of the fractions by this procedure. The distribution is easily interpreted if the marker is uniquely associated with a subcellular organelle that the fractionation procedure separates in good yield and purity. However, as we have seen, bimodal distributions are to be expected with many synaptosomal components, and a minor fraction that contains a relatively large amount of a component, i.e., contains it in high specific concentration (or activity), always merits further study.

Thus, in addition to the recovery and distribution of a marker, we also need to know its specific concentration. This, if expressed in absolute values (e.g., units of enzyme activity per milligram of protein), will be affected by errors and losses in both numerator and denominator. Thus, to compare experiments it is often more useful to calculate the relative specific concentration (or activity), which is the percentage of the total recovered component present in the fraction divided by the percentage of total recovered protein present in the fraction. Values of this ratio greater than unity represent an enrichment of the marker in this fraction relative to the homogenate (or other parent fraction).

A useful method of presenting percentage distributions and relative specific concentrations in a single figure is to plot, as a histogram, the relative specific concentrations of the fractions as ordinates against their cumulative percentage protein distributions as abscissae. The area of each block is then proportional to the percentage distribution of the marker in each fraction. This is sometimes known as a de Duve diagram, from its originator.[6]

Another measure of relative specific concentration is the concentration ratio or enrichment factor, which is the specific concentration of a marker in a fraction divided by the corresponding quantity for the homogenate or parent fraction. This is a useful procedure if analytical data are not available for all fractions, but random losses and analytical errors will lead to a larger variance in the enrichment factor between experiments than in the relative specific concentration.

Examples of these methods of assessment of subcellular fractions are given in Sections 4 and 5 and can be found in numerous papers and in an article on practical methods.[53]

2.3.2. *Morphological Evaluation*

2.3.2a. Preparation of Material for Electron Microscopy. Synaptosome fractions can be prepared for electron microscopy by any of the standard methods of fixing, staining, and embedding of nervous tissue that give good preservation of nerve terminals in whole-tissue blocks. As a routine, glutaraldehyde–osmium–lead–uranium[34] is suitable, but osmium–phosphotungstic acid,[3] permanganate–lead,[56] and aldehyde–phosphotungstic acid[95] are useful for special purposes. Because of the particulate nature of the material, penetration of fixative is rapid, and a fixative such as permanganate that does not give good results with whole brain tissue is satisfactory with subcellular fractions.

Fixation can be carried out on pellets[3] or suspensions. The latter may be infiltrated after being sealed into agar cups[96] or mixed with fixative and subsequently centrifuged to form a pellet that is dehydrated and embedded without further disturbance.[34,38,56] The permanganate method, when used with suspensions, is particularly convenient because the manganese dioxide formed during fixation seems to act as a cushion during centrifuging, and a preparation is obtained in which the structures are well separated without the packing distortions often visible in pellets.[56,88]

Synaptosomes fixed in osmium are fragile—much more so than mitochondria—and will readily break up if the pellet is handled carelessly. This and the use of methacrylate rather than the epoxy resins as an embedding medium probably explain why synaptosomes remained undetected for so long;

inadequately preserved synaptosomes can indeed be seen in electron micrographs of brain mitochondria published before their existence became generally known.[97]

Subfractions of synaptosomes (Section 4) present no special difficulties except for the usual one of identifying the provenance of membrane fragments in the absence of the landmarks of an ordered morphology.

Synaptosomes and their subfractions can also be examined in negative staining.[98] The problem here is to remove the solute of the suspension medium (which would otherwise interfere with the even deposition of negative stain) and to secure sufficient penetration of the stain to show up the characteristic internal organelles (synaptic vesicles and small mitochondria) without bringing about damage so extensive as to make the structures unrecognizable. Fixation with aqueous formaldehyde or osmium tetroxide followed by 1–2% sodium phosphotungstate (pH 7.4) has been the most frequently used method; this permits some penetration of stain without osmotic disruption. If isosmotic negative stain (e.g., 2% ammonium molybdate) is used without fixation, little or no penetration occurs; if the strongly hypoosmotic phosphotungstate is used without fixation, extensive disruption and release of cytoplasmic organelles ensue. Some of the clearest pictures of individual synaptosomes that have been obtained were selected images in preparations exposed to mild hypoosmotic disruption.

By contrast, synaptic vesicles isolated from brain synaptosomes may be sustained without prior fixation even with hypoosmotic stains such as 2% sodium phosphotungstate, since they are considerably more resistant to hypoosmotic disruption than synaptosomes. Even after fixation, there is little penetration of negative stain, possibly because of the presence of a core substance.[99] Prior treatment with sodium phosphate buffer permits the ingress of stain, and the synaptic vesicle membrane then becomes clearly visible.[38] Because of the small size of the vesicles, staining conditions must be optimal in terms of solute removal, particle concentration, and evenness and thickness of stain deposition.

Fractions containing synaptosomal plasma membranes are difficult to stain because of the presence of a macromolecular substance (possibly a mucopolysaccharide) that interferes with the even deposition of stain.[9,77] The staining patterns obtained when this material is present resemble those obtained with Ficoll and can vary from that characteristic of a sol (electron-lucent particles of Ficoll embedded in electron-dense stain) to that characteristic of a gel (an electron-lucent network of Ficoll enclosing patches of electron-dense stain). The high extracellular volume associated with synaptosome pellets also suggests the presence of material penetrated by extracellular markers.[100]

There are few reports in the literature of the use of freeze-fracture or scanning electron microscopy for investigating synaptosomes. Freeze-fracture has been applied to synaptosomes derived from *Torpedo* electromotor nerve terminals (Section 3).[101] In scanning electron microscopy (W. B. Essman, personal communication), little of interest is resolved; the synaptosomes look like potatoes scattered on the specimen grid.

2.3.2b. Sampling Problems. When subcellular particles are sedimented before or after fixation, stratification results; the lighter particles accumulate at the top of the pellet, the heavier at the bottom. To avoid misinterpretation, random sampling for electron microscopy throughout the pellet is essential[102]; alternatively, by proper orientation of a small pellet before sectioning, the whole thickness of the pellet may be examined in a single section.[103–105]

2.3.2c. Morphological Characteristics of Synaptosomes. Synaptosomes in section appear[3] as membrane-bounded oval profiles about 0.5 μm in diameter containing numerous vesicles about 42 nm (S.D. ±5 nm) in diameter and, sometimes, small mitochondria (figures relate to permanganate-fixed material). Not infrequently, lengths of thickened membrane can be seen attached to the synaptosomal plasma membrane, with a cleft between about 20 nm wide that is filled with lightly stained striated material. In addition, most profiles contain a few larger vesicles (mean diameter 70 nm, S.D. ±5 nm); these are particularly clearly seen in permanganate-fixed material. "Coated" and dense-cored vesicles are also seen within some profiles.

These morphological features occur together, as far as is known, in only one structure in the central nervous system. Taken in conjunction with the high concentration of putative transmitter substances in the synaptosome fraction, they provide the basis for the identification of these structures as detached nerve terminals. It is assumed that on homogenization the terminal region of the neuron is "pinched off" and seals up to form a discrete subcellular particle, which carries with it a length of the characteristically thickened postsynaptic membrane. Although the contents of the synaptic cleft are only lightly stained, it is evident that the two membranes are securely bonded together; possibly the bonding material is coded for the particular contact made. In cholinergic endings, the surface antigen CHOL-1 is tentatively assigned this role.[106]

In assessing sectioned material, it must be borne in mind that even if all synaptosomes possess postsynaptic attachments and mitochondria as well as synaptic vesicles, only certain planes of section will contain all three structural components. Moreover, serial sections show that some synaptosomes do not contain mitochondria[107]; whether these represent mitochondrion-free lobes or larger nerve endings or whether they are derived from endings that are themselves devoid of mitochondria is not known. The presence of mitochondrion-free synaptosomes has also been deduced from a statistical examination of synaptosome profiles.[102] Although no quantitative studies have been carried out, the number of postsynaptic attachments observed seems less than one would expect if all synaptosomes possessed them. Attachments are more frequently seen in preparations made from regions of the brain (e.g., cerebral cortex and cerebellar cortex) where synaptic contacts are made with presumably easily detached dendritic "spines"; occasionally a complete spine is seen. Fewer attachments are seen in preparations that have been exposed to hyperosmotic sucrose, suggesting that high concentrations of sucrose have a solvent action on the cement in the synaptic cleft.

Sometimes profiles are seen with obvious postsynaptic attachments but few or no synaptic vesicles. These are probably synaptosome "ghosts."

In stereomicrographs of negatively stained synaptosomes that were lightly fixed to permit penetration of stain, the region containing mitochondria is seen to stand higher on the grid than the parts containing only synaptic vesicles.[98] Many synaptosomes are seen to contain only one elongated, coiled mitochondrion, showing that the adjacent oval mitochondrial profiles often seen in thin section may well represent cross sections of a single mitochondrion winding in and out of the plane of section. Sometimes short lengths of axon are seen attached to synaptosomes; these would be hard to identify in thin section because of their small diameter and random orientation in the block. On the other hand, postsynaptic adhesions are difficult to identify with certainty in negative staining, and the synaptic cleft is rarely seen, presumably because the requisite orientation is seldom attained.

The morphological descriptions given above relate primarily to fractions derived from guinea pig, coypu, and, to a lesser extent, rat brain. However, the morphology of the primary subcellular fractions and especially of intact synaptosomes from a wide variety of vertebrate species appears to be very similar, regional differences being greater than species differences.

As far as fractions of disrupted synaptosomes are concerned, the range of species studied is smaller. However, dog, rat, cat, ox, and sheep give similar results to guinea pig and coypu. In the rat, the synaptic vesicles appear to be slightly denser than in the guinea pig; this may be because of the adsorption of more soluble protein onto them at the low ionic strength of the preparation. Vesicle preparations from cat and sheep are considerably less homogeneous than those from the rodents, probably because of the greater difficulty in removing myelin.

In spite of these similarities, it is inadvisable to apply methods worked out with one species to another without adequate morphological controls. Indeed, there is sufficient inherent lack of reproducibility in all subcellular fractionation techniques to make morphological controls essential when a new method is being applied, even when the species used is the same as in the original publication.

2.3.2.d. Quantitative Evaluation of Fractions. An important test of the basic assumption that the vesiculated synaptosome profiles are sections of pinched-off presynaptic nerve terminals would be to compare the number of synaptosomes in brain homogenate or the fractions derived from it with the number of nerve terminals in the intact tissue from which the homogenate is made. Clearly, if the number of synaptosomes were found to be greatly in excess of the number of nerve endings, doubt would be cast on the whole synaptosome concept, and if considerably less, the synaptosome fraction could not be regarded as representative of nerve endings as a whole. Either way, the validity of using synaptosomes for *in vitro* preparations for investigating synaptic function would be questionable.

The application of the test is fraught with difficulty, partly because of the problem of deducing the number of three-dimensional objects from two-dimensional sections and partly because of the sampling problems involved (Section 2.3.2b). However, in a careful study,[56] it has been estimated that guinea pig cerebral cortex homogenates contain about 4×10^{11} synaptosomes/g and that the number of nerve terminals in guinea pig cortex is about equal to this.

Statistical analysis[108] of synaptosome profiles provides quantitative information regarding the size of the various compartments in the "average" synaptosome. A synaptosome of mean volume (0.1 μm^3) contains about 73 synaptic vesicles occupying about 4% of the volume; another 24% is taken up by mitochondria, and 8% by the external membrane, leaving a cytoplasmic volume of 64% of the total volume.

As prepared according to the standard method shown in Fig. 1, the synaptosome fraction (B) is relatively free from myelin whorls or mitochondria but is contaminated with considerable numbers of membrane fragments. Some of these are imperfectly preserved synaptosomes, others are grazing sections of synaptosomes, yet others are of unknown provenance. They range in size from fragments of microsomal dimensions (0.1–0.2 μm in diameter) to large (0.5–1 μm diameter), empty, oval profiles that might be fragments of dendrites or glial processes. Some may be small fragments of myelin, but the low content of the preparation in cerebroside, a myelin marker (Table I), suggests that there is relatively little myelin contamination.

The presence of nonsynaptosomal material is indicated by the considerable amount of nondiffusible dry solids that cannot be accounted for by the morphological estimate of the synaptosomal volume. Thus, the total morphological synaptosomal volume in fraction B from guinea pig cortex has been estimated to be about 31.5 μl/g of original tissue.[29,56] At an equilibrium density of about 1.15, this is 36 mg/g of original tissue. The nondiffusible dry solids in this fraction are about 22 mg/g of original tissue. This implies a water content of only about 29% if all dry solids are assumed to originate from synaptosomes. Although synaptosomes in equilibrium with 0.8–1.2 M sucrose may well have a relatively low water content (62% assuming a dry density of 1.4), it seems clear that about half of the material in fraction B could be nonsynaptosomal, though not necessarily membranous.

With continuous (0.8–1.6 M) sucrose density gradients,[68,102] a very much purer synaptosome fraction may be obtained. Myelin contamination is insignificant below about 1.0 M; the larger membrane fragments remain in the upper part of the gradient (down to about 1.17 M sucrose), and mitochondrial contamination does not commence until a sucrose concentration of about 1.24 M is reached. Thus, the zone between 1.17 and 1.24 M, containing about 10% of the protein of the parent material, constitutes a relatively pure synaptosome preparation, free from myelin and mitochondria and containing only a minor amount of contamination from small membrane fragments.

2.4. Atypical Brain Synaptosomes

Some presynaptic nerve terminals in brain differ considerably from the norm in size or morphology. They nevertheless generate synaptosomes on homogenation, and these deserve specific comment.

The mossy-fiber nerve terminals in the cerebellum are very large (2–5 μm, 4–10 times larger than the average). The synaptosomes generated from them[8,30] by the application of mechanical shear are also large and sediment in the P_1 (crude nuclear) fraction; these retain some of their postsynaptic attachments to shorn-off granule cell dendrites and can be separated from the other constituents of the fraction (mainly nuclei and large myelin fragments) by a simple isosmotic density gradient procedure utilizing Ficoll to stabilize the nuclei. They are fragile, and the vesicles they discharge on rupture may contaminate the microsomal fraction. Thus, the cerebellum is routinely excluded when synaptosomes are prepared from brain.

A more gentle homogenization results in the isolation of almost complete glomerular "islets," mossy fiber terminals surrounded by their glial capsule with numerous attachments to stubby granule cell dendrites, which also retain axodendritic synapses with Golgi nerve terminals.[109] That such complex structures can be isolated in a good state of preparation indicates how versatile tissue fractionation techniques can be.

By contrast, numerous parallel-fiber terminals are recovered in the normal synaptosome fraction from cerebellum; they can be readily identified by their deeply invaginated small spines.[8]

Other types of rather atypical nerve terminals are those forming dendrodendritic synapses in the olfactory bulb and those of the vasopressin- and oxytocin-secreting cells in the neurohypophysis. Both types of nerve terminals can be isolated.[36,110–112] The synaptosomes derived from the neurohypophysis retain their stock of large secretory granules.[36,112] The granules themselves have also been isolated.[113]

3. *THE PREPARATION OF SYNAPTOSOMES FROM PERIPHERAL TISSUES*

As mentioned in Section 1, the mechanical conditions that promote synaptosome formation when brain tissue is homogenized do not seem to be met in peripheral tissues. However, synaptosome preparations have been obtained from the retina,[114,115] superior cervical ganglion,[116,117] myenteric plexus of intestine,[118] and the electric organ of *Torpedo*.[119,120] The myenteric plexus synaptosomes appear to represent pinched off varicosities, the *Torpedo* electric organ synaptosomes resealed portions of electromotor nerve terminals.

That synaptosomes were generated in low yield when electric organ was homogenized was observed during early attempts to submit this collagen-rich and difficultly homogenizable tissue to subcellular fractionation.[57] It was found that a high proportion of terminals were not pinched off but simply torn open, and the synaptic vesicles released. Later, the yield of synaptosomes was improved when juvenile tissue, with a lower collagen content, was used.[119] The synaptosomes obtained were small and contained few vesicles and little cytoplasm; their "atypical" form led to their being called terminal sacs (abbreviated to T-sacs). However, larger synaptosomes can also be obtained from electric tissue.[120] These too, as originally prepared, do not perform well metabolically, but their ability to extrude Na^+ and to take up K^+ and choline can be greatly improved by preparing them in a low-Na^+, high-K^+ medium containing ATP, some of which becomes incorporated into the synaptosomal cytoplasm when the synaptosomes reseal.[121]

It seems clear that as electromotor nerve terminals are torn apart during homogenization much terminal cytoplasm is lost, but a proportion of the terminal fragments do reseal and show a limited metabolic and transmitter-releasing capacity in spite of a depleted cytosol. The exchange of contents between cytoplasm and medium makes it possible to vary the composition of the cytoplasm at will. However, the real significance and usefulness of this preparation lie in the fact that electromotor nerve terminals are purely cholinergic and that these synaptosomes are the first purely cholinergic synaptosomes to be obtained. Furthermore, they completely lack postsynaptic adhesions, so that the plasma membrane preparations obtainable from them are uncontaminated with elements of these membranes.

Analysis[122] of these plasma membranes has provided useful information about the composition of the terminal plasma membrane and has pointed up differences between the plasma membrane and the synaptic vesicles that support the "transient exocytosis" theory of transmitter release.[85,123] By injecting these membranes into sheep, it has proved possible to raise an antiserum to a surface antigen of the presynaptic plasma membrane, which has proved to be a specific component of all mammalian central and peripheral cholinergic terminals tested but not of noncholinergic terminals (Section 6).[106,124] Such sera should prove valuable as a basis for the histochemical identification of cholinergic connections in the central nervous system.

4. *SUBFRACTIONATION OF SYNAPTOSOMES INTO THEIR COMPONENT ORGANELLES*

4.1. Disruption of Synaptosomes

Brain synaptosomes are relatively labile structures and are readily disrupted by a variety of treatments, e.g., warming, vigorous mechanical agitation, supersonic vibrations, detergents, freezing and thawing, incubation with cobra venom, and hypoosmotic treatment.[5] Most of these treatments result in the prompt liberation of soluble cytoplasmic components (e.g., K^+, lactate dehydrogenase, choline acetyltransferase), but some of them (shaking, supersonic vibration, incubation with cobra venom, hypoosmotic treatment, freezing and thawing) liberate only 50% of the transmitter acetylcholine. This led[9,52] to the concept of labile-bound and stable-bound forms of transmitter.

Morphological and radiochemical studies have permitted the identification of the labile-bound fraction of acetylcholine released during hypoosmotic treatment as that fraction of synaptosomal acetylcholine that is present in soluble form in the cytoplasm of the synaptosome and of the stable-bound acetylcholine as the fraction that is associated with synaptic vesicles, probably in a local concentration as high as 0.2 M.[38] Since choline acetyltransferase, the enzyme that synthesizes acetylcholine, has a purely cytoplasmic location, one would expect to find a cytoplasmic pool of acetylcholine. This pool is bound only to the extent that the external membrane provides a barrier to the outward diffusion of free cytoplasmic acetylcholine, as it does to that of any soluble cytoplasmic constituent.

Hypoosmotic disruption provides[9] the most suitable procedure so far found for the isolation of the constituent organelles of brain synaptosomes. Large numbers of synaptic vesicles survive intact; external membranes and intraterminal mitochondria are not unduly fragmented and can thus be readily separated from vesicles by centrifuging. However, as noted below, the conditions for disruption must be carefully chosen for optimum separation.

Relatively little study has been made of disruptive procedures that do not involve hypoosmotic treatment as one step. However, it is known that disruption by freezing and thawing does not yield significant amounts of vesicles[88] and that disruption by supersound gives a vesicle fraction heavily contaminated with small fragments of presynaptic plasma membrane.

4.2. Fractionation of Osmotically Lysed Synaptosomes

4.2.1. By Moving-Boundary (Differential) Centrifugation

In one procedure,[125] a crude synaptosome preparation (approximately equivalent to fraction P_2 in Fig. 1) is suspended in 10 μM $CaCl_2$ (10 ml/g of original tissue) and centrifuged at 11,500 *g* for 20 min to sediment mitochondria and myelin fragments (fraction M_1). The supernatant is centrifuged at 100,000 *g* for 30 min to give a pellet (M_2) said to consist of free synaptic vesicles mixed with some "curved membranes" and a supernatant (M_3) consisting of soluble cytoplasmic constituents diluted with suspension medium.

Although fraction M_2 undoubtedly contains free synaptic vesicles, it is contaminated with considerable amounts of other membranous material and even partly disrupted synaptosomes.[9] This is shown by morphological examination using methods that give adequate preservation and sampling of the material and also by the presence in the fraction of cholinesterase,[125] Na^+,K^+-activated adenosine triphosphatase,[126] and ganglioside,[127] all markers for external membranes but absent from more homogeneous synaptic vesicle preparations.[77,84,85,126,128–132]

Perhaps an even more serious criticism of this preparation is that at the very low ionic strength and relatively low pH of the hypoosmotic suspension, soluble proteins may be precipitated nonspecifically onto the membranes present in the fraction, leading to incorrect conclusions[41,133] regarding the composition of the fractions. This is particularly evident with the soluble enzyme choline acetyltransferase in certain species, notably rat, rabbit, and sheep.[86]

4.2.2. By Density-Gradient Centrifugation

For the adequate separation of the various components of lysed synaptosomes, the ionic strength and pH of the lysate should not be too low, and some form of centrifugal separation using density gradients is essential. In the method developed in the author's laboratory (Fig. 3), crude (fraction P_2)[9] or

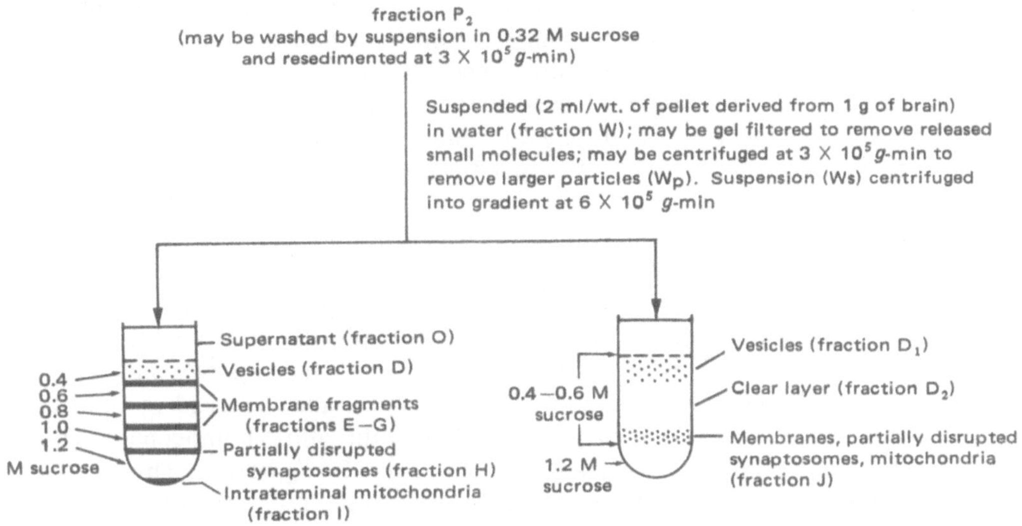

Fig. 3. Separation of synaptosomes prepared from scraped guinea pig[9] or coypu[38] cortex into their constituent components using step[9] or continuous[38] gradients.[53] A purified synaptosome preparation may be substituted[77] for fraction P_2.

purified (fraction B)[77] synaptosomes are lysed in only 2 ml of water/g of original tissue by sucking the suspended synaptosome pellet up and down in a pipette and layering on a discontinuous[9,77] or continuous[38] sucrose density gradient. After centrifuging for 2 h at 53,000 g, a hazy layer is formed in 0.4 M sucrose that consists of a morphologically very pure preparation of synaptic vesicles; larger membrane fragments, incompletely disrupted synaptosomes, and mitochondria are carried further down the gradient. The preparation can be scaled up using a zonal rotor (Fig. 4)[13] The higher ionic strength and pH of our lysed synaptosomes compared to those of others result in less nonspecific adsorption of cytoplasmic proteins onto membranes.

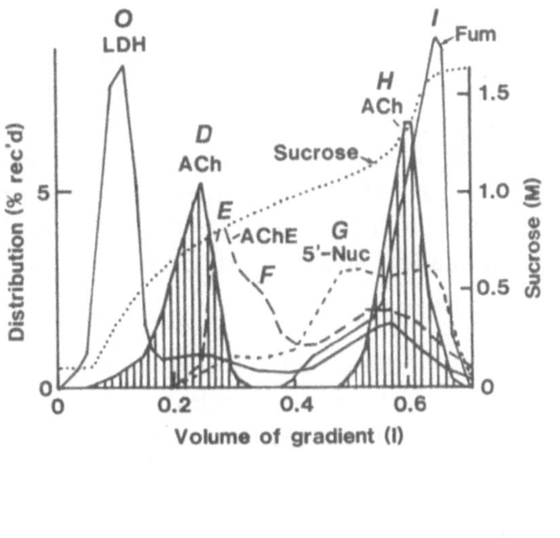

Fig. 4. Separation of synaptosomal components by density-gradient centrifuging in a zonal rotor.[13] The peaks are lettered after the corresponding fractions in Fig. 3. 5'-Nuc, 5'-nucleotidase; LDH, lactate dehydrogenase; other abbreviations as in legend to Fig. 2.

Synaptic vesicle preparations obtained in this way do, however, contain soluble cytoplasmic constituents that have diffused into the hazy layer from the clear band (O) immediately above. These contaminants and any residual contamination by microsomes or vesiculated fragments of plasma membranes may be removed by chromatography on columns of porous glass beads.[132] This leads to a striking increase in the specific concentration of the cholinergic vesicle marker acetylcholine. Calculations making allowances for the dilution of the cholinergic vesicles by noncholinergic ones suggest that the specific concentration of acetylcholine attained is close to the limiting value.[132]

The yield of synaptic vesicles in fraction D_1 (Fig. 3) has been evaluated using a "tagging" procedure with polystyrene beads and negative staining.[38] A volume of fraction D_1 equivalent to 1 g of original tissue (guinea pig cortex) was estimated to contain about 3.8×10^{12} vesicles, and the yield of vesicles was estimated to be 15–16% of those present in the initial homogenate. This suggests a total number of vesicles of about 2.4×10^{13}/g cortex. This is consistent with the number of synaptosomes produced per gram of cortex and the number of vesicles per synaptosome.

Besides the monodispersed synaptic vesicles isolated in fractions D or D_1, appreciable numbers remain attached to synaptosomal plasma membranes. In the scheme shown in Fig. 3, these are largely discarded in fraction W_p. This sedimentation step, by reducing the number of myelin fragments, mitochondria, and incompletely disrupted synaptosomes in the synaptic vesicle suspension before submitting this to density-gradient separation, undoubtedly improves the purity of the monodispersed vesicle fraction D.

However, the vesicles remaining within disrupted synaptosomes may have considerable functional significance. The experiments that demonstrate this[134,135] were ones in which the yield of vesicles attached to plasma membranes was maximized by submitting the entire fraction W to step-gradient centrifugation; the membrane-associated vesicles were recovered in fraction H (Fig. 3). When transmitter stores of acetylcholine were labeled *in vivo* by injecting radioactive choline[134] or choline analogues that are acetylated to form false transmitters,[135] the label appeared preferentially in the vesicles of fraction H. On restimulation, the specific activity of the released acetylcholine and the ratio in which true and false transmitters were released agreed with the values for fraction H and not with those of fraction D.[135] In preparations in which labile-bound (cytoplasmic) transmitter had been stabilized by inhibiting cholinesterases, the values for released transmitter also differed from those of the cytoplasmic pool (recovered in fraction O, Fig. 3).[135] It is concluded that the vesicles in fraction H, not those in fraction D, are the ones directly involved in transmitter release and vesicle recycling. Attempts to separate the vesicles of fraction H from the plasma membranes have not been very successful. The significance of these findings is discussed in greater detail in Section 5.2.1 and in Chapter 2 in this volume.

The discontinuous density gradient shown in Fig. 3 and its continuous-density-gradient counterpart (Fig. 4) yield four other fractions (E, F, G, and I) besides those of cytoplasm (O), monodispersed vesicles (D), and vesicles attached to synaptosomal plasma membranes (H) already mentioned.

Fraction E is a mixed fraction; it contains some synaptic vesicles and numerous vesiculated membrane fragments of microsomal dimensions (0.1–0.2 μm in diameter) and is fairly rich in acetylcholinesterase and Na^+,K^+-activated ATPase.[9,77] It probably consists mainly of smaller fragments of the synaptosomal plasma membrane and endoplasmic reticulum.

Fraction I is rich in mitochondrial markers and is seen in the electron microscope to consist of fairly pure mitochondria.[9] Many of these are small and of the size range of intraterminal mitochondria. Such mitochondria have been systematically investigated.[59]

Fractions F, G, and H contain acetylcholinesterase, Na^+,K^+-activated ATPase, and 5'-nucleotidase, all markers for plasma membranes.[9,77] Fraction H consists of disrupted synaptosomes[9] and, as already mentioned, contains numerous synaptic vesicles, many of which appear in close proximity to the plasma membranes. The disrupted synaptosomes that make up this fraction have lost almost all their soluble cytoplasm, so the highly labeled acetylcholine that can be detected in this fraction after labeling the acetylcholine pools *in vivo* is not in the cytoplasm of undisrupted synaptosomes, as has been erroneously suggested.[136,137]

Fractions F and G can be described as "synaptosome ghosts."[9] They are rich in plasma membrane markers[9,77] and contain large (0.5-μm diameter) membranous sacs but few if any of the cytoplasmic organelles characteristic of synaptosomes. They have quite a different composition from synaptic vesicles, being rich in cholesterol[84] and ganglioside,[85] possessing several enzymes not present in synaptic vesicles, and giving a different pattern of polypeptides on gel electrophoresis.[138] The significance of this difference in composition for notions of vesicle exocytosis is discussed in Chapter 2.

Plasma membranes have also been prepared from the purely cholinergic synaptosomes that can be made from *Torpedo* electric organ. The technique—osmotic shock followed by fractionation on a step gradient—was similar to that used for brain synaptosomes.[122] Again, considerable differences in composition were noted between electromotor synaptic vesicles and presynaptic plasma membranes. The polypeptide patterns of both membranes were simpler than those of the mixture of synaptosomes of diverse provenance from mammalian brain.

4.3. Separation of Postsynaptic Membranes

As mentioned in Section 2.3.2b, many if not all brain synaptosomes have lengths of postsynaptic membranes adhering to their peripheries. This has been recognized since synaptosomes were first isolated,[2,3] and, indeed, such membranes were one of the four main morphological features (the other three being the synaptic vesicles, intraterminal mitochondria, and presynaptic plasma membrane) that led to the identification of the synaptosome as a pinched-off nerve terminal. Thus, there is no justification for distinguishing synaptosomes without postsynaptic adhesions as "synaptosomes" and those with such adhesions as "synaptic complexes" as some authors have attempted to do.[139]

The postsynaptic adhesions are readily recognized by their heavily stained appearance in conventional electron microscopy. This is particularly obvious if preparations are poststained with phosphotungstic acid, when some cleft material also stains.[140] Most methods for isolating postsynaptic membranes or postsynaptic densities as they are now usually called utilize dilute Triton X-100, which solubilizes the lipids in synaptosome and synaptic vesicle membranes and causes them to disintegrate without extracting much protein (so it is thought) from the postsynaptic densities. These may then be further purified on a step gradient.[141–146]

These densities, when analyzed by polyacrylamide gel electrophoresis in sodium dodecyl sulfate, contain numerous proteins including those recognized, by molecular mass and immunochemically, as α- and β-tubulin, actin, intermediate filament proteins, and "synapsin I." There is now reason to believe that the main protein of postsynaptic densities isolated from cortex has a molecular weight of 50,000 and is not to be identified with any of the proteins already mentioned; a corresponding protein from cerebellum has M_r 76,000.[147] Moreover, the immunochemical identification of glial intermediate filament protein as a constituent of isolated postsynaptic densities but not, of course, of these densities *in situ* has suggested that most if not all of the cytoskeletal proteins present in isolated postsynaptic densities (but not the 50- or 76-kilodalton components) are adsorbed onto them during isolation.[147] This redistribution of proteins during subcellular fractionation is an ever-present problem when one attempts to define the protein composition of subcellular fractions and was first encountered with nervous tissue when species variations were reported in the subcellular distribution of choline acetyltransferase (see Sections 2.3.1 and 5.2.1).

5. THE SYNAPTOSOME AS A FUNCTIONAL ENTITY

5.1. The Synaptosome as a Miniature Cell

5.1.1. Techniques

The synaptosome, as a pinched-off cell process, is bounded by a sealed plasma membrane which encloses a cytoplasm containing a full complement of glycolytic enzymes and, usually, one or more small mitochondria. It is therefore in effect a miniature, nonnucleated cell. When warmed and provided with oxidizable substrates, synaptosomes actively respire, synthesize ATP and creatine phosphate,[148] take up K^+,[148–150] extrude Na^+,[151] and also take up large numbers of other substances (amino acids,[152–157] biogenic amines[19,158,159]) by saturable carrier-mediated, Na^+-dependent processes. As already mentioned (Section 2.2.3), such activity is accompanied by a marked improvement in synaptic morphology, but which if any, of the metabolic processes specifically bring about such changes is unknown.[160]

The high local metabolic requirements of the nerve terminal are reflected by the active glucose uptake by synaptosomes. This is inhibited by 2-deoxyglucose, which is itself taken up, however, by the glucose carrier with a K_m of 0.24 mM; *3-O*-methylglucose and phloretin inhibit this uptake with K_is of 0.65 mM and 0.75 μM, respectively.[18,161]

Two types of synaptosome preparation have been used to study metabolic activities, the suspension and the superfused pellet or "synaptosome bed."[21] In suspension, synaptosomes autolyse rapidly at 37°C and are best studied at 20–30°C. This preparation is useful for studying ion movements,[14,148–151,160] the uptake of amines,[19,158,159] amino acids,[152–157] and oxidizable substrates,[148] and the behavior of fluorescent probes.[15] In sterile tissue culture media, function may survive up to 5 days.[162] Uptake is usually terminated by collecting the synaptosomes on a Millipore® filter and washing briefly to remove "extracellular" metabolites, but rapid sedimentation[163] (e.g., in an air-driven miniature ultracentrifuge such as the Beckman Airfuge®) or filtration through a column of Sephadex G-50[29] or similar material may also be used. The last technique has proved particularly useful in studies of passive permeability (Section 5.1.3).

Synaptosome beds[21] are usually prepared as a sandwich between protecting layers to avoid damage to the outer layer of synaptosomes during superfusion. Such beds may be treated very much like tissue slices, which must, indeed, owe much of their metabolism to the nerve terminals contained within them, since these, with their high mitochondrial content, are regions of intense metabolic activity. Thus, synaptosome beds can be electrically stimulated in suitably designed vessels.[21,164] They are particularly useful for studying transmitter release (Section 5.2.2) and the effects of drugs on this process, since reuptake of transmitter (a complication with the amino acids, catecholamines, and 5-hydroxytryptamine) is minimized by its prompt removal by superfusion.[165,166] As compared to tissue slices, other advantages are the absence of neuronal interactions and of the diffusion barriers imposed by the structure of the tissue.

Membrane transport functions may additionally be studied in vesiculated fragments of synaptosomal plasma membranes derived from synaptosomes by hypoosmotic shock and density gradient centrifugation followed by sonication in the presence of materials it is wished to seal up inside the membrane vesicles.[167,168] Usually, this is K^+. The effect of imposed ionic gradients, ionophores, and analogues on the kinetics of uptake can then be studied by transfer of these vesicles to suitable media.

5.1.2. Metabolic Properties

As might be expected from their cytoplasmic content, brain synaptosomes respire much more effectively than mitochondria with glucose as a substrate (Table II, cols. 4–6); the optimum substrate is, however, an equimolar mixture of glucose and succinate. Glucose (Table II, col. 3), succinate, and malate alone give rates only about half that obtained with glucose and succinate together; pyruvate, α-oxoglutarate, and glutamate are relatively poor substrates. Under optimum conditions, the oxygen uptake is from one-third to one-half that of cortex slices and about half of that of whole brain on a protein basis (Table II,

Table II
Glucose and Succinate by Brain Preparations under Varying Conditions

Preparation	Respiration ($\mu l/O_2 \cdot min^{-1} \cdot mg\ protein^{-1}$)	Rate (% of control)	Rate (% of total[e] respiration) in: Glucose	Glucose + succinate	Succinate
Whole brain[a]	400–500	—	—	—	—
Slices[b,c]	600	—	—	—	—
Synaptosomes[b,d]	225 ± 52 (S.D.) (20)	—	76 ± 2	43 ± 16	56 ± 13
Mitochondria[b,d]	—	—	11 ± 5	44 ± 14	35 ± 13
Myelin[b,d]	—	—	13 ± 3	13 ± 5	8 ± 2
Synaptosomes[f]	—	100	—	—	—
No succinate	—	53	—	—	—
No succinate or glucose	—	18	—	—	—
Plus 100 mM K^+	—	121	—	—	—
Plus 100 mM K^+, 6 mM Ca^{2+}	—	77	—	—	—

[a] Cat, assuming 100 mg protein per gram of wet tissue.
[b] Guinea pig cortex.
[c] In Krebs–Ringer solution containing glucose, pyruvate, and fumarate.
[d] The incubation medium for synaptosomes contained 100 mM NaCl, 1.0 mM P, 10 mM $MgCl_2$, 10 mM glucose, 10 mM succinate, and Tris buffer, pH 7.4, together with catalytic amounts of ADP, CoA, NAD, NADP, glutathione, lipoic acid, pyridoxal, and thiamine. Measurements were made at 30°C.
[e] Total respiration is the sum of those of the three components indicated. Values are means ± S.D. of three determinations.
[f] Control as in *d*; others with modifications indicated.

col. 2). The respiratory rate is relatively insensitive to Ca^{2+} or Mg^{2+}. Inclusion of 100 mM Na^+ or K^+ caused a small increase in respiration (about 20% in the case of K^+); this is smaller than with slices (40–60%); however, as with slices, the stimulatory effect of K^+ was abolished by Ca^{2+} (Table II, col. 3). The concentration of high-energy phosphates (ATP and creatine phosphate) rises during incubation in the presence of substrates.

5.1.3. The Passive Permeability of the Synaptosomal Plasma Membrane: Osmotic Properties and Evidence for Sealing

In an early study,[29] synaptosomes were transferred to solutions containing K^+ or other ions and equilibrated for up to 16 h at 0–4°C, then passed quickly through a column of Sephadex G-50 gel, and the distribution of K^+ (or other ion) monitored in successive fractions of effluent (Fig. 5a). Part of the K^+ appeared in the void volume, the rest in the nonoccluded volume. Since the concentration of K^+ inside and out at equilibrium must be equal, the amount of K^+ in the void volume divided by the external K^+ concentration gives the occluded volume occupied by K^+. This is about 30μl/g of tissue, in close agreement with the morphological data. The Na^+ and D-[^{14}C]galactose spaces were

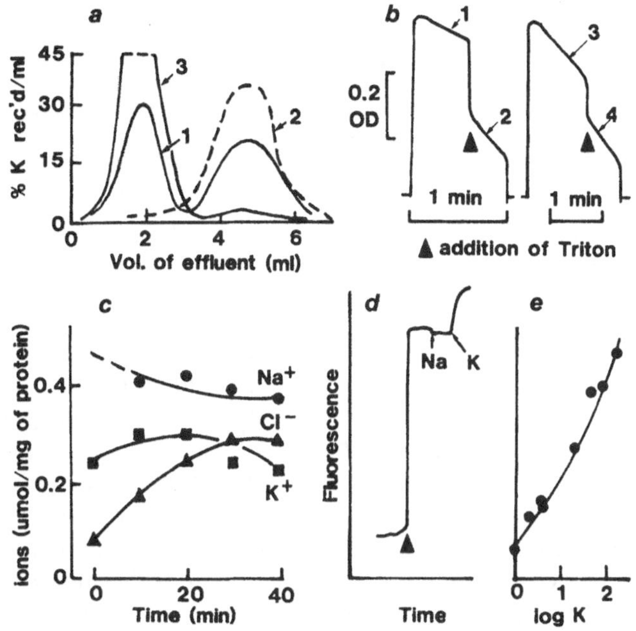

Fig. 5. Evidence that synaptosomes are sealed structures. a: Filtration of synaptosomes through a Sephadex G-50 gel column equilibrated with an isosmotic medium (1) or a strongly hypoosmotic medium (2). Curve 3 is the distribution of a void-volume marker.[13,29] Note the disappearance of the void-volume (synaptosomal) K^+ in 2. b: Occluded state of lactate dehydrogenase under isosmotic assay conditions (1) and after deocclusion under hypoosmotic assay conditions (3) or after treatment with 1% Triton X-100 (2,4); the slope of the tracing shows the rate at which $NADH_2$ is oxidized by pyruvate (decrease in E_{340} with time), and this increases with deocclusion.[13,29] c: Accumulation of K^+ and Cl^- and extrusion of Na^+ by synaptosomes incubated under metabolizing conditions as a function of time.[14] Under steady-state conditions (30 min), the membrane potential calculated from the Goldman equation is -27 ± 3 (S.E.M.; $n = 5$) mV. d,e: Detection of synaptosomal membrane potential by means of a fluorescent probe.[15] Addition of synaptosomes (arrow) to the fluorochrome 3,3'-dipentyl-2,2'-oxacarbocyanine caused (d) an increase in fluorescence attributable to a membrane potential. Reversible changes in fluorescence (d,e) induced by K^+ but not by Na^+ show that the membrane potential primarily reflects a potassium diffusion potential.

close to the K^+ space; by contrast, the space measured by a nonpenetrating macromolecule, ^{131}I-albumin, was only 5% of the K^+ space. In other studies, gel filtration was carried out at varying time intervals, and the rate of diffusion of K^+ and other substrates was measured (Table III). These were as expected from what is known of the properties of neuronal membranes.

In a more recent study, the short-term passive permeability of the synaptosomal plasma membrane to a variety of substances was measured by testing their ability to prevent the osmotic swelling of synaptosomes as detected by changes in light scattering.[169] In this way it was shown that the plasma membrane is impermeable (in the short term) to Na^+, K^+, Ca^{2+}, Mg^{2+}, $SO_4{}^{2-}$, $PO_4{}^{3-}$, and oxalate but permeable to $NH_4{}^+$ and acetate and also to small organic molecules in the expected order: glycerol < thiourea = formamide < propylene oxide = dimethylsulfoxide.

Table III
Permeability of the Synaptosomal Plasma Membrane[29]

Substance	Permeability (cm/sec) × 10^9	Volume occupied (% of K^+ volume)
K^+	10.1	1.00
Na^+	6.1	1.07
[^{14}C]galactose	26.0	0.99
^{131}I-albumin	—	0.05

The ability of synaptosomes to behave as osmometers and to respond by swelling and shrinking, at any rate over a limited range, to changes in external osmotic pressure is clear evidence that they are sealed structures. Further evidence is provided by the fact that the activity of lactate dehydrogenase and other enzymes, when measured under conditions that respect synaptosomal integrity, are "occluded"; i.e., their activity is only a small fraction of that recorded when the synaptosomal plasma membrane is disrupted by detergents or hypoosmotic shock (Fig. 5b).[29] For this reason, detergent (e.g., Triton X-100, Nonidex) is routinely added when synaptosomal enzymes are measured.

5.1.4. *Ion Movements and the Resting Membrane Potential of Synaptosomes*

Freshly prepared synaptosomes have a relatively high K^+ and relatively low Na^+ content. On incubation in a suitable medium (e.g., Krebs–Henseleit buffer), Na^+ is extruded,[151] and K^+ accumulated[148–150]; the Na^+ and K^+ concentrations of the synaptosomal cytoplasm assume values closer to those of intact neurons (Fig. 5c).[14] The observed ion gradients imply the existence of a resting membrane potential; calculations show that this must be close to -30 mV. Recently, the synaptosomal membrane potential has been estimated from the uptake of triphenylphosphonium and Rb^+ ions measured simultaneously in respiring synaptosomes.[170] A value of -45 ± 2.4 (S.D.) mV was obtained.

Another way in which a synaptosomal membrane potential has been detected is by experiments with fluorescent probes that report such potentials (Fig. 5d).[15] The probe also shows that elevated K^+, or Na^+ plus veratridine, causes depolarization. A value for the resting potential of -55 to -60 mV was deduced. Synaptosomes also take up Ca^{2+} by several mechanisms, the main one of which is stimulated by K^+.[171,172]

5.1.5. *Can Synaptosomes Synthesize Lipids and Proteins?*

To what extent are nerve terminals dependent for their maintenance on the cell body to which they are functionally linked by axonal transport, and to what extent are they autonomous? The metabolic versatility of synaptosomes noted in the previous sections suggests that these questions can be answered

Table IV
Properties of Systems Incorporating Labeled Amino Acids into Protein in Synaptosome and Microsome Fractions of Brain[12]

Condition	Incorporation by	
	Synaptosomes	Microsomes
RNase treatment	stable	destroyed
Na^+	required	not required
Ouabain	inhibits	does not inhibit
hypoosmotic shock	inhibits irreversibly	unaffected
hyperosmotic media	inhibit reversibly	unaffected
ATP	not required	required
Cyclohexamide	75% inhibition	100% inhibition
Chloramphenicol	25% inhibition	unaffected

by investigating the synthetic abilities of synaptosome preparations *in vitro*. However, it must be admitted that synaptosomes may lose cofactors or intermediates on incubation and that the contamination with other organelles found in even the purest synaptosome preparation may also be a problem when assessing the synaptosome's anabolic ability.

Morphologically, nerve terminals and synaptosomes appear to be devoid of rough endoplasmic reticulum. However, when nerve terminals in amphibian muscle are depleted of vesicles by means of black widow spider venom, small amounts of smooth reticulum remain.[173] Further, synaptosome profiles nearly always contain one or more large vesicles distinct from synaptic vesicles, which have been variously interpreted as endoplasmic reticulum[174] or lysosomes,[175] and in a careful study, residual amounts of endoplasmic reticulum marker enzymes were found in synaptosome preparations that could not be accounted for by microsomal contamination.[176] On the other hand, occasional profiles are seen in synaptosome preparations that cannot be identified as synaptosomes and appear to consist of pinched-off bags of rough-surfaced endoplasmic reticulum.[177]

If synaptosomes are prepared at varying time intervals after the injection of labeled leucine into the brain, the rate at which synaptosomal protein is labeled is consistent with its acquisition by axonal transport.[24] Synaptosome preparations incorporate radioactive amino acids into protein *in vitro*; this incorporation is small compared with that shown by whole-brain homogenates or microsomes but clearly does not represent microsomal contamination.[178–181] First, it has totally different properties (Table IV), consistent with the notion that incorporation is taking place within a sealed membrane, which protects the protein synthesizing system from RNase, through which labeled amino acids must be transported, and within which ATP must be generated by metabolism. Secondly, if microsomes containing tritium-labeled protein are added to synaptosome preparations whose proteins have been labeled with ^{14}C, and fractionation is carried out using osmotic shock as shown in Fig. 3, the labeled synaptosomal proteins are mainly recovered in the soluble cytoplasmic (O) and intraterminal mitochondrial (I) fractions, whereas the labeled microsomal protein is recovered mainly in fraction E. This nonmicrosomal protein synthesis is partly by a cycloheximide-sensitive 80 S cytoribosomal system and partly by a chloramphenicol-sensitive 70 S mitoribosomal system; after isolation, intraterminal mitochondria retain their protein-synthesizing ability, but this is now fully chloramphenicol sensitive.

All of these properties of the synaptosomal system could be accounted for by the cytomembrane-containing bags as well as by synaptosomes. Autoradiography, in fact, shows considerable accumulation of grains over such profiles.[181] There are some grains over synaptosome profiles, but these are not reduced in puromycin-treated controls, suggesting that the incorporation is not via a normal protein-synthesizing system.

To sum up, it is probably fair to say that most if not all of the observed small incorporation of radioactive amino acids into protein in synaptosome preparations represents contaminating structures other than synaptosomes. Some incorporation into intraterminal mitochondrial protein may, however, occur.[180] There is certainly very little evidence for local protein synthesis at the terminal, and it seems likely that the needs of the terminals are almost entirely provided for by axonal transport.

Similar conclusions may apply to lipid synthesis. Synaptosome preparations are able to convert choline to phosphorylcholine[182] and to incorporate [*N-Me*-^{14}C]choline into phospholipids[183]; however, the latter reaction is via a Ca^{2+}-dependent non-energy-requiring base-exchange reaction. Fractionation of synaptosomes labeled in this way showed that the label was chiefly in mitochondrial phospholipid. Synaptosome preparations could synthesize phosphatidylcholine slowly from CDP-[*Me*-^{14}C]choline; this was a property of the synaptic vesicles and synaptosomal endoplasmic reticulum and not of the external membranes or intraterminal mitochondria.

Since microsomes have the capacity both to synthesize labeled phosphatidylcholine from CDP-[*Me*-^{14}C]choline and to carry out the base-exchange reaction,[183] it is possible that the synaptosomal contaminants consisting of bags of endoplasmic reticulum are responsible for much of the incorporation of radioactive precursors in synaptosome preparations not accounted for by microsomal contamination.

5.2. The Synaptosome as a Surviving Presynaptic Nerve Terminal

The excellent metabolic performance and state of preservation of synaptosomes incubated in a suitable medium suggest that their synaptic function should also remain intact. To a considerable extent this seems to be true, though it is doubtful to what extent synaptosomes can perform vesicle recycling.[22]

5.2.1. Compartmentation and Recycling of Transmitters

Acetylcholine is synthesized from choline and acetylcoenzyme A, and this reaction is catalyzed by the enzyme choline acetyltransferase. Subcellular distribution studies[52] show that both acetylcholine and choline acetyltransferase

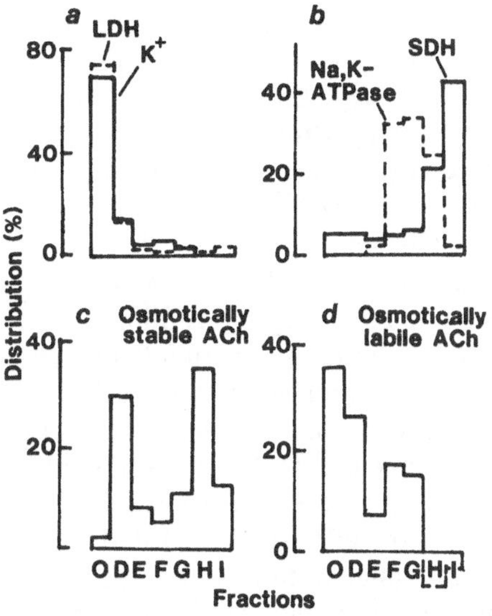

Fig. 6. Characterization of synaptosomal subfractions.[13] For nomenclature, see text and Fig. 3. Abbreviations: Na,K-ATPase, sodium- and potassium-activated adenosine triphosphatase; SDH, succinate dehydrogenase; other abbreviations as in legends to Figs. 2 and 4. Note that osmotically labile ACh (d) is predominantly recovered in the cytoplasmic fraction O, in which the cytoplasmic markers LDH and K^+ are also mainly recovered (a); also note that osmotically stable ACh (c) is bimodally distributed in the two vesicle fractions D and H. The recovery of appreciable amounts of the plasma membrane marker Na^+,K^+-ATPase and of the mitochondrial marker SDH in fraction H together with the morphological evidence shows that the vesicles of fraction H are associated with the plasma membranes and mitochondria of disrupted synaptosomes.

are mainly localized in synaptosomes, although in brain homogenates, about 30% of the enzyme is recovered in free, nonoccluded form in the high-speed supernatant (cell sap fraction). This statement is also true for acetylcholine provided the cholinesterase of the homogenate is blocked by a suitable inhibitor (e.g., 10 μM eserine). Since transmitters and the enzymes that synthesize them are known to be distributed throughout the neuron (although they occur in highest concentrations in the terminals), the supernatant fraction of choline acetyltransferase and acetylcholine is likely to represent mainly the perikaryal and axonal fractions of these components, although some may be contributed by defective (i.e., leaky) synaptosomes. The synaptosomal acetylcholine is the classical "bound" acetylcholine of brain minces, and synaptosome formation adequately explains the phenomenon of bound transmitter. This fraction of tissue acetylcholine escapes hydrolysis by cholinesterases because it is sealed inside synaptosomes and so is protected. A similar fraction of ATP can also be detected.[184]

Analysis of the subfractions of lysed synaptosomes (Section 4.2.2) shows that choline acetyltransferase is a cytoplasmic enzyme since, as we have seen, particulate transferase results from its artifactual adsorption onto membranes at low ionic strength and pH. Acetylcholine is thus synthesized in the cytoplasm, and when radioactive choline is added to a respiring synaptosome preparation, this is the first fraction to become labeled.[134,185] Osmotic lysis of synaptosomes releases this fraction, and for this reason it has been designated the "labile-bound" fraction. However, it cannot be excluded that part of the labile-bound fraction is derived from leakage from vesicles. In the absence of an anticholinesterase, the released labile-bound acetylcholine is destroyed by the cholinesterases of the preparation, but in its presence, labile-bound acetylcholine is recovered mainly in the cytoplasmic (O) fraction (Fig. 6).

The analyses further show that the stable-bound fraction of acetylcholine is to be identified with vesicle-bound acetylcholine (Fig. 6) and, as previously mentioned, is associated with two fractions, D and H, in our original fractionation scheme, consisting of monodispersed and membrane-associated vesicles, respectively. Labeling experiments *in vivo* both with labeled choline and with false transmitter precursors show that included among the membrane-associated vesicles are the actively recycling vesicles, and the monodispersed vesicles represent the reserve population.[134,135] In experiments in which synaptosomes are labeled *in vitro*, the distinction between these two vesicle pools is more difficult to establish, probably because of the lack of transmitter turnover and vesicle emptying and refilling that accompanies the impulse traffic in the living awake animal. In unstimulated synaptosomes, the contrast between the readily labeled cytoplasmic and poorly labeled vesicular pools is particularly striking.

Metabolizing synaptosomes take up choline by a carrier-mediated uptake mechanism.[20,186] Work with highly cholinergic synaptosome preparations from squid optic lobes[20] revealed two uptake systems, a high-affinity ($K_t \sim 1$–$2\ \mu$M) and a low-affinity ($K_t \sim 1$ mM) system. The high-affinity system was also found in mammalian brain synaptosomes[186] but was less easy to spot because of the much lower concentration of cholinergic neurons. The localization of the high-affinity system in synaptosomes derived from cholinergic neurons is evident from degeneration studies in which sectioning of the cholinergic input to a specific brain region (sectioning of the fimbrial input to the hippocampus) destroyed the capacity of the synaptosomes subsequently isolated from this region to take up choline by the high-affinity route[187] and from the similar Na^+ dependence of the high-affinity choline uptake system and of choline acetylation by synaptosomes.[188,189]

One peculiarity of acetylcholine as a transmitter is the rapidity with which its synaptic action is terminated by hydrolysis. Acetylcholinesterase, the enzyme catalyzing this, exists in numerous forms, some of which are generated by the postsynaptic (cholinoceptive) cell and others by the presynaptic (cholinergic) neuron. The contribution of the two parts of the synapse varies with different cholinergic synapses. Thus, in muscle, the postsynaptic (muscle) cell contributes a considerable amount.[190] In the CNS, it has been plausibly argued that the main contribution is made by the cholinergic cell,[191] and acetylcholinesterase is widely used as a cholinergic marker.[192]

Among subcellular fractions, acetylcholinesterase is found in highest specific concentration in the microsome fraction[193] and in greatest amount in the synaptosome fraction.[52] There is no evidence in any fraction for occlusion of the enzyme. These findings are consistent with the electron-microscopic histochemical evidence that acetylcholinesterase is present within the lumen of the perikaryal and axonal endoplasmic reticulum and in high concentration around the outside of the presynaptic plasma membrane.[194] Other evidence suggests that the enzyme is conveyed from the cell body by axonal transport to the terminal, where it is secreted onto the outer surface of the terminal.[194,195] Thus, functionally, acetylcholinesterase is always external to the cytoplasm. Released acetylcholine is rapidly hydrolyzed and is salvaged as choline and acetate.

Our conclusions concerning the acetylcholine pools in cholinergic endings and the manner of their recycling have been greatly strengthened by work with

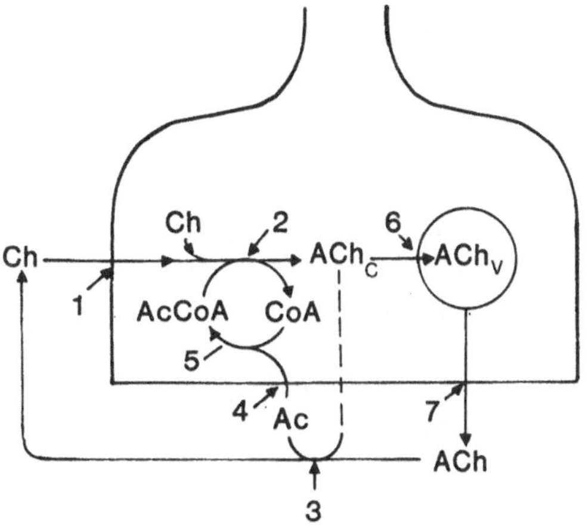

Fig. 7. Pools of acetylcholine in the cholinergic electromotor nerve terminal of *Torpedo marmorata*. Cytoplasmic acetylcholine (ACh_c) is constantly recycling (1–5) and becomes rapidly labeled; it accounts for 20% of the total ACh at rest. The remaining 80% is vesicular (ACh_v) and does not readily acquire ACh_c (6) unless vesicle emptying is induced by stimulation (7). Recycling vesicles rapidly become labeled at the expense of the cytoplasmic pool and can be separated physically from the reserve population because of their lower water content, smaller size, and greater density. Control may be exerted via an inhibitory effect of ACh_c on choline uptake (2); this is reduced when ACh_c falls (as a result of vesicular uptake) on stimulation. Reaction 2 is catalyzed by choline acetyltransferase, an exclusively cytoplasmic enzyme, and reaction 3 by acetylcholinesterase. The synthesis of acetylcoenzyme A (AcCoA) from coenzyme A (CoA) (5) is a complex reaction involving mitochondria and endogenous sources of acetyl groups as well as salvaged acetate.

a purely cholinergic terminal, the electromotor terminal, and are summarized in Fig. 7. In *Torpedo*, both types of vesicle can be extracted from frozen and crushed tissue without first isolating synaptosomes, and the two types can be separated by density-gradient centrifugation.[196,197] Further details are given in Chapter 2 on synaptic vesicles.

The compartmentation and metabolism of catecholamine and amino acid transmitter candidates have also been studied by means of synaptosome preparations. Uptake systems exist for norepinephrine, dopamine, glutamate, γ-aminobutyrate, and glycine with fairly broad specificities.[198,199] Although catecholamines can be inactivated by O-methylation, under normal conditions they are thought to be inactivated by being rapidly taken up by the presynaptic nerve terminal. As with acetylcholine, cytoplasmic and vesicular pools of catecholamines exist.

5.2.2. *Vesicle Recycling and Transmitter Release*

Vesicle recycling has been extensively investigated morphologically in nerve terminals *in situ*.[200–210] Stimulation under conditions in which exocytosis is stimulated and/or endocytosis is blocked by toxins, cold, or strong stimulation causes vesicle depletion as well as massive transmitter release.[201,205,206] In cold-blocked or excessively stimulated preparations, lifting of the block[206] or discontinuation of the stimulus[201,207–209] restores vesicle numbers. Under conditions of vesicle depletion, the increase in the area of the plasma membrane and any invaginations attached to it correspond to the decrease in the area of vesicle membrane.[200,210] The recycling of vesicles can be followed by means of electron-dense markers such as horseradish peroxidase[22,200,201] or dextran particles,[196,202] which enter the vesicle lumen during exocytosis and are trapped when the vesicles reform. Under conditions of rapid freezing and subsequent fracture to expose the outer surface of the presynaptic plasma membrane in the cleft region, exocytotic pits are observed, which increase in number with the amount of transmitter release.[203,204]

The question now arises: can synaptosomes release transmitter on stimulation, and if so is vesicle recycling taking place?

Synaptosomes can indeed be stimulated by high K^+ concentrations or electrically to release transmitters, and this release is largely Ca^{2+} sensitive.[211,212] Usually superfused synaptosome beds are used. However, only a small proportion of total transmitter is releasable. Horseradish peroxidase is taken up.[22] Freeze–etch studies with *Torpedo* synaptosomes have revealed characteristic subunits on the external surface of the plasma membranes, which increase in number as stimulation and transmitter release continue; these may represent imperfectly preserved exocytotic pits.[101]

Thus, it seems clear that synaptosomes retain the basic mechanism for stimulus–secretion coupling, but it is not clear how far vesicles within synaptosomes can be reconstituted in a functional state, restocked with transmitter, and recycled.

5.2.3. Synaptosomes as a Source of New Transmitters

Just how many transmitters are utilized by the mammalian brain is not known; the list of putative transmitters has greatly lengthened to over 20 with the discovery of many pharmacologically active peptides in brain including some long known from nonnervous sources. Subcellular fractionation studies have shown, as we have seen, that most putative central transmitters are present in two pools in brain homogenates, a free pool, which is recovered on fractionation in the high-speed supernatant (cell-sap) fraction, and a bound pool, which is recovered in synaptosomes. Localization of bound putative transmitter in synaptosomes should be a good test for transmitter function. We can therefore expect that synaptosomes contain numerous unidentified transmitters; we cannot, however, expect that there will always be a convenient test system available for them based on receptors in peripheral neuromuscular preparations.

It should be possible, as a means of identifying new central transmitters, to isolate specific types of nerve endings as synaptosomes from different parts of the CNS and apply extracts of them to neurons with which they normally make synaptic contact. The electrical response of the neuron would then serve as the detection and assay system. This approach has been shown to be feasible; extracts of cortical synaptosomes applied to cortical pyramidal cells evoked responses that could be accounted for by the glutamate and γ-aminobutyrate content of the synaptosome extracts,[23] but this approach has not been developed further.

5.2.4. Analysis of the Synaptic Actions of Drugs and Toxins

Many centrally acting drugs and toxins are believed to exert their pharmacological effect by interfering, in one way or another, with transmitter uptake, storage, release, and postsynaptic action. The various types of action can be studied very conveniently with synaptosome preparations. Synaptosomes provide a powerful tool both for identifying the type of synaptic effect exerted by a drug and for screening drugs of a given type and assessing their potency and potential usefulness. Drugs may thus:

1. Block uptake of transmitters or transmitter precursors.
2. Enter the terminal and block vesicular uptake.
3. Enter the terminal and vesicle and act as a false transmitter; if the postsynaptic action of the false transmitter is small, synaptic blockade may ensue.
4. Block stimulus–secretion coupling.
5. Interact with presynaptic receptors, usually to inhibit synaptic activation.
6. Interact with postsynaptic receptors either as agonists or as blocking agents.

Binding of ligands to pre- and postsynaptic receptors is usually studied by means of Scatchard plots, which are very conveniently constructed with intact or lysed synaptosomes. An intriguing recent example is the discovery of a peptide in the urine of schizophrenic patients that powerfully blocks dopamine uptake into synaptosomes and induces a behavioral syndrome in rats with features similar to that of schizophrenics.[213]

6. *SOME LIMITATIONS AND FUTURE DEVELOPMENTS IN SYNAPTOSOME TECHNOLOGY*

Synaptosomes are such versatile structures, show such good preservation of synaptic properties *in vitro*, and have so many applications that it is well to remember that they also have several limitations. First and foremost among these is the heterogeneous or mixed origin of brain synaptosomes, derived, as they are, from many different terminals utilizing many different transmitters. For some purposes this does not matter—much has, for example, been learned about the organization of the cholinergic nerve terminal from cortical synaptosomes of mixed provenance—and may, indeed, be a positive advantage as in the search for new transmitters. For other purposes—e.g., for the identification of proteins and other macromolecules specific for certain types of nerve terminal—it may be a severe disadvantage.

Synaptosomes have proved too heterogeneous in density to allow more than a slight enrichment in any one transmitter type to be achieved by density-gradient centrifugation. As far as cholinergic nerve terminals are concerned, the synaptosome preparation from *Torpedo* electromotor terminals provides a purely cholinergic preparation. For other transmitters, pure sources are not available. Moreover, the degree of purification of membrane markers reported is often not consistent with the estimated contribution of the synaptosomal membrane to total brain volume. One possible solution may be to use the monoclonal antibody technique to derive, from a heterogeneous population of synaptosomal plasma membranes, antibodies selective for surface antigens characteristic of particular subpopulations of synaptosomes. For screening, selective complement lysis and immunohistochemical techniques can be used. Such antibodies could then provide the basis for immunoaffinity separation of the selected population or for attaching a fluorescent "tag," which could be used to generate a signal in an electrostatic cell sorter. In a feasibility study, it was shown that polystyrene beads of the size range of synaptosomes could be separated into different size classes utilizing signals generated by light scattering.[214]

Counting synaptosomes by the only published method[56] is tedious and perhaps not very accurate. Attempts by the author to count synaptosomes by means of a Coulter counter failed, probably because of their small size (V. P. Whittaker, unpublished data). However, if a quick and accurate method could be devised, this might have considerable application in studies of normal and pathological synaptic development. A cell sorter can not only separate cells, it can also count them, and if specific synaptosomal subpopulations could be labeled with different fluorescent markers, the numbers of cholinergic, adrenergic, etc. synaptosomes could be estimated. Correlations could then be attempted with behavior; e.g., variations in the number of cholinergic terminals might correlate with maze performance.

Synaptosomes, as we have seen, may possess only a limited capacity for vesicle recycling. Nevertheless, the processes and macromolecules that bring about exocytosis and endocytosis must be present in synaptosomes, and a careful analysis of synaptosomal cytoplasm and the interaction of cytoplasmic components with vesicles may throw light on these processes. Specific questions can be posed and answered. Thus, synaptic vesicles do not seem to contain lysolecithin,[215] which has been thought to be a fusogen for exocytosing chromaffin granules, nor is there an increased turnover in the fatty acids of synaptosomal phospholipids on depolarization,[216] as would be expected if a small, hard to detect, but functionally important pool of lysolecithin were turning over during vesicle exocytosis. Modification of the cytoplasmic environment of synaptic vesicles while still *in situ* could be a useful way of studying cytoplasm–vesicle interaction in a relatively intact preparation.

Complement fixation utilizing antisera to surface antigens of erythrocytes that cross react with determinant groups on the synaptosome membrane have enabled changes in synaptosomal cytoplasmic Ca^{2+} concentration to be made without loss of macromolecular cytoplasmic constituents.[217] Possibly, the electrostatic technique for rendering membranes permeable to small molecules, as worked out for adrenal medullary cells,[218] could also be applied to synaptosomes.

These concluding remarks may serve to show that in spite of the large volume of work that has been done on the synaptosome and the mass of information that has been accumulated about the molecular organization of the synapse as a result of it, a considerable potential for further work with this preparation still remains.

ACKNOWLEDGMENTS. I am most grateful to Dr. F. Hajós for reading the manuscript, making many helpful suggestions, and providing me with numerous references. I am also grateful to Drs. G. Levi and M. Raiteri for providing preprints of their chapters and to Mrs. Irene Fried for typing the manuscript using the word-processing facility of our PDP-11 computer.

REFERENCES

1. Whittaker, V. P., 1963, *Regional Neurochemistry: The Regional Chemistry, Physiology and Pharmacology of the Nervous System* (S. S. Kety and J. Elkes, eds.), Pergamon Press, Oxford, pp. 259–263.
2. Gray, E. G., and Whittaker, V. P., 1960, *J. Physiol. (London)* **153**:35–37P.
3. Gray, E. G., and Whittaker, V. P., 1962, *J. Anat.* **96**:79–88.
4. Hebb, C. D., and Whittaker, V. P., 1958, *J. Physiol. (Lond.)* **142**:187–196.
5. Whittaker, V. P., 1959, *Biochem. J.* **72**:694–706.
6. de Duve, C., Pressman, B. C., Gianetto, R., Wattiaux, R., and Applemans, F., 1955, *Biochem. J.* **60**:604–617.
7. Novikoff, A. B., and Essner, E., 1962, *Fed. Proc.* **21**:1130–1142.
8. Whittaker, V. P., 1963, *Biochem. Soc. Symp.* **23**:109–126.
9. Whittaker, V. P., Michaelson, I. A., and Kirkland, R. J. A., 1964, *Biochem. J.* **90**:293–305.
10. De Robertis, E., de Iraldi, A. P., Arnaiz, G. R., de L., and Salganicoff, L., 1962, *J. Neurochem.* **9**:23–35.
11. Gray, E. G., and Whittaker, V. P., 1981, *Curr. Contents* **24**:16.
12. Whittaker, V. P., 1973, *Naturwissenschaften* **60**:281–289.
13. Whittaker, V. P., 1972, *Structure and Function of Synapses* (G. D. Pappas and D. P. Purpura, eds.), Raven Press, New York, pp. 87–100.
14. Campbell, C. W. B., 1976, *Brain Res.* **101**:594–599.
15. Blaustein, M. P., and Goldring, J. M., 1975, *J. Physiol. (Lond.)* **247**:589–615.
16. Gordon-Weeks, P. R., Burgoyne, R. D., and Gray, E. G., 1982, *Neuroscience* **3**:739–749.
17. Hajós, F., Csillag, A., and Kálmán, M., 1979, *Exp. Brain Res.* **35**:387–393.
18. Heaton, G. M., and Bachelard, H. J., 1973, *J. Neurochem.* **21**:1099–1108.
19. Bogdanski, D. F., Blaszkowski, T. P., and Tissari, A. H., 1970, *Biochim. Biophys. Acta* **211**:521–532.
20. Dowdall, M. J., and Simon, E. J., 1973, *J. Neurochem.* **21**:969–982.
21. de Belleroche, J. S., and Bradford, H. F., 1972, *J. Neurochem.* **19**:585–602.
22. Fried, R. C., and Blaustein, M. P., 1978, *J. Cell Biol.* **78**:685–700.
23. Krnjević, K., and Whittaker, V. P., 1965, *J. Physiol. (Lond.)* **197**:288–322.
24. Barondes, S. H., 1966, *J. Neurochem.* **13**:721–727.
25. Meldolesi, J., 1982, *J. Neurochem.* **38**:1559–1569.
26. Baba, A., and Cooper, J. R., 1980, *J. Neurochem.* **34**:1369–1379.
27. Yoshida, K., and Imura, H., 1979, *Brain Res.* **172**:453–459.
28. Terenius, L., 1973, *Acta Pharmacol. Toxicol. (Kbh.)* **32**:317–320.
29. Marchbanks, R. M., 1967, *Biochem. J.* **104**:148–157.
30. Israël, M., and Whittaker V. P., 1965, *Experientia* **21**:325–326.
31. Lemkey-Johnson, M., and Larramendi, L. M. H., 1968, *Exp. Brain Res.* **5**:326–340.
32. del Cerro, M. P., Snider, R. S., and Oster, M. L., 1969, *Exp. Brain Res.* **8**:311–320.
33. Laverty, R., Michaelson, I. A., Sharman, D. F., and Whittaker, V. P., 1963, *Br. J. Pharmacol.* **21**:482–490.
34. Ross, L. L., Andreoli, V. M., and Marchbanks, R. M., 1971, *Brain Res.* **25**:103–119.
35. Michaelson, I. A., Whittaker, V. P., Laverty, R., and Sharman, D. F., 1963, *Biochem. Pharmacol.* **12**:1450–1453.
36. LaBella, F. S., and Sanwal, M., 1965, *J. Cell Biol.* **25**:(3.2):179–194.
37. Weinstein, H., Roberts, E., and Kakefuda, T., 1963, *Biochem. Pharmacol.* **12**:503–509.
38. Whittaker, V. P., and Sheridan, M. N., 1965, *J. Neurochem.* **12**:363–372.
39. Chakrin, L. W., and Whittaker, V. P., 1969, *Biochem. J.* **113**:97–107.
40. Tuček, S., 1967, *J. Neurochem.* **14**:531–545.

41. Metzger, H. P., Cuénod, M., Grynbaum, A., and Waelsch, A., 1961, *J. Neurochem.* **14**:99–104.
42. Dodd, P. R., Hardy, J. A., Oakley, A. E., and Stroyf, A. J., 1981, *Brain Res.* **224**:419–425.
43. Oestreicher, A. B., and Leeuwen, C. van, 1975, *J. Neurochem.* **24**:257–259.
44. Abood, L. G., Kurahashi, K., and del Cerro, M., 1967, *Biochem, Biophys. Acta* **136**:521–532.
45. Newkirk, R. F., Ballou, E. W., Vickers, G., and Whittaker, V. P., 1976, *Brain Res.* **101**:103–111.
46. Breer, H., 1981, *Neurochem. Int.* **3**:155–163.
47. Dowdall, M. J., and Whittaker, V. P., 1973, *J. Neurochem.* **20**:921–935.
48. Jones, D. G., 1967, *J. Cell Sci.* **2**:573–586.
49. Florey, E., and Winesdorfer, F., 1968, *J. Neurochem.* **15**:169–177.
50. Simon, E. J., Whittaker, V. P., Meilman, H., Sher, H., Vickers, G., and Couch, E., 1976, *Neurochem. Res.* **1**:83–92.
51. Whittaker, V. P., and Greengard, P., 1971, *J. Neurochem.* **18**:173–176.
52. Whittaker, V. P., 1965, *Prog. Biophys. Mol. Biol.* **15**:39–96.
53. Whittaker, V. P., and Barker, L. A., 1972, *Methods of Neurochemistry*, Volume 2 (R. Fried, ed.), Marcel Dekker, New York, pp. 1–52.
54. Coakley, W. T., 1974, *Brain Res.* **70**:281–284.
55. Whittaker, V. P., and Dowe, G. H. C., 1965, *Biochem. Pharmacol.* **14**:194–196.
56. Clementi, F., Whittaker, V. P., and Sheridan, M. N., 1966, *Z. Zellforsch.* **72**:126–138.
57. Sheridan, M. N., Whittaker, V. P., and Israël, M., 1966, *Z. Zellforsch.* **74**:291–307.
58. Michaelson, I. A., and Dowe, G. H. C., 1963, *Biochem. Pharmacol.* **12**:949–956.
59. Lai, J. C. K., Walsh, J. M., Dennis, S. C., and Clark, J. B., 1977, *J. Neurochem.* **28**:625–631.
60. Wood, M. D., and Wyllie, M. G., 1981, *J. Neurochem.* **37**:705–797.
61. Leskawa, K. C., Yohe, H. C., Matsumoto, M., and Rosenberg, A., 1979, *Neurochem. Res.* **4**:483–504.
62. Dodd, P. R., Hardy, J. A., Oakley, A. E., Edwardson, J. A., Perry, E. K., and Delaunoy, J.-P., 1981, *Brain Res.* **226**:107–118.
63. Tamir, H., Rapport, M. M., and Roizin, L., 1974, *J. Neurochem.* **23**:943–949.
64. Abdel-Latif, A. A., 1966, *Biochim. Biophys. Acta* **121**:403–406.
65. Joó, F., and Karnushina, I., 1975, *J. Neurochem.* **24**:839–840.
66. Delaunoy, J.-P., Hog, F., DeFeudis, F. V., and Mandel, P., 1979, *J. Neurochem.* **33**:611–612.
67. Henn, F. A., Anderson, D. J., and Rustad, D. G., 1976, *Brain Res.* **101**:341–344.
68. Fonnum, F., 1968, *Biochem. J.* **106**:401–412.
69. Bretz, V., Baggiolini, M., Hauser, R., and Hodel, C., 1974, *J. Cell Biol.* **61**:466–480.
70. Laduron, P., 1977, *Int. Rev. Neurobiol.* **20**:251–281.
71. Kuhar, M. J., Green, A. J., Snyder, S. H., and Gfeller, E., 1970, *Brain Res.* **21**:405–417.
72. Lagercrantz, H., and Pertoft, H., 1972, *J. Neurochem.* **19**:811–823.
73. de Duve, C., Berthet, J., and Beaufay, H., 1959, *Prog. Biophys. Biophys. Chem.* **9**:325–369.
74. Baldessian, P. J., and Vogt, M., 1971, *J. Neurochem.* **18**:951–962.
75. Hannig, K., 1967, *Electrophoresis* (M. Bier, ed.), Academic Press, New York, pp. 423–471.
76. Ryan, K. J., Kalant, H., and Thomas, E. L., 1971, *J. Cell Biol.* **49**:235–246.
77. Hosie, R. J. A., 1965, *Biochem. J.* **96**:404–412.
78. Aldridge, W. N., and Johnson, M. K., 1959, *Biochem. J.* **73**:270–276.
79. Toschi, G., 1957, *Exp. Cell Res.* 232–235.
80. Koenig, H., Gaines, D., McDonald, T., Gray, R., and Scott, J., 1965, *J. Neurochem.* **11**:729–743.
81. Koelle, G. B., 1954, *J. Comp. Neurol.* **100**:211–228.
82. Bülbring, E., Philpot, S. J., and Bosquanet, F. D., 1953, *Lancet* **1**:865–866.
83. Cavanagh, J. B., Thompson, R. H. S., and Webster, G. R., 1954, *Q. J. Exp. Physiol.* **39**:185–197.
84. Eichberg, J., Whittaker, V. P., and Dawson, R. M. C., 1964, *Biochem. J.* **92**:91–100.
85. Whittaker, V. P., 1966, *Ann. N.Y. Acad. Sci.* **137**:982–998.

86. Fonnum, F., 1967, *Biochem. J.* **103**:262–270.
87. Fonnum, F., 1968, *Biochem. J.* **109**:389–398.
88. Johnson, M. K., and Whittaker, V. P., 1963, *Biochem. J.* **88**:404–409.
89. Tanaka, R., and Abood, L. G., 1963, *J. Neurochem.* **10**:571–576.
90. Stahl, W. L., Smith, J. C., Napolitano, L. M., and Basford, R. E., 1963, *J. Cell Biol.* **19**:293–307.
91. Skou, J. C., 1960, *Biochim. Biophys. Acta* **42**:6–23.
92. Harwood, J. L., and Hawthorne, J. N., 1969, *J. Neurochem.* **16**:1377–1387.
93. Ryall, R. W., 1964, *J. Neurochem.* **11**:131–145.
94. Mangan, J. L., and Whittaker, V. P., 1966, *Biochem. J.* **98**:128–131.
95. Jones, D. G., and Revell, E., 1970, *Z. Zellforsch.* **111**:195–208.
96. Whittaker, V. P., 1962, *Discovery* **23**:7–13.
97. Petruschka, E., and Giuditta, D., 1959, *J. Biochem. Biophys. Cytol.* **6**:129–132.
98. Horne, R. W., and Whittaker, V. P., 1961, *Z. Zellforsch.* **58**:1–16.
99. Whittaker, V. P., 1965, *Pharmacol. Rev.* **18**:401–412.
100. Heaton, G. M., and Bachelard, H. S., 1974, *J. Neurochem.* **22**:561–564.
101. Israël, M., Manaranche, R., Morel, N., Dedieu, J. C., Gulik-Krzywicki, T., and Lesbats, B., 1981, *J. Ultrastruct. Res.* **75**:162–178.
102. Whittaker, V. P., 1968, *Biochem. J.* **106**:412–417.
103. Baudhuin, P., and Berthet, J., 1967, *J. Cell Biol.* **35**:631–648.
104. Cotman, C. W., and Flansburg, D. A., 1970, *Brain Res.* **22**:152–156.
105. Grove III, W. E., Bondareff, W., and Veis, A., 1973, *J. Neurochem* **21**:703–704.
106. Richardson, P. J., Walker, J. H., Jones, R. T., and Whittaker, V. P., 1982, *J. Neurochem.* **38**:1605–1614.
107. Jones, D. G., and Brearley, R. F., 1973, *Z. Zellforsch.* **140**:481–496.
108. Whittaker, V. P., 1968, *Structure and Function of Inhibitory Neuronal Mechanisms* (C. von Euler, S. Skoglund, and U. Söderberg, eds.), Pergamon Press, Oxford, pp. 487–504.
109. Balázs, R., Hajós, F., Johnson, A. L., Reynierse, G. L. A., Tapia, R., and Wilkin, G. P., 1975, *Brain Res.* **86**:17–30.
110. Kornguth, S., Fuhl, U., Knoblock, L., Sutherland, L., Johnson, T., and Scott, G., 1976, *Brain Res.* **105**:423–435.
111. Quinn, M. R., and Cagan, R. H., 1980, *J. Neurochem.* **35**:583–590.
112. Barer, R., Heller, H., and Lederis, K., 1963, *Proc. R. Soc. Lond.* [*Biol.*] **158**:388–416.
113. Nordmann, J. J., Louis, F., and Morris, S. J., 1979, *Neuroscience* **4**:1367–1379.
114. Atterwill, C. K., and Neal, M. J., 1976, *J. Neurochem.* **27**:529–537.
115. Thomas, T. N., and Redburn, D. A., 1978, *J. Neurochem.* **31**:63–68.
116. Giacobini, E., Hökfelt, T., Kerpel-Fronius, S., Koslow, S. H., Mitchard, M., and Noré, B., 1971, *J. Neurochem.* **18**:223–231.
117. Wilson, W. S., and Cooper, J. R., 1972, *J. Neurochem.* **19**:2779–2790.
118. Jonakait, G. M., Gintzler, A. R., and Gershon, M. D., 1979, *J. Neurochem.* **32**:1387–1400.
119. Dowdall, M. J., and Zimmermann, H., 1977, *Neuroscience* **2**:405–421.
120. Morel, N., Israël, M., Manaranche, R., and Mastour-Frachon, P., 1977, *J. Cell Biol.* **75**:43–55.
121. Richardson, P. J., and Whittaker, V. P., 1981, *J. Neurochem.* **36**:1536–1542.
122. Stadler, H., and Tashiro, T., 1979, *Eur. J. Biochem.* **101**:171–178.
123. Whittaker, V. P., 1977, *Naturwissenschaften* **64**:606–611.
124. Jones, R. T., Walker, J. H., Richardson, P. J., Fox, G. Q., and Whittaker, V. P., 1981, *Cell Tissue Res.* **218**:355–373.
125. De Robertis, E., Arnaiz, G. R. de L., Salganicoff, L., de Iraldi, A. P., and Zieher, L. M., 1963, *J. Neurochem.* **10**:225–235.
126. Germain, M., and Proulx, P., 1965, *Biochem. Pharmacol.* **14**:1815–1816.
127. Burton, R. M., Howard, R. E., Baer, S., and Balfour, Y. M., 1964, *Biochim. Biophys. Acta* **84**:441–447.
128. Wiegandt, H., 1967, *J. Neurochem.* **14**:671–674.
129. Mellanby, J., and Whittaker, V. P., 1968, *J. Neurochem.* **15**:205–208.
130. Breckenridge, W. C., Gombos, G., and Morgan, I. G., 1972, *Biochim. Biophys. Acta* **266**:695–707.

131. Breckenridge, W. C., Morgan, I. G., Zanetta, J. P., and Vincendon, G., 1973, *Biochim. Biophys. Acta* **320**:681–686.
132. Nagy, A., Baker, R. R., Morris, S. J., and Whittaker, V. P., 1976, *Brain Res.* **109**:285–309.
133. McCaman, R. E., Arnaiz, G. R. de L., and De Robertis, E., 1965, *J. Neurochem.* **12**:927–935.
134. Barker, L. A., Dowdall, M. J., and Whittaker V. P., 1972, *Biochem. J.* **130**:1063–1080.
135. Schwarzenfeld, I. von, 1979, *Neuroscience* **4**:477–493.
136. Marchbanks, R. M., 1975, *Handbook of Psychopharmacology*, Volume 3 (L. L. Iversen, S. D. Iversen, and S. H. Snyder, eds.), Plenum Press, New York, pp. 247–326.
137. Marchbanks, R. M., 1978, *Trends Neurosci.* **1**:168.
138. Morris, S. J., Ralston III, H. J., and Shooter, E. M., 1971, *J. Neurochem.* **18**:2279–2290.
139. Kornguth, S. E., Anderson, J. W., and Scott, G., 1969, *J. Neurochem.* **16**:1017–1024.
140. Cotman, C. W., Levy, W., Banker, G., and Taylor, D., 1971, *Biochim. Biophys. Acta* **249**:406–418.
141. Cotman, C. W., and Matthews, D. A., 1971, *Biochim. Biophys. Acta* **249**:380–394.
142. Cotman, C. W., Banker, G., Churchill, L., and Taylor, D., 1974, *J. Cell Biol.* **74**:181–203.
143. Kelly, P. T., and Cotman, C. W., 1978, *J. Cell Biol.* **79**:173–183.
144. Cohen, R. S., Blomberg, F., Berzins, K., and Siekevitz, P., 1977, *J. Cell Biol.* **74**:181–203.
145. Matus, A. I., and Taff-Jones, D. H., 1978, *Proc. R. Soc. Lond.* [*Biol.*] **203**:135–151.
146. Thieren, H. M., and Mushynki, W. E., 1975, *J. Cell Biol.* **71**:807–822.
147. Matus, A. I., Pehling, G., Ackermann, M., and Maeder, J., 1980, *J. Cell Biol.* **87**:346–359.
148. Bradford, H. F., 1969, *J. Neurochem.* **16**:675–684.
149. Appel, S. H., Autilio, L., Festoff, B. W., and Escueta, A. V., 1969, *J. Biol. Chem.* **244**:3166–3172.
150. Escueta, A. V., and Appel, S. H., 1969, *Biochemistry* **8**:725–733.
151. Ling, C.-M., and Abdel-Latif, A. A., 1968, *J. Neurochem.* **15**:721–729.
152. Peterson, N. A., and Raghupathy, E., 1972, *J. Neurochem.* **19**:1423–1438.
153. Navon, S., and Lajtha, A., 1969, *Biochim. Biophys. Acta* **173**:518–531.
154. Bennett, J. P., Logan, W. T., and Snyder, S. H., 1972, *Science* **178**:997–999.
155. Weinstein, H., Varon, S., Muhlemann, D. R., and Roberts, E., 1965, *Biochem. Pharmacol.* **14**:273–288.
156. Kuriyama, K., Weinstein, H., and Roberts, E., 1969, *Brain Res.* **16**:470–492.
157. Grahame-Smith, D. G., and Parfitt, A. G., 1970, *J. Neurochem.* **17**:1339–1353.
158. White, T. D., and Keen, P., 1970, *Biochim. Biophys. Acta* **196**:285–295.
159. Colburn, R. W., Goodwin, F. K., Murphy, D. L., Bunney Jr., W. E., and Davis, J. M., 1968, *Biochem. Pharmacol.* **17**:957–964.
160. Hajós, F., and Csillag, A., 1976, *Brain Res.* **112**:207–213.
161. Diamond, I., and Fishman, R. A., 1973, *J. Neurochem.* **20**:1533–1542.
162. Bradford, H. F., Jones, D. G., Ward, H. K., and Booker, J., 1975, *Brain Res.* **90**:245–259.
163. Bradford, H. F., Ward, H. K., and Thomas, A. J., 1978, *J. Neurochem.* **30**:1453–1459.
164. Bradford, H. F., 1972, *Methods of Neurochemistry*, Volume 3 (R. Fried, ed.), Marcel Dekker, New York, pp. 155–202.
165. Raiteri, M., Angelini, F., and Levi, G., 1974, *Eur. J. Pharmacol.* **25**:411–414.
166. Raiteri, M., and Levi, G., 1978, *Rev. Neurosci.* **3**:77–130.
167. Kanner, B. I., and Sharon, I., 1978, *Biochemistry* **17**:3949–3953.
168. Marvizón, J. G., Mayn, F., Jr., Aragón, M. C., Giménez, C., and Valdivieso, F., 1981, *J. Neurochem.* **37**:1401–1406.
169. Keen, P., and White, T. D., 1970, *J. Neurochem.* **17**:565–571.
170. Scott, I. D., and Nicholls, D. G., 1980, *Biochem. J.* **186**:21–33.
171. Blaustein, M. P., 1975, *J. Physiol.* (*Lond.*) **247**:617–655.
172. Åkerman, K. E. P., and Nicholls, D. G., 1981, *Eur. J. Biochem.* **117**:491–497.
173. Clark, A. W., Hurlbut, W. P., and Mauro, A., 1972, *J. Cell Biol.* **52**:1–14.
174. McGraw, C. F., Soyo, A. V., and Blaustein, M. P., 1980, *J. Cell Biol.* **85**:228–241.
175. Gordon, M. K., Bench, K. G., Deanin, G. G., and Gordon, M. W., 1968, *Nature* **217**:523–527.
176. Miller, E. K., and Dawson, R. M. C., 1972, *Biochem. J.* **126**:805–821.

177. Morgan, I. G., 1970, *FEBS Lett.* **10**:273–275.
178. Morgan, I. G., and Austin, L., 1968, *J. Neurochem.* **15**:41–51.
179. Autilio, L. A., Appel, S. H., Pettis, P., and Gambetti, P., 1968, *Biochemistry* **7**:2615–2622.
180. Hernández, A. G., 1974, *Biochem. J.* **142**:7–17.
181. Gambetti, P., Autilio-Gambetti, A., Gonatas, N. K., and Shafer, B., 1972, *J. Cell Biol.* **52**:526–535.
182. Dowdall, M. J., Barker, L. A., and Whittaker, V. P., 1972, *Biochem. J.* **130**:1081–1094.
183. Miller, E. K., and Dawson, R. M. C., 1972, *Biochem. J.* **126**:823–835.
184. Nyman, M., and Whittaker, V. P., 1963, *Biochem. J.* **87**:248–255.
185. Marchbanks, R. M., 1969, *Biochem. Pharmacol.* **18**:1763–1766.
186. Yamamura, H., and Snyder, S. H., 1973, *J. Neurochem.* **21**:1355–1374.
187. Kuhar, M. J., Sethy, V. H., Roth, R. H., and Aghajanian, G. K., 1973, *J. Neurochem.* **20**:581–593.
188. Haga, T., 1971, *J. Neurochem.* **18**:781–798.
189. Haga, T., and Noda, H., 1973, *Biochim. Biophys. Acta* **291**:564–575.
190. Crone, H. D., and Freeman, S. E., 1972, *J. Neurochem.* **19**:1207–1208.
191. Lewis, P. R., and Shute, C. C. D., 1966, *J. Cell Sci.* **1**:381–390.
192. Lewis, P. R., Shute, C. C. D., and Silver, A., 1967, *J. Physiol.* (*Lond.*) **191**:215–224.
193. Aldridge, W. N., and Johnson, M. K., 1959, *Biochem. J.* **73**:270–276.
194. Tennyson, V. M., and Brzin, M., 1970, *J. Cell Biol.* **46**:69–80.
195. Csillik, B., 1975, *Int. Rev. Neurobiol.* **18**:69–140.
196. Zimmermann, H., and Whittaker, V. P., 1977, *Nature* **267**:633–635.
197. Zimmermann, H., and Denston, C. R., 1977, *Neuroscience* **2**:715–730.
198. Iversen, L. L., and Johnston, G. A. R., 1971, *J. Neurochem.* **18**:1939–1950.
199. Logan, W. F., and Snyder, S. H., 1972, *Brain Res.* **42**:413–431.
200. Heuser, J. E., and Reese, T. S., 1973, *J. Cell Biol.* **57**:315–344.
201. Ceccarelli, B., Hurlbut, W. P., and Mauro, A., 1973, *J. Cell Biol.* **54**:30–38.
202. Ceccarelli, B., Hurlbut, W. P., and Mauro, A., 1973, *J. Cell Biol.* **57**:499–524.
203. Ceccarelli, B., Grohovaz, F., and Hurlbut, W. P., 1979, *J. Cell Biol.* **81**:178–192.
204. Heuser, J. E., Reese, T. S., Dennis, M. J., Jan, Y., Jan, L., and Evans, L., 1979, *J. Cell Biol.* **81**:275–300.
205. Model, P. G., Highstein, S. M., and Bennett, M. V. L., 1975, *Brain Res.* **98**:209–228.
206. Bennett, M. V. L., Model, P. G., and Highstein, S. M., 1976, *Cold Spring Harbor Symp. Quant. Biol.* **40**:25–35.
207. Zimmermann, H., and Whittaker, V. P., 1974, *J. Neurochem.* **22**:435–450.
208. Zimmermann, H., and Whittaker, V. P., 1974, *J. Neurochem.* **22**:1109–1114.
209. Suszkiw, J. B., 1980, *Neuroscience* **5**:1341–1349.
210. Boyne, A. F., Bohan, T. P., and Williams, T. H., 1975, *J. Cell Biol.* **67**:814–825.
211. De Belleroche, J., and Bradford, H. F., 1972, *J. Neurochem.* **19**:1817–1819.
212. Osborne, R. H., Bradford, H. F., and Jones, D. G., 1973, *J. Neurochem.* **21**:407–419.
213. Hole, K., Bergslien, H., Jørgensen, H. A., Berge, O.-G., Reichelt, K. L., and Trygstad, O. E., 1979, *Neuroscience* **4**:1883–1893.
214. Jovin, T. M., Morris, S. J., Striker, G., Schultens, H. A., Digweed, M., and Arndt-Jovin, D. J., 1976, *J. Histochem. Cytochem.* **24**:269–283.
215. Baker, R. R., Dowdall, M. J., and Whittaker, V. P., 1975, *Brain Res.* **100**:629–644.
216. Baker, R. R., Dowdall, M. J., and Whittaker, V. P., 1976, *Biochem. J.* **154**:65–75.
217. Schweitzer, E. S., and Blaustein, M. P., 1980, *Exp. Brain Res.* **38**:443–453.
218. Baker, P. F., and Knight, D. E., 1981, *Phil. Trans. R. Soc.* (*Lond.*) [*Biol.*] **298**:83–103.

2

The Synaptic Vesicle

V. P. Whittaker

I. INTRODUCTION

Synaptic vesicles are the characteristic organelles of the presynaptic nerve terminals of chemical synapses (i.e., those utilizing the release of a specific chemical transmitter substance to bring about synaptic transmission). They are normally 45–50 nm in diameter and must be among the most homogeneous and uniform lipoprotein membrane organelles known. Exceptionally, the synaptic vesicles of the electromotor nerve terminals in the electric organ of *Torpedo* are larger (85–90 nm in diameter); these vesicles can be isolated relatively easily in bulk and high purity and have thus provided an important source material for most of the recent work on the biochemistry and biophysics of synaptic vesicles. By contrast, *Torpedo* motor nerve terminals have normal-sized vesicles. The subject has recently been comprehensively reviewed,[1–3] and only a selection from the vast literature can be given here.

Synaptic vesicles are too small to be resolved by light microscopy; they were discovered independently by several workers as soon as high-resolution electron microscopy, using reliable methods of fixation and embedding, began to be applied to nervous tissue, peripheral and central. At least four publications can be cited[4–7] that represent presumably independent work and appeared almost simultaneously. The most significant of these[7] suggested that the vesicles might be the morphological basis for the recently discovered packaging or quantization of transmitter release. Difficulties in estimating accurately either the size of a quantum or the number of transmitter molecules in a vesicle have made it difficult to apply a stringent test to this idea, but the calculations that have been made (Section 2.1.1) show that there is no serious inconsistency in the magnitude of these two quantities.

The general appearance of the neuron—the cell body tightly packed with polysomes, rough- and smooth-walled cisternae of endoplasmic reticulum, and vesicles, the terminals packed with synaptic vesicles—strongly suggests that this is a secreting cell in which the release zone is separated from the cell body

---

V. P. Whittaker • Abteilung Neurochemie, Max-Planck-Institut für Biophysikalische Chemie, Göttingen, Federal Republic of Germany.

by a narrow tube, the axon, and in which, possibly in consequence, the secretory mechanism has become miniaturized. Early attempts to demonstrate synaptic vesicle depletion on stimulation were usually unsuccessful,[8] but later experiments in which exocytosis was stimulated by prolonged excitation[9–11] or toxins[12] or in which endocytosis was inhibited by cold[13] clearly showed that transmitter and vesicle depletion occurred *pari passu* and that without vesicles the transmitter stores required for synaptic transmission cannot be regained. Studies with high-molecular-mass substances unable to cross lipoprotein membranes[10,11,14] and others with an antibody to a vesicle core antigen[15,16] have shown that synaptic vesicles recycle and have suggested that failure to observe vesicle and transmitter depletion at physiological rates of stimulation reflects a rapid rate of vesicle reformation and reutilization coupled with rapid transmitter resynthesis (or reuptake). In this respect, synaptic vesicles have a higher degree of autonomy than normal storage granules, which are probably recovered in a metabolically depleted form and need to be recycled through the Golgi apparatus to be fully reconstituted.

The concepts that transmitter storage and release are an adaptation of cell secretion and that synaptic vesicles are the storage sites of transmitter and the source of the transmitter released on stimulation (sometimes referred to as the vesicle theory of transmitter storage and release) accommodate a large number of experimental results with many different synapses and have commanded general acceptance. The anomaly that the specific radioactivity of the total vesicle transmitter pool is lower than that of released transmitter in synapses whose transmitter stores have been labeled with a radioactive precursor[17–20] has been satisfactorily explained by the discovery of vesicle metabolic heterogeneity, i.e., the finding that only a proportion of the total vesicle population under normal conditions of stimulation is actively recycling.[10,11,21] Nevertheless, a few, better known for their copious reviews[22–25] than for their original contributions, have provided a dissenting view, which is discussed below (Section 4.4). They feel that the known facts of transmitter release are best described by a model in which transmitter is directly released from a cytoplasmic pool via a carrier in the external membrane. The vesicles are held to provide either a buffer system for maintaining cytoplasmic levels of transmitter or a calcium-sequestering system.[26]

Although the synaptic vesicles of nerve terminals using different transmitters are usually very similar in morphology, peripheral and some central terminals containing biogenic amines normally contain a population of dense-cored vesicles larger on average than synaptic vesicles. The dense core is thought to be a condensation product of norepinephrine and osmium since it becomes more prominent when terminals are pretreated with norepinephrine. Dense-cored vesicles are also seen in 5-hydroxytryptamine-containing endings and in endings secreting oxytocin and vasopressin. Here, the dense core is the neurophysin–oxytocin or neurophysin–vasopressin complex. Occasional dense-cored vesicles are also seen in most other types of terminal, as are larger vesicles, cisternae, "coated vesicles," microtubules, neurofilaments, and small mitochondria. All of these structures are potential contaminants of isolated synaptic vesicles, as are fragments of the presynaptic and postsynaptic plasma membranes. Although dense-cored vesicles and synaptic vesicles are undoubtedly homologous structures, the literature on the former is large and for reasons of length has been largely excluded from this chapter. The reader is referred to other, more comprehensive reviews.[27,28]

2. *ISOLATION*

2.1. *Mammalian Synaptic Vesicles*

2.1.1. *From Central Synapses*

For a decade after the first morphological descriptions, synaptic vesicles were cytoplasmic entities known only from electron micrographs. Speculations about their function remained unsupported by experimental evidence. One pioneer even doubted his initial discovery and went to great trouble to prove by means of a tilting stage that what he had originally described as vesicles really were indeed vesicles and not cross sections of tubules.[29] Progress, especially for our understanding of the molecular basis of transmitter storage and release, required the isolation of synaptic vesicles, and, emboldened by the recent success of Blaschko[30] and Hillarp[31] and their co-workers in isolating chromaffin granules, I attempted to isolate synaptic vesicles from brain cortex in 1957–1958 using "bound acetylcholine" as a marker.[32]

What I actually succeeded in isolating, as described in the previous chapter, were synaptosomes,[33] i.e., the vesicles *in situ* in the detached nerve terminals. Having noted that synaptosomes are readily disrupted by exposure to hypoosmotic media with the release of synaptic vesicles[34,35] and cytoplasmic components but that about 50% of the acetylcholine remained stable under these conditions,[32] I thought that this osmotically "stable bound acetylcholine" might be vesicular acetylcholine and decided to purify the particles containing it.[36] This was done by step-gradient fractionation of the supernatant (W_s) from a hypoosmotically disrupted synaptosome–mitochondrial (P_2) fraction after removal of the larger particles by centrifuging at an intermediate speed (W_p).

The vesicles (and the stable bound acetylcholine) distributed themselves bimodally, partly in a fraction of monodispersed synaptic vesicles (D) in a layer of 0.4 M sucrose and partly in a fraction of disrupted synaptosomes (H) lying at the interface between 1.0 and 1.2 M sucrose. Some vesicles are lost in the discarded W_p; these augment those in fraction H if the whole fraction W (for this purpose preferably prepared from fraction B rather than fraction P_2) is separated on the step gradient. Fraction D was remarkably free from nonvesicular membrane fragments, as was shown morphologically and by the absence of the nonvesicular membrane markers acetylcholinesterase and Na^+,K^+-activated ATPase. We were therefore surprised when, during the course of this work, electron micrographs of a fraction (M_2) equivalent to all the particulate material in W_s were published showing a homogeneous field of synaptic vesicles.[37] We knew that the vesicles in W_s were heavily contaminated with membrane fragments, partially disrupted synaptosomes and mitochondria and offered our own explanation of the discrepancy.[36,38] Subsequent work by several groups confirmed that density-gradient fractionation of W_s is needed to remove extensive contamination from the monodispersed vesicles.[39–41]

The D fraction or a similar fraction (D_1) prepared on a simpler gradient provided an excellent preparation for our initial attempt to estimate the number of acetylcholine molecules per vesicle.[42,43] This gave the rather low number of 300. However, it was appreciated that only a small proportion of the vesicles could have been derived from cholinergic terminals; if the proportion of cholinergic vesicles is x%, then the acetylcholine content per cholinergic vesicle would be $300 \times 100/x$. We assumed that $x = 15$, giving a value of 2000 molecules/vesicle, which is a little more than the amount (1500 molecules) of acetylcholine that a structure the size of the core of a synaptic vesicle could store as a solution isosmotic with plasma.

How does this compare with the size of a quantum? Although our measurements were done with brain vesicles and measurements of quantal size were done with frog muscle, the question is not meaningless because brain and motor nerve terminal vesicles do not differ much in size. If the size of the quantum were as high as 10^5, as one early estimate suggested,[44] then one could abandon the notion that vesicles are the morphological basis of quantized transmitter release, because it is extremely unlikely that vesicle cores could be made of crystalline acetylcholine. The then popular estimate of 10^4 also seemed rather high, but another estimate was much lower (900),[45] and the most reliable estimate that has been made, using snake neuromuscular junction and electrophoretic application of acetylcholine, is 6000.[46] A less direct estimate[46] based on the ratio of the potential generated by one molecule of acetylcholine (the so-called "unit event")[47,48] to that generated by one quantum (the miniature end-plate potential) is 2000. Recent work using selective complement lysis of cholinergic synaptosomes[49] indicates (on the assumption that the yield of vesicles from cholinergic nerve terminals is the same as that from noncholinergic terminals) that x is nearer 6 than 15%, which would raise the estimated number of acetylcholine molecules per cholinergic vesicle to about 5000. Thus, although the agreement is not as precise as one would like, it is not inconsistent with the vesicle-as-quantum theory.

In view of the high concentration of acetylcholinesterase in the synaptic cleft, people sometimes wonder how the acetylcholine escapes destruction before reaching the receptor. In the concentration in which acetylcholine appears to be stored in the vesicle, it is an excellent anticholinesterase; it produces marked autoinhibition down to about 10 mM. Not until the acetylcholine concentration in the cleft has fallen to 1 mM is it hydrolyzed at a maximum rate. The enzyme is also inhibited at low pHs. Thus, the sudden release of a puff of acetylcholine in a local concentration of 0.16 M or more and at a pH of 6.5 (Section 3.2), as predicted by the vesicle theory, may be an essential feature of the mechanism of cholinergic transmission.

Although fraction D (or D_1) is a highly homogeneous preparation of monodispersed vesicles as judged by morphological criteria (contamination by nonvesicular membrane fragments is as low as 0.3% in good preparations,[42]), its content of acetylcholine relative to protein is rather low, suggesting the pres-

Table I
Separation of Cholinergic from Noncholinergic Synaptic Vesicles from Guinea Pig Brain[51]

Parameter	Vesicle class: Cholinergic	Vesicle class: Adrenergic	Ratio cholinergic : adrenergic
Acetylcholine (pmol/mg protein)	2500	60	42
Norepinephrine (pmol/mg protein)	3.2	75	0.043
Dopamine (pmol/mg protein)	7.8	203	0.038
Dopamine-β-hydroxylase (nmol/h per mg protein)	0	5	0
Diameter (nm)	50	60	—
Purification factor (relative to W_s)	21	11	—

ence of large amounts of extraneous (nonvesicular) protein as well as noncholinergic vesicles. This has been confirmed by the 70-fold enrichment of acetylcholine relative to protein obtained by zonal fractionation and gel filtration.[50]

Assuming that mammalian brain cholinergic vesicles are similar in composition to *Torpedo* vesicles, but making allowance for the lower core/membrane ratio and assuming that the acetylcholine is present in the core in isosmotic concentration, it has been calculated[50] that the specific concentration of acetylcholine in pure cholinergic synaptic vesicles must be about 100 nmol/mg of protein. The highest concentration actually attained by zonal centrifugation and porous-glass-bead chromatography is 14 nmol/mg protein.

This will represent a limiting concentration if the cholinergic vesicles comprise only 15% of the total, as originally assumed. Thus, this preparation may well be a pure preparation of synaptic vesicles (i.e., free from extraneous protein) though not homogeneous with respect to transmitter type. Some success in separating cholinergic from noncholinergic vesicles has been attained by using D_2O gradients followed by column chromatography.[51] That the lengthy purification procedure resulted in some loss of transmitter is shown by the fact that the specific concentration of acetylcholine in the cholinergic fraction was lower (2.5 nmol/mg of protein) than previously attained[50]; however, there was a clear separation (Table I) of vesicles with high acetylcholine content and low or negligible norepinephrine, dopamine, and dopamine β-hydroxylase from those with relatively low acetylcholine and high catecholamine content. The latter fraction also had appreciable dopamine β-hydroxylase activity and contained vesicles of about 20% greater diameter, many with dense cores. The cross contamination of each fraction by the other was estimated to be about 10%.

2.1.2. From Peripheral Synapses

The only attempt to prepare synaptic vesicles from a mainly cholinergic mammalian source is that utilizing bovine superior cervical ganglion.[52] The specific concentration of acetylcholine in this preparation ranged from 3 to 14 nmol/mg protein, but estimates of the number of acetylcholine molecules/vesicle gave a figure of 1600, in satisfactory agreement with those obtained for brain vesicles. Thus, most of the protein in this preparation must have been extravesicular.

The myenteric plexus of guinea pig, although far from being purely cholinergic, is very rich in acetylcholine: 150 nmol/g tissue compared to 15 nmol/g tissue for brain; by contrast the norepinephrine, dopamine, and 5-hydroxytryptamine contents are much lower (3.3, 0.4, 0.6 nmol/g, respectively). Vesicles have been isolated from the low-speed supernatant of a homogenate of myenteric plexus–longitudinal muscle with a concentration of 20 nmol acetylcholine/mg protein, an enrichment factor of 36, and a yield of 50% in a single step by zonal gradient separation.[53] The vesicle fraction has one main contaminant: actomyosin fibrils. Calculations suggest that the cholinergic vesicles are 10–25% pure, the contaminants being actomyosin and some noncholinergic vesicles. This preparation should be an excellent starting point for further work.

2.2. Vesicles from *Torpedo* Electromotor Nerve Terminals

Most recent work on synaptic vesicles has concentrated on those from the *Torpedo* electromotor nerve terminals so abundantly present in the electric organs of these fish, and several reviews[2,54,55] have appeared. The advantages of this source[56] are these: it is purely cholinergic; the tissue contains about 1000 times the amount of nerve tissue than another enbryologically related and purely cholinergic source, muscle; and large-scale preparations yield essentially pure cholinergic vesicles in milligram quantities.

It must be emphasized that *Torpedo* vesicles are larger (90 nm diameter) than those in other endings (including *Torpedo* neuromuscular synapses) (45–50 nm diameter), and vesicle recycling is probably slower than at the neuromuscular junction; thus, the electromotor synapse is not a typical cholinergic synapse. However, the electric organ originates from myotubes,[57–61] and there is no reason to believe that the evolutionary adaptation that has occurred renders the electromotor synapse invalid as a model for other cholinergic synapses. Indeed, recent immunochemical studies indicate a high degree of cross reactivity between antigens specific for mammalian cholinergic nerve terminals and antibodies raised to the corresponding *Torpedo* electromotor antigens.[15,16,62–64]

Torpedo vesicles were first isolated soon after brain cortex vesicles,[65,66] but difficulties in supply precluded much further work for a time. The first vesicle preparations[65] were reasonably pure morphologically and gave a figure of 40,000 for the number of acetylcholine molecules/vesicle. This value, over 100 times greater than that for mammalian brain vesicles, was explained[66] on the basis that the vesicles were purely cholinergic (i.e., uncontaminated by noncholinergic vesicles), had much larger cores and might be expected to store acetylcholine in concentrations roughly isosmotic with elasmobranch plasma, which has two to three times the osmolarity of mammalian plasma. This figure is probably an underestimate; the most recent value is about 200,000, corresponding to an internal acetylcholine concentration of about 0.9 M.[67] Since miniature potentials similar to those recorded at frog neuromuscular junctions are observed in resting *Torpedo* electric organ,[68,69] it may be wondered why the package is so large: are we dealing with a multiquantum vesicle? The answer may well be found in the low resistance of the membrane of the postsynaptic cell, the electrocyte; to generate a "normal" postsynaptic response, the quantum of acetylcholine must be much larger. Unfortunately, no direct measurement of quantal size has been made on this synapse.

For large-scale isolation of synaptic vesicles from this source,[70] recourse was had to a method of tissue comminution that had previously been successfully used for the isolation of dense-cored vesicles from vas deferens,[71] a tissue equally difficult to homogenize: freezing in liquid nitrogen, pounding to a coarse powder, and extraction of the powder (preferably before thawing[67]) with isosmotic sucrose, sucrose–saline, or saline media. After removal of coarse and medium-sized fragments by squeezing through cheesecloth and centrifuging in a moderately intense centrifugal field (10,000 *g* for 30 min), the supernatant (the "cytoplasmic extract") rich in synaptic vesicles is submitted to density-gradient separation in a zonal rotor. A large peak of bound acetylcholine consisting morphologically of almost pure synaptic vesicles (VP) separates from a less dense peak of soluble protein (SP) and another denser peak of membrane fragments (MP).

The acetylcholine : lipid ratio of VP is close to the limiting value, indicating the essential freedom of the fraction from contaminating membranes, but the fraction still contains extraneous (nonvesicular) protein. To remove this, an additional purification step is necessary. This may take the form of a second zonal run,[67] passage through a column of porous glass beads,[50,67] or a preliminary removal of most of the soluble cytoplasmic protein by concentrating the vesicles on a simple step gradient in the presence of a low concentration of EGTA.[72] The limiting concentration of acetylcholine is 6–7 μmol/mg of protein or 1.8–2.2 μmol/mg of phospholipid. The only value much higher than this is that of 28 μmol/mg of protein reported for a single experiment in a recent paper[26]; this is impossibly high and suggests that the protein content of the fraction was seriously underestimated. In our experience,[67] this is easily done if an insufficiently large vesicle sample is taken for protein determination.

Torpedo synaptic vesicles appear to be more sensitive to hypoosmotic lysis than mammalian cortical vesicles and less easily dispersed when pelleted; the preparative procedure described above is based on a modification of our original isolation method that took these factors into account.[73]

Most of the work with synaptic vesicles has utilized *Torpedo marmorata*, but other Torpedinidae, including *Narcine* give similar preparations.[70,74] The electric organ of another unrelated electric fish, *Electrophorus electricus*, is much less suitable.[75]

2.3. Vesicles from Squid Optic Lobes

Synaptic vesicles have been isolated in relatively pure form from the brain of squid, which has a greater acetylcholine content than the electric organ of *Torpedo* but is not purely cholinergic.[76] This preparation has so far not been exploited.

3. VESICLE COMPOSITION

3.1. Vesicles Derived from Mammalian Central and Peripheral Synapses

Analysis of mammalian cortical vesicles necessarily yields average values for a mixed vesicle population derived from many different types of nerve terminals, which may also contain, even when carefully prepared, small amounts of coated vesicles and vesiculated fragments of a Ca^{2+}-sequestering endoplasmic reticulum as contaminants. Enzymes and other components of the cytosol may be adsorbed during isolation or may be present as unadsorbed contaminants because of inadequate fractionation methods. Vesiculated fragments of the presynaptic plasma membrane may also contaminate the preparations.

In general, unadsorbed contaminants do not copurify with vesicle markers and are fairly easily identified. Adsorbed components or components of other vesicles of similar size and density may copurify through several fractionation steps and may thus prove more difficult to spot as contaminants, especially when sensitive methods exist for their detection. One should always be suspicious of assigning an enzyme or other activity to synaptic vesicles if other types of subcellular particles possess this activity in a markedly greater concentration. These remarks may seem obvious, but the literature abounds in errors.

One clear result emerges: the lipid, protein, enzyme, and, most recently, antigen composition of adequately purified synaptic vesicles differs in several important respects from that of presynaptic plasma membranes.[77] This means that even if the surface area of the presynaptic plasma membrane can be increased under conditions that promote exocytosis at the expense of endocytosis, not only does endocytosis restore the *status quo ante* during the recovery phase, it does so by retrieving vesicle membrane. As described in Section 4, there is much evidence that the vesicles so retrieved retain their storage capacity as well as their composition. I have proposed the term "transient exocytosis" to denote this highly conservative type of exocytosis.[78]

Tables II–IV summarize the main information available concerning the composition of synaptic vesicles from mammalian brain: Na^+,K^+-ATPase,[39,80] cholinesterase,[36] and ganglioside[40,50,77,79] are absent from synaptic vesicles; if present in a preparation, they indicate contamination with plasma membranes, which are relatively rich in these components.[36,77,80,82,83] In their low and high cholesterol content, synaptic vesicles and presynaptic plasma membranes conform to other internal and external cell membranes, respectively.[50,77,79,82]

Table II
Lipid Content of Brain Synaptic Vesicles

Species	Tissue	Ref.	Acetylcholine[a] (nmol/mg protein)	Lipid/ protein (w/w)	Molar ratios Acetylcholine: ATP	Cholesterol: phospholipid	Sialic acid: phospholipid
Guinea pig	Cortex	50	14	1.11	1.2	0.51	<0.001
Rat	Brain	79	—	1.44	—	0.58	0.007[b]

[a] Present in cholinergic subpopulation.
[b] Derived value.

In spite of some promising work[51] (see Section 2.1.1), it is doubtful whether mammalian brain synaptic vesicles have yet been purified sufficiently to make reliable identifications of vesicle proteins possible. In one study[86] using step-gradient vesicles from ox brain further purified on columns of Sepharose 6B to remove cytoplasmic and plasma membrane contamination, but with only a modest enrichment ratio as defined by transmitter content, proteins of $M_r \times 10^{-3}$ of 200, 160, 120, 58, 55, 46, 42, 34, 31, and 27 were identified; the 200-, 55-, 46-, and 42-kilodalton components were tentatively identified as myosin heavy chains, tubulin, filamin, and actin, respectively. Similar components have been observed in another study,[87] and an association of contractile proteins with vesicles has been postulated to be involved in the mechanism of exocytosis.[88]

Torpedo electromotor synaptic vesicles rigorously purified in the presence of Ca^{2+}-sequestering agents have, by contrast, a much simpler protein composition and, of the various components of the cytoskeleton, retain only actin (Section 3.2); it is therefore permissible to conclude that the relatively complex protein composition reported for mammalian synaptic vesicles is partly a result of their heterogeneous synaptic origin and partly of the adsorption of cytoplasmic components. That such adsorption can occur is illustrated by the controversy as to whether choline acetyltransferase is a constituent of synaptic

Table III
Phospholipid Content of Brain Synaptic Vesicles

Phospholipid component	Component as percent of total phospholipid Ref. 79	Ref. 50
Phosphatidylcholine	42.2	38.1
Phosphatidylethanolamine	36.3	32.3
Phosphatidylserine	11.8	15.4
Sphingomyelin	4.9	7.0
Phosphatidylinositide	2.9	5.2
Phosphatidic acid	1.8	1.1
Lysophosphatidylcholine	1.5	0.8

Table IV
Differences in Composition of Brain Synaptic Vesicle and Presynaptic Plasma Membranes

Component	Vesicles[a]	Ref.	Presynaptic plasma membrane[a]	Ref.
Enzymes				
Na^+,K^+-ATPase	—	39,80	+	80
Acetylcholinesterase	—	36	+	36
Choline acetyltransferase	—	81	−	81
Lipids				
Cholesterol	low (0.5)	50,77,79	high (0.8)	77,82
Ganglioside	—	40,50,77,79	+	77,82,83
Proteins	Different patterns in gel electrophoresis			84,87
Antigens				
CHOL-1[c]	−	62–64	+	62–64
Vesicle-specific proteoglycan[c]	+	15,16,62	−	15,16,62
Receptor for α-bungarotoxin[d]	−	85	+	85

[a] Symbols: +, present; −, absent.
[b] Figures in parentheses are cholesterol:phospholipid molar ratios.
[c] Cholinergic subpopulation only.
[d] Frog sartorius motor nerve terminals.

vesicles: this was resolved by the discovery that the enzyme may be adsorbed onto vesicles and contaminating fragments of presynaptic plasma membrane at low ionic strength[81]; this effect is more obvious with some species than with others because of species variations in the amount of positive charge on the choline acetyltransferase molecule.[89]

Calcium uptake has been detected in synaptic vesicle preparations but has not been included because it can easily be explained by contamination with endoplasmic reticulum.[90]

Although acetylcholine is commonly used as a synaptic vesicle marker, only a small proportion of vesicles (Section 2.1.1) can be derived from cholinergic terminals. We should therefore expect other putative central transmitters to be present in this fraction.

As far as amino acids are concerned, the large cytoplasmic pool of free amino acid tends to swamp any vesicular pool,[91] making it difficult to decide if one exists[92] and, if it does, how large it is.[93] However, the best evidence suggests that there is a vesicular pool and that, though small compared with the free pool, it is quite large compared with that of acetylcholine.[93] Somewhat disturbing is the alleged presence,[94] along with accepted putative excitatory and inhibitory[95] amino acid transmitters, of smaller amounts of others such as glutamine, lysine, and cysteic acid, which do not have the appropriate pharmacological effects on neurons.

Substance P is an example of a brain peptide that is highly localized in the synaptosome[96,97] fraction. Much more needs to be done on the isolation of neuropeptide-containing vesicles.

In contrast to the various brain vesicle preparations studied, the relatively highly purified cholinergic vesicles from the myenteric plexus[53] have a rather simple protein composition, several components of which have the same electrophoretic behavior as the main components of cholinergic electromotor vesicles. The main contaminant is actomyosin. Lipid and other analyses have not so far been made on these vesicles.

3.2. Torpedo Vesicles

The composition of electromotor nerve terminal synaptic vesicles from *Torpedo marmorata* is summarized in Table V. Less complete data are available for *Narcine* vesicles (for references see ref. 98). The vesicle appears to consist of two parts, a lipoprotein limiting membrane and an aqueous core in which the small-molecular-weight constituents are stored. In addition to acetylcholine, there are considerable amounts of ATP[103]; in this respect, this cholinergic vesicle resembles chromaffin granules and noradrenergic vesicles, which also contain considerable amounts of ATP. The ATP content relative to acetylcholine is much lower than that of brain vesicles (Table II), but some of the ATP of brain vesicles may be in noncholinergic vesicles. The concentrations of both acetylcholine and ATP are high enough to be detected by nuclear magnetic resonance (NMR)[104,105]; the main resonances of acetylcholine show a chemical shift, which has been attributed to the shielding effect of the membrane, whereas the β- and γ-P resonances of the ATP of vesicles freshly isolated in K^+-glycine buffer ("K^+ vesicles") show both a chemical shift and peak broadening, which can be duplicated in solutions containing acetylcholine and Mg^{2+} ions at pH 6.5.

From this information it has been inferred that the vesicle may contain considerable amounts of Mg^{2+} and an acidic core. The chemical shift and peak broadening are seen in whole tissue, from which two sets of β and γ resonances are obtainable, one sharp and in the normal position, presumably originating from cytoplasmic ATP, the other showing the same chemical shifts and peak broadening as are seen in the K^+ vesicles, presumably generated by vesicular ATP *in situ*. The chemical shift and peak broadening are not seen in vesicles isolated in Na^+-containing solutions or in K^+ vesicles treated with nigericin, a lipid-soluble ionophore that exchanges K^+ for H^+ ions and would cause the collapse of any pH gradient across the membrane of such vesicles. They also disappear from K^+ vesicles slowly on storage.

Measurements of water spaces[101] show that the *Torpedo* vesicle is a highly hydrated structure. A comparison of its water space as measured by glycerol and dimethylsulfoxide with that of vesicle ghosts (vesicles that have resealed, with loss of acetylcholine and ATP, on exposure to hypoosmotic solutions) shows that the membrane occupies about 25% of the vesicle volume, consistent with a diameter of 90 nm and a membrane thickness of 4 nm. The small-molecular-weight solutes occupy about 3% of the vesicle volume in their nonhydrated form and 7% in their hydrated form. The water not bound to these solutes can be largely removed by hyperosmotic suspension media, i.e., is "osmotically active."

Table V
Composition of Cholinergic Synaptic Vesicles from Torpedo Electromotor Nerve Terminals

Component	Amount/vesicle		Amount/mg of protein		Percent of total	Mol per mol phospholipid
	Molecules	Mass (mg)	pmol	mg		
Small molecules[2,67,72]			[c]4 to 7			
Acetylcholine	2.0×10^5	—	—	—	—	—
ATP	2.7×10^4	—	—	—	—	—
Lipids[a2,50,67]	—	1.4×10^{-13}	—	1.7 to 2.9	—	—
PC	—	—	—	—	46.6	—
PE	—	—	—	—	29.5	—
PS	—	—	—	—	12.6	—
PI	—	—	—	—	5.1	—
SM	—	—	—	—	5.1	—
PA	—	—	—	—	0.6	—
LPC	—	—	—	—	0.4	—
Cholesterol	—	—	—	—	—	0.42
Ganglioside	—	—	—	—	—	0.01
Protein[b2,99]	—	$4.8–8.1 \times 10^{-14}$	—	—	—	—
0_1 (220)	—	—	—	—	3	—
0_2 (200)	—	—	—	—	3	—
1 (160) } 2 (147) }	—	—	—	—	24	—
4 (115)	—	—	—	—	2	—
6_1 (85)	—	—	—	—	1	—

7_1 (76)	—	—	—	—	4	—
7_2 (52)	—	—	—	—	3	—
8 (42)	—	—	—	—	23	—
10 (35)	—	—	—	—	4	—
11 (34)	—	—	—	—	23	—
13 (25)	—	—	—	—	10	—
GAG[a2,100]	—	10^{-14}	—	—	—	—
Water[c2,100]						
Osmotically active[d]	—	—	—	—	78	—
Bound to small molecules[e]	—	—	—	—	10	—
Bound to membrane[f]	—	—	—	—	12	—
Ions[g102]						
Na^+	—	—	0.18	—	—	—
K^+	—	—	0.04	—	—	—
Mg^{2+}	—	—	0.09	—	—	—
Ca^{2+}	—	—	0.18	—	—	—

[a] Abbreviations: PC, phosphatidylcholine; PE, phosphatidylethanolamine + plasmalogen; PS, phosphatidylserine; PI, phosphatidylinositide; SM, sphingomyelin; PA, phosphatidic acid; LPC, lysophosphatidylcholine; GAG, glycosaminoglycan.

[b] Figures in parentheses are apparent molecular masses ($\times 10^{-3}$) after gel electrophoresis in sodium dodecylsulfate.

[c] Water content of vesicles (D_2O space) is 83% v/v; hydrated membrane space, 25% v/v.[101]

[d] Measured by glycerol space.

[e] Measured by difference between dimethylsulfoxide and glycerol space.

[f] Measured by difference between D_2O and dimethylsulfoxide space.

[g] After Millipore® filtration.

The membrane is a relatively low-protein,[50,67] highly hydrated structure. Its water content (defined as the difference between the deuterium oxide and dimetylsulfoxide spaces) is about 40%. Its lipid : protein ratio is 1.7 to 2.9 depending on the degree of exposure of the vesicles to Ca^{2+} chelators during preparation[98]; these values are consistent with the observed membrane density of 1.11 to 1.13 g/ml.[67,98,101] The amount of lipid found per vesicle is just sufficient to form a bilayer limiting membrane of the observed diameter and thickness.[67]

Contrary to earlier conclusions based on the analysis of vesicles that, though morphologically pure, we now know still contained much extraneous protein,[70] there is calculated to be little or no core protein[67]; in this respect, the *Torpedo* vesicle is quite unlike the chromaffin granule with its large concentration of the core protein chromogranin A. Most of the membrane protein is accounted for by five main components,[72] designated 1, 2, 8, 11, and 13. Component 8 is actin, and 11 closely resembles (but is not identical to) the ADP/ATP translocase of mitochondria[106]; thus, both bind the ligands palmitoylcoenzyme A and atractylate have molecular weights of 30,000–34,000, and are positively charged. In addition to lipid and protein, these vesicles contain appreciable amounts of a glycosaminoglycan of the heparan type, much of which appears to be conjugated to protein and is set free on prolonged dialysis.[100] However, since the protein content of proteoglycans is quite low, this protein probably only accounts for a few percent of total vesicle protein. Immunochemical studies[15,16] indicate that the proteoglycan is on the inside of the vesicle membrane (see Section 4.3).

Exactly how the vesicle acquires the extraordinarily high concentrations of acetylcholine and ATP (amounting to 0.9 and 0.12 M, respectively) is not known, but there is evidence from work with isolated vesicles that the membrane contains saturable uptake systems for both acetylcholine[107,108] and ATP,[109] and the driving force is likely to be a combination of the pH gradient and an electrogenic ion gradient generated perhaps by the Ca^{2+},Mg^{2+}-activated ATPase that copurifies with vesicles[110] and is thought to be an integral vesicle membrane protein. Presumably, this ATPase is of the proton-translocating type. Imposition of a Na^+ gradient[111,112] across the vesicle membrane provides energy *in vitro* for acetylcholine (or choline) uptake, but a K^+ gradient is ineffective.[112] Conceivably, the uptake of Na^+ from the extracellular medium in exchange for acetylcholine cations during exocytosis creates a Na^+ gradient across the membrane of the reformed vesicle, the dissipation of which generates at least part of the energy for acetylcholine uptake via an electroneutral Na^+/acetylcholine antiport system which discriminates between acetylcholine and other cations in the cytosol.

Although there is much to be done on vesicle structure and function, the studies shown in Table V have enabled a self-consistent biochemical and biophysical model of vesicle structure to be worked out that is also consistent with the morphology.

Table VI summarizes the differences that have been found in the composition of the electromotor synaptic vesicles and its presynaptic plasma membrane. As with brain synapses, there are considerable differences in compo-

Table VI
Comparison of the Composition of Torpedo Electromotor Synaptic Vesicles with That of Presynaptic Plasma Membranes

Component	Vesicles	Presynaptic plasma membrane
Proteins[a 2,99]		
0_1 (220)	±	−
0_2 (200)	±	±
1 (160), 2 (147)	+ + + +	±
4 (115)	±	±
5 (100)	−	+ + + + + +
6_1 (90)	±	−
6_2 (85)	−	±
7_1 (76)	±	+ + +
7_2 (52)	±	±
8 (42)	+ + + +	+ +
10 (35)	±	−
11a (35)	−	+ + +
11 (34)	+ + + +	−
12 (32)	±	−
13 (25)	+ + + +	−
13a (25)	−	±
Enzymes[b 110]		
Na^+,K^+-ATPase (= 5)	−	+
Mg^{2+},Ca^{2+}-ATPase	+	−
Cholinesterase (= 7_1)	−	+
Permeases[b 106–109,113,114]		
Choline	−	+
Acetylcholine	+	−
Adenosine	−	+
ATP (= 11)	+	−
Antigens[b 62,64]		
I (= 11a)	−	+
VI (= 7_1)[c]	−	+
CHOL-1	−	+
Proteoglycan	+	−

[a] Symbols: −, less than 1%; ±, 1–5%; +, 6–10%; + +, 11–15%; + + +, 16–20%; + + + +, 21–25%; + + + + + +, 31–35% of total protein in gel measured by densitometric scanning of Coomassie-blue-positive peptides. Components 3, 9, and 14 as originally described[72] are present in less than 1% amounts in either membrane and have been omitted.
[b] Symbols: +, present; −, absent.
[c] This antigen was numbered V in ref. 2.

sition, which again supports the notion of the conservation of vesicle membrane composition through cycles of exo- and endocytosis. The core proteoglycan has been postulated as a possible stabilizer of vesicle composition during such recycling.[16]

The inorganic cation contents of vesicles listed in Table V were obtained[102] on preparations of less than maximum purity and need confirmation. It is clear from this and other work[105,111,112] that the vesicle membrane is fairly permeable to univalent ions and that the univalent ion content can be influenced by the preparative procedure. The divalent ion content is more stable.

As far as proteins are concerned,[99] the only major component in common is actin (no. 8). Like the vesicle, the presynaptic plasma membrane has a relatively simple protein pattern. The major components (apart from actin) are 5 (probably the Na^+,K^+-ATPase), 7_1 (cholinesterase), and 11a. The last two are antigenic, so much so that antisera to highly purified vesicles contain antibodies to them[62] even though contamination of the vesicle preparation with fragments of plasma membrane must have been, in biochemical terms, negligible. Such antibodies can be removed by adsorption with plasma membranes. Antisera to presynaptic plasma membranes do not usually contain antibodies to vesicle antigens but do have antibodies to highly antigenic components of the basement membrane and noninnervated face, fragments of which are presumably contaminants of the presynaptic plasma membrane preparations. When these are removed, the main antigens recognized are the protein components 7 and 11a and a glycolipid, probably a pentasialoganglioside, which has been designated CHOL-1.[64] This is not present in vesicles but is present as a specific component in all peripheral and central cholinergic presynaptic plasma membranes so far tested, but not in other types of synapse.

The difference in the antigens of the two membranes has thus made it possible to develop antisera specific for the vesicles and for the presynaptic plasma membranes, respectively, which cross react with immunochemically identical antigens in other cholinergic neurons. Such sera are expected to have numerous applications, some of which are mentioned below.

4. VESICLE RECYCLING AND METABOLIC HETEROGENEITY

4.1. The Preferential Release of Newly Synthesized Transmitter and the Concept of Vesicle Metabolic Heterogeneity

A large number of experiments with radiolabeled transmitter (norepinephrine)[17] or transmitter precursor (acetylcholine)[18] show that the recently taken up (norepinephrine) or synthesized (acetylcholine) transmitter is preferentially released. The released transmitter is invariably found to have a higher specific radioactivity than the main fraction of isolated versicles.[19,115] These findings can quite simply be explained on a two-compartment model of transmitter storage[116,117] combined with the concept of vesicle metabolic heterogeneity.[10,11,118]

The two compartment model assumes that in a resting terminal there are two pools of transmitter, a relatively small cytoplasmic pool, which is subject to futile recycling and rapidly comes into equilibrium with extracellular label, and a much larger vesicular pool, which, being fully charged with transmitter, does not appreciably exchange with the cytoplasmic pool. A recent estimate puts the size of the cytoplasmic pool at 20% of the total transmitter store in the electromotor synapse.[117]

On stimulator, the small fraction of vesicles near the presynaptic plasma membrane preferentially discharge their content of transmitter into the synaptic cleft and are refilled from the cytoplasm. They thus come into equilibrium with

Table VII
Recycling of Vesicular GAG[16a]

Treatment	Automatic exposure time (s) (mean ± S.D.)	P
No stimulation	49 ± 12	<0.01%
Stimulation (5 Hz, 10 min)	29 ± 10	
Recovery (5 h)	42 ± 11	<0.01%
Restimulated	35 ± 8	

[a] The figures are for the mean automatic exposure times, in seconds (± SD, n = 60), for the ventral surface of electrocytes in sections of *Torpedo* electric organ treated with an antiserum specific for vesicular GAG that was subsequently labeled with FITC-labeled IgG. The exposure times are a reciprocal measure of fluorescence intensity. Note the increase in fluorescence (reduction in exposure time) as a result of stimulation (at 5 Hz for 10 min) via the lobe, the return to control, unstimulated values on recovery (5 h), and the reappearance of fluorescence on restimulation. The exposure times are adjusted for varying section thicknesses between the coembedded bracketed pairs. The results are taken to indicate exterioration of a previously inaccessible vesicle core antigen by stimulus-induced recovery and its reexternalization on restimulation. When the antigen was rendered fully accessible by acetone-treatment, exposure times for all sections was equal at 54 ± 9 s (n = 16), showing conservation of total vesicular GAG.

the cytoplasm and with extracellular label. As stimulation proceeds, more of the "reserve" population of vesicles are recycled and become labeled.[10,11] The varying degrees of recycling and labeling through the vesicle population are expressed by the term "vesicle metabolic heterogeneity."

Under steady-state conditions and for short stimuli that do not recruit significant numbers of reserve vesicles with unlabeled transmitter, the specific activity of released transmitter should equal that of the recycling vesicle population and thus that of the cytoplasmic pool, and this prediction has been experimentally verified.[116]

Evidence for this model has mainly come from the *Torpedo* electromotor synapse, where recycling causes a change in the physical properties of vesicles large enough to enable the recycling and reserve vesicles to be separated in a zonal density gradient (Fig. 1).[11,116] The peak of reserve vesicles (VP_1) has the same density and water spaces as the vesicle peak from resting tissue; the peak of recycling vesicles (VP_2) has a higher density and lower water content.[119] When the innervated blocks from which these fractions are isolated are perfused during stimulation with radioactive precursors, the vesicles in the reserve fraction are only slightly labeled with newly synthesized transmitter, whereas those in the recycling fraction are heavily labeled. Restimulation of the labeled block releases transmitter of approximately the same specific radioactivity as that of the recycling vesicles. That these vesicles have undergone cycles of exo- and endocytosis during transmitter release and refilling is shown by experiments in which dextran particles appear exclusively in the lumen of the VP_2 fraction.

These experiments have been criticized[24] on the grounds that fraction VP_2 is contaminated with vesiculated membrane fragments and/or small synaptosomes, which might have entrapped part of the (highly labeled) cytoplasmic pool. This objection has been disposed of in two ways, first by removing membrane contamination by passing the vesicles through columns of suitably treated

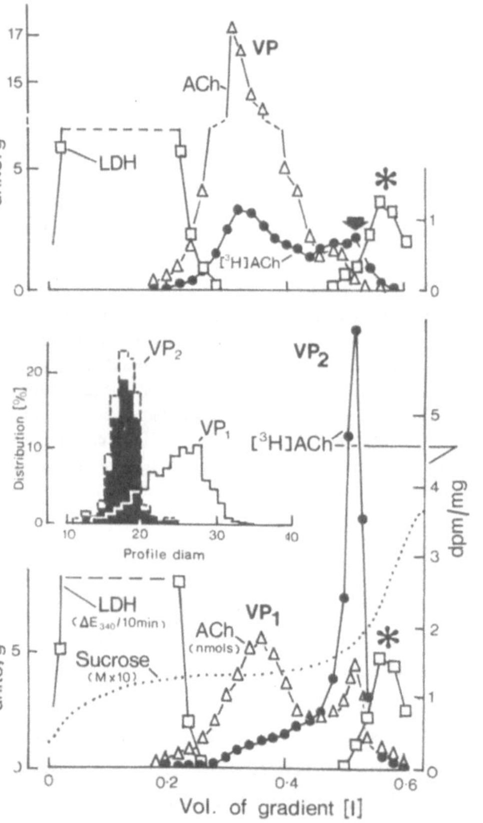

Fig. 1. Distribution[11,118] of the soluble cytoplasmic marker lactate dehydrogenase (LDH, squares) and vesicular acetylcholine (ACh, triangles) in a zonal density gradient after centrifuging cytoplasmic extracts of (top) an unstimulated innervated perfused block of *Torpedo* electric tissue or (bottom) a similar block stimulated through the nerve at 0.1 Hz for 3 h into the gradient until isopycnic conditions had been established. Note (top) the separation of the ACh-rich vesicle peak (VP) from the LDH-rich soluble protein peak (SP) and from a membrane peak (MP) that contains a small amount of occluded LDH, and (bottom) the fall on stimulation in the size of the original ACh peak (here designated VP_1) with the appearance of a second, denser ACh peak (VP_2), usually represented in unstimulated tissue by an inconspicuous shoulder (arrow, top diagram). In this experiment, radioactive acetate and dextran were added to the perfusate. The filled circles show the distribution of labeled ACh. It will be seen that VP in unstimulated tissue and the corresponding VP_1 in stimulated tissue acquire relatively little radioactivity, but the vesicle peak generated by stimulation (VP_2) acquires much. Dextran particles are seen, in intact tissue, to be absent from vesicles in unstimulated tissue, but as stimulation proceeds, they are present in increasing numbers of vesicles whose mean profile diameter is 20–30% less than that of the unlabeled population. The insert diagram shows that this size difference persists in the isolated vesicle and that dextran particles are present exclusively in fraction VP_2. These results are taken to mean that VP_2 is the fraction of recycling vesicles (1) generated by stimulation, (2) discharging transmitters, and (3) avidly taking up newly synthesized transmitter from the cytoplasm, whereas VP_1 represents the original population of reserve, resting, metabolically inert, and fully charged vesicles. All changes induced by stimulation are reversed during a subsequent rest period.

porous glass beads[120]—the purified vesicles retain their radioactivity—and secondly by tagging preparations with differently labeled synaptosomes; the synaptosomes sediment to a denser part of the gradient and are clearly distinguishable from VP_2.[11]

Vesicle metabolic heterogeneity is also a feature of mammalian brain and was, indeed, first observed there.[121] When synaptosomes and, from them, synaptic vesicles are prepared from brain cortex whose acetylcholine has been labeled *in vivo*, the fraction (D) of monodispersed synaptic vesicles is much less labeled than either the cytoplasmic fraction (O) or a fraction (H) of vesicles that remain in close association with the presynaptic plasma membrane after synaptosome disruption. It has been plausibly argued[21] that the vesicles of fraction H are those near the presynaptic plasma membrane and actively engaged in recycling; thus, they can be regarded as analogues of the VP_2 vesicles from *Torpedo*. It should be pointed out that the relative specific radioactivities of the transmitter in fractions D and H are critically dependent on the conditions of vesicle isolation: the fractions must be prepared from the total fraction of water-shocked synaptosomes (W) and not from the supernatant (W_s) obtained after centrifuging this fraction; centrifuging fraction W removes most of the highly labeled vesicles along with synaptosomal debris, as is shown by the relatively high specific radioactivity of the acetylcholine in the discarded pellet (W_p). This accounts for some discrepancies in the literature.[21] The specific radioactivity of released acetylcholine is close to that found in fraction H when this is prepared from total shocked synaptosomes.

4.2. *The Use of False Transmitters*

Although agreement between the specific radioactivity of the transmitter released on stimulation and that of the transmitter stored in recycling vesicles is a necessary condition for the correctness of the vesicle theory, its verification[21,116] does not prove the vesicle theory: the transmitter could have originated from the similarly labeled cytoplasmic pool. To label the cytoplasmic and recycling vesicular pools differently, recourse has been made to false transmitters, that is, transmitter analogues that are taken up into the cytoplasm and from thence into the pool of recycling vesicles and are released, along with the endogenous transmitter, on stimulation. Because of differences in the specificity of the uptake mechanisms for transmitter analogues (or their precursors) in the presynaptic plasma membranes and the vesicle membrane, false transmitters can be expected to distribute themselves differently, relative to the newly synthesized or newly taken up endogenous transmitter, between the cytoplasmic and vesicular pools. One can then compare the ratio in which false transmitters are released with the ratio in which they are present in the vesicular and cytoplasmic pools (Fig. 2). Invariably, the ratio for released transmitters is the same as that for recycling vesicles even when this differs greatly from the ratio for the cytoplasm. Such studies, both in the adrenergic and the cholinergic system,[122–124] have provided strong additional support for the vesicle theory.

4.3. *Immunohistochemical Evidence for Vesicle Exo- and Endocytosis*

As mentioned above, it has recently been possible to prepare[62–64] from *Torpedo* electromotor synaptic vesicles and presynaptic plasma membranes antisera, from rabbit and sheep, respectively, that, after suitable adsorption to remove antibodies to highly antigenic contaminants, are selective for antigens specific for these two structures.

The vesicle antigen is a proteoglycan to which the heparanlike glycosaminoglycan (GAG) previously isolated from these vesicles is conjugated.[62,125] The presynaptic plasma membrane antigen (CHOL-1) has been tentatively identified as a pentasialoganglioside,[64] although an antibody is also present in this

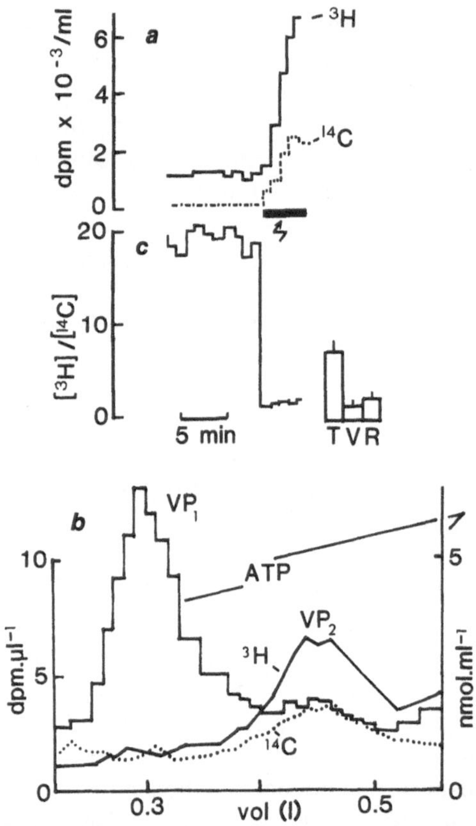

Fig. 2. Experiments[123] illustrating the incorporation of false transmitters into the fraction (VP_2) of recycling vesicles and their release from this fraction. Blocks of electric organ were loaded with [^{3}H]homocholine and [^{14}C]choline after having been depleted of endogenous transmitter by stimulation through the nerve at 1 Hz for 30 min. On restimulation (10 Hz for 5 min) 8 h later in the presence of paraoxon (to stabilize released esters) and hemicholinium-3 (to inhibit reuptake of label), both radioactive labels were released, ^{3}H as a mixture of homocholine and acetylhomocholine, ^{14}C as acetylcholine (a). The labels were incorporated exclusively in the fraction of recycling vesicles (b). The ratio in which the true and false transmitters were released (c, block R) is the same as in the fraction of recycling vesicles (c, block V) and much lower (because of preferential vesicular uptake of acetylcholine relative to acetylhomocholine plus homocholine) than in whole tissue (c, block T). By contrast, the ratio in which the labels were released during the prestimulation resting period (c, left) is much higher than even the tissue ratio, reflecting a preponderance of homocholine and acetylhomocholine relative to acetylcholine in the cytoplasm. The tissue ratio observed is consistent with a 25% cytoplasmic acetylcholine pool and a 77% cytoplasmic homocholine plus acetylhomocholine pool. In other experiments,[107] it was deduced that the proportion of acetylcholine that is cytoplasmic in resting tissue without any loading stimulus is 22%. Blocks are mean values of eight experiments: bars are S.E.M.s.

preparation directed against the protein component 11a (mol. wt. 33,000; Table VI).[62]

Both antisera cross react specifically with both peripheral and central mammalian cholinergic terminals.[15,16,63] The specificity of the anti-CHOL-1 serum has been documented (1) by its ability to initiate selective complement-induced lysis of the cholinergic subpopulation of brain synaptosomes,[49,63,64] (2) by the disappearance of staining in the hippocampus when the cholinergic pathway to this brain region is interrupted,[63] (3) by the similarity of the staining pattern in a variety of peripheral and central regions to that obtained with acetylcholinesterase reagents[63]—the anti-CHOL-1 pattern is, however, more restricted and apparently limited to nerve terminals, and (4) by the complementary nature of the staining to that obtained with fluorescence-labeled α-bungarotoxin in neuromuscular junctions and retina.[62,63]

The specificity of the anti-GAG serum has been shown by the general similarity in its staining reaction to that of anti-CHOL-1; this is, however, less completely confined to nerve terminals and tends to extend along tracts (C. Zimmermann, unpublished data). For a reaction with the anti-GAG serum to occur, Triton treatment is necessary to expose the antigen. This is particularly clear in the electromotor system and indicates that the antigen is internal to the vesicle. On stimulation (Table VII), the antigen is rendered accessible to the antibody even without detergent treatment.[15] This indicates that stimulation has caused vesicle exocytosis and exteriorization of the internal antigen. After recovery, the antigen is reinternalized, as shown by the loss of the tissue's capacity to be stained. That the antigen has been reinternalized and not simply degraded is shown by the reappearance of the reaction on restimulation.[16]

A somewhat similar but apparently less specific vesicle antiserum has been described by another group.[126] This enables the massive irreversible exocytosis induced by lanthanum treatment to be detected, but not that induced by stimulation.[127]

Such anti-GAG sera are proving useful for investigating the biosynthesis and axonal transport of vesicles. This is described in more detail below (Section 5).

4.4. The Cytoplasmic Theory of Transmitter Release

In several recent review articles,[21–25] it has been proposed that acetylcholine might be released on stimulation directly from the cytoplasmic pool, presumably via "gates" in the presynaptic plasma membrane that operate in such a way as to generate quanta. Whereas the vesicle theory of transmitter release, although not conclusively proved in each and every type of synapse does subsume an impressive range of morphological, physiological, and biochemical findings, virtually no positive evidence has ever been offered for the cytoplasmic theory, and several known facts are difficult to reconcile with it.[128,129] These are as follows:

1. Acetylcholine release should be but is not affected by the state of depolarization of the nerve terminal plasma membrane.
2. The release of acetylcholine should itself be electrogenic but is not.
3. Changes in cytoplasmic acetylcholine concentration should affect the size of the quantum and the amount of acetylcholine released but do not.

These difficulties seem to have prompted the formulation of a modified cytoplasmic theory in which the cytoplasmic compartment is considered to be heterogeneous, with a small subpool bound to a structure called an "operator" on the inner surface of the plasma membrane, which binds and releases cytoplasmic acetylcholine.[24,25] To explain the results with false transmitters, this operator must also be credited with the same specificity for acetylcholine analogues as the vesicles. It becomes conceptually difficult to distinguish the "operator" from the population of recycling vesicles, which are, of course, those nearest the membrane.

The protagonists of the cytoplasmic theory also have difficulty in assigning a role for the vesicles. In two recent publications,[26,130] cholinergic vesicles from *Torpedo* are stated to take up Ca^{2+}, and it has been suggested that the function of the vesicles is to act as a Ca^{2+}-sequestering mechanism.[26] Ionized

Ca^{2+} admitted during stimulation is thought to be sequestered within vesicles, and this is thought to result in the intracellular release of transmitter, thereby restoring its concentration in the cytoplasmic compartment after the depletion caused by transmitter release. Later, the Ca^{2+} is released by vesicular exocytosis.

Presumably, such a mechanism would operate at noncholinergic synapses too; it is unlikely that vesicles have one function in cholinergic endings and a different one in adrenergic ones. But intracellular release of vesicular transmitter is not consistent with the known facts about false transmitters in either adrenergic[27] or cholinergic[122] synapses. It must also be appreciated that the observed uptake of Ca^{2+} by synaptic vesicle preparations would be accounted for by as little as 1% contamination by vesiculated fragments of a Ca^{2+}-sequestering endoplasmic reticulum similar to those that have been isolated from muscle. No attempt has been made to prove that such particles were absent from the preparations studied.

There seems no reason to doubt that nonquantized transmitter release—at least from cholinergic terminals—can and does occur as a result of diffusion from the cytoplasmic compartment. The release of transmitter from resting motor terminals in diaphragm *in vitro* is much greater than can be accounted for by the observed rate of miniature potentials and is differently affected by changes in the cation composition of the bathing fluid.[131] Under normal conditions, the acetylcholine released will be hydrolyzed by cholinesterase, and the products taken up by the terminal; this may well be the basis for the "futile recycling" of the cytoplasmic compartment, which is a central feature of the two-compartment model described above.[116] Direct evidence for the release of the cytoplasmic pool at rest has been obtained by means of false transmitters (Fig. 2).[122,123] However, there is no evidence for anything more than a slight increase in the nonquantized release on stimulation, in contrast to the many-hundred-fold increase observed in quantized (i.e., vesicular) release.[132]

5. THE LIFE CYCLE OF THE SYNAPTIC VESICLE

5.1. Genesis and Axonal Transport

All parts of the cell, including the cell organelles, have only a limited life and are constantly being renewed. The synaptic vesicles are no exception. Labeling experiments have suggested a half-life of about 21 days.[133] In adrenergic neurons, there is convincing evidence that dense-cored vesicles are synthesized in the cell-body and conveyed to the terminals by fast axonal transport,[134] undergoing chemical modification on the way.[27,28] In other types of neuron, the evidence is less complete. Possibly dense-cored vesicles lose at least part of their core with each cycle of exo- and endocytosis and do have to be renewed more frequently; if so, axonal transport might play a larger part in their life cycle than in that of synaptic vesicles and so would be more easily detectable. The characteristic morphology of dense-cored vesicles makes them easier to spot in the cytoplasm of the cell body and axon,[135] and the transmitter can be readily detected by fluorescence histochemistry. In electromotor neurons, mRNAs coding for vesicle protein have been isolated from their perikarya,[136] and axonal transport of acetylcholine[137–139] and of a stable, presumed vesicular fraction of ATP[139] has been detected; it can be deduced that both compartments are conveyed by fast transport.[139] The ratio of acetylcholine to stable ATP is, however, much lower[139] than that in vesicles isolated from the nerve terminals; transmitter uptake during transport may occur with cholinergic as well as with adrenergic vesicles.[27,28]

Further evidence from the axonal transport of synaptic vesicles comes from immunofluorescence histochemistry with an antiserum raised to vesicular GAG.[15] After Triton treatment, punctate fluorescence is observed in electromotor perikarya, especially in the region near the axon hillock, and, in confirmation of earlier work with a less well characterized serum,[140,141] the immunoreactive material accumulates above a ligature applied to the electromotor axons. In other work,[142] ^{35}S injected into the region of the cells of origin was found to be incorporated into rapidly transported macromolecules in both guinea pig vagus and *Torpedo* electromotor nerves; on gel electrophoresis, a proportion of the organically bound ^{35}S extracted from the axons was found to migrate as an Alcian-blue-positive band, which was immunochemically identified as vesicular GAG. In guinea pig vagus and *Torpedo* electromotor nerves, the ^{35}S-labeled vesicles eventually reached the nerve terminals and could be isolated in the usual way.

There seems little reason to doubt, therefore, that synaptic vesicles, like dense-cored vesicles, are made in the cell body and transported to the synapse[31] to replace vesicles that have lost their function after undergoing several, perhaps many, cycles of transient exocytosis. Whether these exhausted, functionless vesicles are retrieved by retrograde transport for rehabilitation in the Golgi region of the cell is at present obscure.

5.2. Exo- and Endocytosis

The mechanism of exocytosis itself is not understood. Several processes must contribute to it: movement of vesicles towards the presynaptic plasma membrane; contact with the plasma membrane, often at specialized "release sites," fusion of the vesicle and plasma membranes followed by "fission" (communication of the vesicle interior with the extracellular space via a hole); recovery of the vesicle membrane (endocytosis); and refilling of the vesicle. In some unknown way, the vesicle must interact with elements of the cytoskeleton during the cycle of exo- and endocytosis, and, since the process of transmitter release is Ca^{2+} dependent, Ca^{2+} must play a part in this interaction and/or in membrane fusion. *In vitro*, Ca^{2+} induces vesicle aggregation,[143] a process that is diffusion controlled, indicating that each random collision has a high probability of resulting in the formation of a stable complex. A protein, synexin, has been isolated from the cytoplasm of adrenal medullary glands and enhances the Ca^{2+}-induced coacervation of chromaffin granules, but not the Mg^{2+}-induced coacervation[144,145]; thus, synexin confers Ca^{2+} specificity on the reaction. There is some evidence for the existence of a similar protein in electromotor nerve terminals.[146]

When vesicles are incubated with ^{32}P-labeled ATP in the presence of cyclic AMP and Ca^{2+}, certain vesicle proteins are phosphorylated and show up as dark bands in autoradiographs of gels after electrophoresis.[147,148] Prominent among these are two polypeptides of M_r 80 and 85 kilodaltons, collectively known as protein I or synapsin I. Immunohistochemistry shows that synapsin I is present in all central synapses and, within the synapse, forms a coat around the vesicles. One of three phosphorylation sites is on a collagenlike tail and is catalyzed by an endogenous Ca^{2+}-activated protein kinase. The substrate and kinase copurify with the vesicles through several stages including chromatography on columns of porous glass beads. The binding of the protein to vesicles is favored by low ionic strength; it is eluted by raising the salt concentration, and its affinity for vesicles is reduced by phosphorylation. It is proposed that the protein is involved in some way in the interaction of the vesicle with the terminal cytoskeleton to effect translocation of vesicles within the cytoplasm and that its phosphorylation and dephosphorylation somehow regulate this process.

In other work on Ca^{2+}-stimulated phosphorylation of vesicle proteins, a key protein has been identified as calmodulinlike; the process is thought to be concerned with Ca^{2+}-requiring transmitter release.[149]

6. *FUTURE PERSPECTIVES*

Much progress has been made since synaptic vesicles were first isolated[36] in a morphologically but not biochemically pure state almost two decades ago both in the technology of isolation[50,67,70,72,73,120] and in understanding the composition and function of these organelles. Much of this progress has, however, been achieved with an admittedly somewhat atypical synaptic vesicle, that of the electromotor nerve terminal of *Torpedo*. It is surprising that so little recent work has been done with vesicles derived from mammalian tissues and of conventional size. Perhaps this is because of the inherent unattractiveness in studying preparations heterogeneous with respect to transmitter type. However, it should be possible to overcome this difficulty either by using sources of vesicles especially rich in one type or another or by the use of immunoselective methods based on antibodies able to recognize components specific for synaptosomes or vesicles of defined transmitter type.

Nuclear magnetic resonance promises to provide a noninvasive technique[105] whereby the cytoplasmic and vesicular compartments of transmitters, or components copackaged with transmitters, can be recognized and their fate on stimulation determined. Gas chromatography–mass spectrometry is a powerful analytical technique which enables the metabolism and compartmentation of acetylcholine[117] and other transmitters, and their false-transmitter analogues,[150] to be studied.

The electric organ of *Torpedo*[56] combined with improved methods of comminution,[70] high-resolution centrifugal density-gradient separation,[70] and new column chromatographic techniques[120] have proved ideal for the bulk separation and purification to constant composition of cholinergic synaptic vesicles. Biophysical and biochemical studies have led to a self-consistent and fairly detailed structural model. Such vesicles retain the capacity to take up acetylcholine and ATP *in vitro* by saturable carrier mechanisms, but the precise mechanism whereby vesicles are able to store transmitter and ATP in such high concentrations is not fully understood. The study of recycling vesicles suggests that osmotically induced changes in water content accompany the release and uptake of acetylcholine and ATP,[119] and osmotic forces may play a role in vesicle fusion and exocytosis, as they have been shown to do in model systems.[151,152] However, the molecular basis for the mobilization and recycling of vesicles remains obscure in this as in other systems involving exo- and endocytosis.

The discovery that at least one vesicle component, the proteoglycan,[100] is both a specific component of the vesicle core and highly antigenic[15,62] has made possible the immunochemical identification of a similar substance in other cholinergic nerve terminals. This may provide a powerful tool for studying the formation, transport, recycling, and subsequent fate of the vesicles.

Although many gaps remain in our knowledge of how transmitters are packaged and vesicles transported and recycled, the vesicle theory of transmitter storage and release continues to accommodate more of the known facts and experimental findings than alternatives. Transmitter storage and release can thus be seen as a phylogenetically ancient evolutionary adaption of a widespread and general cellular mechanism to the manifold and specialized needs of synaptic transmission. As has often been true in the history of neurochemistry, the solution of the problems raised by vesicular storage and release may well ultimately depend on developments in other areas of cell biology.

REFERENCES

1. Zimmermann, H., 1979, *Neuroscience* **4**:1773–1804.
2. Whittaker, V. P., and Stadler, H., 1980, *Proteins of the Nervous System*, 2nd ed. (R. A. Bradshaw and D. M. Schneider, eds.), Raven Press, New York, pp. 231–255.
3. MacIntosh, F. C., and Collier, B., 1976, *Handbook of Experimental Pharmacology*, Volume 42 (E. Zaimis, ed.), Springer-Verlag, Heidelberg, pp. 99–228.
4. Sjöstrand, F., 1953, *J. Appl. Physiol.* **24**:1422.
5. Palay, S. L., and Palade, G. E., 1954, *Anat. Rec.* **118**:336.
6. De Robertis, E. D. P., and Bennett, H. S., 1954, *Fed. Proc.* **13**:35.
7. Castillo, J. del, and Katz, B., 1955, *J. Physiol. (Lond.)* **128**:396–411.
8. Mountford, S., 1963, *Exp. Cell Res.* **9**:403–418.
9. Ceccarelli, B., Hurlbut, W. P., and Mauro, A., 1973, *J. Cell Biol.* **57**:499–524.
10. Zimmermann, H., and Denston, C. R., 1977, *Neuroscience* **2**:695–714.
11. Zimmermann, H., and Denston, C. R., 1977, *Neuroscience* **2**:715–730.
12. Clark, A. W., Hurlbut, W. P., and Mauro, A., 1972, *J. Cell Biol.* **52**:1–14.
13. Model, P. G., Highstein, S. M., and Bennett, M. V. L., 1975, *Brain Res.* **98**:209–228.
14. Ceccarelli, B., and Hurlbut, W. P., 1975, *J. Physiol. (Lond.)* **247**:163–168.
15. Jones, R. T., Walker, J. H., Stadler, H., and Whittaker, V. P., 1982, *Cell Tissue Res.* **223**:117–126.
16. Jones, R. T., Walker, J. H., Stadler, H., and Whittaker, V. P., 1982, *Cell Tissue Res.* **224**:685–688.

17. Kopin, I. J., Breese, G. R., Kraus, K. R., and Weise, V. K., 1968, *J. Pharmacol. Exp. Ther.* **161**:271–278.
18. Collier, B., 1969, *J. Physiol. (Lond.)* **205**:341–352.
19. Chakrin, L. W., Marchbanks, R. M., Mitchell, J. F., and Whittaker, V. P., 1972, *J. Neurochem*. **19**:2727–2736.
20. Levy, W. B., Haycock, J. W., and Cotman, C. W., 1976, *Brain Res.* **119**:243–256.
21. Schwarzenfeld, I. von, 1979, *Neuroscience* **4**:477–493.
22. Marchbanks, R. M., 1975, *Int. J. Biochem.* **6**:303–312.
23. Marchbanks, R. M., 1977, *Synapses* (G. A. Cottrell and P. N. R. Usherwood, eds.), Blackie, Glasgow, pp. 81–101.
24. Dunant, Y., and Israël, M., 1979, *Trends Neurosci.* **2**:130–132.
25. Tauc, L., 1979, *Biochem. Pharmacol.* **28**:3493–3498.
26. Israël, M., Manaranche, R., Marsal, J., Meunier, F. M., Morel, F., Franchon, P., and Lesbats, B., 1980, *J. Membr. Biol.* **54**:115–126.
27. Smith, A. D., 1972, *Biochem. Soc. Symp.* **36**:103–131.
28. Lagercrantz, H., 1971, *Acta Physiol. Scand.* [*Suppl.*] **366**:1–44.
29. Robertson, D. J., 1961, *Regional Neurochemistry* (S. S. Kety and J. Elkes, eds.), Pergamon Press, Oxford, pp. 491–534.
30. Blaschko, H., Hagen, P., and Welch, A. D., 1955, *J. Physiol. (Lond.)* **129**:27–49.
31. Hillarp, N.-A., Högberg, B., and Nilson, B., 1955, *Nature* **176**:1032–1033.
32. Whittaker, V. P., 1959, *Biochem. J.* **72**:694–706.
33. Gray, E. G., and Whittaker, V. P., 1962, *J. Anat.* **96**:79–81.
34. Johnson, M. K., and Whittaker, V. P., 1962, *Acta Neurol. Scand.* [*Suppl.*] **160**:60.
35. Johnson, M. K., and Whittaker, V. P., 1963, *Biochem. J.* **88**:404–409.
36. Whittaker, V. P., Michaelson, I. A., and Kirkland, R. J. A., 1964, *Biochem. J.* **90**:293–303.
37. De Robertis, E., Arnaiz, G. R. de L., and de Iraldi, A. P., 1962, *Nature* **194**:794–795.
38. Whittaker, V. P., 1964, *Prog. Brain Res.* **8**:90–117.
39. Germain, M., and Proulx, P., 1965, *Biochem. Pharmacol.* **14**:1815–1819.
40. Wiegandt, H., 1967, *J. Neurochem.* **14**:671–674.
41. Lapetina, E. G., Soto, E. F., and De Robertis, E., 1968, *J. Neurochem.* **15**:437–445.
42. Whittaker, V. P., and Sheridan, M. N., 1965, *J. Neurochem.* **12**:363–372.
43. Whittaker, V. P., 1966, *Mechanisms of Release of Biogenic Amines* (U. S. von Euler, S. Rosell, and B. Uvnäs, eds.), Pergamon Press, Oxford, pp. 147–163.
44. Krnjević, K., and Mitchell, J. F., 1961, *J. Physiol. (Lond.)* **155**:246–262.
45. MacIntosh, F. C., 1959, *Can. J. Biochem.* **37**:343–356.
46. Kuffler, S. W., and Yoshikama, D., 1975, *J. Physiol. (Lond.)* **251**:465–482.
47. Katz, B., and Miledi, R., 1972, *J. Physiol. (Lond.)* **224**:665–699.
48. Anderson, C. R., and Stevens, C. F., 1973, *J. Physiol. (Lond.)* **235**:655–691.
49. Richardson, P. J., 1981, *J. Neurochem.* **37**:258–260.
50. Nagy, A., Baker, R. R., Morris, S. J., and Whittaker, V. P., 1976, *Brain Res.* **109**:285–309.
51. Nagy, A., Várady, G., Joó, F., Rakonczay, Z., and Pilc, A., 1977, *J. Neurochem.* **29**:449–459.
52. Wilson, W. S., Schulz, R. A., and Cooper, J. R., 1973, *J. Neurochem.* **20**:659–667.
53. Dowe, G. H. C., Kilbinger, H., and Whittaker, V. P., 1981, *J. Neurochem.* **35**:993–103.
54. Whittaker, V. P., and Roed, I. S., 1982, *Compartmentation and Transmitter Interaction* (H. F. Bradford, ed.), Plenum Press, New York.
55. Zimmermann, H., Stadler, H., and Whittaker, V. P., 1981, *Chemical Transmission 75 Years* (L. Stjärne, P. Hedqvist, H. Lagercrantz, and A. Wennmalm, eds.), Academic Press, London, pp. 92–104.
56. Whittaker, V. P., and Zimmermann, H., 1976, *Biochemical and Biophysical Perspectives in Marine Biology* (D. C. Malins and J. R. Sargent, eds.), Academic Press, London, pp. 67–116.
57. Fritsch, G., 1890, *Die elektrischen Fische*, Von Veit, Leipzig.
58. Fox, G. Q., and Richardson, G. P., 1978, *J. Comp. Neurol.* **179**:677–697.
59. Fox, G. Q., and Richardson, G. P., 1979, *J. Comp. Neurol.* **185**:293–316.
60. Krenz, W.-D., Tashiro, T., Wächtler, K., Whittaker, V. P., and Witzemann, V., 1980, *Neuroscience* **5**:617–624.

61. Mellinger, J., Belbenoît, P., Ravaille, M., and Szabo, T., 1978, *Dev. Biol.* **67**:167–188.
62. Walker, J. H., Jones, R. T., Obrocki, J., Richardson, G. P., and Stadler, H., 1982, *Cell Tissue Res.* **223**:101–116.
63. Jones, R. T., Walker, J. H., Richardson, P. J., Fox, G. Q., and Whittaker, V. P., 1981, *Cell Tissue Res.* **218**:355–373.
64. Richardson, P. J., Walker, J. H., Jones, R. T., and Whittaker, V. P., 1982, *J. Neurochem.* **38**:1605–1614.
65. Sheridan, M. N., and Whittaker, V. P., 1964, *J. Physiol.* (*Lond.*) **175**:25–26P.
66. Sheridan, M. N., Whittaker, V. P., and Israël, M., 1966, *Z. Zellforsch.* **74**:291–3077.
67. Ohsawa, K., Dowe, G. H. C., Morris, S. J., and Whittaker, V. P., 1979, *Brain Res.* **161**:447–457.
68. Miledi, R., Molinoff, P., and Potter, L. T., 1971, *Nature* **229**:554–557.
69. Erdélyi, L., and Krenz, W.-D., 1984, *Comp. Biochem. Physiol.* (in press).
70. Whittaker, V. P., Essman, W. B., and Dowe, G. H. C., 1972, *Biochem. J.* **128**:833–846.
71. Whittaker, V. P., 1956, *Pharmacol. Rev.* **18**:401–412.
72. Tashiro, T., and Stadler, H., 1978, *Eur. J. Biochem.* **90**:479–487.
73. Israël, M., Gautron, G., and Lesbats, B., 1970, *J. Neurochem.* **17**:1441–1450.
74. Wagner, J. A., Carlson, H. S., and Kelly, R. B., 1978, *Biochemistry* **17**:1199–1205.
75. Zimmermann, H., and Denston, C. R., 1976, *Brain Res.* **111**:365–376.
76. Dowdall, M. J., and Whittaker, V. P., 1973, *J. Neurochem.* **20**:921–935.
77. Whittaker, V. P., 1966, *Ann. N.Y. Acad. Sci.* **137**:982–998.
78. Whittaker, V. P., 1977, *Naturwissenschaften* **64**:606–611.
79. Breckenridge, W. C., Morgan, I. G., Zanetta, J. P., and Vincendon, G., 1973, *Biochim. Biophys. Acta* **320**:681–686.
80. Hosie, R. J. A., 1965, *Biochem. J.* **96**:404–412.
81. Fonnum, F., 1967, *Biochem. J.* **103**:262–270.
82. Breckenridge, W. C., Gombos, G., and Morgan, I. G., 1972, *Biochim. Biophys. Acta* **266**:695–707.
83. Mellanby, J., and Whittaker, V. P., 1968, *J. Neurochem.* **15**:205–208.
84. Mahler, H. R., 1977, *Neurochem. Res.* **2**:119–147.
85. Lentz, T. L., and Chester, J., 1982, *Neuroscience* **7**:9–20.
86. Zisapel, N., and Zurgil, N., 1979, *Brain Res.* **178**:297–310.
87. Richter-Landsberg, C., Neuhoff, V., and Waehneldt, T. V., 1977, *J. Neurosci. Res.* **3**:103–113.
88. Nicklaus, W. J., Puszkin, S., and Berl, S., 1973, *J. Neurochem.* **20**:109–121.
89. Fonnum, F., 1970, *Drugs and Cholinergic Mechanisms in the CNS* (E. Heilbronn and A. Winter, eds.), Försvarets Forskningsanstalt, Stockholm, pp. 83–95.
90. Tsudzuki, T., 1979, *J. Biochem.* **86**:777–782.
91. Mangan, J. L., and Whittaker, V. P., 1966, *Biochem. J.* **98**:128–137.
92. Rassin, D. K., 1972, *J. Neurochem.* **19**:139–148.
93. De Belleroche, J. S., and Bradford, H. F., 1973, *J. Neurochem.* **21**:441–451.
94. Lähdesmäki, P., Karppinen, A., Saarni, H., and Winter, R., 1977, *Brain Res.* **138**:295–308.
95. Kuriyama, K., Roberts, E., and Vos, T., 1968, *Brain Res.* **9**:231–252.
96. Cleugh, J., Gaddum, J. H., Mitchell, A. A., Smith, M. W., and Whittaker, V. P., 1964, *J. Physiol.* (*Lond.*) **170**:69–85.
97. Duffy, M. J., Mulhall, D., and Powell, D., 1975, *J. Neurochem.* **25**:305–307.
98. Morris, S. J., 1980, *Neuroscience* **5**:1509–1516.
99. Stadler, H., and Tashiro, T., 1979, *Eur. J. Biochem.* **101**:171–178.
100. Stadler, H., and Whittaker, V. P., 1978, *Brain Res.* **153**:408–413.
101. Giompres, P. E., Morris, S. J., and Whittaker, V. P., 1980, *Neuroscience* **6**:757–763.
102. Schmidt, R., Zimmermann, H., and Whittaker, V. P., 1980, *Neuroscience* **5**:625–638.
103. Dowdall, M. J., Boyne, A. F., and Whittaker, V. P., 1974, *Biochem. J.* **140**:1–12.
104. Stadler, H., and Füldner, H. H., 1980, *Nature* **286**:293–294.
105. Füldner, H. H., and Stadler, H., 1982, *Eur. J. Biochem.* **12**:519–524.
106. Fenwick, E. M., and Stadler, H., 1981, *Abstracts 8th International Meeting International Society for Neurochemistry*, Nottingham, p. 168.

107. Giompres, P., and Luqmani, Y. A., 1980, *Neuroscience* **5**:1041–1052.
108. Michaelson, D. M., and Angel, I., 1980, *Life Sci.* **27**:39–44.
109. Luqmani, Y. A., 1981, *Neuroscience* **6**:1011–1021.
110. Breer, H., Morris, S. J., and Whittaker, V. P., 1977, *Eur. J. Biochem.* **80**:313–318.
111. Carpenter, R. S., and Parsons, S. M., 1978, *J. Biol. Chem.* **253**:326–329.
112. Suszkiw, J. B., 1981, *Cholinergic Mechanisms, Phylogenetic Aspects Central and Peripheral Synapses and Clinical Significance* (G. Pepeu and H. Ladinsky, eds.), Plenum Press, New York, pp. 313–320.
113. Zimmermann, H., Dowdall, M. J., and Lane, D. A., 1979, *Neuroscience* **4**:979–993.
114. Dowdall, M. J., 1977, *Biochemistry of Characterised Neurons* (N. N. Osborne, ed.), Pergamon Press, Oxford, pp. 177–216.
115. Marchbanks, R. M., and Israël, M., 1971, *J. Neurochem.* **18**:439–448.
116. Suszkiw, J. B., Zimmermann, H., and Whittaker, V. P., 1978, *J. Neurochem.* **30**:1269–1280.
117. Weiler, M., Roed, I. S., and Whittaker, V. P., 1982, *J. Neurochem.* **38**:1187–1191.
118. Zimmermann, H., and Whittaker, V. P., 1977, *Nature* **261**:633–635.
119. Giompres, P. A., Zimmermann, H., and Whittaker, V. P., 1981, *Neuroscience* **6**:775–785.
120. Giompres, P. A., Zimmermann, H., and Whittaker, V. P., 1981, *Neuroscience* **6**:765–774.
121. Barker, L. A., Dowdall, M. J., and Whittaker, V. P., 1972, *Biochem. J.* **130**:1063–1080.
122. Whittaker, V. P., and Luqmani, Y. A., 1980, *Gen. Pharmacol.* **11**:7–14.
123. Luqmani, Y. A., Sudlow, G., and Whittaker, V. P., 1980, *Neuroscience* **5**:153–160.
124. Schwarzenfeld, I. von, Sudlow, G., and Whittaker, V. P., 1979, *Prog. Brain Res.* **49**:1613–174.
125. Jones, R. T., and Walker, J. H., 1982, *Abstracts European Symposium on Cholinergic Transmission, Presynaptic Aspects, Strasbourg*, p. 60.
126. Hooper, J. E., Carlson, S. S., and Kelly, R. B., 1980, *J. Cell Biol.* **87**:104–113.
127. Wedel, R. J. von, Carlson, S. S., and Kelly, R. B., 1981, *Proc. Natl. Acad. Sci. U.S.A.* **78**:1014–1018.
128. Zimmermann, H., 1979, *Trends Neurosci.* **2**:282–285.
129. Zimmermann, H., 1981, *Chemical Transmission 75 Years* (L. Stjärne, P. Hedqvist, H. Lagercrantz, and A. Wennmalm, eds.) Academic Press, London, pp. 179–185.
130. Michaelson, D. M., Ophis, I., and Angel, I., 1980, *J. Neurochem.* **35**:116–124.
131. Mitchell, J. F., and Silver, A., 1963, *J. Physiol. (Lond.)* **165**:117–129.
132. Vizi, E. S., and Vyskočil, F., 1979, *J. Physiol. (Lond.)* **286**:1–14.
133. Hungen, K. von, Mahler, H. R., and Moore, W. T., 1968, *J. Biol. Chem.* **243**:1415–1423.
134. Dahlström, A., 1971, *Phil. Trans. R. Soc. Lond. [Biol.]* **261**:319–323.
135. Grillo, M. A., 1966, *Pharmacol. Rev.* **18**:387–399.
136. Schmid, D., Stadler, H., and Whittaker, V. P., 1982, *Eur. J. Biochem.* **122**:633–639.
137. Zimmermann, H., and Whittaker, V. P., 1973, *Abstracts 4th International Meeting International Society for Neurochemistry*, Tokyo, p. 321.
138. Heilbronn, H., and Pettersson, H., 1973, *Acta Physiol. Scand.* **88**:590–592.
139. Davies, L. P., 1978, *Exp. Brain Res.* **33**:149–151.
140. Ulmar, G., and Whittaker, V. P., 1974, *J. Neurochem.* **22**:451–454.
141. Ulmar, G., and Whittaker, V. P., 1974, *Brain Res.* **71**:155–159.
142. Stadler, H., and Tashiro, T., 1982, *Abstracts European Symposium on Cholinergic Transmission, Presynaptic Aspects, Strasbourg*, p. 92.
143. Haynes, D. H., Lansman, J., Cahill, A., and Morris, S. J., 1979, *Biochim. Biophys. Acta* **557**:340–353.
144. Creutz, C. E., Pazoles, C. J., and Pollard, H. B., 1978, *J. Biol. Chem.* **253**:2858–2866.
145. Morris, S. J., Hughes, J. M. X., and Whittaker, V. P., 1982, *J. Neurochem.* **39**:529–535.
146. Südhof, T. C., Walker, J. H., and Obrocki, J., 1982 *EMBO. J.* **1**:1167–1170.
147. Greengard, P., 1978, *Cyclic Nucleotides, Phosphorylated Proteins, and Neuronal Function*, Raven Press, New York.
148. Huttner, W. B., De Camilli, P., Schiebler, W., and Greengard, P., 1981, *Abstr. Soc. Neurosci.* **7**:441.

149. DeLorenzo, R. J., Freedman, S. D., Yohe, W. B., and Maurer, S. C., 1979, *Proc. Natl. Acad. Sci. U.S.A.* **76**:1838–1842.
150. Weiler, M., and Roed, I. S., 1980, *Hoppe-Seyler's Z. Physiol. Chem.* **361**:1354–1355.
151. Cohen, F. S., Zimmerberg, J., and Finkelstein, A., 1980, *J. Gen. Physiol.* **75**:251–270.
152. Pollard, H. B., Pazoles, C. J., Creutz, C. E., and Linder, O., 1979, *Int. Rev. Cytol.* **58**:159–197.

3

Noradrenergic Vesicles
Molecular Organization and Function

Richard L. Klein and Åsa K. Thureson-Klein

1. NORADRENERGIC VESICLES: OCCURRENCE AND FORMATION

1.1. Large Dense-Cored Vesicles

Morphological and biochemical evidence supports the concept that large dense-cored vesicles (LDVs, 75–90 nm diameter) originate exclusively in the perikaryon, where their membranes are synthesized and they are packaged with specific membrane and matrix components except for norepinephrine (NE). Thus, LDVs can be found along the nerve axons as they are transported toward and accumulated in the terminals of all peripheral noradrenergic nerves (Fig. 1). It follows that the yield of LDVs in isolated fractions will be greater per nerve weight in the proximal to distal direction, and this can be demonstrated experimentally by taking sequential segments of the nerve for vesicle isolation (Section 3.1).

The LDVs accumulate in large numbers proximal to the point of interruption of normal axoplasmic transport by various procedures *in situ*, *in vitro*, and in tissue culture. At a point distal to interruption of axoplasmic flow, there is also an accumulation of particles, although to a lesser extent. Retrograde transport studies using horseradish peroxidase and specific antibodies to nerve growth factor and dopamine β-hydroxylase (DβH) substantiate a considerable movement of particles back toward the cell body. This may account, at least in part, for the number of empty vesicle profiles seen by electron microscopy *in situ* and occurring *in vitro* in less dense regions of a density gradient used to prepare the vesicles (Section 4.3).

1.2. Small Dense-Cored Vesicles

Ultrastructurally, small dense-cored vesicles (SDVs, 45–55 nm diameter) also can be seen in the axons of noradrenergic nerves, but they occur primarily

Richard L. Klein and Åsa K. Thureson-Klein • Department of Pharmacology and Toxicology, University of Mississippi Medical Center, Jackson, Mississippi 39216.

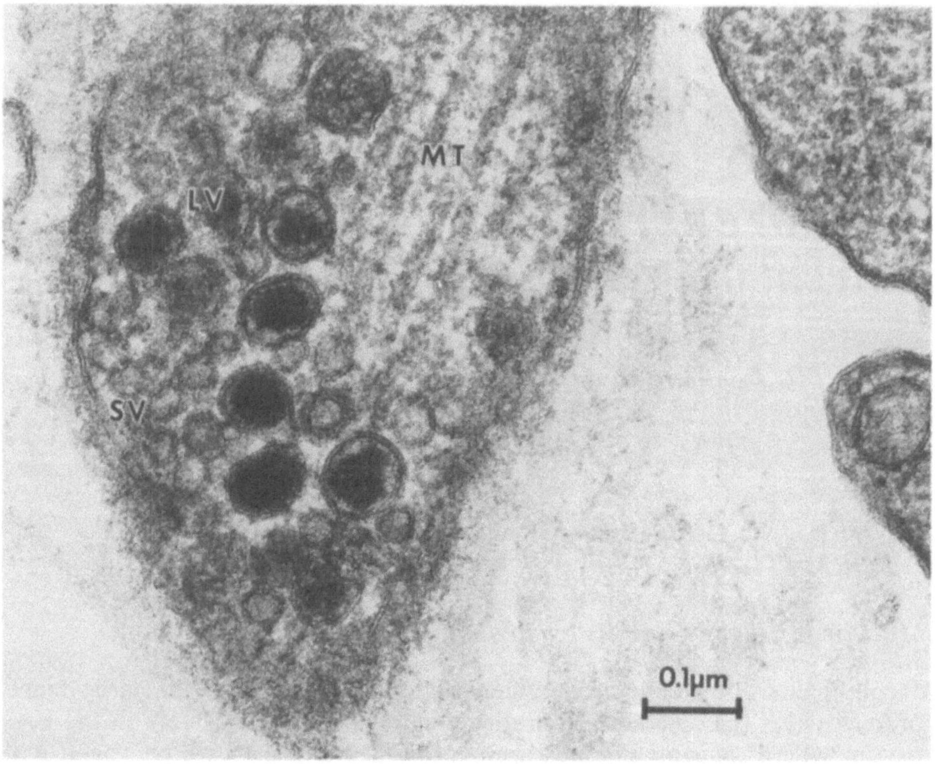

Fig. 1. Large dense-cored vesicles (LV) are aligned along microtubules (MT) in a terminal region of a human vein. SV, small vesicles. Fixation with glutaraldehyde and osmium tetroxide.

in the terminal varicosities. However, the identity of the axonal small vesicles is difficult to prove biochemically, because there are no preparations of SDVs that have been purified primarily from axons rather than from terminals; for example, vas deferens preparations will contain primarily terminal SDVs. Even if such preparations were available, an additional problem is that the axonal SDVs probably contain little NE to use as a marker except for the small amount that could be taken up from the axoplasm during transport; there is no evidence supporting NE synthesis by SDVs *per se*. (Section 4.3).

An axial orientation of SDVs parallel to microtubular elements can be found but does not appear as routinely as profiles of LDVs. However, analogous to the LDVs in Fig. 1, the SDVs in varicose regions often appear to be lined up perpendicular to the tangent of the neuronal membrane, at least in rodent terminals (see Fig. 14). The SDVs accumulate in various numbers depending on the individual varicosity and tissue innervated. They may be loosely packed to fill only a small fraction of the terminal volume or be relatively tightly packed to fill some 40 to 50% of the volume in a terminal. The proportion of SDVs in the axons rarely, if ever, reflects the proportion in the terminals. In fact, there is convincing evidence that small-vesicle formation is primarily a

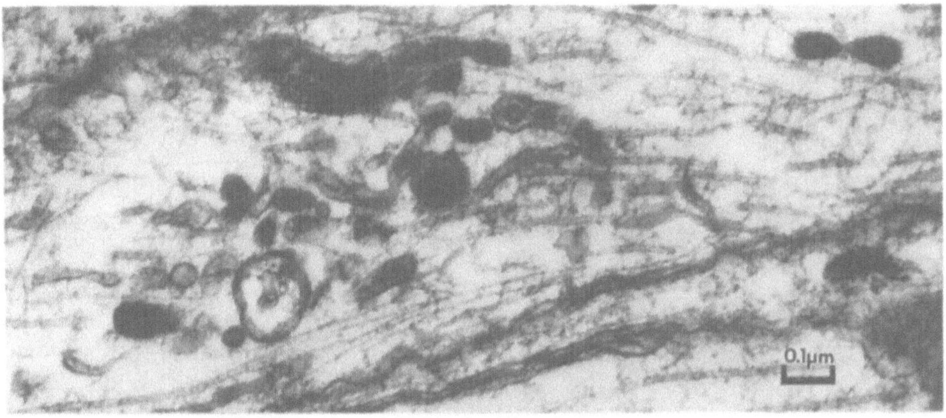

Fig. 2. Tubular elements containing dense material are present among LDVs and SDVs in bovine axons. The diameter of the tubular elements is considerably larger than the diameter of adjacent microtubules. Fixation with glutaraldehyde and osmium tetroxide.

local event in the nerve terminal involving specialized elements of the endoplasmic reticulum (see review[1]). In support of the latter, both random tubular profiles (Fig. 2) and accumulations of tubular elements containing dense material can be found along the axon, as if undergoing axoplasmic transport. In addition, neuronal ligation experiments cause accumulations of similar endoplasmic reticular elements on the proximal side.

Thus, it is reasonable to suggest that the specialized endoplasmic reticular membranes with some stainable contents (ATP, phospholipids?) other than NE also originate in the perikaryon for transport to the terminals. The potential to form SDVs from these specialized membranes is expressed by the axonal occurrence of some small vesicles; however, the primary formation takes place in the terminal regions (Fig. 3). The stimulus for such formation could come from neuronal activity (Section 10.2.1), NE depletion, and/or induction of NE synthesis. Local formation may be necessary to keep pace with a more rapid turnover of SDVs than of LDVs. The lack of NE synthesis capability, absence of matrix proteins, and the probable function of SDVs to release a single quantum of transmitter support the concept of local formation.

1.3. Percentages of LDVs and SDVs in Different Noradrenergic Terminals: An Interpretation

In recent years it has become obvious from ultrastructural studies that the proportion of large and small dense-cored vesicles varies widely not only among terminals in a given tissue but, in particular, according to the tissue and the species examined (Table I). The fact that the percentages of LDVs in the terminals of certain tissues increase in general according to animal size could reflect a similar LDV turnover rate but a greater distance of axoplasmic transport in the larger animals. A larger standing population of LDVs would satisfy

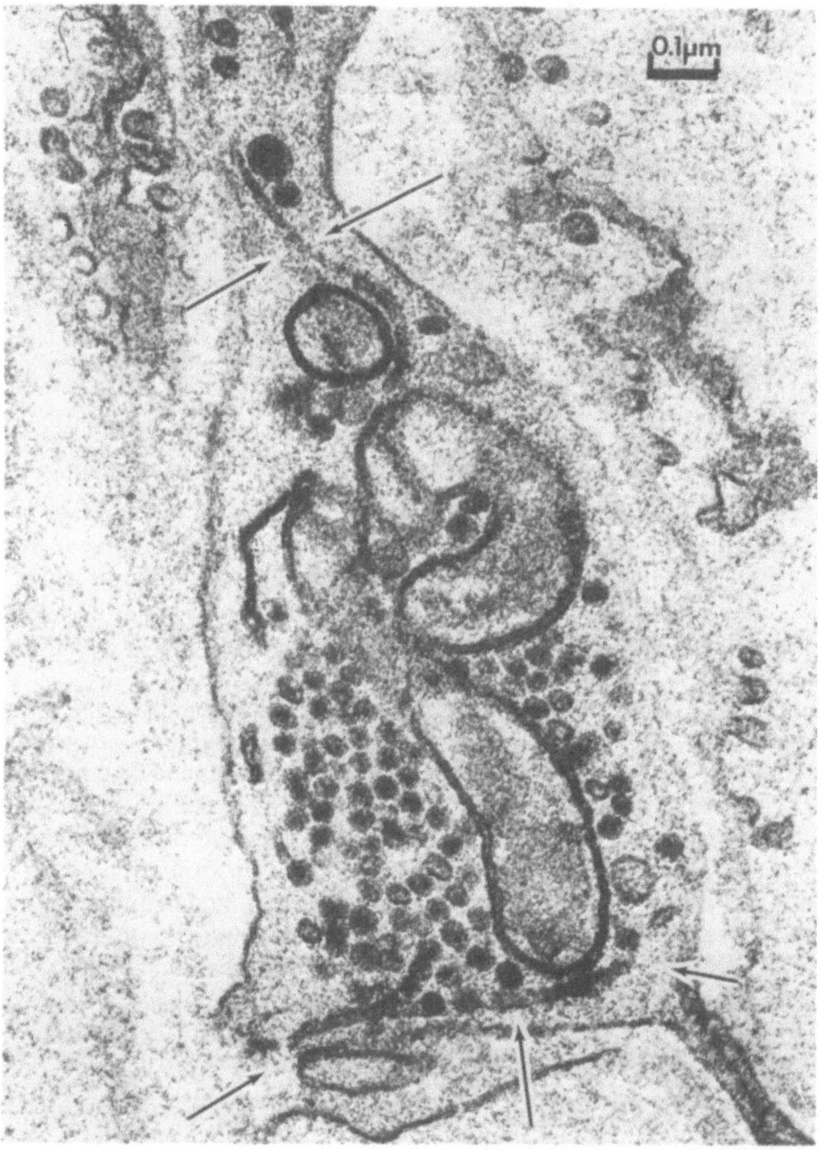

Fig. 3. Tubular elements are present in both an axon and a varicosity (arrows) of a guinea pig noradrenergic fiber, where they appear to contribute to local formation of SDVs by "budding" of vesicles. Microtubules are not preserved in this preparation because of the permanganate fixation used.

Table I
Distribution of Noradrenergic Vesicles in Terminals of Different Tissues and Species[a]

Small dense-cored vesicles	Large dense-cored vesicles	Relative storage capacity in LDV core volume	Animal and tissue
95%	5%	29%	Rat, guinea pig: iris, vas deferens
85–90%	10–15%	46–57%	Rat: brown adipose, heart, spleen Rabbit: spleen, blood vessels
80%	20%	65%	Cat: spleen Frog: heart
65–80%	20–35%	77–81%	Dog, pig, sheep: spleen Rabbit: arteries Human: mesenteric vein and artery, saphenous vein, vas deferens?
50–60%	40–50%	84–88%	Dog: saphenous vein Calf: spleen Cow: spleen, vas deferens, blood vessels

[a] See reviews[2,3] for specific refs.

the requirement to maintain adequate NE synthesis capacity, opioid peptide stores, ATP stores, and others.

The contrast in vesicle distribution is exemplified by comparing terminals in human mesenteric vein (30% LDVs, 70% SDVs and clear vesicles) with those of guinea pig vas deferens (5% LDVs, 95% SDVs and clear vesicles) in Fig. 4.

1.3.1. Determinants of Dense Cores in Small Vesicles

The percentages of SDVs given in Table I include both dense-cored and clear vesicles of 45–55 nm diameter. The proportion with dense cores varies with the physiological state of the nerve but typically is 50–80% of the total small-vesicle population. The appearance of dense cores depends in part on the fixation and staining reactions used for electron microscopy.[3,4] In the terminals of rodents it seems to reflect the NE content of the vesicle.[5] However, there is some evidence that ATP and phospholipids also contribute to the staining reaction, particularly as observed after uranyl acetate.[4] Electrical stimulation of the nerves and pharmacological depletion with tyramine, guanethidine, and reserpine cause the loss of electron-opaque cores *in situ*. Subsequent administration of NE results in neuronal and vesicle uptake, which replenishes the dense-cored appearance in the small-vesicle population. *In vitro*, incubation

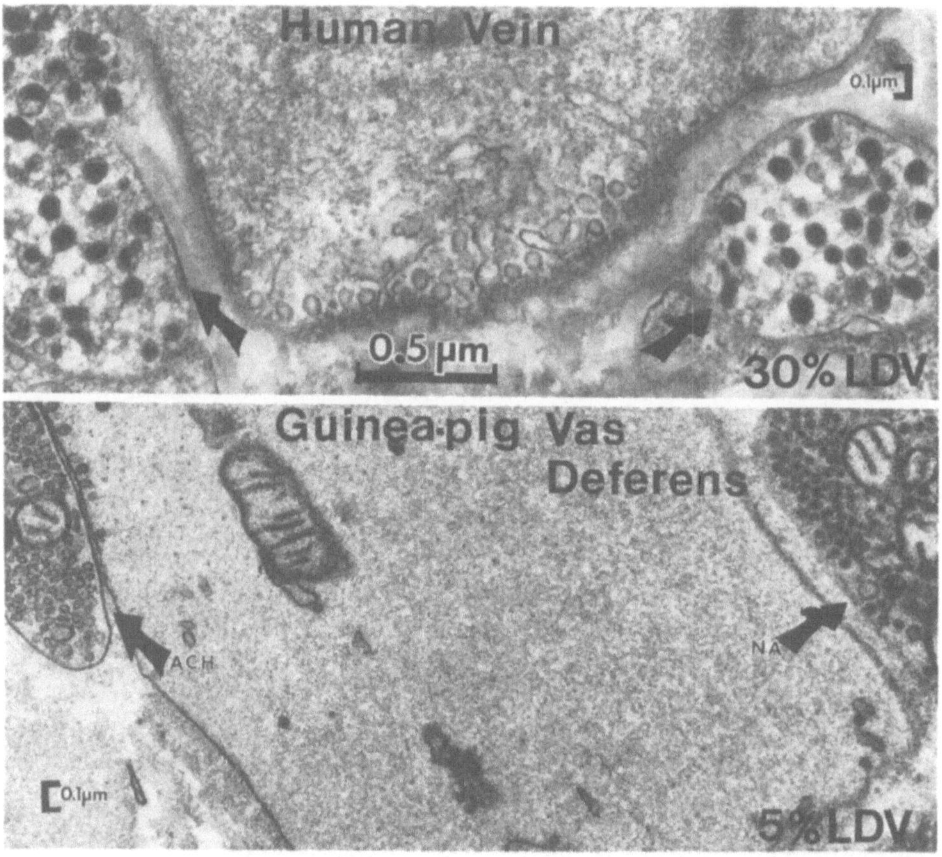

Fig. 4. Two noradrenergic terminals on either side of a smooth muscle cell in human omental vein are shown in the upper panel. The lower panel is from guinea pig vas deferens and shows a portion of a single smooth muscle cell with both a noradrenergic terminal (right) with the occasional LDV and a cholinergic terminal (left).

with NE and Mg^{2+}-ATP produces a similar effect on partially depleted, purified SDV fractions.[6]

The LDVs, on the other hand, can be nearly completely depleted of NE content by spontaneous release or by tyramine displacement but will still retain their dense-cored appearance to a large extent.[4,7]

1.3.2. Interpretation of Data from Rodent Tissues and SDV Preparations: A Cautionary Word

A word of caution is in order when interpreting the data from preparations containing primarily SDVs. Most of the ultrastructural information about the SDVs emanates from studies on rodents and has not been confirmed in larger animals such as cat, dog, or cow or in human material. It is well to remember that the initial observations on noradrenergic terminals at the ultrastructural level were also made in rodent tissues. These resulted quite unintentionally in the misleading generalization that all terminal vesicle populations are predominantly composed of SDVs, and the occasional LDV was relegated to an inconsequential presence. Even now, some tendency in this regard persists in the literature.

A combined physiological, biochemical, and ultrastructural approach is necessary to provide rational interpretations of any studies on vesicle dynamics. It is also just as important to choose an appropriate model to investigate the role of the terminal vesicle population. For example, use of the common rodent vas deferens (95% SDVs, 5% LDVs) as a model has many advantages and can provide informative data on SDV function and participation in exocytosis. It is particularly suitable for electrophysiological studies of quantal release. The rodent vas deferens also has several important disadvantages: it is innervated by rather specialized, short postganglionic fibers; it has relatively confined synaptic spaces atypical of the overwhelming number of postganglionic sympathetic nerves in the body; it is physiologically conditioned to infrequent stimulation for only brief periods; and others. Furthermore, one is at a distinct disadvantage to conclude anything about the function of the LDVs in this preparation at the ultrastructural level. The fact that one typically finds only one to three LDV profiles in an electron microscopic section through a varicosity in rodent vas deferens (see, e.g., Figs. 3,4,14a–c,15b) precludes any meaningful morphometric analyses to determine LDV function in terms of vesicle turnover and exocytosis. Yet, it is known that hydrophilic DβH or opioid peptide release must originate by exocytosis from the large-type vesicles.

Thus, it is strongly recommended that results with the popular rodent vas deferens model be considered in proper perspective and that studies be initiated on other appropriate models such as blood vessels or even vas deferens from larger animals (cat, dog, ox, and human), which may contain 20–40% LDVs in their terminal varicosities.

2. *PURIFIED FRACTIONS OF NORADRENERGIC VESICLES*

Knowledge of the molecular composition and organization of noradrenergic storage vesicles in the peripheral nervous system depends entirely on a practical source from which to purify the vesicles. At present, this is satisfied only by mammalian systems,[8] namely, the bovine splenic nerve for the large-type dense-cored vesicles and the castrated rodent vas deferens for the small-type dense-cored vesicles.

2.1. *Large Dense-Cored Vesicles of 80–90% Purity*

An improved method for the purification of U. S. von Euler's original LDV preparation from the bovine splenic nerve (~98% sympathetic C fibers) was developed in 1969–70.[9] The method was refined further when the significance of the minimal postmortem delay preparation was discovered[10] and the need arose for gradient distribution studies and sampling with minimal arti-

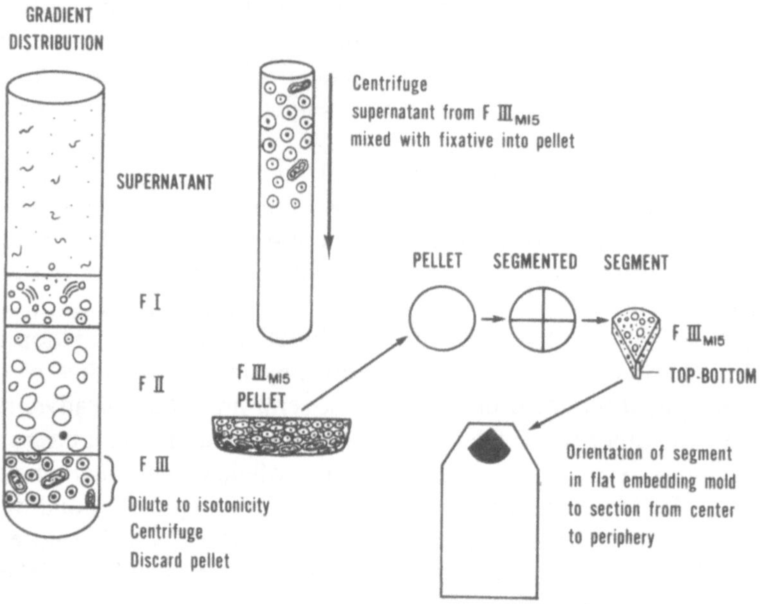

Fig. 5. Procedures for sucrose–D_2O density-gradient sampling and electron microscopic preparation of neuronal particle fractions in order to estimate vesicle purity. (Reproduced with modifications from ref. 4 with permission from Academic Press.)

facts.[11] The overall scheme for purification is available in reviews[7,12] and has recently been discussed in detail.[8]

Procedures for gradient sampling, fixation, and quantitation of vesicle purity for electron microscopy (Fig. 5) were developed during the same period in order to support the biochemical findings and indications of vesicle purification with correlative morphological proof.[13,14] Thus, in the minimal 10- to 12-min postmortem delay preparation of LDVs from bovine splenic nerve, it is possible to achieve an overall purity estimated at 80–90% with the primary contamination (mitochondrial and other large membrane fragments) largely restricted to the bottom 5–10% of the pellet. This contaminated vesicle layer merges into the major layer of nearly pure LDVs above (Fig. 6).

2.2. *Small Dense-Cored Vesicles of 25–40% Purity*

The practical purification of SDVs was not achieved until 1978[15] using rodent vas deferens after a period of castration. The small-vesicle yield and purity were much improved over earlier preparations as measured by NE:protein ratios in the "light" vesicle peak compared to control animals; this was a result of the selective atrophy of smooth muscle but retention of sympathetic innervation. In a correlative morphological study,[6] SDV purity was esimated up to 25–40% in the best preparations (Fig. 7).

2.3. *Comments on Vesicle Purity Estimates*

A correlative estimation of vesicle purity using both biochemical and ultrastructural criteria is highly desirable if not mandatory. It is common to find

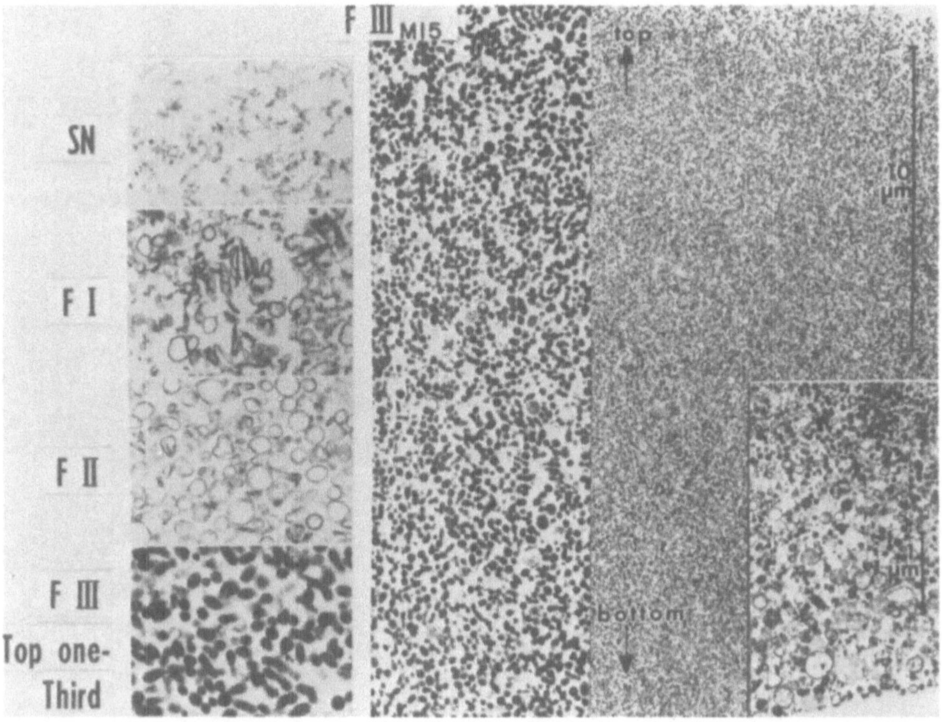

Fig. 6. Purified LDVs from bovine splenic nerve. Left panel shows typical contents of the various fractions: SN, supernatant; FI–FIII, gradient fractions (see refs. 4,8 for more details). An enlarged view of fraction $FIII_{M15}$ (left center), the purest fraction of LDVs, is shown to consist of nearly pure vesicles in the upper third of the pellet depth. The lower-power view (right center) shows ~40% of the pellet depth from the top surface (upper right). The inset (lower right) shows the pellet bottom surface with the transition from nearly pure vesicles above to a mixture with mitochondrial fragments. (Reproduced with modifications from ref. 4 with permission from Academic Press.)

preparations of biochemically "pure" subcellular particles that show insignificant marker activity for contaminants but still contain considerable contamination at the ultrastructural level. Conversely, nearly anything can be shown at the ultrastructural level by selecting an infinitesimally small area of the total. Thus, with subcellular particles, pellets should be prepared in relatively long slender tubes with flat bottoms using a rotor with swing-out buckets (Fig. 5). This minimizes tube wall effects perpendicular to the vector of centrifugal force. Fixation in suspension during centrifugation is often useful to achieve stratification of particles in the pellet. Sections for electron microscopic observation must be chosen in a way that reveals the particle distribution from top to bottom and from center to periphery in order to ascertain that a relatively constant distribution of the purified particle and possible contaminants has been obtained (Fig. 6). Morphometric quantitation of particles must then insure random selection from several preparations and at sites in the pellet both vertically and

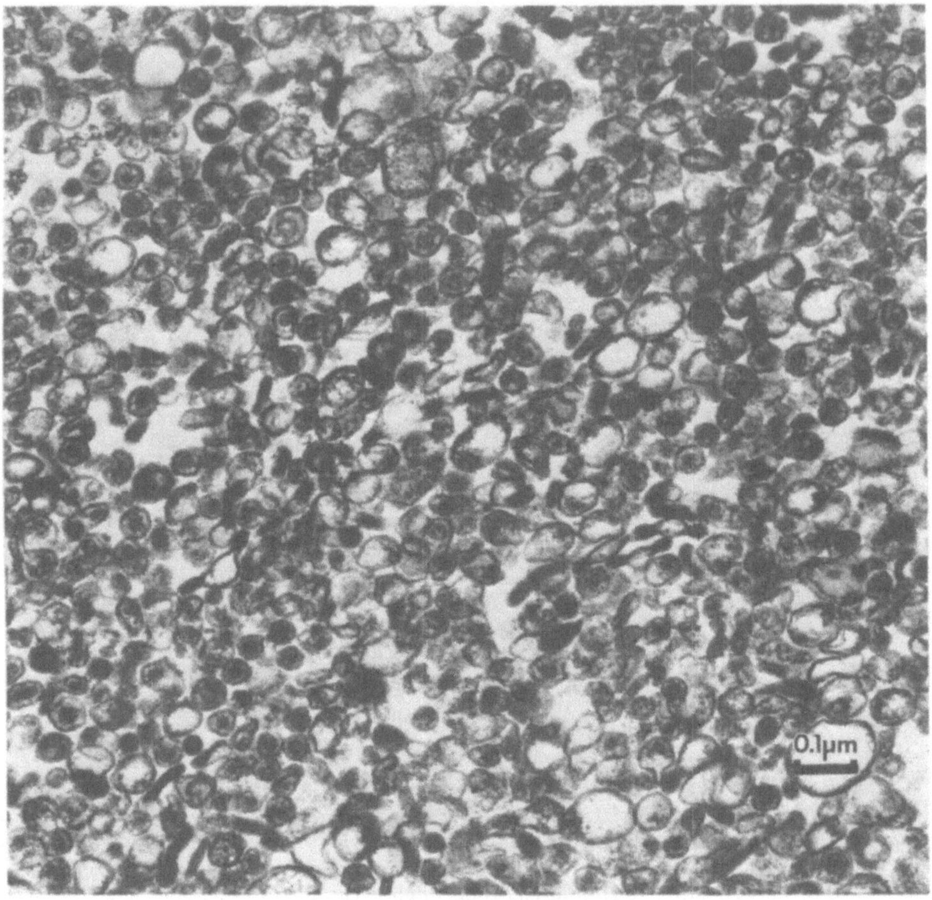

Fig. 7. A portion of the pellet depth in a preparation of SDVs from castrated rat vas deferens with a relatively high purity estimated at 25–40%.

horizontally, such as can be achieved with the use of predetermined discrete areas on an overlying grid.

3. *AXOPLASMIC TRANSPORT OF NEWLY FORMED LARGE DENSE-CORED VESICLES*

3.1. *A Unique Method to Study Aspects of Axoplasmic Transport*

As neuronal axoplasmic transport is covered in detail elsewhere in Volume 5, it is pertinent to the present chapter only to mention that studies of purified LDV fractions prepared identically and simultaneously from equal weights of sequential segments of the bovine splenic nerve also provide a measure of vesicle compositional and functional changes during transport from the cell

Table II
Norepinephrine:ATP Ratios in Large and Small Dense-Cored Vesicles

	Axon segment			
	Extrasplenic 10–12 cm		Distal (intrasplenic 15–30 cm)	Terminal varicose regions
	Proximal	Midportion		
Larger vesicles[a]				
NE:ATP ratio				
(measured)	4.5:1	5.9:1	7.2:1	17.6:1[b]
Corrected[c]	8.1:1	10.6–15.3:1	13–25:1	33–61:1
Small vesicles				20–60:1[d]

[a] From bovine splenic nerve.
[b] Calculation based on the increased NA:DβH activity ratio in the density-gradient fraction of LDVs (1.178 $g \cdot cm^{-3}$) from crude spleen homogenate[12] and assuming ATP and DβH contents continue to remain constant as during axoplasmic transport to the terminals.
[c] Corrections are made for 85% vesicle purity, 45% nonspecifically adsorbed albuminlike protein,[2] and for NE loss during 10 to 12-min postmortem delay in the presence and absence of 5 mM Mg^{2+}-ATP. With Mg^{2+}-ATP, there are two NE pools with $t_{1/2}$ of 4.5 and 36 min for spontaneous net release at 30°C; without Mg^{2+}-ATP, there is a single pool of NE with $t_{1/2}$ of 13 min at 30°.[18]
[d] Ref. 19.

body to the terminals.[16] The data indicate that the DβH, ATP, and opioid peptide contents are prepackaged into the LDV in the nerve cell body and that the content per milligram vesicle protein does not change during axoplasmic transport to the terminals.

3.2. Establishment of a Fast-Release Pool of NE: A Compositional and Functional Change

Norepinephrine is synthesized continuously, and the LDV content increases at least 70% as the vesicle moves down the axon to the distal nerve segment.[17] Further synthesis in the terminal region can be calculated to increase the content another 2.4-fold.[12] An important point, the significance of which was not originally apparent, is that the fast-release pool from which newly synthesized NE originates in LDVs is also established during axoplasmic transport.[10] In fact, the increase in NE content of the LDVs during transport is equivalent to the amount of NE that can be shown kinetically to be stored in the fast-release pool as distinct from the more slowly released, Mg^{2+}-ATP-dependent uptake pool.[18]

3.3. Changes in NE:ATP Ratios and Implications for the Role of ATP

On the basis of the same type of studies, it can be shown that the molar ratios of NE:ATP in LDVs increase markedly during axoplasmic transport to reach 33–61:1 in the terminal varicosities (Table II). This is similar to the ratio of 20–60:1 reported for the purified SDVs from castrated rat vas deferens.

Thus, the prior notion of a fixed 4:1 ratio for NE:ATP complex analogous to that proposed earlier for adrenal chromaffin granules should be discarded in favor of other roles for ATP. Recent NMR studies do not support such a complex, although some charge interaction may occur.

Speculative roles for ATP in LDVs include protection of DβH from H_2O_2 generated by autooxidation of ascorbate, provision of substrate for protein phosphorylation reactions, inhibition of transmitter release in the short term by presynaptic receptor feedback, and stimulation of NE synthesis and enhanced release through increased formation of cyclic AMP in the longer term.[20]

4. NEURONAL SUBCELLULAR LOCALIZATION OF DOPAMINE β-HYDROXYLASE AND THE ORIGIN OF CIRCULATING ENZYME

4.1. The Problem and the Hypothesis

Although one may measure some parallels between DβH release and NE release in perfused organs *in situ* and in isolated preparations *in vitro* under specified stimulation conditions, there is seldom a correlation *in vivo* between circulating DβH and sympathetic activity as reflected by catecholamine levels in the blood.[21] This may in part be because DβH has been measured only by enzymic activity in some studies, although various proportions of the released enzyme are inactive and can only be detected antigenically. However, the lack of correlation during normal levels of sympathetic activity is most likely because of the fact that hydrophilic (water-soluble) DβH is found only in the matrix of the LDVs and not in the SDVs. Thus, it will only be under particular circumstances including stressful stimuli, when exocytosis from LDVs appears to be most frequent (Section 10.2), that a reasonable correlation may exist and can be demonstrated experimentally.

The hypothesis that NE can be released from both SDVs and LDVs but that DβH is released only from LDVs was originally proposed in 1972[22] (see ref. 20 for more detailed discussion).

4.2. Dopamine β-Hydroxylase in LDVs

Because the specific localization of DβH enzymic activity is of primary importance for the origin of newly synthesized NE, we shall discuss some of the more conclusive evidence on the subject. In the most purified fraction of intact LDVs from axons of bovine splenic nerve, two-thirds of the DβH is latent, and one-third is active enzymatically.[23] On vesicle lysis with the French Press, the amount of hydrophilic enzyme (66% polar amino acid residues[16]) released into the water-soluble phase corresponds to the two-thirds that is contained in the matrix. On release, this DβH becomes enzymatically active; however, it can also be activated without release by treatment with a low concentration of Triton X-100 (0.01% v/v), which has negligible effect to disrupt the vesicles. The latter activation of enzymic activity results because enzymic sites become exposed to penetrating substrate, usually tyramine or dopamine.[23] The activation is not caused by a permeability change in the vesicle membrane *per se*, because tyramine exchanges very rapidly across the intact vesicle membrane both to displace NE in a nonstoichiometric manner[18] and, if ascorbate is added, to be converted into octopamine by DβH.

A curious comparison can be made between the membrane distributions of DβH in LDVs and adrenal chromaffin granules (ACGs) (refs. in Tables IV, V). In the LDV, there is calculated to be three or four molecules of DβH associated with the membrane, which is equivalent to the one-third active enzyme measured in intact vesicles. In the ACG, there is calculated to be 208 molecules of DβH associated with the membrane of a total of 344 molecules. However, the enzyme is reported to be only 7–12% active in the intact ACG, which is equivalent to 24–41 molecules. Thus, if the active enzyme is membrane associated, as logic would argue, all of the membrane enzyme in LDVs is in the active state, whereas only 12–20% of the DβH molecules in the ACG membrane are active in the intact isolated granule.

4.3. *Evidence for Dopamine β-Hydroxylase in SDVs?*

There has been considerable earlier controversy on the localization of DβH in SDVs. In our opinion, this can now be largely resolved. In all of the earlier preparations as well as in the most purified SDV fraction from the castrated rat vas deferens (Section 2.2), there has been a notably poor correlation in density gradients between DβH enzymic activity peaks and NE peaks at a density corresponding to the population of SDVs.[24] In fact, several papers have shown that no correlation exists.

In the purest preparation of 25–40% SDVs, there occurs only a plateau of low enzymic activity under a sharp peak of NE and a coincident lesser peak of ATP.[15] In this and in all of the earlier preparations, some 50% of the total DβH activity resided at a density corresponding to the LDV fraction, even though LDVs comprise only 5% of the total terminal vesicle population of the rodent vas deferens. A simple calculation can equate the DβH activity under the SDV peak to about one molecule of enzyme per ten small vesicles.[2,25] From a purely biochemical standpoint, this can only be regarded as contamination. Furthermore, there are several proven probabilities for contamination of the light vesicle peak of SDVs with enzyme from other sources. These include "immature" LDVs from the more proximal segments of the neuron, depleted LDVs either through vesicle rupture during homogenization or leakage of contents, and spent vesicles forming retrograde transport particles. It is also known that retrograde particles[26] and other light particles[27,28] have relatively low enzymic DβH activity compared to DβH antigenicity; i.e., their homospecific activity is low. In an earlier rodent vas deferens preparation of partially purified SDVs, it was calculated that a contamination of only one LDV per 6250 particles could account for all of the DβH activity in the light vesicle peak.[11]

Recent data are also pertinent to this argument. The distribution of NE, DβH, and ATP are shown (Fig. 8A) with marker enzymes for contaminants (Fig 8B) on a sucrose–D_2O density gradient using a rat vas deferens homogenate

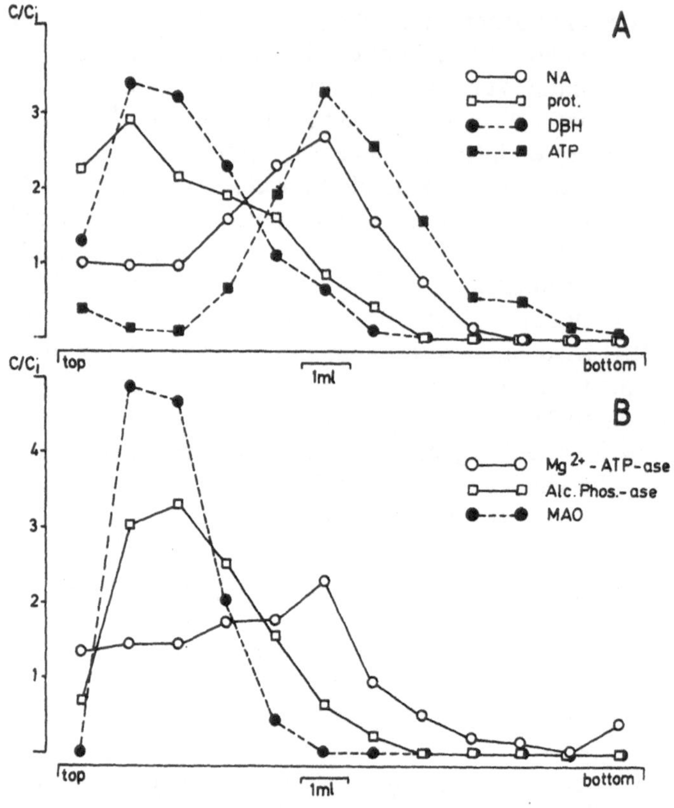

Fig. 8. Distribution of NE (NA), marker enzymes, and proteins in gradients of a vesicle-enriched fraction from rat vas deferens. After differential centrifugation and differential gradient centrifugation, the upper fractions (containing the slow-sedimenting NE vesicles) were layered on a sucrose–D_2O density gradient for isopycnic centrifugation. The relative concentration, c/c, is the amount in each fraction relative to the total amount applied to the gradient.[29] The figure was kindly provided by W. P. DePotter.

fraction (slowly sedimenting particles, i.e., "light" vesicles).[29] No correlation exists between the DβH activity peak and the primary peaks of NE and ATP corresponding to the SDV fraction. The specific activity achieved with the D_2O-containing gradient is 21.2 nmol NE/mg protein, a 2.46-fold increase over the usual sucrose–H_2O gradient. If the castrated rat vas deferens preparation is used, the same separation of NE/ATP peaks from the DβH peak was achieved, but a 3.57-fold increase in specific activity to 75.7 nmol NE/mg protein resulted in the SDV fraction, uncorrected for vesicle purity and NE leakage. (M. F. Williams and W. P. DePotter, unpublished data.) This compares most favorably with the initial reports using castrated rat vas deferens of 33.6 nmol NE/mg protein before corrections.[2]

As a different approach, a density gradient separation was made from a homogenate of bull vas deferens, which contains ~30% LDVs and 70% SDVs. In this case, both chromogranin A, the soluble matrix protein, and DβH, the partially soluble and membrane protein, were identified by specific antibody reactions.[30] Both chromogranin A and DβH were localized at a density coincident with the peak of LDVs and not the peak of SDVs.

4.4. Comparisons with Bovine Adrenal Chromaffin Granules

Intact LDVs and ACGs contain largely latent DβH enzyme amounting to 67% in LDVs and 88–93% in ACGs.[16] About 70% of the fully activated enzyme is in the water-soluble phase of LDVs (53% of the total soluble protein), and this is hydrophilic type.[31] In ACGs, 50–67% of the fully activated enzyme is in the water-soluble phase (~5% of the total soluble protein), and about 50% of the total enzyme is of the hydrophilic-type. In LDVs, the one-third membrane-associated enzyme comprises 16% of the water-insoluble protein (0.45 nmol/mg protein) and is presumed to be of the amphiphilic type, analogous to the membrane enzyme of ACGs. However, the latter is unproven and even may be questionable at this time because preliminary analyses of the total enzyme from LDVs give a ratio of polar to nonpolar amino acids not much lower than the 66% found for the purified hydrophilic enzyme.[31] In ACGs, the amphiphilic enzyme is ~50% of the total and comprises a similar 13–25% of the membrane protein.

Each LDV is calculated to contain a total of ~12 molecules of DβH,[2] and each ACG a total of 344 molecules of DβH (208 in the membrane).[32]

5. COEXISTENCE OF NEUROTRANSMITTER AND NEUROPEPTIDE

5.1. Coexistence at the Cell Level

The coexistence of various peptides and transmitters in neurons and neuronlike cells has been implicit in the APUD[33] and paraneuron[34] concepts for a number of years. Numerous recent studies have utilized histofluorescence and immunohistofluorescence techniques at the light microscopic level to demonstrate the coexistence of peptide and transmitter in the same nerve bundle and cell population.

5.2. Coexistence at the Vesicle Level

The possible coexistence of peptide and neurotransmitter in the same neuronal vesicle population demands much more critical evaluation and usually will involve a combination of biochemical, histochemical, and electron microscopic techniques. Proof of coexistence at the vesicle level has particular physiological implications for the concomitant release of the peptide and transmitter by exocytosis. Only three examples are relatively firm at this date.

5.2.1. Sympathetic Postganglionic Nerve LDVs: Opioid Peptides and NE

The coexistence of opioid peptides (OPs) at 400–800 μM concentration in the LDV core volume together with NE and hydrophilic DβH was first reported

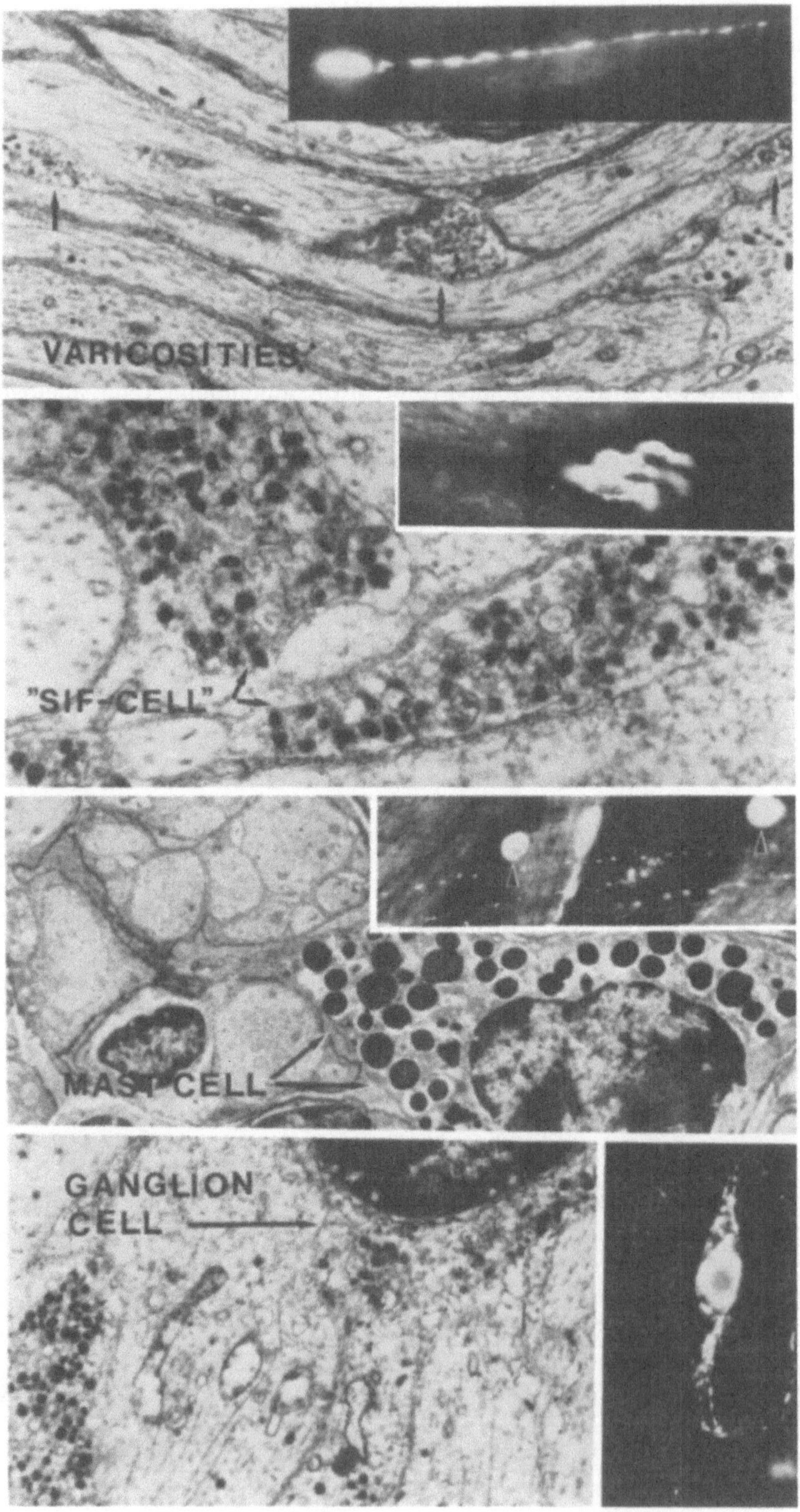
VARICOSITIES
"SIF-CELL"
MAST CELL
GANGLION
CELL

in 1980[2,35] using the highly purified fraction of LDVs from bovine splenic nerve. These studies showed that the ratios of OPs to NE and to DβH activity remained similar throughout the multistep purification procedure. Further, the distribution of NE and OPs in sedimentable particles was essentially identical in the sucrose–D_2O density gradient used as an intermediate step in LDV purification; NE and OPs formed a sharp coincident peak in the LDV fraction. Efforts to separate OPs, NE, DβH, and dopamine (DA) peaks by subjecting semipurified LDVs from the sucrose–D_2O density gradient to modified second gradient separations with and without D_2O did not reveal any minor vesicle contaminant that could account for the high opioid content of the primary LDV fraction.[36]

Further consideration was given to the only other catecholamine-containing cells that occur along the bovine splenic nerve trunk.[36] These could be identified by histofluorescence and electron microscopy (Fig. 9) and include sympathetic growth fibers (a noncontaminant), SIF cells, mast cells, and ganglion cells. The latter three cell types can be eliminated from making a significant contribution to the LDV fraction on the bases of differential frequency of occurrence and distribution along the axon from the celiac ganglion to the terminal varicosities in the spleen and the size and fragility of their vesicle or granule populations.

Recent studies[37] using antibodies specific to dynorphin 1-13 demonstrate the presence of this opioid by RIA in the bovine splenic nerve homogenate where it occurs 90% particulate compared to Leu-enkephalin 85% particulate in the 10,500g for 15 min supernatant. The dynorphin co-purifies ~66-fold similar to Leu-enkephalin (RIA specificity 96% with 4% cross-reactivity to Met-enkephalin) in the sequential steps to produce the purest NA-containing LDV fraction. Dynorphin (2148D) occurs at a molar ratio of ~1:7 Leu-enkephalins and likely comprises a portion of the 50% opioid peptides shown by radioreceptor assay[36] to be less than 5000D exclusive of the Met- and Leu-enkephalins.

5.2.1a. Opioid Peptides in LDVs during Axoplasmic Transport. In the central nervous system, the question has been raised whether the larger opioidlike petides are subject to peptidase breakdown to form enkephalins and other small opioids during axoplasmic transport from the point of formation in the cell body

←

Fig. 9. Cells of the bovine splenic nerve. Upper panel: Varicose fibers terminating in a conelike process are strongly fluorescent and typical of sympathetic C fibers in the growing stage. They occur more frequently than SIF cells and are randomly distributed along the extrasplenic portion and less frequently along the intrasplenic extensions of the nerve bundle. The vesicle population is identical to that in sympathetic C fibers with 70- to 80-nm diameter LDVs. Second panel: SIF (small intensely fluorescent) cells show strong yellow–green fluorescence. They occur infrequently and more often toward the proximal end of the nerve bundle near the celiac ganglion. Their vesicle population is easily recognized and consists of 89% LDVs of 120- to 140-nm diameter, considerably larger than the typical noradrenergic LDVs. Third panel: Mast cells show bright yellow–green fluorescence partly attributable to DA. They are relatively numerous along the nerve bundle. Their granule population is easily recognized by the large 300 to 800-nm diameter. The granules are highly fragile during homogenization, and essentially all histamine and DA are found in the supernatant. Bottom panel: Ganglion cells show strong fluorescence typical of sympathetic fibers. They occur infrequently and mostly at the proximal end of the nerve bundle near the celiac ganglion, similarly to SIF cells. Their vesicle population is low compared to the terminals.

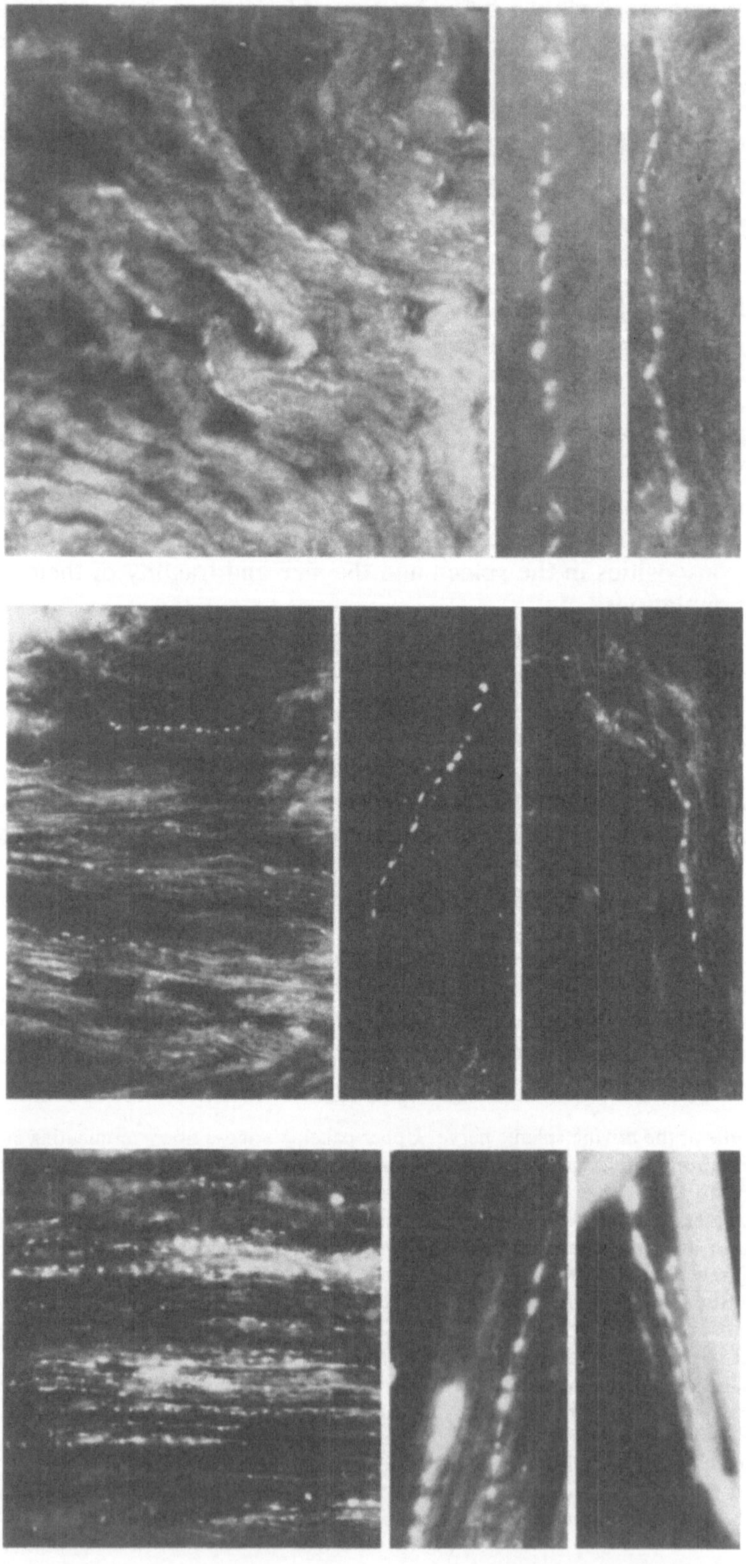

to the terminals. The finding that the OP content as measured by radioreceptor assay and HPLC does not change in the purified LDVs during axoplasmic transport indicates that negligible conversion occurs from larger opioids with cryptic activity to Met- and Leu-enkephalins in the peripheral sympathetic nerves.[35] The finding that about half of the total opioid activity is enkephalin pentapeptides compared to 10–25% in purified adrenal chromaffin granules also suggests primarily a transmitter or local modulator function for the opioids in the peripheral noradrenergic terminals, whereas an endocrinelike function may be speculated for the majority of large opioids in the adrenal chromaffin cells.[36]

5.2.1b. Molar Ratios of OP:NE Should Reflect the Percentage of LDVs: A Hypothesis. Given that OPs and NE are stored in the same LDVs of sympathetic nerve for which DβH can be used as a marker, evidence is being sought to test the working hypothesis that ratios of OP:NE should reflect the percentage of LDVs in the noradrenergic terminals of the particular tissue and species. The experimental design to test this hypothesis must be subject to rigorous critique using correlative ultrastructural, immunofluorescence, fluorescense histochemical, and biochemical techniques. For example, it must be ascertained that NE, DβH, and OPs are associated primarily with sympathetic innervation rather than with other neuronallike cell types.

The initial approach to this problem has been with the bovine splenic nerve. Norepinephrine histofluorescence and DβH and OP immunohistofluorescence show similar patterns in the bovine splenic nerve trunk and in individual varicose fibers (Fig. 10) (Å. K. Thureson-Klein, unpublished data). Preliminary biochemical studies on OP:NE ratios in several bovine tissues including splenic nerve, vein, and artery and mesenteric vein and artery give molar ratios in the range of 1:10–70 (R. L. Klein, B. H. Douglas, II., and K. B. Moorthy, unpublished data). With guinea pig vas deferens, the NE histofluorescence and DβH immunohistofluorescence give typical patterns, but opioid immunohistofluorescence is very low (Å. K. Thureson-Klein, unpublished data). This is supported by preliminary molar ratios of OP:NE that are 10- to 100-fold lower in rodents than in bovine tissues.

5.2.2. Bombesin and Other Bioactive Peptides

Recent studies[37] have also identified bombesin by RIA in the bovine splenic nerve homogenate where it occurs about 95% particulate in the 10,500g for 15 min supernatant. The bombesin co-purifies 77-fold similar to Leu-enkephalin and dynorphin during the sequential steps to produce the purest LDV fraction. Bombesin (1620D) occurs at a molar ratio of about 1:2–4 Leu-enkephalins.

←

Fig. 10. Localization of NE, dopamine β-hydroxylase (DβH), and opioid peptides (OPs) in the bovine splenic nerve. Upper row of panels are sections of nerve trunk showing similar patterns of fluorescence; middle and lower rows of panels show individual varicose fibers with similar patterns of fluorescence. NE, Falck–Hillarp histofluorescence; DβH, FITC–antibody immunohistofluorescence using rabbit antibody to borine adrenal DβH (from M. Goldstein); OPs, FITC-antibody immunohistofluorescence using rabbit antibody to Leu-enkephalin glutaraldehyde bovine thyroglobulin conjugate.

Vasoactive intestinal polypeptide can be detected by RIA in the bovine splenic nerve homogenate at low levels, 5% or less than Leu-enkephalin on a molar basis. Preliminary trials do not indicate a co-purification with the LDV fraction. This low level is in keeping with the minor percentage (1–2%) of cholinergic fibers in the bovine splenic nerve bundle where this peptide is speculated to occur.

Neurotensin and substance P occur just at or below the RIA detectable limits of 0.5 and 1.0 pmol, respectively, in samples of the bovine splenic nerve homogenate used to measure the other peptides.

5.2.3. *Parasympathetic Postganglionic Nerve LDVs: Vasoactive Intestinal Polypeptide and Acetylcholine*

In combined biochemical, histochemical, and electron microscopic studies, VIP, (vasoactive intestinal polypeptide) and ACh (acetylcholine) have been demonstrated in the same population of LDVs.[38] In density gradient separation of cat salivary gland homogenate, two peaks of ACh could be distinguished: a major peak corresponding in density to the predominant small-type synaptic vesicles (50 nm diameter) and a minor ACh peak at a density corresponding to the LDVs (100 nm diameter), which comprise only a small percentage of the total terminal vesicle population in the peripheral cholinergic system. A single large immunoreactive peak of VIP was found coincident with the small peak of ACh and LDVs. This biochemical finding was confirmed by electron microscopic immunocytochemistry showing that VIP-reactive material occurred in the LDV population *in situ* but not in the small synaptic vesicles.[39]

This example differs from that of NE and OPs in the noradrenergic LDVs because exocytosis would result in a large augmentation of NE release with the OPs, whereas relatively minimal release of ACh would occur from the LDVs containing VIP in the cholinergic terminals. Because of the relative impurity of the two cholinergic vesicle populations in this study, it is not known whether the cholinergic LDVs store ACh endogenously or whether the LDVs are able to take up ACh during isolation procedures.

5.2.4. *Central Neuronal LDVs: Substance P and Serotonin*

In an enviably simple but highly technical ultrastructural study, immunoreactive substance P and serotonin were shown to coexist in about 20% of the terminals of raphe nuclei and dorsal horn cells of the spinal cord in exactly the same pattern of LDVs (60–90 nm diameter) with each substance, using adjacent serial sections.[40] Unfortunately, there are no analytical data to quantitate the transmitter and peptide contents of these cells or vesicles.

5.3. *Peptides Are Stored Only in Large-Type Dense-Cored Vesicles: A Tentative Generalization*

A perusal of the literature on this subject[41] suggests that peptides are almost exclusively stored in the large-type granular vesicles of various diameters depending on the cell type throughout the animal kingdom.

Table III
Glycoproteins and Glycosaminoglycans of Bovine Noradrenergic Large Dense-Cored Vesicles Compared with Adrenomedullary Chromaffin Granules

	Large dense-cored vesicles[a] (μmol amino sugar/100 mg lipid-free dry weight)		Adrenal chromaffin granules[b] (μmol amino sugar/100 mg lipid-free dry weight)	
	Soluble fraction[c]	Membrane fraction	Soluble fraction	Membrane fraction
Glycoproteins	6.89 (9.1%)[d]	6.36 (7.1%)	5.0 (74%)	8.2 (10%)
Glycosaminoglycans				
Heparan sulfate	0.45	0.31	0.2	0.3
Chondroitin sulfate	0.18	0.10	3.4	2.6
Hyaluronic acid	trace (?)	—	0	0

[a] R. U. Margolis, R. K. Margolis, B. H. Douglas II., L. T. Callaghan, and R. L. Klein, unpublished data.
[b] Data from refs. 42, 43. Also see refs. 44, 45 for comparisons.
[c] Vesicles lysed by French press method.[8,23]
[d] Percentage N-acetylgalactosamine (of total N-acetylglucosamine + N-acetylgalactosamine).

6. GLYCOPROTEIN AND GLYCOSAMINOGLYCAN CONTENTS OF LARGE DENSE-CORED VESICLES COMPARED TO CHROMAFFIN GRANULES

Preliminary studies on the glycoprotein and glycosaminoglycan contents of highly purified (washed, dialyzed, and lyophilized) preparations of LDVs from bovine splenic nerve provide interesting comparisons with data from bovine ACGs (Table III). The following tentative conclusions seem reasonable:

1. Water-soluble protein. The LDVs contain ~38% compared to 78% soluble proteins in ACGs. Note: highly purified cholinergic synaptic vesicles from *Torpedo* sp. electric organ and SDVs in sympathetic nerves contain little or no soluble protein.
2. Glycoprotein. The LDVs contain 52% compared to ACGs 67% soluble glycoprotein but at a similar concentration of 5–8%. However, in LDVs, 90% is N-acetylglucosamine (mainly N-glycosidically linked), whereas 74% is N-acetylgalactosamine (mainly O-glycosidically linked) in ACGs.
   a. Chromogranin A. The latter hexosamine distribution is in keeping with expectations from the content of chromogranin A, which in LDVs is a relatively minor soluble protein but is a major soluble protein (40–50%) in ACGs; chromogranin A contains 80% N-acetylgalactosamine.
   b. Dopamine β-hydroxylase. The latter hexosamine distribution is also in keeping with expectations from content of hydrophilic DβH, which is a major soluble protein (53%) in LDVs but is a minor soluble protein (5%) in ACGs; DβH contains high N-acetylglucosamine.
3. Glycosaminoglycan. In LDVs the glycosaminoglycans constitute only 14–18% of the amount found in ACGs, of which 61% is soluble in LDVs compared to 80% in ACGs. Of the total hexosamine content, in LDVs

71–76% is N-acetylglucosamine (from heparin sulfate?), whereas it is 94% N-acetylgalactosamine from chondroitin sulfate in ACGs. In this respect, the neuronal LDVs resemble the highly purified *Torpedo* sp. cholinergic vesicle preparation from the electric organ. The latter contains mostly glycosaminoglycans (from heparan sulfate?), of which 50–70% is reported to be a membrane-associated proteoglycan called vesiculin.[46,47]

7. LIPID CONTENT OF NORADRENERGIC LARGE DENSE-CORED VESICLES COMPARED TO CHROMAFFIN GRANULES

The partial lipid composition of the most purified fraction of LDVs from the bovine splenic nerve has been determined[12,23] and compared to that of bovine ACGs.[16] The common phospholipids, including phosphatidylcholine 34.9%, phosphatidylethanolamine 28.9%, sphingomyelin 19.4%, phosphatidylinositol 7.6%, and phosphatidylserine 6.8%, are present in LDVs at a percentage of total phospholipid similar to that found in ACGs. A major quantitative difference occurs in the lipid phosphorus:protein ratio, which is 2.20 μmol/mg protein in LDVs compared to 0.48 in ACGs. There are also two times the relative cholesterol content in LDVs, 0.56 μmol/mg protein compared to 0.29 in ACGs.

The much higher ratio of phospholipids to protein in LDVs can be explained by the knowledge that (1) ACGs contain 78% soluble proteins compared to 38% in LDVs and (2) LDVs contain about 50% of their total phospholipids in the soluble matrix of the core, which is released by purely physical lysis, whereas physical lysis of ACGs releases little or no phospholipids.

An additional intriguing contrast can be made by comparing the lysolecithin content, which is uniquely high in ACGs, comprising 17% of the total phospholipids,[32,48] but is found only as a trace amount in LDVs.[23]

The maximum purity attainable of 25–40% SDVs from the castrated rat vas deferens preparation precludes any meaningful quantitation of SDV lipid content.

8. BINDING OF CYTOPLASMIC PROTEINS TO VESICLES

Compositional studies of adequately purified cell particles are complicated by the possibility that various substances, often proteins and peptides, will bind to the organelle surface. This can have functional significance in the case of specific binding, but when chemical composition is studied, it can be most bothersome in the case of nonspecific adsorption. An example of the latter case is the nonspecific adsorption of albuminlike protein to the LDVs from bovine splenic nerve. Adsorbed protein is carried by the vesicles close to the bottom of a sucrose–D_2O density gradient and remains in the diluted supernatant suspension of the most purified LDVs during the final fractional centrifugation step (Fig. 5). This albuminlike protein can be identified by comparing gel filtration and electrophoresis patterns with crystalline bovine serum albumin and by immunoprecipitation with antibodies to bovine serum albumin. It accounts for 45% of the total vesicle fraction protein content. The adsorbed protein can be completely removed by Affi-Gel Blue® adsorption[2] and mostly removed by washing the purified vesicles one time.

8.1. Calmodulin Binding to LDVs

The ubiquitous and recently widely studied Ca^{2+}-binding protein, calmodulin (~15,000 daltons), will bind specifically to particles in partially purified preparations of mixed synaptic vesicles from brain.[49] A number of Ca^{2+}–calmodulin-dependent protein phosphorylation reactions have been demonstrated and are subject to specific pharmacological antagonism. Certain reactions are suggested to be linked to coincidentally stimulated and antagonized vesicle functions, which include transmitter secretion.

In the density gradient used to purify the LDVs from bovine splenic nerve homogenate, calmodulin as measured by RIA occurs nearly entirely in the soluble fractions of the gradient tube, with no peak of calmodulin or calmodulin specific activity (per milligram protein) in the gradient proper.[50] In the untreated control LDV fraction, calmodulin occurs at 0.4 μg/mg LDV protein and can be reduced to an insignificant value <0.1 μg/mg protein by Ca^{2+} chelation with EGTA. However, calmodulin apparently binds specifically to the LDVs in the presence of 0.1 μM Ca^{2+} and 5 mM ATP (requirement not established) at a markedly elevated level of ~0.8% of the total vesicle protein content (R. L. Klein, B. H. Douglas, II., and K. B. Moorthy, unpublished data). This is equivalent to seven molecules of calmodulin per LDV or 40×10^{10} molecules/cm^2 vesicle membrane surface area.

8.2. Protein I Binding to LDVs or an Endogenous Constituent?

Preliminary reports[16] suggested that protein I (86,000 daltons) was a constituent of the noradrenergic LDVs, and this was recently supported by quantitative data.[51] However, calculations based on these data leave considerable doubt. In the recent report using bovine splenic nerve homogenate, the fraction of semipurified LDVs from the sucrose–D_2O density gradient step gave a NE : protein content of 4.4 nmol/mg protein, which coincided with the peak of protein I measured by detergent-based RIA. The protein I was increased eight times and NE four to five times per milligram protein compared to the normally achieved 30- to 40-fold purification based on NE, DβH, and OPs in the LDV fraction of the sucrose–D_2O density gradient. At the peak concentration, sedimentable protein I was 3.9 pmol/mg protein. Based on corrections for 42% LDV purity in the sucrose–D_2O density gradient fraction and 45% nonspecific adsorption of albuminlike protein, it can be calculated that each LDV contains only 0.24 molecules of protein I. Even if one assumes that the reported purification was proportionately low, e.g., eight times compared to 40 times, the calculation still results in only 1.2 molecules per LDV. The purified SDVs from castrated rat vas deferens contained even less protein I, and it was concluded that bovine adrenal chromaffin granules had negligible amounts.[51]

Thus, one must conclude that protein I is unlikely to be an endogenous constituent of the LDVs. Tentatively, protein I can be considered to have some affinity for the LDV, perhaps analogous to calmodulin. One can speculate that the binding of protein I to LDVs may be favored under phosphorylating conditions.

8.3. Tubulin Binding to LDVs

Tubulin subunits can be shown to bind in an apparently specific manner to chromaffin granule membranes by immunohistochemical and biochemical techniques.[52] No tubulin binding was demonstrated, at least by [^{3}H]colchicine labeling experiments, in the density-gradient fraction of LDVs or in the finally purified vesicle suspension.[12] All tubulin was found in the soluble fractions above the gradient.

8.4. Actin Binding to Neuronal Vesicles

Actin can be shown to bind specifically to chromaffin granule membranes at α-actinin sites.[52,53] Similar experiments have not been performed on the neuronal noradrenergic vesicles.

9. TRANSPORT SYSTEMS

A summary of data that compare transport systems in neuronal LDVs, SDVs, and adrenal chromaffin granules is presented in Table IV.

9.1. H^+-Translocating Mg^{2+}-ATPase

There is considerably more known about the function of the ATPase in ACGs than in neuronal LDVs or SDVs. It is now proven that an electrogenic H^+-translocating Mg^{2+}-ATPase will establish a pH gradient and is linked to catecholamine uptake in ACGs and ACG ghosts.[54,55] At cytoplasmic pH the positively charged amines will exchange with protons, $2H^+$/catecholamine, by an antiport mechanism. In the presence of permeant anions, exogenous ATP will also generate a membrane potential difference ($\Delta\psi$) of +80–100 mV inside, which can drive catecholamine uptake. Thus, an electrochemical gradient provides the driving force for catecholamine uptake: a ΔpH of 1.5 will generate a 10^3 gradient, a $\Delta\psi$ of 60 mV will account for another 10:1 gradient, and these will be augmented to a lesser extent by the equilibrium from some intragranular interaction between negative charges and catecholamines.[54] The maintenance of a 10^4 to 10^5 concentration gradient for catecholamines in ACGs seems entirely feasible on experimental grounds.

On the basis of kinetic and binding studies with solubilized and reconstituted extracts from ACG ghosts,[56] it has been estimated that there are 75 mol-

Table IV
Transport Systems and Related Information

	Large dense-cored vesicles	Small dense-cored vesicles	Adrenal chromaffin granules
Catecholamine gradient	~10^5 (terminals)	~5×10^4	10^4 to 10^5
Mg^{2+},(Ca^{2+})-ATPase (37°, 5 mM ATP)	0.22 μM P_i·min^{-1} mg protein^{-1}	Unk[a]	0.1–0.2 μM P_i·min^{-1} mg protein^{-1}
Reserpine, ouabain, oligomycin	Insensitive	Unk	Insensitive
Proton pump, 2H$^+$/CA	Unk	Unk	Yes
ΔpH	Unk	Unk	~pH 5.5 inside
Δψ	Unk	Unk	+80–100 mV inside
Mg^{2+}-ATP-dependent CA uptake; K_m	1.5 μM	22 μM	~ 30 μM (10–100 μM)
V_{max}, CA uptake	~0.01 μmol·min^{-1} mg protein^{-1}	Unk	0.01 μmol·min^{-1} mg protein^{-1}
NE/DA compete	Yes	Unk	Yes
Reserpine inhibition; ~K_i	0.05–0.1 μM[62]	Yes, K_i unk	~30 μM[56]
NE reverses	Yes	Unk	Yes
Oxid-phosph. uncouplers inhibit	Yes	Unk	Yes
NE synthesis from DA	Yes	No	Yes
Rate for intact vesicle (optimal)	30 molec NE·sec^{-1} per molec DβH		Unk
V_{max} purified DβH	~20 Units[63]		~20 Units
Theoretical	30–40 Units		
Ascorbic acid			
Gradient	Unk	Unk	Yes[65,66]
For half-max DβH activity	~1.3 mM		Unk
Cytochrome b_{561}	~20 molec/LDV[64] (8.5% memb. prot.)	~16 molec/SDV (14% total prot.)	1750 molec/ACG (10–20% membr. prot.)
Relationship to proton pump	Independent (?)		Unk
Mg^{2+}-ATP required	No		Unk
NE/DA Compete	No		Unk
References	2, 12, 16, 18	24, 57, 58	59–61

[a] Abbreviations: Unk, unknown; CA, catecholamine; Oxid-phosph, oxidative-phosphorylation; molec, molecule; Unit of DβH activity is 1.0 μmol product formed·min^{-1}·mg protein^{-1} at 37°; memb. prot., membrane protein.

ecules of catecholamine carrier protein, each of which turns over 35 molecules of NE per minute. A considerably lower estimate of ~10 molecules of Mg^{2+}-ATPase has been calculated for each ACG.[32]

9.1.1. Comparisons of the Mg^{2+}-ATPase and the Mg^{2+}-ATP-Dependent Uptake

The characteristics of the Mg^{2+}-ATPase and the Mg^{2+}-ATP-dependent uptake in neuronal LDVs are compared with those of ACGs on the bases of the incomplete data available (Table IV). Although a proton pump mechanism and pH have not been measured in the LDVs, a number of properties based on comparisons with ACGs support their probable existence:

1. Mg^{2+}-ATPase. There is a similar rate of ATP hydrolysis; reserpine does not inhibit the enzyme; and cytochrome b_{561} and NADH (acceptor) oxidoreductase are present.
2. Mg^{2+}-ATP-Dependent Catecholamine Uptake. The V_{max} for catecholamine uptake is similar; there is competition between NE and DA for uptake; and reserpine inhibition is competitively reversed by NE.

The LDVs show certain quantitative differences from ACGs in that the apparent K_m for NE uptake is much lower, and the approximate K_i for resperine inhibition of NE uptake is correspondingly lower. At 30°, the k_1 for influx of NE is 0.85 s^{-1}, and the absolute flux is 80 $fmol \cdot s^{-1} \cdot cm^{-2}$ at ~1.0 μM NE_o in purified LDVs in the presence of 2.5–5.0 mM Mg^{2+}-ATP.[18]

9.2. *Norepinephrine Synthesis from Dopamine*

9.2.1. *A Rapidly Released NE Synthesis Pool in LDVs*

There is considerably more known about NE synthesis from the immediate precursor DA in the highly purified fraction of neuronal LDVs than in ACGs. In fact, there are data that suggest that DA uptake related to NE synthesis in LDVs is by a mechanism independent of the carrier for Mg^{2+}-ATP-dependent uptake and the Mg^{2+}-ATPase proton pump.

In LDVs there are two pools of NE, one characterized by rapid exchange with $t_{½}$ of 2.7 min at 30° and one characterized by Mg^{2+}-ATP-dependent uptake with $t_{½}$of 19 min.[18] The addition of exogenous ascorbic acid is required to activate DβH, thereby inducing synthesis of NE from DA. In double-labeling experiments, it can be demonstrated that newly synthesized [^{14}C]NE from [^{14}C]DA occurs exclusively in the fast-release pool, which quickly saturates, resulting in overflow of [^{14}C]NE into the medium as if into the neuroplasm. Measurements of [^{3}H]NE exchange and calculations of unidirectional fluxes in the same experiment show that the two pools do not mix in the presence of Mg^{2+}-ATP. Newly synthsized [^{14}C]NE can be taken up readily into the more slowly released NE pool and will exchange with [^{3}H]NE in that pool subsequent to overflow into the medium.

9.2.2. *Differences between the Uptake of DA for NE Synthesis and for the Mg^{2+}-ATP-Dependent Carrier*

In contrast to the competition between NE and DA for uptake into the Mg^{2+}-ATP-dependent pool, [^{14}C]NE synthesis from [^{14}C]DA is unaffected by the level of exogenous NE at least up to 10 μM, at which concentration Mg^{2+}-ATP-dependent uptake can no longer be measured. Synthesis is also unaffected by the level of NE inside the vesicles from the initial 70 nmol/mg protein to near complete depletion of transmitter. Furthermore, the synthesis of [^{14}C]NE from [^{14}C]DA does not require the addition of exogenous Mg^{2+}-ATP.[18] Note: in the presence of Mg^{2+}-ATP, the K_m for uptake of NE is 1.5 μM under these conditions, and the maximum rate of uptake will occur between ~0.5 and 1.0 μM experimentally.

9.2.3. Rate of NE Synthesis by Intact LDVs

The rate of NE synthesis by intact LDVs is very rapid.[2,18] (Section 9.2.4b) and is half-maximal at ~1.3 mM exogenous ascorbic acid, which concentration coincidentally produces ~75% reduction of cytochrome b_{561}.[64] It is not proven in which process(es) cytochrome b_{561} functions as an electron translocator, i.e., NE synthesis and/or the Mg^{2+}-ATPase proton pump.

9.2.4. Provocative Hypotheses Related to NE Synthesis by LDVs

9.2.4a. Dopamine Uptake for NE Synthesis Is Independent of Mg^{2+}-ATP-Dependent Uptake and the Proton Pump. Although it has been presumed by most investigators that the proton pump mechanism in ACGs provides DA uptake for NE synthesis, in our opinion the evidence to date from studies of LDVs (Section 9.2.2) does not appear to be in support of any generalization in this regard. Rather, the data suggest that a separate mechanism may exist in LDVs with a much higher affinity for DA binding (uptake) than for Mg^{2+}-ATP-dependent uptake and is operable only when DβH is activated by ascorbic acid. For some time now, we have speculated about the existence of a membrane macromolecule with a DA receptor or uptake site on the outside of the vesicle membrane directed to a DβH enzymatic site on the inside of the membrane.

9.2.4b. Noradrenergic LDVs Provide Newly Synthesized NE to Fill the Small Synaptic Vesicles. We have recently proposed that LDVs provide newly synthesized NE to fill locally formed and depleted small vesicles.[2,25] This is based on the following considerations: (1) SDVs seem to lack NE synthesis capacity; (2) LDVs possess an outstanding potential to synthesize NE. At 37° and optimal substrate concentration, each LDV can synthesize 300 molecules of NE per second. Thus, one LDV has the potential to fill one SDV completely in 2–3 s. Since only a few vesicles actually participate in exocytosis during a given stimulus (Section 10.2), the refilling of small vesicles can occur instantaneously. This is in keeping with *in vivo* evidence that NE depletion can not be demonstrated after usual physiological stimulation, even though substantial release of transmitter can be analyzed biochemically. (3) Initially, newly synthesized NE appears exclusively in the fast-release pool of the LDVs. This pool saturates rapidly, resulting in NE overflow into the medium (neuroplasm) at a continuous linear rate. (4) The K_m for Mg^{2+}-ATP-dependent uptake of NE by the LDV and the neuronal membrane is about 1 μM compared to ~20 μM for SDVs (Table IV). Thus, LDVs will always be preferentially filled with NE from the neuroplasm. At intermediate K_m values, a combination of low neuroplasmic NE concentration, tyrosine substrate, cofactors, and ascorbate avail-

Table V
Potential for Exocytotic Release of Soluble Contents from Noradrenergic Vesicles Compared to Adrenal Chromaffin Granules

	Numbers of molecules		
	Small dense-cored vesicles[2]	Large dense-cored vesicles[2,16]	Adrenal chromaffin granules
Norepinephrine			
Axonal		3,700–6,600[a]	
Terminal	700–1,000	9,000–16,000[b]	2–3 × 10^6
ATP	12–50	150–500	6.7 × 10^5
Opioid peptides			
Met-, Leu-enk.	0	58–100	9,000
Larger opioids	0	Equal activity[c]	4–8 times activity[c]
DβH	0	~9	~136
Chromogranin A	0	24 (?)[d]	5,000
Ascorbic acid	Unk	Unk	1 × 10^5
Phospholipids	Unk	14,600[e]	Little if any
Others (see text)			

[a] Corrected for 85% vesicle purity and 45% nonspecifically absorbed albuminlike protein.[2]
[b] Corrected for NE loss during 10 to 12-min postmortem delay and synthesis in the terminal region.[2]
[c] Activation by trypsin or others increased the opioid radioreceptor activity two-fold in LDVs and to a much greater extent in ACGs.
[d] This tentative calculation is based on the value of 32 μg chromogranin A per mg protein in semipurified LDV fraction III from the sucrose–D_2O density gradient after sedimentation.[67] It has been corrected for an estimated purity of 42% and 45% nonspecifically absorbed albuminlike protein. (See also text Section 10.1.3)
[e] Approximately 50% of the total phospholipid content can be released from the most purified LDVs by purely physical lysis with the French press.[16]

ability will activate tyrosine hydroxylase and DβH to perpetuate synthesis of NE by the LDVs until the SDVs are filled from the overflow of newly synthesized NE. (5) With SDV filling, a subsequent buildup of neuroplasmic NE will result in feedback (end product) inhibition of tyrosine hydroxylase and related induction of DβH activity. (6) the K_m for MAO is ~100 μM and is not expected to compete with SDV uptake for neuroplasmic NE.

10. EXOCYTOSIS: BIOCHEMICAL, PHYSIOLOGICAL, AND ULTRASTRUCTURAL EVIDENCE

10.1. Potential for Exocytotic Release of Soluble Contents

The estimated numbers of molecules of soluble components available for exocytotic release by SDVs, LDVs, and ACGs are compared in Table V. Perhaps some help to appreciate the data may be gained by comparing the relative sizes of these catecholamine storage particles (Fig. 11) and the calculated volume and surface area ratios given in the legend to this figure.

There are major qualitative and quantitative differences among the three catecholamine storage particles relating to the potential to release soluble contents. The neuronal SDVs probably release only NE and ATP and, specula-

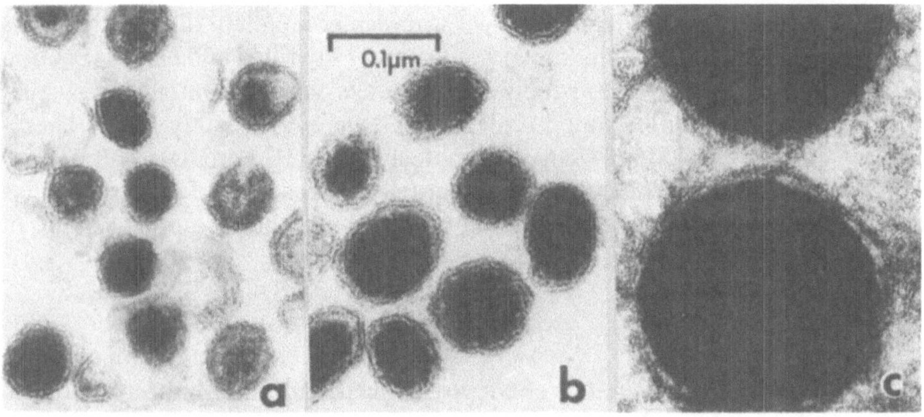

Fig. 11. Purified preparations of (a) SDVs from castrated rat vas deferens, (b) LDVs from bovine splenic nerve, and (c) chromaffin granules from bovine adrenal medulla. Using average diameters of 45, 75, and 350 nm *in situ*, respectively, and 70 nm for membrane thickness, the following ratios can be calculated for SDVs:LDVs:ACGs (LDVs:ACGs): total volume, 1:4.6:471 (1:102); core volume, 1:7.6:1273 (1:167); membrane volume, 1:3.2:81 (1:25); surface area, 1:2.8:61 (1:22).

tively, some phospholipids. Comparisons of the LDVs with ACGs indicate that the LDVs release relatively smaller amounts of nucleotides and chromogranin A on the basis of total particle protein. The LDVs also release proportionately much less of the larger opioid peptides with cryptic enkephalin activity. On the other hand, the LDVs appear to release considerable phospholipids. Only the catecholamine, opioid peptide, and chromogranin A (?) (see Section 10.1.3) contents occur at close to the ratio expected from the relative core volumes, indicative of similar soluble-phase concentrations in the two particles.

10.1.1. Norepinephrine Release by SDVs: An Estimate of Quantal Size

Under normal physiological conditions, it is generally assumed that SDVs are responsible for the primary release of NE from sympathetic nerve terminals. This would preserve an analogy with the neuromuscular junction and autonomic cholinergic nerves and provide for quantal packaging of the transmitter, norepinephrine. However, it is abundantly clear that the specific mechanisms involved in exocytosis and vesicle recycling will be different from the predominant thinking, which has resulted from the elegant ultrastructural and freeze fracture studies on the skeletal neuromuscular junction.

To estimate the NE content of each completely filled SDV is difficult. With direct analytical data of the purest available preparations from castrated rat vas deferens, there is no need to correct for postmortem delay. However, there is loss of NE into the soluble phase as a result of tissue homogenization and purification procedures. In the calculations, this loss has been treated as if all soluble NE originates from the SDVs, even though there occur ~5% LDVs in this preparation. A somewhat exaggerated NE content results. On the other hand, it is well known that the small-vesicle population in the typical varicosity is never entirely filled with NE, as reflected by potassium permanganate staining reactions and others; typically, 50–80% appear completely filled with dense-cored material, and the remainder vary between partly filled and empty (Section 10.2). This situation tends to produce an average vesicle content below that which potentially could be found if all vesicles were completely filled. Therefore, the calculated value of 700–1000 molecules of NE per SDV (~0.078 M) as a proposed estimate of quantal size actually represents a fair compromise between the exaggerated estimate using total NE and the incomplete filling of some vesicles.

10.1.2. Concomitant Release of NE and DβH by LDVs

That LDVs do participate in exocytosis even under normal physiological conditions is conclusively demonstrated ultrastructurally, biochemically, and physiologically. However, exocytotic release from LDVs is much enhanced under stressful stimuli (Section 10.2), when *a priori* NE release will be greatly augmented (9,000–16,000 molecules/vesicle). Exocytotic release must be accompanied by concomitant release of hydrophilic DβH (Section 4.2) and opioid peptides (Section 5.2.1), which originate exclusively from the LDVs. The latter can be demonstrated by sympathetic nerve stimulation *in situ* and *in vitro*.

10.1.3. Chromogranin A Release from LDVs

Only small amounts of chromogranin A have been reported in LDV fractions in the earlier literature. Quantitation involved the use of antibodies from purified proteins that are now suspect of DβH contamination. There is also a minor electrophoretic band on SDS-polyacrylamide gels from the water-soluble fraction of semipurified LDVs with a mobility corresponding to chromogranin A.[68] Thus, at most, a very tentative estimate of 24 molecules of chromogranin A released from each LDV can be made at present. If correct, this is equivalent to ~14% of the total vesicle protein content and 37% of the soluble protein.

Antibodies prepared to rigorously purified chromogranin A recently have been used to detect this acidic protein in the LDV fraction but not in the SDV fraction on density-gradient separations of bull vas deferens homogenate (~30% LDVs and 70% SDVs).[30]

In another recent study, RIA measurements have been made using antibodies to chromogranin A purified electrophoretically from bovine ACGs. (D. T. O'Connor and R. L. Klein, unpublished data.) The sucrose-D_2O density gradient distribution of chromogranin A is remarkably similar to that of total NA, i.e., both in particulate and soluble fractions. In the initial experiments <40% of the chromogranin was particulate. This contrasts to 89–95% particulate bioactive peptides (Section 5.2.1 and 5.2.2). There is co-purification of chromogranin with particulate NE at each step in the preparation of the purest LDV fraction. There is a 30-fold enrichment of chromogranin A specific activity and preliminary calculations indicate 1.4 ± 0.1 μg/mg vesicle protein (uncorrected for vesicle purity and non-specifically adsorbed albumin-like protein). The preliminary data suggest only ~1 molecule of chromogranin/LDV at a molar ratio of 1:2000 NE by direct analysis of the purest LDV fraction. The LDV chromogranin molecular weights were estimated using SDS gels with anti-chromogranin A immunoblotting. A major band at 67kD was found identical to chromogranin A from ACGs and a minor band of immunoreactivity occurred at 97kD.

10.1.4. Ascorbic Acid Release

The ascorbic acid content of the LDVs has not been measured. Tentatively, it can be assumed that ascorbic acid is a very labile constituent of the LDVs and is rapidly lost during tissue homogenization and vesicle purification procedures, as occurs from ACGs.[66] However, it is likely that under appropriate conditions *in vitro* an ascorbate gradient can be demonstrated. Ascorbic acid is the natural cosubstrate for DβH, and it has to enter the vesicle in order to reach the active site on the enzyme, possibly via the proton pump.[69] With purified vesicles, ascorbic acid must be added exogenously, but it can serve alone to initiate synthesis of NE from DA by LDVs at a maximal rate (Table IV).

10.1.5. Release of Other Soluble Components

The potential release of soluble complex carbohydrates from LDVs compared to ACGs can be inferred from the data discussed in Section 6. The LDVs also contain in the water-soluble phase a nonchromogranin glycoprotein (100–120,000 dalton range), a colored protein (22-55,000 dalton range), and a proteoglycan (? <14,000).[16] The colored protein has not been identified, but it is in the size range of cytochrome b_{561}. It may be pertinent to point out that the earlier positive identification of cytochrome b_{561} in LDV fractions[64] can not be conclusively restricted to the LDV membrane, as the methods of freeze–thaw, hypoosmotic shocks, and dialysis that were used do not produce efficient vesicle lysis.[16]

10.2. Ultrastructural Evidence for Exocytosis from Noradrenergic Terminals: A Human Blood Vessel Model

It has been demonstrated ultrastructurally that both LDVs and SDVs can fuse with the neuronal membrane of noradrenergic varicosities to release their contents by exocytosis.[3,70–73] This is true for bovine tissues, which may contain up to 40–50% LDVs (Fig. 12), for human tissues, which contain 30–35% LDVs (Fig. 13), and for guinea pig tissues, which contain only 5% LDVs (Figs. 14,15).

10.2.1. Dynamic Changes in Large- and Small-Vesicle Populations per Varicosity

A model that can be used to study the dynamics of both SDVs and LDVs is the human mesenteric blood vessel.[73] A summary of results from recent studies is given in Table VI, and a list of interpretations follows.

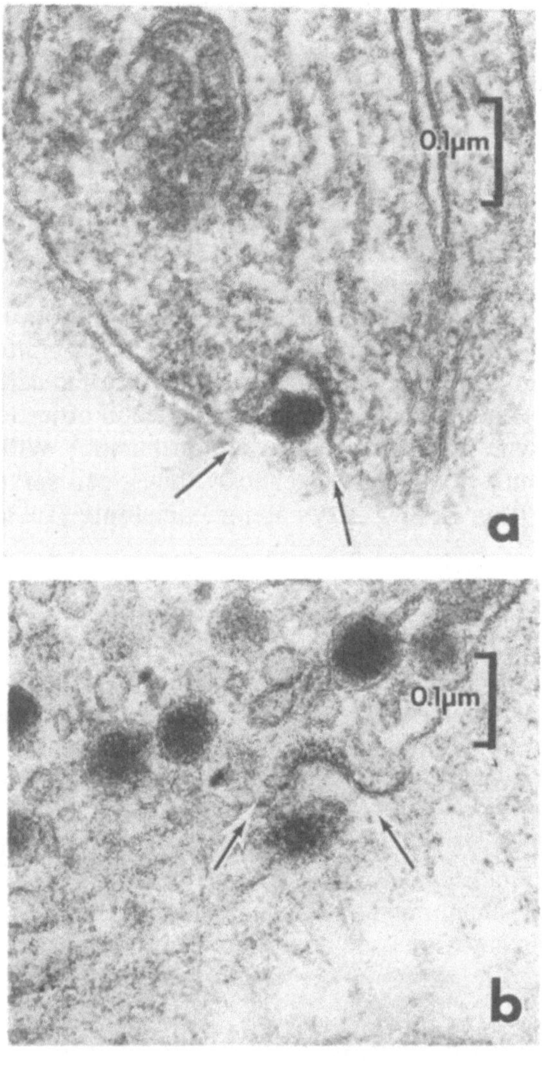

Fig. 12. a: An exocytotic profile in a bovine splenic vein terminal demonstrates the core of an LDV still within its membrane fused with the terminal membrane to form an omega figure. The vesicle membrane shows a coat of dense material, which may indicate reuptake of membrane at the site of release. Fixed during stimulation with glutaraldehyde. b: An exocytotic profile in a bovine vas deferens terminal. The expelled core appears to be dispersing outside the coated omega profile. (Reproduced from ref. 73 with permission from Pergamon Press.)

In human mesenteric vessels fixed immediately after surgical removal, a count of varicosity sections ($n = 275$) gives a terminal vesicle population of ~300 with a distribution of 33% LDVs, 39% SDVs, and 28% SCVs (small clear vesicles). These occupy an average of 3–4% of the varicosity volume.

After a period of 1 h of superfusion to achieve equilibration of the tissue, there is an increased accumulation of 25–40% in total vesicles in the average varicosity but no significant change in their ratios. This is thought to result from continued axoplasmic transport *in vitro*.

On field stimulation for 1 min at 10 Hz, there is a further 38% increase in the total vesicle population, which is caused primarily by a 49% increase in small vesicles accompanied by a significant shift from SDVs to SCVs. This

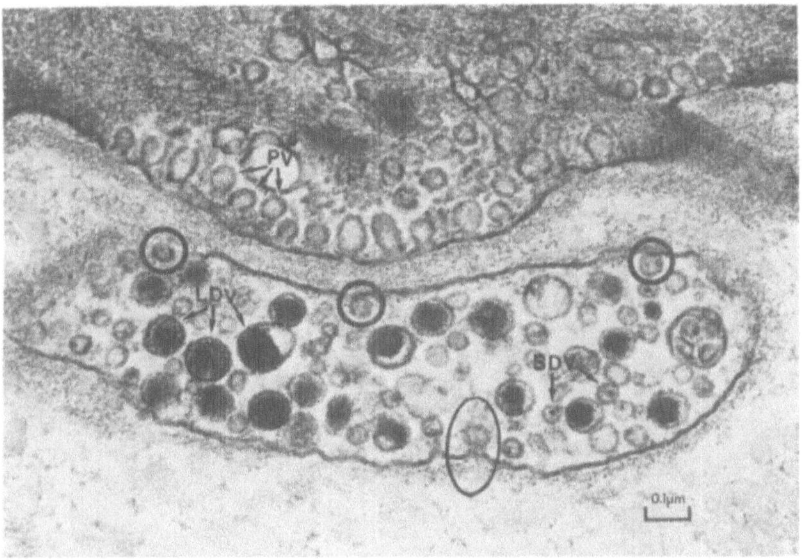

Fig. 13. A terminal in a human omental vein subjected to electrical stimulation in the presence of phentolamine. Large (LDV) and small (SDV) vesicles are scattered in the neuroplasm. Small vesicles (encircled) are interpreted to be in various stages of exocytosis, from fusing with the terminal membrane to expelling of contents. PV, pinocytotic vesicles. (Reproduced from ref. 73 with permission from Pergamon Press.)

short-term response to repetitive electrical stimulation is interpreted to reflect a coincident induction of local small-vesicle formation and of NE synthesis to fill SCVs. A smaller increase in LDVs of 11% is not statistically significant; however, the relative decrease in LDVs compared to the total vesicle population is significant.

With phentolamine present during stimulation, there is a significant decrease in the LDVs and SDVs compared to stimulation alone, giving a population of vesicles close to the control values, but there is also a significant shift from SDVs to SCVs. The most obvious ultrastructural difference, compared to stimulation in the absence of phentolamine, is an increased number of both LDVs and SDVs that occur closely associated or fused with the neuronal membrane.[73] Thus, phentolamine may affect some process that enhances the availability of vesicles for potential exocytotic release.

The decrease in total vesicles and the shift to SCVs could reflect an inability to accommodate the drug-induced augmented overflow of NE. The increased exocytosis known to occur with phentolamine results in both NE and DβH release. Thus, there may be insufficient time for refilling of empty vesicles and replacement of used vesicles in the presence of drug. It could also indicate that the drug somehow interferes with the process of local formation of small vesicles normally occurring during electrical stimulation. It should be noted that there is no proof that all of the small vesicles counted in the varicosities have the potential to form SDVs; it cannot be excluded that some of these small vesicles are pinocytotic in nature and form from the neuronal membrane.

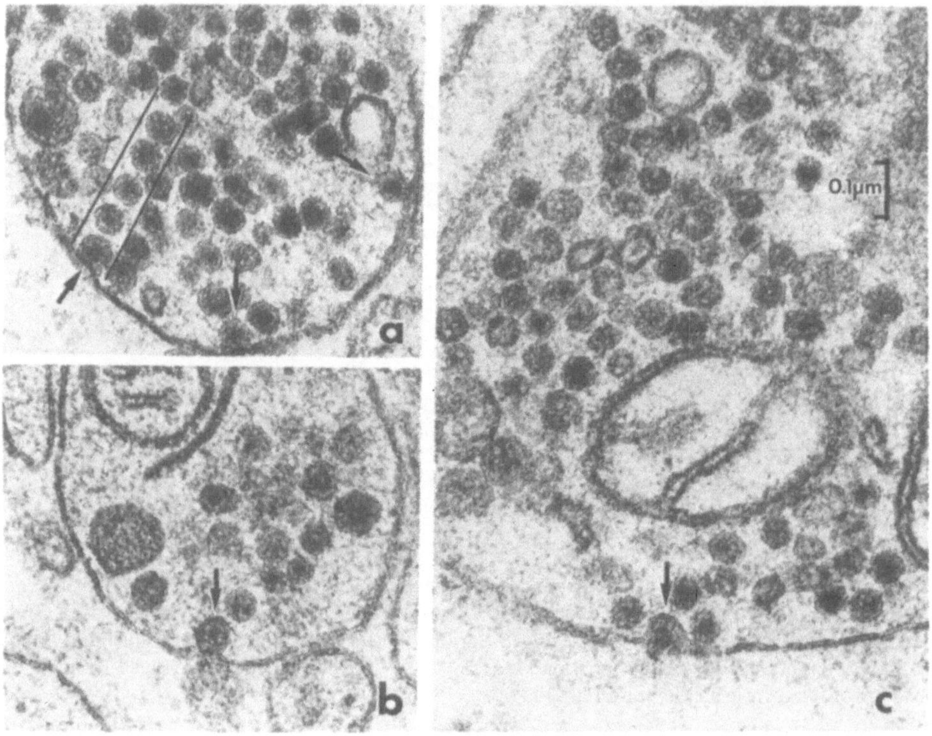

Fig. 14. Noradrenergic terminals in guinea pig vas deferens containing predominantly SDVs. a: Three SDVs (arrows) are in position to fuse with the terminal membrane and release their contents by exocytosis. A row of vesicles (between lines) appear to be lined up perpendicular to the membrane as if in position for sequential release. b: A small vesicle (arrow) is in position to release its contents. c: A small vesicle (arrow) is fusing with the terminal membrane, and its core is exposed.

There is no indication under any circumstances studied that an increased incorporation of vesicle membranes occurs into the neuronal membrane that would be reflected by an average increase in varicosity circumference. Then again, the small numbers of vesicles interpreted to participate in exocytosis with a given stimulus would not be expected to have a dramatic effect in this regard. In fact, electron microscopic observations suggest that there is immediate retrieval of vesicle membrane at the site of exocytosis. This is supported by the fact that one can find omega profiles in which the dense core has been caught in the act of release and on which a typical coated membrane is already formed (Figs. 12a,b; refs. 3,70,71). The coated membrane is accepted by many as an indication of membrane reuptake.

Electrical stimulation alone and stimulation with phentolamine present increase the incidence of omega profiles from one per 37–43 varicosity sections in controls (immediately fixed or equilibrated) to one in 8–10 varicosity sections (includes sections with two or three omega profiles). On an average, the increased incidence is equivalent to a little more than one presumed exocytotic event per varicosity (Table VI).

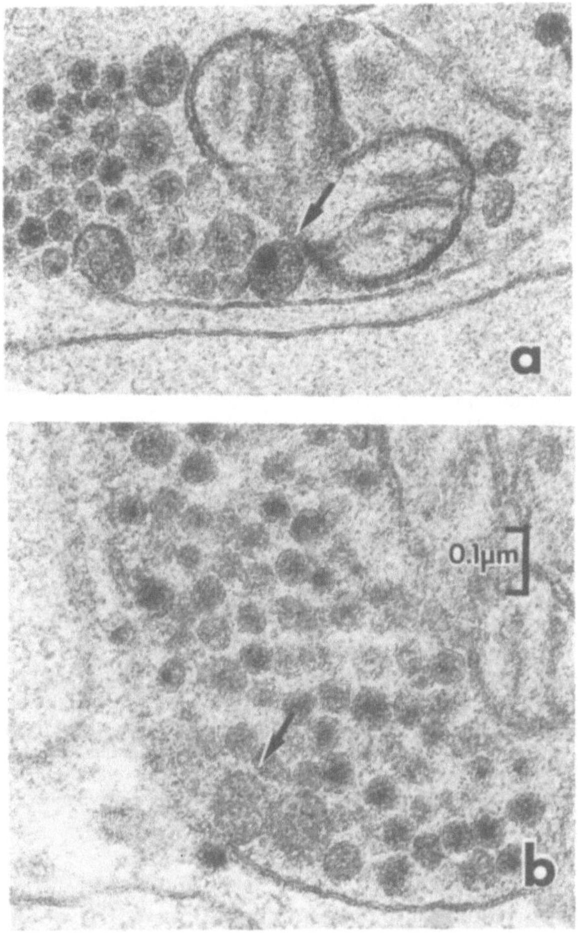

Fig. 15. Terminals in guinea pig vas deferens. a: An LDV (arrow) adjacent to the terminal membrane is in a position to release its contents by exocytosis. b: Core material is present just outside the LDV, which has fused with the terminal membrane.

After electrical stimulation alone or with phentolamine present, there is not only an increased incidence of individual varicosity sections with single omega profiles, but there is also an increased tendency for two or more omega profiles in a single section. This could indicate that the grading of exocytotic secretion in response to repetitive stimulation may be caused both by invasion of more varicosities ("recruitment") and by an increase in the number of vesicles involved per varicosity.

It should be noted that because of the rapidity of the exocytotic event, omega profiles counted probably represent only those events that were in progress when the superfusate containing fixative reached the membrane. In this sense, the results represent the dynamics of a much shorter period than the 1-min duration of stimulation. It is easier to visualize exocytosis by LDVs than by SDVs in this type of experiment and after other stressful stimuli,[70] even though the incidence of quantal NE release from SDVs is expected to be much more frequent. This could reflect a longer duration for LDVs to release their

Table VI
Number of Vesicles per Terminal Varicosity of Human Mesenteric Vein[a]: Effects of Stimulation and Phentolamine[72]

Conditions (varicosity sections)	LDVs	Percent of total vesicles	SDVs + SCVs	Percent SCVs of total SCVs + SDVs	Total vesicles per varicosity	LDV omega profiles per varicosity
Equilibrated (n = 300)	80	30.5	183[b]	43.4	263	1:3.5–4.1
Stimulated 1 min, 10 Hz (n = 400)	89	24.4	274[b]	48.5[b]	363[b]	1.3:1
Stimulated + phentolamine (0.75 μM) (n = 282)	68[b]	24.2	214[b]	52.0[b]	282[b]	1.1:1

[a] Total varicosities counted 982; total vesicles counted 8036 LDVs, 29,930 SDVs + SCVs; LDVs avg. 87.5 nm diameter; SDVs avg. 51 nm diameter.
[b] Statistically significant change from number above at $P < 0.001$ level. The noradrenergic varicosities are considered to have a shape resembling an oblate spheroid, and the electron microscopic sections estimated to be 70 nm in thickness by interference color. Measurements of sections through varicose regions (n = 150) averaged 0.129 ± 0.031 μm long axis and 0.642 ± 0.36 μm short axis giving a varicosity volume of 0.36 μm^3 and a surface area of 1.98 μm^2. The varicosity volume:section volume ratio is 9.8:1, and the varicosity surface area:section perimeter surface area ratio is 10.54:1. The former ratio was used to convert numbers of vesicles per section to per varicosity, and the latter ratio to convert exocytotic profiles (omega figures) per section perimeter to total varicosity surface area. Note: the omega figure diameter from an LDV is essentially the same as the section thickness. After the appropriate time during treatment, the tissues were fixed by adding glutaraldehyde to the superfusion medium. This was followed by postosmification. Phentolamine was used at a concentration that produced near optimal NE overflow but was too low to inhibit neuronal amine uptake.[72]

larger molecular content, including DβH, compared to the presumed millisecond release of a single quantum of NE from an SDV.

In general, the changes in vesicle numbers should all be viewed as part of a neuronal response to maintain homeostasis, and any increase or decrease in vesicle numbers as a temporary adjustment to the experimental conditions.

The data are in keeping with the concept that not all varicosities are invaded and/or activated by a given train of repetitive stimuli and that only a few vesicles of the total actually participate in exocytosis when a varicosity is activated. For more comprehensive treatment of the electrophysiological parameters in sympathetic varicose fibers, see refs. 74,75. On the basis of from one to three omega profiles per varicosity section, which is not an uncommon finding in species with a high ratio of LDVs to SDVs, it can be speculated that under the influence of repetitive stimuli and drugs, as many as 10–32 of the total 90 LDVs may be released from an individual varicosity.

Acknowledgments. This work was supported in part by the American Heart Association and its Mississippi Affiliate, Grant 82–604.

REFERENCES

1. Holtzman, E., 1977, *Neuroscience* **2**:327–355.
2. Klein, R. L., and Lagercrantz, H., 1981, *Chemical Neurotransmission—75 Years. Second Nobel Conference, Stockholm* (L. Stjärne, P. Hedqvist, H. Lagercrantz, and Å. Wennmalm, eds.), Academic Press, London, pp. 69–83.
3. Thureson-Klein, Å., 1982, *Neurotransmitter Vesicles* (R. L. Klein, H. Lagercrantz, and H. Zimmermann, eds.), Academic Press, London, pp. 65–87.
4. Thureson-Klein, Å., 1982, *Neurotransmitter Vesicles* (R. L. Klein, H. Lagercrantz, and H. Zimmermann, eds.), Academic Press, London, pp. 119–132.
5. Fillenz, M., 1979, *The Release of Catecholamines from Adrenergic Neurons* (D. M. Paton, ed.), Pergamon Press, New York, pp. 17–37.
6. Fried, G., Thureson-Klein, Å., and Lagercrantz, H., 1981, *Neuroscience* **6**:787–800.
7. Klein, R. L., and Thureson-Klein, Å., 1974, *Fed. Proc.* **33**:2195–2206.
8. Lagercrantz, H., and Klein, R. L., 1982, *Neurotransmitter Vesicles* (R. L. Klein, H. Lagercrantz, and H. Zimmermann, eds.), Academic Press, London, pp. 89–118.
9. Lagercrantz, H., Klein, R. L., and Stjärne, L., 1970, *Life Sci.* **9**:639–650.
10. Yen, S. S., Klein, R. L., and Chen-Yen, S. H., 1973, *J. Neurocytol.* **2**:1–12.
11. Klein, R. L., Thureson-Klein, Å., Chen-Yen, S. H., Baggett, J. McC., Gasparis, M. S., and Kirksey, D. F., 1979, *J. Neurobiol.* **10**:291–307.
12. Lagercrantz, H., 1976, *Neuroscience* **1**:81–92.
13. Klein, R. L., and Thureson-Klein, Å., 1971, *J. Ultrastruct. Res.* **34**:473–491.
14. Thureson-Klein, Å., Klein, R. L., and Lagercrantz, H., 1973, *J. Neurocytol.* **2**:13–27.
15. Fried, G., Lagercrantz, H., and Hökfelt, T., 1978, *Neuroscience* **3**:1271–1291.
16. Klein, R. L., 1982, *Neurotransmitter Vesicles* (R. L. Klein, H. Lagercrantz, and H. Zimmermann, eds.), Academic Press, London, pp. 133–173.
17. Klein, R. L., 1973, *Frontiers in Catecholamine Research* (E. Usdin and S. Snyder, eds.), Pergamon Press, New York, pp. 423–425.
18. Klein, R. L., 1982, *Neurotransmitter Vesicles* (R. L. Klein, H. Lagercrantz, and H. Zimmermann, eds.), Academic Press, London, pp. 189–217.
19. Fredholm, B. B., Fried, G., and Hedqvist, P., 1982, *Eur. J. Pharmacol.* **79**:233–244.
20. Klein, R. L., and Lagercrantz, H., 1982, *Neurotransmitter Vesicles* (R. L. Klein, H. Lagercrantz, and H. Zimmermann, eds.), Academic Press, London, pp. 219–239.
21. Weinshilboum, R. M., 1979, *Pharmacol. Rev.* **30**:133–166.
22. Smith, A. D., 1972, *Pharmacol. Rev.* **24**:435–457.
23. Kirksey, D. F., Klein, R. L., Baggett, J. McC., and Gasparis, M. S., 1978, *Neuroscience* **3**:71–81.
24. Lagercrantz, H., and Fried, G., 1982, *Neurotransmitter Vesicles* (R. L. Klein, H. Lagercrantz, and H. Zimmermann, eds.), Academic Press, London, pp. 175–188.
25. Klein, R. L., and Thureson-Klein, Å., 1981, *Function and Regulation of Monoamine Enzymes—Basic and Clinical Aspects* (E. Usdin, N. Weiner, and M. B. Youdim, eds.), Macmillan, London, pp. 443–451.
26. Nagatsu, I., Kondo, Y., Kato, T., and Nagatsu, T., 1976, *Brain Res.* **116**:277–285.
27. Helle, K. B., and Serck-Hanssen, G., 1981, *Chemical Neurotransmission—75 Years, Second Nobel Conference, Stockholm* (L. Stjärne, P. Hedqvist, H. Lagercrantz, and Å. Wennmalm, eds.), Academic Press, London, pp. 85–90.
28. Yang, W. H., 1981, *Function and Regulation of Monoamine Enzymes—Basic and Clinical Aspects* (E. Usdin, N. Weiner, and M. B. Youdim, eds.), Macmillan, London, pp. 453–460.
29. Willems, M. F., and DePotter, W., 1982, *Arch. Int. Pharmacodyn.* **258**:333–334.
30. Lingg, G., Neuman, B., Schmidt, W., Zangerle, R., Fischer-Colbrie, R., and Winkler, H., 1982, *Molecular Neurobiology of Peripheral Catecholamine Systems. An International Conference, Ibiza, Spain.* The Faculty of Medicine Universidad Autónoma de Madrid, Madrid, p. 121A.
31. Gasparis, M. S., Yang W.-H., and Klein, R. L., 1983, *Neurochem. Res.* **8**:1417–1439.

32. Winkler, H., and Westhead, E., 1980, *Neuroscience* **5**:1803–1823.
33. Pearse, A. G. E., 1980, *Mikroskopie* **36**:257–269.
34. Fujita, T., and Kobayashi, S., 1979, *Trends Neurosci.* **2**:27–30.
35. Wilson, S. P., Klein, R. L., Chang, K. J., Gasparis, M. S., Viveros, O. H., and Yang, W. H., 1980, *Nature* **288**:707–709.
36. Klein, R. L., Wilson, S. P., Dzielak, D. J., Yang, W. H., and Viveros, O. H., 1982, *Neuroscience* **7**:2255–2261.
37. Klein, R. L., Day, R., and Lemaire, S., 1983, *Soc. Neurosci. Abstr.* **9**:575.
38. Lundberg, J. M., Fried, G. Fahrenkrug, J., Holmstedt, B., Hökfelt, T., Lagercrantz, H., Lundgren, G., and Anggard, A., 1981, *Neuroscience* **6**:1001–1010.
39. Johansson, O., and Lundberg, J. M., 1981, *Neuroscience* **5**:847–862.
40. Pelletier, G., Steinbusch, H. W. M., and Verhofstad, A. A. J., 1981, *Nature* **293**:71.
41. Fried, G., 1982, *Neurotransmitter Vesicles* (R. L. Klein, H. Lagercrantz, and H. Zimmermann, eds.), Academic Press, London, pp. 361–374.
42. Geissler, D., Martinek, A., Margolis, R. U., Margolis, R. K., Skrivanek, J. A., Ledeen, R., Konig, P., and Winkler, H., 1977, *Neuroscience* **2**:685–693.
43. Kiang, W. L., Krusius, T., Finne, J., Margolis, R. U., and Margolis, R. K., 1982, *J. Biol. Chem.* **257**:1651–1659.
44. Blaschke, E., 1979, *Acta Physiol. Scand* [*Suppl.*] **466**:7–42.
45. Uvnäs, B., and Åborg, C. H., 1977, *Acta Physiol. Scand.* **99**:476–483.
46. Stadler, H., and Whittaker, V. P., 1978, *Brain Res.* **153**:408–413.
47. Jones, R. T., Walker, J. H., Stadler, H., and Whittaker, V. P., 1982, *Cell Tissue Res.* **224**:685–688.
48. Winkler, H., 1976, *Neuroscience* **1**:65–80.
49. DeLorenzo, R. J., 1982, *Fed. Proc.* **41**:2265–2272.
50. Klein, R. L., 1982, *Molecular Neurobiology of Peripheral Catecholamine Systems. An International Conference, Ibiza, Spain*, The Faculty of Medicine Universidad Autónoma de Madrid, Madrid, p. 107A.
51. Fried, G., Nestler, E. J., DeCamilli, P., Stjärne, L., Olson, L., Lundberg, J. M., Hökfelt, T., Ouimet, C. C., and Greengard, P., 1982, *Neurobiology* **79**:2717–2721.
52. Aunis, D., Bader, M. F., Bernier-Valentin, F., Ciesielski-Treska, J., and Rousset B., 1982, *Molecular Neurobiology of Peripheral Catecholamine Systems. An International Conference, Ibiza, Spain*, The Faculty of Medicine Universidad Autónoma de Madrid, Madrid, pp. 165A–166A.
53. Trifaro, J. M., 1982, *Molecular Neurobiology of Peripheral Catecholamine Systems. An International Conference, Ibiza, Spain*, The Faculty of Medicine Universidad Autónoma de Madrid, Madrid, pp. 167A–168A.
54. Knoth, J., Zallakian, M., and Njus, D., 1982, *Fed. Proc.* **41**:2472–2745.
55. Johnson, R. G., Carty, S., and Scarpa, A., 1982, *Fed. Proc.* **41**:2746–2754.
56. Henry, J. P., Isambert, M. F., Roisin, M. P., and Scherman, D., 1982, *Molecular Neurobiology of Peripheral Catecholamine Systems. An International Conference, Ibiza, Spain*, The Faculty of Medicine Universidad Autónoma de Madrid, Madrid, p. 87A.
57. Fried, G., 1978, *Biochim. Biophys. Acta* **507**:175–177.
58. Fried, G., 1980, *Acta Physiol. Scand.* [*Suppl.* 493] **111**:1–28.
59. Apps, D. K., 1982, *Fed. Proc.* **41**:2775–2780.
60. Phillips, J. H., 1982, *Neuroscience* **7**:1595–1609.
61. Winkler, H., and Carmichael, S. W., 1982, *The Secretor Granule* (A. M. Poisner and J. M. Trifaro, eds.), Elsevier Biomedical Press, New York, pp. 3–79.
62. Klein, R. L., and Lagercrantz, H., 1971, *Acta Physiol. Scand.* **83**:179–190.
63. Klein, R. L., Kirksey, D. F., Rush, R. A., and Goldstein, M., 1977, *J. Neurochem.* **28**:81–86.
64. Flatmark, T., Lagercrantz, H., Terland, O., Helle, K. B., and Stjärne, L., 1971, *Biochim. Biophys. Acta* **245**:249–252.
65. Ingebretsen, O. C., Terland, O., and Flatmark, T., 1980, *Biochim. Biophys. Acta* **628**:182–189.
66. Daniels, A. J., Dean, G., Viveros, O. H., and Diliberto, E. J. Jr., 1982, *Science* **216**:737–739.

67. Lagercrantz, H., 1971, *Acta Physiol. Scand.* [*Suppl.* 366] **82**:1–44.
68. Bartlett, S. F., Lagercrantz, H., and Smith, A. D., 1976, *Neuroscience* **1**:339–344.
69. Njus, D., 1982, *Molecular Neurobiology of Peripheral Catecholamine Systems. An International Conference, Ibiza, Spain*, The Faculty of Medicine Universidad Autónoma de Madrid, Madrid, pp. 83A–84A.
70. Thureson-Klein, Å., Klein, R. L., and Johansson, O., 1979, *J. Neurobiol.* **10**:309–324.
71. Bevan, J. A., Bevan, R. D., and Duckles, S. P., 1980, *Handbook of Physiology*, Volume II (D. F. Bohr, A. P. Somlyo, and H. V. Sparks, Jr., eds.), American Physiological Society, Bethesda, pp. 515–566.
72. Thureson-Klein, Å., and Stjärne, L., 1981, *Chemical Neurotransmission—75 Years, Second Nobel Conference, Stockholm* (L. Stjärne, P. Hedqvist, H. Lagercrantz, and Å. Wennmalm, eds.), Academic Press, London, pp. 152–164.
73. Thureson-Klein, Å., 1983, *Neuroscience* **10**:245–252.
74. Stjärne, L., 1982, *Advances in Pharmacology and Therapeutics II*, Volume II (H. Yoshida, Y. Hagibara, and S. Ebashi, eds.), Pergamon Press, Oxford, New York, pp. 111–120.
75. Cunnane, T. C., and Stjärne, L., 1982, *Neuroscience* **7**:2565–2576.

4

Synaptic Proteins

Henry R. Mahler†

1. INTRODUCTION

The highly specialized regions of contact (d = 1–2 μm) between individual neurons, including their axons and dendrites, are known as synapses. They provide sites of communication between such cells not only for the transmission of electric impulses[1] but also for mutual recognition for the construction and refashioning of neural networks in the course of ontogeny and in consequence of insults (ablation) and experiential stimulation. The synapse proper therefore consists of a junction between two closely apposed membranes separated by a cleft some 15 nm wide[2]: a presynaptic membrane derived from the terminal bouton of the axon or dendrite of a transmitting neuron and a postsynaptic membrane usually derived from the cell body or a dendrite of the receiving neuron. Both membranes exhibit the ultrastructure and other properties of a typical plasma (cell surface or unit) membrane with the important proviso that they also contain—or are attached to—distinct and highly characteristic specializations on their internal faces (or aspects) capable of interaction and communication with the respective cytosolic compartments, the presynaptic web, or dense projections, providng sites of attachment for transmitter storage granules (synaptic vesicles) and filaments of the synaptoplasm, and the postsynaptic densities attached to but discrete from the postsynaptic membrane (see refs. 2–6 for useful reviews and actual and diagrammatic representations of these structural elements).

Of particular relevance to the neurochemist is the observation that under appropriate conditions all of these structures are retained in a satisfactory state of structural and functional presentation in preparations of synaptosomes (nerve-ending particles; V. P. Whittaker, Chapter 1, this volume) and membranes derived from them, which makes possible at least a beginning towards their molecular characterization.

This chapter is devoted to a brief but reasonably comprehensive summary of the current state of the art (early 1982). It is focused on proteins for which

Henry R. Mahler • Deceased. Department of Chemistry and Cellular, Developmental and Molecular Biology Program, Indiana University, Bloomington, Indiana 47405.

there is some reasonable certainty with regard to both their chemical nature and their synaptic localization. The reader is referred to other recent reviews concerning earlier work and additional details.[7–10]

2. METHODOLOGICAL ASPECTS

2.1. Criteria for Inclusion

2.1.1. Synaptic Localization

For a protein, or any other component of a synaptic membrane, one can devise criteria of decreasing rigor to justify their inclusion in a catalogue of such entities. The easiest and most rigorous is uniqueness, based on a demonstration of exclusive localization and association with synaptic structures by cytochemical (immunologic or enzymatic) techniques. If this criterion is met, then it is obvious that there exists a biochemical correlate, namely, that there must be an increase in the specific content (amount per milligram protein, analogous to specific activity in the case of an enzyme) of the entity as a function of the extent of purification of the synaptic (or subsynaptic) structure with which it is associated. Operationally, specific content must increase in the progressions:

whole homogenate (WH) → total membrane fraction [P_1 (crude nuclear fraction), P_2, P_3 (microsomal fraction)] → P_2 (crude synaptosomes and mitochondria) (1)

P_2 → lysed P_2 → crude membranes → synaptic membranes (SM)

P_2 → purified synaptosomes (P_2B) → lysed P_2B → synaptic membranes (SM) ↑; ↓ synaptic vesicles (SV); ↓ synaptic mitochondria (SMt) (2)

SM → synaptic junctions (SJ) → postsynaptic membrane (PSM) → postsynaptic densities (PSD) (3)
SM ↓, SJ ↓ presynaptic membranes (PreSM)

Once such unique entities have been identified, they constitute biochemical or immunochemical markers for the structure and the subcellular or subsynaptosomal fraction in question. This well-recognized postulate forms the basis for most biochemical fractionation techniques and constitutes the second operational criterion useful for our purposes: copurification or parallel increases in specific content with an authentic marker. Of equal importance and utility is the converse: purification must accompany removal of negative markers, i.e., those unique to and characteristic for structures and fractions likely to contaminate the one of interest. It should be noted that this set of criteria is the only operationally useful one in case the entity in question, although specific for a particular membrane, is not unique to it. The application of these criteria to the design of fractionation schemes for the various subsynaptic membranes is described in Section 2.2, and possible markers for them, based on the observations described in Section 3, are summarized in Table I.

2.1.2. Molecular Characterization

In this section, the emphasis is on proteins that have been or are being subjected to rigorous characterization in order to establish their identity as discrete molecules. For unique proteins, this means their extraction and purification to a state approaching homogeneity together with, or followed by, the establishment of their characteristic physical and chemical properties; among the former are apparent molecular weight by gel exclusion chromatography and electrophoresis in polyacrylamide gels with denaturing (SDS) and nondenaturing detergents (such as low concentrations of Triton X-100) and isoelectric points (pI); among the latter are protein sequence or fingerprint (proteolytic fragment) data. For proteins that are assigned to a particular membrane by virtue of copurification with a marker appropriate to this membrane, identification entails a careful comparison of these properties of the protein in the membrane with those of the authentic pure molecule obtained from a standard source.

2.1.3. Some Quantitative Aspects

It may be well to anticipate some of the findings developed in the next section by providing reasonable estimates of the amounts of protein in the various membranes present in a homogenate of rat brain cortex, the object of most of the studies described in this chapter. These amounts are in milligrams for a single cortex, with percentages relative to WH in parenthesis (all corrected, wherever possible, for incomplete recoveries): WH, 150 (100); $P_1P_2P_3$, 115 (77); P_2 membranes, 58 (39); P_2B, 12 (8); SM, 10–16 (7–12); SJ, 2–5 (1.5–3.3); PreSM, 1–3 (0.7–2); PSD (Triton X-100), 0.5–1.5 (0.1–0.3); PSD (sodium lauroyl sarcosinate), 0.04–0.12 (0.03–0.08); SV, 0.2–0.5 (0.12–0.3); SMt, ~5 (~3).

It may also be helpful to calculate the actual molar amounts of a protein (its specific content) present in a membrane fraction to the extent of 1% of its total protein: assuming a molecular weight equal to 50,000, such a protein would be present to the extent of 200 pmol/mg membrane protein. Conversely, a specific content of such a protein of 1 pmol/mg would correspond to 5×10^{-3} mg%. If the same protein is enzymatically active, with, let us say, a specific activity of 1 μmol $\cdot$ min^{-1} $\cdot$ mg^{-1}, and the pure enzyme exhibits a turnover number equal to 1.0×10^3 min^{-1}, then its specific content will be equal to 1.00 nmol/mg or 5 mg%.

Table I
Possible Markers for Synaptic and Subsynaptic Membranes

Marker	SM: Pre	SM: Post (incl. PSD)	SJ	SV
Enzymes				
Alkaline phosphatase				+[j]
Galactosyl transferase, sialidase			+	
3′,5′(Cyclic)nucleotide phosphodiesterase		+	+	
Filamentous proteins				
PSD protein		+	+	–
Receptors				
Aspartate, glutamate		+(?)	+	
Spiroperidol (high affinity)		+(?)	+	
α-BTX[a]		+(?)	+	
QNB[b]		+(?)	+	
Neurotransmitters				
ACh[c], GABA[d]				+
Catecholamines (esp. *nor*epinephrine, DA[e])				+
Channel proteins				
Na^+ channel (STX[f], ScTX[g])	+		+	
Thiamine[h]	+(?)		+	
Glycoproteins				
gP112.5 (±2.5)		+	+	
gP100 (±5)		+	+	
Phosphoproteins				
B50	+		+	
Proteins Ia, Ib		+	+	+
Protein antigens				
Synaptin				+
P65				+
D1	+(?)		+	
D2, Thy1		+(?)	+	
D3	+		+	
P95		+	+	
P43[i]				

[a] α-Bungarotoxin (see text).
[b] Quinuclidylbenzilate (see text).
[c] Acetylcholine.
[d] γ-Aminobutyrate.
[e] Dopamine.
[f] Saxitoxin.
[g] Scorpion toxin.
[h] Matsuda and Cooper.[13]
[i] Associated with nicotinic ACh receptor.
[j] Probably coated vesicles rather than SV proper; clathrin (see text) is an unambiguous marker for coated vesicles.

2.2. Isolation and Characterization of Synaptic and Subsynaptic Membranes

2.2.1. General Considerations

All preparations use synaptosomes as a starting material, either directly in relatively crude form as a combined synaptosomal–mitochondrial (P_2) fraction or after their prior purification in which they are separated from the principal contaminants provided by myelin and somatic (cell body) mitochondria by isopycnic (density-gradient) centrifugation or discontinuous (step) gradients, usually of sucrose or Ficoll–sucrose (cf. 11,12; V. P. Whittaker, Chapter 1, this volume). Synaptosomes, which, when intact, behave as osmometers, are then subject to hypoosmotic shock by exposure to a medium of low ionic strength (10 mM Tris or HEPES) at an alkaline pH (usually 8.0–8.5; however, a recent study by Matsudo and Cooper[13] suggests that contamination by adhering mitochondrial membranes can be reduced substantially by raising the pH to 9.5). This treatment liberates the content of the presynaptic cytosol (synaptoplasm), including the SV and SMt. The former can then be isolated by sedimentation at high speed and further purified by appropriate means.[14–17] The latter can be separated from the SM by rate and isopycnic sedimentation or flotation on continuous or discontinuous gradients of sucrose[11,12,18,19] or other media (e.g., sodium diatrizoate[20]).

The extent of retention of intact junctions, PSM, and its attached specializations, especially PSD, is a function of the presence of low concentration (~50 μM) of Ca^{2+} in the isolation and lysis media as well as the extent of mechanical shear accompanying lysis.[11,18,21] Following the pioneering studies of de Robertis *et al.*,[22] SJs are commonly prepared by exposure of SM to low concentrations of Triton X-100,[21,23,24] which solubilizes contaminating membranes including that part of the PreSM not protected by cleft material. The PSDs in various stages of functional integrity are then obtained either from the SJ or, more usually, directly from SM preparations by their exposure to detergents [such as Triton X-100, Na-cholate or deoxycholate, Na-laur(o)yl sarcosinate] at relatively high concentrations followed by isopycnic sedimentation.[4,25–27] A preparation of PSDs still attached to PSM requires extraction of SM with salt and EGTA followed by sonication and sedimentation in a discontinuous sucrose gradient.[28,29]

The principal source of contamination in all of these preparations comes from mitochondria and their membranes and fragments, which even after careful separation on density gradients can reach levels >10%. To obviate this difficulty, Cotman and Taylor[23] have introduced the use of iodonitrotetrazolium (INT) plus succinate; mitochondria loaded with this mixture convert it to iodonitroformazan and in consequence increase in bouyant density, which facilitates their removal. The disadvantage of the method is the potential inactivation of many enzymes or other biologically active proteins by oxidation with INT. Very recently, Matsuda and Cooper[13] have shown that osmotic shock of synaptosomes at pH 9.5 for 10–15 min also serves to reduce mitochondrial contamination to <2% in SM obtained subsequently from the 0.8–1.0 M sucrose interface of a five-step sucrose gradient.

2.2.2. Some Specific Examples

2.2.2a. Synaptic Membranes (without INT). This procedure was developed by Salvaterra and Matthews.[18] A 10% (w/v) homogenate of rat forebrains was prepared in 0.32 M sucrose, pH 7.4, and centrifuged at 1000 *g* for 10 min. The supernatant together with a wash of the pellet (P_1) was centrifuged at 10,000 *g* for 20 min to yield a crude P_2. The latter was lysed by resuspension in 5 mM Tris, pH 8.0, and kept at 0°C for 30 min; 48% sucrose was added to a final concentration of 34% and placed at the bottom of a gradient of 28.5% sucrose followed by 10% sucrose in a 30 Beckman rotor. The gradient was centrifuged at 80,000 *g* for 30 min, and the material banding between the 28.5 and 34% sucrose layer was removed (SM), diluted 1:3, pelleted at 150,000 *g* for 60 min in the 60 Ti rotor, resuspended in 50 mM Tris, 50 mM NaCl, 0.02% NaN_3, and stored frozen; yield was 14 mg (12.7% of homogenate particulate) per rat brain.

2.2.2b. Synaptic Membranes and Synaptic Junctions (with INT). This procedure[24] is a modification of the method of Cotman and Taylor.[23] The forebrains of 26 rats (30–45 days) were homogenized (12.5% w/v) in 0.32 M sucrose containing 50 μM $MgCl_2$ and 200 μM HEPES buffer, pH 7.4, and fractionated. The crude P_2 (14,500 *g* for 10 min in the 30 rotor) was shocked by homogenization with 200 μM HEPES, 0.05 mM $CaCl_2$ and incubated with one-fifth volume of INT (2.5 mg/ml), Na succinate (65 mg/ml), 0.2 M Na phosphate (pH 7.4), pelleted (8720 *g* for 7 min), washed, and subfractionated into myelin (0.32 M–0.8 M sucrose interface), light SM (0.8 M–1.0 M), SM (1.0 M–1.3 M), and mitochondria (pellet). The SM fractions were suspended in 0.32 M sucrose (2–4 mg/ml) and treated with 2 volumes of 0.4% Triton X-100 containing 2 mM EDTA and 10 mM HEPES, pH 7.4. The mixtures were then placed over 10 ml of 1.0 M sucrose and centrifuged at 40,000 *g* for 60 min in an SW25 rotor, with the pellet constituting the SJ fraction. Yield was ~3 mg of SM or 0.3 mg of SJ protein per gram wet weight brain.

2.2.2c. Postsynaptic Densities (Using Na-Lauroyl Sarcosinate). Synaptic membranes (40–80 mg) obtained as above in 5–8 ml buffer (2 mM BICINE, pH 7.5) were treated with 20 ml of NLS (3.9% w/v in BICINE, final detergent concentration ~3%) for 10 min at 4°. The mixture (8–9 ml) was applied to a discontinuous gradient (SW25.1 rotor) consisting of 7 ml each of 1.0, 1.4, and 2.2 M sucrose, all in 0.05 mM $CaCl_2$, pH 7.4, and centrifuged at 63,600 *g* for 75 min. The material collected from the 1.4–2.2 M sucrose interface was diluted 1:3 with 0.1 mM EDTA and, after pelleting at 78,500 *g* for 20 min, constitutes the PSD fraction, obtained in a yield of ~12.5 μg/g wet weight.[25]

2.2.2d. Postsynaptic Densities (Using Triton X-100). The following procedure, originally developed by Cohen *et al.*[26] and Carlin *et al.*[27] for dog brain, works equally well for rat brain. Brain regions (10 g) were homogenized in 40 ml of solution A (0.32 M sucrose, 1 mM $NaHCO_3$, 1 mM $MgCl_2$, 0.5 mM $CaCl_2$), made 10% (w/v) in solution A, and centrifuged at 710 *g* for 10 min. The pellet was washed by resuspension in 10 volumes of solution A and centrifuged as before. The supernatants were pooled and centrifuged at 13,800 *g* for 10 min, the pellet resuspended in solution B (0.32 M sucrose, 1 mM $NaHCO_3$), and 8 ml of the suspension placed on a discontinuous gradient consisting of 10 ml each of 0.85, 1.0, and 1.2 M sucrose, all in 1.0 M $NaHCO_3$, and centrifuged at 82,500 *g* for 120 min. The band at the 1.0/1.2 M sucrose interface was removed and diluted with solution B (60 ml per 10 g of tissue), an equal volume of 1% (v/v) Triton X-100 and 0.32 M sucrose in 12 mM Tris HCl (pH 8.1) was added, and the suspension was stirred for 15 min at 4°C. After centrifugation at 32,800 *g* for 20 min, the pellet was resuspended in 2.5 ml of solution B per 10 g of tissue, and 2 ml of this was layered on gradients composed of 4 ml of 2.0 M sucrose and 3 ml each of 1.5 M and 1.0 M sucrose, both in 1 mM $NaHCO_3$, and centrifuged at 201,800 *g* for 120 min (SW40 rotor). The band at the 1.5/2.0 M interface was removed, diluted to a final volume of 6 ml with 3 ml of solution B and an equal volume of 1% Triton, 150 mM KCl, and pelleted at 201,800 *g* for 20 min. The PSD was then resuspended by homogenization; yield was ~200 μg protein per gram wet weight.

Alternatively, SM are isolated first by one of the methods described above and treated with Triton X-100 in the manner described.

2.2.2e. Synaptic Vesicles. This procedure, developed by J. Near, is an adaptation of that of Burke and de Lorenzo[17] for use with slaughterhouse tissue (e.g., corpus striatum from bovine brain). Tissue was homogenized immediately at 4°C in 0.32 M sucrose, 0.3 mM PMSF, transported to the laboratory on ice, rehomogenized by ten additional strokes using a motor-driven Teflon® pestle at 500 rpm, and centrifuged at 2000 g_{max} for 10 min. The supernatant was centrifuged at 12,000 g_{max} for 20 min, and the pellet resuspended in 0.32 M sucrose, 0.3 mM PMSF and repelleted at 30,000 g_{max} for 30 min. The resultant pellet was resuspended (5 ml/g original tissue) in distilled water containing 0.3 mM PMSF by means of five homogenization strokes, and to the suspension was added 3 ml of a buffer consisting of 160 mM KCl, 5 mM NaCl, 10 mM Tris maleate (pH 6.5), and 260 μM pargyline. The suspension was centrifuged at 12,200 g_{max} for 25 min, and the supernatant centrifuged at 64,000 g_{max} for 40 min to pellet coated vesicles. The supernatant was then made 5 mM in $MgCl_2$ and centrifuged at 186,000 g_{max} for 60 min to pellet synaptic vesicles. Approximate yields (per gram wet weight of tissue) are 200 μg for synaptic vesicles and 400 μg of coated vesicles.

3. PROTEIN CONSTITUENTS OF SYNAPTIC MEMBRANES

3.1. General Considerations

This section provides a summary of the evidence for the presence of a number of classes of well-characterized proteins associated with either (or both) the membranes or their attachments found in synaptic junctions of vertebrate CNS. The topics discussed include a brief description of their molecular properties, their uniqueness—if any—to the CNS, and their presence in subsynaptic membranes and their specializations.

Table II
Amounts (in Weight %) of Filamentous Proteins in Synaptic and Subsynaptic Membranes[a]

	Actin	α-Tubulin	β-Tubulin	PSD protein
SPM	5.4 ± 0.5	5.9 ± 0.1	5.5 ± 0.3	3.9 ± 0.2
PSM–PSD	4.1 ± 0.1	5.4 ± 0.4	5.7 ± 0.5	7.4 ± 0.3
PSD	2.9 ± 0.2	6.2 ± 0.5	5.7 ± 0.9	11.6 ± 1.2
SJ	5.6 ± 0.5	7.9 ± 0.8		9.8 ± 0.9
PSD*	6.0 ± 0.2	14.2 ± 0.8		43.0 ± 1.5
SV	11.2	25.0		None
Mol. wt.	45,000	57,000	53,000	50,000

[a] All data from Mahler *et al.*[32] and Ratner and Mahler[29] except for SJ and PSD*[109]; PSD prepared with Triton X-100 according to Cohen *et al.*[26]; PSD* with Na-laurylsarcosinate (3%). SV data from Zisapel and Zurgil[16] and Zisapel *et al.*[45] for beef brain; confirmed qualitatively for rat brain cortex by Kleine and for beef brain *corpus striatum* by Near in the author's laboratory. Similar data have also been published for chick brain by Babitch and Benavides.[37]

3.2. Filamentous Proteins

The presence of the subunits of certain filamentous proteins such as G-actin and tubulin in close association with isolated SM and in synapses *in situ* was suggested relatively early on the basis of fingerprint and immunohistochemical evidence.[5,30,31] The more recent history of the problem of identification and appropriate references are to be found in papers by Mahler[7] and Gurd.[10] The actual amounts present in SM and its subfractions are summarized in Table II. With the question of their molecular identity having been settled, the major remaining problem is whether these proteins are integral constituents of the membrane and its characteristic specializations or are peripheral and provide sites for attachment to, or other interactions with, their polymerized filamentous cytoskeletal counterparts (8 to 10-nm microfilaments composed of F-actin and microtubules).

3.2.1. Actin

The actin species found in SM, SJ, and PSDs have been identified, on the basis of position on denaturing two-dimensional gels, fingerprints after proteolytic digestion, and immunochemical techniques, as identical to the β- and γ-actins found in brain cytoplasm and synaptoplasm.[10,32–34] These molecules are similar to the α form found in skeletal and heart muscle in containing 374 amino acid (aa) residues but differ from it in 25 (β) and 24 (γ) of these positions, many of them in the N-terminal sequence; they also differ in the genes required for their specification.[35] All three species have blocked (N-acetyl) N termini and exhibit the same actual (41,800) and apparent molecular weights (43,000–45,000 depending on the gel system used) but differ in their isoelectric points. The actin associated with the postsynaptic density has been shown to be in the form of F-actin and localized at its exterior surface in contact with the microfilaments of the postsynaptic cytosol.[28,36] Actin also forms part of the membrane of SV from brain[37] and electrocytes of *Torpedo*[38] as well as adrenal chromaffin granules.[39] In beef brain, actin has been estimated to account for 11.2% of the SV protein.[16]

3.2.1a. α-Actinin. α-Actinin, a protein with a mol. wt. of 100,000, is present in brain in association with coated vesicles[40] and chromaffin granules.[41] Kleine has some unpublished fingerprint evidence for its presence in synaptic and subsynaptic membranes.

3.2.1b. Actin-Depolymerizing Factor. Such a protein with mol. wt. ~19,000 has been isolated and partially purified from chick embryo brain.[42] As indicated by its name, this molecule has the ability to convert F- into G-actin. Nothing is known about its presence at the synaptic junction.

3.2.1c. Myosin. There have been persistent but somewhat controversial reports of the presence of myosin on preSM. Recent studies by Beach *et al.*[43] have provided somewhat more rigorous identification of this molecule in SJs and PSDs. The molecular properties of brain myosin and the regulation of its interaction with myosin—requiring phosphorylation of one of its light chains by means of a CaM-dependent kinase—are probably similar to those established for other nonmuscle cells.[43a]

3.2.1d. Tropomyosin. Although tropomyosin is present in brain cytosol, a protein with a similar mobility in SM and PSD differs from it in pI and proteolytic fingerprints (L. Kleine, unpublished observations).

3.2.1e. Troponin. There is now no longer any good evidence for the presence of troponin in synaptic junctions. In its stead, they contain calmodulin, a structurally and functionally analogous protein (see Section 3.3.1).

3.2.1f. Spectrin. This protein constitutes a major filamentous structural protein on the interior aspect of erythrocyte membranes and provides anchoring sites for the actin attached to these structures. It is an extended rodlike molecule present as a tetramer formed by head-to-head association of nonidentical (α,β) dimers, both in the mol. wt. range of 260,000. The β subunit is attached to the cytosolic domain of band 3, the major transmembrane protein, by means of yet another protein called ankyrin. An immunoreactive analogue of spectrin (exhibiting ~25% homology) has recently been identified in crude (not necessarily synaptic) membrane preparations from pig brain[43a,b] and purified from beef brain microsomes.[43c] It is capable of strong interactions with actin, and it is similar to the erythrocyte protein in morphology, mol. wts. of the tetramer and its constituents (265,000 or 240,000 and 260,000 or 235,000), but not in their peptide maps. Its precise localization, relation to other actin-binding proteins and other proteins in the same mol. wt. range [e.g., MAP2 (cf. Section 3.2.2) and a CaM-binding protein (cf. Section 3.3.2f)] remain to be established. It has also been referred to as calspectin[43c] and fodrin.[43a,b,d]

3.2.2. Tubulin

There is unambiguous evidence for the presence of α- and β-tubulin, i.e., the subunits of cytoplasmic microtubules, in SM, SJ, and PSD[10] as well as in SV from rat,[17] mouse,[44] chick,[37] and beef brain.[45] These two molecules contain 450 (α) and 445 (β) aa residues, corresponding to actual molecular weights of 49,400 and 50,000, respectively[46,47] In contrast, their mol. wts. in different gel systems are found to equal 55–60,000 and 52–55,000 for α and β forms, respectively. In addition, there is evidence for a number (≥2) of subspecies of both α- and β-tubulin.[34,48,49] Furthermore, α-tubulin is subject to the addition of tyrosine at its C-termini. Tubulin is probably an integral protein of the synaptic junction,[10,50] including the PSM and PSD*[32,50] as well as the PreSM[52] and the SV.[17,45,53]

Assembled cytoplasmic microtubules contain a number of tightly bound proteins present in a constant stoichiometry of ⩽1 molecule per molecule of α- or β-tubulin. Among these are two microtubule-associated protein molecules (MAP1 and MAP2), with mol. wts. variously reported as 300,000 (or 370,000) and 250,000 (or 350,000), respectively,[55,55a] also known collectively as high-molecular-weight proteins (HMWPs). They facilitate assembly of microtubules and contribute characteristic side arms to their morphology. Recent immunohistochemical studies by Matus *et al.*[55] have demonstrated the presence of these proteins in dendrites of neurons to the exclusion of their axons and other (glial) cell types. One would therefore expect that in SM these proteins should be concentrated on the postsynaptic side, and they have in fact been found to be present in PSD preparations.

3.2.3. Postsynaptic Density Protein

A characteristic protein of synaptic membranes and synaptic junctions that appears to be concentrated exclusively in its prominent postsynaptic specialization, the postsynaptic density (PSD) (Table II), has been studied extensively by Cotman and his collaborators.[24,51] This protein appears to contain a single polypeptide chain of mol. wt. 50,000 ± 2,000 with an N-terminal tyrosine, and its isoelectric point and tryptic fingerprint maps distinguish it from all other filamentous proteins including the neurofilament triplet.[19] It is specific to the central nervous system and is concentrated in asymmetric (inhibitory, type 1) axodendritic postsynaptic structures, as found in the cerebrum (≥80%) and midbrain but is present at low concentrations or absent where such structures are relatively sparse or modified, such as the cerebellum.[27,27a]

By the use of immunochemical techniques Sampedro *et al.*[56] have identified a protein with mol. wt. 95,000 specifically associated with, and immunocytochemically localized at, PSDs of bovine and rodent brain.

Similarly, Froehner *et al.*[57] have provided evidence for the presence of a protein with mol. wt. 43,000 tightly and functionally associated with the ace-

* A very recent report[50a] suggests that tubulin may become attached to the PSD in the course of its isolation.

tylcholine receptor of *Torpedo* electrocytes in both its (postsynaptic) membrane-bound and purified forms. However, membranes can be stripped of the protein at pH 11 without affecting their binding of cholinergic ligands or the control of ion permeability. The receptor proper consists of four subunits: α (mol. wt. 40,000), β (mol. wt. 50,000), γ (mol. wt. 60,000), and δ (mol. wt. 65,000).

3.2.4. Other Filamentous Proteins

Other filamentous proteins, especially the constituents of intermediate filaments,[58] which are prominent in other parts of the CNS (such as the neurofilament triplet), in glia (such as the acidic glial fibrillar protein), and in other cell types (such as desmin and vimentin), do not appear to be present at synaptic junctions.

3.3. Calcium-Binding Proteins

Calcium ions play an essential role in triggering the release of neurotransmitters from their storage sites in synaptic vesicles and liberating them into the synaptic cleft. It might therefore be anticipated that a number of proteins capable of binding Ca^{2+} and depending on this ion for their activity are concentrated at the synaptic junction, and this expectation has been amply borne out by recent studies.

3.3.1. Calmodulin

This small (actual molecular weight 17,000, apparent mol. wt. 17–22,000), ubiquituous, heat- and acid-stable Ca^{2+}-binding protein is an essential modulator or regulator of a large number of Ca^{2+}-dependent processes.[59–61] Calmodulin (CaM) contains a single polypeptide chain of 140 residues and is related in its primary structure and the nature of its Ca^{2+}-binding sites (four in the case of CaM) to a number of other proteins with similar functions[62] such as myosin light chains, troponin C, and parvalbumin. Its concentration in brain is high (10^{-5} M or 682 μg/g wet weight in rabbit brain,[63]) and it appears to be present there in both soluble, and membrane-associated forms. Its concentration in synaptic membranes is in the range from 25 (cortex) to 11 (cerebellum) μg/mg protein. These values are increased approximately threefold in isolated PSDs to which the protein has been localized by both biochemical and immunocytochemical techniques.[64–67] The affinity of the protein for Ca^{2+} is high ($K_D^{av} = 1–10 \times 10^{-7}$ M, depending on ionic strength), and its interactions with CaM-requiring enzymes is blocked by phenothiazines in the presence of Ca^{2+} ($K_D = 1$ μM) (also see Section 3.5.1).

3.3.2. Proteins Requiring CaM for Activity (CaM-Binding Proteins)

3.3.2a. Calcineurin. A protein capable of forming tight complexes with CaM has been isolated together with brain cyclic nucleotide phosphodiesterase, which it inhibits.[68] This protein appears specific to the central nervous system and has been called calcineurin.[69,70] It contains two subunits: the first, A (mol. wt. 61,000), is responsible for the binding of CaM in a reaction requiring Ca^{2+} ions, which are bound to the second (B) subunit (mol. wt. 15,000). This subunit has four Ca^{2+}-binding sites ($K_D^{av} \leq 10^{-6}$ M) and requires both K^+ (0.1 M) and Mg^{2+} (1 mM) for the reaction. These properties of CaM and calcineurin regulate the kinetic and thermodynamic buffering of the intraterminal and postsynaptic concentration of both Ca^{2+} and CaM. In addition, recent studies[110] have revealed that in the presence of Ca^{2+} and CaM, calcineurin is a potent protein phosphatase and may thus constitute part of a Ca^{2+}/CaM-dependent system for the regulation of a variety of proteins for this form of covalent modification (see Section 3.3.5).

3.3.2b. Cyclic Nucleotide Phosphodiesterase. This enzyme is known to be concentrated in the postsynaptic density and, perhaps, in the postsynaptic region of the synaptic cleft.[71] The purified enzyme consists of two identical subunits with mol. wt. 58,000. Both the monomer and the dimer are catalytically active, but whereas the former does not require any cofactors, the latter is completely ineffective unless it is supplemented with Ca^{2+} and CaM. Both cyclic AMP and cyclic GMP can function as substrates, but although the affinity for cyclic GMP is higher, maximal activity requires cyclic AMP: the K_m (cyclic GMP) is 2.5–3.0 μM, and V_{max} in the presence of Ca^{2+} and CaM is 30–40 μmol/min per mg with 200 μM and 150–250 μmol/min per mg with 2 mM cyclic AMP, respectively.[72,72a] The K_m and maximal activities for cyclic AMP are affected not only by CaM but also by NH_4^+ ions: the value of this parameter is raised from 3.1 to 4.1 × 10^{-4} M by 0.2 M NH_4^+ in the absence and from 1 to 2.2 × 10^{-6} M in the presence of CaM.[69]

The activity associated with PSDs exhibits similar properties.[73] It can be extracted by sonication, and the solubilized protein, with a mol. wt. of 215,000, is activated by CaM, which lowers the K_m and raises the V_{max}; the ratios of K_m (cyclic AMP)/K_m (cyclic GMP) and V_{max} (cyclic AMP)/V_{max} (cyclic GMP) are ≥5 and 3.5, respectively. The activities of the enzyme in isolated SM and SJ (23 nmol/min per mg) can be used to estimate its concentration in these preparations; it accounts for approximately 1% of their total protein.

3.3.2c. S-100 Proteins. This group of small, highly acidic Ca^{2+}-binding proteins are now known to be capable of forming complexes with CaM. They are commonly believed to be predominantly of glial origin; however, they are also present in neuronal nuclei and can be bound strongly and relatively specifically to synaptic membranes.[74] Two such proteins exist: S-100a or PAP1a, with a mol. wt. of 20,000, and S-100b or PAP1b, with a mol. wt. of 21,000. Whereas the former contains two closely related subunits, α (mol. wt. 10,400) and β (mol. wt. 10,500), the latter is a homodimer of two α subunits.[77,78] These α chains have been sequenced: they consist of 91 amino acid residues with a molecular weight of 10,507 and contain a single Ca^{2+}-binding domain similar in structure to the canonical Ca^{2+}-binding regions of related proteins such as

CaM (see above), troponin C (four domains), and parvalbumin* (two domains).[62] These domains contain 12 residues and, for S100b, troponin C, CaM, and parvalbumin (Parv), consist of (conserved residues in italics):

S100b: ^{61}*Asp*-Ser-*Asp*-*Gly*-*Asp*-*Gly*-Glu-Cys-*Asp*-*Phe*-Gln-*Glu*-(^{73}*Phe*)-
Troponin C: ^{65}*Asp*-Glu-*Asp*-*Gly*-Ser-*Gly*-Thr-Val-*Asp*-*Phe*-Asp-*Glu*-(^{77}*Phe*)-
CaM: ^{56}*Asp*-Ala-*Asp*-*Gly*-*Asp*-*Gly*-Thr-Ile-*Asp*-*Phe*-Pro-*Glu*-(^{68}*Phe*)-
Parv (carp): ^{50}*Asp*-Glu-*Asp*-Lys-Ser-*Gly*-Phe-Ile-Glu-Glu-Asp-*Glu*-(^{64}Leu)-

3.3.2d. Ca^{2+}-ATPase. This important CaM-dependent enzyme is predominantly localized in the presynaptic membrane.[77,78] It acts as a Ca^{2+} pump to restore the intraterminal concentration of Ca^{2+} ions to their resting value subsequent to their influx during depolarization, which results in release of neurotransmitters. Its specific activity in such membrane preparations is of the order of 0.18 μmol/min per mg. Although the enzyme has not been purified from this source, the closely related protein from sarcolemma is a single polypeptide with mol. wt. 150,000 and an activity of 1.1 μmol/min per mg.[81] These values suggest that the ATPase may contribute 15% or more of the proteins of the presynaptic membrane.

3.3.2e. Other CaM-Binding Proteins of Unknown Functions. Studies by Grand and Perry[63] indicate that CaM-binding proteins are present to the extent of 1 mg/g wet weight of rabbit brain. Among them, proteins with mol. wts. 140,000 and 58,000 are the most prominent and form the tightest complexes with CaM. On the basis of the earlier discussion, they may represent the Ca^{2+}-ATPase and the cyclic nucleotide phosphodiesterase, respectively. Immunocytochemical localization experiments[65,66] have demonstrated a tight association of CaM with a heat-labile protein of mol. wt. 80,000 and with postsynaptic specializations (PSDs and microtubules) of dendrites in various regions of the rodent CNS such as caudate–putamen, deep cerebellar nuclei, and retina. This protein is most likely identical with calcineurin (Section 3.3.2a).

Furthermore, Carlin *et al.*,[80] using iodinated CaM to probe complex formation with PSD proteins separated on denaturing polyacrylamide gels, were able to demonstrate that the PSD protein (see above) as well as proteins with mol. wt. 60,000 (phosphodiesterase?), 114,000 (Ca^{2+}-ATPase?), and 230,000 (MAP2?) were susceptible to this form of interaction. Using [^{125}I]azidocalmodulin as a photoaffinity analogue, Andreasen *et al.*[80a] found labeled products that, corrected for the covalently linked CaM analogue, exhibited mol. wts. of 44,000 (PSD protein), 57,000 (phosphodiesterase), 95,000, and 155,000, respectively. In contrast, Pelfrey *et al.*,[80b] using a filter-transfer technique for the identification of proteins capable of binding to [^{125}I]CaM, found the most prominent component in brain (and other membranes) at mol. wt. 240,000 (MAP2 or brain spectrin?). The latter is a serious possibility since this molecule exhibits strong CaM-binding activity.[43b]

* A recent report has provided immunohistochemical evidence for the localization of this protein in a neuronal (GABAergic?) subpopulation.

3.3.2f. Binding of CaM to Peptide Transmitters and Model Substrates. In a systematic investigation of the Ca^{2+}-dependent binding of a number of oligo- and polypeptides to CaM, using a covalent fluorescent derivative prepared by condensation with 5-(dimethylamino)-1-naphtalene sulfonyl (dansyl) chloride, Malencik and Anderson[80c] showed that ACTH, β-endorphin, glucagon, and substance P but not myelin basic protein were capable of such interactions with K_Ds in the range of 2.5 μM. The sequences responsible for this binding (see below) resemble those of substrates for cyclic-AMP-dependent protein kinase (Section 3.5.1) and for CaM-binding proteins (above). The hydrophobic core sequence in one of the CaM domains is given for comparison. Competition for CaM by neuroactive polypeptides may therefore modulate the state of phosphorylation and thereby the activity of the latter.

Substance P:	*Arg-Pro-Lys*-Pro-Gln-Gln-*Phe-Phe*-Gly-Leu-Met-
ACTH:	*Lys*^{16}*-Arg-Arg*-Pro-Val-Lys-*Val-Tyr*-Pro-Asn-Gly-
β-Endorphin:	*Lys*^{19}*-Lys-His*-Ala-Asn-Lys-*Ile-Ile*-Ala-Asn-Lys-
Synth. substr. for cyclic AMP kinase:	*Lys*^{9}*-Arg-Lys*-Glu-Ile-Ser*-*Val-Ala*-Gly-Leu-
Myelin basic protein:	*Gln*^{8}*-Arg-His*-Gly-Ser*-Lys-*Tyr-Leu-Ala*-
CaM:	Thr-Glu^{45}-Ala-Glu-Leu-Gln-Asp-Met-Ile-Asn-Glu-Val-Phe^{65}-Pro-Glu-Phe-Leu-Thr-Met-Met-Ala-Arg-Lys-

Likely residues involved in binding to CaM are in italics, and the phosphorylatable serine is indicated by an asterisk.

3.4. Glycoproteins

Many of the proteins in SM preparations carry carbohydrate prosthetic groups[81] (R. K. Margolis and R. U. Margolis, Volume 5, this *Handbook*). This is not surprising, since polypeptides with an exterior orientation, i.e., those that penetrate into the synaptic cleft, would be expected to, and are in fact, glycosylated.[7,10,44]

Recently, the focus of attention has shifted to those glycoproteins of definite junctional, and specifically subsynaptic, localization. Operationally, these proteins are concentrated in SJ and PSD preparations, and their elaboration coincides with or just precedes synaptogenesis. Four (sets of) species stand out; all of them are capable of binding the lectin concanavalin A, either on separating gels or on columns, and are referred to as Con A BPs or GPs I, II, III (doublet IIIa + IIIb) and IV; their mol. wts. have been reported as 160 or 180,000, 123 or 130,000, 110,000, and 95,000, respectively.[10,24,82,83] On the basis of tryptic digest maps, the three largest exhibit considerable amino acid sequence homology, whereas GP IV appears unique. The number of Con A binding sites per molecule is in the order II > I > III. These sites appear to be principally (60–70%) polymannose chains linked to Asn residues of the polypeptides. The proteins also bind to the lectins from *Lotus tetragonolobulus* (specific for α-L-fucose), wheat germ (N-acetyl-D-glucosamine), and lentils (mannose, *N*-acetylglucosamine). Since isolated GPI and III can be fractionated into separate components on the basis of their retention on wheat germ lectin–sepharose columns, both of these entities must exhibit some additional heterogeneity in their carbohydrate side chains.

3.5. Protein Kinases and Phosphoproteins

3.5.1. Protein Kinases

Regulation of the activity of proteins by controlling the level of phosphorylation of susceptible residues (serine, threonine, and tyrosine) by affecting the relevant protein kinases and/or phosphatases is a pervasive and characteristic property of eukaryotic cells. It has been postulated to be operative at synaptic junctions in affecting both presynaptic (e.g., neurotransmitter synthesis and release) and postsynaptic (receptor function and control of ion permeability) events (e.g., refs. 7,32,84–86; R. Rodnight, Volume 4, this *Handbook*).

The presence of protein kinases capable of acting on specific proteins in the two apposed membranes constitutes an essential requirement of the hypothesis, and the demonstration of a number of membrane-associated enzymes exhibiting this capability provides powerful support for it. Among them are protein kinases with requirements for (1) cyclic AMP (two forms with identical catalytic but different regulatory subunits), (2) cyclic GMP (principally in the cerebellum, (3) Ca^{2+} plus CaM (mol. wt. 120,000,[87] probably more than one form,[87a] with an enzyme for tubulin located in synaptic vesicles[17,17a]), (4) Ca^{2+} plus acidic phospholipids, further activated by unsaturated fatty acids (mol. wt. 90,000),[88,88a–c] (5) Mg^{2+} (basal), and (6) a specific MAP2 kinase. It is of interest to note that, contrary to earlier views, certain neuroleptics such as trifluoroperazine constitute effective inhibitors not only of system 3 (by binding to CaM) but also of system 4 and that myelin basic protein is a good substrate for system 4[88b] but not for system 3[80b] (see Section 3.3.2f). Another synapse-specific protein phosphorylated by system 4 is a soluble protein with apparent mol. wt. 87,000.[88c]

3.5.2. Nature of the Substrates

A large number (>25) of the proteins in synaptic membranes are susceptible to phosphorylation by one or more of the kinases just described. In the summary discussion to follow, emphasis is placed on proteins for which there exists evidence concerning the identify of both the nature of the protein molecule and the kinase responsible for its phosphorylation (Table III). Many of the proteins, particularly in the range of mobilities in SDS gels corresponding to apparent mol. wt. between 45 and 60,000, move to identical or overlapping positions on one-dimensional gels. Therefore, their unambiguous identification has required their comparison with authentic standards by two-dimensional electrophoresis and fingerprint techniques.

Besides a number of the major filamentous proteins present in the junction, two additional classes of proteins are subject to ready phosphorylation. The

Table III
Phosphoproteins in Junctional Membranes

		Nature of kinase			Localization[a]		
Mol. wt. ($\times 10^{-3}$)	Protein	Cyclic AMP	CA^{2+}–CaM	Other	PreSM	PSD	SV
250	MAP2[b]	+		b		+	
80	Ia	+	+[c]		+	+	+
75	Ib	+	+[c]		+	+	+
60	Cyclic nucleotide phosphodiesterase		+			+	
57	α-Tubulin (pI = 5.8)		+			+[g]	+[h]
57	R^B_{II} (pI = 5.2)	+			+		
55	IIIB (pI = 6.6–7.3)	+					
54	R^B_{II} (pI = 5.1)	+			+		
52[d]	B50			e	+	−	−
53	β-Tubulin		+			+[g]	+[h]
51	R^B_I (?)	+			+		
50	PSD protein		+		−	+[g]	−
45	Actin		+			+[g]	
43	α-PDH[f]				f	−	−

[a] A plus indicates definite evidence for, a minus definite evidence against, presence of the entity in question; an empty space shows either that the question has not been addressed or that the answer is ambiguous.
[b] Probably by MAP2 kinase; brain spectrin, which exhibits a similar mobility, is phosphorylated by a Ca–CaM-dependent kinase with mol. wt. 800,000 and a Ca-stimulated kinase (mol. wt. 80,000).[112]
[c] Two kinases: one phosphorylates sites overlapping, the other different from, those phosphorylated by cyclic-AMP-dependent kinase.
[d] In other gel systems the apparent mol. wt. is 48–50,000.
[e] Ca^{2+}-phospholipid-dependent kinase.
[f] α-Subunit of pyruvate dehydrogenase, a mitochondrial protein phosphorylated by a specific kinase; mitochondria contaminate PreSM.
[g] The phosphorylation of these PSD proteins is greatly stimulated by exposure to nonionic detergents at low concentrations and may also be catalyzed by the Ca^{2+}-phospholipid-dependent kinase.
[h] Probably catalyzed by a specific kinase.

first is constituted by the brain-specific form of regulatory subunits of the cyclic-AMP-dependent protein kinase. The predominant species is R^B_{II}, the subunit of type II kinase found in brain, which, like R_{II} subunits in general, is subject to autophosphorylation by the catalytic (C) subunit of the same molecule according to the reaction sequence

$$R_2C_2 + 2n \text{ cyclic AMP} \rightleftarrows \underset{B}{R_2} (\text{cyclic AMP})_{2n} + 2C \qquad (n = 2)$$

At least two phosphorylated forms of R_{II} differing in their mobilities and isoelectric point, probably because of the number of sites phosphorylated, have been identified in SM fractions.[32,89]

The other class consists of kinase substrates apparently specific for and restricted to the nervous system. The first of these is a basic (pI = 10.2), highly elongated (frictional ratio = 2.2), collagenase-sensitive protein called synapsin I (formerly protein I).[90,91] It consists of a mixture of two closely related polypeptide chains, Ia with mol. wt. in the range 73–86,000 and Ib in the range of 68–80,000, depending on the gel system used, in an approximate ratio of 1 to 1. Using radioimmunoassay in detergent extracts, Goelz *et al.*[92] have determined the concentration of the protein in several mammalian species and in various regions of cat brain: the protein accounts for 0.4% of total cortex protein and is present (concentration in pmol/mg, shown in parentheses) in frontal cortex (51), occipital cortex (36), hippocampal cortex (54), posterior amygdala (37), caudate (31), putamen (23), thalamus (50), hypothalamus (32), posterior pituitary (9), anterior pituitary (<0.2), substantia nigra (25–33), medial geniculate (30), superior colliculus (15), pons (1–4), medulla (7), and cerebellar cortex (28). In synapses, the protein has been localized principally at PSDs and synaptic vesicles.[93,94] The protein is subject to phosphorylation at distinct and separate sites by cyclic-AMP- and Ca^{2+}–CaM-dependent protein kinases.[95]

The second such protein is an acidic protein (pI = 4.5; mol. wt. 48,000–52,000) identified and characterized by Gispen, Zwiers, and their collaborators.[96] It is specific for neurons[96a] and their synaptic junctions, where it appears to be concentrated in the presynaptic membrane.[97,98] Its concentration, based on the amount of phosphate transferred from ATP under comparable conditions, is of the same magnitude as that of synapsin I. The kinase responsible for this reaction requires Ca^{2+} but not CaM and may be identical with the acidic phospholipid-dependent enzyme; the reaction is inhibited by ACTH or the N-terminal fragment $ACTH_{1-24}$ at micromolar concentrations.

The third protein is now referred to as synapsin IIIb (formerly protein IIIB) and has been partially purified 660-fold to a purity of 86% from calf brain homogenates by Huang *et al.*[99] Its mol. wt. is 55,000 by gel electrophoresis, and it is heterogeneous on isoelectric focusing with pIs between 6.6 and 7.3. Like synapsin I, it is highly elongated (frictional ration of 1.5) and rich in proline; it differs from it by being collagenase resistant. It is subject to phosphorylation (0.82 mol/mol protein) by the catalytic subunit of cyclic-AMP-dependent protein kinase. Its function, localization, specificity, and possible relationship to other proteins in this mol. wt. range (Table III), especially R_{II}^{B}, is as yet unknown.

Finally, the α subunit of the sodium channel (mol. wt. 270,000; see Section 3.6.1) is susceptible to phosphorylation by the cyclic-AMP-dependent kinase to the extent of 3–4 mol phosphoserine/mol protein.[99a]

3.6. *Proteins Involved in Transmitter Function*

In this rubric belong proteins that are directly linked to the transmission of the action potential, i.e., those constituting the presynaptic Na^{+} channel and the postsynaptic receptors for neurotransmitters, which in turn alter the ion permeability of that membrane.

3.6.1. Na^+-Channel Proteins

The polypeptides constituting the voltage-sensitive Na^+ channel are operationally defined in terms of their ability to interact with and be functionally altered by neurotoxins at three distinct sites.[100,100a]

Binding to site I by ligands, of which the water-soluble heterocyclic guanidines tetrodotoxin (TTX) and saxitoxin (STX) are examples, inhibits Na^+ transport in a voltage-independent fashion. Both of these ligands are believed to bind to a common site located near the extracellular opening of the ion-conducting channel that is unaffected by occupancy of the other two sites. The density of site I equals 12,000 μm^{-2} in the node of Ranvier or 1 pmol/mg of purified axolemma,[100b] and binding by one molecule of ligand per channel ($K_{0.5}$ ~2 + 10^{-10} M in axolemma and 3 × 10^{-9} M in synaptic membranes) is sufficient to block it.

Binding to sites II and III is voltage sensitive and interactively coupled, so that occupancy of sites of one class facilitates binding to those of the other. It leads to hyperexcitability and depolarization. Specific ligands ($K_{0.5}$ in parentheses) for site II are the lipid-soluble alkaloids veratridine (80 μM), batrachotoxin (0.4 μM), aconitine (20 μM), and grayanotoxin (50 μM); their binding, which is unaffected by occupancy of site I, causes persistent activation of the Na^+ channels at the resting potential by blocking their inactivation, with one molecule of ligand sufficient for the opening of that channel. It also shifts its voltage dependence to more negative membrane potentials and affects ion selectivity, particularly with regard to the ability of Cs^+ to substitute for Na^+. Site III, which interacts with toxic polypeptides from sea anemone nematocysts ($K_{0.5}$ ≃2 × 10^{-7} M) and scorpions (*Leirus quinquestriatus* or *Androctonus anstratis* Hector) (ScTX, basic polypeptides with mol. wt. 7000; $K_{0.5}$ = 0.5–15 μM for the former and 1–3 × 10^{-9} M for the latter), appears to constitute the voltage (action potential)-dependent "gating protein" or "voltage sensor." The affinity for these ligands increases by a factor of 10 for each 31 mV of depolarization, and their binding slows or blocks inactivation of the channel and enhances the persistent activation brought about by occupancy of site II.

The complete Na^+ channel and the receptors for the three ligand classes have been solubilized from crude synaptosomal and microsomal fractions from rat brain by means of detergents such as Triton X-100 and sodium cholate; the channel, with properties similar to those exhibited by intact excitable membranes, can then be reconstituted by incorporation into proteoliposomes.[101,101a] The most highly purified (~50%), reconstitutively active preparations of the STX receptor[102] exhibit a specific activity of 1500 pmol/mg protein, a mol. wt. of 316,000 ± 63,000, and consists of at least three subunits: one molecule of α (mol. wt. 270,000 ± 10,000) and two (?) of β (β_1, 39,000; β_2, 37,000). The same polypeptides are also labeled by exposure of intact synaptosomes to photoaffinity labeling with ^{125}I derivatives of ScTX. When reconstituted into phosphatidylcholine vesicles, the preparation exhibits Na^+ flux and inhibition parameters similar to those of the intact membrane. Thus, these polypeptides are probably both necessary and sufficient to define all three binding sites and the functional channel in the (pre)synaptic membrane, where they are present to the extent of ~5 pmol (or 1.6 μg)/mg protein.

Gill[102a] has been able to demonstrate the localization of the Na^+-channel protein(s), of a Ca^{2+} channel activated by veratridine-sensitive Na^+ efflux, as well as the sodium pump (see next section) in a highly purified preparation of inverted (pre?)synaptic membranes.

3.6.2. Na^+, K^+-ATPase

Subsequent to the depolarization-induced influx of Na^+ ions into the terminal, the ionic balance is restored to resting conditions by operation of a sodium pump in the presynaptic membrane, constituted by its ouabain (strophantidin)-sensitive Na^+, K^+-ATPase. These molecules (see J. D. Robinson, Volume 4, this *Handbook*) consist of two subunits: α (mol. wt. 106,000 ± 3,000), containing both the catalytic and the ouabain-binding sites, and β (a glycoprotein with mol. wt. 52,000), present in a probable stoichiometry of $(\alpha\beta)_n$, with a mol wt. $n \times 170{,}000$ where $n = 1$ or 2.[103] The pure enzyme exhibits a specific activity of 25 μmol/min per mg protein. Since its specific activity in synaptic membranes is of the order of 1.5 μmol/min per mg, its concentration must equal ~60 μg/mg protein.

3.6.3. Neurotransmitter Receptors

These molecules (see Peck and Kelner, Volumes 2 and 6, this *Handbook*) have traditionally been considered to be located in and thus constituting markers for postsynaptic membranes. However, ambiguities in this interpretation have been introduced by the demonstration of the existence of presynaptic forms (autoreceptors). In either case, they are concentrated in junctional membranes. However, other, related molecules may also be present in light microsomal membranes, perhaps constituting precursors in receptor biosynthesis. More severe are the problems of their actual molecular properties because of instability or other difficulties involved in their solubilization and purification.

Concentrations in rat SM have been reported[18] to equal 24.1 fmol/mg for α-bungarotoxin (αBTX, probably identical with nicotinic AChR) and 2.0 pmol/mg for quinuclidyl benzylate (muscarinic AChR); the corresponding values (all in pmol/mg[24,104]) are ~3 for L-glutamate; ~1 for L-aspartate, and 0.002 for kainate. Conversion of SM to SJ increases these values by a factor of seven- to eightfold. The putative nicotinic AChR in brain synaptic membranes is associated with three polypeptides with apparent mol. wts. 57,000, 35,000, and 25,000 located at or close to its αBTX binding site.[104a] However, the subunits of the receptor proper isolated from fish electrocytes and neuromuscular junctions are known to be polypeptides with mol. wts. 65,000 (δ), 60,000 (γ), 50,000 (β), and 40,000 (α).

3.7. Synaptic Antigens

A number of protein antigens located in synaptosomal and SM preparations have been investigated, principally by Bock and Jørgensen (see E. Bock, Chapter 9). They are constituted by the proteins referred to as synaptin and proteins D1, D2, and D3 and are included here mainly for the sake of completeness.

Synaptin is an antigen identified by an antibody (Ab) raised against SV. It is concentrated in SV at a concentration ten times, and in SM three times, that found in whole homogenate, and its presence has been demonstrated in ACh- as well as in catecholamine-containing storage vesicles. It appears to be localized on the interior face of the presynaptic and the exterior of the SV membrane. The Abs used for the identification of D1, D2, and D3 were all raised against SM. These three proteins appear to be present in about equal amounts in adult rat brain. D1 is a protein consisting of two subunits with mol. wts. 50,300 and 116,000; one or both of these polypeptides exhibit a transmembrane orientation in the SM, where it is enriched two- to fourfold over the homogenate. It is present in both synapses and neurons but absent from glia. In all of these properties as well as its ontogeny it is similar to the Thy-1 antigen[10] (R. K. Margolis and R. U. Margolis, Volume 5, this *Handbook*).

Protein D2, a glycoprotein with mol. wt. 139,000, is localized unambiguously on the exterior aspect of the postsynaptic membrane of adult rat brain. It is immunochemically related to the adhesion molecule of chick retina and is probably involved in adhesion phenomena among neurites and in early steps of synaptogenesis. At that stage of development, the molecule is present as a sialylated form called D2p, whereas the form present in adult brain ($d \geq 25$ in cortex, ≥ 35 in cerebellum) is desialylated and is referred to as D2a. This molecule exhibits a lower electrophoretic mobility than D2p and can be obtained *in vitro* by treating the latter with neuraminidase at pH 5. Finally, protein D3, consisting of three polypeptide subunits with mol. wts. 14,100, 23,500, and 34,400, respectively, appears to be localized on the internal (synaptoplasmic) face of the presynaptic membrane and may form part of its dense projections.

3.8. Synaptic Vesicle Proteins

In agreement with the exocytosis–fusion hypothesis for depolarization-induced transmitter release, highly purified preparations of *bona fide* storage vesicles and presynaptic membranes exhibit closely similar profiles of major proteins. This is particularly so with respect to the filamentous proteins discussed in Section 3.2 with the important and necessary exception of the postsynaptic density protein (see Table II). Conversely, certain proteins, such as the antigens synaptin and a second protein with mol. wt. 65,000,[105] appear to be concentrated in SV; the latter is a specific antigen for the external surface of vesicles in neuronal and neurosecretory tissue in shark, amphibia, birds, and mammals. Synapsin I (Section 3.5.2) may also be associated with the same (synaptoplasmic) surface of SV.

In contrast, the protein clathrin, with a mol. wt. of 170,000, is characteristic for coated vesicles, and its presence can be taken as a diagnostic for the presence of these organelles in a preparation (e.g., ref. 106). Recent experiments by Near suggest that alkaline phosphatase, a protein with mol. wt. 125,000, appears to be present in vesicle fractions rich in clathrin rather than in SV proper, as claimed by Zisapel and Haklai.[107]

4. OUTLOOK

Although it is apparent from the material presented in this chapter that considerable progress has been made recently in defining the protein molecules that constitute the essential building blocks of the synapse, even more remains to be done. The most difficult, and at the same time an indispensable, methodological hurdle consists of the development of convenient and rapid means for the separation and isolation of pure pre- and postsynaptic membranes with and without their characteristic attachments from specific regions of the vertebrate brain. Once this barrier has been breached, perhaps by use of immunochromatography,[111] using some of the specific markers described as antigens, a number of more penetrating studies becomes possible.

Among these are determinations of the topography (lateral organization) and topology (cross-sectional organization) of membrane proteins, the nature and significance of their interaction with gangliosides and phospholipids in these organizational features, the dynamics of changes in them during synaptogenesis, turnover, and function, especially depolarization and the resulting release of neurotransmitter from its presynaptic storage sites into the cleft and its interaction with pre- and postsynaptic receptors, exocytosis of synaptic vesicles and fusion with the presynaptic membrane and their retrieval from it from distal sites, the occurrence and possible participation in all of these processes of covalent modification of proteins within the confines of the synaptic junction, not just by phosphorylation/dephosphorylation but by methylation/demethylation, sulfation of tyrosine residues, nucleotidyl transfer, transglycosylation, etc.

Once studies of this sort have generated reproducible results, perhaps by the time the next edition of this *Handbook* goes to press, neurochemistry will have begun to lay a firm molecular foundation to permit an approach to the pervasive role of the synapse in all of neuroscience.

REFERENCES

1. Sherrington, C. S., 1906, *The Integrative Action of the Nervous System*, Yale University Press, New Haven.
2. Pfenninger, K. H., 1978, *Annu. Rev. Neurosci.* **1**:445–471.
3. Jones, D. G., 1978, *Adv. Anat. Embryol. Cell Biol.* **55**:7–65.
4. Matus, A. I., and Taff-Jones, D. H., 1978, *Proc. R. Soc. Lond.* [*Biol.*] **203**:135–151.
5. Matus, A. I., Walters, B. B., and Jones, D. H., 1975, *J. Neurocytol.* **4**:369–375.
6. Gray, E. G., 1975, *J. Neurocytol.* **4**:315–339.
7. Mahler, H., 1977. *Neurochem. Res.* **2**:119–147.
8. Mahler, H. R., 1979, *Complex Carbohydrates of Nervous Tissue* (R. U. Margolis and R. K. Margolis, eds.), Plenum Press, New York, pp. 165–184.
9. Smith, A. P., and Loh, H. H., 1979, *Life Sci.* **24**:1–20.
10. Gurd, J. W., 1982, *Molecular Approaches to Neurobiology* (I. R. Brown, ed.), Academic Press, New York, 99–130.
11. Gurd, J. W., Jones, L. R., Mahler, H. R., and Moore, W. J., 1974, *J. Neurochem.* **22**:281–290.
12. Babitch, J. A., Breithaupt, T. B., Chiu, T.-C., Garadi, R., and Helseth, D. L., 1976, *Biochim. Biophys. Acta* **433**:75–89.

13. Matsuda, T., and Cooper, J. R., 1981, *Proc. Natl. Acad. Sci. U.S.A.* **78**:5886–5889.
14. Morgan, I. G., Breckenridge, W. C., Vincendon, G., and Gombos, G., 1973, *Proteins of the Nervous System* (D. M. Schneider, ed.), Raven Press, New York, pp. 171–192.
15. deLorenzo, R. J., and Freedman, S. D., 1978, *Biochem. Biophys. Res. Commun.* **80**:183–192.
16. Zisapel, N., and Zurgil, N., 1979, *Brain Res.* **178**:297–310.
17. Burke, B. E., and de Lorenzo, R. J., 1982, *J. Neurochem.* **38**:1205–1218.
17a. Goldenring, J. R., Gonzalez, B., and de Lorenzo, R. J., 1982, *Biochem. Biophys. Res. Commun.* **108**:421–428.
18. Salvaterra, P. M., and Matthews, D. A., 1980, *Neurochem. Res.* **5**:181–195.
19. Matus, A., Pehling, G., Ackermann, M., and Maeder, J., 1980, *J. Cell Biol.* **87**:346–359.
20. Tamir, H., Mahadik, S. P., and Rapport, M. M., 1976, *Anal. Biochem.* **76**:634–647.
21. Cotman, C. W., Levy, W., Banker, G., and Taylor, D., 1971, *Biochim. Biophys. Acta* **249**:406–418.
22. De Robertis, E., Azcurra, J. M., and Fiszer, S., 1967, *Brain Res.* **5**:45–56.
23. Cotman, C. W., and Taylor, D., 1972, *J. Cell Biol.* **55**:696–711.
24. Mena, E. E., Foster, A. C., Fagg, G. E., and Cotman, C. W., 1981, *J. Neurochem.* **37**:1557–1566.
25. Cotman, C. W., Banker, G., Churchill, L., and Taylor, D., 1974, *J. Cell Biol.* **63**:441–455.
26. Cohen, R. S., Blomberg, F., Berzins, K., and Siekevitz, P., 1977, *J. Cell Biol.* **74**:181–203.
27. Carlin, R. K., Grab, D. J., Cohen, R. S., and Siekevitz, P., 1980, *J. Cell Biol.* **86**:831–843.
27a. Flanagan, S. D., Yost, B., and Crawford, G., 1982, *J. Cell Biol.* **94**:743–748.
28. Ratner, N., 1982, Ph.D. Thesis, Indiana University, Bloomington.
29. Ratner, N., and Mahler, H. R., 1983, *Neuroscience.* **9**:631–644.
30. Blitz, A. L., and Fine, R. E., 1974, *Proc. Natl. Acad. Sci. U.S.A.* **71**:4472–4476.
31. Walters, B. B., and Matus, A. I., 1975, *Nature* **257**:496–498.
32. Mahler, H. R., Kleine, L. P., Ratner, N., and Sorensen, R. G., 1982, Volume 56, *Progress in Brain Research*, W. H. Gispen and A. Rauttenberg, Elsevier/North-Holland Biomedical Press, Amsterdam, pp. 27–48.
33. Hall, Z. W., Lubit, B. W., and Schwartz, J. H., 1981, *J. Cell Biol.* **90**:780–792.
34. Marotta, C. A., Stocchi, P., and Gilbert, J. M., 1978, *J. Neurochem.* **30**:1441–1451.
35. Vandekerchkove, J., and Weber, K., 1978, *Eur. J. Biochem.* **90**:451–562.
36. Gulley, R. L., and Reese, T. S., 1981, *J. Cell Biol.* **91**:298–302.
37. Babitch, J. A., and Benavides, L. A., 1979, *Neuroscience* **4**:603–613.
38. Stadler, H., and Tashiro, T., 1979, *Eur. J. Biochem.* **101**:171–178.
39. Lee, R. W. H., Mushynski, W. E., and Trifaro, J. M., 1979, *Neuroscience* **4**:843–852.
40. Schook, W., Ores, C., and Puszkin, S., 1978, *Biochem. J.* **175**:63–72.
41. Jockusch, B. M., Burger, M. M., da Prada, M., Richards, J. G., Chaponnier, C., and Gabbiani, G., 1977, *Nature* **270**:628–629.
42. Bamburg, J. R., Harris, H. E., and Weeds, A. G., 1980, *FEBS Lett.* **121**:178–182.
42a. Adelstein, R. S., 1982, *Cell* **30**:349–350.
43. Beach, R. L., Kelly, P. T., Babitch, J. A., and Cotman, C. W., 1981, *Brain Res.* **225**:75–93.
43a. Bennett, V., Davis, J., and Fowler, W. E., 1982, *Nature* **299**:126–131.
43b. Glenney, J., Glenny, P., and Weber, K., 1982, *Cell* **28**:843–854.
43c. Shimo-Oka, T., and Watamake, Y. J., 1981, *J. Biochem.* **90**:1297–1307.
43d. Levine, J., and Willard, M., 1981, *J. Cell Biol.* **90**:631–643.
44. Smith, A. P., and Loh, H. H., 1981, *J. Neurochem.* **36**:1749–1757.
45. Zisapel, N., Levi, M., and Gozes, I., 1980, *J. Neurochem.* **34**:26–32.
46. Ponstingl, H., Krauhs, E., Little, M., and Kempf, T., 1981, *Proc. Natl. Acad. Sci. U.S.A.* **78**:2757–2761.
47. Krauhs, E., Little, M., Kempf, T., Hofer-Warbinek, R., Ade, W., and Ponstingl, H., 1981, *Proc. Natl. Acad. Sci. U.S.A.* **78**:4156–4160.
48. Little, M., 1979, *FEBS Lett.* **108**:283–286.
49. Gozes, I., de Baetselier, A., and Littauer, U. Z., 1980, *Eur. J. Biochem.* **103**:13–20.
50. Babitch, J. A., 1981, *J. Neurochem.* **37**:1394–1400.
50a. Carlin, R. K., Grab, D. J., and Siekevitz, P., 1982, *J. Neurochem.* **38**:94–100.

51. Kelly, P. T., and Cotman, C. W., 1978, *J. Cell Biol.* **75**:173–183.
52. Gozes, I., and Littauer, U. Z., 1979, *FEBS Lett.* **99**:86–90.
53. Burke, B. E., and deLorenzo, R. J., 1981, *Proc. Natl. Acad. Sci. U.S.A.* **78**:991–995.
54. Sloboda, R. D., Rudolph, S. A., Rosenbaum, J. L., and Greengard, P., 1975, *Proc. Natl. Acad. Sci. U.S.A.* **72**:177–181.
55. Matus, A., Bernhardt, R., and Hugh-Jones, T., 1981, *Proc. Natl. Acad. Sci. U.S.A.* **78**:3010–3014.

55a. Davis, J., and Bennett, V. J., 1982, *J. Biol. Chem.* **256**:5816–5820.

56. Sampedro, M. N., Bussineau, C. M., and Cotman, C. W., 1981, *J. Cell Biol.* **90**:675–686.
57. Froehner, S. C., Gulbrandsen, V., Hyman, C., Jeng, A. Y., Neubig, R. R., and Cohen, J. B., 1981, *Proc. Natl. Acad. Sci. U.S.A.* **78**:5230–5234.
58. Lazarides, E., 1981, *Cell* **23**:649–650.
59. Cheung, W. Y., 1980, *Science* **207**:19–27.
60. Klee, C. B., Crouch, T. H., and Richman, P. G., 1980, *Annu. Rev. Biochem.* **49**:489–515.
61. Means, A. R., and Dedman, J. R., 1980, *Nature* **285**:73–77.
62. Kretsinger, R. H., 1980, *CRC Crit. Rev. Biochem.* **8**:119–174.
63. Grand, R. J. A., and Perry, S. V., 1979, *Biochem. J.* **183**:285–295.
64. Grab, D. J., Berzins, K., Cohen, R. S., and Siekevitz, P., 1979, *J. Biol. Chem.* **254**:8690–8696.
65. Wood, J. G., Wallace, R. W., Whitaker, J. N., and Cheung, W. Y., 1980, *Ann. N.Y. Acad. Sci.* **356**:75–82.
66. Wood, J. G., Wallace, R. W., Whitaker, J. N., and Cheung, W. Y., 1980, *J. Cell Biol.* **84**:66–76.
67. Lin, C.-T., Dedman, J. R., Brinkley, B. R., and Means, A. R., 1980, *J. Cell Biol.* **85**:473–480.
68. Klee, C. B., Crouch, T. H., and Krinks, M. H., 1979a, *Biochemistry* **18**:722–729.
69. Klee, C. B., Crouch, T. H., and Krinks, M. H., 1979b, *Proc. Natl. Acad. Sci. U.S.A.* **76**:6270–6273.
70. Klee, C. B., and Haiech, J., 1980, *Ann. N.Y. Acad. Sci.* **356**:43–54.
71. Therien, H.-M., and Mushynski, W. E., 1979, *Biochim. Biophys. Acta* 585:201–209.
72. Kincaid, R. L., Manganiello, V. C., and Vaughan, M., 1981, *J. Biol. Chem.* **256**:11345–11350.

72a. Kincaid, R. L., Kempner, E., Manganiello, V. C., Osborne, J. C., Jr., and Vaughan, M., 1981, *J. Biol. Chem.* **256**:11351–11355.

73. Grab, D. J., Carlin, R. K., and Siekevitz, P., 1981, *J. Cell Biol.* **89**:433–439.
74. Donato, R., 1981, *J. Neurochem.* **36**:532–537.
75. Endo, T., Tanaka, T., Isobe, T., Kasai, H., Okuyama, T., and Hidaka, H., 1981, *J. Biol. Chem.* **256**:12485–12489.
76. Isobe, T., and Okuyama, T., 1978, *Eur. J. Biochem.* **89**:378–388.
77. Sorensen, R. G., and Mahler, H. R., 1981, *J. Neurochem.* **37**:1407–1418.
78. Sorensen, R. G., and Mahler, H. R., 1984, *J. Neurochem.* **40**:1349–1365. (in press).
79. Caroni, P., and Carafoli, E., 1981, *J. Biol. Chem.* **256**:9371–9373.
80. Carlin, R. K., Grab, D. J., and Siekevitz, P., 1981, *J. Cell Biol.* **89**:449–455.

80a. Andreasen, T. J., Keller, C. H., LaPorte, D. C., Edelman, A. M., and Storem, D. R., 1981, *Proc. Natl. Acad. Sci. U.S.A.* **78**:2782–2785.

80b. Palfrey, H. C., Schiebler, W., and Greengard, P., 1982, *Proc. Natl. Acad. Sci. U.S.A.* **79**:3780–3784.

80c. Malencik, D. A., and Anderson, S. R., 1982, *Biochemistry* **21**:3480–3486.

81. Mahler, H. R., 1979, *Complex Carbohydrates of Nervous Tissue* (R. U. Margolis and R. K. Margolis, eds.), Plenum Press, New York, pp. 165–184.
82. Gurd, J. W., 1980, *Can. J. Biochem.* **58**:941–951.
83. Mena, E. E., and Cotman, C. W., 1982, *Science* **216**:422–424.
84. Greengard, P., 1976, *Nature* **260**:101–108.
85. Greengard, P., 1978, *Science* **199**:146–152.
86. Williams, M., and Rodnight, R., 1977, *Prog. Neurobiol.* **8**:183–250.
87. Kennedy, M. B., and Greengard, P., 1981, *Proc. Natl. Acad. Sci. U.S.A.* **78**:1293–1298.

87a. Miyamoto, E., Fukunaga, K., Matsui, K., and Iwasa, Y., 1981, *J. Neurochem.* **37**:1324–1330.

88. Takai, Y., Kishimoto, A., Iwasa, Y., Kawahara, Y., Mori, T., and Nishizuka, Y., 1978, *J. Biol. Chem.* **254**:3693–3695.
88a. Wise, B. C., Raynor, R. L., and Kuo, J. F., 1982, *J. Biol. Chem.* **257**:8481–8488.
88b. Wise, B. C., Glass, D. B., Chou, C.-H. J., Raynor, R. L., Katoh, N., Schatzman, R. C., Turner, R. S., Kibler, R. F., and Kuo, J. F., 1982, *J. Biol. Chem.* **257**:8489–8495.
88c. Wu, W. C.-S., Walaas, S. I., Nairn, A. C., and Greengard, P., 1982, *Proc. Natl. Acad. Sci. U.S.A.* **79**:5249–5254.
89. Rubin, C. S., Fleischer, N., Sarkar, D., and Erlichman, J., 1982, *Protein Phosphorylation*, Volume 8, Cold Spring Harbor Laboratory, New York, pp. 1333–1346.
90. Ueda, T., and Greengard, P., 1977, *J. Biol. Chem.* **252**:5155–5163.
91. Forn, J., and Greengard, P., 1978, *Proc. Natl. Acad. Sci. U.S.A.* **75**:5195–5199.
92. Goelz, S. E., Nestler, E. J., Chehrazi, B., and Greengrad, P., 1981, *Proc. Natl. Acad. Sci. U.S.A.* **78**:2130–2134.
93. Ueda, T., Greengard, P., Berzins, K., Cohen, R. S., Blomberg, F., Grab, D. J., and Siekevitz, P., 1979, *J. Cell Biol.* **83**:308–319.
94. Bloom, F. E., Ueda, T., Battenberg, E., and Greengard, P., 1978, *Proc. Natl. Acad. Sci. U.S.A.* **76**:5982–5986.
95. Huttner, W. B., DeGennaro, L. J., and Greengard, P., 1981, *J. Biol. Chem.* **256**:1482–1488.
96. Zwiers, H., Schotman, P., and Gispen, W. H., 1980, *J. Neurochem.* **34**:1689–1699.
96a. Kristjansson, G. I., Zwiers, H., Oestreicher, A. B., and Gispen, W. H., 1982, *J. Neurochem.* **39**:371–378.
97. Oestreicher, A. B., Zwiers, H., Schotman, P., and Gispen, W. H., 1981, *Brain Res. Bull.* **116**:145–153.
98. Sorensen, R. G., Kleine, L. P., and Mahler, H. R., 1981, *Brain Res. Bull.* **7**:57–61.
99. Huang, C.-K., Browning, M. D., and Greengard, P., 1982, *J. Biol. Chem.* **257**:6524–6528.
99a. Costa, M. R. C., Casnellie, J. E., and Catterall, W. A., 1982, *J. Biol. Chem.* **257**:7918–7921.
100. Catterall, W. A., 1980, *Annu. Rev. Pharmacol. Toxicol.* **20**:15–43.
100a. Catterall, W. A., 1982, *Cell* **30**:672–674.
100b. DeVries, G. H., and Lazdunski, M., 1982, *J. Biol. Chem.* **257**:11684–11688.
101. Tamkun, M. M., and Catterall, W. A., 1981, *J. Biol. Chem.* **256**:11457–11463.
101a. Talvenheimo, J. A., Tamkun, M. M., and Catterall, W. A., 1982, *J. Biol. Chem.* **257**:11868–11871.
102. Hartshorne, R. P., and Catterall, W. A., 1981, *Proc. Natl. Acad. Sci. U.S.A.* **78**:4620–4624.
102a. Gill, D. L., 1982, *J. Biol. Chem.* **257**:10986–10990.
103. Craig, W. S., 1982, *Biochemistry* **21**:2667–2674.
104. Foster, A. C., Menga, E. E., Fagg, G. E., and Cotman, C. W., 1981, *J. Neurosci.* **1**:620–625.
104a. Betz, H., Graham, D., and Rehm, H., 1982, *J. Biol. Chem.* **257**:11390–11394.
104b. Froehner, S. C., Gulbrandsen, V., Hyman, C., Jeng, A. Y., Neuberg, R. R., and Cohen, J. B., 1981, *Proc. Natl. Acad. Sci. U.S.A.* **78**:5230–5234.
105. Matthews, W. D., Tsavaler, L., and Reichardt, L. F., 1981, *J. Cell Biol.* **91**:257–269.
106. Rubenstein, J. L. R., Fine, R. E., Luskey, B. D., and Rothman, J. E., 1981, *J. Cell Biol.* **89**:357–361.
107. Zisapel, N., and Haklai, R., 1980, *Neuroscience* **5**:2297–2303.
108. Lohmann, S. M., Walter, U., and Greengard, P., 1980, *J. Biol. Chem.* **255**:9985–9992.
109. Kelly, P. T., and Cotman, C. W., 1977, *J. Biol. Chem.* **252**:786–793.
110. Yang, S.-D., Tallant, E. A., and Cheung, W. Y., 1982, *Biochem. Biophys. Res. Commun.* **106**:1419–1425.
111. Miljanich, G. P., Brasier, A. R., and Kelly, R. G., 1982, *J. Cell Biol.* **94**:88–96.
112. Sobue, K., Kanda, K., and Kakiuchi, S., 1982 *FEBS Lett.* **150**:185–190.

5

Brain Mitochondria

J. B. Clark and W. J. Nicklas

1. INTRODUCTION

As previous reviews on brain mitochondrial properties appeared in the first edition of the *Handbook in Neurochemistry* (Volumes 2 and 3, 1969–70), we confine ourselves mainly to those reports appearing in the last 10–12 years. Certain aspects of what may legitimately be considered as mitochondrial activity have been left out, as they form subjects in their own right which are being treated elsewhere, e.g., mitochondrial protein synthesis, monoamine oxidase activity. The emphasis of this review is therefore, in the main, concerned with the properties of mitochondria derived from the main cell types, neurons and glia; the metabolic properties of these preparations particularly with respect to energy metabolism; the role of the mitochondria in the development of brain function; and clinical disorders that are considered to be the result of a primary mitochondrial lesion in the brain.

2. PREPARATION AND PROPERTIES OF BRAIN MITOCHONDRIA

2.1. Preparative Techniques

Investigations in this area have been primarily concerned with achieving preparations of mitochondria from the two main cell types of brain, i.e., neurons and glia, that are relatively uncontaminated and also metabolically active. Two approaches have been followed: (1) the preparation of whole-cell fractions enriched in neurons or glia followed by the isolation of mitochondria from the two cell types and (2) the preparation of mitochondrial populations that are derived from synaptosomes by lysis (synaptic mitochondria) as compared to those isolated free of synaptosomes (free or nonsynaptic mitochondria). Al-

J. B. Clark • Department of Biochemistry, The Medical College of St. Bartholomews Hospital, University of London, London EC1M 6BQ, England. *W. J. Nicklas* • Department of Neurology, University of Medicine and Dentistry of New Jersey, Rutgers Medical School, Piscataway, New Jersey 08854.

though in theory the first approach should yield a better comparison, in practice, the impurity of cell populations prepared from brain by current techniques and the time involved have led to mitochondrial populations that are not in a very active metabolic state as judged by such indices as respiratory control ratios. Further progress in preparing metabolically active brain mitochondria has been achieved by using Ficoll–sucrose gradients rather than gradients consisting of sucrose alone, thus decreasing possible osmotic damage caused by the high concentration of sucrose used in the gradients.

Hamberger *et al.*,[1] using neuronal-enriched or glial-enriched cell fractions derived either from beef brain or rabbit cerebral cortex, were able to prepare mitochondria from both cell types on discontinuous sucrose gradients. The mitochondria from the neuronally enriched fraction had a higher rate of amino acid incorporation into proteins, a higher cytochrome oxidase activity, and a greater buoyant density than those derived from the glial enriched fraction. The glial-enriched fraction, however, showed a higher monoamine oxidase and Na^+,K^+-ATPase activity.

2.1.1. Gradient Separation of Brain Homogenates

A number of studies have been carried out with continuous sucrose gradients using both differential and zonal centrifugation techniques with a view to establishing the heterogeneity of the mitochondrial population of the brain. Studies from Cotman's laboratory[2] established the gradient conditions for the optimal separation of mitochondria, synaptosomes, and lysosomes from brain homogenates by differential centrifugation in sucrose solutions. Neidle *et al.*,[3] using sucrose gradients, reported high activities of both acetyl-CoA synthetase and glutamate dehydrogenase in the mitochondrial fraction with the highest density and, in the population of a lighter density, found high activities of monoamine oxidase, succinate dehydrogenase, and glutaminase.

Later studies by Van den Berg and colleagues[4,5] using continuous gradients (0.8–1.6 M sucrose) and centrifuging for 1 h, reported subpopulations of brain mitochondria with different activities of glutamate dehydrogenase and 4-aminobutyrate:2-oxoglutarate aminotransferase (GABA transaminase). This was interpreted as further support for the compartmentation of glutamate metabolism in the brain. Further studies[6] using similar techniques showed that the short-chain fatty acid synthases; acetyl-CoA, propionyl-CoA, and butyryl-CoA synthase, were located in a mitochondrial population of higher density than those in which NAD-isocitrate dehydrogenase (ICDH) was located but of similar density to those in which glutamate dehydrogenase (GDH) was enriched.

A number of studies on the heterogeneity of brain mitochondria have been carried out by using zonal centrifugation. Blokhuis and Veldstra,[7] employing a linear sucrose gradient, detected several mitochondrial populations including two main groups, one that peaked at 1.53 M sucrose and contained high specific activities of glutamate dehydrogenase and NADP-linked isocitrate dehydrogenase and the other, peaking at a lower sucrose concentration, which contained high activities of NAD-linked isocitrate dehydrogenase, monoamine oxidase, and citrate synthase. Further studies on conditions to achieve optimal separation of both mitochondrial and synaptosomal populations from rat brain homogenates have been reported by Churchill and Cotman.[8] Use of rate zonal centrifugation as a means of establishing the purity or otherwise of mitochondrial populations prepared from whole brain or synaptosomal fractions has also been reported[9] and is considered below.

2.1.2. Synaptic and Nonsynaptic ("Free") Populations of Brain Mitochondria

In order to achieve mitochondrial populations of minimal contamination, most procedures employed currently involve the use of gradients, and although a number of different compounds are used in gradients, a combination of Ficoll–sucrose appears to be the most popular. An early report of the use of such a gradient procedure was made by the authors, which led to the preparation of a relatively pure preparation of essentially nonsynaptic (free) mitochondria exhibiting good respiratory control and the oxidation of a number of substrates (see ref. 10). Further modification of the Ficoll–surcose gradient procedures used allowed the isolation of a relatively pure synaptosomal fraction, which was subsequently lysed by a dilute Tris buffer (see ref. 9), and further separation permitted the isolation of a metabolically active mitochondrial population derived from synaptosomes (synaptic mitochondria). When the synaptic mitochondrial fraction was compared by rate zonal centrifugation to a "nonsynaptic" free mitochondrial population prepared as before,[10] the two separated at distinct densities, the nonsynaptic mitochondria having the higher buoyant density, and whereas the free mitochondria ran essentially as a single band, the synaptic mitochondrial fraction showed signs of heterogeneity.

Further refinement of the gradient conditions permitted the isolation from a single homogenate of rat brain of a "free" mitochondrial population and two distinct synaptic mitochondrial subpopulations, all in a metabolically active, well-coupled, and relatively uncontaminated state.[11,12] The capability of these various brain mitochondrial populations to oxidize various substrates and their complement of enzyme activities are shown in Tables I and II. Wilson and his group[13] have reported a preparation of "free" rat brain mitochondria involving a 10–18% gradient of sodium diatrizoate, which is centrifuged for 15 min at 53,000 *g* in a swing-out rotor. The purity of the preparation as judged by lactate dehydrogenase contamination is good, as is the metabolic integrity of the preparation. However, some modification of the outer membrane appears to have occurred, as this preparation has no hexokinase activity, unlike most other brain mitochondrial preparations (see Section 2.2.3).

Nicholls and his group[14,15] have used Ficoll gradients in the preparation of both "free" and "synaptic" mitochondria from guinea pig cortex. In their lysis of synaptosomes to release the synaptic mitochondria, they used digitonin treatment rather than osmotic lysis used by other workers.[9–12]

2.1.3. Purity of Brain Mitochondrial Preparations

The relative purity of the various mitochondrial preparations has been assessed by electron microscopy and measurement of enzyme markers, par-

Table I
Substrate Oxidation by "Free" and Synaptic Mitochondria from Rat brain[a]

Substrate	K^+ concn. (mM)	O_2 uptake rate (atoms/min per mg protein)							
		Synaptic mitochondria				Nonsynaptic mitochondria			
		Light fraction (SM)		Heavy fraction (SM_2)		A		B	
		State 3	RCR	State 3	RCR	State 3	RCR	State 3	RCR
2.5 mM Malate + 5 mM pyruvate	5	73 ± 9	8.6	55 ± 15	6.7	66 ± 15	5.8	93 ± 10	6.7
	100	161 ± 6	12.8	135 ± 19	11.6	158 ± 15	5.4	183 ± 13	8.0
2.5 mM Malate + 10 mM glutamate	5	54 ± 12	5.8	54 ± 6	9.3	40 ± 2	14.5	97 ± 8	4.8
	100	89 ± 15	7.7	67 ± 11	5.2	107 ± 20	5.3	158 ± 11	4.1
2.5 mM Malate + 10 mM glutamine	5	—	—	—	—	—	—	47 ± 3	2.6
	100	—	—	43 ± 3	—	—	—	55 ± 2	2.4
2.5 mM Malate + 5 mM DL-3-hydroxybutyrate	5	57 ± 5	2.5	46 ± 0.5	2.5	30 ± 3	3.0	30 ± 6	3.0
	100	66 ± 1	1.7	55 ± 0.7	2.0	53 ± 5	2.5	60 ± 7	3.0
2.5 mM Malate + 5 mM citrate	5	59 ± 1	3.7	52 ± 1	3.6	54 ± 1	10	66 ± 5	3.2
	100	131 ± 1	2.4	105 ± 1	2.7	60 ± 12	11	92 ± 9	2.0
2.5 mM Malate + 5 mM 2-oxoglutarate	5	46 ± 4	4.0	52 ± 9	2.7	66 ± 6	9.4	46 ± 5	3.0
	100	135 ± 8	3.8	113 ± 17	3.9	135 ± 9	5.3	95 ± 10	2.8
2.5 mM Malate + 10 mM succinate	5	106 ± 2	2.7	80 ± 8	2.3	93 ± 3	1.8	122 ± 10	3.3
	100	135 ± 2	2.9	140 ± 17	2.6	138 ± 8	2.7	150 ± 10	2.0

[a] Results are expressed as atoms O consumed/min per mg mitochondrial protein and are means ± S.D.s of a minimum of three experiments. Experiments were carried out at 28°, and synaptic mitochondria and the A preparation of free (nopsynaptic) mitochondria were prepared according to ref. 11, and the B preparation of free mitochondria according to ref. 10. State 3 is the respiration rate in the presence of ADP, and the respiratory control ratio (RCR) is defined as the respiration rate in the presence of ADP divided by the rate in its absence. These data are taken from refs. 11, 26, 27.

Table II
Cytochrome Content of "Free" (Nonsynaptic) and Synaptic Rat Brain Mitochondria[a]

Cytochrome (nmol/mg protein)	Synaptic mitochondria		Nonsynaptic mitochondria	
	Light fraction (SM)	Heavy fraction (SM_2)	A	B
b	0.14	0.13	0.19	0.20
c_1	0.27	0.23	0.31	0.31
c	0.45	0.46	0.54	0.51
a	0.28	0.27	0.33	0.36
a_3	0.2	0.21	0.24	0.23

[a] Synaptic mitochondria (SM and SM_2) and free mitochondria (population A) were prepared by the method of Lai *et al.*,[11] and the cytochrome data are taken from ref. 33. Free mitochondria (population B) were prepared by the method of Clark and Nicklas,[10] and the cytochrome data are taken from ref. 29. Cytochromes have been calculated from spectra taken at liquid N_2 temperatures and using absorption coefficients as detailed in ref. 10. The results are expressed in nmol/mg protein and represent the mean of at least two estimations.

ticularly lactate dehydrogenase (cytosolic contamination), rotenone-insensitive NADH-cytochrome *c* reductase (microsomal contamination), and acetylcholinesterase (membranal contamination). Such measurements, however, only provide information on nonmitochondrial contaminants and do not help with the problem of heterogeneity of whole-brain mitochondria from a functional or regional aspect. Generally, free brain mitochondria prepared by the Ficoll–sucrose gradient procedures[9–12,14,15] are relatively free of cytosolic and membranal contamination as judged by electron microscopy and enzyme marker studies. For the free, nonsynaptic mitochondria, lactate dehydrogenase levels are less than 0.3% of the homogenate values, and acetylcholinesterase less than 0.5%,[9–12] and EM studies suggest that 90–95% of the particles present are mitochondrial.[10] The method using Na diatrizoate gradients[13] showed a lactate dehydrogenase activity of >0.05% of the homogenate value. This, however, must be taken in conjunction with the observation that the hexokinase activity normally associated with the outer mitochondrial membrane is absent, suggesting that some modification/removal of the outer membrane may have occurred during preparation.

In the case of synaptic mitochondria, the general level of cytosolic and membranal contamination is similar to that for free mitochondria. However, a further level of contamination has to be considered, that of the presence of free, nonsynaptic mitochondria in the original synaptosomal preparation from which the synaptic mitochondria are prepared. This has been variously assessed by using the latency of marker enzymes such as the rotenone-insensitive NAD(P)H cytochrome *c* reductase and put at a level of 5% or less on a protein basis.[9,11,15] The advent of methods using flotation on Ficoll–sucrose gradients rather than precipitation,[16] which leads to purer synaptosomal fractions, will clearly assist in further reducing this aspect of contamination. In addition, it is likely that mitochondria derived from a cholinergic terminal as opposed to those from a GABinergic or adrenergic system will show different properties. This aspect may be approached by preparing synaptic mitochondria from specific areas of the brain and is currently in progress in one of the author's laboratories (S. F. Leong and J. B. Clark, unpublished data). These studies together with improved methods for lysing the synaptosome[17] to improve the yield and decrease the damage to the mitochondria will undoubtedly yield some interesting information in this direction.

The Ficoll–sucrose gradient methods referred to earlier have also been utilized successfully to prepare mitochondria from developing rat brain of several ages that exhibit good respiratory control and minimal contamination.[18–20] Techniques have also been established for the removal of the outer membrane of brain mitochondria by the selective use of digitonin[21] and also a study of the glycoprotein content of intrasynaptosomal mitochondria.[22] In the latter case, synaptosomes prepared on a Ficoll gradient were lysed osmotically to yield synaptic mitochondria. Investigations showed that there were no glycoprotein- or concanavalin-A-binding glycoproteins intrinsic to the membranes of the synaptic mitochondria. A comparison of the glycoprotein content of the microsomal, synaptosomal, and mitochondrial fractions of brain using the sodium diatrizoate gradient procedure[13] has also indicated the low glycoprotein content of "free" brain mitochondria.[23]

A detailed survey of the polypeptide components of rat brain mitochondria prepared on discontinuous sucrose gradients has also been carried out on polyacrylamide gels.[24] Seventeen polypeptide bands in the mol. wt. range 22,000–50,000 were recorded, with no bands of mol. wt. greater than 100,000 detectable. These latter reports are of interest in the light of the electron microscope investigations of Gray's group[25] in which they have reported that under certain conditions in brain slices the inner membrane of intrasynaptosomal mitochondria becomes exposed and attracts aggregations of synaptic vesicles to the exposed area. These observations are dependent on the presence of Ca^{2+}/Mg^{2+} in the incubation medium, and the possibility is suggested that they may be associated with vesicle exocytosis.

2.2. Properties of Brain Mitochondria

2.2.1. Respiratory Studies

Table I summarizes some of the data concerning the oxidation of various substrates by nonsynaptic and synaptic mitochondria derived from whole brain. These have been prepared essentially as detailed in references 9–12. Brain mitochondria, whether derived from synaptic or nonsynaptic sources, showed a marked stimulation of respiration when incubated in a medium containing a high K^+ concentration (>100 mM) as compared with medium containing only 5 mM K^+. Although the respiratory stimulation varies from substrate to substrate, it is often at least a twofold increase and sometimes more. Both pyruvate and succinate are good substrates for oxidation in the presence of malate as judged by the high rates of O_2 uptake but do not vary markedly from synaptic to nonsynaptic mitochondria. Pyruvate oxidation, however, is much more markedly stimulated by high K^+ concentrations than is succinate oxidation, and this may be related to a specific K^+ effect on the pyruvate dehydrogenase complex.[31] Similarly, 3-hydroxybutyrate and 2-oxoglutarate in the presence of malate show similar rates of oxidation by all types of mitochondria, although 2-oxoglutarate is oxidized in the high-K^+ medium at twice the rate of 3-hydroxybutyrate. Glutamine is also oxidized at only marginally higher rates by nonsynaptic mitochondria. In contrast, glutamate and malate are oxidized by "free" mitochondria at rates that are almost twice those of "synaptic" mitochondria. This correlates clearly with the activities of the glutamate–oxaloacetate transaminase (Table III) and supports the suggestion that most of glutamate metabolism in brain mitochondria occurs via transamination and not via glutamate dehydrogenase.[32] Citrate, however, is marginally better oxidized by synaptically derived mitochondria.

Table II indicates the cytochrome spectrum and concentration in the different types of mitochondria. As can be seen, all populations of mitochondria contain a normal spectrum for mammalian mitochondria with possibly a slightly higher concentration on a protein basis in the nonsynaptic mitochondria. This, however, may not be of any real significance and may merely reflect the fact that the nonsynaptic mitochondria are generally somewhat less contaminated with nonmitochondrial protein than the synaptic preparations.

2.2.2. *Enzymes*

Table III lists the specific activities of a number of enzymes that have been measured in the different populations of mitochondria. Although there have been a number of reports in which "free" mitochondrial enzyme activity has been compared with synaptosomal activity of certain mitochondrial enzymes,[34–38] there are relatively few comprehensive surveys of enzyme activity of isolated purified free and synaptic mitochondria.[10–12] Pyruvate dehydrogenase, a key enzyme in brain and mitochondrial metabolism, appears to have a higher specific activity in the free mitochondrial preparations.[12] More recent investigations on a regional basis have confirmed this, indicating that in some brain regions the free mitochondria may have four to five times the activity of the synaptic populations on a protein basis (S. F. Leong and J. B. Clark, 1984). The specific activity quoted in Table III is the active portion of the pyruvate dehydrogenase (PDH), which on extraction is about 70% of the total available activity (112 $\pm$ 13 nmol/min per mg protein, n = 10).

Studies on the control of this enzyme[39] by the satellite PDH kinase and phosphatase have indicated that the brain enzyme may be controlled in much the same way as the PDH from other tissues, full activation being achieved by the appropriate concentrations of Ca^{2+} and Mg^{2+} (cf. ref. 40). The rat brain PDH complex has been purified, and a detailed analysis has been carried out on its mechanism and the metabolites that inhibit it.[41] The data are consistent with a nonclassical, three-site ping–pong mechanism.[41] When the activity of the pyruvate dehydrogenase is considered on a whole-tissue basis[20] and compared with estimates of the rate of whole brain glycolysis *in vivo*,[42] it is clear that, under most circumstances, the brain pyruvate dehydrogenase is working very close to its maximum capacity. More recently,[43,44] evidence has been put forward involving Ca^{2+}-mediated changes in the phosphorylation of the subunit

Table III
Enzyme Activities in "Free" (Nonsynaptic) and Synaptic Rat Brain Mitochondria[a]

Enzyme	Specific activity (nmol/min per mg mitochondrial protein)				Reference
	Synaptic mitochondria		Nonsynaptic mitochondria		
	Light fraction (SM)	Heavy fraction (SM_2)	A	B	
Pyruvate dehydrogenase	73 ± 4.8 (3)	57 ± 5.5 (3)	72 ± 12.4 (3)	72 ± 6.1 (14)	12
Citrate synthase	621 ± 0 (2)	771 ± 0 (2)	1,242 ± 0 (2)	1,070 ± 104 (14)	26, 29
NAD^+–isocitrate dehydrogenase	142 ± 12 (6)	127 ± 5.9 (6)	161 ± 18 (3)	141 ± 13 (4)	12
$NADP^+$–isocitrate dehydrogenase	24 ± 6.2 (7)	47 ± 8.3 (7)	27.5 ± 2.2 (3)	34 ± 6.8 (4)	12
2-Oxoglutarate dehydrogenase	45 ± 8.5 (3)	36 ± 3.4 (3)	62 ± 4 (3)	53 ± 3.3 (3)	12
Fumarase	323 ± 38 (4)	281 ± 42 (4)	364 ± 37 (5)	371 ± 24 (3)	12
NAD^+–malate dehydrogenase	8,625 ± 1,143 (5)	7,955 ± 720 (5)	10,357 ± 950 (3)	7,919 ± 1,055 (4)	12
NAD^+–glutamate dehydrogenase	550 ± 38 (6)	992 ± 99 (6)	534 ± 71 (7)	578 ± 44 (9)	12
$NADP^+$–glutamate dehydrogenase	469 ± 66 (3)	786 ± 41 (3)	454 ± 34 (3)	492 ± 39.4 (4)	12
Glutamate–oxaloacetate transaminase	1,592 ± 141 (7)	1,216 ± 168 (7)	1,794 ± 142 (6)	1,752 ± 83 (4)	12
Glutaminase, phosphate dependent	—	294 ± 37 (2)	—	384 ± 65 (2)	28
Glutaminase, phosphate independent	—	70 (1)	—	70 (1)	28
4-Aminobutyrate transaminase	21 ± 1.2 (4)	26 ± 4.6 (4)	20 ± 1 (3)	38 ± 3.3 (7)	12
Succinic semialdehyde dehydrogenase	23 ± 4 (7)	29 ± 5 (7)	27 ± 2 (3)	57 ± 0.6 (7)	12
Adenylate kinase	—	—	—	21.5 ± 2.5 (5)	26
Creatine kinase	208 ± 18 (3)	215 ± 13 (3)	774 ± 43 (3)	—	29
Hexokinase	138 ± 9 (3)	131 ± 6 (3)	300 ± 21 (3)	—	29
3-Hydroxybutyrate dehydrogenase	26.3 ± 3 (3)	45.5 ± 3.2 (3)	17.8 ± 1.2 (3)	16.3 ± 1.2 (5)	30
3-Oxoacid transferase	180 ± 12.2 (50)	97.6 ± 5.1 (3)	98.6 ± 8.2 (3)	98.3 ± 6.2 (5)	30
Acetoacetyl-CoA thiolase	52.7 ± 3 (3)	69.4 ± 3 (3)	66.3 ± 4 (3)	66.3 ± 3 (5)	30
Acetoacetyl-CoA deacylase	—	—	—	38.5 ± 1.6 (3)	30
3-Hydroxy-3-methylglutaryl-CoA synthase	14.4 ± 1.2 (3)	17.9 ± 1 (3)	17.8 ± 1.2 (3)	17.7 ± 1.1 (5)	30
Acetyl-CoA synthetase	2 ± 0.6 (3)	7 ± 1.4 (3)	3.1 ± 0.4 (3)	3 ± 0.3 (3)	12

[a] Synaptic mitochondria (SM, SM_2) and free mitochondria (population A) were prepared by the method of Lai *et al.*,[11] and free mitochondria (population B) by the method of Clark and Nicklas.[10] All values are expressed as means ±S.D. with the number of determinations in parentheses. Determinations were carried out at 25°C.

of pyruvate dehydrogenase, which may be related through alterations in internal synaptosomal Ca^{2+} levels to the passage of a nerve impulse.

Pyruvate carboxylase and phosphoenolpyruvate carboxykinase activity have been reported in rat brain mitochondrial preparations.[35,45,46] The relative importance of these enzymes in brain metabolism is still unclear, particularly in view of the relatively low activities, but the pyruvate carboxylase could clearly act in an anaplerotic sense during periods of rapid turnover. Such a role would fit in with the fact that the activity of pyruvate carboxylase is higher in the young developing brain than in the adult.[47] Citrate synthase activity appears to be higher in the nonsynaptic mitochondrial populations (Table III), with only marginal differences in the activities of the NAD^+ and $NADP^+$ isocitrate dehydrogenases. Conflicting reports exist in the literature regarding isocitrate dehydrogenase; Salganicoff and Koeppe[35] reported a decrease in the free mitochondrial fraction, but Wilson[36] reported an increase. 2-Oxoglutarate dehydrogenase activity is generally lower in the synaptic mitochondrial fractions, but fumarase and malate dehydrogenases show essentially the same order of activity in all mitochondrial populations.

Because of the extensive glutamate compartmentation studies in brain (see Section 3), considerable interest has been shown in the mitochondrial enzymes associated with glutamate, glutamine, and GABA metabolism. As was previously pointed out, the glutamate oxidation studies on the various mitochondrial preparations correlate best with the measured glutamate–oxaloacetate transaminase activities, i.e., higher in the free mitochondrial populations than in the synaptic. There seems to be a general consensus[34–38,48] that the glutamate dehydrogenase activity is higher in the free mitochondrial populations than in the synaptic. Lai *et al.*[11] have reported very similar glutamate dehydrogenase activities between the mitochondrial populations with the exception of the heavy synaptic population, which had a specific activity approximately 50% higher. Further analysis on a regional basis has shown that this is not the case for all regions and that the true picture is complex (S. F. Leong and J. B. Clark, 1984).

The two reports on glutaminase activity from rat brain[28,38] both concur that phosphate-dependent glutaminase is higher in the free mitochondrial population, together with a higher GABA transaminase and succinic semialdehyde dehydrogenase activity.[12,35] Buu *et al.*,[49] working with mouse brain GABA transaminase, have confirmed this, showing that free mitochondrial fractions have five times more activity and that the apparent K_m for GABA for the synaptic enzyme (2.6 mM) is twice that of the free mitochondrial enzyme. Further kinetic analyses on the GABA transaminase from synaptic and free mitochondrial fractions of rat brain have been carried out,[50] indicating differences in their response to the inhibitors diaminobutyrate and aminoxyacetate, and a detailed analysis of the inhibitory effects of the branched-chain fatty acids 2-methyl-2-ethyl caproate and 2,2-dimethyl valerate on free mitochondrial GABA transaminase has also been reported.[51] Succinate dehydrogenase has been reported to be both higher in free mitochondrial populations[38] and lower.[35]

A number of kinases have been reported to be associated with brain mitochondrial fractions: adenylate kinase,[29] hexokinase,[20,29,36,38] and creatine kinase.[29,52,53] Data from our laboratory suggest that there is a considerably higher activity of both hexokinase and creatine kinase associated with the free mitochondrial population than the synaptic. This, however, conflicts with the evidence from other laboratories,[36,38] which suggests the opposite. It should, however, be stressed that these latter laboratories[36,38] did not purify synaptic mitochondria but measured mitochondrial enzymes in lysed synaptosomal fractions. The problems of measuring accurately the specific activity of a mitochondrial enzyme, particularly in the case of hexokinase, which is bimodally distributed, are considerable under these circumstances.

Studies have also been made on the enzymes of ketone body metabolism, 3-hydroxybutyrate dehydrogenase, 3-oxo-acid transferase, and acetoacetyl-CoA thiolase. Only minor variations are seen of the 3-oxo-acid transferase and the acetoacetyl-CoA thiolase between the various mitochondrial populations except that the light synaptic mitochondrial fraction has a 3-oxo-acid transferase activity twice as high as the other populations. However, the 3-hydroxybutyrate dehydrogenase, which has a lower activity than either the 3-oxo-acid transferase or acetoacetyl-CoA thiolase, shows considerable variation between the populations, with the free mitochondrial activity being one-half or one-third of the synaptic activity. There is some evidence to suggest that the brain enzyme may be a variant from that present in other tissues such as the liver.[54] Although the brain is not ketogenic, no 3-hydroxy-3-methylglutaryl-CoA lyase being detectable,[30,55] a mitochondrial 3-hydroxy-3-methylglutaryl-CoA synthase and reductase have been reported.[30,55] There does not appear to be any major difference in the specific activity of the latter enzyme in the various mitochondrial populations.

Substrate specificity studies have been carried out on the aromatic aminotransferase activity of the rat brain aspartate aminotransferase, which is bimodally distributed.[56] The mitochondrial form of aspartate aminotransferase is mainly active towards tyrosine and phenylalanine with 2-oxoglutarate. There is also a fatty acid elongation system present in brain mitochondria as well as the microsomal fraction. This is primarily active towards fatty acyl-CoAs of C_{16} length (10 nmol/mg protein per h) but will also interact with higher chain lengths but at a slower rate.[57] This enzyme activity is particularly high in the developing brain but levels off in the adult. More recently, Srere and his colleagues,[58] using two-dimensional polyacrylamide gel electrophoresis to analyze the enzymatic protein composition of mitochondrial systems, have compared "free" brain mitochondrial preparations with those from other tissues. In the case of free brain mitochondria, 120–150 polypeptide components were detectable, among which glutamate dehydrogenase, glutamate oxaloacetate transaminase, pyruvate carboxylase, pyruvate dehydrogenase, citrate synthase, fumarase, and 2-oxoglutarate dehydrogenase could be identified. Comparison with other tissues suggests that brain mitochondrial pyruvate carboxylase content was intermediate between those of liver and muscle, whereas the brain citrate synthase content was higher than that in other tissues. Comparisons of the different brain mitochondrial populations by this technique would be valuable.

2.2.3. Mitochondrially Bound Kinases

For many years brain preparations were reported to possess a particulate bound hexokinase activity. There has been considerable controversy over whether this bound hexokinase represented a true reflection of the brain cell or whether it was a preparative artifact of the mitochondrial preparations employed. Over the last 10 years it has become increasingly clear that brain cells possess not only a mitochondrially bound hexokinase but possibly other bound kinase enzymes as well. Furthermore, this situation may be an important part of the control and efficiency of energy transfer that are necessary requirements for normal brain function.

The studies of Basford and his co-workers[59] using immunohistochemical techniques with a purified antibody to hexokinase showed this enzyme to be bound to mitochondria in brain cortex slices. Further studies[60] involving the removal and isolation of the outer mitochondrial membrane of purified brain mitochondria suggested that more than half of the total hexokinase was located on the outer mitochondrial membrane in the brain cortex. However, other work[61] localized the hexokinase on the outside of the inner membrane. What is generally agreed, however, is that the hexokinase is bound on the cytosolic side of the atractylate-sensitive barrier of the brain mitochondria,[20,61] since the production of glucose-6-phosphate by the mitochondrially bound hexokinase in the presence of glucose + ADP and inorganic phosphate is sensitive to both oligomycin and atractylate. Subsequent work from the Wilson group[62] has also localized a specific hexokinase binding protein on the outer mitochondrial membrane.

Wilson[36] reported, on the basis of studies with sucrose density gradients, that synaptosomal mitochondria have a relatively high hexokinase activity as compared to that of the free mitochondria, and similar conclusions have been reached on the basis of developmental studies.[38] However, studies by Booth[29] (see Table III) have suggested that the free mitochondria may have a slightly higher activity of bound hexokinase than the synaptic fraction. More recent studies by Wilson's group using immunofluorescence and histochemical techniques[63] in a regional study of the localization of hexokinase in rat brain have lead to the view that neuronal and glial cells may contain variable amounts of hexokinase. Thus, their previous suggestion that, generally, there is a higher hexokinase activity in synaptic mitochondria may not be true for all regions. This is in accord with recent investigations on the hexokinase bound to mitochondria purified from specific brain regions (cortex, striatum, and medulla). In all cases, there were comparable specific activities of bound hexokinase associated with both synaptic and nonsynaptically derived mitochondria.[143] There have been a number of studies that concern the ability of certain agents and/or conditions to solubilize the mitochondrially bound hexokinase and the relevance that this may have for the control of metabolism. Hockman and Sacktor[64] showed that ATP generated during the course of oxidative phosphorylation caused the release of mitochondrially bound hexokinase and that in brain the solubilized hexokinase had an apparent K_i for glucose-6-phosphate severalfold lower than the bound enzyme. Glucose-6-phosphate itself may also act to release bound hexokinase,[20,64] a factor that may be important in the regulation of glycolytic flux in the brain.[20] Brain mitochondrial hexokinase released by glucose-6-phosphate rebinds to mitochondrial outer membrane fractions in the presence of Mg^{2+} but not to inner mitochondrial membranes or to other membrane fragments, e.g., microsomes or myelin.[64]

Further studies on the binding of hexokinase to the mitochondrial membrane in the presence of various cations suggest that the interaction between the two is primarily electrostatic and that neutral salts, e.g., K^+, Na^+, or Li^+, will promote rebinding at concentrations <0.2 M but will cause solubilization of the membrane bound enzyme at concentrations >0.2 M.[65] Further support for the electrostatic interaction has been obtained by studying the effects of free fatty acids on this system, since these cause solubilization of the hexokinase,[66] as do certain lipophilic drugs and anaesthetics such as barbiturates, which cause hexokinase release both *in vitro* and *in vivo*.[67] It has been found in experimental diabetes that there is an increase in cytosolic hexokinase that is reversible by administration of insulin.[68] Further studies on the binding of hexokinase to brain mitochondria have suggested that there is a specific binding protein associated with the mitochondrially bound enzyme.[69] A protein of 31,000 daltons has been isolated from mitochondrial outer membranes that, when reincorporated into lipid vesicles, confers on the vesicles the ability to bind hexokinase in a glucose-6-phosphate-dependent fashion.[62] A number of studies on other metabolic systems have suggested that *in vivo* an equilibrium between soluble and mitochondrially bound hexokinase may exist, which is responsive to the energy state of the cell and could be important in the energy homeostasis of the brain.[20,70,71]

Evidence exists also for the existence of a mitochondrially bound creatine kinase on brain mitochondria,[53,72] although this only represents about 5% of the total cell creatine kinase[53] as compared, for example, with the 80% of the total hexokinase bound to the adult brain mitochondria. Furthermore, the bound creatine kinase is not solubilized by increasing the ionic strength of the medium , as happens in the case of the hexokinase.[53] Reports have also been made that phosphofructokinase is bound to the mitochondrial membrane in brain and that this binding may be enhanced by ADP[73] and that adenylate kinase is also associated with the mitochondrial membranes.[2,71,72]

3. *INTEGRATED METABOLISM OF INTACT BRAIN MITOCHONDRIA*

3.1. *Introduction*

Mitochondria obtained from nervous tissue are, in many ways, similar to those prepared from other tissues, as can be seen from the previous section. Because of factors outlined above, it is probably not worthwhile to use brain mitochondrial preparations to examine general problems of mitochondrial biophysics and biochemistry. Conversely, it is also not always justifiable to use a more easily obtained mitochondrial preparation, e.g., from rat liver, to study a condition or drug that primarily interacts with the CNS. This has been done frequently in the past and poses serious questions of the relevance of such studies to the problem being investigated. In such cases, comparative studies on preparations from different organs would be more useful.

However, the very factors that render biochemical studies of brain mitochondria difficult also impart a uniqueness to such studies so that they cannot be duplicated by using mitochondria from other sources. The heterogeneous nature of the CNS is a fact that must be taken into account in all neurobiological studies. The brain is composed of heterogeneous structures, each of which contains heterogeneous populations of neurons and glial cells. At the subcellular level, heterogeneity also abounds. For example, the synaptic apparatus is quite different in its metabolic machinery from the perikaryon, and the same is probably true of other processes such as dendrites. The process of glial cells, e.g., astrocytic end feet, are also highly differentiated and no doubt perform different metabolic functions than their corresponding cell bodies.

The biochemical heterogeneity of the CNS has been amply documented and reviewed.[74–77] It is not surprising, then, that mitochondria isolated from brain tissue also evidence a heterogeneity. This is most easily observed in the distributions of specific mitochondrial enzymes on analytical ultracentrifugation of subcellular preparations. Van den Berg and his colleagues have shown a disparate distribution of GABA transaminase, glutamate dehydrogenase, and short-chain fatty acyl-CoA synthetase as compared to NAD^+-dependent isocitrate dehydrogenase and monoamine oxidase in such preparations,[4,6] which are in general accord with the proposed models of glutamate and GABA metabolism in brain.[5,78] Similar measurements of the activity of tricarboxylate cycle and related enzymes have been done on Ficoll-gradient-prepared mitochondria—so called "free" or nonsynaptic mitochondria and synaptic mitochondrial fractions from synaptosomal preparations.[9,79,80] These latter studies also indicate a "compartmentation" of enzymes involved in carbohydrate and amino acid metabolism.

As has been pointed out previously,[79] one must be cautious with interpretation of this type of data, which usually involve measurements of maximal enzyme activities. The control of actual metabolism within the mitochondria can be quite complex in that it is regulated by such factors as substrate availability and the phosphate and redox potentials. Therefore, conclusions based on optimal enzyme activities may not reflect the true metabolic functions of these organelles *in vitro* or *in vivo*.

In recent years, studies have been done on the metabolism of substrates important to brain function in intact, metabolically active brain mitochondria. The following sections review these investigations.

3.2. Pyruvate and Ketone Body Metabolism

The central role of aerobic glycolysis as the prime energy source in adult CNS is reflected in the fact shown in Table I that under normal physiological conditions pyruvate is the best substrate for oxygen utilization by brain mitochondria. However, under certain conditions such as during early development or in adult brain following prolonged starvation, other substrates such as ketone bodies may also be utilized. This section deals with various aspects of metabolism of these substrates by brain mitochondria.

3.2.1. Transport

It has been generally accepted only recently that some mitochondria have a specific translocase for pyruvate transport.[81] This has been demonstrated in isolated rat brain mitochondria.[82] Moreover, in brain mitochondria, 3-hydroxybutyrate and acetoacetate seem to be transported by the same carrier system. The latter point is deduced from the fact that the pyruvate translocase inhibitor, α-cyanocinnamate, inhibits both the oxidation[83,84] and the actual uptake[82] of both pyruvate and the ketone bodies by rat brain mitochondria. It has been suggested that the ability of the mitochondria to accumulate pyruvate against a concentration gradient may allow pyruvate oxidation to proceed at a high enough rate to account for the observed glycolytic rate of rat brain.[82] The latter authors also proposed that this system allows ketone body oxidation to occur when pyruvate availability or utilization is limiting (e.g., starvation or in young animals). In the developing brain, when both ketone bodies and pyruvate are contributory to energy provision,[42] it has been shown that pyruvate utilization may be inhibited by the presence of 3-hydroxybutyrate,[85] a phenomenon that may be mediated in part by competition at the level of the translocase. Phenylpyruvate and α-ketoiscaproate (keto acids that accumulate in phenylketonuria and maple syrup urine disease, respectively) have also been found to be inhibitors of the pyruvate/3-hydroxybutyrate translocase[82] at concentrations that have no direct effect on pyruvate dehydrogenase activity. This may explain the known inhibition of pyruvate metabolism by these compounds and may be relevant to the mechanism of the disease processes involved.

3.2.2. Pyruvate Metabolism and Pyruvate Dehydrogenase Activity

Although some pyruvate can be metabolized by brain mitochondria via pyruvate carboxylase,[45,46] the overwhelming pathway is oxidation via pyruvate dehydrogenase and the tricarboxylate cycle. In isolated rat brain mitochondria, exogenous malate appears to be mandatory for maximal pyruvate oxidation.[10,31] Presumably, this allows optimal levels of oxalacetate to drive citrate synthesis. Metabolism of pyruvate by rabbit brain mitochondria does not appear to have this exogenous malate requirement.[86] Oxidations of pyruvate/malate are comparable in "free" and synaptic mitochondria.[9] The activity and regulation of the pyruvate dehydrogenase complex in isolated rat brain mitochondria is similar to that observed in other mammalian tissues, as described in Section 2.2.2. In summary, pyruvate metabolism in brain mitochondria appears to be controlled at the levels of the translocase, pyruvate dehydrogenase, and (in some cases) oxalacetate availability.

3.3. Glutamate Transport, Metabolism, and Synthesis

3.3.1. Transport

There are several proposed mitochondrial translocators for glutamate, three of which are the glutamate–OH^- antiporter, the glutamate–aspartate antiporter, and the glutamine–glutamate antiporter (see ref. 87 for discussion). Brand and Chappell[88] initially reported that brain and heart mitochondria, unlike liver mitochondria, possess only the glutamate–aspartate translocator. Subsequent studies by other groups have indicated that the glutamate–OH^- antiporter also exists in brain mitochondria,[87,89,90] but with the latter system working at a much slower rate than the glutamate–aspartate translocator.[87] The studies using the swelling technique for anion transport[89] also provided evidence for the presence of the malate–phosphate and succinate–phosphate translocases. No evidence has been found of a glutamine–glutamate translocator.[87,88]

3.3.2. Metabolism of Glutamate

Glutamate, whether presented exogenously to brain mitochondria[27,32,91] or endogenously via glutamine and glutaminase,[32] is primarily metabolized via transamination rather than oxidative deamination. This may reflect the fact that the glutamate–aspartate antiporter is more rapid in these mitochondria than the glutamate–OH^- system.[87] It also explains the kinetics observed for glutamate oxidation in the presence of malate.[27] In both synaptic and "free" mitochondria, glutamate oxidation in the presence of malate was biphasic, one system with an apparent K_m for glutamate of 0.25 mM, the other 1.6 mM. Aspartate production followed a single kinetic process with an apparant K_m for glutamate of 2 mM. This is close to the reported K_m of glutamate for the glutamate–aspartate antiporter.[90] Thus, the low-K_m system can perhaps be equated to the non-aspartate-mediated glutamate transport.

3.3.3. Synthesis of Glutamate by Brain Mitochondria

The high maximal activities of glutamate dehydrogenase (GDH) in brain and its distribution among various subpopulations of brain mitochondria (see Section 2.2.2) have generated some studies into glutamate synthesis via reductive amination of 2-oxoglutarate in brain mitochondria. Nonsynaptic mitochondria were found to synthesize glutamate via GDH at twice the rate of synaptic mitochondria when pyruvate was the precursor for endogenous 2-oxoglutarate but with similar rates when exogenous 2-oxoglutarate was present.[92] This is in spite of the 70% higher maximal activity of GDH in the synaptic mitochondria. In all cases, however, the synthesis of glutamate via reductive amination of 2-oxoglutarate in intact preparations was <10% of the available GDH activity[27,92] Glutamate dehydrogenase is a highly regulated enzyme, and its activity *in situ* would be controlled by many factors. Indeed, the studies with isolated rat brain mitochondria cast into doubt the use of the GDH redox couple in determining brain mitochondrial redox states in whole tissue (see ref.

92 for discussion). In any case, the role of GDH in glutamate metabolism in the brain is still unclear.

3.4. Transport and Metabolism of 4-Aminobutyrate (GABA)

The mitochondrial enzyme 4-aminobutyrate:2-oxoglutarate aminotransferase (GABA transaminase) is primarily localized in the inner membrane (not matrix space) of brain mitochondria.[93] Although the distribution of this enzyme in rat brain mitochondria appears to be heterogeneous in analytical ultracentrifugation studies, bulk preparations of "free" and "synaptic" mitochondria both contain transaminase activity (see Section 2.2.2.). There is, perhaps, greater activity in the "free" mitochondria. Succinic semialdehyde dehydrogenase, on the other hand, appears to be twice as active in "free" as in the "synaptic" mitochondria.[94] This is in accordance with other subcellular distribution studies with rat brain, which show a major enrichment of this oxidizing enzyme in "free" mitochondria compared to the synaptosomal fraction.[95] It is generally conceded that mitochondrial oxidation of GABA is controlled by the activity of the transaminase, not the dehydrogenase,[51,94] but it may be that this enzyme could play a regulatory role in some mitochondria.

3.4.1. Transport of GABA

There is, apparently, no specific transport system for GABA.[94,96] It probably enters as a neutral species at rates that are able to sustain maximal activity of the transaminase–dehydrogenase couple. This is consistent with the lack of latency of transaminase activity in brain mitochondria[96] and the accessibility it possesses because of its localization in the inner membrane.[93]

3.4.2. Metabolism of GABA

GABA can be metabolized by intact brain mitochondria with net oxygen uptake[51] and concomitant formation of glutamate (and aspartate).[51,94] The K_m for GABA for its oxidation, ~1 mM,[51] is comparable to the K_m for the GABA transaminase[94] in these same mitochondria. Oxygen uptake is tightly coupled to the formation of glutamate and aspartate; the ratio of oxygen utilization to rate of amino acid formation is close to the theoretical value of 3 when exogenous malate is supplied, indicating complete oxidation to oxaloacetate of the carbon structure of the succinic semialdehyde generated from the transaminase.[51] Approximately 50% more glutamate is found from GABA oxidation in "free" compared to synaptic mitochondria.[94] GABA oxidation apparently interacts directly with the normal tricarboxylate cycle of brain mitochondria since it also decreases the rate of substrate level phosphorylation.[97]

3.5. Mitochondrial/Cytosolic Transport Systems

In addition to the various translocases described above for mitochondrial metabolism of pyruvate, ketone bodies, and amino acids, a number of other mitochondrial/cytoplasmic shuttles have been studied with brain mitochondria. These transport systems are important to the integrative functions of brain cells.

3.5.1. Citrate Transport

Although brain mitochondria do not oxidize citrate as well as pyruvate (see Table I), nevertheless a citrate–malate translocase has been demonstrated[98] in nonsynaptic mitochondria that exhibits sensitivity to the classical inhibitors of this system, butylmalonate and benzene-1,2,3-tricarboxylate. Further studies[99] have demonstrated efflux of citrate from brain mitochondria that is malate and energy dependent. In addition, with certain substrates, and under some conditions, synaptic mitochondria show significantly lower rates of citrate efflux than do nonsynaptic mitochondria.[99]

3.5.2. Redox Shuttles

Of great importance are the shuttle systems by which reducing potential may be transferred from the cytosol to the mitochondrial compartment and *vice versa*; of these, two are likely to be operative in the brain: the malate–aspartate (or Borst) shuttle and the α-glycerophosphate shuttle (see ref. 100 for review). Both of these have been studied by isolation of brain mitochondria and the reconstitution of the cytosolic part of the system to study a functional shuttle.[88,90] It is clear that, in particular, the malate–aspartate shuttle may play an important role in brain energy metabolism.

3.5.3. Problem of Acetylcholine Synthesis

Studies in mammalian systems have shown that aerobic glycolysis is the source of the acetyl moiety of acetylcholine.[101] Since acetyl CoA does not readily traverse the mitochondrial membrane, a central question is which mitochondrial/cytosolic carbon transport system is responsible for the cytoplasmic acetyl-CoA used for acetylcholine synthesis in cholinergic terminals.[102–105] Efflux of citrate and cytosolic acetyl-CoA formation via ATP:citrate lyase is usually considered to be the prime mechanism for this process.[106] Citrate lyase is active in the cytoplasm of synaptosomal fractions,[104] and, as described in Section 3.5.1., citrate is transported by a specific carrier in brain mitochondria. Although some studies have indicated that this system could operate at rates consistent with acetylcholine synthesis,[107] others have suggested it may not be the only system operative.[102,103,105] N-Acetyl-L-aspartate formation by rat brain mitochondria and its efflux via the dicarboxylic acid translocase have been demonstrated and suggested to be an alternative route of generating acetyl-CoA in the cytoplasm.[102,108] Studies using brain slices and synaptosomal preparations[103,105] have indicated that citrate is not the sole intermediate between pyruvate oxidation and the acetyl moiety of acetylcholine. On the basis of these studies, it was suggested that acetylcarnitine and N-acetylaspartate might also be involved.

3.6. Calcium Transport in Brain Mitochondria

The transport of Ca^{2+} by various brain preparations and its importance in regulating such diverse functions as protein phosphorylation and exocytosis have been amply reviewed elsewhere in this series. Mitochondria generally show energy-dependent uptake (and storage systems) of calcium ion.[14,109] It has been suggested that Ca^{2+} accumulation by rat brain mitochondria is different from that found in mitochondria from rat liver or kidney; the affinity appears to be somewhat higher, as is the total number of uptake sites.[110] Furthermore, the latter authors reported that electron microscopy indicated that all of the Ca^{2+} was retained on the inner membrane. It was remarked that this prevents massive loading and resultant swelling, which in other mitochondria usually result in Ca^{2+}-induced membrane disintegration.

Others have also reported comparative differences in the uptake of Ca^{2+} by mitochondria. Crompton *et al.*[111] suggested that heart, brain, and adrenal cortical mitochondria have in common a capacity to catalyze a recycling of Ca^{2+}, Na^{+}, and H^{+} across the inner membrane. This appeared as a high rate of Na^{+}-induced Ca^{2+} efflux. These authors further proposed that this recycling may be related to the functions of these particular tissues in muscle contraction and neuro- or homorosecretion. The possibility that synaptic mitochondria play a role in the Ca^{2+}-stimulated efflux of neurotransmitters is an intriguing one,[112] but the evidence is not entirely convincing.[113]

4. DEVELOPMENTAL AND AGING STUDIES ON BRAIN MITOCHONDRIA

4.1. Developmental Studies

Interest in developmental studies on brain mitochondrial activity has mainly centered around facets of energy metabolism in view of the transition in the utilization of substrate fuels from the developing to adult brain. It is now well established in the rat[42,114] and human[115] that the young developing brain utilizes a mixture of glucose and ketone bodies, whereas the adult brain has an obligatory requirement for glucose except under conditions of long-term starvation. Baquer *et al.*,[116] in an extensive study on enzyme development during the first 20 days of life in rats, showed a closely coordinated development of enzyme activity between enzymes of glycolysis and the tricarboxylate cycle, particularly hexokinase and citrate synthase. However, enzymes associated with glutamate metabolism, notably glutamate–oxaloacetate transaminase, glutamate dehydrogenase, and glutamine synthetase, although their activity increased as the brain developed, did not show a close correlation with the tricarboxylate cycle enzymes. Previous studies[117,118] had shown that glutamate dehydrogenase developed during the period from 8 days after birth to 21 days, when the adult activity was attained. Furthermore, a study on synaptic and nonsynaptic mitochondria from 15-day-old and adult rats[38] has suggested that certain enzymes, notably glutaminase and succinate dehydrogenase, may develop differently in the two types of mitochondria.

Of particular interest has been the development of pyruvate metabolism, since earlier reports[42] suggested on the basis of studies with rat brain homogenates that considerable quantities of lactate were effluxed from the young rat brain. Cremer and Teal[119] showed that activity of pyruvate dehydrogenase increased fivefold from 2 days to 20 days after birth. This observation was further extended[20] by studies on purified rat brain mitochondria from developing animals 1 to 30 days old: it was shown that at about 5 days only 5–10% of the adult activity of the pyruvate dehydrogenase was present. The major rate of increase in pyruvate dehydrogenase activity occurred in the late suckling period (10–21 days after birth). The pattern of pyruvate dehydrogenase development was also paralleled by the hexokinase development, the major increase of which occurred in the mitochondrially bound form.[20]

Studies on other tricarboxylate cycle enzymes suggest a similar pattern of development (i.e., an increase of severalfold during development to adult values), but in the case of citrate synthase the maximal rate of enzyme activity increase occurs slightly earlier than that of pyruvate dehydrogenase.[20] Similar observations to those previously mentioned on pyruvate dehydrogenase but assayed in rat brain homogenates rather than purified mitochondria have recently been reported.[114] There have also been reports on the development of other enzymes of the citric acid cycle, e.g., aconitase, NAD^+–malate dehydrogenase, and the transaminases, aspartate and alanine aminotransferase.[116,120] All of these show increases of three- to tenfold during the period 10–35 days after birth in the rat. In a crude rat brain mitochondrial preparation, similar increases have been reported[120] for phosphoenolpyruvate carboxykinase and pyruvate carboxylase. However, in the latter case, in studies utilizing a different assay procedure,[47] it has been shown that although the activity of the brain pyruvate carboxylase increased some tenfold over the period 10–21 days, the activity subsequently fell to approximately half the maximal level in the adult. 2-Oxoglutarate dehydrogenase has also been shown to increase its activity during development, increasing from 0.2 unit/g at 2 days and reaching the adult value (~0.9 unit/g) at the end of the third postnatal week.[121] This enzyme has been shown to be inhibited by the branched-chain 2-oxoacids (2-oxo-4-methylvalerate, 2-oxo-3-methylvalerate, 2-oxo-isovalerate), which may be related to the pathogenesis of maple syrup urine disease.

Similar studies on the development of mitochondrially bound hexokinase, pyruvate dehydrogenase, and citrate synthase have also been carried out on the guinea pig brain.[122] In all cases, it was found that these enzymes were already present in the newborn guinea pig at a level of activity closely approaching that of the adult. Thus, the newborn guinea pig brain is potentially capable of utilizing glucose via glycolysis and the citric acid cycle at almost the same rate as the adult. This contrasts with the rat, which needs to utilize ketone bodies as well as glucose during the early development of the brain. This also suggests that a fully active aerobic glycolysis is a necessary requirement for neurological competence, since the guinea pig (a precocial species) is born neurologically mature, in contrast to the rat (a nonprecocial species), which does not show signs of neurological maturity until the latter end of the suckling period.[122]

A number of studies have been carried out on the development of the mitochondrial enzymes of ketone body utilization in rat brain. A general pattern for these enzymes has emerged: they develop during the suckling period, reach their optimal activities at or around weaning (21 days after birth in the rat), and then decline in activity in the adult. Page *et al.*[117] reported that 3-hydroxybutyrate dehydrogenase and 3-oxoacid transferase were approximately two-thirds of the adult activity just after birth, rising to a value that was three times the adult value at weaning, followed by a decline to the adult value. Middleton[123] showed that the acetoacetyl-CoA thiolase enzyme of rat brain was bimodally distributed and that whereas the cytosolic form gradually declined from a specific activity of 4 units/g at birth to 1.3 units/g in the adult, the mitochondrial form showed a similar profile to the other enzymes of ketone body utilization, having a specific activity of 1 unit/g at birth, 5 units/g at 25 days, and 2 units/g in the adult. Furthermore, Williamson *et al.*[124] were unable to detect any reversal in the decline in activity of these enzymes in the adult by starvation or fat-fed diets or a diabetic condition.

The developmental patterns of the enzymes of ketone body utilization have been recently confirmed[122] in the rat brain with purified brain mitochondrial preparations. However, similar studies in the guinea pig brain[122] have shown that these enzymes are very low in activity and do not change significantly from birth to adulthood. Metabolic studies using mitochondria derived from developing rat brain have also been carried out and have shown that these mitochondria may oxidize acetoacetate[84] and also that 3-hydroxybutyrate may inhibit the utilization of pyruvate.[85] Both of these studies imply the presence of a monocarboxylate translocase in neonatal brain mitochondria, which may transport 3-hydroxybutyrate, acetoacetate, and pyruvate (see Section 3.2.1). Evidence has also been provided for an acetoacetyl-CoA deacylase in rat brain mitochondria,[125] which more than doubles its activity during suckling and then maintains its activity into adulthood (1.1 units/g). However, 3-hydroxy-3-methylglutaryl-CoA synthase is bimodally distributed,[105] and whereas there is little change in activity up until weaning, after that time the mitochondrial enzyme approximately doubles in activity, but the cytosolic form decreases by a similar amount. The ability of neonatal brain mitochondria to efflux citrate and acetoacetate when utilizing different substrates has been studied as a function of development and related to the ability of glucose or ketone bodies to act as precursors of acetyl groups in the cytosol.[99]

The effect of hypothyroidism has been studied in the development of ketone-body-utilizing enzymes.[126] This state appears to retard the development of these enzymes, which is normally seen in the neonatal rat brain. Creatine kinase has also been shown to develop in a pattern very similar to that exhibited by the total brain hexokinase[53] although, as the mitochondrial form represents only 5% of the total, its activity probably resides mainly in the cytosolic form. Equally, the alanine aminotransferase in brain mitochondria does not appear to change during development, but the cytosolic form increases.[127] The effects of copper deficiency[128] and inorganic lead[129] have also been investigated on neonatal rat brain mitochondria. Copper deficiency led to a generalized decrease in oxidation rates characterized by a marked deficiency in cytochromes a_1, a_3, and cytochrome oxidase. Inorganic lead showed marked disturbances of mitochondrial function from several brain regions characterized by a decreased respiratory control and cytochrome oxidase activity.

Mitochondrial synthesis of macromolecules including DNA, RNA, and proteins has been studied in several brain regions as a function of development. In particular, purified mitochondria from cerebral hemispheres, cerebellum, and brainstem have been investigated in 10-day and 30-day-old brains.[130,131] The results indicate that DNA and protein synthesis in the 10-day-old rat brain is 70% higher than in the 30-day-old brain in all regions, with very little change in the RNA synthesis rate. This suggests that there is a rapid biogenesis of the mitochondria during this development period.

4.2. Aging Studies

Although aging of the brain and, in particular, senile dementia represents a growing area of investigation, the number of papers concerning specifically mitochondrially related functions is relatively small. No doubt this is an area of expansion. Indeed, some studies on rats with defective lung function have suggested that under conditions in which the inspired oxygen is decreased the mitochondrial cytochrome redox state in the cerebral cortex of both normal adult and aged animals is affected similarly.[132] However, recent studies[133] on isolated synaptic and nonsynaptic mitochondria from 12- and 24-month-old rats have shown that a significant reduction in the state 3 mitochondrial respiration rate occurs with pyruvate and malate as substrates in the aged animals as compared to the normal adults. No changes occur in the activity of the pyruvate dehydrogenase complex under similar conditions. On the other hand, significant reductions occur in the activities of the ketone-body-utilizing enzymes, 3-hydroxybutyrate dehydrogenase, 3-oxoacid-CoA transferase, and acetoacetyl-CoA thiolase, when brain mitochondrial populations from 12- and 24-month-old rats are compared to 3-month-old animals, but significant reductions in the oxidation of DL-3-hydroxybutyrate occur only in the 24-month-old animals.

Regional brain studies on enzymes in aged animals have been investigated.[134,135] No changes in a number of brain regions were seen in the activities of hexokinase and pyruvate dehydrogenase when 3-month- and 2-year-old animals were compared.[135] In contrast, significant decreases in all areas of D-3-hydroxybutyrate dehydrogenase activity were observed. Furthermore, marked decreases were also recorded in the activity of aldolase, lactate dehydrogenase, citrate synthase, NAD–isocitrate dehydrogenase, and acetylcholinesterase.[135] Studies on 1-year-old animals showed that glutamate dehydrogenase and malate dehydrogenase also decreased in activity in certain brain regions on aging.[134] Surprisingly, however, these workers[134] reported that succinate dehydrogenase activity increased on aging in the frontal cortex. Clearly, further work is required to resolve these points in this very important area.

5. NEUROPATHOLOGICAL STUDIES WITH BRAIN MITOCHONDRIA

There are many disease states in which morphological alterations are seen in the mitochondria of brain on microscopic examination of autopsy or biopsy samples. In most cases it is not known whether these changes are primary or secondary to the disease process. In a very few of these cases, mitochondria have been isolated from the brains of experimental animals and examined. As described in Section 4.1, the sequelae of copper deficiency have been studied in mitochondria from developing rat brain.[128] Thyroid deficiency, which has profound effects on neurodevelopment, has also been studied with isolated mitochondria.[126,136,137] Oxidative phosphorylation was deleteriously altered by thyroid deficiency in rat liver, kidney, and brain preparations.[137] There were several differences among the organs studied that suggested that the effects were specific to the target tissue. The results are also a good illustration of the dictum that when a disease or deficiency state has multiple system effects, comparative studies are a necessity.

One pathological condition with profound CNS effects that has been extensively studied in experimental systems is that of ischemia/hypoxia.[138] Ginsberg *et al.*[139] have studied brain mitochondrial metabolism following bilateral cerebral ischemia in the gerbil. Respiration of these mitochondria appeared relatively resistant to irreversible impairment following the ischemic insult. Furthermore, alterations were observed in respiratory activity only after 1–2 h of ischemia, whereas significant cerebral edema was observed within 30 min of cerebral ischemia. Alterations in brain mitochondrial enzymes have been measured in mitochondrial and synaptosomal fractions of rat brain after postdecapitative normothermic ischemia of various duration.[140] Malate dehydrogenase and total NADH–cytochrome *c* reductase activity were decreased in mitochondria, as was acetylcholinesterase in synaptosomes. These enzyme decreases were variably affected by pretreatment with vincamine, trimetazidine, or suloctidil. Thus, although mitochondrial changes are observed in the hypoxic/ischemic condition, both microscopically and in isolated mitochondria, the relationship of these changes to the irreversible sequelae of stroke remains to be established. Similar studies may also elucidate the role of brain mitochondrial changes in hypoglycemic brain injury[141] and in patients dying from shock and trauma.[142]

REFERENCES

1. Hamberger, A., Blomstrand, C., and Lehninger, A. L., 1970, *J. Cell Biol.* **45:**221–234.
2. Cotman, C., Brown, D. H., Farrell, B. W., and Anderson, N. G., 1970, *Arch. Biochem. Biophys.* **136:**436–447.
3. Neidle, A., Van den Berg, C. J., and Grynbaum, A., 1969, *J. Neurochem.* **16:**225–234.
4. Reijnierse, G. L., Veldstra, H., and Van den Berg, C. J., 1975, *Biochem. J.* **152:**469–475.
5. Van den Berg, C. J., Matheson, D. F., Ronda, G., Reijnierse, G. L., Blokhuis, G. G., Kroon, M. C., Clarke, D. D., and Garfinkel, D., 1975, *Metabolic Compartmentation and Neurotransmission* (S. Berl, D. D. Clarke, and D. Schneider, eds.), Plenum Press, New York, pp. 515–543.

6. Reijnierse, G. L., Veldstra, H., and Van den Berg, C. J., 1975, *Biochem. J.* **152**:477–484.
7. Blokhuis, G. G., and Veldstra, H., 1970, *Febs. Lett.* **11**:197–199.
8. Churchill, L., and Cotman, C. W., 1973, *Neurobiology* **3**:311–319.
9. Lai, J. C., and Clark, J. B., 1976, *Biochem. J.* **154**:423–432.
10. Clark, J. B., and Nicklas, W. J., 1970, *J. Biol. Chem.* **245**:4724–4731.
11. Lai, J. C., Walsh, J. M., Dennis, S. C., and Clark, J. B., 1977, *J. Neurochem.* **28**:625–631.
12. Lai, J. C., and Clark, J. B., 1979, *Methods Enzymol.* **55**:51–60.
13. Manthorpe, C. M., Jr., Nettleton, D. O., and Wilson, J. E., 1976, *J. Neurochem.* **27**:1547–1549.
14. Nicholls, D. G., 1978, *Biochem. J.* **170**:511–522.
15. Scott, I. D., and Nicholls, D. G., 1980, *Biochem. J.* **186**:21–33.
16. Booth, R. F., and Clark, J. B., 1978, *Biochem. J.* **176**:365–370.
17. Booth, R. F., and Clark, J. B., 1979, *Febs. Lett.* **107**:387–392.
18. Holtzman, D., and Moore, C. L., 1973, *Biol. Neonate* **22**:230–242.
19. Holtzman, D., and Moore, C. L., 1975, *J. Neurochem.* **24**:1011–1015.
20. Land, J. M., Booth, R. F., Berger, R., and Clark, J. B., 1977, *Biochem. J.* **164**:339–348.
21. Watanabe, H., 1971, *J. Biochem. (Tokyo)* **69**:275–281.
22. Zanetta, J. P., Reeber, A., Ghandour, M. S., Vincendon, G., and Gombos, G., 1977, *Brain Res.* **125**:386–389.
23. Krusius, T., Finne, J., Margolis, R. U., and Margolis, R. K., 1978, *Biochemistry* **17**:3849–3854.
24. Mahadik, S. P., Korenovsky, A., and Rapport, M. M., 1976, *Anal. Biochem.* **76**:615–633.
25. Gray, E. G., Jones, D. H., and Barron, J., 1979, *J. Neurocytol.* **8**:675–685.
26. Land, J. M., 1974, Ph.D. Thesis, University of London, London.
27. Dennis, S. C., Lai, J. C., and Clark, J. B., 1977, *Biochem. J.* **164**:727–736.
28. Lai, J. C., 1975, Ph.D. Thesis, University of London, London.
29. Booth, R. F., 1978, Ph.D. Thesis, University of London, London.
30. Patel, T. B., 1978, Ph.D. Thesis, University of London, London.
31. Nicklas, W. J., Clark, J. B., and Williamson, J. R., 1971, *Biochem. J.* **123**:83–95.
32. Dennis, S. C., and Clark, J. B., 1977, *Biochem. J.* **168**:521–527.
33. Dennis, S. C., 1976, Ph.D. Thesis, University of London, London.
34. Balázs, R., Dahl, D., and Harwood, J., 1966, *J. Neurochem.* **13**:897–905.
35. Salganicoff, L., and Koeppe, R. E., 1968, *J. Biol. Chem.* **243**:3416–3420.
36. Wilson, J. E., 1972, *Arch. Biochem. Biophys.* **150**:96–104.
37. Gurd, J. W., Jones, L. R., Mahler, H. R., and Moore, W. J., 1974, *J. Neurochem.* **22**:281–290.
38. Dienel, G., Ryder, E., and Greengard, O., 1977, *Biochim. Biophys. Acta* **496**:484–494.
39. Booth, R. F., and Clark, J. B., 1978, *J. Neurochem.* **30**:1003–1008.
40. Jope, R., and Blass, J. P., 1975, *Biochem. J.* **150**:397–403.
41. Ngo, T. T., and Barbeau, A., 1978, *J. Neurochem.* **31**:69–75.
42. Cremer, J. E., and Heath, D. F., 1974, *Biochem. J.* **142**:527–544.
43. Magilen, G., Gordon, A., Au, A., and Diamond, I., 1981, *J. Neurochem.* **36**:1861–1864.
44. Browning, M., Baudry, M., Bennett, W. F., and Lynch, G., 1981, *J. Neurochem.* **36**:1932–1940.
45. Patel, M. S., and Tilghman, S. M., 1973, *Biochem. J.* **132**:185–192.
46. Patel, M. S., 1974, *J. Neurochem.* **22**:717–724.
47. Land, J. M., and Clark, J. B., 1975, *Normal and Pathological Development of Energy Metabolism* (F. A. Hommes and C. J. Van den Berg, eds.), Academic Press, London and New York, pp. 155–167.
48. Wilkin, G. P., Reijnierse, G. L., Johnson, A. L., and Balázs, R., 1979, *Brain Res.* **164**:153–163.
49. Buu, N. T., and Van Gelder, N. M., 1974, *Can. J. Physiol. Pharmacol.* **52**:674–680.
50. Tunnicliff, G., Ngo, T. T., Rojo-Ortega, J. M., and Barbeau, A., 1977, *Can. J. Biochem.* **55**:479–484.
51. Cunningham, J., Clarke, D. D., and Nicklas, W. J., 1980, *J. Neurochem.* **34**:197–202.
52. Jacobus, W. E., and Lehninger, A. L., 1973, *J. Biol. Chem.* **248**:4803–4810.

53. Booth, R. F., and Clark, J. B., 1978, *Biochem. J.* **170**:145–151.
54. Dombrowski, G. J., Cheung, G. P., and Swiatch, K. R., 1977, *Life Sci.* **21**:1821–1829.
55. Patel, T. B., and Clark, J. B., 1981, *J. Neurochem.* **36**:1281–1284.
56. King, S., and Phillips, A. T., 1978, *J. Neurochem.* **30**:1399–1407.
57. Morad, S., and Kishimoto, Y., 1978, *Arch. Biochem. Biophys.* **185**:300–306.
58. Henslee, J. G., and Srere, P. A., 1979, *J. Biol. Chem.* **254**:5488–5497.
59. Craven, P. A., and Basford, R. E., 1969, *Biochemistry* **8**:3520–3525.
60. Craven, P. A., Goldblatt, P. J., and Basford, R. E., 1969, *Biochemistry* **8**:3525–3532.
61. Vallejo, C. G., Marco, R., and Sebastian, J., 1970, *Eur. J. Biochem.* **14**:478–485.
62. Felgner, P. L., Messer, J. L., and Wilson, J. E., 1979, *J. Biol. Chem.* **254**:4946–4949.
63. Wilkin, G. P., and Wilson, J. E., 1977, *J. Neurochem.* **29**:1039–1051.
64. Hochman, M. S., and Sacktor, B., 1973, *Biochem. Biophys. Res. Commun.* **54**:1546–1553.
65. Felgner, P. L., and Wilson, J. E., 1977, *Arch. Biochem. Biophys.* **182**:282–294.
66. Domanska-Janik, K., Broniszewska-Ardelt, B., and Wroblewski, J. T., 1978, *J. Neurochem.* **30**:1157–1161.
67. Hanke, J., Hofeler, H., Krieclstein, J., and Wever, K., 1979, *Naunyn-Schmiedebergs Arch Pharmacol.* **307**:171–176.
68. Ouchi, M., Dohmoto, C., Kaminashi, T., Murakami, K., and Ishibashi, S., 1975, *Brain Res.* **98**:410–414.
69. Kurokawa, M., Kimura, J., Tokooka, S., and Ishibashi, S., 1979, *Brain Res.* **175**:169–173.
70. Knull, H. R., Taylor, W. F., and Wells, W. W., 1973, *J. Biol. Chem.* **248**:5414–5417.
71. Inui, M., and Ishibashi, S., 1979, *J. Biochem. (Tokyo)*, **85**:1151–1156.
72. Lapin, E. P., Maker, H. S., and Lehrer, G. M., 1974, *J. Neurochem.* **23**:465–469.
73. Craven, P. A., and Basford, R. E., 1974, *Biochim. Biophys. Acta* **354**:49–56.
74. Balázs, R., and Cremer, J. E. (eds.), 1973, *Metabolic Compartmentation in the Brain*, Macmillan, London.
75. Berl, S., Clarke, D. D., and Schneider, D. (eds.), 1975, *Metabolic Compartmentation and Neurotransmission*, Plenum Press, New York.
76. Fonnum, F. (ed.), 1978, *Amino Acids as Chemical Transmitters*, Plenum Press, New York.
77. Bradford, H. F. (ed.), 1982, *Neurotransmitter Interaction and Compartmentation*, Plenum Press, New York.
78. Garfinkel, D., 1970, *Brain Res.* **23**:387–406.
79. Lai, J. C., Walsh, J. M., Dennis, S. C., and Clark, J. B., 1975, *Metabolic Compartmentation and Neurotransmission* (S. Berl, D. D. Clarke, and D. Schneider, eds.), Plenum Press, New York, pp. 487–496.
80. Lai, J. C., and Clark, J. B., 1978, *Biochem. Soc. Trans.* **6**:993–995.
81. Halestrap, A. P., Scott, R. D., and Thomas, A. P., 1980, *Int. J. Biochem.* **11**:97–105.
82. Land, J. M., Mowbray, J., and Clark, J. B., 1976, *J. Neurochem.* **26**:823–830.
83. Land, J. M., and Clark, J. B., 1974, *Febs. Lett.* **44**:348–351.
84. Patel, T. B., Booth, R. F. G., and Clark, J. B., 1977, *J. Neurochem.* **29**:1151–1153.
85. Booth, R. F. G., and Clark, J. B., 1981, *J. Neurochem.* **37**:179–185.
86. Beck, D. P., Broyles, J. L., and Von Korff, R. W., 1977, *J. Neurochem.* **29**:487–493.
87. Dennis, S. C., Land, J. M., and Clark, J. B., 1976, *Biochem. J.* **156**:323–331.
88. Brand, M. D., and Chappell, J. B., 1974, *Biochem. J.* **140**:205–210.
89. Minn, A., Gayet, J., and Delome, P., 1975, *J. Neurochem.* **24**:149–156.
90. Minn, A., and Gayet, J., 1977, *J. Neurochem.* **29**:873–881.
91. Gayet, J., Minn, A., and Lehr, P., 1970, *Brain Res.* **18**:368–371.
92. Dennis, S. C., and Clark, J. B., 1978, *J. Neurochem.* **31**:673–680.
93. Schousboe, I., Bro, B., and Schousboe, A., 1979, *Biochem. J.* **162**:303–307.
94. Walsh, J. M., and Clark, J. B., 1976, *Biochem. J.* **160**:147–157.
95. Sims, K. L., and Davis, G. A., 1973, *Eur. J. Biochem.* **35**:450–453.
96. Brand, M. D., and Chappell, J. B., 1974, *J. Neurochem.* **22**:47–51.
97. Rodichok, L. D., and Albers, R. W., 1980, *J. Neurochem.* **34**:303–307.
98. Patel, M. S., 1975, *Brain Res.* **98**:607–611.
99. Patel, T. B., and Clark, J. B., 1981, *Biochim. Biophys. Acta* **677**:373–380.
100. Meijer, A. J., and Van Dam, K., 1974, *Biochim. Biophys. Acta* **346**:213–244.

101. Tucek, S., and Cheng, S. C., 1974, *J. Neurochem.* **22**:893–914.
102. Patel, T. B., and Clark, J. B., 1979, *Biochem. J.* **184**:539–546.
103. Gibson, G. E., and Shimada, M., 1980, *Biochem. Pharmacol.* **29**:167–174.
104. Hafalowska, U., and Ksiezak, H., 1976, *J. Neurochem.* **27**:813–815.
105. Clark, J. B., Booth, R. F. G., Harvey, S. A. K., Leong, S. F., and Patel, T. B., 1982, *Neurotransmitter Interaction and Compartmentation* (H. Bradford, ed.), Plenum Press, New York p. 431–460.
106. Sterling, G. H., and O'Neill, J. J., 1978, *J. Neurochem.* **31**:525–530.
107. Szeutowicz, A., Lysiak, W., and Angielski, S., 1977, *J. Neurochem.* **29**:375–378.
108. D'Adamo, A. F., Gidez, L. I., and Yatzu, F. M., 1968, *Exp. Brain Res.* **5**:267–273.
109. Carafoli, E., and Lehninger, A. L., 1971, *Biochem. J.* **122**:681–690.
110. Mela, L., and Wrobel-Kuhl, K., 1978, *Ann. N.Y. Acad. Sci.* **307**:242–245.
111. Crompton, M., Moser, R., Ludi, H., and Carafoli, E., 1978, *Eur. J. Biochem.* **82**:25–31.
112. Sandoval, M. E., 1980, *Brain Res.* **181**:357–367.
113. Blaustein, M. P., Ratzlaff, R. W., Kendrick, N. C., and Schweitzer, E. S., 1978, *J. Gen. Physiol.* **72**:15–41.
114. Stumpf, B., and Kraus, H., 1979, *Pediatr. Res.* **13**(1):585–590.
115. Kraus, A., Schlenker, S., and Schwedesky, D., 1974, *Hoppe-Seylers Z. Physiol. Chem.* **355**:164–170.
116. Baquer, N. Z., McLean, P., and Greenbaum, A. L., 1975, *Normal and Pathological Development of Energy Metabolism* (F. A. Hommes and C. J. Van den Berg, eds.), Academic Press, London, pp. 109–132.
117. Page, M. A., Krebs, H. G., and Williamson, D. H., 1971, *Biochem. J.* **121**:49–53.
118. Prosky, L., and O'Dell, R. G., 1972, *J. Neurochem.* **19**:1405–1407.
119. Cremer, J. E., and Teal, H. M., 1974, *Febs. Lett.* **39**:17–20.
120. Wilbur, D. O., and Patel, M. S., 1974, *J. Neurochem.* **22**:709–715.
121. Patel, M. S., 1974, *Biochem. J.* **144**:91–97.
122. Booth, R. F., Patel, T. B., and Clark, J. B., 1980, *J. Neurochem.* **34**:17–25.
123. Middleton, B., 1973, *Biochem. J.* **132**:731–737.
124. Williamson, D. H., Bates, M. W., Page, M. A., and Krebs, M. A., 1971, *Biochem. J.* **121**:41–47.
125. Patel, T. B., and Clark, J. B., 1978, *Biochem. J.* **176**:951–958.
126. Patel, M. S., 1979, *Biochem. J.* **184**:169–172.
127. Orlicky, J., Ruscak, M., Ruscakova, D., and Hager, H., 1979, *J. Neurochem.* **32**:1551–1558.
128. Prohaska, J., and Wells, W. W., 1975, *J. Neurochem.* **25**:221–228.
129. Holtzman, D., and Shen Hsu J., 1976, *Pediatr. Res.* **10**:70–75.
130. Gadaleta, M. N., Giuffrida, A. M., Renis, M., Serra, I., Del Prete, G., Geremia, E., and Saccone, C., 1979, *Neurochem. Res.* **4**:25–35.
131. Giuffrida, A. M., Gadaleta, M. N., Serra, I., Renis, M., Geremia, E., Del Prete, G., and Saccone, C., 1979, *Neurochem. Res.* **4**:37–52.
132. Sylvia, A. L., and Rosenthal, M., 1978, *Brain Res.* **146**:109–122.
133. Deshmukh, D. R., Owen, O. E., and Patel, M. S., 1980, *J. Neurochem.* **34**:1219–1224.
134. Ryder, E., 1980, *J. Neurochem.* **34**:1550–1552.
135. Leong, S. F., Lai, J. C., Lim, L., and Clark, J. B., 1981, *J. Neurochem.* **37**:1548–1556.
136. Sterling, K., Milch, P. O., Brenner, M. A., and Lazarus, J. H., 1977, *Science* **197**:996–999.
137. Katyare, S. S., Joshi, M. V., Fatterpaker, P., and Sreenivasan, A., 1977, *Arch. Biochem. Biophys.* **182**:155–163.
138. Maker, H. S., and Nicklas, W. J., 1978, *Extrapulmonary Manifestations of Respiratory Disease* (E. D. Robin, ed.), Marcel Dekker, New York, pp. 107–149.
139. Ginsberg, M. D., Mela, L., Wrobel-Kuhl, K., and Reivich, M., 1977, *Ann. Neurol.* **1**:519–527.
140. Villa, R. F., Benzi, G., and Curti, D., 1981, *Biochem. Pharmacol.* **30**:2399–2408.
141. Kalimo, H., Agardh, C. D., Olsson, Y., and Siesjo, B. K., 1980, *Acta Neuropathol.* **50**:43–52.
142. Cowley, R. A., Mergner, W. J., Fisher, R. S., Jones, R. T., and Trump, B. F., 1979, *Am. Surg.* **45**:255–269.
143. Leong, S. F., Lai, J. C. K., Lim, L., and Clark, J. B., 1984, *J. Neurochem.* (in press).

6

The Smooth Endoplasmic Reticulum

Gérard Alonso

1. INTRODUCTION

The concept of endoplasmic reticulum (ER) was introduced by electron microscopists to designate a well-developed system of reticular membrane occurring in the internal regions of most cells. This specific intracellular membranous system is particularly important in neurons, where it forms a complex network of intraneuronal channels isolated from the surrounding neuroplasm. Three main components have generally been attributed to this intracellular membranous compartment: the rough endoplasmic reticulum (RER), the Golgi apparatus, and the smooth endoplasmic reticulum (SER). The particular aspect of the neuronal ER results from its highly regionalized organization: whereas the RER and Golgi complexes are rigorously confined to the perikarya and the proximal part of the dendrites, the SER is present in all neuronal sections, especially all along the dendritic and axonal processes, which may extend over distances thousands of times greater than the cell body diameter.

A series of studies based on autoradiography and cell fractionation have clearly established that proteins as well as phospholipids are essentially synthetized in the RER.[1–3] The Golgi apparatus, an aggregation of smooth membrane saccules and their associated vesicles, was identified as an active site of post-translational modification—mainly by glycosylation, sulfatation, and proteolytic processing—of proteins previously synthetized in the RER.[4,5] Until now, however, little was known about the role of the SER in neuronal metabolism. Because of the SER's extensive distribution within the neuron and close association with the RER, the first hypothesis proposed was that it might constitute an adequate structure for the diffusion of substances from their sites of synthesis. Clearly, the perikaryal SER is now acknowledged to play such a role and is generally considered to be the preferential pathway for the transfer of material from the RER to the Golgi.[6,7] Recently, because of its anatomic

Gérard Alonso • Laboratoire de Neuroendocrinologie, Université de Montpellier II, 34100 Montpellier, France.

organization within neuronal processes, a similar role has been proposed for SER in the somatofugal transfer of macromolecules within dendrites and axons.

Because of the complex geometry of the branching organization of the dendrites and their partial autonomy for protein synthesis, fairly little information is available on the dendritic transport of macromolecules. On the other hand, a considerable body of literature has accumulated on macromolecule transport in the axonal processes.[8–10] The present chapter therefore deals essentially with recent developments relating to the structure of the SER and its role in axonal processes.

2. *STRUCTURAL ORGANIZATION OF AXONAL SMOOTH ENDOPLASMIC RETICULUM*

Axonal SER has been identified in all neurons so far investigated. It generally forms an extensive system of smooth membrane cisternae or elongated tubules visible all along the axon.[11,12] As long as observation of the SER was restricted to ultrathin sections, the general organization of this highly polymorphous organelle was hard to visualize in conventional electron microscopy. Recently, however, better insight into the three-dimensional organization of axonal SER was gained by the discovery of new staining techniques that are specific for the intracellular membrane system and permit observation of 0.5- to 2-μm-thick sections with both conventional (100 kV) and high-voltage (200–1000 kV) electron microscopes.

2.1. *Three-Dimensional Organization of Axonal Smooth Endoplasmic Reticulum*

Two main techniques have been developed for specific staining of the axonal SER of central and peripheral nerves. They consist of *en bloc* staining of the nervous tissue either with heavy metal impregnation[13] or with a cytochemical technique initially developed to stain the sarcoplasmic reticulum and that follows the procedure for demonstrating endogenous peroxidase activity.[14] Although the nature of those stainings is not yet known, they generally provide a good contrast of intraneural membrane systems and permit the three-dimensional organization of the axonal SER to be visualized by stereoscopic observation of thick sections.[14–16]

It was striking to observe that in the various types of neuron investigated this organization was fairly similar in appearance in the different parts of the axon, in which it displays the following characteristics:

1. In the axon hillock and the proximal parts of the axons, SER tubules form a complex, irregularly meshed, three-dimensional network whose organization is roughly comparable to that of the peripheral RER of the perikaryon. However, this type of network gradually disappears along the axonal route, and the SER tubules then tend to run parallel to the axonal axis.

2. All along the axon, the axonal SER forms a homogeneous system composed of variously dilated tubules, frequently interconnected by oblique anastomoses.
3. In the terminal axon sections, the SER tubules lose their polarity and again form a complex three-dimensional network extending into the whole axoplasm. Finally, terminals of these tubules develop into thinner tubules, which are rarely observed to be in direct contact with the terminal axolemma but are often closely connected with synaptic vesicles.

Furthermore, the study of thick sections of selectively stained material provided strong support for Palay's original suggestion[17] that axonal SER formed a continuous structure from the perikarya to the axon terminal. Although no direct continuity has been established with the Golgi apparatus, axonal SER has been shown to be connected to the peripheral RER of certain neuronal perikarya.[18] Moreover, stereoscopic examination of 1- to 2-μm-thick sections always led to the observation of continuous interconnected intra-axonal structures. This concept of anatomic continuity within the axonal SER may appear to contradict the findings of Kreutzberg and Gross[19] and Porter *et al.*,[20] who reported clear-cut discontinuities in this membrane system. However, the apparent discrepancy between the two series of observations may reflect methodological differences in the preparation of the materials. In this connection, the conventional tissue fixation and staining techniques used by these last two groups of authors have been shown to be less efficient than heavy metal impregnation techniques for preserving the structural integrity of axonal SER.[21]

2.2. *Relationships between Axonal Smooth Endoplasmic Reticulum and Other Axonal Structures*

In material treated by conventional ultrastructural techniques or selective staining, it has often been stressed that in a variety of neurons, axonal SER exhibits preferential relationships with various axonal structures.

2.2.1. Axolemma

In addition to the SER–axolemma relationships frequently observed,[11,12] SER tubules may develop, all along the axons, local specializations such as dilated cisternae or fenestrated subaxolemmal plates, which are either closely apposed to[22,23] or in direct contact with the axolemma.[15,24]

2.2.2. Mitochondria

Close relationships between mitochondria and axonal SER have often been observed in conventional ultrastructural studies.[25] Obviously, these associations are even more clearly detectable on thick sections of selectively stained tissues, where the axonal SER frequently appears as a meshed network enclosing the mitochondria.[14–16] However, whatever the technique, the most intriguing pictures are those showing direct connections between outer mitochondrial membranes and specific elements of the axonal SER.[25]

2.2.3. Secretory Vesicles

Secretory vesicles are visualized as dense-cored, membrane-delimited structures with a diameter of 100 to 200 nm. They are observed in a large variety of neurons and are generally assumed to concentrate specific secretory substances. Direct continuity between axonal SER and the limiting membranes of secretory vesicles has been described in several types of aminergic and peptidergic secretory neurons of both adult and embryonic rats.[16,21,26–28]

2.2.4. Synaptic Vesicles

Synaptic vesicles consist of small, electron-lucent membrane-bound structures (30 to 50 nm in diameter) essentially located in presynaptic axonal regions. It is generally agreed that they contain specific neurotransmitters and associated proteins, which are discharged by exocytosis during synaptic activity. Large amounts of similar organelles known as synapticlike vesicles are also seen in the perivascular regions of neurosecretory axonal endings, although their role is still controversial.

Continuity between axonal SER and synaptic vesicles has been reported in many types of neurons on both ultrathin sections of conventionally treated material[17,26,29,33] and thick sections of selectively impregnated tissue.[14,15,34] In neurosecretory neurons, synapticlike vesicles were also frequently observed to be directly connected with thin elements of axonal SER.[16,21]

2.2.5. Cytoskeleton

The axonal cytoskeleton is known to comprise three main elements: microtubules and neurofilaments, which form straight elongated structures parallel to the axonal axis; microfilaments, consisting mainly of actin filaments; and microtrabeculae, forming a matrix of spongelike material whose composition is still unknown.[35]

Axonal SER was, in fact, shown to establish close relationships with actin filaments[36,37] and with microtrabeculae, which themselves connect to microtubules or neurofilaments.[35]

3. SMOOTH ENDOPLASMIC RETICULUM AND AXONAL TRANSPORT

In all types of axons so far studied, three main types of transport have been distinguished, depending on their velocity and polarity[9]: fast anterograde transport (hundreds of millimeters per day), slow anterograde transport (less than 100 mm/day), and retrograde transport (50 to 250 mm/day). It is now universally admitted that whatever the variety of molecules transported, the different components of axonal transport are related to the motility of different cytological structures rather than to individual molecules.[38] Cytoskeletal materials were thus selectively associated with slow anterograde transport, and membranous cellular structures with fast axonal transport in both the orthograde and retrograde directions. Because of the widespread distribution of the SER throughout all the axonal processes, it may, indeed, be a good candidate for a vehicle in fast axonal transport.

3.1. Evidence for Smooth Endoplasmic Reticulum Participation in Axonal Transport

3.1.1. Anterograde Transport

The concept of possible axonal SER involvement in the mechanisms of fast transport first arose from morphological studies of axons after experimental blockade of the axonal flow. After sectioning or local compression of various nerves, the accumulation of polymorphous tubules and vesicles derived from the SER was observed at the proximal end of the axons.[15,39–41] The objection that such mechanical lesions might have caused local structural alterations was recently refuted by Tsukita and Ishikawa,[42] who clearly showed that SER-like structures accumulated on the proximal side of axons when the axonal flow had been interrupted by local cooling *in situ*.

A connection between axonal SER and fast anterograde transport was also strongly suggested by the results of autoradiographic studies in which SER structures were labeled by rapidly transported proteins.[15,34,43,44] Finally, it was shown that in experimentally induced neuropathy, labeled macromolecules carried by fast transport were retained in distal parts of the axons, at SER accumulation sites located beneath the axolemma.[45]

However, the concept that SER may be the essential vehicle of fast anterograde transport is far from being generally accepted. According to Schwartz[8]: "the existing evidence that membranes of the endoplasmic reticulum contain materials moved by fast transport is strong; the evidence that the reticulum mediates transport is weak." As suggested by this author, the visualization of labeled macromolecules in SER profiles might simply be related to the fact that radioactive materials have reached their final destination. In this connection, morphometric analysis of autoradiographs designed to detect rapidly transported radioactive proteins indicated that in several types of axons, the specific activity of the vesicles was higher than that of the SER.[46,47] The alternative hypothesis was therefore proposed that fast axonal transport of various materials would in fact imply the budding off of rapidly moving vesicles and their subsequent renewed fusion with a relatively static SER. Further, Ellisman and Lindsey[48] recently reported that unlike other axonal membrane structures, SER displayed no apparent movement between two local cooling points on the saphenous nerve.

A different interpretation of this last series of data may, however, be suggested: (1) a higher specific activity of axonal vesicles may be explained by the concentration in these vesicles of specific radioactive proteins transported by the SER; and (2) the absence of motility of axonal SER between two cooling points may indicate that the propulsion of this organelle depends on the integrity of its functional connection with the perikaryon. According to the numerous data, which certainly provided very strong arguments in favor of active anterograde propulsion of the axonal SER (see above), a more accurate view of the axonal transport of membrane-associated molecules might therefore be to consider that it consists of a rapid intra-axonal propelling of both the tubular system of SER and of vesicles, which may themselves bud off and fuse with the SER at various points along the axon.

3.1.2. Retrograde Transport

Several authors have stressed that retrograde transport of exogenous material[5] such as nerve growth factor, viruses, toxins, and horseradish peroxidase takes place within axonal tubules of the SER system.[49–53] However, after interruption of the axonal flow by local cooling, Tsukita and Ishikawa[42] observed that retrograde transport of horseradish peroxidase accumulated distal to the point at which the axonal flow was interrupted, within spherical or oblong membranous structures different from the usual SER profiles. Furthermore, a recent investigation by LaVail *et al.*[54] clearly demonstrated that the tubular structures containing retrogradely transported horseradish peroxidase were always discontinuous and that their limiting membrane was morphologically different from that of the axonal SER. Finally, the intense retrograde transport of particles observed at light microscope level in living axons obviously did not involve SER tubules but larger organelles such as mitochondria or lamellar and multivesicular bodies.[55]

It is therefore now generally accepted that retrograde transport probably involves the participation of membranous structures other than those of the axonal SER system.

3.2. Nature and Role of Materials Transported by Axonal Smooth Endoplasmic Reticulum

The fact that fast axonal transport consists mainly of particulate material implies that rapidly transported macromolecules are composed of both membrane components and substances enclosed in membranous structures. Both types of transport have been associated with axonal SER.

3.2.1. Membrane Components

Much neuronal metabolism is related to the steady-state turnover of membranes. As both proteins and phospholipids are essentially synthesized in the perikarya, the development of an efficient system of membrane maintenance is particularly important in axonal processes.

Since phospholipids migrate within the axon at the same velocity as proteins,[56,57] and since inhibition of protein synthesis also blocks axonal transport of phospholipids,[57,58] the axonal migration of newly formed membrane components was suggested to occur via constituted membranes. In neurons as in other cell types, membrane maintenance is therefore generally thought to be ensured by the bulk transfer of membrane components from newly formed membranes towards older ones.[59] This transfer may occur by two main mechanisms: (1) the lateral flow of molecules via direct continuity or contacts between membranes and (2) the budding of vesicles from one membrane and their fusion with another. In this respect, the frequent observation, all along the axons, of pictures showing continuity or contacts between the SER and the axolemma or other membranous organelles, as well as the repeated visualization of vesicles budding off from the SER, strongly support the proposition that axonal SER constitutes a preferential system for membrane maintenance in the axons.

The transfer of material from the SER to other membranous structures was also strongly suggested by the results of a series of autoradiographic studies.[43] A few hours after central administration of radioactive precursors, the SER contained most of the radioactivity when the front of labeled material moved along the axons. Only later was radioactivity seen to increase in the axolemma, the synaptic vesicles, and, to a lesser extent, the mitochondria. The possibility that axonal SER ensures maintenance of the axolemma is further supported by the finding that fast axonal transport of glycoproteins—an essential component of plasma membranes—is closely associated with the SER[34,43]

According to Palay,[17] axonal SER might also be involved in the maintenance of synaptic vesicles. This idea has been strengthened by repeated observation of continuity between these two structures (see above) and was recently reinforced by the results of specific staining of axonal SER tubules with zinc–iodide–osmium, known to stain the synaptic vesicles in various neurons selectively.[27,60–62]

In addition to simple structural maintenance, axonal SER may participate in the maintenance of specific membrane components. Several biochemical findings clearly indicate that a large part of material carried by fast axonal transport consists of enzymes known to be bound to membranes.[8,9] Furthermore, acetylcholinesterase, an enzyme intended for the external surface of the plasma membrane, was histochemically demonstrated to be associated with SER tubules all along axons.[63,64] Similarly, dopamine-β-hydroxylase, which converts dopamine into norepinephrine, was immunocytochemically detected in the membranes of both axonal SER and synaptic vesicles.[65]

3.2.2. The Intraluminal Components

In all cell types, a number of specific macromolecules are known to be transported throughout the cytoplasm within membrane delimited structures. This is the case for acid hydrolases and for various secretory substances that are abundantly produced by neurons.

Although normally restricted to the perikarial lysosomes, acid hydrolases were in fact shown to be transported along neuronal processes, at least under special conditions such as nerve injuries or extreme functional stimulation.[66–68] Consequently, such axonal transport does not occur within isolated lysosomes but within the lumen of elongated tubules apparently belonging to the axonal

SER system and frequently appearing in connection with autophagic vacuoles containing various degraded axonal structures.[69]

According to classical scheme, most secretory materials are transported to excretion sites isolated inside the lumen of membrane-limited vesicles. In neurons, a similar role is assigned to the secretory and/or synaptic vesicles, which are transported down to the axonal endings, where they eventually fuse with the axolemma. In various types of neuron, the observation of connections between axonal SER and both secretory and synaptic vesicles led to the idea that similar secretory substances may also be conveyed along the axons within the lumen of the axonal SER. The arguments favoring this view have been essentially formulated in relation to the hypothalamo–neurohypophysial neurosecretory neurons, where secretory proteins form electron-dense cores within the lumen of neurosecretory vesicles.[70,71] Earlier suggestions conforming to this theory were derived from ultrastructural observations of (1) the accumulation of electron-dense material within SER tubules at the proximal end of sectioned axons[72,73] and (2) direct connections between SER tubules and neurosecretory vesicles.[16,72,73]

Recent studies under various experimental conditions have reinforced this hypothesis. It is well known, for instance, that in rats, osmotic stimulation increases both the synthesis and axonal transport of vasopressin and oxytocin in these neurosecretory neurons.[74,75] In these conditions, the axonal SER system displays an extraordinary development, which coincides with the striking disappearance of the axonal secretory vesicles.[21] On the other hand, the sudden suppression of the osmotic stimulus, which quickly stopped hormonal excretion and induced the gradual hormonal reloading of the axon terminals,[76] was observed to be correlated first with extensive development of the axonal SER[76] and later with the accumulation of electron-dense material within SER tubules.[21] Despite the present limits of ultrastructural immunochemical techniques, even more direct arguments have been gathered from the study of the intraaxonal location of vasopressin. In such studies, immunocytochemical labeling was detected within intra-axonal tubular structures whose organization definitely resembled that of axonal SER.[77–79]

A similar mode of transport within axonal SER tubules has also been considered in nonendocrine neurons. In adrenergic neurons, monoamines were visualized within axonal SER tubules.[80,81] The role of calcium transport and sequestration, now generally admitted to be axonal SER functions,[82,85] may also be related to the intraluminal transport by the SER of specific calcium-binding proteins known to be present within the neurons.[86] The last and most provocative hypothesis in this series postulated that axonal SER even mediates the axonal transport of mitochondria.[25] According to this theory, the inner mitochondrial membrane and its enclosed matrix move within SER channels, which themselves supply the external membrane of the mitochondria. However intriguing the continuity observed between axonal SER and the external mitochondrial membrane, such an interpretation obviously requires clarification.

Since no pictures of fusion between axonal SER tubules and terminal axolemma have been observed, extraneuronal excretion of intraluminal SER secretory proteins is not likely to occur via direct branching of the axonal SER

Table I
Presumed Origins of Some of the Specific Macromolecules Transported by the Axonal Smooth Endoplasmic Reticulum

Source	Macromolecules	References
Rough ER	Glucose-6-phosphatase	26,100
	Acetylcholinesterase	101,102
All saccules of the Golgi apparatus	Thiamine pyrophosphatase	103–106
	Dopamine-β-hydroxylase	65
Preferentially saccules of the *trans* side of the Golgi apparatus and associated vesicles	Macromolecule-associated carbohydrates	80
	ZIO-stainable material	60–62,80
GERL[a]	Acid hydrolases	106,107

[a] Acronym introduced by Novikoff[108] to designate a system of smooth membrane cisternae adjacent to the saccules of the *trans* side of the Golgi apparatus and giving rise to lysosomes.

into the extracellular spaces. A more tenable hypothesis is therefore that excretion occurs by inclusion of the material in secretory or synaptic vesicles, which eventually bud off from the terminal segments of the axonal SER and fuse with the axolemma. Indeed, this idea is strengthened, both in neurosecretory and in adrenergic neurons, by the observation of frequent connections between axonal SER tubules and secretory or synaptic vesicles in the terminal portions of axons.[16,21,80,81] In axon terminals, a category of small vesicles was also shown to store calcium,[87,88] which again fits the concept that calcium-binding proteins might, like several others, ultimately may be incorporated into vesicles.

3.3. Heterogeneity of Axonal Smooth Endoplasmic Reticulum

Although the term SER is generally used for membranous tubules and cisternae extending throughout the axons, it is far from certain that all structures belong to the same morphological and functional system.

First of all, it is worth noting that the various enzymatic systems cytochemically investigated within axons were always shown to be present in only some of the SER tubules. As regards the protein transported within the axonal SER lumen, it may be considered *a priori* that hydrolytic enzymes are processed separately from other intraluminal proteins. Hence, the recently evolved concept that axonal SER is divided into several structurally and functionally different compartments.

Since axonal SER components essentially originate in the perikaryon, it seems reasonable to assume that, if true, such compartmentalization would already occur at the perikaryal level, when the different components are elaborated. In this line of research, it has been demonstrated that all cisternae of the RER system are not identical but comprise distinct individual microenvironments synthesizing different protein species.[89] That axonal SER may directly derive from the RER has further been suggested by the cytochemical demonstration that the two structures sometimes share the same enzymatic system (Table I). The recent visualization of direct connections between the

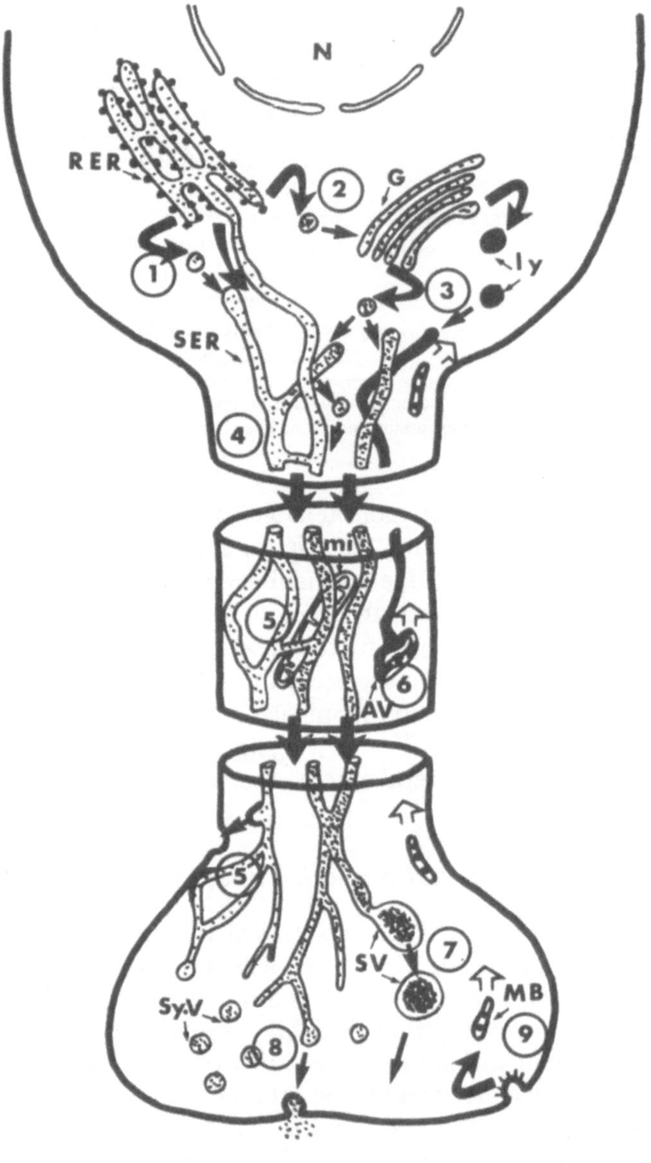

Fig. 1. Diagrammatic representation of the possible functions of axonal smooth endoplasmic reticulum (see also refs. 43, 45). (1) The membrane components, i.e., phospholipids and membrane proteins, are synthetized in the RER. The newly formed membranes give rise to SER, which is transported to the axon hillock either by channels directly connected to the RER or by vesicle shuttle. (2) Nonmembranous proteins are discharged in the lumen of the RER. Most of them (i.e., secretory proteins of lysosomal enzymes) are transported within vesicles to the Golgi apparatus (G), where they undergo posttranslational modifications. (3) Secretory proteins are then transferred from the Golgi apparatus to the axonal SER by vesicle shuttle. Lysosomal enzymes are either stored in lysosomes (ly), which remain in the perikaryon, or transported via vesicles to special axonal SER channels. (4) The interconnected tubular elements of the axonal SER are propelled down to the axon terminal by fast transport mechanisms. During this transport, some vesicles may bud off from axonal SER tubules and fuse with them again later. (5) Specific components of the

Table II
Presumed Destinations of Some of the Specific Macromolecules Transported by the Axonal SER

Macromolecules	Destination	References
Acid hydrolases	Autophagic vacuoles	69
Dopamine-β-hydroxylase	Synaptic vesicles	65
ZIO-stainable material	Synaptic vesicles	60,61,80
Catecholamines	Synaptic and secretory vesicles	80,81
Secretory proteins	Synaptic and secretory vesicles	16,21,28,79,105
Calcium-binding protein	Synaptic vesicles	87,88
Macromolecule-associated carbohydrates	Axolemma and synaptic vesicles	34,43
Acetylcholinesterase	Axolemma	103,109

axonal SER and the RER also strengthened this view considerably.[18] It may therefore be tentatively proposed that different regions of the RER generate differentiated axonal channels, which themselves may be driven towards the axons as soon as their membrane components are synthesized.

Another kind of perikaryal compartmentalization may be related to the elaboration of secretory proteins. Unlike the proteins intended for the components of membrane structures, secretory proteins are vectorially discharged into the lumen of the RER immediately after synthesis.[6] They are then transferred via SER elements to the Golgi apparatus before being propelled towards their excretion sites (Table 2).[90] In addition to the modifications of the chemical structure of secretory proteins, one of the main functions of the Golgi apparatus is to provide a membrane container that is competent for their excretion.[5] Since axonal SER seems to be involved in the transport of such secretory material (see above), it would be logical to consider that there are also Golgi-associated axonal SER channels. In this connection, the cytochemical demonstration of concomitant presence in axonal SER and Golgi complexes of various specific macromolecules (Table I) strongly favors a close relationship between the two structures. However, in the absence of clear evidence of direct tubular continuity, the mechanisms of the structural connections between the Golgi ap-

---

SER membrane are transferred to the axolemma or to the mitochondrial membrane (mi) by transient contact or by the fusion of vesicles budding off from SER elements. (6) Lysosomal enzymes are transferred to autophagic vacuoles (AV), which are then transported back to the perikaryon via retrograde axonal transport. (7) Secretory material is concentrated in portions of the SER tubules that dilate and eventually break down, giving rise to secretory vesicles (SV). (8) The terminal segments of axonal SER tubules become thinner and break down into synaptic vesicles (Sy.V). These vesicles provide specific macromolecules for local mechanisms such as neurotransmitter synthesis or calcium uptake. By fusion with the terminal axolemma, these vesicles further insure the maintenance of the terminal axolemma and the release of the SER intraluminal material, including neurotransmitters, secretory proteins, etc. (9) The retrograde transport of various substances including specific macromolecules destined to be hydrolyzed by perikaryal lysosomes or exogenous material taken up by microendocytosis occurs within polymorphous membranous bodies (MB), which form a discontinuous axonal system.

paratus and axonal SER remain purely speculative. One possible hypothesis is that vesicles sprouting from the Golgi then fuse to form a continuous tubular system.[91]

On the other hand, the possibility that various materials may, in fact, intermingle during their axonal transport in the SER cannot be completely ruled out. It is admitted that a single membrane system may indeed serve for the maintenance of very different types of membranes and that the bulk transfer of membrane components by membrane contacts or by vesicle shuttling results not in the mere mixing of all components but rather in the transfer of specific components from one membrane to the other.[59] Therefore, it appears that the different membrane proteins synthesized by different RER subsections may be conveyed in intermingled form in the membrane of a single axonal SER system.

In the same way, it may be assumed that the secretory proteins and their packaging membranes originating from the Golgi apparatus may be transported to the axonal ending together with SER deriving from RER structures by the fusing of vesicles from the Golgi saccules with preexisting axonal SER tubules.

The nature of the quantity of material transported by the axonal SER understandably differs among specific neuron species. However, as already mentioned, the presence of acid hydrolases was demonstrated in the axonal SER tubules of stressed neurons only.[66–68] In neurosecretory neurons, osmotic stimulation was shown to induce an intensive development of the axonal SER system, whereas its tubular elements thinned markedly.[21] The amount and nature of the material transported by axonal SER therefore clearly depend on the physiological state of the neurons. So, in full agreement with Cardell's theory,[92] neurons as well as other cell types may respond to specific stimuli by forming SER that differs quantitatively and qualitatively from the SER of unstimulated neurons.

3.4. Mechanisms of Axonal Smooth Endoplasmic Reticulum Transport

Microtubules and contractile proteins have both been reported to be involved in the mechanisms of fast anterograde axonal transport,[93] and these structures certainly help to propel the axonal SER along the axon. As already mentioned, the observation of actin filaments attached to the membrane of SER profiles[36,37] as well as the constant linking of axonal SER and microtubules by the microtrabeculae[35] provide morphological evidence for such involvement. Furthermore, in neurosecretory neurons, administration of colchicine, which is toxic to microtubules, results in very drastic accumulation of SER tubules in the proximal sections of axons and in the perikarya.[94]

Existing concepts of the molecular mechanisms governing the axonal transport of membrane organelles are still conjectural, although ATP and calcium are certainly of prime importance.[95,96] In this respect, it is significant that ATPase has been biochemically and cytochemically associated with microtubules throughout the entire length of neuronal processes.[97,98] In addition, there is a major calcium-dependent stage of the propelling machinery which appears to be restricted to the perikarya, since suppression of intra- and extracellular calcium reduces rapid axonal transport and causes accumulation of SER elements in the Golgi region.[99] In this connection, the idea that axonal SER propulsion depends partly on perikaryal mechanisms may find support in the experiment of Ellisman and Lindsey[48] in which they demonstrated the absence of SER motility in axon segments isolated from the perikarya by local cooling.

A last highly speculative question concerns the possible existence of distinct transport mechanisms for membrane of the axonal SER and for intraluminal materials. The intraluminal accumulation by axonal SER of electron-dense substances observed when axonal flow of SER is impaired[39,40,94,97] might in fact result from the existence of partly independent propelling mechanisms.

4. *CONCLUSIONS*

If the statement by Schwartz[8] that "axonal transport is a specialization of some universal process" is true, axonal organelles should not be considered fundamentally different from the corresponding structures in a large variety of nonneuronal cells. Consequently, the concept that the SER system mainly serves to transport specific molecules synthesized in the RER would apply to axonal SER (Figure 1). Similarly, axonal SER should not be considered as a static network but rather as a very dynamic system, mostly continuous from the perikaryon to the axonal ending, which may vary quantitatively and qualitatively according to the physiological state of the neuron. Active diffusion of part of the newly formed SER in the axon, or in other neuronal processes, would therefore be an efficient way for the neurons to ensure both their structural and functional maintenance in close adaptation to the fluctuations of their environment.

REFERENCES

1. Droz, B., and Koenig, H. L., 1970, *Protein Metabolism in the Nervous System* (A. Lajtha, ed.), Plenum Press, New York, pp. 93–108.
2. Roberts, S., Zomzely, C. E., and Bondy, S. C., 1970, *Protein Metabolism of the Nervous System* (A. Lajtha, ed.), Plenum Press, New York, pp. 3–37.
3. Droz, B., and Boyenval, J., 1975, *J. Microsc. Biol. Cell.* **23:**45a–46a.
4. LeBlond, C. P., and Bennett, G., 1977, *International Cell Biology* (B. R. Brinkley and K. R. Porter, eds.), Rockefeller University Press, New York, pp. 326–336.
5. Farquhar, M. G., and Palade, G. E., 1981, *J. Cell Biol.* **91:**77s–103s.
6. Palade, G., 1975, *Science* **189:**347–358.
7. Whalley, W. G., 1975, *The Golgi Apparatus*, Volume 2, Springer, Vienna.
8. Schwartz, J. H., 1979, *Annu. Rev. Neurosci.* **2:**467–504.
9. Grafstein, B., and Forman, D. S., 1980, *Physiol. Rev.* **60:**1167–1283.
10. Thoenen, H., and Kreutzberg, G. W. (eds.), 1981, *Neurosci. Res. Prog. Bull.* **20:**1–138.
11. Peters, A., Palay, S. L., and Webster, H. F., 1976, *The Fine Structure of the Nervous System*, Saunders, Philadelphia.
12. Palay, S. L., and Chan-Palay, V., 1977, *Handbook of Physiology* (E. R. Kandel, ed.), American Physiological Society, Bethesda, pp. 5–37.
13. Thiery, G., and Rambourg, A., 1976, *J. Microsc. Biol. Cell* **26:**103–106.
14. Tsukita, S., and Ishikawa, H., 1976, *J. Electron Microsc.* **25:**141–149.

15. Droz, B., Rambourg, A., and Koenig, H. L., 1975, *Brain Res.* **92**:1–13.
16. Alonso, G., and Assenmacher, I., 1978, *Biol. Cell* **32**:205–208.
17. Palay, S. L., 1958, *Exp. Cell Res.* [*Suppl.*] **5**:275–293.
18. Beaudet, A., and Rambourg, A., 1979, *Biol. Cell* **35**:14a.
19. Kreutzberg, G. W., and Gross, G. W., 1977, *Cell. Tissue Res.* **181**:443–457.
20. Porter, K. R., Byers, H. R., and Ellisman, M. H., 1979, *The Neurosciences, 4th Study Program* (F. O. Schmitt and F. G. Worden, eds.), MIT Press, Cambridge, Massachusetts, pp. 703–722.
21. Alonso, G., and Assenmacher, I., 1979, *Cell. Tissue Res.* **189**:415–429.
22. Rosenbluth, J., 1962, *J. Cell Biol.* **13**:207–217.
23. Palay, S. L., and Chan-Palay, V., 1973, *Metabolic Compartimentation in the Brain* (R. Bálazs and J. E. Cremer, eds.), Macmillan, London, pp. 187–207.
24. Reiter, W., 1966, *Z. Zellforsch. Mikrosk. Anat.* **72**:446–461.
25. Spacek, J., and Lieberman, A. R., 1980, *J. Cell. Sci.* **46**:129–147.
26. Holtsman, E., Teichberg, S., Abrahams, S. J., Citkowitz, E., Crain, S. M., Kawai, N., and Peterson, E. R., 1972, *J. Histochem. Cytochem.* **21**:349–385.
27. Teichberg, S., and Holtzman, E., 1973, *J. Cell Biol.* **57**:88–108.
28. Dellman, H. D., Castel, M., and Linner, J. G., 1978, *Gen. Comp. Endocrinol.* **36**:477–486.
29. Robertson, J. D., 1970, *The Neurosciences Second Study Program* (F. O. Schmidt, ed.), Rockefeller University Press, New York, pp. 715–728.
30. Korneliussen, H., 1972, *Z. Zellforsch.* **130**:28–57.
31. Hernandez-Nicaise, M. C., 1973, *J. Neurocytol.* **2**:249–263.
32. Taxi, J., and Sotelo, C., 1973, *Brain Res.* **62**:431–437.
33. Couteaux, R., 1974, *Advances in Cytopharmacology*, Volume 2 (B. Ceccarelli, F. Clementi, and J. Meldolesi, eds.), Raven Press, New York, pp. 369–379.
34. Markov, D., Rambourg, A., and Droz, B., 1976, *J. Microsc. Biol. Cell.* **25**:57–60.
36. Lebeux, Y. J., and Willemot, J., 1975, *Cell. Tissue Res.* **160**:1–36.
37. Alonso, G., Gabrion, J., Travers, E., and Assenmacher, I., 1981, *Cell. Tissue Res.* **214**:323–341.
38. Tytell, M., Black, M. M., Garner, J. A., and Lasek, R. J., 1981, *Science* **214**:179–181.
39. Pellegrino de Iraldi, A., and De Rogriguez, E., 1968, *Z. Zellforsch.* **87**:330–344.
40. Kapeller, K., and Mayor, D., 1979, *Proc. R. Soc.* (*Lond.*) [*Biol.*] **172**:39–51.
41. Smith, R. S., 1980, *J. Neurocytol.* **9**:39–65.
42. Tsukita, S., and Ishikawa, H., 1980, *J. Cell Biol.* **107**:315–395.
43. Rambourg, A., and Droz, B., 1980, *J. Neurochem.* **35**:16–25.
44. Schonbach, J., Schonbach, C., and Cuenod, M., 1971, *J. Comp. Neurol.* **141**:485–498.
45. Droz, B., Koenig, H. L., Di Giamberardino, L., Couraud, J. Y., Chretien, M., and Souyri, F., 1979, *Progress in Brain Research*, Volume 49 (S. Tucek, ed.), Elsevier, Amsterdam, New York, pp. 23–44.
46. Hendrickson, A. E., 1972, *J. Comp. Neurol.* **144**:381–397.
47. Thompson, E. B., Schwartz, J. H., and Kandel, E. R., 1976, *Brain Res.* **112**:251–281.
48. Ellisman, M. H., and Lindsey, J. D., 1981, *J. Cell Biol.* **91**:91a.
49. Sotelo, C., and Riche, D., 1974, *Anat. Embryol.* **146**:209–218.
50. Nauta, H. J. W., Kaiserman-Ambramof, I. R., and Lasek, R. J., 1975, *Brain Res.* **85**:373–384.
51. Schwab, M., and Thoenen, H., 1977, *Brain Res.* **122**:459–474.
52. Lasek, R. J., and Hoffman, P. N., 1976, *Cell Motility*, Volume 3 (R. Goldman, J. Pollard, and J. Rosenbaum, eds.), Cold Spring Harbor Laboratory, New York, pp. 1021–1049.
53. Dolivo, M., Beretta, E., Bonifas, F., and Foroglou, C., 1978, *Brain Res.* **140**:111–123.
54. LaVail, J. H., Rapisardi, S., and Sugino, I. K., 1980, *Brain Res.* **121**:3–20.
55. Breuer, A. C., Christian, C. N., Henlart, M., and Nelson, P. G., 1975, *J. Cell. Biol.* **65**:562–576.
56. Grafstein, B. J., Miller, J. A., Ledeen, R. W., Haley, J., and Specht, S. C., 1975, *Exp. Neurol.* **46**:261–281.
57. Rostas, J. A. P., Austin, L., and Jeaffrey, P. L., 1979, *J. Neurochem.* **32**:1461–1466.
58. Abe, T., Haga, T., and Kurokawa, M., 1973, *Biochem. J.* **136**:731–740.

59. Holtzman, E., and Mercurio, A. M., 1980, *Int. Rev. Cytol.* **67**:1–67.
60. Stelzner, D. J., 1971, *Z. Zellforsch.* **120**:332–335.
61. Lane, N. J., and Swales, L. S., 1976, *J. Cell Sci.* **22**:435–453.
62. Reinecke, M., and Walther, C., 1968, *J. Cell Biol.* **78**:839–855.
63. Brzin, M. V., Tennuson, M., and Duffy, P. E., 1966, *J. Cell Biol.* **31**:215–242.
64. Kaza, P., Mann, S. P., Karsch, S., Toth, L., and Jordan, S., 1973, *J. Neurochem.* **21**:431–436.
65. Cimarusti, D. L., Saito, K., Vaughn, J. E., Baker, R., Roberts, E., and Thomas, R. E., 1979, *Brain Res.* **162**:55–67.
66. Holtzman, E., 1977, *Neuroscience* **2**:327–355.
67. Broadwell, R. E., 1980, *J. Histochem. Cytochem.* **28**:87–89.
68. Whitaker, S., and Labella, F. S., 1972, *Z. Zellforsch.* **125**:1–15.
69. Boudier, J. A., Marchi, D., Cataldo, C., Massacrier, A., and Cau, P., 1981, *Biol. Cell.* **40**:33–40.
70. Sachs, H., 1970, *Handbook of Neurochemistry*, Volume 4 (A. Lajtha, ed.), Plenum Press, New York, pp. 373–428.
71. Mason, C. A., and Bern, H. A., 1977, *Handbook of Physiology*, Volume I (E. R. Kandel, ed.), American Physiological Society, Bethesda, pp. 651–689.
72. Palay, S. L., 1957, *Ultrastructure and Cellular Chemistry of Neural Tissue* (H. Waelsch, ed.), Cassels, London, pp. 31–49.
73. Dellman, H. D., and Rodriguez, E. M., 1970, *Experientia* **26**:414–415.
74. Norstrom, A., 1971, *Brain Res.* **28**:131–142.
75. Norstrom, A., and Sjostrand, J., 1972, *J. Endocrinol.* **52**:87–105.
76. Rougon-Rappuzzi, G., Cau, P., Boudier, J. A., and Cupo, A., 1978, *Neuroendocrinology* **27**:46–62.
77. Alonso, G., Gabrion, J., Lutz, B., and Assenmacher, I., 1980, *J. Physiol.(Paris) 76:*1b.
78. Krisch, B., 1980, *J. Histochem. Cytochem.* **28**:472–474.
79. Dreyfuss, F., Burlet, A., Chateau, M., and Czernichow, P., 1979, *Biol. Cell.* **35**:141–164.
80. Quatacker, J., 1981, *Histochem. J.* **13**:109–124.
81. Tranzer, J. P., 1972, *Nature* **237**:57–58.
82. Hammerschlag, R., Dravid, A. R., and Chiu, A. Y., 1975, *Science* **188**:273–275.
83. Blaustein, M. P., Ratzlaff, R. W., Kendrick, N. C., and Schweitzer, E. S., 1978, *J. Gen. Physiol.* **72**:15–41.
84. Duce, I. R., and Keen, P., 1978, *Neuroscience* **3**:837–848.
85. Henkart, M. P., Reese, T. S., and Brinley, F. J., 1978, *Science* **202**:1300–1303.
86. Iqbal, Z., and Ochs, S., 1978, *J. Neurochem.* **31**:409–418.
87. Blitz, A. L., Fine, R. E., and Toselli, P. A., 1977, *J. Cell Biol.* **75**:135–147.
88. Shaw, F. D., and Morris, J. F., 1980, *Nature* **287**:56–58.
89. Svardal, A. M., and Pryme, I. F., 1980, *Sub. Biochem.* **7**:117–170.
90. Tartakoff, A., and Vassalli, I. P., *J. Cell Biol.* **79**:694–707.
91. Hammerschlag, R., and Stone, G., 1982, *Axoplasmic Transport* (D. G. Weiss, ed.), Springer-Verlag, Berlin, Heidelberg, New York,pp. 406–413.
92. Cardell, R. R., 1977, *Int. Rev. Cytol.* **48**:221–279.
93. Puszkin, S., and Schook, W., 1979, *Methods and Achievements in Experimental Pathology*, Volume 9 (G. Jasmin and M. Cantin, eds.), S. Karger, Basel, pp. 87–111.
94. Alonso, G., and Arancibia, S., (to be published).
95. Ochs, S., 1982, *Science* **176**:252–260.
96. David, A. R., and Hammerschlag, R., 1975, *J. Neurochem.* **24**:711–718.
97. Droz, B., 1975, *The Nervous System*, Volume 1 (D. B. Tower, ed.), Raven Press,New York, pp. 111–127.
98. Sharp, G. A., Fritzsimons, J. T. R., and Kerkut, G. A., 1980, *Comp. Biochem. Physiol.* **66A**:415–429.
99. Lindsey, J. D., Hammerschlag, R., and Ellisman, M. H., 1981, *Brain Res.* **205**:275–287.
100. Stephens, H. R., and Sandbrown, E. B., 1976, *Brain Res.* **113**:127–132.
101. Gisiger, V., Venkov, L., and Gautron, J., 1975, *J. Neurochem.* **25**:737–748.
102. Somogyi, P., and Chubb, I. W., 1976, *Neuroscience* **1**:413–421.

103. Griffith, D. L., and Bondareef, 1973, *Am. J. Anat.* **136**:549–556.
104. Csillik, B., Knyihar, E., Laszlo, I., and Boncz, I., 1974, *Brain Res.* **70**:179–183.
105. Castel, M., and Dellman, H. D., 1980, *Cell Tissue Res.* **210**:205–221.
106. Broadwell, R. D., and Oliver, C., 1981, *J. Cell Biol.* **90**:474–484.
107. Novikoff, A. B., Mori, M., Quintana, N., and Hauw, J. J., 1971, *J. Cell Biol.* **50**:859–886.
108. Novikoff, A. B., 1967, *The Neuron* (H. Hyden, ed.), Elsevier, Amsterdam, pp. 255–318.
109. Kreutzberg, G. W., Schubert, P., Toth, L., and Rieske, E., 1973, *Brain Res.* **62**:399–404.

7

Lysosomes

Harold Koenig

1. INTRODUCTION

Lysosomes ("lytic bodies") are membrane-enclosed cytoplasmic organelles that have a low internal pH and contain numerous hydrolytic enzymes with acid pH optima in a latent state. Lysosomes are in fact the primary component of an extraordinarily dynamic and variegated membrane system, the lysosomal–vacuolar or lysosomal system, which serves as the intracellular digestive tract in virtually all eucaryotic cells. Lysosomes are unique among cell organelles in their morphological and enzymatic polymorphism and in the multiplicity of cell processes, both physiological and pathological, in which they participate. There have been many important advances in our knowledge of lysosomes since the appearance of the chapter on this subject in the first edition of the *Handbook*. In recent years, interest has focused on a number of new topics, including the chemical structure, biosynthesis, and intracellular transport of lysosomal enzymes; the role of the phosphomannosyl recognition marker and its receptor in lysosomal enzyme targeting; the regulation of lysosomal proteolysis and its role in intracellular protein degradation; receptor-mediated endocytosis of transport proteins, hormones , and other cell agonists; phagosome–lysosome fusion; lysosomotropic agents; and drug-induced lysosomal storage disorders.

2. LYSOSOMAL ENZYMES

2.1. Biochemical Composition

More than 60 enzymes have been identified as lysosomal enzymes. The great majority of these enzymes have pH optima in the acid range. The deduction that lysosomal enzymes are glycoproteins[1–3] appears to have held up as an increasing number of the enzymes have been sufficiently purified for chemical analysis. We originally found that an enzyme protein fraction pre-

Harold Koenig • Neurology Service, Veterans Administration Lakeside Medical Center; and Department of Neurology, Northwestern University Medical School, Chicago, Illinois 60611.

pared from purified rat kidney and liver lysosomes contained approximately 10% carbohydrate, consisting of mannose, glucose, N-acetylglucosamine, N-acetylneuramic acid, galactose, and fucose.[2,3] Nearly all of the many protein components resolved from this enzyme fraction by polyacrylamide gel electrophoresis stained for carbohydrate by the periodic acid–Schiff method, indicating that lysosomal hydrolases are largely glycoproteins.[2–4] Neuraminidase digestion decreased the mobility of the acidic forms (also some basic forms) of seven lysosomal hydrolases[5] and increased their isoelectric points,[6] thereby corroborating that these lysosomal enzymes are glycoproteins and demonstrating that their electronegative charge partly reflects the presence of N-acetylneuraminic acid residues. Further confirmation for the glycoprotein nature of lysosomal enzymes came from the finding that these enzymes bind to a concanavalin A–Sepharose matrix and can be eluted at different pH values in the presence of α-methylglucoside.[7] Finally, a growing number of purified lysosomal enzymes have been shown to contain carbohydrate residues, including β-glucuronidase,[8,9] acid DNase,[10] acid phosphatase,[11] arylsulfatase A,[12] hexosaminidase,[13,14] cathepsin D,[15] β-galactosidase,[16] α-L-iduronidase,[17] and β-glucosidase.[18] The carbohydrate residues of lysosomal enzymes are linked to asparagine residues of the peptide chain.[19,20]

2.2. *Biosynthesis of Lysosomal Enzymes*

Lysosomal enzymes resemble secretory proteins in that they apparently are synthesized in membrane-bound ribosomes and initially follow the secretory route. This view was originally based on cytochemical observations showing the presence of hydrolases in the lumen of the endoplasmic reticulum (ER), Golgi complex, and the GERL membrane systems.[21,22] We isolated a special population of rough microsomes enriched 10- to 30-fold in lysosomal hydrolases from the bulk of rough microsomes in rat kidney because of their greater density on isopycnic centrifugation.[23] The cisternal content of these special rough microsomes resembled the matrix of lysosomes in that it was electron dense, osmiophilic, and plumbophilic, reacted positively for protein-bound carbohydrate by the PAS procedure, and stained for acid phosphatase activity. Kinetic studies of isotope uptake showed that amino acids, N-acetylglucosamine, mannose, and glucose are incorporated into lysosomal glycoprotein in these special rough microsomes; shortly thereafter, the labeled proteins appeared in the Golgi-enriched fraction, and at 1–3 h the lysosomal fraction became maximally labeled.[24] N-Acetylneuraminic acid was initially incorporated in the Golgi-enriched fraction, and labeled protein rapidly appeared in the lysosomes.[24]

An electron microscopic autoradiographic study in rat kidney proximal tubule cells generally confirmed these radiochemical findings, showing that [^{3}H]lysine is initially (5 min) incorporated in the rough ER.[25,26] Radioactivity appeared in the Golgi apparatus at 15–30 min and in lysosomes by 30–60 min. [^{3}H]Mannose, [^{3}H]glucose, and N-acetyl-[^{3}H]glucosamine were incorporated into the rough ER, and N-acetyl-[^{3}H]mannosamine and additional N-acetyl-[^{3}H]glucosamine were incorporated into the Golgi apparatus.[25,26] Essentially all dense bodies (secondary lysosomes) were labeled by the various precursors, indicating that lysosomes continually receive newly synthesized glycoproteins, presumably through fusion with primary lysosomes generated in the Golgi apparatus.

2.2.1. Translation and Cotranslational Processing of Lysosomal Enzymes

The mRNAs for β-glucuronidase and cathepsin D have recently been shown to be segregated intracellularly in membrane-bound polysomes.[27] The nascent β-glucuronidase[27] and cathepsin D[27,28] polypeptides contain amino-terminal insertion signals, which apparently determine the cotranslational insertion of the polypeptides into the cisternae of the ER. These signal peptides are ~2000 daltons in size and are removed at the time of insertion by a peptidase in the ER. Cotranslational glycosylation of the nascent enzymes involves the transfer of a preformed $glucose_3$ $mannose_9$ N-$acetylglucosamine_2$ species from a lipid carrier to the asparagine residues of the nascent protein.[29] The decrease in polypeptide length resulting from the removal of the insertion signal is normally masked by the increase in apparent molecular weight resulting from cotranslational glycosylation. This glycosylation can be prevented by tunicamycin and leads to quantitative secretion of unglycosylated lysosomal enzymes from cultured cells into the extracellular medium.[27]

2.2.2. Synthesis of Phosphorylated Recognition Marker in Lysosomal Enzymes

There is a considerable body of evidence indicating that phosphomannosyl residues on acid hydrolases are an essential component of the recognition signal necessary for the targeting of these enzymes to lysosomes.[30,31] The mannose-6-phosphate moiety is synthesized as a phosphodiester covered with N-acetylglucosamine via a novel biosynthetic pathway. UDP-N-Acetylglucosamine serves as the donor of N-acetylglucosamine-1-phosphate to a high-mannose oligosaccharide on the nascent lysosomal enzyme, giving rise to an oligosaccharide substituted with a phosphodiester or "covered" phosphate. The N-acetylglucosamine cover is then removed, exposing a mannose 6-phosphomonoester or "uncovered" phosphate. Phosphorylation is preceded by the removal of glucose residues from the core oligosaccharides. Glucose processing is a posttranslational event mediated by two specific glucosidases, glucosidase I, located in the rough ER, and glucosidase II, located in smooth membranes.[29]

Mannose processing apparently occurs in several stages. In the final stage of mannose processing, the neutral, high-mannose-type oligosaccharides are gradually converted to complex-type oligosaccharides, and this presumably involves the α1,2 mannosidase and the α1,3, α1,6-specific mannosidase present in the Golgi apparatus.[29] Phosphorylation of mannose residues is achieved soon after completion of glucose processing in the early stages of mannose processing. This inference is consistent with the finding that the enzymes mediating mannose phosphorylation, N-acetylglucosamine-1-phosphotransferase and α-N-acetylglucosaminylphosphodiesterase, are localized in the *cis* elements of the Golgi apparatus. The Golgi apparatus is differentiated into proximal (*cis*) and distal (*trans*) elements in relation to the endoplasmic reticulum. In the modification of carbohydrate side chains of glycoproteins, the *cis* and *trans* elements are the sites of the early and late reactions, respectively.[33] In β-glucuronidase, the most thoroughly studied of the lysosomal enzymes, phosphorylation of enzyme oligosaccharides is a posttranslational event, with each of the three glycosylation sites per subunit being partially phosphorylated.[34] The completion of complex-type oligosaccharides involves the addition of the terminal sugars, fucose, galactose, N-acetylneuraminic acid, and N-acetylglucosamine, from their corresponding sugar nucleotides. The transferase enzymes that catalyze this reaction are located in the *trans* (late) elements of the Golgi apparatus.[33] The newly completed lysosomal enzymes are transported to lysosomes in the form of very acidic sialoglycoproteins with pIs of 2 to 4.9.[39–42]

2.2.3. Proteolytic Processing of Lysosomal Enzymes

Lysosomal enzymes are usually synthesized as precursors with higher molecular weights than those of the mature forms.[30] This has been shown for a number of hydrolases, including β-galactosidase,[16] hexosaminidase,[17,35] cathepsin D,[17,28] α-glucosidase,[17] and α-L-iduronidase.[18] The larger enzyme precursors are trimmed to their eventual size over the course of several hours to several days, apparently as a consequence of limited proteolysis after they have arrived in their lysosomal target.

2.3. Role of Mannose-6-Phosphate Recognition Marker

Adsorptive endocytosis of lysosomal enzymes by fibroblasts requires the presence of mannose-6-phosphate recognition markers on the enzymes and high-affinity lysosomal enzyme receptors on the cell surface of fibroblasts.[31] Thus, lysosomal enzyme uptake is a saturable process that occurs largely by means of receptor-mediated endocytosis. However, most of the phosphomannosyl–enzyme receptors in rat liver are intracellular and face the interior of the ER, Golgi apparatus, and lysosomes, with the highest specific activity of these receptors occurring in these organelles.[36] The intracellular receptors are occupied by endogenous enzymes, which can be displaced by mannose-6-phosphate and present an occupancy gradient that runs downhill from ER to lysosomes. Sly and associates have therefore proposed that the delivery of newly synthesized acid hydrolases to lysosomes depend on these specific receptor binding sites. The receptor-bound hydrolases are thought to be collected in coated vesicles, which bud off the Golgi or GERL to become primary lysosomes. Primary lysosomes fuse with secondary lysosomes, and the resultant fall in pH permits a pH-dependent dissociation of receptors from enzymes and recycling of free receptors to the Golgi apparatus. Fibroblasts express these receptors on the cell surface, which allows recapture of extracellular enzymes by receptor-mediated endocytosis.

2.4. *Disease- and Drug-Induced Defects in Lysosomal Enzyme Processing*

In I-cell disease, a fatal Hurlerlike disease resulting from a single gene defect, the fibroblasts are deficient in numerous acid hydrolases as a result of a quantitative secretion of the newly synthesized enzymes into the extracellular medium. These secreted forms are recognition defective; i.e., they are not pinocytosed by fibroblasts.[30] Several lysosomal enzymes, β-hexosaminidase, α-glucosidase, and cathepsin D, have been shown to be synthesized as precursors of apparently normal molecular weight but are not phosphorylated. Thus, although deficient phosphorylation of lysosomal enzymes is a cause of I-cell disease, it could be secondary to another abnormality, for example, a defect in the synthesis of a suitable oligosaccharide intermediate.[30] Unexplained as yet is the fact that the I-cell mutation affects predominantly connective tissue cells but spares the parenchymal cells, e.g., leucocytes, neurons, hepatocytes, commonly involved in other lysosomal storage diseases.

Treatment of fibroblasts with lysosomotropic agents such as chloroquine and NH_4^+ results in an almost quantitative secretion of newly formed precursor hydrolases into the medium, and the residual intracellular hydrolases fail to undergo proteolytic modification.[30] The lysosomotropic amines also inhibit the uptake of extracellular, high-uptake hydrolases.[37] These findings have been explained on the grounds that the amines accumulate in lysosomes and raise the intralysosomal pH, thereby inhibiting the pH-dependent dissociation of receptor-bound enzymes so that ultimately all receptors are saturated.[37] Although this drug effect superficially resembles the effect of the I-cell mutation in inducing the quantitative secretion of lysosomal hydrolases, it differs in that the secreted enzymes are phosphorylated. The antibiotic tunicamycin inhibits the first step of oligosaccharide–lipid synthesis, the addition of N-acetylglucosamine-phosphate to dolichol phosphate to yield N-acetylglucosaminylpyrophosphoryldolichol, and thus blocks the synthesis of asparagine-linked oligosaccharide. Consequently, tunicamycin induces the synthesis of unglycosylated, and therefore unphosphorylated, hydrolases, which are secreted rather than targeted to lysosomes.[27,30]

2.5. *Physicochemical Modifications of Lysosomal Enzymes during Intracellular Transport*

We have investigated five acid hydrolases, acid phosphatase, β-galactosidase, β-N-acetylhexosaminidase, β-glucuronidase, and arylsulfatase, in rat kidney subcellular fractions with respect to solubilization by freeze-thawing, electrophoretic mobility in polyacrylamide gels,[4] and pI values as measured by column isoelectric focusing.[38] The following subcellular fractions were prepared and characterized by ultrastructural and cytochemical criteria and by biochemical and enzymatic markers: (1) an ordinary rough microsomal fraction; (2) a special rough microsomal fraction enriched seven- to ninefold in acid hydrolases over the homogenate[23]; (3) a smooth microsomal fraction; (4) a Golgi membrane fraction enriched 2.5-fold in acid hydrolases and 10-, 15-, and 20-

fold in sialyl transferase, N-acetyllactosamine synthetase, and galactosyl transferase, respectively; and (5) a lysosomal fraction enriched 15- to 25-fold in acid hydrolases. In the special rough microsomal fraction, the five lysosomal enzymes occurred largely or exclusively as a single bound basic form associated with stainable glycoprotein in gel electrophoretograms[4,23] and with pI values of about 8.[38] In the lysosomal fraction, these hydrolases were largely present as a soluble acidic form with pI values of 4.4–5.0 and a smaller bound basic form with pI values of 6–8.[38] In the Golgi membrane fraction, nearly all of the various enzymes occurred in the acidic form, with pI values between 4.0 and 4.8. The solubility, anodic mobility, and pI values of the hydrolases in the smooth microsomal fraction were intermediate between those of the special rough microsomal fraction and those of the Golgi membrane fraction.

2.6. *Synthesis and Turnover of Lysosomal Glycoproteins in Relation to pIs*

Lysosomes in rat kidney[24] and liver (tritosomes)[39] feature an active incorporation of newly synthesized sialoglycoproteins. Under steady-state conditions, this synthesis is counterpoised by an equally active biodegration of these constituents. However, the turnover rates of the peptide and N-acetylneuraminic acid (NANA) residues, estimated from semilogarithmic plots of the radioactive decay of ^{14}C and ^{3}H as a function of time after administration of [^{14}C]lysine and N-acetyl-[^{3}H]mannosamine, were dissimilar. Thus, NANA residues in the soluble constituents of rat kidney lysosomes turned over twice as rapidly as the polypeptide, with half-lives of 3.2 and 6.8 days, respectively.[40] In the soluble extract (tritosol) of rat liver lysosomes (tritosomes), NANA turned over seven times faster than the polypeptide, with half-lives of 1 and 7 days, respectively.[40,41] The rapid catabolic loss of NANA is attributable to the terminal position of this residue in the carbohydrate side chain of glycoproteins, which renders it readily accessible to cleavage by lysosomal neuraminidase during physiological autolysis. A rapid autolytic cleavage of NANA residues from lysosomal glycoproteins has also been observed when kidney lysosomal extracts are incubated *in vitro* at an acid pH.[5,42]

As previously indicated, the lysosomal hydrolases in the Golgi apparatus have a decidedly acidic character[4,38] because of the attachment of NANA residues to the nascent glycoenzymes as they flow through the Golgi cisternae.[24–26] Accordingly, the new synthesized hydrolases are packaged in lysosomes of rat kidney,[40,42] liver, and preputial gland[39,41] in the form of acidic sialoglycoproteins. This was established by fractionating soluble extracts of purified lysosomal fractions doubly labeled with [^{14}C]lysine and N-acetyl-[^{3}H]mannosamine by column isoelectric focusing. The radioactivity profiles of ^{14}C and ^{3}H label in the gradient fractions 1.5 h after isotope administration were essentially congruent, indicating that the labeled constituents were mostly glycoproteins. Further, the great bulk of the labeled sialoglycoproteins at 1.5 h were very acidic, with pIs between 2 and 4.9; less than 10 and 5% of the ^{14}C and ^{3}H labels occurred in the more basic glycoproteins (pI 5.9) at this time.

A marked change in the ^{14}C and ^{3}H radioactivity profiles developed at later intervals after a single administration of the isotopes. This consisted of a time-dependent shift of ^{14}C and ^{3}H labels from the glycoprotein fractions of low pIs, which contained most of the label at 1.5 h, into more basic glycoprotein fractions that were unlabeled or sparsely labeled at this time. Evidently, during biodegradation, the lysosomal glycoproteins become progressively more basic because the NANA and, inferentially, other carbohydrate residues as well turn over more rapidly than the peptide moieties. Similar observations have been made during autolysis *in vitro* of rat kidney lysosomal glycoproteins prelabeled in the NANA and polypeptide portions with N-acetyl-[^{3}H]mannosamine and [^{14}C]lysine.[42] When a lysosomal extract was incubated at pH 5, labeled NANA was cleaved twice as rapidly as peptide, and the pI values of the labeled glycoproteins and of two representative hydrolases, β-glucuronidase and arylsulfatase, markedly increased.

These findings provide additional direct evidence for the view that each newly completed acid hydrolase is packaged in lysosomes in a highly acidic form. The observation that the more basic glycoprotein fractions, which contain mostly the basic forms of the various lysosomal hydrolases, are essentially unlabeled at 1.5 h after isotope injection but acquire substantial peptide-associated [^{14}C]lysine at a later time supports the view that the basic forms are formed during biodegradation through a partial autolytic cleavage of NANA, carbohydrate, and glycopeptide residues.

Lysosomal enzymes are relatively resistant to autolytic inactivation.[43] That the resistance of these hydrolases to autolytic inactivation may be related to their glycoprotein nature is shown by the following *in vitro* incubation experiments on lysosomal glycoproteins prelabeled in the peptide ([^{14}C]lysine) and NANA (N-acetyl-[^{3}H]mannosamine) or N-acetylglucosamine (N-acetyl-[^{14}C]glucosamine) moieties.[43] *p*-Nitrophenyloxamic acid, an inhibitor of bacterial neuraminidase,[44] protected labeled NANA from autolytic cleavage and retarded protein digestion. Further, galactono-, glucono-, and mannonolactones, inhibitors of the corresponding glycosidases, in combination inhibited the release of labeled N-acetylglucosamine, presumably by blocking the cleavage of the distal monosaccharides on the side chains, and thus slowed protein digestion. Cathepsin D seems to be largely responsible for the autolytic digestion of lysosomal protein, as it is prevented by pepstatin (A. Goldstone and H. Koenig, unpublished data), a potent inhibitor of this enzyme.[45] These findings support the view that the carbohydrate side chains protect the protein core of the lysosomal hydrolases from catheptic attack.

3. LYSOSOMAL LIPOGLYCOPROTEINS

Soluble acidic lipoglycoproteins (SALGP) are the major nonenzymatic component of secondary lysosomes isolated from rat liver and kidney. Ultracentrifugal flotation in KBr solutions was used to separate the SALGP from the enzyme glycoproteins in Triton X-100 extracts of purified lysosomal fractions. The SALGP account for about 50% of the soluble lysosomal protein and contain 15–60% phospholipid (depending on buoyant density) and 5% carbohydrate (hexose, hexosamine, NANA). On polyacrylamide gel electrophoresis, SALGP migrate as a single fast anodic band, which stains for protein, carbohydrate, lipid (Sudan black B), and anionic groups (acridine orange metachromasia).[46–48] The SALGP are rapidly labeled *in vivo* by precursors of phospholipid (^{32}Pi), [^{3}H]choline, [^{3}H]*myo*-inositol), aminosugars and NANA ([^{14}C]glucosamine, N-acetyl-[^{3}H]mannosamine), and polypeptides ([^{14}C]leucine, [^{14}C]lysine). The SALGP have an apparent molecular weight of approximately 15,000 as determined by gel filtration and SDS polyacrylamide gel electrophoresis and pIs of 3–4.5.[47,48] The SALGP probably play an important role in maintaining the low internal pH of lysosomes required for acid hydrolysis and account, in part at least, for the accumulation of basic dyes and other organic cations within lysosomes *in vivo* and *supra vitam*.[46–48] The SALGP tend to form nonspecific aggregates or complexes with lysosomal enzymes and may modulate the catalytic activities of these enzymes. This subject has been previously reviewed.[47,48]

4. *ENDOCYTOSIS*

Heterophagy is the complex process whereby the lysosomal system digests extracellular material taken up into the cell by endocytosis. The contents of the internalized vesicles are transported to the lysosomes by fusion of their respective membranes (phagolysosome fusion). In this section we briefly consider the process of endocytosis. For a more detailed account, the reader is referred to a recent review.[49]

4.1. *Types of Endocytosis*

Endocytosis encompasses phagocytosis, the uptake of large insoluble particles visible in the light microscope, and pinocytosis, the vesicular uptake of smaller submicroscopic components ranging from small particles (immune complexes, lipoproteins, colloids) to soluble macromolecules (hormones, enzymes, toxins, transport proteins), low-molecular-weight solutes, and water. The terms fluid endocytosis and adsorptive endocytosis or pinocytosis have been used to distinguish the uptake of substances in the extracellular fluid (fluid-phase or bulk-phase pinocytosis) from uptake of substances that bind to the cell surface and subsequently appear on the internal aspect of vesicle membrane (adsorptive pinocytosis). Adsorptive endocytosis is a more selective and concentrating process, which allows cells to internalize large amounts of specific solutes without ingesting large amounts of fluid. Ligands that bind to specific receptors in the cell surface are interiorized by the process of receptor-mediated endocytosis, a special form of adsorptive endocytosis. Neurons and exocrine and endocrine gland cells display an enhanced rate of fluid-phase pinocytosis following a burst of secretory activity. This secretory endocytosis provides for membrane recycling by retrieval of the delimiting membrane of synaptic vesicles or other storage organelles, which is inserted into the plasma membrane at time of exocytosis. A fourth form of pinocytosis, "constitutive" pinocytosis, refers to a relatively active state of pinocytosis that is sustained for long periods of time in certain cells cultured *in vitro*, e.g., growing fibroblasts, nondividing macrophages, and *in vivo* in some epithelial cells such as in kidney tubules and vas deferens.[49]

In some instances, endocytic vesicles may traverse the cell without fusing with lysosomes and exocytose their contents elsewhere at the cell surface. This route, designated vesicular transport or diacytosis (endoexocytosis), has been well studied in the endothelium of blood vessels, where pinocytic vesicles shuttle plasma from the capillary lumen to the tissue space.[50] Diacytosis also serves to transport materials across epithelial cell layers, e.g., immunoglobulin A across mammary epithelium[51] and proteins across fetal and neonatal gut.[52] Endocytosis appears to play an important role in the turnover of the plasma membrane and some of its individual components, e.g., the acetylcholine receptor.[49,53]

Pinocytosis has been divided into macropinocytosis, characterized by the formation of macropinosomes 0.3–3 μm in diameter, and micropinocytosis, with micropinosomes about 0.07 μm in diameter.[49] Macropinocytosis probably occurs when surface ruffles fall back and fuse with adjacent areas of the cell surface and trap medium. Micropinocytosis involves the invagination and pinching off of small identations or caveolae in the cell surface.[49,54,55]

4.2. Receptor-Mediated Endocytosis

In the last few years a special internalization pathway, termed receptor-mediated endocytosis, has been recognized and characterized. This pathway, which serves to convey hormones and a number of biologically active proteins from surface to the cell interior, has recently been reviewed.[54–56] Embedded in the plasma membrane of fibroblasts and other cells are a wide variety of specific receptors, including those for the hormones insulin, nerve growth factor, epidermal growth factor, chorionic gonadotropin, triiodothyronine, and somatomedin C; the transport proteins low-density lipoprotein (LDL), transcobalamin II, transferrin, and yolk proteins (lipovitellin, phosvitin) and for other proteins such as lysosomal enzymes, asialoglycoproteins, α-2-macroglobulin, maternal immunoglobulins (IgG), thrombin, diphtheria toxin, *Salmonella* toxin, *Pseudomonas* toxin, and various viruses.

The LDL receptors occur largely preclustered in coated pits, but most other receptors are diffusely distributed on the cell surface.[54–56] After ligands bind to their specific receptors, the ligand–receptor (L–R) complexes move to specialized regions of the plasma membrane and form clusters. These membrane regions possess or acquire a bristle coat consisting of the protein clathrin, which has subunits of 185,000 daltons that are polymerized in a basket configuration around the region and are designated (bristle-) coated pits.[57] These coated pits constitute the portal of entry for various ligands binding to physiologically important receptors. It has been widely assumed that the L–R complexes are internalized by invagination and pinching off of the coated pits to form coated vesicles, which then rapidly lose their clathrin coat.[56] However,

Pastan, Willingham, and associates have recently presented evidence that clathrin-coated vesicles are not formed during the endocytosis process and that coated pits are stable elements.[54,55] Instead, they propose that a smooth-walled vesicle 0.25–0.4 μm in diameter, termed a "receptosome," which forms in an as yet unknown manner from the coated pit, is used by all ligands. Receptosomes bearing ligand arrive at the Golgi zone in the perinuclear cytoplasm shortly (10–30 min) after entry, and subsequently (15–60 min) the ligand begins to accumulate in lysosomes and undergo degradation. According to these workers, receptosomes probably fuse selectively with certain portions of the Golgi–GERL system or with each other but do not fuse with lysosomes, ER, or other mature organelles.[54,55] The usual fate of ligands internalized by receptor-mediated endocytosis is lysosomal degradation. However, in certain instances, some ligand may also be directed to specific subcellular organelles, such as the nucleus, or else may be exocytosed without further change.

4.3. Cell Regulation of Endocytosis

The molecular mechanisms that regulate fluid-phase, adsorptive, and receptor-mediated endocytosis are poorly understood. There is evidence that extracellular Ca^{2+} is necessary for ligand binding to specific surface receptors,[56] L–R clustering in coated pits,[58] and fluid-phase and adsorptive pinocytosis.[59–62] Recruitment of clathrin coats to the membrane opposite L–R clusters is sensitive to the calmodulin-directed drug trifluoperazine.[62] Fluid-phase endocytosis is more markedly inhibited than receptor-mediated endocytosis by trifluoperazine.[62] Transglutaminase has been implicated in receptor-mediated endocytosis on the grounds that a diverse group of compounds, including primary alkylamines, dansylcadaverine, and bacitracin, which inhibit L–R clustering and internalization, also inhibit transglutaminase.[63] These workers postulate that a ligand-mediated activation of membrane-associated transglutaminase covalently links receptor proteins by forming an isopeptide bond between a lysine residue of one protein and a glutamine residue of another, leading to clustering.[63] Cytoskeletal components have been implicated in receptor-mediated and fluid endocytosis. Thus, the antimicrofilament agent cytochalasin B inhibits coated vesicle formation,[62] receptor-mediated endocytosis,[62,64] and fluid endocytosis.[59–65] Colchicine, an antimicrotubular agent, inhibited fluid endocytosis[65] and receptor-mediated endocytosis of LDL, apparently by decreasing high-affinity surface binding of LDL to fibroblasts.[66]

Studies in amebae indicate that considerable amounts of extracellular Ca^{2+} are bound to anionic sites of the mucous layer (glycocalyx) of the plasma membrane and stabilize it.[67,68] Pinocytosis-inducing agents range in size from ions to macromolecules, are frequently cationic, and bind electrostatically to anionic sites on the cell surface. Pinocytosis-inducing substances bind to these anionic sites, release the surface-bound Ca^{2+}, and destabilize the surface membrane, thereby increasing ion influx, including Ca^{2+}, into the cell. Membrane invagination and vesiculation involve a Ca^{2+}-mediated stimulation of the local contractile response of the submembranous microfilament (actomyosin) system and membrane flow process.[67] The Ca^{2+} ionophore A23187 directly stimulates endocytosis in amebae without the use of an inducer, presumably by increasing cytosolic Ca^{2+}.[69] Cyclic AMP stimulates fluid-phase pinocytosis in amebae[68] and mediates the pinocytosis induced by vasopressin in toad bladder.[70] Pinocytosis in brain capillaries is stimulated by cyclic AMP,[72,73] norepinephrine,[73] and serotonin.[73] Enhanced fluid-phase endocytosis is a characteristic feature of denervated,[74] dystrophic,[75] and protamine-treated muscle.[76] Enhanced endocytosis and vesicular transport are generally believed to be the principal mechanisms by which plasma protein and tracers are conveyed across the blood–brain barrier, i.e., the endothelium of the cerebral microvessels, in pathologic states associated with vasogenic brain edema.

5. *AUTOPHAGY*

The first well-defined step in the autophagic–lysosomal pathway of intracellular protein degradation is the sequestration of a region of cytoplasm by one or several membranes, which eventually form a closed vacuole, the autophagosome. The second step involves the juxtaposing of this autophagosome and a lysosome by movement of these organelles through the cytoplasm, followed by fusion of their membranes and the formation of a common organelle, the autolysosome or autophagic vacuole. The third step is the digestion of the sequestered cell components by the hydrolases, followed by diffusion or transport of the monomeric digestion products across the lysosomal membrane and/or by exocytotic extrusion of the lysosome from the cell. The morphological and cytochemical aspects of autophagy have been the subject of a recent review.[77]

5.1. *Role of Lysosomal Autophagy in Intracellular Protein Degradation*

Three general classes of cellular proteins have been distinguished according to their degradation rates. (1) Long-half-life proteins comprise the bulk of cellular proteins and display relatively slow turnover rates. (2) Short-half-life proteins constitute a minor portion of cellular proteins and are characterized by unusually rapid turnover rates. (3) Abnormal proteins are produced by the incorporation of amino acid analogues or puromycin or by certain mutations and turn over even more rapidly than short-half-life proteins.[78] It is generally thought that the degradation of short-half-life proteins and abnormal proteins occurs mostly by nonlysosomal mechanisms, possibly involving a soluble energy- and ATP-dependent proteolytic system. The degradation of these short-lived proteins is little affected by hormones, nutritional deprivation, or inhibitors of protein synthesis or of lysosomal proteolysis. With respect to the degradation of the long-lived proteins, two states have been distinguished, an accelerated or enhanced state under conditions of nutritional privation and a basal state in the presence of adequate nutrients, growth-promoting factors, or insulin. It is now reasonably well established that enhanced protein degradation in mammalian cells is largely mediated by the autophagic–lysosomal pathway.

However, there has been considerable debate concerning the role of the lysosomal system in the basal degradation of long-lived proteins.

Evidence for the participation of lysosomes in protein degradation has come chiefly from studies in perfused rat liver and in isolated or cultured cells. In all instances, general protein breakdown increases markedly during nutrient deprivation or glucagon treatment, and this accelerated proteolysis is promptly suppressed by insulin, amino acids, and various growth-promoting factors. The conclusion that this deprivation-induced proteolysis occurs through augmented lysosomal autophagy is supported by several lines of evidence.[78–82]

5.1.1. *Studies in Perfused Liver*

Early experiments by Mortimore and associates[83] revealed that lysosomes prepared from livers perfused with an unsupplemented medium, which enhances proteolysis, show an increased sensitivity to lysis in hypotonic sucrose, reflecting an increased accumulation of autophagic vacuoles. Addition of insulin, amino acids, or cycloheximide to the perfusion fluid prevented the increase in osmotic sensitivity of the lysosomes as well as the accelerated proteolysis. The osmotic alterations were rapidly reversed by amino acids or insulin with a half-life of about 8 min.[83] Ultrastructural observations of liver subsequently confirmed that perfusion with unsupplemented medium induced the appearance of numerous autophagic vacuoles, and this was suppressed by amino acid supplementation.[84] The amount of endogenous proteolysis in lysosomes incubated *in vitro* correlated well with protein breakdown in intact livers under various physiological conditions (perfusion with unsupplemented medium or with amino acids, insulin, or glucagon). Proteolysis occurred only in particulate suspensions and not in particle-free supernatants.[87]

More recently, fractional volumes of lysosomal–vacuolar elements and long-lived protein degradation were quantitatively correlated in rat livers perfused with varying levels of amino acids.[86] The turnover rate ($t_{\frac{1}{2}}$) of autophagic vacuoles, estimated from their regression rate after addition of amino acids or insulin, was found to be 7.5 min, close to the value of 8 min found by the osmotic sensitivity method.[83] The rates of lysosomal proteolysis calculated from the fractional volumes of the lysosomal vacuoles and the rate constant of the turnover of these vacuoles agreed quantitatively with the rates of overall protein degradation over a range of amino acid concentrations.[86]

In a third approach, Mortimore and Ward[87] estimated the degradable intralysosomal protein pool by prolonged *in vitro* incubation of liver homogenates and subcellular fractions thereof to allow proteolysis of endogenous proteins to go to completion. The rates of lysosomal proteolysis calculated by this method were in good agreement with rates of protein degradation obtained in various physiological states, e.g., perfusion with unsupplemented medium or with glucagon, insulin, or amino acids. These experiments provide strong evidence that essentially all of deprivation-accelerated degradation of intracellular proteins is mediated by lysosomal autophagy. In addition, the demonstration of a distinct intralysosomal pool of degradable protein under basal conditions (in untreated liver and after perfusion with insulin or amino acids) indicates that basal protein degradation is also mediated by the lysosomal system.[87]

5.1.2. *Effects of Lysosomal Inhibitors*

Important evidence for the role of lysosomes in intracellular protein catabolism comes from studies involving the use of agents that specifically inhibit lysosomal proteolysis. Wibo and Poole[88] first demonstrated that chloroquine inhibits protein turnover in cultured fibroblasts by inhibiting lysosomal cathepsin B. Chloroquine is one of a number of weakly basic substances, designated lysosomotropic amines,[89] that accumulate within lysosomes, thereby raising the intralysosomal pH. The plasma and lysosomal membranes are highly permeable to the unionized form of the weak bases but are relatively impervious to the protonated form of the bases. Since the pH inside lysosomes is considerably lower than that in the ambient cytoplasm, the weak bases are protonated and trapped inside lysosomes. As the concentration of the base inside the lysosome approaches isotonicity, water enters the lysosome osmotically, and the lysosomes swell to form large vacuoles.[89–93] Thus, chloroquine is concentrated about 1000-fold within lysosomes, and direct measurement of intralysosomal pH in living macrophages with a fluorescence probe (fluorescein-labeled dextran) technique disclosed a marked increase in pH (from *ca.* 4.7 to 6.5), an increase sufficient to block most lysosomal protease activity.[91] Other lysosomotropic amines, e.g., neutral red, acridine orange, NH_4Cl, methylamine, and tributylamine, accumulate in lysosomes, induce lysosomal vacuolation, increase intralysosomal pH, and inhibit lysosomal proteolysis.[81,90–94]

Dean[95] first used a specific lysosomal protease inhibitor to investigate the role of lysosomes in protein catabolism. He found in perfused rat liver that pepstatin, an inhibitor of lysosomal cathepsin D, when administered enclosed in phagocytosable liposomes to facilitate uptake by hepatocytes substantially inhibited the turnover of long-lived protein.[95] Leupeptin, an inhibitor of lysosomal cathepsin B and some other serine proteases, and antipain, an inhibitor of cathepsins A, B, papain, and trypsin,[96] were subsequently found to inhibit protein degradation in hepatoma cells[97] and hepatocytes.[98,99] Leupeptin decreased the degradation rates of average protein in denervated and dystrophic skeletal muscle by 35–50% and inhibited lysosomal cathepsin B,[100] thus possibly accounting for the observation of McGowan *et al.*[101] that leupeptin, antipain, and pepstatin retard the degeneration of dystrophic muscle cells *in vitro*. In cultured myotubes, leupeptin, chymostatin, and chloroquine decreased the degradation rate of the acetylcholine receptors by two-to 11-fold but had a much smaller effect on the catabolism of average muscle protein.[102] These findings provide support for a lysosomal role in intracellular protein catabolism, although it should be noted that these peptide inhibitors may inhibit intracellular proteinases other than lysosomal proteases (except for pepstatin, which apparently inhibits only cathepsin D).

5.2. Hormonal Regulation of Lysosome-Mediated Protein Degradation

As already mentioned, insulin inhibits protein degradation in liver, skeletal and heart muscle, and various other cell types.[103–105] Insulin decreases lysosome-mediated protein catabolism in liver by rapidly suppressing the formation of autophagic vacuoles.[83,85,86] Conversely, glucagon increases endogenous protein degradation in liver by enhancing lysosomal autophagy.[106,107] Cyclic AMP also stimulates lysosomal autophagy in liver *in vivo*.[107] In cultured hepatocytes, glucagon apparently stimulates proteolysis by increasing cyclic AMP levels.[104] β-Adrenergic agonists enhance intracellular protein breakdown in cultured hepatocytes by increasing cyclic AMP, whereas in skeletal muscle they increase cyclic AMP but inhibit protein degradation.[104,105]

Thyroidectomy decreases protein degradation, and thyroxine or triiodothyronine enhances protein degradation in skeletal muscle and liver but not in heart or kidney.[103–105] Thyroidectomy also decreases the levels of several lysosomal proteases in liver and muscle, and thyroid hormones increase these enzymes.[108] These findings suggest that thyroid hormones influence intracellular protein degradation by regulating lysosomal proteolysis.[108] Thyroid hormones also regulate the activities of various other lysosomal enzymes in rat liver and muscle.[108] It has therefore been suggested that the two characteristic features of human hypothyroidism, the accumulation of mucopolysaccharides, i.e., myxedema, and the increase in serum cholesterol esters, may result from the reduced activities of the lysosomal hydrolases that are rate limiting in the catabolism of hyaluronic acid and cholesterol esters.[103,108] The muscle wasting and weakness of thyrotoxic myopathy may be related, in part, to enhanced degradation of muscle proteins.[103] Hypophysectomy decreases protein turnover in muscle and liver as a result of the decrease in circulating thyroid hormones.[103]

Glucocorticoids are considered catabolic in muscle but anabolic in liver. Hence, the effect of these steroid hormones on protein degradation in different cell types is of considerable interest. Glucocorticoids apparently decrease muscle proteolysis in fed rats but increase muscle proteolysis in fasted rats. These contrasting effects of glucocorticoids suggest that another hormone dependent on nutritional status, probably insulin, influences the muscles response to glucocorticoids.[103] Although they are generally assumed to promote proteolysis, the effects of glucocorticoids on protein catabolism in muscle remain controversial in view of the fact that a number of studies have yielded contradictory results.[103,105] Studies in the author's laboratory have revealed that glucocorticoids decrease the levels of several lysosomal hydrolases in rodent brain, skeletal muscle, heart, and kidneys, suggesting that these steroid hormones regulate (decrease) the activity of the lysosomal–vacuolar system.[109] Furthermore, glucocorticoids antagonize the long-term effects of androgens in mouse kidney and several other organs, including the androgen-induced increase in lysosomal enzymes (see below).[110]

Testosterone generally exerts an overall anabolic effect on nonsexual target organs such as kidney, heart, skeletal muscle, and aorta. In addition, testosterone regulates (increases) the activity of the lysosomal–vacuolar system in these tissues. We have described a testosterone-mediated sexual dimorphism in the ultrastructure of mouse kidney proximal tubule cells involving the lysosomes, the tissue activities of numerous lysosomal hydrolases, and the urinary excretion of lysosomal enzymes.[111] In proximal tubules of male mice, autophagic vacuoles and lysosomes (predominantly membrane-filled myeloid bodies) are more numerous and voluminous, and exocytosed intraluminal lysosomes are more common than in female mice. Males have higher kidney activities of lysosomal hydrolases and excrete larger amounts of hydrolases and protein into the urine. Orchidectomy evokes the feminine pattern, whereas testosterone administration induces the male pattern, indicating that endogenous testosterone regulates (enhances) the activity of the lysosomal–vacuolar system in proximal tubule cells.[111]

We have also observed a testosterone-mediated sex difference in lysosomal enzyme levels in mouse and rat skeletal muscle,[112,113] aorta,[114] heart,[113,115] and brain.[116] Lysosomes from mouse and rat kidney,[117] brain,[117] and heart[115,117] display a characteristic testosterone-dependent sex difference in that the enzyme latency and membrane stability of these lysosomes in males are substantially lower than in females, consistent with a greater degree of autophagy in the male tissues. Testosterone induces an acute (<5 min) decrease in enzyme latency and membrane stability of kidney lysosomes and coincidentally reduces the equilibrium density of a large proportion of lysosomes in a sucrose gradient.[117] These acute changes reflect the properties of an enlarged population of heterolysosomes resulting from testosterone-stimulated pinocytosis in proximal tubule cells.[117,118]

5.3. Lysosome-Mediated Protein Degradation in Muscle Pathology

Enhanced lysosomal autophagy and proteolysis also occur in various pathological states. Thus, an increase in lysosome-mediated proteolysis plays a major role in the development of skeletal muscle atrophy resulting from denervation,[100] muscular dystrophy,[100] and endocrinopathies.[103,108] The degradation of acetylcholine receptors involves endocytosis and lysosomal digestion of the membrane-associated receptor protein[98,119] and is markedly accelerated in myasthenia gravis because of the formation of immune complexes with humoral antibodies directed toward the receptor.[120] Furthermore, in innervated muscle, the acetylcholine receptors are localized in the neuromuscular junctions; after denervation there is a great increase of extrajunctional acetylcholine receptors, and the turnover rate of extrajunctional and junctional receptors increases sharply because of enhanced lysosomal degradation.[119]

As mentioned earlier (Section 4.1.2), the lysosomal inhibitors leupeptin, chymostatin, and chloroquine reduce the degradation rate of the acetylcholine receptors to a much greater extent than that of the average muscle protein in cultured myotubes.[102] In denervated and dystrophic muscle, leupeptin decreased the degradation rates of average protein by 35–50% coincident with inhibition of lysosomal cathepsin B.[100] Accordingly, leupeptin, antipain, and pepstatin decrease the degeneration of dystrophic muscle cells grown in culture.[101] We found that dexamethasone and aspirin suppress the denervation-induced increase in acid hydrolases and attenuate the muscle atrophy, suggesting that these agents may inhibit lysosomal proteolysis, possibly by interfering with prostaglandin synthesis.[109,121] This subject has been discussed in a recent review.[122]

5.4. *Molecular and Cellular Regulation of Lysosomal Proteolysis*

The molecular signals that regulate lysosomal proteolysis in normal and diseased tissues are largely unknown. It seems likely that there are specific control points for autophagosome formation, phagolysosome fusion, and lysosomal hydrolysis. Continuous protein synthesis appears to be necessary both for autophagosome production and lysosomal digestion. Thus, cycloheximide, given in a dose that abolishes protein synthesis, blocks deprivation-induced lysosomal proteolysis partly by suppressing the formation of autophagosomes and partly by decreasing lysosomal protease (cathepsins B and D) activities.[123,124] A rapidly turning over protein could be important in autophagosome production.

Phagolysosome fusion probably involves microtubule-mediated translocations and contiguities of autophagosomes and lysosomes, since microtubular poisons such as vinblastine inhibit protein degradation while inducing a paradoxical accumulation of autophagosomes.[125] Leupeptin and the lysosomotropic amine propylamine also elicit an accumulation of autophagosomes coincident with inhibition of lysosomal proteolysis, suggesting that such accumulation may be secondary to the inhibition of lysosomal digestion.[125] In contrast, amino acid mixtures, particularly leucine and asparagine, inhibit protein degradation in various biological systems but suppress the formation of autophagic vacuoles.[84,86,99] The methylated adenosine derivatives 6-dimethylaminopurine riboside and puromycin aminonucleotide[127] inhibit lysosomal degradation to a much greater extent than protein synthesis, apparently by suppressing autophagosome formation.[126] 3-Methyladenine is a particularly potent inhibitor of autophagosome formation and lysosome-mediated proteolysis.[127] Low temperature (20°C) selectively inhibits fusion of heterophagosomes and lysosomes in perfused hepatocytes, possibly by reducing membrane fluidity.[128] An analogous inhibition of autophagosome–lysosome fusion may account for the decrease in the degradation of long-lived proteins observed in cultured hepatocytes at low temperature (17°C).[98]

5.4.1. *Calcium and Prostaglandins as Regulatory Signals*

Ionic calcium and prostaglandins appear to be important regulatory signals for lysosomal proteolysis in skeletal and cardiac muscle. Goldberg and associates[129,130] showed that prostaglandins E_2 and $F_{2\alpha}$ stimulate protein degradation in skeletal and cardiac muscle. The stimulation of protein degradation by PGE_2 and $PGF_{2\alpha}$ and by arachadonic acid, the precursor of prostaglandins, could be blocked by leupeptin and Ep475, inhibitors of lysosomal thiol proteases, as well as by indomethacin, an inhibitor of prostaglandin synthetase (cyclooxygenase), indicating that prostaglandins probably activate intralysosomal proteolysis.[129]

The calcium ionophores A23187 and ionomycin in the presence of extracellular calcium stimulate protein degradation, presumably by increasing intracellular Ca^{2+}.[130,131] High extracellular potassium (100 mM), which can depolarize the muscle membrane and increase cytosolic Ca^{2+} by mobilizing calcium in the sacroplasmic reticulum, also enhances muscle proteolysis.[130] This effect seems to involve lysosomal proteases, as it is blocked by leupeptin and Ep475. A23187 and high potassium also stimulate prostaglandin synthesis in muscle. Moreover, indomethacin and aspirin, cyclooxygenase inhibitors, and mepacrine, an inhibitor of phospholipase A_2, suppress prostaglandin synthesis and the proteolysis induced by A23187 and high potassium. Thus, Ca^{2+} appears to promote muscle protein degradation by activating phospholipase A_2 and increasing the arachidonic acid pool, thereby stimulating prostaglandin synthesis and enhancing lysosomal proteolysis.[130] Prostaglandins are increased in denervated muscle.[132] The ability of dexamethasone and aspirin, which are known to inhibit prostaglandin synthesis, to suppress the denervation-induced increase in lysosomal enzymes and atrophy in skeletal muscle lends strong support to the view that prostaglandins are involved in stimulating intralysosomal proteolysis.[109,121]

5.4.2. The Role of Polyamines as Regulatory Signals

Increased synthesis of the polyamines putrescine, spermidine, and spermine and of their rate-regulatory synthetic enzyme ornithine decarboxylase (ODC) is one of the earliest events that occur during tissue growth, differentiation, replication, and transformation. The polyamines are thought to be involved in regulating nucleic acid and protein synthesis, although their precise physiological roles and mechanism of action are still unknown. Recent studies in the author's laboratory have disclosed that the polyamines are involved in the regulation of certain aspects of lysosomal function.

5.4.2a. Polyamines in Muscle Denervation. Polyamine synthesis, measured by ODC activity and the concentrations of putrescine, spermidine, and spermine, is increased in denervated[133,134] and dystrophic muscle.[133,135] We recently confirmed that polyamine synthesis is enhanced in denervated muscle and showed that this synthesis is indispensible to the postdenervation induction of lysosomal enzymes and muscle atrophy in mouse gastrocnemius.[136,137] Administration of the specific, irreversible ODC inhibitor α-difluoromethylornithine (DFMO) to mice inhibited the postdenervation increase in ODC and polyamines, suppressed the induction of cathepsin B, β-galactosidase, and several other lysosomal enzymes, and decreased the loss in gastrocnemius wet weight and protein by 65% 4 days following sciatic nerve section. Administration of exogenous putrescine nullified the effects of DFMO, thereby verifying that putrescine (and polyamine) depletion is responsible for DFMO suppression of the denervation-induced increase in lysosomal enzymes and loss of muscle mass and protein. These data implicate enhanced polyamine synthesis in the regulation of lysosomal enzyme levels and lysosome-mediated proteolysis in skeletal muscle during denervation atrophy.

We previously reported that aspirin and dexamethasone attenuate the denervation-induced increase in lysosomal enzymes and muscle atrophy in mouse gastrocnemius.[109,121] We have since found that these antiinflammatory agents exert these effects by suppressing the increase in ODC activity and polyamine concentrations in denervated mouse gastrocnemius.[136,137] Aspirin inhibits cyclooxygenase, whereas dexamethasone inhibits the phospholipase-A_2-mediated release of arachidonic acid, the precursor of prostaglandins. Therefore, these findings suggest that enhanced prostaglandin synthesis may be mandatory for the denervation-mediated increase in ODC activity, possibly via activation of a prostaglandin-linked adenylate cyclase. Cyclic AMP and adenylate cyclase levels are increased in denervated muscle,[134] as are prostaglandins.[132]

5.4.2b. Polyamines in Androgenic Hormone Action. In an earlier section (5.2), we described the impact of testosterone on the lysosomal–vacuolar system of a number of extragenital tissues in rodents. These tissues also exhibit a testosterone-mediated sex difference in polyamine synthesis. Male mice display higher levels of kidney ODC and polyamine and excrete larger amounts of polyamines into the urine than female mice.[138] Orchidectomy evokes the female pattern, and testosterone induces the male pattern of polyamine synthesis and excretion. Furthermore, DFMO suppresses the testosterone-induced increase in kidney ODC and polyamines and urinary excretion of polyamines and also suppresses the testosterone-induced increment in kidney lysosomal hydrolases and lysosomal enzymuria resulting from exocytosis of lysosomes. Administration of exogenous putrescine augmented kidney polyamine levels and nullified the inhibitory effect of DFMO. These data indicate that enhanced ODC activity and polyamine synthesis are indispensible in mediating the effects of testosterone on the lysosomal–vacuolar system of proximal tubules, including the increase in lysosomal enzymes, exocytosis of lysosomes, and lysosomal enzymuria.[138] Utilizing a similar approach, we demonstrated that polyamines are involved in the testosterone-mediated increase in lysosomal enzymes in skeletal muscle,[139,140] heart,[139] brain,[141] and aorta (H. Koenig, A. Goldstone, and C. Y. Lu, unpublished data).

Polyamine synthesis is also mandatory for the mediation of early androgenic responses. We recently found that testosterone induces an early (<1 min), Ca^{2+}- and receptor-dependent stimulation of endocytosis (also hexose and amino acid transport) in mouse kidney cortex.[142,143] This response involves Ca^{2+} fluxes apparently leading to an increase in cytosolic Ca^{2+}.[144,145] Testosterone induced a rapid (<30s), transient threefold increase in ODC activity and an early (<2 min), sustained, 1.5- to twofold increase in polyamines in mouse kidney cortex slices,[146,147] whereas DFMO blocked the testosterone-induced increment in polyamines and endocytosis (also hexose and amino acid transport) and also abolished the early (<30 s) testosterone-induced increase in the influx and efflux of $^{45}Ca^{2+}$ and the mobilization of mitochondrial ^{45}Ca.[147] Exogenous putrescine nullified DFMO inhibition and restored the increment in polyamines, Ca^{2+} fluxes, and endocytosis. These findings indicate that enhanced polyamine synthesis is obligatory for the testosterone-mediated stimulation of Ca^{2+} fluxes, endocytosis, and hexose and amino acid transport.[147]

The early (<5 min) testosterone-induced activation of the lysosomal system in mouse kidney *in vivo* reflects an augmented formation of phagolysosomes and autolysosomes, which are characterized by decreased lysosomal enzyme latency and increased osmotic sensitivity resulting from enhanced endocytosis and probably autophagy.[117] When administered just before testosterone, DFMO blocked lysosomal activation, whereas putrescine itself evoked an activation of the lysosomal system.[117] These observations strongly suggest that polyamines are implicated in the mediation of androgenic effects on the lysosomal system in target cells, including endocytosis, autophagy, and phagolysosome formation. The prior administration of the calcium channel blocker verapamil or the prostaglandin synthesis inhibitors aspirin and dexamethasone also abolished the testosterone-induced lysosomal activation, indicating that this activation is dependent on Ca^{2+} influx and prostaglandin synthesis. Androgenic stimulation of endocytosis and membrane transport in cortex slices is also dependent on Ca^{2+} influx, as it is blocked by Ca^{2+}-free medium, EGTA, and verapamil.[144,145] Testosterone-mediated stimulation of endocytosis and hexose and amino acid transport also requires prostaglandin synthesis, as aspirin and dexamethasone suppress this response in a dose-dependent manner, and this inhibition is reversed by micromolar concentrations of the prostaglandins PGA_2 and PGE_2 (H. Koenig, A. Goldstone, and C. Y. Lu, unpublished data). The testosterone-induced increase in ODC activity and polyamine levels is also dependent on extracellular Ca^{2+} and prostaglandin synthesis. (H. Koenig, A. Goldstone, and C. Y. Lu, unpublished data).

5.4.2.c. Isoproterenol-Induced Endocytosis. In kidney, β-adrenergic stimulation generally enhances cyclic AMP formation and has been implicated in increasing tubular reabsorption of Na^+ and the production and secretion of renin and kallikrein. We found that the β-adrenergic agonist isoproterenol (1 μM) induces an early stimulation of endocytosis and hexose and amino acid transport in mouse kidney cortex slices.[148] In histochemical preparations stained for the endocytosable marker horseradish peroxidase, isoproterenol-induced endocytosis was restricted to the proximal tubules. This rapid membrane response is receptor and Ca^{2+} dependent, as it is blocked by the β-adrenergic antagonist propranolol and by deleting Ca^{2+} from the medium. This β-adrenergic response also involves an early (<30 s) increase in $^{45}Ca^{2+}$ influx and efflux and a mobilization of mitochondrial ^{45}Ca.[148] Isoproterenol (1 μM) induced a rapid, transient increase (up to sevenfold) in ODC activity and an early sustained increase in putrescine, spermidine, and spermine concentrations in cortex slices. Small doses of isoproterenol in mice elicit a rapid (<2 min) increase in kidney and heart polyamine concentrations *in vivo*; DFMO suppressed the isoproterenol-induced increment in polyamines, Ca^{2+} fluxes, endocytosis, and hexose and amino acid transport, whereas exogenous putrescine reversed DFMO Inhibition.[148a] These data show that enhanced polyamine synthesis is an early event following β-adrenoreceptor activation and is indispensible for the mediation of the early membrane response.

Many surface agonists are known to increase ODC and polyamine levels in responsive tissues after a lag period of ~4 h. We found that a number of different surface agonists and ligands induce a rapid (<5 min), Ca^{2+}-dependent stimulation of endocytosis and hexose and amino acid transport in mouse kidney cortex slices.[149] The following agonists were also found to induce a rapid (<1.5 min) transient increase in ODC activity and an early (2–4 min), sustained increase in putrescine, spermidine, and spermine concentrations in cortex slices *in vitro*[150]: the tumor promotor phorbol myristate acetate (PMA, 15 nM); the mitogen concanavalin A (con A, 140 nm); the prostaglandins PGA_2 and PGE_2 (1.5–5 μM); the Ca^{2+} ionophore A23187 (5 μM); and dibutyryl cyclic AMP (3 mM). The ODC activation and polyamine accumulation apparently precede the early stimulation of membrane transport functions induced by these diverse ligands. Studies involving the use of the ODC inhibitor DFMO with PMA and con A indicate that enhanced polyamine synthesis is mandatory for the mediation of the plasma membrane transport response to these agonists.[150] These data suggest that a rapid activation of a latent ODC (probably associated with the plasma membrane or submembranous cytosol) and an early, polyamine-mediated stimulation of plasma membrane transport functions, probably triggered by a cytosolic Ca^{2+} signal, may be a general feature of surface receptor activation.[147-150]

5.4.2d. Mechanism of Action of Polyamines. Our data indicate that polyamine synthesis is mandatory for the increase of Ca^{2+} fluxes induced in kidney cortex by testosterone[146,147] and β-adrenergic agonists.[148a] On the basis of these and other findings, we have proposed that the polyamines serve as an intracellular signal to increase cytosolic Ca^{2+} concentrations and stimulate Ca^{2+}- (or Ca^{2+}- and calmodulin-) mediated processes. The postulated mechanism is that polyamines, the major organic cations of cells, enhance Ca^{2+} fluxes and mobilize intracellular calcium via a cation-exchange reaction. With respect to denervation-induced muscle proteolysis, the available evidence is consistent with the hypothesis that nerve section initiates a cascade of membrane reactions involving sequential increases in Ca^{2+} influx, prostaglandin synthesis, adenylate cyclase activity, and cyclic AMP levels and leads to an increase in ODC activity.[109,121,136,137] The resultant accumulation of cellular polyamines evokes a rise in cytosolic Ca^{2+}, which in turn activates the lysosomal–vacuolar system (and associated cytoskeletal components), stimulating endocytosis, autophagocytosis, phagolysosome fusion, and lysosomal enzyme synthesis. A similar scheme appears to be compatible with the testosterone-induced activation of the lysosomal–vacuolar system[117,138,142-147] and the isoproterenol-mediated stimulation of endocytosis in mouse kidney.[148a]

6. LYSOSOMAL STORAGE DISORDERS

Lysosomal storage generally results from incomplete degradation of material that has entered the lysosomal–vacuolar system via heterophagic or autophagic pathways. The mechanisms for such storage are manifold but can usually be reduced to defective enzymatic activity (enzyme deficiency; enzyme inhibition; suboptimal conditions for enzymatic activity) or a substrate abnormality (foreign, inert, or indigestible substrate; surfeit of normal substrate). Lysosomal storage may result from endogenous causes or may be induced by external factors.

6.1. Endogenous Lysosomal Storage

The inherited storage diseases are the best known form of pathological lysosomal storage. These number at least 23 distinct disorders that result from a genetic deficiency of one or more lysosomal enzymes and include such familiar disorders as Tay–Sachs, Niemann–Pick, and Gaucher's diseases as well as rare diseases such as Fabry, Scheie, Moroteaux–Lamy, and Morquio syndromes. These diseases feature the lysosomal accumulation of indigestible residues derived from the incomplete lysosomal degradation of glycoproteins or mucopolysaccharides as a result of the deficiency of a specific acid hydrolase necessary for their digestion. Eventually, these engorged lysosomes come to occupy a large fraction of the cell cytoplasm, interfering with cell function and leading to premature death of the cells.

These diseases have been the subject of several excellent reviews.[151,152] An interesting recent development is the demonstration that patients with the AB form of Tay–Sachs disease (in which hexosaminidase A and B activities are normal but there is nevertheless a pathological accumulation of GM_2 ganglioside) have a decreased amount of a heat-stable "activator" necessary for glycolipid catabolism.[153] In one hereditary storage disease, cystinosis, the lysosomal accumulation has been shown to be caused by a decreased maximum rate of egress of cystine from lysosomes as a result of a defective carrier-mediated transport system for the amino acid.[154] Neuronal ceroid lipofuscinosis exemplifies endogenous storage resulting from substrate indigestibility. This group of hereditary neurological diseases features an accumulation of autofluorescent ceroid lipofuscin pigments in residual lysosomes in brain and other tissues. These lysosomal granules contain greatly increased amounts of lipid-insoluble retinoic acid complexes[155] and lipid-soluble dolichols, and increased amounts of dolichols appear in the urine of patients with the disease.[156]

6.2. Lysosomal Storage Induced by Exogenous Agents

Abnormal lysosomal storage can be induced by introducing agents that interfere with enzyme activity, e.g., enzyme inhibitors such as inhibitory lipid analogues,[157] conduritol β-epoxide,[158] antibodies to lysosomal enzymes,[159] and lysosomotropic amines that raise lysosomal pH.[89–93] Alternatively, the administration of inert or indigestible substances, e.g., Triton WR-1339, thorotrast, heavy metals, sucrose, mannitol, dextrans, or polyvinylpyrrolidine, leads to severe overloading of lysosomes in cells that endocytize these substances. Various aspects of this subject have been presented in several recent reviews.[160–162]

6.2.1. Lysosomal Storage Induced by Cationic Amphiphilic Drugs

A wide variety of drugs with diverse therapeutic actions, when given for prolonged periods, induce a generalized "phospholipidosis" characterized by an intracellular accumulation of enlarged, membrane-filled lysosomes (myeloid bodies, cytoplasmic membranous bodies) resembling those seen in human cases of inborn lipidoses such as Tay–Sachs and Niemann–Pick disease. These agents differ widely in structure but share common features, namely, a hydrophobic moiety and a hydrophilic residue, usually a primary or substituted amine that is positively charged. These agents are therefore amphiphilic (amphipathic) cations and include the following examples: the anticholesterolemic agents triparanol and 20,25-diazacholesterol; the antimalarial and antirheumatic chloroquine; the coronary vasodilator 4,4-diethylaminoethoxyhexestrol; the anorectic drugs chlorphentermine and fenfluramine; the tricyclic antidepressants iprindole, imipramine, and amitriptyline; the antihistamine drugs chlorcyclizine and meclizine; the neuroleptic chlorpromazine; and the aminoglycoside antibiotics gentamicin, erthromycin, and clindamycin. This subject has been covered in a detailed review.[163]

These drugs cause an accumulation of lipids, mainly glycerophospholipids, that is largely confined to lysosomes. In contrast to the genetic lipidoses, in which sphingolipids, particularly glycosphingolipids, are the principal lipids accumulating in tissues, in drug-induced lipidoses, there are increased proportions of acidic phospholipids (phosphatidylinositol, phosphatidylglycerol, phosphatidic acid, lysobisphosphatidic acid) and decreased proportions of the major neutral phospholipids (phosphatidylcholine, phosphatidylethanolamine) and triacylglycerol. Several amphiphilic cations have been shown to accumulate in lysosomes, and it is likely that all such agents that induce lipid accumulation undergo sequestration in lysosomes. One explanation for drug-induced phospholipidoses is that phospholipids are normally degraded in lysosomes by acid phospholipases. When complexed with amphiphilic cations, however, the phospholipids are protected from phospholipase attack, with the consequence that phospholipid–drug complexes accumulate and produce lysosomal hypertrophy.[163] An alternative theory holds that the lipid accumulation results from a drug-induced increase in the biosynthesis of acidic phospholipids. Many amphiphilic cations decrease the rates of synthesis of triacylglycerol and the neutral phospholipids phosphatidylcholine and phosphatidylethanolamine while increasing the synthesis of the acidic phospholipids, e.g., phosphatidic acid, phosphatidylinositol, phosphatidylglycerol, and CDP-diacylglycerol.[164,165] These agents inhibit phosphatidate phosphohydrolase, the enzyme catalyzing the conversion of phosphatidate to 1,2-diglyceride.[166] This enzyme lies at the main branch point in phospholipid synthesis and promotes the formation of the neutral phospholipids and triacylglycerol from 1,2-diglyceride. Its inhibition leads to the conversion of phosphatidate to CDP-diglyceride and a resultant increase in the synthesis of the acidic phospholipids.[166] It is of course possible that both of these mechanisms operate jointly to bring about the accumulation of phospholipids.

The cellular and molecular events that initiate the storage phenomenon have not been clearly defined. Moreover, the mode of entry of most of these agents is unknown. In the case of certain drugs, e.g., chloroquine, neutral red, and acridine orange, an extensive intralysosomal accumulation occurs within minutes, and it is generally assumed that this lysosomotropism involves a passive permeation of the plasma and lysosomal membrane by the unprotonated molecules and a protonation of these molecules in the acid interior of the lysosome, causing them to become "trapped."[89–93] However, most of the amphiphilic cations are protonated at physiological pH and cannot diffuse through the cell membrane but are able to bind to anionic residues in the cell surface. Therefore, adsorptive endocytosis and phagolysosome fusion probably constitute the major pathways for the transmembrane transport and lysosomal accumulation of these agents.

There is evidence that the aminoglycoside gentamicin gains entrance into kidney proximal tubule cells by endocytosis at the luminal surface and subsequently accumulates in lysosomes[167] after first binding to specific membrane "receptors" that have been characterized as anionic phospholipids.[168] Other aminoglycosides and organic cations, e.g., polylysine, spermine, tetralysine, and spermidine, inhibit gentamicin uptake by competing for the same transport system with relative affinities that correlate well with their net cationic charge.[169] Amphiphilic cations probably enhance self-uptake into cells by stimulating endocytotic activity because of their cationic nature (see Section 4.3). In keeping with this inference, we found that gentamicin stimulates fluid-volume endocytosis as well as hexose and amino acid transport in kidney cortex slices *in vitro*.[170] This uptake appears to be a polyamine-dependent process, as the ODC inhibitor DFMO suppresses gentamicin-stimulated endocytosis and hexose and amino acid transport (H. Koenig, A. Goldstone, and C. Y. Lu, unpublished data). Increased intracellular autophagy has been observed in various cell types treated with gentamicin,[171] chloroquine,[172,173] and other amphiphilic cations,[173] and this response has been invoked in the pathogenesis of myeloid body formation.[173]

We found that chloroquine and iprindole induce an early (<1 h) activation of the lysosomal system in rat brain, manifested as a decrease in lysosomal enzyme latency and an increased osmotic sensitivity of the lysosomes.[109] Pretreatment with dexamethasone or aspirin suppressed the lysosomal activation evoked by these agents, suggesting that prostaglandin synthesis is involved in this response,[109,174] and DFMO also inhibited this drug effect, thereby implicating polyamine synthesis in its mediation (H. Koenig and C. Y. Lu, unpublished data). The cellular basis for the lysosomal activation has not been established, but it probably reflects a drug-mediated stimulation of endocytosis and/or autophagocytosis of brain cells.

On the basis of these and other observations, we advance the following model for drug-induced phospholipidosis. Amphiphilic cations bind by electrostatic interaction forces to anionic phospholipids on the plasma membrane and initiate a cascade of membrane reactions that involves an activation of phospholipase A_2 and enhanced prostaglandin and polyamine synthesis and leads to enhanced endocytosis and autophagocytosis (see Section 5.4.2 for a detailed discussion). The interiorization and lysosomal compartmentation of cationic drug–anionic phospholipid complexes by "receptor"-mediated (adsorptive) endocytosis and phagolysosome fusion and the resistance of the complexed phospholipids to lysosomal phospholipases would account for the selective storage of acidic phospholipids. The enhanced synthesis of acidic phospholipids is compensatory and would serve to replace depleted membrane phospholipids.

ACKNOWLEDGMENTS. The research conducted in the author's laboratory was supported by the Research Service of the Veterans Administration and NIH grants (NS 14700, NS 18047, and HL 26835).

REFERENCES

1. Goldstone, A., and Koenig, H., 1968, *J. Histochem. Cytochem.* **16:**511–512.
2. Goldstone, A., and Koenig, H., 1970, *Life Sci.* **9:**1341–1350.
3. Koenig, H., 1980, *The Biochemistry of Brain*, Volume 2 (S. Kumar, ed.), Pergamon Press, Oxford, pp. 559–578.
4. Goldstone, A., and Koenig, H., 1973, *Biochem. J.* **132:**259–266.
5. Goldstone, A., Konecny, P., and Koenig, H., 1971, *FEBS Lett.* **13:**68–72.
6. Needleman, S. B., and Koenig, H., 1975, *Biochim. Biophys. Acta* **379:**43–56.
7. Bishayee, S., and Bachhawat, B. K., 1974, *Biochim. Biophys. Acta* **334:**378–388.
8. Plapp, B. V., and Cole, R. D., 1966, *Arch. Biochem. Biophys.* **116:**193–266.
9. Tulsiani, D. R. P., Keller, R. K., and Touster, O., 1975, *J. Biol. Chem.* **246:**5398–5406.
10. Bernardi, G., Appella, E., and Zito, R., 1965, *Biochemistry* **4:**1725–1729.
11. Derechin, M., Ostrowski, W., Galka, M., and Barnard, E. A., 1971, *Biochim. Biophys. Acta* **250:**143–154.
12. Graham, E. R. B., and Roy, A. B., 1973, *Biochim. Biophys. Acta* **329:**88–92.
13. Banerjee, D. K., and Basu, D., 1975, *Biochem. J.* **145:**113–118.
14. Aruna, R. M., and Basu, D., 1975, *J. Neurochem.* **25:**611–617.
15. Barrett, A. J., and Heath, M. F., 1977, *Lysosomes: A Laboratory Handbook*, (J. T. Dingle, ed.), Elsevier/North-Holland, Amsterdam, pp. 19–145.
16. Skudlarek, M. D., and Swank, R. T., 1979, *J. Biol. Chem.* **254:**9939–9942.
17. Hasilik, A., and Neufeld, E. F., 1980, *J. Biol. Chem.* **255:**4937–4945.
18. Myerowitz, R., and Neufeld, E. F., 1981, *J. Biol. Chem.* **256:**3044–3048.
19. Hickman, S., Shapiro, L. J., and Neufeld, E. F., 1974, *Biochem. Biophys. Res. Commun.* **57:**55–61.
20. Tabas, I., and Kornfeld, S., 1980, *J. Biol. Chem.* **255:**6633–6639.
21. Bainton, D. F., and Farquhar, M. G., 1970, *J. Cell Biol.* **45:**54–73.
22. Novikoff, A. B., 1976, *Proc. Natl. Acad. Sci. U.S.A.* **73:**2781–2787.
23. Goldstone, A., Koenig, H., Nayyar, R., Hughes, C., and Lu, C. Y., 1973, *Biochem. J.* **132:**259–266.
24. Goldstone, A., and Koenig, H., 1972, *Life Sci.* **11:**511–523.
25. Nayyar, R., and Koenig, H., 1972, *J. Cell Biol.* **55:**187a.
26. Nayyar, R., Goldstone, A., and Koenig, H., 1972, *Proceedings 4th International Congress of Histochemistry and Cytochemistry (Kyoto)*, Japanese Society of Histochemistry and Cytochemistry, Kyoto,p. 335.
27. Rosenfeld, M. G., Kreibich, S., Popov, D., Kato, K., and Sabatini, D. C., 1982, *J. Cell Biol.* **93:**135–143.
28. Erickson, A. H., and Blobel, G., 1979, *J. Biol. Chem.* **154:**11771–11774.
29. Hubbard, S. C., and Ivatt, R. J., 1981, *Annu. Rev. Biochem.* **50:**555–583.
30. Neufeld, E. F., 1981, *Lysosomes and Lysosomal Storage Diseases* (J. W. Callahan and J. A. Lowden, eds.), Raven Press, New York, pp. 115–129.
31. Sly, W. S., Natowicz, M., Gonzalez-Noriega, A., Grubb, J. H., and Fischer, H. D., 1981, *Lysosomes and Lysosomal Storage Diseases* (J. W. Callahan and J. A. Lowden, eds.). Raven Press, New York, pp. 131–146.

32. Pohlmann, R., Waheed, A., Hasilik, A., and von Figura, K., 1982, *J. Biol. Chem.* **257**:5323–5325.
33. Rothman, J. E., 1981, *Science* **213**:1212–1219.
34. Goldberg, D. E., and Kornfeld, S., 1981, *J. Biol. Chem.* **256**:13060–13067.
35. Frisch, A., and Neufeld, E. F., 1981, *J. Biol. Chem.* **256**:8242–8246.
36. Fischer, H. D., Gonzalez-Noriega, A., Sly, W. S., and Morré, D. J., 1981, *J. Biol. Chem.* **255**:9608–9615.
37. Gonzales-Noriega, A., Grubb, J. H., Talkod, V., and Sly, W. S., 1980, *J. Cell Biol.* **85**:839–852.
38. Needleman, S. B., Koenig, H., and Goldstone, A., 1975, *Biochim. Biophys. Acta* **379**:57–73.
39. Sanghavi, P., and Koenig, H., 1976, *Biochem. J.* **155**:725–728.
40. Goldstone, A., and Koenig, H., 1974, *FEBS Lett.* **39**:176–181.
41. Koenig, H., Sanghavi, P., and Goldstone, A., 1975, *Proceedings 5th International Meeting of the International Society for Neurochemistry (Barcelona)*, p. 217.
42. Goldstone, A., and Koenig, H., 1974, *Biochem. J.* **141**:527–535.
43. Aronson, N. N., and de Duve, C., 1968, *J. Biol. Chem.* **243**:4564–4573.
44. Edmond, J. D., Johnston, R. G., Kidd, D., Rylance, H. J., and Sommerville, R. G., 1966, *Br. J. Pharmacol. Chemother.* **27**:415–426.
45. Barrett, A. J., and Dingle, J. T., 1972, *Biochem. J.* **127**:438–441.
46. Goldstone, A., Szabo, E., and Koenig, H., 1970, *Life Sci.* **9**:607–616.
47. Koenig, H., 1974, *Advances in Cytopharmacology*, Volume 2 (B. Ceccarelli, F. Clementi, and J. Meldolesi, eds.), Raven Press, New York, pp. 273–301.
48. Koenig, H., 1980, *The Biochemistry of Brain* (S. Kumar, ed.), Pergamon Press, Oxford, pp. 559–578.
49. Silverstein, S. C., Steinman, R. M., and Cohen, Z. A., 1977, *Annu. Rev. Biochem.* **46**:669–722.
50. Bruns, R. R., and Palade, G. E., 1968, *J. Cell Biol.* **37**:277–299.
51. Hennings, W. A., 1976, *Maternofetal Transmission of Immunoglobulins*, Cambridge University Press, Cambridge, England.
52. Tomasi, T. B., 1976, *The Immune System of Secretions*, Prentice-Hall, Englewood Cliffs, New Jersey.
53. Libby, P., and Goldberg, A. L., 1981, *J. Cell. Physiol.* **107**:185–194.
54. Pastan, I. H., and Willingham, M. C., 1981, *Annu. Rev. Physiol.* **43**:239–250.
55. Pastan, I. H., and Willingham, M. C., 1981, *Science* **214**:504–509.
56. Goldstein, J. L., Anderson, R. G. W., and Brown, M. S., 1979, *Nature* **279**:679–685.
57. Pearse, B. M. F., 1976, *Proc. Natl. Acad. Sci. U.S.A.* **73**:1255–1259.
58. Maxfield, F. R., Willingham, M. C., Davies, P. J. A., and Pastan, I., 1979, *Nature* **277**:661–663.
59. Pratten, M. K., Duncan, R., and Lloyd, J. B., 1980, *Coated Vesicles* (C. J. Ockleford and A. Whyte, eds.), Cambridge University Press, Cambridge, pp. 179–218.
60. Prusch, R. D., 1980, *Science* **209**:691–692.
61. Goldstone, A., Koenig, H., and Lu, C. Y., 1983, *Biochem. Biophys. Acta* 1983, **762**:366–371.
62. Salisbury, J. L., Condeelis, J. S., and Satir, P., 1980, *J. Cell Biol.* **87**:132–141.
63. Davies, P. J., Davies, D. R., Levitzski, A., Maxfield, F. R., Milhaud, P., Willingham, M. C., and Pastan, I. H., 1980, *Nature* **283**:162–166.
64. Miller, N. E., and Yin, J. Y., 1978, *Biochim. Biophys. Acta* **530**:145–150.
65. Leake, D. S., Muir, E. M., and Bowyer, D. E., 1982, *Exp. Mol. Pathol.* **36**:262–275.
66. Miller, N. E., and Yin, J. A., 1979, *Biochim. Biophys. Acta* **552**:428–437.
67. Gawlitta, W., and Stockem, W., 1980, *Cell Tissue Res.* **213**:9–20.
68. Josefsson, J.-O., 1975, *Acta Physiol. Scand.* [*Suppl.*] **432**:1–65.
69. Prusch, R. D., 1980, *Science* **209**:691–692.
70. Masur, S. K., Holtzman, E., Schwartz, J. L., and Walter, R., 1971, *J. Cell Biol.* **49**:582–594.
71. Joá, F., 1972, *Experientia* **28**:1470–1471.
72. Joá, F., Rakonczay, Z., and Wollemann, M., 1975, *Experientia* **30**:582–584.
73. Westergaard, E., 1975, *J. Ultrastruct. Res.* **50**:383.
74. Libelius, R., Lundquist, I., Templeton, W., and Thesleff, S., 1978, *Neuroscience* **3**:641–647.

75. Libelius, R., Jirmanová, I., Lundquist, I., and Thesleff, S., 1978, *J. Neuropathol. Exp. Neurol.* **37**:387–400.
76. Libelius, R., and Lundquist, I., 1978, *Cell Tissue Res.* **186**:1–11.
77. Holtzman, E., 1976, *Lysosomes: A Survey*, Springer-Verlag, Berlin, Heidelberg, New York.
78. Goldberg, A. L., and St. John, A. C., 1976, *Annu. Rev. Biochem.* **45**:747–803.
79. Segal, H. L., and Doyle, D. J., eds., 1978, *Protein Turnover and Lysosome Function*, Academic Press, New York.
80. Wiedenthal, K., ed., 1980, *Degradative Processes in Heart and Skeletal Muscle*, Elsevier/North-Holland, Amsterdam.
81. Amenta, J. S., and Brocher, S. C., 1981, *Life Sci.* **28**:1195–1208.
82. Hershko, A., and Ciechanover, A., 1982, *Annu. Rev. Biochem.* **51**:335–364.
83. Neely, A. N., Nelson, P. B., and Mortimore, G. E., 1974, *Biochim. Biophys. Acta* **338**:458–472.
84. Mortimore, G. E., and Schworer, C. M., 1977, *Nature* **270**:174–176.
85. Mortimore, G. E., Ward, W. F., and Schworer, C. M., 1978, *Protein Turnover and Lysosome Function*, Academic Press, New York, pp. 67–87.
86. Schworer, C. M., Schiffer, K. A., and Mortimore, G. E., 1981, *J. Biol. Chem.* **256**:7652–7658.
87. Mortimore, G. E., and Ward, W. F., 1981, *J. Biol. Chem.* **256**:7659–7765.
88. Wibo, M., and Poole, B., 1974, *J. Cell Biol.* **63**:430–440.
89. de Duve, C., de Barsy, T., Poole, B., Trouet, A., Tulkens, P., and van Hoof, F., 1974, *Biochem. Pharmacol.* **23**:2495–2531.
90. Koenig, H., 1968, *Prog. Brain Res.* **29**:87–121.
91. Okhuma, F., and Poole, B., 1978, *Proc. Natl. Acad. Sci. U.S.A.* **75**:3327–3331.
92. Poole, B., and Ohkuma, S., 1981, *J. Cell Biol.* **90**:665–669.
93. Ohkuma, S., and Poole, B., 1981, *J. Cell Biol.* **90**:656–664.
94. Seglen, P. O., and Gordon, P. B., 1980, *Mol. Pharmacol.* **18**:468–475.
95. Dean, R. T., 1975, *Nature* **257**:414–416.
96. Umizawa, H., and Aoyagi, T., 1977, *Proteinases in Mammalian Cells and Tissues* (A. J. Barrett, ed.), North-Holland, Amsterdam, pp. 637–662.
97. Knowles, S. E., and Ballard, F. J., 1976, *Biochem. J.* **156**:609–617.
98. Neff, N. I., de Martino, G. N., and Goldberg, A. L., 1979, *J. Cell. Physiol.* **101**:439–458.
99. Grinde, B., and Seglen, P. O., 1980, *Biochim. Biophys. Acta* **632**:73–86.
100. Libby, P., and Goldberg, A. L., 1978, *Science* **199**:534–536.
101. McGowan, E. B., Shafiq, S. A., and Stracher, A., 1976, *Exp. Neurol.* **50**:649–657.
102. Libby, P., and Goldberg, A. L., 1981, *J. Cell. Physiol.* **107**:185–194.
103. Goldberg, A. L., Tischler, M., de Martino, G., and Griffin, G., 1980, *Fed. Proc.* **39**:31–36.
104. Ballard, F. J., 1980, *Biochem. Actions Horm.* **7**:91–117.
105. Tischler, M. R., 1981, *Life Sci.* **28**:2569–2576.
106. Deter, R. L., Baudhuin, P., and de Duve, C., 1967, *J. Cell Biol.* **35**:C11–C16.
107. Shelburne, J. D., Arstila, A. U., and Trump, B. F., 1973, *Am. J. Pathol.* **72**:521–540.
108. de Martino, G. N., and Goldberg, A. L., 1978, *Proc. Natl. Acad. Sci. USA* **75**:1369–1373.
109. Koenig, H., Goldstone, A., Lu, C. Y., Blume, G., and Hughes, C. T., 1980, *Neurochemistry and Clinical Neurology*, Alan R. Liss, New York, pp. 275–290.
110. Goldstone, A., Koenig, H., Blume, G., and Lu, C. Y., 1981, *Biochim. Biophys. Acta* **677**:133–139.
111. Koenig, H., Goldstone, A., Blume, G., and Lu, C.Y., 1980, *Science* **209**:1023–1026.
112. Koenig, H., Goldstone, A., and Lu, C. Y., 1980, *Biochem. J.* **192**:349–353.
113. Koenig, H., and Goldstone, A., 1980, *Trans. Am. Soc. Neurochem.* **11**:98.
114. Goldstone, A., Koenig, H., and Lu, C. Y., 1981, *Biochim. Biophys. Acta* **673**:170–176.
115. Koenig, H., Goldstone, A., and Lu, C. Y., 1982, *Circ. Res.* **50**:782–787.
116. Koenig, H., and Lu, C. Y., 1980, *Trans. Am. Soc. Neurochem.* **11**:99.
117. Koenig, H., Lu, C. Y., and Goldstone, A., 1981, *Proceedings 8th Meeting of the International Society for Neurochemistry (Nottingham)*, p. 318.
118. Koenig, H., Goldstone, A., and Lu, C. Y., 1982, *Biochem. Biophys. Res. Commun.* **106**:346–353.

119. Fambrough, D. M., 1979, *Physiol. Rev.* **59**:165–227.
120. Appel, S. H., Blosser, J. C., McManaman, J. L., and Ashizawa, T., 1982, *Am. J. Physiol.* **243**:E31–E36.
121. Koenig, H., Goldstone, A., Blume, G., and Lu, C. Y., 1980, *Neurology (N.Y.)* **30**:400.
122. Libby, P., and Goldberg, A. L., 1980, *Degradative Processes in Heart and Skeletal Muscle* (K. Wiedenthal, ed.), Elsevier/North Holland, Amsterdam, pp. 201–222.
123. Kovács, A. L., and Seglen, P. O., 1981, *Biochim. Biophys. Acta* **676**:213–220.
124. Baccino, F. M., Tessitore, L., Cecchini, G., Messina, M., Zuretti, M. F., Bonelli, G., Gabriel, L., and Amenta, J. S., 1982, *Biochem. J.* **206**:395–405.
125. Kovács, A. L., Reith, A., and Seglen, P. O., 1982, *Exp. Cell Res.* **137**:191–201.
126. Kovács, A. L., Molnár, K., and Seglen, P. O., 1981, *FEBS Lett.* **134**:194–196.
127. Seglen, P. O., and Gordon, P. B., 1982, *Proc. Natl. Acad. Sci. U.S.A.* **79**:1889–1892.
128. Dunn, W. A., Hubbard, A. L., and Aronson, N. N., Jr., 1980, *J. Biol. Chem.* **255**:5971–5978.
129. Roedemann, H. P., and Goldberg, A. L., 1982, *J. Biol. Chem.* **257**:1632–1638.
130. Roedemann, H. P., Waxman, L., and Goldberg, A. L., 1982, *J. Biol. Chem.* **257**:8716–8723.
131. Kameyama, T., and Etlinger, J. D., 1979, *Nature* **279**:344–346.
132. Jaweed, M. M., Alam, A., Herbison, S. J., and Ditunno, J. F., Jr., 1981, *Neuroscience* **6**:2787–2792.
133. Kremzner, L. T., Tennyson, V. M., and Miranda, A. F., 1978, *Adv. Polyamine Res.* **2**:241–256.
134. Hopkins, D., and Manchester, K. L., 1981, *Biochem. J.* **197**:603–610.
135. Rudman, D., Kutner, M. H., Chawla, R. K., and Goldsmith, M. A., 1980, *J. Clin. Res.* **65**:95–102.
136. Koenig, H., Goldstone, A., and Lu, C. Y., 1982, *Trans. Am. Soc. Neurochem.* **13**:207.
137. Koenig, H., Goldstone, A., and Lu, C. Y., 1982, *J. Neuropathol. Exp. Neurol.* **41**:376.
138. Goldstone, A., Koenig, H., and Lu, C. Y., 1982, *Biochem. Biophys. Res. Commun.* **104**:165–172.
139. Goldstone, A., Koenig, H., and Lu, C. Y., 1981, *Trans. Am. Soc. Neurochem.* **12**:89.
140. Koenig, H., Goldstone, A., and Lu, C. Y., 1981, *Neurology* (N.Y.) **31**:46.
141. Koenig, H., Goldstone, A., and Lu, C. Y., 1981, *Trans. Am. Soc. Neurochem.* **12**:88.
142. Koenig, H., Goldstone, A., and Lu, C. Y., 1982, *Fed. Proc.* **41**:1598.
143. Koenig, H., Goldstone, A., and Lu, C. Y., 1982, *Biochem. Biophys. Res. Commun.* **106**:346–353.
144. Goldstone, A., Koenig, H., and Lu, C. Y., 1983, *Fed. Proc.* **42**:295.
145. Goldstone, A., Koenig, H., and Lu, C. Y., 1983, *Biochim. Biophys. Acta* **762**:366–371.
146. Koenig, H., Goldstone, A., and Lu, C. Y., 1983, *Fed. Proc.* **42**:299.
147. Koenig, H., Goldstone, A., and Lu, C. Y., 1983, *Nature* **305**:530–534.
148. Goldstone, A., Koenig, H, Lu, C. Y., and Trout, J., 1983, *Biochem. Biophys. Res. Commun.* **114**:913–921.
148a. Koenig, H., Goldstone, A. D., and Lu, C. Y., 1983, *Proc. Nat. Acad. Sci. U.S.A.* **80**:7210–7214.
149. Lu, C. Y., Goldstone, A., and Koenig, H., 1983, *Fed. Proc.* **42**:294.
150. Koenig, H., Goldstone, A., and Lu, C. Y., 1983, *Trans. Am. Neurochem. Soc.* **14**:198.
151. Brady, R. O., 1978, *Annu. Rev. Biochem.* **47**:687–713.
152. Brady, R. O., 1983, *Annu. Rev. Neurosci.* **5**:33–56.
153. Conzelmann, E., and Sandhoff, K., 1978, *Proc. Natl. Acad. Sci. U.S.A.* **75**:3979–3983.
154. Gahl, W. A., Bashan, N., Tietze, F., Bernardini, I., and Schulman, J. D., 1982, *Science* **217**:1263–1265.
155. Wolfe, L. S., Ng Ying Kin, N. M. K., Baker, C. R., Carpenter, S., and Andermann, F., 1977, *Science* **195**:1360–1362.
156. Wolfe, L. S., Ng Ying Kin, N. M. K., Palo, J., and Haltea, M., 1983, *Neurology (N.Y.)* **33**:103–106.
157. Radin, N. S., 1976, *Current Trends in Sphingolipidoses and Allied Disorders* (B. W. Volk and L. Schneck, eds.), Plenum Press, New York, pp. 453–472.
158. Kanfer, J. N., Raghaven, S. S., Mumford, R. A., Sullivan, J., Spielvogel, C., Legler, G., Labow, R. S., Williamson, D. G., and Layne, D. S., 1976, *Current Trends in Sphingolipidoses*

and Allied Disorders (B. W. Volk and L. Schneck, eds.), Plenum Press, New York, pp. 77–98.

159. Dingle, J. T., Poole, A. R., Lazarus, G. S., and Barrett, A. J., 1973, *J. Exp. Med.* **137**:1124–1141.
160. Werb, G., and Dingle, J. T., 1976, *Lysosomes in Biology and Pathology*, Volume 5 (J. T. Dingle and R. T. Dean, eds.), North-Holland, Amsterdam, pp. 127–156.
161. Sternlieb, I., and Goldfischer, S., 1976, *Lysosomes in Biology and Pathology* (J. T. DIngle and R. T. Dean, eds.), North-Holland, Amsterdam, pp. 185–200.
162. Lloyd, J. B., 1973, *Lysosomes and Storage Diseases* (H. G. Hers and F. Van Hoof, eds.), Academic Press, New York, pp. 173–195.
163. Lüllman-Rauch, R., 1979, *Lysosomes in Biology and Pathology*, Volume 6 (J. T. Dingle, P. J. Jacques, and I. H. Shaw, eds.), North-Holland, Amsterdam, pp. 49–130.
164. Hauser, G., and Eichberg, J., 1975, *J. Biol. Chem.* **250**:105–112.
165. Michell, R. H., Allen, D., Bowley, M., and Brindley, D. N., 1976, *J. Pharm. Pharmacol.* **28**:331–332.
166. Bowley, M., Cooling, J., Burditt, S. L., and Brindley, D. N., 1977, *Biochem. J.* **165**:447–454.
167. Silverblatt, F. J., and Kuehn, C., 1979, *Kidney Int.* **15**:335–345.
168. Sastrasinh, M., Knaus, T. C., Weinberg, J. M., and Humes, H. D., 1982, *J. Pharmacol. Exp. Ther.* **222**:350–358.
169. Josepovitz, C., Pastoria-Munoz, E., Timmerman, D., Scott, M., Feldman, S., and Kaloyanides, G. J., 1982, *J. Pharmacol. Exp. Ther.* **223**:314–321.
170. Lu, C. Y., Koenig, H., and Goldstone, A., 1983, *Fed. Proc.* **42**:294.
171. Kosek, J. C., Mazze, R. I., and Cousins, M. J., 1974, *Lab. Invest.* **30**:48–57.
172. Fedorko, M. E., Hirsch, J. G., and Cohn, Z. A., 1968, *J. Cell Biol.* **38**:377–391.
173. Hruban, Z., Slesers, A., and Hopkins, E., 1972, *Lab. Invest.* **27**:62–70.
174. Koenig, H., and Lu, C. Y., 1980, *Proceedings 6th International Histochemical Cytochemistry Congress*, Brighton, England, Royal Microscopical Society, Oxford, p. 210.

8

Brain Nuclei

Yasuo Takahashi

1. INTRODUCTION

In the 1970s great progress was made in the study of the mechanisms of gene expression, although the bulk of the existing information is from tissue other than the brain.[1–4] However, since reliable data on brain nuclei also became available after the publication of Rappoport's work, this chapter primarily describes the data from brain. Reports on brain nuclei published before 1970 have been reviewed by Rappoport *et al.*[5] and McEwen and Zigmond.[6]

2. ISOLATION OF NUCLEI

Rappoport *et al.*[5] described the following criteria for isolated nuclei before these can be used for analytical or enzymological investigations: (1) the nuclei must be free of adherent contaminants such as cytoplasm; (2) they must retain their morphological and biochemical integrity; (3) the isolation procedure must yield a representative sample of nuclei from a heterogeneous cell population from such organs as the brain; (4) the isolation must be rapid. Rappoport *et al.* and McEwen and Zigmond[6] gave a brief summary of the isolation methods used prior to 1970. Basic methods for isolating brain nuclei were established at that time by several investigators, for example (1) isolation of brain nuclei by centrifugation of brain homogenate through dense sucrose (2 M or greater) solutions and (2) the use of detergents (Triton X-100, deoxycholate, Nonidet P-40, and Tween-40) in isolating clean brain nuclei. It is important that sucrose solutions contain divalent cations. Furthermore, two groups of investigators[7,8] have attempted to isolate neuronal and glial nuclei. Since then, the following isolation procedures have been devised.

1. Fractionation of nuclei from brain by zonal centrifugation into five fractions.[9] Zone I contained neuronal nuclei (70%) and astrocytic nuclei (23%); zone II contained astrocytic nuclei (81%) and neuronal nuclei

Yasuo Takahashi • Department of Neuropharmacology, Brain Research Institute, Niigata University, Niigata 951, Japan.

(15%); zone III contained astrocytic nuclei (84%) and oligodendrocytic nuclei (15%); zone IV contained oligodendrocytic nuclei (92%) and zone V contained oligodendrocytic nuclei.

2. A modification[10] of 1 by using metrizamide instead of sucrose. It has advantages over sucrose in certain circumstances because it can form solutions of higher density and lower viscosity.
3. Isolation of nuclei[11] on sucrose density gradient centrifugation from neuronal perikarya that were separated from rat brain cortex and cerebellum and pigeon forebrain by the method of Sellinger and Azcurra.[12]
4. Thompson's procedure,[13] which is a simplified modification of the method of Løvtrup-Rein and McEwen.[7]
5. Isolation of oligodendrocytic nuclei[14] by sucrose density gradient centrifugation from sheep brain white matter.
6. Preparation of nuclei from cultured neural cells[15]: neuroblastoma, C-6 glioma, primary cultured cells, and cloned cells. This procedure mainly consists of the use of nonionic detergents such as Triton X-100 and Nonidet P-40.

Figure 1 shows a phase-contrast micrograph of the rat neuronal and glial nuclei isolated by H. Ozawa, E. Kushiya, and Y. Takahashi (unpublished data).

Several reports in relation to the isolation procedure must be referred to. McEwen *et al.*[6] examined the nuclei isolated from several brain regions and found that the nuclei from the cerebellum were small and that their ratio of RNA, histone, and nonhistone proteins to DNA was low. Further, 70% of cerebellar cell nuclei were found to be the nuclei of granule cells. This finding was recently confirmed and examined in detail by Zagon and McLaughlin.[16] Olpe *et al.*[17] compared the morphological character of the isolated brain nuclei and *in situ* nuclei and pointed out that the criteria of size and number of nucleoli and staining pattern were not sufficient. H. Higa, K. Araki, and Y. Takahashi (unpublished data) isolated nuclei from the motor neurons separated from bovine ventral horn and found a high RNA-to-DNA ratio. In spite of the above-described efforts, it must be noted that isolation of astroglial nuclei is still difficult, and specific enzymic markers for each of the neuronal and glial nuclei are not known.

3. *BRAIN NUCLEOLUS*

The morphological and cytochemical studies of the nucleoli of nerve cells have been reviewed in detail by Busch and Smetana[18] and briefly by Rappoport *et al.*,[5] although brain nucleoli had not been isolated until 1973. Takahashi *et al.*[19,20] and Banks and Johnson[21] first isolated brain nucleoli. Since the latter did not report the morphological and analytical data on isolated nucleoli, the characteristics of the isolated nucleoli are briefly described according to the data of the former group. The isolation procedure consists of sonication of the isolated brain nuclei in isotonic sucrose containing Ca^{2+} and sedimentation through 0.88 M sucrose solution. The conditions for sonication appear to be important. Under phase-contrast microscopy and electron microscopy, the iso-

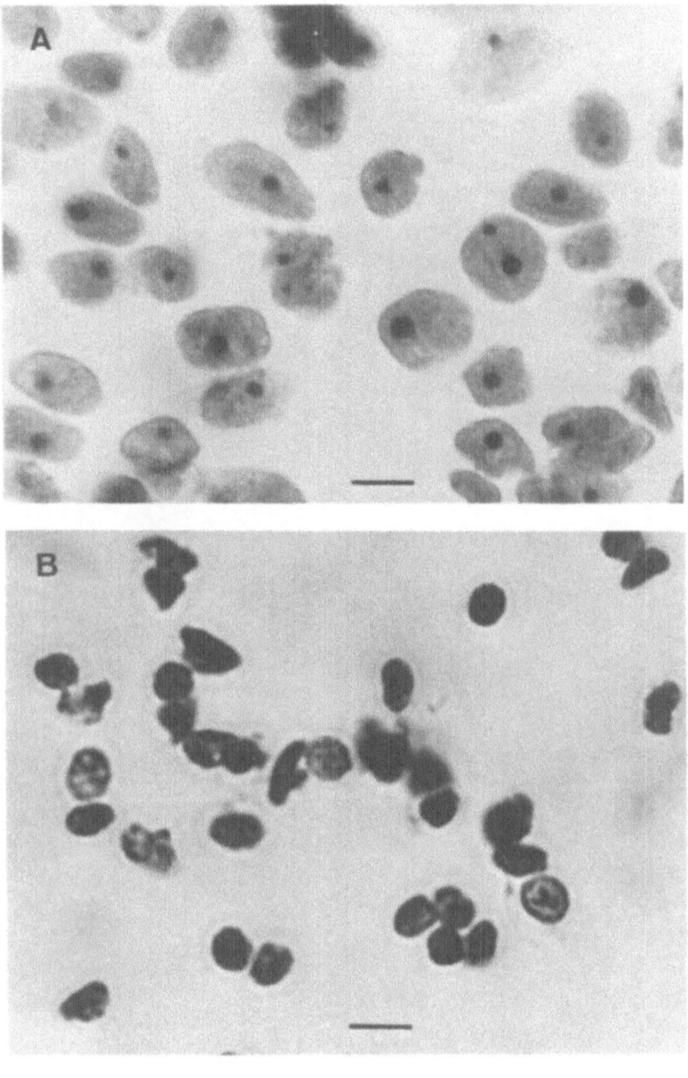

Fig. 1. Phase-contrast micrographs of isolated neuronal and oligodendroglial nuclei of rat brain (H. Ozawa, E. Kushiya, and Y. Takahashi, unpublished data) A: Neuronal nuclei. B: Oligodendroglial nuclei. Bars indicate 10 μm.

lated nucleoli are intact and pure, and they show the existence of nucleolonema and pars amorpha, resembling the ultrastructural morphology of liver nucleoli. The identification of the cellular origin of nucleoli seems to be difficult.

Figure 2 shows an electron micrograph of the isolated brain nuclei (H. Yamamoto and Y. Takahashi, unpublished data) and nucleoli.[19,20] Chemical analyses of brain and liver nucleoli are presented in Table I. Nucleolar RNA extracted from rat brain and analyzed on polyacrylamide gel electrophoresis (PAGE) gave a pattern of 45 S, 36 S, and 28 S RNA, different from that of

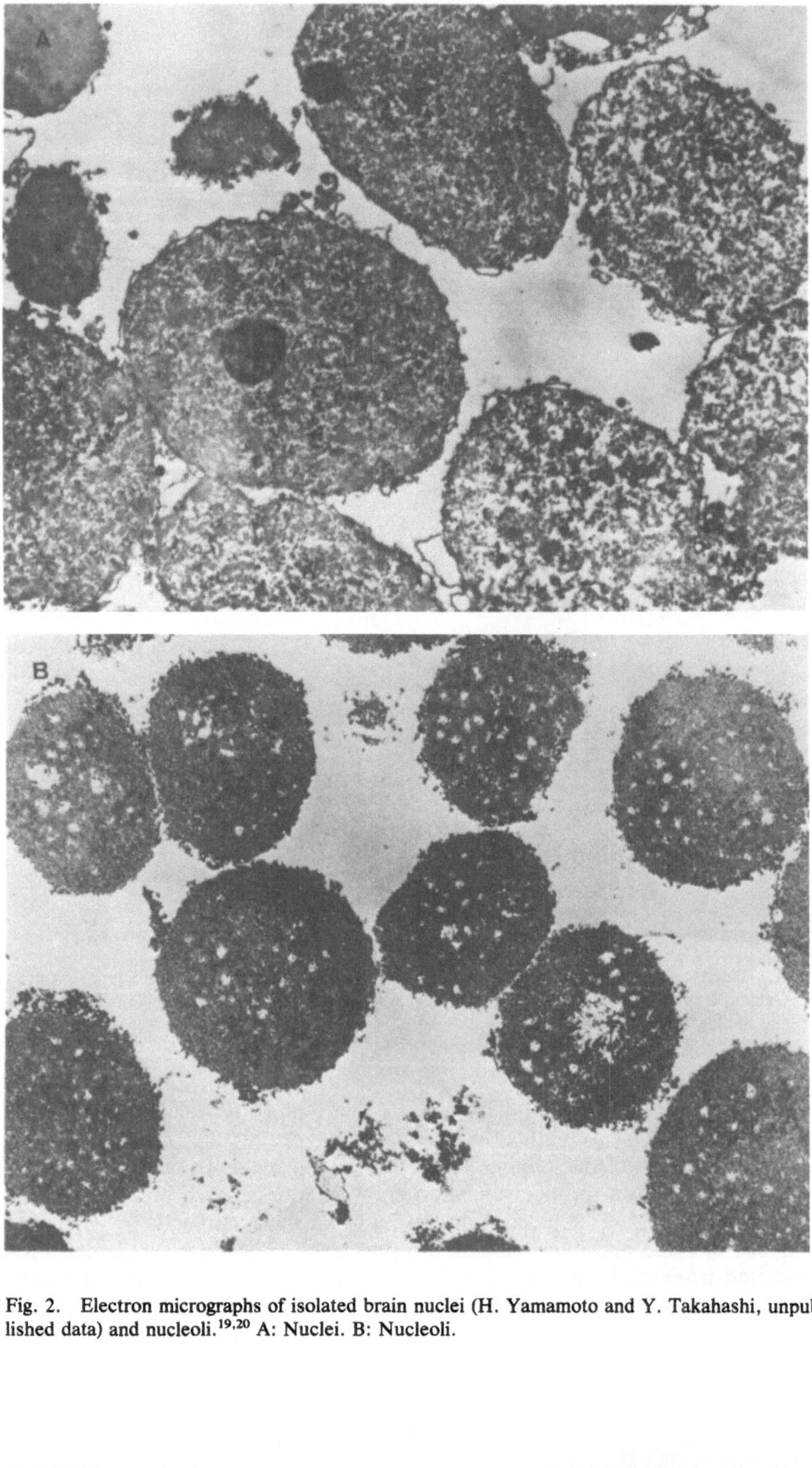

Fig. 2. Electron micrographs of isolated brain nuclei (H. Yamamoto and Y. Takahashi, unpublished data) and nucleoli.[19,20] A: Nuclei. B: Nucleoli.

Table I
Composition of Isolated Nucleolus from Brain and Liver

	Percent dry weight			
Source	RNA	DNA	Protein	RNA/DNA
Pig brain	13.3	10.7	76.0	1.2
Rat liver	8.4	6.1	85.5	1.4

extranucleolar RNA, which gave 28 S, 18 S, 6 S, and 4 S. Biosynthesis and processing of these RNA species will be described later.

Nucleolar RNA is the precursor of ribosomal RNA (rRNA). Nucleolar proteins can be separated into classes that are (1) soluble in diluted saline solution, (2) soluble in acid but insoluble in dilute saline solution, and (3) soluble in neither acid nor in dilute saline solution.[18]

Takahashi *et al.* found remarkable differences between the nucleolar and extranucleolar proteins extracted with Tris-HCl 0.01 M (pH 7.6), $MgCl_2$ 0.01 M, EDTA 1 mM, β-mercaptoethanol 1 mM, in 8 M urea. It is very characteristic that extranucleolar proteins show high-molecular-weight bands in PAGE; these results will be referred to in Section 4. Banks and Johnson and Takahashi *et al.*[19–22] observed RNA polymerase activity in brain nucleolus, and the former described the characteristics of the enzyme in detail; the brain enzymic reaction does not need an energy donor system. The effects of ions and α-amanitin were examined in nucleolar RNA polymerase. α-Amanitin did not give any effect in either newborn or adult mice.[22] There are no analytical reports on brain rRNA genes.

4. *CHROMATIN ANALYSIS*

The extranucleolar fraction of the eucaryotic nucleus contains mainly chromatin and ribonucleoprotein granules. Chromatin is organized in repeating nucleoprotein particles termed nucleosomes. Each nucleosome consists of a core region and a linker region. Histones H2A, H2B, H3, and H4, which are associated with DNA, are found in the core region; H1 is associated with DNA of the linker region (see reviews[23]).

Brain chromatin was isolated by a modification of the procedure of Marushige and Bonner. The analytical data on the chromatin are shown in Table II (K. Ikeda and Y. Takahashi, unpublished data). Isolated chromatin, including brain chromatin, can function as a template for purified *E. coli* RNA polymerase and RNA polymerase B. The template activity of brain chromatin for purified brain RNA polymerase B and *E. coli* polymerase was about 20% of that of pure DNA.[24] RNA synthesis on the brain chromatin template was greater in newborn and fetal rats than in adults.[25] Singh and Sung[24] found that brain RNA polymerase B was three to four times more active with dehistonized chromatin than with pure DNA, whereas the *E. coli* enzyme was almost equally active with

Table II
Chemical Compositions of Various Chromatins

Source of chromatin	Content (relative to DNA)			
	DNA	RNA	Histone	Nonhistone protein
Rat brain	1.00	0.045	1.36	0.75
Rat liver	1.00	0.107	0.92	0.43
Bovine cerebral cortex	1.00	0.100	0.99	0.69
Bovine white matter	1.00	0.03	0.96	0.24

either of these two templates, reflecting the specificity of the transcriptional control mechanism in brain cells. Smith *et al.*[26] reported that specific brain transcription occurred in transcription systems with brain chromatin and *E. coli* polymerase using competitive RNA–DNA hybridization procedures. This finding was almost confirmed by K. Ikeda and Y. Takahashi (unpublished data). However, since such hybridization conditions allowed only the reaction of repetitive DNA sequences, no precise conclusions could be drawn concerning the overall fidelity of *in vitro* transcription.

A more specific approach was devised that involved measuring the specific sequence concentration in the transcript by hybridization to a probe complementary to isolated specific mRNA.[27,28] However, it was claimed that the early works on chromatin transcription suffered from contamination with endogenous RNA. The use of mercurated nucleotides to label *de novo* synthesized transcripts seemed to overcome this difficulty,[29] but this technique was later shown to be only inconsistently effective in separating elongated endogenous RNA because of RNA-dependent transcription.[30] This contamination of endogenous RNA was found to be removed by heating the transcripts.[31] Furthermore, it was shown that a useful method of quantitating specific transcripts seems to be hybridization of labeled transcripts with excess DNA probe that was provided by the technique of cloning a specific genome.[32]

Parker and Roeder[33] reported that homologous RNA polymerase was essential for the accurate transcription of 5 S RNA genes in isolated chromatin. Although homologous RNA polymerase B has been used as a chromatin template since then, the important problem of tissue-specific transcription of chromatin as template has still been left unsolved. In the case of brain chromatin, such studies have not been performed because a brain-specific probe has not yet been obtained. Recently, Brown[34] reported the fractionation of transcriptionally active brain chromatin. Since they used *E. coli* RNA polymerase and did not use any brain-specific DNA probe, the above-described problems still remain in their experiment.

Sarkandar and Dulce[35,36] found that a double number of RNA initiation sites was measured on neuronal when compared to glial chromatin, independently of whether the neuronal or the glial RNA polymerase preparation was used. Further, they fractionated neuronal and glial brain chromatin into transcribable and repressed portions. Active neuronal and glial chromatin fractions

Table III
DNA Repeat Length[a] of Nuclear Preparations from Neuron (N) and Glia (G) of Three Different Brain Regions and Liver

Age (days)	Cerebral cortex		Hypothalamus		Cerebellum		Liver
	N	G	N	G	N	G	
Rat[42,43]							
−2	195	200	179	—	—	—	185
7	173	212	174	—	204	205	
25	175	207	174	214	201	207	
60	174						207
Rabbit[40]							
Adult	162	197			200		200

[a] Mean repeat length (base pairs) measured by linear regression.

are associated with an increased content of nonhistone proteins compared to repressed fractions. Histone H1 is almost absent in active neuronal chromatin and is reduced in glial chromatin.

Transcriptionally, more repressed neuronal and glial fractions demonstrated a minimal ability to initiate RNA synthesis, unlike the template-active neuronal and glial fractions, which are differently enriched in RNA initiation sites. Brown *et al.*[37] produced chromatin subunits by digestion of brain nuclei with micrococcal nuclease and observed the presence of transcribed DNA in these subunits. Currently, studies of tissue specificity on gene expression are being carried out from two directions: (1) chromatin study and (2) transcription factors, as Roeder *et al.*[38] and Manley *et al.*[39] have reported.

Recent developments on brain chromatin concern the chromatin repeat length. Thomas and Thompson[40] found a reduced chromatin repeat length of about 160 base pairs of DNA in cortical neurons from adult rabbits and rats, as compared with about 200 base pairs for glial cells and liver. Subsequently, Brown[41] and Ermini and Kuenzle[42] demonstrated that the short DNA repeat length in cortical neurons is not present at birth but appears during postnatal development of the brain between $2\frac{1}{2}$ and $3\frac{1}{2}$ days in the rabbit and between 4 and 7 days in the mouse. The length of DNA associated with the linker region of the nucleosome appears to decrease from 60 base pairs to 20 base pairs during the developmental change in chromatin of cortical neurons. The short chromatin repeat length of hypothalamic neurons was found by 2 days before birth, indicating that hypothalamic neurons differentiate earlier than cortical neurons during brain development.[43] However, cerebellar neurons showed 203 base pairs 7 days after birth (Table III). Thomas and Thompson[40] ascribed this feature to the high transcriptional activity prevailing in these cells. However, no general correlation appears to exist between transcriptional activity and chromatin repeat length. Neuronal nuclei were found to contain less histone H1 per milligram DNA compared with glial or kidney nuclei.[44] Isolated neuronal perikaryon showed a maximum incorporation of labeled lysine and arginine into histones at 42 h.[45] This period may correlate with the conversion of neuronal chromatin to a short DNA repeat length. The significance of chromatin shortening in the cortical neuron is not yet clear.

Tashiro and Kurokawa[46] observed changes in circular dichroism spectra and thermal melting profiles of DNA reassociated with histones and/or nonhistone proteins from the cerebral or liver chromatin, suggesting an organ-specific modification of chromatin by nonhistone proteins.

Recently, several authors[47,48] published tissue-specific exposure of chromatin structure at the 5′ terminus of specific genes such as the preproinsulin II gene, heat-shock genes, and the globin gene. These exposed sites were revealed by gently digesting chromatin in isolated nuclei with DNase I and mapping the sites of cleavage by a simple indirect end-labeling technique that involves Southern blotting. If we obtain a neuron-specific or glia-specific probe, similar experiments should be carried out on brain chromatin.

5. *NUCLEAR COMPONENTS*

Nuclear components can be conventionally classified into DNA, RNA, histone, and nonhistone proteins. It is considered that DNA in the brain is qualitatively indistinguishable from DNA of other tissues. Heizman *et al.*[49] claimed that brain cortex neurons from adult mammals contain amounts of DNA elevated to between the diploid and tetraploid level. This extra DNA of brain cortex neurons was also qualitatively indistinguishable from other somatic DNA. It is well known that DNA consists of unique DNA and repetitive DNA.[1] Ranjekar and Murthy[50] reported that 10% of the total DNA from the brain exhibited rates of reassociation characteristic of repetitive DNA, and differences were observed in the pattern of denaturation of brain and liver DNAs of newborn and adult rats. However, such phenomena may reflect a contaminant in their DNA preparations. Wintzerith *et al.*[51] reported on satellite DNA of mouse brain nuclei.

Brain nuclear RNA consists of nucleolar and nucleoplasmic RNA, as shown in other tissues. As described in Section 3, rRNA precursors (45 S, 36 S, 32 S and 28 S) were found using PAGE and sucrose density gradient centrifugation. Further, evidence of heterogeneous RNA (hnRNA) as precursor of mRNA was also observed using PAGE, sucrose density gradient centrifugation, and radioisotope labeling. The existence and processing of these nuclear RNAs were demonstrated by analysis with PAGE and sucrose density gradient centrifugation after incorporation of ^{32}P, [^{14}C]orotic acid, [^{3}H]uridine, and [^{3}H]methionine into RNA[52–54] (Fig. 3). In addition to tRNA, brain nuclei contain the small nuclear RNAs that are designated as $A(U_3)$, $C(U_2)$, $D(U_1)$, G(5 S III), and H(4.5 S).[55] The nucleotide sequences of such RNAs and their genes in other tissue nuclei were determined, and at present, some are presumed to participate in splicing of hnRNA. Recently, Grunning *et al.*[55] investigated the changes of these RNAs in brain nuclei during development. Several investigators studied the biosynthesis of RNA *in vivo* in neuronal and glial brain nuclei. Løvtrup-Rein[56] found that the rate of incorporation of label into nuclear RNA was four times higher in astroglial and neuronal nuclei than in the other glial

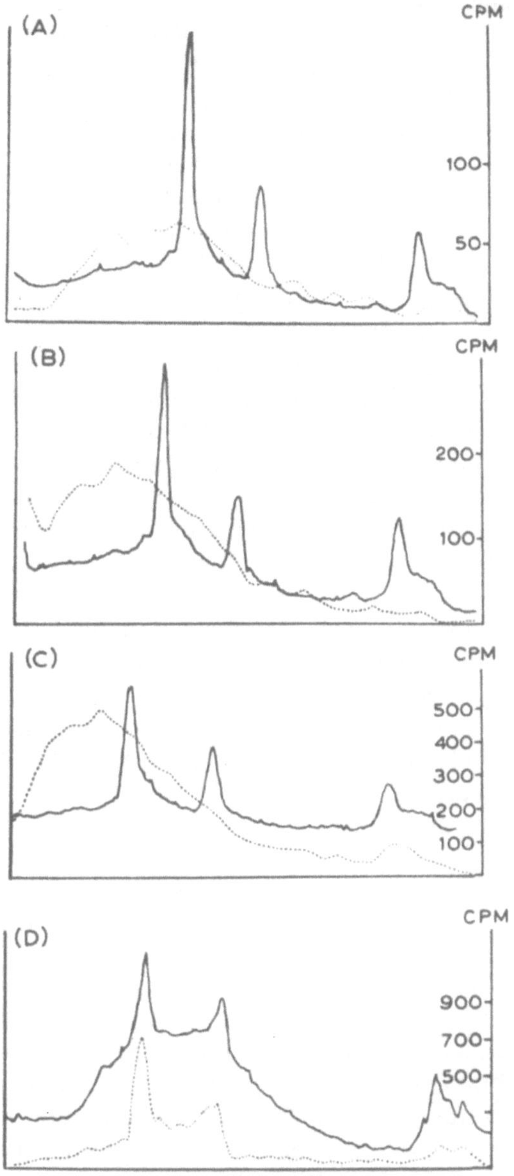

Fig. 3. Labeling pattern of rat brain nuclear RNA by electrophoresis.[52] 10 μCi (A), 10 μCi (B), 6.5 μCi (C), and 2 μCi (D) of [^{14}C]orotic acid were injected into the subarachnoidal space of rats. Nuclear RNA was isolated 15 min (A), 30 min (B), 60 min (C), and 20 h (D) after injection of [^{14}C]orotic acid. Solid line, densitometry; dashed line, radioactivity.

nuclei. Giuditta *et al.*[57] reported that radioactive RNA is almost three times higher in the large than in the small nuclei 4 h after [^{14}C]orotate injection. According to Austoker *et al.*,[58] with [^{3}H]uridine, the astrocytes in zone II contain the highest radioactivity except at the beginning of the experiment, when the neuronal nuclei of zone I are more highly labeled. With [^{14}C]orotic acid, zone I utilizes more than the nuclei of other zones.

Sedimentation and electrophoretic analysis of labeled RNA in neuronal and glial nuclei after incorporation of precursors revealed some differences among cellular nuclei. According to Løvtrup-Rein,[59] all three types of nuclei contained a 45 S and a 38 S RNA. Further, a 32 S component could be identified in astrocytic nuclei, a 35 S in neuronal nuclei, and both 32 S and 35 S RNA in glial nuclei. Giuditta *et al.* also reported that heterogeneous and preribosomal RNAs are present in both nuclear classes. After 30 min of incorporation, preribosomal species can be detected in the large nuclei, whereas hnRNA prevails in the small nuclei. On the other hand, mature radioactive 18 S and 28 S rRNAs appear after 60 min in the small nuclei but only after 180 min in the large nuclei. Recently, however, Stoykova *et al.*[60] reported that the patterns of nucleolar preribosomal RNA and rRNA species in neuronal and glial nuclei are identical.

Messenger RNA and hnRNA generally contain the caps $M^7G(5')$ $PPPN_1(m)$ $PN_2(m)$ at the 5′ terminus. Since Furuichi and Miura[61] discovered this in 1975, many investigations have been carried out on the structure, distribution, and physiological roles of the cap. However, to my knowledge there is no information about the cap structure in brain nuclear RNA, although very recently the presence of the cap in brain mRNA was reported.[62]

Mammalian brain contains nuclear and cytoplasmic RNA containing poly(A) at the 3′ end as in other tissues.[63,64] Cytoplasmic poly(A) RNA acts as mRNA. It was found that about 10% of nuclear RNA contains poly(A) as shown by poly(U) sepharose or oligo(dT) cellulose affinity chromatography. The proportion of RNA that is polyadenylated appears to decrease during brain development, although Lim[65] reported that throughout development of the rat forebrain the poly(A) content of the cellular and of the cytoplasmic high-molecular-weight RNA remains constant. Newborn brain nuclei synthesize a greater amount of poly(A) RNA than adult mouse brain nuclei.[66] The mean size distribution of poly(A) RNA species in the forebrain nucleus of young rats is larger than corresponding adult values. Nuclear RNA from rat neurons appears to contain more poly(A) than that isolated from glial cells.[67]

The RNA sequence complexity of mammalian brain nuclei has been studied by several investigators. Recently, Grouse *et al.*[68] and W. E. Hahn, J. V. Van Ness, and N. Chaudhari (personal communication) have written excellent reviews, and Kaplan reviews this problem in another volume. Therefore, I shall describe it here only briefly. Several studies on the sequence complexity of brain nuclear RNA indicated that about 15–20% of the unique DNA sequence is transcribed in the adult rodent brain, whereas the unique DNA expression values in the nuclear RNA from other tissues were much lower than those in the brain.[69–72] When brain nuclear poly(A) hnRNA was used to drive the reaction with unique DNA, about 12% of the DNA probe reacted.[73] A similar situation was observed in the complexity of cytoplasmic RNA of the brain and other tissues.[74] The reason for its high complexity has not been clarified. Very recently, H. Ozawa, E. Kushiya, and Y. Takahashi (unpublished data) found that this high complexity in the brain nuclei results only from the neuronal nuclei and that the complexity increases about 20% during postnatal brain development (Tables IV, V). The RNA complexity of isolated oligodendroglia was considerably lower. These results may be related to previous findings of differences between neuronal and glial nuclear RNA using hybridization of nuclear RNA in vast DNA excess.[75,76] Hahn *et al.*[77] reported that most of the

Table IV
Saturation Hybridization of Nuclear RNA to Unique [^3H]DNA

Nuclear RNA	Unique DNA hybridized at saturation (%) Observed value[a] (mean ± S.E.)	Unique DNA hybridized at saturation (%) Corrected value[b] (mean)	Complexity[c] in nucleotides
Neuron (Method 1)	13.5 ± 0.42	16.5	5.9×10^8
Neuron (Method 2)	13.1 ± 0.59	16.0	6.1×10^8
Glia (Method 1)	9.2 ± 0.48	11.2	4.3×10^8
Glia (Method 4)	9.0 ± 1.4	10.9	4.1×10^8
Granule cells (Method 3)	12.9 ± 0.82	15.7	6.0×10^8
Brain	12.8 ± 0.36	15.6	5.9×10^8
Liver	7.8 ± 0.51	9.5	3.6×10^8
Spleen	8.2 ± 0.59	10.0	3.8×10^8
Neuron + glia[d]	12.9 ± 0.32	15.7	6.0×10^8
Neuron + liver[e]	13.0 ± 0.43	15.9	6.0×10^8

[a] Data obtained from hybridization reaction mixtures incubated to Cot > 25,000. S.E., standard error.

[b] Observed values were corrected for terminal reactivity of unique (^{3}H)DNA probe, which was maximally reassociated to Cot > 20,000.

[c] Percentage of unique DNA hybridized is expressed in terms of nucleotides assuming that DNA transcription is asymmetric and that a unique genome is 1.9×10^9 nucleotides.

[d] Combination contained equal amounts of nuclear RNA from neuronal nuclei (Method 2) and glial nuclei (Method 1).

[e] Combination contained equal amounts of nuclear RNA from neuronal nuclei (Method 2) and liver nuclei.

3′-proximal sequences of large nuclear poly(A) hnRNA are homologous with mRNA in the mouse brain.

Nuclear proteins are composed of histones, nonhistone chromosomal proteins, soluble proteins, and residual proteins. Recently, references on chromosomal proteins were reviewed,[78] and advances in the analytical and preparative methods for nuclear proteins were also reviewed in detail.[3] By 1970, Graziano and Huang[79] had carried out chromatographic separtation and polyacrylamide electrophoretic analysis of histones and nonhistone proteins in the

Table V
Developmental Changes in Saturation Hybridization of Cerebral Cortical Neuronal Nuclear RNA to Unique DNA

Days after birth	Unique DNA hybridized at saturation (%) Corrected value (mean ± S.E.)	Unique DNA hybridized at saturation (%) Percent of adult value
1	13.2 ± 0.66	81.5
3	13.7 ± 0.38	84.6
6	14.1 ± 0.40	87.0
21	16.2 ± 0.34	100.0

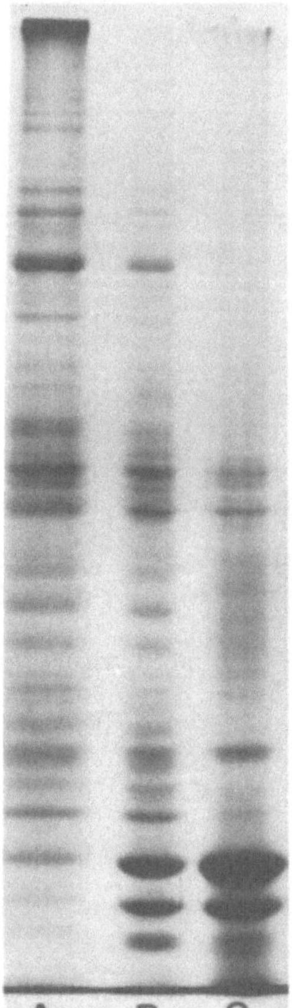

Fig. 4. Electrophoretic pattern of chromatin proteins. A, neuronal chromatin; B, oligodendroglial chromatin; C, liver chromatin. (K. Araki, E. Kushiya, and Y. Takahashi, unpublished data.)

brain nuclei. Histones consist of H1, H2A, H2B, H3, and H4, as in the nuclei of other tissues. This finding was observed in the early works of Shaw and Huang, 1970[79a] and was also confirmed in the recent report by Greenwood *et al.*[80] However, Greenwood *et al.* observed a quantitative change in histone H1 in the neuronal nuclei in agreement with the results of Sarkander *et al.*[36]

A number of investigators[79,81–90] (K. Araki, E. Kushiya, and Y. Takahashi, unpublished data) reported the analysis of nonhistone chromosomal proteins by one-dimensional and two-dimensional gel electrophoresis and column chromatography. The composition of nonhistone chromosomal proteins is very complex, and these proteins currently are considered to be the major regulators of gene expression. However, there is a high degree of homology between the nonhistone chromosomal proteins of different organs as well as of different vertebrates.[82] The brain nonhistone protein fraction shows a number of unique high-molecular-weight proteins different from those of liver and kidney[83–86] (K. Araki, E. Kushiya, and Y. Takahashi, unpublished data), as shown in Fig. 4. Such brain specificity may be related to the role of this fraction in the control of specific gene expression. Furthermore, it was reported that high-molecular-weight proteins mainly exist in the neuronal nuclei.[84,86] (K. Araki, E. Kushiya, and Y. Takahashi, unpublished data). However, Tsitilou *et al.*[87] recently contradicted these reports and found that the only major difference was the existence of a specific protein (pI 8.5, molecular weight 10,000) in zone I (neuronal nuclei). Tsitilou *et al.* also observed the electrophoretic patterns of buffer-saline-soluble and 0.35 M NaCl-soluble proteins. The early work of Dravid and Burdman[88] did not show any differences among acidic proteins isolated from the nuclei of neurons, astrocytes, and oligodendrocytes. Fujitani and Holonbek[89] did not find any differences in nonhistone proteins among several brain regions.

Heizmann *et al.*[90] analyzed nonhistone chromosomal proteins of neuronal nuclei by labeling *in vitro* with iodo[1-^{14}C]acetamide and two-dimensional gel electrophoresis followed by fluorography. Comparison of nuclear protein patterns of cortex neurons, cerebellar neurons, cortex glia, and liver revealed significant differences among the four tissues, some proteins seeming to be specific for certain tissues and others appearing to be present universally. Among the low-mobility-group proteins, the two basic proteins (mol. wt. 35,000 and 38,000) showed marked developmental changes in cortex and cerebellar neurons. The discrepancies of chromatin protein patterns may have resulted from differences in preparation procedures of brain chromatin or in extraction and analytical methods for these proteins. High-mobility-group proteins were detected in neuronal and glial nuclei, but no differences were observed among the high-mobility-group proteins isolated from brain, liver, and neuronal and glial nuclei.[91]

6. *RNA SYNTHESIS*

In the 1960s, several investigations showed RNA synthesis using an aggregate enzyme from rat brain nuclei.[92,93] Then studies of RNA polymerase branched out in two directions: (1) purification and characterization of this enzyme and (2) *in vivo* and *in vitro* determination of RNA polymerase activity using isolated neuronal and glial nuclei.[94]

Purification and characterization of RNA polymerase were carried out in liver and thymus by Roeder and Rutter[95] and Chambon and colleagues.[96,97] Similar experiments were performed in brain nuclei. Singh and Sung[98] partially separated RNA polymerase A, which participates in synthesis of rRNA, and B, which synthesizes mRNA, from bovine brain nuclei. Yamamoto and Takahashi[99] isolated and characterized RNA polymerase A, B_I, B_{II}, C_I, C_{II}, and C_{III} from rat brain nuclei (Fig. 5). C enzymes participate in the synthesis of tRNA. Three classes of the enzymes are different in α-amanitin sensitivity, ion dependence, and template dependence [DNA and poly d(A–T)]. Furthermore, they[100] succeeded in complete purification of RNA polymerase B_{II} and analyzed the subunit structure, which is slightly different from that of calf thymus enzyme. Stokes *et al.*[101] partially separated RNA polymerases from mouse brain small nuclei. A is localized in nucleolus; B and C are in nucleoplasm.

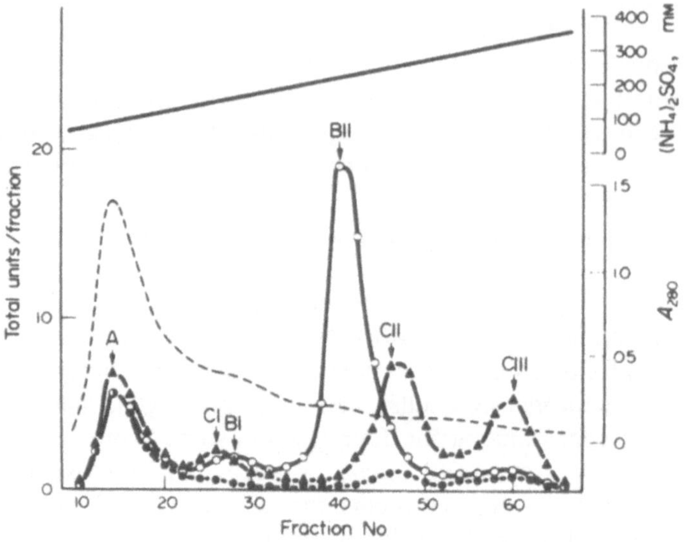

Fig. 5. Multiple RNA polymerases of rat brain nuclei.[99] Fraction IV (482 mg protein) from rat brain nuclei was subjected to DEAE Sephadex A-25 chromatography. An aliquot of sample was assayed with DNA in the absence (○—○) or presence (●—●) of α-amanitin (1 μg/ml) and with poly[d(A–T)] in the presence of α-amanitin (▲—▲). Ammonium sulfate was added to the assay mixture: 12.5 μmol with DNA template; 2.5 μmol with poly [d(A–T)]. ----, absorbance at 280 nm; —, ammonium sulfate concentration.

RNA polymerase activity was determined in isolated neuronal and glial nuclei by a number of investigators.[8,13,102] Enzyme activity per nucleus was much higher in neuronal nuclei than in glial and liver nuclei, although higher poly(A) synthesis activity was found in liver than in neuronal nuclei.[102] Austoker *et al.*[9] described the highest RNA polymerase activity in the zone II fraction (81% astrocytic, 15% neuronal nuclei) and the lowest in zones IV and V (oligodendrocytic nuclei as shown in Table VI). The high activity of neuronal and astrocytic nuclei may be accounted for by difference of RNA polymerase itself or the chromatin structure of each cell nucleus. However, according to Yamamoto and Takahashi,[99] properties and activity (500 units/mg protein) of purified RNA polymerase B were not very much different from those of calf thymus. Austoker *et al.*[9] related the high enzyme activity to highly stained euchromatin in neuronal and astroglial nuclei. Mizobe *et al.*[103] reported evidence that in neuronal chromatin, a part of the steric restrictions imposed on RNA synthesis appears to be intrinsically weakened. RNA synthesis is not increased by trypsin treatment of neuronal chromatin but is increased by the same treatment of liver chromatin. This result may be related to the observation that the circular dichroism spectrum of neuronal chromatin as compared with that of liver chromatin is characterized by its higher mean residue ellipticity at 260–300 nm. However, the above results may be related to evidence that acetylation or phosphorylation activities are higher in neuronal chromatin than in glial chromatin, as described later.

Table VI

RNA Polymerase Activity in Fractionated Rat Brain Nuclei in the Absence and in the Presence of α-Amanitin with Different Labeled Nucleotides[a]

	Incorporation of nucleotide (pmol/20 min per 100 μg of DNA)												
	$[^3H]UTP$[b]				$[^3H]ATP$			$[^3H]CTP$			$[^3H]GTP$		
	Mg^{2+}		Mn^{2+} – $(NH_4)_2SO_4$		Mg^{2+}	Mn^{2+} – $(NH_4)_2SO_4$		Mg^{2+}	Mn^{2+} – $(NH_4)_2SO_4$		Mg^{2+}	Mn^{2+} – $(NH_4)_2SO_4$	
Zone	+[c]	–[c]	+	–	+	+	–	+	+	–	+	+	–
I	59	352	55	777	33	29	367	56	54	679	33	40	377
II	67	364	65	902	26	25	407	45	44	947	50	59	277
III	36	333	37	521	23	20	327	38	48	650	39	39	227
IV	15	139	24	394	22	22	211	33	33	389	27	27	166
V	25	87	34	375	25	27	199	31	31	412	17	19	171

[a] From Austoker *et al.*,[9] with their permission.

[b] In the columns headed Mg^{2+}, 6 mM $MgCl_2$ and 0.15 M KCl were used. In the columns headed Mn^{2+} – $(NH_4)_2SO_4$, 3 mM $MnCl_2$ and 0.3 M $(NH_4)_2SO_4$ were used. $[^3H]$-labeled UTP, ATP, CTP, and GTP were each 1 μCi (specific radioactivity 2 mCi/mmol).

[c] + indicates the presence, and – the absence, of 1.3 μg of α-amanitin.

In relation to the above *in vitro* studies, it is noteworthy that the *in vivo* incorporation of [^{3}H]uridine or [^{14}C]orotic acid into RNA of neuronal nuclei is higher than that into glial nuclei.[56–60] The effects of development, physiological changes, and various drugs on brain RNA synthesis were also examined. Data on RNA synthesis during brain development were reviewed by Johnson and Weck[104] in 1976 and by Mandel and Wintzerith [105] in 1980. Some results are, however, contradictory. Banks and Johnson[106] reported that glial nuclear RNA polymerase decreased and that neuronal enzyme increased during development. But Giuffrida *et al.*[107] observed that RNA polymerases from all classes of nuclei decreased during the growth of the rat. Developmental changes of three types of RNA polymerase have not been simultaneously examined. According to Guiditta *et al.*,[108] changes in brain RNA synthesis in sleep were found only in rRNA and hnRNA of neuronal nuclei. Dietary protein restriction introduced a decrease of RNA polymerase B but not A.[109] Sunde *et al.*[110,111] found that a dehydration stimulus resulted in a generalized and selective increase in the labeling of the various RNA species of rat neural lobes (neuroglial cells). Long-lasting effects of electroconvulsive shock on the pattern of poly(A) RNA synthesis in rabbit cerebral cortical nuclei were also observed.[112] Intravenous injection of LSD into rabbits increased RNA synthesis of brain nuclei, especially of the nucleoplasm.[113] After chronic morphine treatment, specific activity of RNA polymerase A decreased, and ion sensitivity of B and C changed in brain nuclei.[114] Intracerebral injection of α-amanitin 6 h before training was followed by impaired performance of rats on retesting after 7 days.[115]

Nuclear–cytosol interactions were also examined in the brain by Weck and Johnson,[116,117] Lim,[65] and others.[6] Takahashi *et al.*[52] used dialyzed purified brain cytosol fractions to prepare undegraded brain nuclear RNA. That the addition of cytosol resulted in a significant increase in the radioactivity and size of the RNA with isolated 10-day-old and adult brain nuclei was also reported by Weck and Johnson.[116,117] Such results were also observed in both neuronal and glial nuclei. The addition of cytosol to isolated brain nuclei stimulated an ATP-dependent release of mRNA from the nuclei, as described for liver. Undernutrition increased this release. Lim[65] observed higher transfer of synthesized nuclear RNA into cytoplasm in newborn than in adult rat forebrain. In the brain, some classes of histones are also synthesized in the cytoplasm and then enter the nuclei to form a complex with DNA. The hormones bound to the cytoplasmic receptor are transferred to the hypothalamic nucleus, where the hormones alter transcription. Such hormonal nucleocytoplasmic interaction is discussed below.

7. DNA SYNTHESIS

It is well known that the key enzyme for DNA synthesis in eucaryotic cells is DNA polymerase. Three different forms, α, β, and γ, were found in cells of brain, as in other eucaryotic cells. DNA polymerase α, which is located in the nucleus, is involved in DNA replication. The location of DNA polymerase

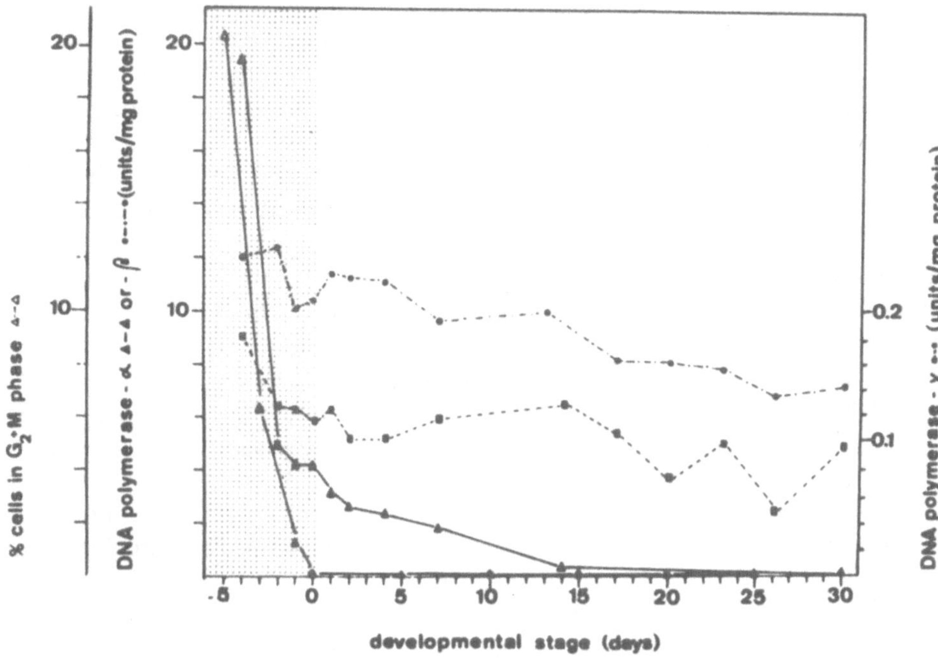

Fig. 6. Developmental profiles of the three DNA polymerases of brain: DNA polymerase α (▲), DNA polymerase β (●), and DNA polymerase γ (■). Note the parallel decline of DNA polymerase α and of the percentage of cells in G_2 + M phase (△). (From Hübscher *et al.*,[121] with their permission.)

β is also nuclear, and the role of this enzyme may be DNA repair. DNA polymerase γ is located in mitochondria and nucleus and undertakes replication in the mitochondria, but its nuclear role is still unknown.[1] A part of the studies on DNA polymerase in brain nuclei is the fractionation and purification of this enzyme. Shimada and Terayama[118] reported a stimulating factor of infant rat brain cytosol on DNA synthesis in rat brain. Chiu and Sung[119,120] fractionated the enzyme to A (α) and B (β) forms and observed developmental changes of subcellular localization: the 10-day-old brain contained more enzyme activity in the soluble fraction than in the nuclear fraction, which is the reverse of the adult brain. It was reported that DNA polymerase A (α) exists in the brain soluble fraction and that most of the B (β) form is in the chromatin. However, the experiments of Kuenzle, Spadari, *et al.*[121–123] showed evidence for the existence of DNA polymerase, α, β, and γ in brain as in other tissues, and they proposed that DNA polymerase α is the replicating nuclear enzyme, DNA polymerase β is able to repair UV-damaged nuclear DNA, and DNA polymerase γ is responsible for DNA replication in mitochondria from the changes of DNA synthesis in the developing brain, measurement of UV-induced repair-type DNA synthesis, and the studies of *in vitro* DNA synthesis in synaptosomes (Fig. 6). They isolated DNA polymerase from adult rat cortex neuronal nuclei, which contained no α polymerase, 99.2% β polymerase, and only 0.8% γ polymerase. Its properties were similar to those of other mammalian β polymerase. They purified DNA polymerase γ from rat brain synaptosomal mitochondria and brain nuclei. Polymerases from these sources may be identical by several criteria, such as molecular weight (180,000), ion requirement, and template preferences. To my knowledge, α-polymerase has never been purified from the brain.

According to Stambolova *et al.*,[124] DNA polymerase activity was the highest in the neuronal nuclei from zone I and exhibited a progressive decline in II, III, IV, and V, which were fractionated by zonal centrifugation of rat brain nuclei. Nuclei from adults had a much higher activity than those from 10-day-old rats and expressed a preference for native DNA as template. These results were in contrast with those on the *in vivo* incorporation of [^{3}H]thymidine into nuclear DNA, in which the specific radioactivity of the DNA in the infant rats was highest in zone V. In the nuclei of adult rats, which exhibited a comparatively low incorporation, the highest specific radioactivity was associated with zones I and V. The results of Stambolova *et al.*[124] indicate that the measurement of DNA polymerase *in vitro* may represent the total replication and repair activity of the nuclei and that the high activity of the adult brain nuclei could be associated with a process of repair. This view is in general agreement with the recent data on α and β polymerases by Hübscher *et al.*[121] In fact, Stambolova *et al.*[125] found later that the two forms of DNA polymerase are present in all types of nuclei from both 10-day-old and adult rats and that the larger part of total DNA polymerase activity of all nuclear fractions, with the exception of zone IV and V nuclei from infant rats, was mediated by the action of enzyme B (β).

The results of Szijan and Burdman[126] in which the total DNA polymerase activity of glial nuclei is higher than that of neuronal nuclei in the adult rat were in contrast with those of the abovementioned authors. Inoue *et al.*[127–129] observed results similar to those of Stambolova *et al.* in mature neuronal nuclei and, furthermore, showed that DNA ligase is contained in neuronal nuclei in greater amounts than in glial and liver nuclei isolated from an adult guinea pig. Norton and Viola[130] also reported the presence of α-polymerase activity in neuronal nuclei only and β-polymerase activity in all nuclear fractions of adult rats, which did not agree with data of previous authors. It is difficult to interpret the differences of these studies concerning DNA polymerase, although the methods for determination of this enzyme activity and the purity of subclasses of brain nuclei were variable. Thymidine incorporation, and the effects of various conditions on it, including autoradiographic investigations, were reviewed in detail by Mandel and Wintzerith.[105]

8. *PROTEIN SYNTHESIS AND OTHER ENZYMES*

The great difficulties in localizing the enzymes in the nuclei that Rappoport *et al.*[5] described are still present: (1) cross contamination of enzymes between fractions and (2) leaking of enzymes from nuclei during fractionation. DNA polymerase and RNA polymerase are described in the preceding pages.

A number of investigators[131–141] observed protein biosynthesis in isolated brain nuclei, although inconsistencies were evident on several points such as the effect of hypertonic sucrose for preparation of brain nuclei, the effect of sodium ion, the effects of some inhibitors, the involvement of an energy-yielding process, and the differences between neuronal and glial nuclei. Some of the reasons for such controversy may be cross contamination of nuclei by other subcellular fractions, bacterial contamination, or removal of endoplasmic reticulum with detergents. However, it is difficult to understand the discrepancies in the results with neuronal and glial nuclei. Haglid[140] reported that the isolated nuclei of astrocytoma incorporated 40 times more [^{3}H]leucine into proteins than the nuclei of a normal human brain. Recently, Gozes *et al.*[142] observed that nuclei prepared from either enriched neuronal or glial populations synthesized similar proportions of tubulin and actin *in vitro*. In their nuclear preparation, cycloheximide inhibited and the cytosol pH 5 factor stimulated amino acid incorporation.

Acidic and alkaline RNases were found in brain nuclei and nucleoli.[19,143] Ittel *et al.*[144–147] reported the purification and characterization of two brain nuclei RNases, one an alkaline RNase, and the other an acidic RNase, and further determined the change of RNase activities during postnatal development of rat brain nuclei. They reported[147] that the acidic and alkaline RNase activities of the neuronal nuclei were much higher than those of the glial nuclei, although the RNase activity per nucleus was the same in all five fractionated zones according to Austoker *et al.*[9] Some of these RNases may be involved in the processing of the RNA precursors. The DNase activity of brain nuclei was rather low, one-tenth of that of liver nuclei. Stambolova *et al.*[124] reported that DNase activity was markedly higher in zone I, decreased gradually in the next three zones, and was slightly increased in zone V. Inoue *et al.*[127] observed a similar pattern.

Modification of histone or nonhistone protein has also been found in brain nuclei and was implicated in the regulation of gene expression. The higher *in vitro* RNA synthesis in the neuronal nuclei corresponded to a higher nuclear acetylation rate of neuronal histones compared to glial histones.[148] Developmental changes in histone acetylation were also compared with the changes of RNA synthesis in brain nuclei.[149] Furthermore, Sarkander *et al.*[150] found that the phosphorylation of neuronal nonhistone proteins was considerably higher than that of glial ones. Most nonhistone proteins of high molecular weight were highly phosphorylated. Acetylation of brain chromatin protein was also studied by Kanungo *et al.*[151,152] Yanagihara and co-worker[153,154] reported a difference in phosphorylation of the chromatin protein between liver and brain and the changes in phosphorylation with cerebral anoxia and ischemia. The specific activities of chromatin protein kinase and histone methyltransferase were higher in neuronal than in oligodendroglial nuclei.[155]

Hollenbeck *et al.*[156] found the translocation of cytosol protein kinase into nuclei with the induction of tyrosine hydroxylase in neuroblastoma cells. The preparation and properties of brain histone lysine methyltransferase, histone arginine methyltransferase, and protein methylase II were recently described in detail by Paik and associates[157,158] and also by Duerre and Quick.[159] The

ADP-ribosylation of histones and nonhistone proteins by the nuclear enzyme poly(ADP-ribose) synthetase has also been found in the brain,[160] where the activity was considerably higher. Recently, Bilen *et al.*[161] reported that the poly(ADP-ribose) synthetase activity is similar in neuronal and glial nuclear suspensions, but the neuronal soluble enzyme activity is higher than that of the glial soluble enzyme. This system may be implicated in the regulation of gene expression and chromatin condensation, although its difinite biological function is still obscure.[162]

Since Rappoport *et al.*[5] reviewed the references on the glycolytic and respiratory enzymes in brain nuclei, only a few investigators have reported information concerning these enzymes. Respiratory activity of guinea pig brain nuclei has been observed.[163] Kato and Lowry[164] determined the distribution of several enzymes between nucleus and cytoplasm of single dorsal root ganglion cells. Of the seven enzymes from major energy-yielding systems, three were higher in concentration in the nucleus than in the cytoplasm (hexokinase, 6-P-gluconic dehydrogenase, isocitric dehydrogenase). Glutamic dehydrogenase was found in the nucleus as well as in the cytoplasm. ATP:NMN adenyltransferase was found to be localized almost exclusively in the cytoplasm, in contrast to findings of Kato and Kurokawa,[8] who examined nuclei isolated by the usual bulk fractionation procedures. The reason for this discrepancy was not clear, but it may reflect a combination of better avoidance of artifacts and differences among the various nuclear types. Thompson[165,166] reported the synthesis of dCDP-diglyceride in the neuronal nuclei. This may reflect a generally high specific activity of liponucleotide synthesis in the endoplasmic reticulum of excitable cells.

9. RECEPTOR ACTIVITY

According to current theories,[167] in both neuronal and nonneuronal tissues, steroid hormones act by (1) binding to cytosol receptors, (2) translocation of the receptor to the cell nucleus and association with chromatin, and (3) alteration of transcription and subsequent protein synthesis. Recently, McEwen[168] reviewed endocrine effects on the brain. In this Handbook gonadal hormones are also reviewed by McEwen. I describe them only briefly here.

Autoradiographic methods utilizing [^{3}H]-labeled steroid were useful for the mapping of target cells in various brain regions. Further, cell nuclei isolated with the aid of Triton X-100 were always used for the study of [^{3}H]steroid hormone binding. First, the receptor for estradiol was examined. Autoradiography[169] revealed that the neurons of the hypophyseotropic area and amygdala contained the highest concentrations of [^{3}H]estradiol receptor sites. The estradiol bound to cytosol receptor is transported into the cell nuclei. In fact, cell nuclear retention of [^{3}H]estradiol was demonstrated in various brain regions.[170] The purified nuclear receptor has a sedimentation coefficient of 4.5 S in the calf uterus[171]; that of brain nuclei may have a similar coefficient. The postnatal development of estrogen receptor systems was observed,[172] and the presence of unoccupied nuclear receptors for estradiol in the rat hypothalamus was also found.[173] Nuclear and cytosol receptors for testosterone and its metabolite (5α-dihydrotestosterone) also were demonstrated in the hypothalamus, amygdala, and pituitary tissue.[174,175]

The highest concentrations of binding sites for [^{3}H]corticosterone were found in the hippocampus, amygdala, and septum.[170,176] This binding also was to cytosol and nuclear binding sites. About 16,000 molecules of corticosterone are bound per cell nucleus in the hippocampus. It should be noted that glial cells contain putative receptor sites for [^{3}H]corticosterone.

[^{3}H]Progesterone and its metabolite accumulated in midbrain region, hypothalamus and cerebral cortex. Recently, demonstrations of the receptor for this steroid have been successfully carried out using a synthetic progestin, 17α, 21-dimethyl-19-norpregna-4,9-dione-3,20-dione.[177] The receptors were found in the brain regions described above, but only those in hypothalamus and the preoptic area were induced by estrogen treatment.[178] There is evidence that progestin is bound by brain cell nuclei. There are reports[179] that rat brain contains proteins capable of binding [^{3}H]deoxycorticosterone mineralocorticoid.

Studies of the relationship between gonadal steroids and sexual differentiation of brain have been extensively carried out by a number of investigators. Several important findings were observed, for example, (1) androgen-dependent changes in the volume of the cell nucleus and in RNA synthesis in hypothalamus and amygdala during the perinatal period,[168] (2) blockage of androgen-induced sexual differentiation of brain by inhibitors of DNA and RNA synthesis,[180] (3) significant differences between male and female rats in the retention of [^{3}H]estradiol-17 by purified hypothalamic chromatin.[181]

Receptors for thyroid hormones such as triiodothyronin also exist in brain nuclei.[182,183] Nerve growth factors bind to the receptor in the nuclear membrane.[184] There are also reports that S-100 protein[185] and myelin basic protein[186] are able to bind to the brain chromatin.

10. CONCLUSION

Factors that affect the yield and purity of the nuclei isolated from nervous tissue were discussed in the reviews of Rappoport *et al.*[5] and McEwen *et al.*[6] However, the separation procedures for the subclasses of brain nuclei were not extensively discussed. To my knowledge, a satisfactory procedure has not yet been devised for isolation of neuronal and astroglial nuclei, particularly for the latter, although the method of Mathias *et al.* provided astroglial nuclei of 81% purity. It may be desirable to develop a new procedure, such as application of antibody-affinity chromatography, cell sorter, centrifugal elutriation, or others. Another problem with the isolated brain nuclei is the absence of any marker enzyme or marker protein for neuronal, astroglial, or oligodendroglial nuclei. The isolation of the special nucleus-specific proteins (nonhistine proteins) and production of the antibodies against them may possibly provide the marker for each isolated specific nucleus. These antibodies may be useful together with morphology for identification of each subclass of brain nuclei. At present, the study of specific gene expression is one of the most challenging problems of general biology. In order to develop this project in the area of brain research or neurochemistry, a brain-specific or nervous-cell-specific DNA probe should be obtained.

After this manuscript had been completed, two monographs about molecular genetic neurobiology were published.[187,188]

REFERENCES

1. Lewin, B., 1980, *Gene Expression*, 2nd ed., John Wiley & Sons, New York.
2. Busch, H. (ed.), 1974–1981, *The Cell Nucleus*, Volumes 1–8, Academic Press, New York.
3. Stein, G., Stein, J., and Kleinsmith, L. J., 1977–1978, *Chromatin and Chromosomal Protein Research*, Volumes 1–4 (D. Prescott, ed.), Academic Press, New York.
4. Adams, R. L. P., Burdon, R. H., Campbell, A. M., Leader, D. P., and Smellie, R. M. S., 1981, *The Biochemistry of the Nucleic Acids*, 9th ed., Chapman & Hall, London.
5. Rappoport, D. A., Maxcy, P., Jr., and Daginawala, H. F., 1969, *Handbook of Neurochemistry*, Volume 2 (A. Lajtha, ed.), Plenum Press, New York, pp. 241–254.
6. McEwen, B. S., and Zigmond, R. E., 1972, *Research Methods in Neurochemistry*, Volume 1 (N. Marks and R. Rodnight, eds.), Plenum Press, New York, pp. 139–161.
7. Løvtrup-Rein, H., and McEwen, B. S., 1966, *J. Cell Biol.* **30**:405–416.
8. Kato, T., and Kurokawa, M., 1967, *J. Cell Biol.* **32**:649–662.
9. Austoker, J. L., Cox, D., and Mathias, A. P., 1972, *Biochem. J.* **129**:1139–1155.
10. Mathias, A. P., and Wynter, C. V. A., 1973, *FEBS Lett.* **33**:18–22.
11. Knüsel, A., Lehner, B., Kuenzle, C. C., and Kistler, G. S., 1973, *J. Cell Biol.* **59**:762–765.
12. Sellinger, O. Z., and Azcurra, J. M., 1974, *Research Methods in Neurochemistry*, Volume 2 (N. Marks and R. Rodnight, eds.), Plenum Press, New York, pp. 3–38.
13. Thompson, R. J., 1973, *J. Neurochem.* **21**:19–40.
14. Slagel, D. E., and Akers, R. D., 1972, *Brain Res.* **44**:245–260.
15. Kaplan, B. B., Schachter, B. S., Osterburg, H. H., de Vellis, J. S., and Finch, C. E., 1978, *Biochemistry* **17**:5516–5524.
16. Zagon, I., and McLaughlin, P. J., 1979, *Brain Res.* **170**:443–457.
17. Olpe, H. R., Honegger, C. G., Leuba, G., and Rabinowicz, T., 1974, *Exp. Brain Res.* **21**:131–138.
18. Busch, H., and Smetana, K., 1970, *The Nucleolus*, Academic Press, New York.
19. Takahashi, Y., Araki, K., Ikeda, K., and Oyanagi, S., 1973, *Brain Res.* **73**:189–203.
20. Takahashi, Y., Araki, K., Ikeda, K., and Oyanagi, S., 1973, *Proc. Jpn. Acad.* **49**:765–770.
21. Banks, S. P., and Johnson, T. C., 1973, *Biochim. Biophys. Acta* **294**:450–460.
22. Banks, S. P., and Johnson, T. C., 1973, *Brain Res.* **62**:201–212.
23. McGhee, J. D., and Felsenfeld, G., 1980, *Annu. Rev. Biochem.* **49**:1115–1156.
24. Singh, V. K., and Sung, S. N., 1972, *Biochem. J.* **130**:1095–1099.
25. Bondy, S. C., and Roberts, S., 1969, *Biochem. J.* **115**:341–349.
26. Smith, K. D., Church, R. B., and McCarthy, B. J., 1969, *Biochemistry* **8**:4271–4277.
27. Gilmour, R. S., and Paul, J., 1973, *Proc. Natl. Acad. Sci. U.S.A.* **70**:3440–3442.
28. Harris, S. E., Schwartz, R. J., Tsai, M.-J., and O'Malley, B. W., 1976, *J. Biol. Chem.* **251**:524–529.
29. Dale, R. M. K., and Ward, D. C., 1975, *Biochemistry* **14**:2458–2469.
30. Zasloff, M., and Felsenfeld, G., 1977, *Biochemistry* **16**:5135–5145.
31. Pays, E., Donaldson, D., and Gilmour, R. S., 1979, *Biochim. Biophys. Acta* **562**:112–130.
32. Jaequet, M., Levy, S. B., Robert, B., and Gros, F., 1977, *Gene* **1**:373–383.
33. Parker, C. S., and Roeder, R. G., 1977, *Proc. Natl. Acad. Sci. U.S.A.* **74**:44–48.
34. Brown, I. R., 1977, *J. Neurochem.* **28**:1389–1391.
35. Sarkandar, H. I., and Dulce, H.-J., 1978, *Exp. Brain Res.* **31**:317–327.
36. Sarkandar, H. I., and Dulce, H.-J., 1979, *Exp. Brain Res.* **35**:109–125.
37. Brown, I. R., Heikkila, J., Silver, J., and Straus, N., 1977, *Biochim. Biophys. Acta* **477**:288–294.

38. Segall, J., Matsui, T., and Roeder, R. G., 1980, *J. Biol. Chem.* **255**:11986–11991.
39. Manley, J. L., Fire, A., Cano, A., Sharp, P. A., and Gefter, M. L., 1980, *Proc. Natl. Acad. Sci. U.S.A.* **77**:3855–3859.
40. Thomas, G. O., and Thompson, R. J., 1977, *Cell* **10**:633–640.
41. Brown, I. R., 1978, *Biochem. Biophys. Res. Commun.* **84**:285–292.
42. Ermini, M., and Kuenzle, C. C., 1978, *FEBS Lett.* **90**:167–172.
43. Whatley, S. A., Hall, C., and Lim, L., 1981, *Biochem. J.* **196**:115–119.
44. Greenwood, P. D., Silver, J. C., and Brown, I. R., 1981, *J. Neurochem.* **37**:498–505.
45. Brown, I. R., 1980, *Dev. Biol.* **80**:248–252.
46. Tashiro, T., and Kurokawa, M., 1975, *Eur. J. Biochem.* **60**:569–577.
47. Wu, C., and Gilbert, W., 1981, *Proc. Natl. Acad. Sci. U.S.A.* **78**:1577–1580.
48. McGhee, J. D., Wood, W. I., Dolan, M., Engle, J. D., and Felsenfeld, G., 1981, *Cell* **27**:45–55.
49. Heizman, C. W., Hobi, R., Winkler, G. C., and Kuenzle, C. C., 1981, *Exp. Cell Res.* **135**:331–339.
50. Ranjekar, P. K., and Murthy, M. R. V., 1973, *J. Neurochem.* **20**:1257–1264.
51. Wintzerith, M., Mori, K., and Mandel, P., 1973, *J. Neurochem.* **21**:1341–1343.
52. Takahashi, Y., Araki, K., and Suzuki, Y., 1971, *Brain Res.* **32**:179–188.
53. Tencheva, Z. S., and Hadjiolov, A. A., 1969, *J. Neurochem.* **16**:769–776.
54. Vesco, C., and Giuditta, A., 1967, *Biochim. Biophys. Acta* **142**:385–402.
55. Grunning, P. W., Shooter, E. M., Austin, L., and Jeffrey, P. L., 1981, *J. Biol. Chem.* **256**:6663–6669.
56. Løvtrup-Rein, H., 1970, *J. Neurochem.* **17**:853–863.
57. Giuditta, A., Rutigliano, B., Casola, L., and Romano, M., 1972, *Brain Res.* **46**:313–328.
58. Austoker, J., Cox, D., and Mathias, A. P., 1973, *Biochem. J.* **132**:813–819.
59. Løvtrup-Rein, H., and Grahn, B., 1970, *J. Neurochem.* **17**:845–852.
60. Stoykova, A. S., Dabeva, M. D., Dimova, R. N., and Hadjiolov, A. A., 1979, *J. Neurochem.* **33**:931–937.
61. Furuichi, Y., and Miura, K., 1975, *Nature* **253**:374–375.
62. Murthy, M. R. V., 1982, *J. Neurochem.* **38**:28–40.
63. DeLarco, J., Abramowitz, A., Bromwell, K., and Guroff, G., 1975, *J. Neurochem.* **24**:215–222.
64. Mahoney, J. B., and Brown, I. R., 1975, *J. Neurochem.* **25**:503–507.
65. Lim, L., 1977, *Biochemical Correlates of Brain Structure and Function* (A. N. Davison, ed.), Academic Press, New York, pp. 15–41.
66. Banks, S. P., and Johnson, T. C., 1973, *Science* **181**:1064–1065.
67. Soga, K., Kushiya, E., Araki, K., and Takahashi, Y., 1975, *Annual Report Brain Research Institute, Niigata University*, Volume 8, Niigata University, Niigata, Japan, p. 27.
68. Grouse, L. D., Schrier, B. K., Letendre, C. H., and Nelson, P. G., 1980, *Curr. Top. Dev. Biol.* **16**:381–397.
69. Hahn, W. E., and Laird, C. D., 1971, *Science* **173**:158–161.
70. Brown, I. R., and Church, R. B., 1972, *Dev. Biol.* **29**:73–84.
71. Grouse, L. D., Chilton, M.-D., and McCarthy, B. J., 1972, *Biochemistry* **11**:798–805.
72. Chikaraishi, D. M., Deeb, S. S., and Sueoka, N., 1978, *Cell* **13**:111–120.
73. Bantle, J. A., and Hahn, W. E., 1976, *Cell* **8**:139–150.
74. Chikaraishi, D. M., 1979, *Biochemistry* **18**:3249–3256.
75. Soga, K., and Takahashi, Y., 1975, *Nature* **256**:233–234.
76. Soga, K., and Takahashi, Y., 1976, *J. Neurochem.* **26**:89–94.
77. Hahn, W. E., Van Ness, J., and Maxwell, I. H., 1978, *Proc. Natl. Acad. Sci. U.S.A.* **75**:5544–5547.
78. Elgin, S. C. R., and Weintraub, H., 1975, *Annu. Rev. Biochem.* **44**:725–774.
79. Graziano, S. L., and Huang, R. C. C., 1971, *Biochemistry* **10**:4770–4777.
79a. Shaw, L. M. J., and Huang, R. C. C., 1970, *Biochemistry* **9**:4530–4542.
80. Greenwood, P. D., Silver, J. C., and Brown, I. R., 1981, *J. Neurochem.* **37**:498–505.
81. MacGillivray, A. J., and Rickwood, D., 1974, *Eur. J. Biochem.* **41**:181–190.
82. Elgin, S. C. R., and Bonner, J., 1970, *Biochemistry* **9**:4440–4447.

83. MacGillivray, A. J., Cameron, A., Krauze, R. J., Rickwood, D., and Paul, J., 1972, *Biochim. Biophys. Acta* **227**:384–402.
84. Tashiro, T., Mizobe, F., and Kurokawa, M., 1974, *FEBS Lett.* **38**:121–124.
85. Wu, F. C., Elgin, S. C. R., and Hood, L. E., 1973, *Biochemistry* **12**:2792–2797.
86. Olpe, H.-R., Von Hahn, H. P., and Honegger, C. G., 1973, *Brain Res.* **58**:453–464.
87. Tsitilou, S. G., Cox, D., Mathias, A. P., and Ridge, D., 1979, *Biochem. J.* **177**:331–346.
88. Dravid, A. R., and Burdman, J. A., 1968, *J. Neurochem.* **15**:25–30.
89. Fujitani, H., and Holoubek, V., 1974, *J. Neurochem.* **23**:1215–1224.
90. Heizmann, C. W., Arnold, E. M., and Kuenzle, C. C., 1980, *J. Biol. Chem.* **255**:11504–11511.
91. Greenwood, P. D., Silver, J. C., and Brown, I. R., 1981, *Neurochem. Res.* **6**:673–679.
92. Barondes, S. H., 1964, *J. Neurochem.* **11**:663–669.
93. Bondy, S. C., and Waelsch, H., 1965, *J. Neurochem.* **12**:751–756.
94. Mahler, H. R., 1981, *Basic Neurochemistry*, 3rd ed. (G. J. Siegel, R. W. Albers, B. W. Agranoff, and R. Katzman, eds.), Little, Brown, Boston, pp. 371–440.
95. Roeder, R. G., and Rutter, W. J., 1970, *Proc. Natl. Acad. Sci. U.S.A.* **65**:675–682.
96. Kedinger, C., and Chambon, P., 1972, *Eur. J. Biochem.* **28**:283–290.
97. Chambon, P., 1978, *Annu. Rev. Biochem.* **44**:613–638.
98. Singh, V. K., and Sung, S. C., 1971, *Brain Res.* **25**:677–679.
99. Yamamoto, H., and Takahashi, Y., 1978, *J. Neurochem.* **31**:449–456.
100. Yamamoto, H., and Takahashi, Y., 1980, *J. Neurochem.* **34**:255–260.
101. Stokes, K. B., Lee, N. M., and Loh, H. H., 1978, *Biochem. Pharmacol.* **27**:1787–1792.
102. Kato, T., and Kurokawa, M., 1970, *Biochem. J.* **116**:599–609.
103. Mizobe, F., Tashiro, T., and Kurokawa, M., 1974, *Eur. J. Biochem.* **48**:25–33.
104. Johnson, T. C., and Weck, P. K., 1976, *Neurochem. Res.* **1**:557–572.
105. Mandel, P., and Wintzerith, M., 1980, *Biochemistry of Brain* (S. Kumar, ed.), Pergamon Press, Oxford, pp. 241–282.
106. Banks-Schlegel, S. P., and Johnson, T. C., 1975, *J. Neurochem.* **24**:947–952.
107. Giuffrida, A. M., Cox, D., and Mathias, A. P., 1975, *J. Neurochem.* **24**:749–755.
108. Giuditta, A., Rutigliano, B., and Vitale-Neugebauer, A., 1980, *J. Neurochem.* **35**:1259–1266.
109. Kubát, B., Morén, G. M., and von der Decken, A., 1978, *J. Neurochem.* **31**:1143–1148.
110. Sunde, D., Osinchak, J., and Sachs, H., 1972, *Brain Res.* **47**:195–216.
111. Sunde, D., McKelvy, J., and Sachs, H., 1972, *Brain Res.* **47**:237–253.
112. Cupello, A., Ferrillo, F., and Rosadini, G., 1981, *Neurochem. Res.* **6**:175–182.
113. Brown, I. R., 1975, *Proc. Natl. Acad. Sci. U.S.A.* **72**:837–839.
114. Stokes, K. B., Lee, N. M., and Loh, H. H., 1980, *J. Neurochem.* **34**:1058–1064.
115. Montanaro, N., Norrello, F., and Stripe, F., 1971, *Biochem. J.* **125**:1087–1090.
116. Weck, P. K., and Johnson, T. C., 1978, *J. Neurochem.* **30**:1057–1065.
117. Weck, P. K., and Johnson, T. C., 1978, *Neurochem. Res.* **3**:325–343.
118. Shimada, H., and Terayama, H., 1972, *Biochim. Biophys. Acta* **287**:415–426.
119. Chiu, J. F., and Sung, S. C., 1971, *Biochim. Biophys. Acta* **246**:44–50.
120. Chiu, J. F., and Sung, S. C., 1973, *J. Neurochem.* **20**:617–620.
121. Hübscher, U., Kuenzle, C. C., Limacher, W., Scherrer, P., and Spadari, S., 1978, *Cold Spring Harbor Symp. Quant. Biol.* **43**:625–629.
122. Hübscher, U., Kuenzle, C. C., and Spadari, S., 1977, *Eur. J. Biochem.* **81**:249–258.
123. Waser, J., Hübscher, U., Kuenzle, C. C., and Spadari, S., 1979, *Eur. J. Biochem.* **97**:361–368.
124. Stambolova, M. A., Cox, D., and Mathias, A. P., 1973, *Biochem. J.* **136**:685–695.
125. Stambolova, M. A., Cox, D., and Mathias, A. P., 1974, *Biochem. J.* **140**:65–71.
126. Szijan, I., and Burdman, J. A., 1974, *Brain Res.* **73**:563–567.
127. Inoue, N., Suzuki, O., and Kato, T., 1976, *J. Neurochem.* **27**:113–119.
128. Inoue, N., Ono, T., and Kato, T., 1979, *Biochem. J.* **180**:471–480.
129. Inoue, N., and Kato, T., 1980, *J. Neurochem.* **34**:1574–1583,
130. Norton, P., and Viola, M. V., 1977, *J. Neurochem.* **29**:299–303.
131. Mase, K., 1966, *Adv. Neurol. Sci.* **10**:198–203 [in Japanese].
132. Inamura, H., Kato, T., and Kurokawa, M., 1968, *Proc. Jpn. Acad.* **44**:974–979.
133. Burdman, J. A., and Journey, L. J., 1969, *J. Neurochem.* **16**:493–500.

134. Løvtrup-Rein, H., 1970, *Brain Res.* **19**:433–444.
135. Burdman, J. A., 1970, *J. Neurochem.* **17**:1555–1562.
136. Burdman, J. A., Haglid, K., and Dravid, A. R., 1970, *J. Neurochem.* **17**:669–676.
137. Burdman, J. A., 1972, *J. Neurochem.* **19**:1459–1469.
138. Dravid, A. R., and Wong, E., 1972, *J. Neurochem.* **19**:2709–2725.
139. Burdman, J. A., 1972, *Brain Res.* **41**:413–421.
140. Haglid, K. G., 1972, *J. Neurochem.* **19**:19–25.
141. Fleischer-Lambropoulos, H., and Reinsch, I., 1975, *Brain Res.* **88**:120–126.
142. Gozes, I., Walker, M. D., Kaye, A. M., and Littauer, U. Z., 1977, *J. Biol. Chem.* **252**:1819–1825.
143. Suzuki, Y., 1969, *Bull. Jpn. Neurochem. Soc.* **8**:41–50.
144. Niedergang, C., Okazaki, H., Ittel, M. E., Munoz, D., Petek, F., and Mandel, P., 1974, *Biochim. Biophys. Acta* **358**:91–94.
145. Ittel, M. E., Niedergang, C., Munoz, D., Petek, F., Okazaki, H., and Mandel, P., 1975, *J. Neurochem.* **25**:171–176.
146. Ittel, M. E., and Mandel, P., 1977, *J. Neurochem.* **28**:1355–1358.
147. Ittel, M. E., and Mandel, P., 1979, *J. Neurochem.* **33**:521–525.
148. Sarkander, H.-I., Fleischer-Lambropoulos, H., and Brade, W. P., 1975, *FEBS Lett.* **52**:40–43.
149. Sarkander, H.-I., and Knoll-Köhler, E., 1978, *FEBS Lett.* **85**:301–304.
150. Fleischer-Lambropoulos, H., Sarkander, H.-I., and Brade, W. P., 1974, *FEBS Lett.* **45**:329–332.,
151. Kanungo, M. S., and Thakur, M. K., 1979, *Biochem. Biophys. Res. Commun.* **87**:266–271.
152. Supakar, P. C., and Kanungo, M. S., 1981, *Biochem. Biophys. Res. Commun.* **100**:73–78.
153. Oh'Hara, I., and Yanagihara, T., 1977, *J. Neurochem.* **29**:1065–1073.
154. Yanagihara, T., 1980, *J. Neurochem.* **35**:1209–1215.
155. Lee, N. M., and Loh, H. H., 1977, *J. Neurochem.* **29**:547–550.
156. Hollenbeck, R. A., Chuang, D. M., and Costa, E., 1979, *Brain Res.* **171**:481–487.
157. Durban, E., Lee, H. W., Kim, S. and Paik, W. K., 1978, *Methods Cell Biol.* **19**:59–68.
158. Nochumson, S., Kim, S., and Paik, W. K., 1978, *Methods Cell Biol.* **19**:69–78.
159. Duerre, J. A., and Quick, D. P., 1979, *Transmethylation* (E. Usdin, R. T. Borchart, and C. R. Creveling, eds.), Elsevier/North-Holland, New York, pp. 583–592.
160. Nishizuka, Y., Ueda, K., Nakazawa, K., and Hayaishi, O., 1967, *J. Biol. Chem.* **242**:3164–3171.
161. Bilen, J., Ittel, M. E., Niedergang, C., Okazaki, H., and Mandel, P., 1981, *Neurochem. Res.* **6**:1253–1263.
162. Hayaishi, O., and Ueda, K., 1977, *Annu. Rev. Biochem.* **46**:95–116.
163. Mukherjee, S. K., and Narayanaswami, A., 1972, *J. Neurochem.* **19**:631–640.
164. Kato, T., and Lowry, O. H., 1973, *J. Biol. Chem.* **248**:2044–2048.
165. Thompson, R. J., 1977, *J. Neurochem.* **29**:383–385.
166. Thompson, R. J., 1977, *J. Neurochem.* **29**:387–391.
167. Thrall, C., Webster, R. A., and Spelsberg, T. C., 1978, *The Cell Nucleus* (H. Busch, ed.), Academic Press, New York, pp. 461–529.
168. McEwen, B. S., 1981, *Basic Neurochemistry*, 3rd ed. (G. J. Siegel, R. W. Albers, B. W. Agranoff, and R. Katzman, eds.), Little, Brown, Boston, pp. 775–799.
169. Pfaff, D. W., and Keiner, M., 1973, *J. Comp. Neurol.* **151**:121–158.
170. McEwen, B. S., Zigmond, R. E., and Gerlach, J. L., 1972, *Structure and Function of Nervous Tissue*, Volume 5 (G. H. Bourne, ed.), Academic Press, New York, pp. 205–291.
171. McEwen, B. S., 1978, *Hormone Receptors, Steroid Hormones*, Volume 1 (B. O'Malley and L. Birnbaumer, eds.), Academic Press, New York, pp. 353–400.
172. MacLusky, N. J., Chaptal, C., and McEwen, B. S., 1979, *Brain Res.* **178**:143–160.
173. Thrower, S., Neethling, C., White, J. O., and Lim, L., 1981, *Biochem. J.* **194**:667–671.
174. Barley, J., Ginsburg, M., Greenstein, B. D., MacLusky, N. J., and Thomas, P. J., 1975, *Brain Res.* **100**:383–393.
175. Lieberburg, I., MacLusky, N. J., and McEwen, B. S., 1977, *Endocrinology* **100**:598–607.
176. McEwen, B. S., Gerlach, J. L., and Micco, D. J., 1975, *The Hippocampus*, Volume 1 (R. L. Isaacson and K. H. Pribram, eds.), Plenum Press, New York, pp. 285–322.

177. Kato, J., and Onouchi, T., 1977, *Endocrinology* **101**:920–928.
178. MacLusky, N. J., and McEwen, B. S., 1978, *Nature* **274**:276–278.
179. Ermisch, A., and Rühle, H. J., 1978, *Brain Res.* **147**:154–158.
180. Salaman, D. F., and Birkett, S., 1974, *Nature* **247**:109–112.
181. Whalen, R. E., and Olsen, K. L., 1978, *Brain Res.* **152**:121–131.
182. Dozin-van Roye, B., and De Nayer, P., 1979, *Brain Res.* **177**:551–554.
183. Schwartz, H. L., and Oppenheimer, J. H., 1978, *Endocrinology* **103**:943–948.
184. Yanker, B. A., and Shooter, E. M., 1979, *Proc. Natl. Acad. Sci. U.S.A.* **76**:1269–1273.
185. Donato, R., and Michetti, F., 1981, *J. Neurochem.* **36**:1698–1705.
186. Ganbatz, J. W., 1981, *FEBS Lett.* **127**:179–182.
187. Brown, I. R., 1982, *Molecular Approaches to Neurobiology*, Academic Press, New York.
188. Schmitt, F. O., Bird, S. J., and Bloom, F. E., 1982, *Molecular Genetic Neuroscience*, Raven Press, New York.

9

Membrane Markers of the Nervous System

Elisabeth Bock

1. INTRODUCTION

In recent years a series of neural cell membrane markers has been demonstrated. The study of membrane markers has been undertaken in order to define surface components that are specific for a particular cell type, for example, neurons, astroglia, oligodendroglia, and Schwann cells. Cell–cell interactions in the nervous system are assumed to have developed to a high degree of precision because of the many cellular connections diverging and converging from one cell to another. The molecular mechanisms underlying these phenomena remain largely obscure, but it may be assumed that specific cell surface molecules play an important role. The study of neural cell membrane markers is therefore expected to improve our understanding of cellular communications.

The majority of procedures available for characterization of a given protein require the protein to be present in pure form and in suitable amounts. Such conditions are difficult to fulfill for many proteins of the nervous system. However, immunochemical procedures are not limited by these requirements and have proven valuable tools in the analysis of neural markers. By immunization with purified antigens or mixtures of more or less characterized antigens, a series of markers of the nervous system have been demonstrated by means of polyclonal antibodies.[1,2] Not only can neurons, astroglia, and oligodendroglia be distinguished by the different antigens, but these markers can also be used to trace the development of individual cell types.

Important tools in these studies have been electroimmunoprecipitation in gel, immunocytochemical techniques, and cytotoxicity tests. Immunochemical recognition of an antigen combined with characterization by physicochemical techniques allows the antigen to be characterized in molecular terms. However, progress introducing antibodies against any but major classes of neural cell types has, until recently, been hindered by the heterogeneity of the immune response generated against the available antigens. The introduction of mono-

Elisabeth Bock • The Protein Laboratory, Sigurdsgade 34, Denmark.

clonal antibody technology was therefore a major advance and has already led to the recognition of many new cell-type-specific antigens. By this technology it is possible to obtain specific antibodies using preparations containing many antigens. Immunoglobulin-producing cells are adapted to continuous culture by fusion to a neoplastic derivative of the immune system. The fused hybrid cells, termed hybridomas, combine continuous viability in cell culture with the ability to produce antibody molecules with specificity for only one epitope. Immunochemical assays for characterization and quantification of nervous-system-specific antigens have recently been reviewed.[3]

In this chapter a series of membrane markers of the nervous system is described with special emphasis on the immunochemical approach employed for the demonstration.

2. NEURONAL SURFACE MARKERS

2.1. Neuronal Plasma Membrane Markers

2.1.1. D2-Cell Adhesion Molecule

The D2 protein was originally described as a nervous-system-specific membrane protein enriched in fractions of synaptosomal plasma membranes.[4] An immunochemical relationship between D2 and the chick neuronal cell adhesion molecule (see below) has been demonstrated, and it has been shown that antibodies against D2 can inhibit fasciculation of neurites from cultured rat sympathetic ganglia.[5] Fractionation procedures for rat brain D2 and human brain D2 have recently been published,[6–8] and specific antisera have been produced in rabbits. Purified D2 from adult rat brain or human brain was found to consist of two polypeptides with apparent mol. wt. 150,000 and 125,000, respectively. Purified D2 from human fetal brain was revealed in sodium dodecyl sulfate–polyacrylamide gel electrophoresis (SDS–PAGE) as one blurred band in the mol. wt. range 150,000–190,000. D2 binds to several lectins.[7] Lectin-positive and lectin-negative forms of D2 are present in varying proportions during development, indicating a heterogeneity of the carbohydrate moiety of D2 at any stage of development. Although an amphiphilic nature of D2 has been demonstrated,[9,10] a soluble form of D2 has been found in cerebrospinal fluid,[11] in amniotic fluid,[12] and in serum.[7] It therefore seems as if the epitopes of D2 exist on both a membrane-bound amphiphilic molecule and on a hydrophilic molecule. The relationship between the two forms is unclear.

Quantification of D2 has been performed by means of immunoelectrophoretic techniques.[11,13] Recently, an enzyme-linked immunosorbent assay has been established.[14] Antigens immunochemically identical or partially identical to the D2 glycoprotein have been demonstrated in mammalian, avian, amphibian, and fish brain.[15] D2 has been demonstrated in all investigated areas of the central nervous system.[13] Recent studies have shown that D2, in contrast to what was previously assumed, is not restricted to the nervous system. It has also been demonstrated in other tissues (heart, skeletal muscle, kidney, ventricle, and liver), although in much lower amounts.[7] The concentration of D2 is higher at early stages of development than in adult tissues. Thus, the level of D2 expressed relative to total protein content decreases with age.[16,17] In mouse, the level of D2 at postnatal day 40 is approximately 50% of that on day 12. The electrophoretic migration of D2 has been observed to decrease during development,[17,18] and it has been suggested that the decrease is probably caused by removal of sialic acids during development.[18]

D2 was introduced as a neuronal membrane marker. The immunochemical relationship to the chick neuronal cell adhesion molecule indicates that D2 may be involved in neuronal adhesion phenomena. However, the recent demonstration of D2 in other organs than the nervous system indicates that D2 may be a general cell adhesion molecule or a member of a family of cell adhesion molecules present on many organs at various times of development.

2.1.2. Neural Cell Adhesion Molecule (N-CAM)

Studies of the *in vitro* aggregation of neural cells from chick embryos have led to the characterization of a molecule referred to as a neural cell adhesion molecule (N-CAM).[19–20] This protein has been shown *in vitro* to be involved in cell–cell binding,[21,22] neurite fasciculation,[23] histogenesis of the neural retina,[24] and nerve–muscle interaction.[25] The original procedure for isolation of N-CAM[19,20] began with culture supernatants from retinal tissue and yielded only microgram quantities of the purified molecule. By means of monoclonal antibodies, it has now been possible to purify the molecule in milligram quantities from chick brain,[21] mouse brain,[26] and human brain.[27] In embryonic brain, N-CAM exists as a large cell-surface sialoglycoprotein. After SDS–PAGE, it appears as an indistinct, broad band corresponding to material with apparent mol. wt. ranging between approximately 140,000 and 250,000.[21] The embryonic molecule has a higher content of sialic acid (up to 30% by weight) than the adult form, apparently in part as a polysialic acid and coupled to the polypeptide in an unusual but as yet undetermined linkage.[21,28] Removal of sialic acid converts the broad bands seen in gel electrophoresis into several more discrete bands around mol. wt. 140,000. The N-CAM occurs naturally in partially desialylated form in adult brain.[28] Conversion to this state is seen in late fetal and neonatal development. After incubation at 37° in 10 mM ammonium bicarbonate, the molecule partially degrades to a polypeptide of mol. wt. 65,000.[21] Amino acid composition of N-CAMs from mouse, rat, chicken, and human brain is similar.[26,27]

An immunochemical relationship between the chick N-CAM and the rat D2 glycoprotein has been demonstrated.[5] Furthermore, it has been shown that antibodies against the D2 glycoprotein are able to interfere with neurite fasciculation.[5] Thus, it is reasonable to assume that the D2 glycoprotein is the mammalian correlate of the chick N-CAM.

2.1.3. BSP-2

A monoclonal antibody, anti-BSP-2, has been produced against glycoproteins extracted from neonatal mouse brain.[29] Immunoprecipitates prepared with the BSP-2 antibody contained a triplet of high-molecular-weight glycoproteins with apparent mol. wts. of 180,000, 140,000, and 120,000. In primary cultures of dissociated cerebellar cells, the antibody bound to neuronal cell types but not to astrocytes or to fibroblasts.[29,30] In tissue sections at either the light or the electron microscope levels, immunocytochemical staining indicated the presence of BSP-2 on the surface of Purkinje cell somas and their dendritic tree. In the electron microscope, this surface labeling was seen to extend beneath synaptic boutons on their perykarya. The cytoplasm and nuclei of these cells were free of label.

On macroneurons in the deep cerebellar nuclei, staining was also limited to the plasma membrane. In contrast, the cytoplasm of granule cells was heavily immunolabeled, as were their plasma membranes. Astrocytes situated in the granular layer and their thin processes covering the Purkinje cell soma and its dendritic tree in the molecular layer were found to be labeled, predominantly over their plasma membranes. Oligodendrocytes were never found to be labeled. Endothelial cells were likewise free of staining. The absence of demonstrable BSP-2 in astrocytes in culture although the antigen is clearly demonstrable in the astrocytes of adult mouse brain cerebellum is puzzling: either the antigen is not yet expressed on the surface of the cells in culture or it is not accessible on the external surface of the living cells.[30] Attention has been drawn to common properties of D2 glycoprotein and BSP-2.[30] Further work will be necessary before a definite relationship between the D2 and BSP-2 antigen may be established.

2.1.4. NS-4.

The NS-4 (nervous system antigen 4) antigen was originally detected by an antiserum raised against particulate fractions of cerebellum from 4-day-old mice.[31,32] NS-4 is present in neuroectodermally derived tissues at all ages tested, 12-day-old embryos and 12-month-old adults being the earliest and latest stages. The antigen is expressed not only on cerebellum but also on all parts of the normal central nervous system including retina. NS-4 is not restricted to neuroectodermal tissues but is also present on sperm from vas deferens and epididymis and on preimplantation mouse embryos when tested by the complement-mediated cytotoxicity test.[33] NS-4 was found on all cells of cleavage-stage embryos and on cells of the trophoblast and inner cell mass of the mouse blastocyst.[34] In a study on cerebellar cultures, two major surface components of mol. wt. 200,000 and 145,000 were identified as equivalent to the antigen NS-4.[35] A relationship among NS-4, BSP-2, and D2 has been discussed[30] but not established.

2.1.5. D1

The D1 protein was defined by means of a polyspecific polyclonal rabbit antiserum raised against fractions of rat brain synaptosomal plasma membranes.[4] By immunoprecipitation, D1 was shown to be composed of two polypeptide bands of apparent mol. wts. 116,000 and 50,300.[36] D1 has been found in all investigated areas of the central nervous system.[13] The membrane topography of D1 has been determined, and the results indicate that the protein probably spans the plasma membrane.[37] In a developmental study on mice, D1 was expressed relative to the total protein and was shown to rise from day 8 to day 40 postnatally. At day 40, a steady adult plateau was reached.[16] D1 has been determined in the cerebellar mutants reeler, staggerer, and weaver.[38] The specific concentration was unchanged, but the amount of D1 was decreased.

2.1.6. D3

D3 was demonstrated by means of polyclonal polyspecific rabbit antibodies raised against rat synaptosomal plasma membranes.[4] By immunoprecipitation, D3 has shown to be composed of three polypeptide bands of apparent mol. wts. 34,400, 23,500, and 14,100.[36] Antigens immunochemically identical and partially identical to D3 have been demonstrated in mammalian, avian, amphibian, and fish brain.[15] D3 has been demonstrated in all investigated areas of the central nervous system.[13] Moderate enrichments of D3 have been demonstrated in fractions of synaptic plasma membranes.[4,37] However, D3 has also been demonstrated in the glioma C6 concentrations close to the amount found in adult brain.[37] In the cerebellar mouse mutants weaver and reeler, increased specific concentrations of D3 have been demonstrated.[38] In a developmental study on mice, the level of D3 expressed relative to the total protein content was shown to rise from day 8 to day 40 postnatally. On day 40, a steady adult plateau was reached.[16]

2.1.7. P-400

P-400 is a protein with an apparent mol. wt. of 400,000.[39,40] It is membrane bound, present in normal cerebellum, absent in cerebral cortex, and lost as a consequence of both the nervous and staggerer mutations. Purified Purkinje cell somas from rat contained significant amounts of the protein. P-400 has been studied in homogenates of molecular layer, granular layer, and white matter from cerebellum and was found exclusively in the molecular layer, even if the molecular layer was freed from Purkinje cell somas, indicating that the protein is present in the dendritic arborization of the Purkinje cells.[41]

2.1.8. Thy-1

Thy-1 is one of a number of cell surface components referred to as differentiation antigens. Thy-1 antigen has been purified from rat and mouse lymphoid cells[42,43] and brain.[44–46] In all cases, the molecule is a glycoprotein with an apparent mol. wt. of 25,000 determined by SDS–PAGE. By sedimentation equilibrium, the mol. wt. was determined as 17,500 for rat brain Thy-1 and 18,700 for thymocyte Thy-1.[47] In both cases, the polypeptide has a mol. wt. of 12,500, the rest of the molecule being carbohydrate. In the mouse, two genetic alleles designated Thy-1^a and Thy-1^b code for the specificities Thy-1.1

and Thy-1.2, respectively.[48,49] Only the Thy-1.1 specificity has been demonstrated on rat cells.

Thy-1 is probably the most abundant cell surface glycoprotein of rodent brain and thymus, yet its function is unknown. By means of immunocytochemistry on short-term cultures, Thy-1 was demonstrated to be present on fibroblasts and some neurons but not on the majority of leptomeningeal cells or on oligodendrocytes or astrocytes. However, it was expressed on some astrocytes in longer-term cultures.[50] Localization studies in rat brain and human brain have yielded similar results.[51,52] By use of a monoclonal antibody to human Thy-1, the antigen was demonstrated by immunocytochemical procedures on the membranes of some neuronal cell bodies and their processes, particularly the Purkinje cell of the cerebellar cortex. Staining was also observed on cell bodies of satellite cells in areas of gray matter and on what appeared to be fiber tracts in the basal ganglia and thalamus, the tracts appearing duller than the surrounding gray matter of the nuclei. Finally, staining was observed of only some fibers in the sciatic nerve.[52] No staining of the human adrenal gland was obtained.

Structural similarities between Thy-1 from rat brain and immunoglobulin have been demonstrated.[53] This result suggests that Thy-1 has a structure resembling an Ig domain, which would not be consistent with an enzymatic function. Thy-1 has never been shown to associate with another polypeptide but appears to bind directly to the membrane. The function of Thy-1 is unknown, but it could exist to mediate cell–cell interactions, and a structure like the Ig domain is particularly suited for this role. The versatility of the domain structure has already been proven by the large number of effector functions mediated by Ig constant-region domains.[53]

2.1.9. Invertebrate Neuronal Membrane Antigens

A large number of precise connections have to be established between neurons. Very little is known about the molecular mechanisms that generate these connections. It has been postulated that there must exist molecules, differing in kind or quantity from cell to cell, that mediate the necessary recognition. Because the markers of interest differ in each neuron with different connections, these molecules must somehow be identified in structured material, where the same neuron can be easily identified in repeated experiments.

The leech nervous system was chosen by Zipser and McKay[54] not only for the identification of specific neuronal markers but also for analyzing the rules that establish neuronal connections. As the putative antigens could not in principle be purified, the hybridoma technique was chosen. Of 475 hybridomas tested, approximately 300 bound to the leach nervous system. From these, 41 different antibodies were obtained, which bound to specific subsets of neurons. Twenty of these antibodies were analyzed in detail, all giving different staining patterns when immunocytochemical procedures were employed. Some antibodies (Lan 3-1, Lan 3-3, Lan 3-4) recognized pairs of neurons in the midbody ganglion. In addition, Lan 3-1 bound to two extra pairs of large cells in the fifth and sixth midbody ganglia. These cells are known to be involved in the control of sexual behavior. Several antibodies were isolated that labeled the mechanosensory cells mediating the motor responses to the environmental stimuli of touch, pressure, and noxious mechanical stimulation. The immunochemical identification of leech neurons is a convenient tool to help map variations in specific cell body distributions along the entire leech nerve cord, a task that in the past was done solely with electrophysiological techniques. The findings showed that groups of electrophysiologically distinct neurons were biochemically heterogeneous.

2.1.10. Other Neuronal Membrane Markers Defined by Monoclonal Antibodies

The discriminating power of monoclonal antibodies makes the hybridoma technique the method of choice for defining neuronal cell surface antigens specific for neuronal subpopulation. In recent years, many cell surface antigens defined by monoclonal antibodies have been described. A neuronal surface antigen present on neurons from the CNS but not from the PNS has thus been demonstrated.[55]

A monoclonal antibody (M2)[56] has been produced that, on cerebellar sections, stains the surface of the cell bodies of granule and Purkinje cells by immunocytochemistry. In monolayer cultures, the antigen is detected on the cell surface of all astrocytes and on immature oligodendrocytes. After 3 days in culture, neurons also begin to express this antigen. This antibody was obtained by immunization with particulate fractions from early postnatal mouse cerebellum. Others[57] immunized with hypothalamus and thereby generated more than 100 antibody-producing hybridomas, 60 of which were brain specific. Among these, 54% reacted with glial elements, pituitary cells, or basal lamina of intracerebral capillaries, with little variation among individual hybridomas in each of these groups. Forty-six percent of brain specific antibodies reacted with neuronal structures.

Neuron-specific hybridomas could be classified into groups that localized antigens in anatomically definable overall patterns. Within these patterns, individual hybridomas exhibited extensive quantitative localization diversity. The authors suggested that the genetic message from a common "proantigen" within an overall pattern may be slightly modified during differentiation of a neuron, thus leading to minor variability in antigenic expression. During antibody formation, similar minor changes occur in the differentiation of the genetic message for the antibody-variable region.

Several other authors have reported the production of monoclonal antibodies against neuronal surface antigens. These antigens are presently being characterized (for further information, see ref. 58).

2.2. Retinal Cell Membrane Markers

The retina is a convenient model system in which to study the differences in cell surface molecules of different subclasses of neural cells, and this system has the advantage that many of the cells involved in the transmission of visual information from the photoreceptors to the primary visual cortex have been identified and reasonably well characterized with respect to their physiological responses and synaptic interactions. The hybridoma technology has also been applied successfully here.[59,60] By use of a membrane preparation of adult rat retina, seven monoclonal antibodies have been produced.[59] Three antibodies reacted with particular regions of rat photoreceptor cell surfaces: RET-P1 labeled the cell bodies and outer and inner segments (rods but not cones); RET-P2 labeled only outer segments; and RET-P3 labeled only the cell bodies. Three antibodies reacted with glial cells: RET-G2 and RET-G3 were specific for Müller cells; RET-G1 also labeled glia elsewhere in brain. The seventh antibody, RET-N1, reacted with many types of neuronal cells.

A monoclonal antibody has been obtained that binds to cell membrane molecules distributed in a topographic gradient in avian retina.[60] Thirty-five times more antigen was detected in dorsoposterior retina than in ventroanterior retina. Most of the antigen was associated with the synaptic layers of the retina. Less antigen was detected in cerebrum, thalamus, cerebellum, and optic tectum. Little or none was found in nonneuronal tissues tested. The antigen was found on most or all cell types in retina, and the concentration of antigen found is a function of the square of the circumferential distance from the ventroanterior pole of the gradient towards the dorsoposterior pole. Thus, the antigen can be used as a marker of cell position along the ventroanterior–dorsoposterior axis of the retina.

2.3. Markers of the Synapse

2.3.1. Synaptin

Synaptin was originally demonstrated by means of polyspecific polyclonal rabbit antisera raised against fractions of rat brain synaptic vesicles.[4,15] Synaptin is composed of one polypeptide of apparent mol. wt. 44,000. The isoelectric pH has been determined to be 4.1. The protein has been found in brains from several mammalian species, including man, dog, swine, ox, guinea pig, and mouse.[15] Synaptin has been found in all investigated areas of the central nervous system.[13] It has also been found in subcellular fractions from peripheral nerves, adrenal medulla, and the neurohypophysis. In rat brain synaptic vesicle fractions, enrichments of synaptin of 10–15 times have been found when compared to the starting homogenate. Synaptin has also been demonstrated in fractions of synaptosomal plasma membranes. Determinations of membrane topography have indicated that the protein is localized on the inside of the synaptic plasma membrane and on the cytoplasmic surface of brain synaptic vesicles and chromaffin granules of the adrenal medulla.[37,61]

In a developmental study on mice, the level of synaptin expressed relative to the total protein content was shown to rise from day 8 to day 40 postnatally. At day 40, a steady adult level was reached.[16] Studies on axonal transport of synaptin indicate that the protein is subject to rapid axonal transport.[15] The function of synaptin is unknown, but the membrane topography has led to the suggestion that this protein might be involved in the exoendocytosis of synaptic vesicles during neurotransmission.[61]

2.3.2. *Markers of Electric Organ Membranes and Synaptic Vesicles*

The electric organ from electric fish is used for studying cholinergic neurotransmission, since the organ is derived embryologically from muscle, having lost the ability to contract but not the ability to produce an action potential when stimulated by its purely cholinergic nerve supply. It thus presents a system exclusively cholinergic, extremely rich in both pre- and postsynaptic structures, and it is available in large quantities. By means of polyclonal polyspecific rabbit antibodies raised against *Torpedo* electric organ membranes, antibodies were obtained against three membrane proteins: T1, T2, and T3.[62] By charge-shift immunoelectrophoresis, T1 and T2 were found to be amphiphilic, whereas T3 was a borderline case. The three proteins bound to hydrophobic ligands, indicating that they possessed apolar regions. They also bound to concanavalin A, indicating that they all had a carbohydrate moiety.

T1 was composed of several polypeptides in the mol. wt. range 37,000–98,000. T2 and T3 were each found to be composed of one polypeptide band of apparent mol. wt. 110,000 and 99,000 respectively. T1, T2, and T3 are probably plasma membrane proteins as they copurified with Na^+, K^+-ATPase. T2 was specific for the electric organ, whereas T1 was found in small amounts in all studied tissues. Trace amounts of T3 were present in heart and muscle as well as the electric organ. With respect to the *Torpedo* nervous system, T2 and T3 could only be demonstrated in the electric organ, whereas T1 was also found in the central nervous system.

An isolation procedure for T3 has been developed.[63] Monoclonal antibodies against all three antigens have been produced, and, by immunocytochemistry on frozen sections of rat cerebellum, all antibodies were shown to react with Purkinje cells and to a varying extent with granule cells. No glial staining was obtained.[64]

Polyclonal rabbit antibodies have been generated against purified synaptic vesicles from the marine ray *Narcine brasiliensis*, and a radioimmunoassay was set up to assay quantitatively synaptic vesicle antigens.[65] The antibody cross reacted with the mammalian central and peripheral nervous systems. It bound selectively to nerve terminals and not to axons or cell bodies.[66,67] The antigens reacting with the antibody seemed to be localized on the inside of the synaptic vesicles and were only exposed to the antibodies when synaptic vesicles were inserted into the plasma membrane of the presynaptic terminal during exocytosis.[68]

2.3.3. *Synaptic Membrane Markers Defined by Monoclonal Antibodies*

By immunization with synaptic membranes and junctions, an antibody has been obtained, antibody 30, which identifies a vesicle-specific protein that is probably on every neurosecretory vesicle.[69,70] In sections of retina, antibody 30 stains the two synaptic regions, the inner and outer plexiform layers. In cerebellum, antibody 30 stains the large glomerular terminals in the granule cell layer and large amounts of small, punctate areas in the molecular layer. With the possible exception of the neuromuscular junction, quantitative radioimmunoassays have detected the antigen in every neuronal or neurosecretory tissue tested, including posterior pituitary and adrenal medulla. The antigen cannot be detected in nonsecretory tissues such as extrajunctional diaphragm, muscle, and adrenal cortex, but it is present at lower levels in at least one secretory gland, the anterior pituitary, that is not derived in ontogeny from the neural crest and the neural tube. Quantitative assays also demonstrated the presence of comparable levels of the antigen in every vertebrate species tested.

Electron microscope examination showed that antibody binding is primarily restricted to synaptic vesicles. The molecular weight of the antigen has been determined to be 65,000. Trypsin destroys the antigen. The antigen is released by detergent but not by high or low salt. Therefore, the antigen appears to be an integral membrane protein. Since the antigen is exposed on the cytoplasmic surface on the vesicles, the antibody has been used for purification of secretory vesicles and their internal components from nerve terminals and cell lines.

In an attempt to obtain monoclonal antibodies to molecules in synaptic membranes that affect the formation and/or function of synapses, 50 hybridoma cell lines synthesizing antibodies to molecules in rat cerebral cortex synaptic membrane preparations have been obtained.[71] Two antibodies were found with regional specificity in the nervous system. These antibodies bound to molecules in the cerebral cortex and cerebellum membrane fractions, but few or no antigenic sites were detected in retinal fractions. Antigenic differences between embryonic and adult cerebral cortex membranes also were detected with certain antibodies. Antibodies were found that affected synaptic transmission presynaptically by increasing $^{45}Ca^{2+}$ uptake and acetylcholine secretion at synapses, monitored indirectly by determining MEPP responses of myotubes. Postsynaptic effects of antibodies were also found that resulted in myotube hyperpolarization or depolarization.

3. GLIAL SURFACE MARKERS

3.1. NS-1: An Oligodendroglial Marker

Nervous system antigen 1 (NS-1) was defined by antiserum raised against a methylcholanthrene-induced mouse tumor, the glioma G26.[31,72] The antigen is brain specific and found only in brain and neural tumors. NS-1 is present in several mammalian species. Barely detectable at early postnatal age, the antigen reaches adult levels 3 weeks after birth. NS-1 is detectable by immunocytochemistry in white-matter tracts of adult cerebellum. It is not detectable at birth but can be detected in presumptive white matter tracts by postnatal day 8, before the onset of myelin formation.

NS-1 antiserum can also be used to label specifically one particular cell type in monolayer cultures of early postnatal cerebellum. This particular cell type occurs with a frequency of 3–5% of the total cells and has a distinctive morphology characterized by an extensive network of fine cellular processes that radiate concentrically from the cell soma. Both the cell body and processes are antigen positive. Astroglia are completely NS-1 antigen negative.

3.2. Ran-1: A Schwann Cell Marker

Rat neural antigen 1 (Ran-1)[74] is a cell surface antigen defined by a mouse antiserum made against a rat neural cell line.[75] The antigen is found on a large number of different neural tumors of the rat and normal rat brain nerve but not in significant amounts on nonneural tumors or tissues.[75] It is expressed only on Schwann cells in primary cultures of newborn rat nerve[74] and dorsal root ganglia.[76]

By immunocytochemical staining of cells in dissociated cell cultures (neonatal rat sciatic nerve, dorsal root ganglia, optic nerve, cerebellum, corpus callosum, cerebellar cortex, and leptomeninges), only Schwann cells were labeled with antibodies against Ran-1. Although Ran-1 is expressed on glial and neuronal tumors, it was not found on any normal astrocytes, oligodendrocytes, or neurons.[50]

3.3. Ran-2

By immunization with cultured rat astrocytes, a monoclonal antibody has been obtained that reacts with an antigen called the rat neural antigen 2 (Ran-2).[77] The antibody binds to the surface of rat ependymal cells, retinal Müller cells, and leptomeningeal cells as well as to astrocytes but not to cultured neurons, oligodendrocytes, Schwann cells, microglia, or various nonneural cells. The antigen is protease sensitive and rat specific.

3.4. Other Glial Membrane Markers Defined by Monoclonal Antibodies

Four antibodies have been raised against antigens, designated 01, 02, 03, and 04, that are expressed on the surfaces of oligodendrocytes.[78,79] The antibodies were obtained by immunization of mice with white matter from bovine corpus callosum. In primary cultures of neonatal mouse nervous system (cerebellum, cerebrum, spinal cord, and optic nerve), all four antibodies stained surfaces of cells that were positive for galactocerebroside and were negative for tetanus toxin, fibronectin, and glial fibrillary acidic protein. Two of the antibodies, 03 and 04, reacted with the surface of cells that were negative for galactocerebroside as well as for tetanus toxin, fibronectin, and glial fibrillary acidic protein. These cells had a very similar morphology to oligodendrocytes in electron micrographs of immunoperoxidase-stained cultured cells.[80]

By immunization of mice with rat hypothalami, 135 antibody-producing hybridomas were obtained, of which 54% reacted with glial elements, although only six of the antibodies were specific for glial cells. They stained oligodendrocytes. Each of these hybridomas reacted identically, without variation in regional distribution of immunocytochemical staining.[57]

REFERENCES

1. Bock, E., 1978, *J. Neurochem.* **30**:7–14.
2. Bock, E. (ed.), 1982, *Nervous-System-Specific Proteins*, Blackwell, Oxford.
3. Bock, E., 1984, *Immunology of Brain* (M. Adinolfi and A. Bignami, eds.), Blackwell, Oxford (in press).
4. Bock, E., and Jørgensen, O. S., 1975, *FEBS Lett.* **52**:37–39.
5. Jørgensen, O. S., Delouvée, A., Thiery, J.-P., and Edelman, G. M., 1980, *FEBS Lett.* **111**:39–42.
6. Rasmussen, S., Ramlau, J., Axelsen, N. H., and Bock, E., 1980, *Scand. J. Immunol.* **15**:179–185.
7. Bock, E., Berezin, V., and Rasmussen, S., 1983, *Protides of the Biological Fluids*, Volume 30 (H. Peeters, ed.), Pergamon Press, Oxford pp. 75–78.
8. Rasmussen, S., Berezin, V., Nørgaard-Pedersen, B., and Bock, E., 1983, *Protides of the Biological Fluids*, Volume 30 (H. Peeters, ed.), Pergamon Press, Oxford pp. 83–86.
9. Albeck, M. J., and Bock, E., 1982, *Protides of the Biological Fluids*, Volume 29 (H. Peeters, ed.), Pergamon Press, Oxford, pp. 151–154.
10. Jørgensen, O. S., 1977, *FEBS Lett.* **79**:42–44.
11. Jørgensen, O. S., and Bock, E., 1975, *Scand. J. Immunol.* **4**(Suppl. 2):25–30.
12. Jørgensen, O. S., and Nørgaard-Pedersen, B., 1981, *Prenat. Diagn.* **1**:3–6.
13. Bock, E., and Bræstrup, C., 1978, *J. Neurochem.* **30**:1603–1606.
14. Ibsen, S., Berezin, V., Nørgaard-Pedersen, B., and Bock, E., 1983, *J. Neurochem.* pp. 356–366.
15. Bock, E., Divac, I., Norrild, B., Thorn, N. A., Torp-Pedersen, C., and Treiman, M., 1982, *Nervous-System-Specific Proteins* (E. Bock, ed.), Blackwell, Oxford, pp. 223–240.
16. Jacque, C. M., Jørgensen, O. S., Baumann, N. A., and Bock, E., 1976, *J. Neurochem.* **27**:905–909.
17. Bock, E., Rasmussen, S., Albeck, M., Sensenbrenner, M., Pettmann, B., and Louis, J. C., 1983, *Electroimmunochemical Analysis of Membrane Proteins* (O. J. Bjerrum, ed.), Elsevier, Amsterdam pp. 275–286.
18. Jørgensen, O. S., and Møller, M., 1980, *Brain Res.* **194**:419–429.
19. Brackenbury, R., Thiery, J.-P., Rutishauser, U., and Edelman, G. M., 1977, *J. Biol. Chem.* **252**:6835–6840.
20. Thiery, J.-P., Brackenbury, R., Rutishauser, U., and Edelman, G. M., 1977, *J. Biol. Chem.* **252**:6841–6845.
21. Hoffman, S., Sorkin, B. C., White, P. C., Brackenbury, R., Mailhammer, R., Rutishauser, U., Cunningham, B. A., and Edelman, G. M., 1982, *J. Biol. Chem.* **257**:7720–7729.
22. Rutishauser, U., Hoffman, S., and Edelman, G. M., 1982, *Proc. Natl. Acad. Sci. U.S.A.* **79**:685–689.
23. Rutishauser, U., Gall, W. E., and Edelman, G. M., 1978, *J. Cell Biol.* **79**:382–393.
24. Buskirk, D. R., Thiery, J.-P., Rutishauser, U., and Edelman, G. M., 1980, *Nature* **285**:488–489.
25. Grumet, M., Rutishauser, U., and Edelman, G. M., 1982, *Nature* **295**:693–695.
26. Chuong, C.-M., McClain, D. A., Streit, P., and Edelman, G. M., *Proc. Natl. Acad. Sci. U.S.A.* **79**:4234–4238.
27. McClain, D. A., and Edelman, G. M., 1982, *Proc. Natl. Acad. Sci. U.S.A.* **79**:6380–6384.
28. Rothbard, J. B., Brackenbury, R., Cunningham, B. A., and Edelman, G. M., 1982, *J. Biol. Chem.* **257**:11064–11069.
29. Hirn, M., Pierres, M., Deagostini-Bazin, H., Hirsch, M., and Goridis, C., 1981, *Brain Res.* **214**:433–439.
30. Langley, O. K., Ghandour, M. S., Gombos, G., Hirn, M., and Goridis, C., 1982, *Neurochem. Res.* **7**:349–362.

31. Schachner, M., 1982, *Nervous-System-Specific Proteins* (E. Bock, ed.), Blackwell, Oxford, pp. 201–222.
32. Goridis, C., Joher, J. A., Hirsch, M., and Schachner, M., 1978, *J. Neurochem.* **31**:531–539.
33. Schachner, M., Wortham, K. A., Carter, L. D., and Chaffee, L. K., 1975, *Dev. Biol.* **44**:313–325.
34. Solter, D., and Schachner, M., 1976, *Dev. Biol.* **52**:98–104.
35. Rohrer, H., and Schachner, M., 1980, *J. Neurochem.* **35**:792–803.
36. Jørgensen, O. S., 1979, *Biochim. Biophys. Acta* **581**:153–162.
37. Bock, E., Bjerrum, O. J., Gombos, G., Jørgensen, O. S., Reeber, A., Vincendon, G., Wechsler, W., Yavin, E., and Yavin, Z., 1980, *Synaptic Constituents in Health and Disease* (M. Brzin, D. Sket, and H. Bachelard, eds.), Mladinska Knjiga, Ljubljana, Pergamon Press, London, pp. 210–223.
38. Jørgensen, O. S., and Mikoshiba, K., 1978, *FEBS Lett.* **93**:185–188.
39. Mallet, J., Huchet, M., Pougeois, R., and Changeux, J. P., 1975, *FEBS Lett.* **52**:216–220.
40. Mallet, J., Huchet, M., Pougeois, R., and Changeux, J. P., 1976, *Brain Res.* **103**:291–312.
41. Mikoshiba, K., Mallet, J., and Changeux, J. P., 1977, *Proc. Int. Soc. Neurochem.* **6**:283.
42. Letarte-Muirhead, M., Barclay, A. N., and Williams, A. F., 1975, *Biochem. J.* **151**:685–697.
43. Zwerner, R. M., Barstad, P. A., and Acton, R. T., 1977, *J. Exp. Med.* **146**:986–1000.
44. Barclay, A. N., Letarte-Muirhead, M., and Williams, A. F., 1975, *Biochem. J.* **151**:699–706.
45. Letarte, M., and Meghji, G., 1978, *J. Immunol.* **121**:1718–1725.
46. McClain, L. D., Tomana, M., and Acton, R. T., 1978, *Brain Res.* **159**:161–171.
47. Kuchel, P. W., Campbell, D. G., Barclay, A. N., and Williams, A. F., 1978, *Biochem. J.* **169**:411–417.
48. Itakura, K., Hutton, J. J., Boyse, E. A., and Old, L. J., 1972, *Transplantation* **13**:239–243.
49. Blankenhorn, E. P., and Douglas, T. C., 1972, *J. Hered.* **63**:259–263.
50. Raff, M. C., Fields, K. L., Hakomori, S., Mirsky, R., Pruss, R., and Winter, J., 1979, *Brain Res.* **174**:283–308.
51. Barclay, A. N., and Hyden, H., 1978, *J. Neurochem.* **31**:1357–1391.
52. McKenzie, J. L., and Fabre, J. L., 1981, *Brain Res.* **230**:307–316.
53. Campbell, D. G., Williams, A. F., Bayley, P. M., and Reid, K., B. M., 1979, *Nature* **282**:341–342.
54. Zipser, B., and McKay, R., 1981, *Nature* **289**:549–554.
55. Cohen, J., and Selvendran, S. Y., 1981, *Nature* **291**:421–423.
56. Lagenaur, C., and Schachner, M., 1981, *J. Supramol. Struct.* **15**:335–346.
57. Sternberger, L. A., Harwell, L. W., and Sternberger, N. H., 1982, *Proc. Natl. Acad. Sci. U.S.A.* **79**:1326–1330.
58. Barnstable, C. J., 1980, *Nature* **286**:231–235.
59. Trisler, G. D., Schneider, M. D., and Nirenberg, M., 1981, *Proc. Natl. Acad. Sci. U.S.A.* **78**:2145–2149.
61. Bock, E., and Helle, K. M., 1977, *FEBS Lett.* **52**:175–178.
62. von Gerstenberg, A. C., Ramlau, J., Nyholm, L., Torp-Pedersen, C., and Bock, E., 1983, *Immunoelectrophoretic Analysis of Membrane Proteins* (O. J. Bjerrum, ed.), Elsevier, Amsterdam,pp. 287–299.
63. Bock, E., and von Gerstenberg, A. C., 1982, *Protides of the Biological Fluids* Volume 29, (H. Peeters, ed.), Pergamon Press, Oxford, pp. 155–158.
64. von Gerstenberg, A. C., 1983, *Protides of the Biological Fluids* (H. Peeters, ed.) Volume 30, Pergamon Press, Oxford pp. 91–94.
65. Carlson, S. S., and Kelly, R. B., 1980, *J. Cell Biol.* **87**:98–103.
66. Sanes, J. R., Carlson, S. S., von Wedel, R. J., and Kelly, R. B., 1979, *Nature* **280**:403–404.
67. Hooper, J. E., Carlson, S. S., and Kelly, R. B., 1980, *J. Cell Biol.* **87**:104–113.
68. Kelly, R. B., Carlson, S. S., von Wedel, R. J., Hooper, J. E., Miljanich, G. P., and Brasier, A. R., 1981, *Monoclonal Antibodies to Neural Antigens* (R. McKay, M. C. Raff, and L. F. Reichardt, eds.), Cold Spring Harbor Laboratory, New York, pp. 153–161.
69. Matthew, W. D., Tsavaler, L., and Reichard, L. F., 1981, *J. Cell Biol.* **91**:257–269.
70. Reichardt, L. F., and Matthew, W. D., 1982, *Trends Neurosci.* **5**:24–31.

71. De Blas, A. L., Busis, N. A., and Nirenberg, M., 1981, *Monoclonal Antibodies to Neural Antigens* (R. McKay, M. C. Raff, and L. F. Reichardt, eds.), Cold Spring Harbor Laboratory, New York, pp. 181–191.
72. SundarRaj, N., Schachner, M., and Pfeiffer, S. E., 1975, *Proc. Natl. Acad. Sci. U.S.A.* **72:**1927–1931.
73. Schachner, M., and Willinger, M., 1979, *The Menarini Symposia for Immunopathology* (P. Miescher, ed.), Karger, Basel, pp. 37–60.
74. Brockes, J. P., Fields, K. L., and Raff, M. C., 1977, *Nature* **266:**364–366.
75. Fields, K. L., Gosling, C., Megson, M., and Stern, P. L., 1975, *Proc. Natl. Acad. Sci. U.S.A.* **72:**1286–1300.
76. Fields, K. L., Brockes, J. P., Mirsky, R., and Wendon, L. M. B., 1978, *Cell* **14:**43–51.
77. Bartlett, P. F., Noble, M. D., Pruss, R. M., Raff, M. C., Rattray, S., and Williams, C. A., 1981, *Brain Res.* **204:**339–351.
78. Schachner, M., Kim, S. K., and Zenle, R., 1981, *Dev. Biol.* **83:**328–338.
79. Sommer, I., and Schachner, M., 1981, *Dev. Biol.* **83:**311–327.
80. Schachner, M., Sommer, I., Lagenaur, C., and Schnitzer, J., 1981, *Monoclonal Antibodies to Neural Antigens* (R. McKay, M. C. Raff, and L. F. Reichardt, eds.), Cold Spring Harbor Laboratory, New York, pp. 15–19.

10

Microtubules in the Nervous System

David Soifer and Kathyrn Mack

1. INTRODUCTION

Microtubules are constituents of all eucaryotic cells. As such, they are composed of proteins that are highly conserved from the point of view of evolution. The machinery of the mitotic apparatus is formed of microtubules. The movement of cilia, flagella, and sperm tails depends on axonemal arrays of microtubules. Microtubule proteins have been implicated in a variety of cellular functions such as intracellular translocation, lysosome function, cell motility, hormone release, and the lateral movement of proteins in plasma membranes. These have been reviewed in several monographs in recent years.[45,61,187,217]

Since the discovery that the main protein of microtubules is a major component of the central nervous system,[21] brain has been used as the source of material for most studies of the properties of microtubules and their proteins. Many of the methods developed for the purification of microtubules from brain are only of limited value when applied to other tissues. Consequently, more is known about the microtubules of the nervous system than about microtubules of other tissues.

Microtubules constitute one class of the fibrous structures that are perceived as forming the cytoskeleton. Within cells of the nervous system, cytoskeletal structures include intermediate filaments, microfilaments, and various cross-bridging microtrabecular[63] and filamentous or granular[203] structures. Some of the elements of the neuronal cytoskeleton are discussed in other chapters in this series. In Volume 5, Wallin and Deinum describe many of the properties of tubulin, the main protein of microtubules. In this chapter, we discuss the structure and distribution of microtubules in the nervous system, the assembly of microtubules from constituent proteins, and some aspects of the regulation of microtubule dynamics. This chapter is not meant to be an exhaustive review of these subjects but to serve as an introduction to micro-

David Soifer and Kathryn Mack • Laboratory of Cell Biology, Institute for Basic Research in Developmental Disabilities, Staten Island, New York 10314.

tubules of the nervous system and as a critical discussion of aspects of currently held concepts of microtubule biology.

2. MICROTUBULES IN NERVOUS TISSUE

Although numerous investigators have used the term "neurotubules" to describe microtubules from brain, a large percentage of the microtubules of the central nervous system are constituents of glia. As the biochemical heterogeneity of microtubules has become evident (see below), it has become important to realize that populations of microtubules or microtubule proteins from the brain, from any given region of the brain, and even from any given cell are not a homogeneous group of structures. Nevertheless, examination of the fine structure of nervous tissue reveals microtubules to be prominent structures of nearly all cells studied.[180]

2.1. Distribution in Neurons

2.1.1. Axons and Dendrites

2.1.1a General Features. Axially arrayed microtubules are common features of axons and dendrites. These microtubules may be nearly constant in number for a given population of axons, such as those of the olfactory nerve of the pike.[129] They may often be seen in discrete groups within an axon[100,109,216,222,223,269] (Fig. 1). Freeze–fracture studies using rapid freezing and limited etching confirm that most of the axoplasm is organized into axial domains characterized either by neurofilaments or by microtubules.[203] In dendrites and in a limited number of axons, such as those of olfactory nerves of fish and amphibians, microtubules are the main axial structures, neurofilaments being present only in small numbers. In most axons, neurofilaments appear to be more numerous than microtubules. There have been many reports of close associations between microtubules and vesicular components of the axon such as mitochondria[183] and synaptic vesicles.[216] As a result, in part, of their axial orientation and their apparent relationships to organelles undergoing fast transport, microtubules have been leading candidates for a major role in the transport of substances along axons and dendrites.

2.1.1b. Microtubule Density in Axons. The number of microtubules in a given axon remains approximately constant along its length, at least proximal to peripheral branching.[24,109,235] Although there has been a report that microtubule numbers are increased slightly in the region of the nodes of Ranvier,[14] such changes, if present, are small and not consistent with other reports.[24,109,235] The distribution of neurofilaments contrasts with that of microtubules: the number of neurofilaments per unit cross-sectional area remains constant through the node. Thus, as the diameter of the axon decreases, the number of neurofilaments decreases, whereas the number of microtubules, now packed closely together, does not change significantly.[235]

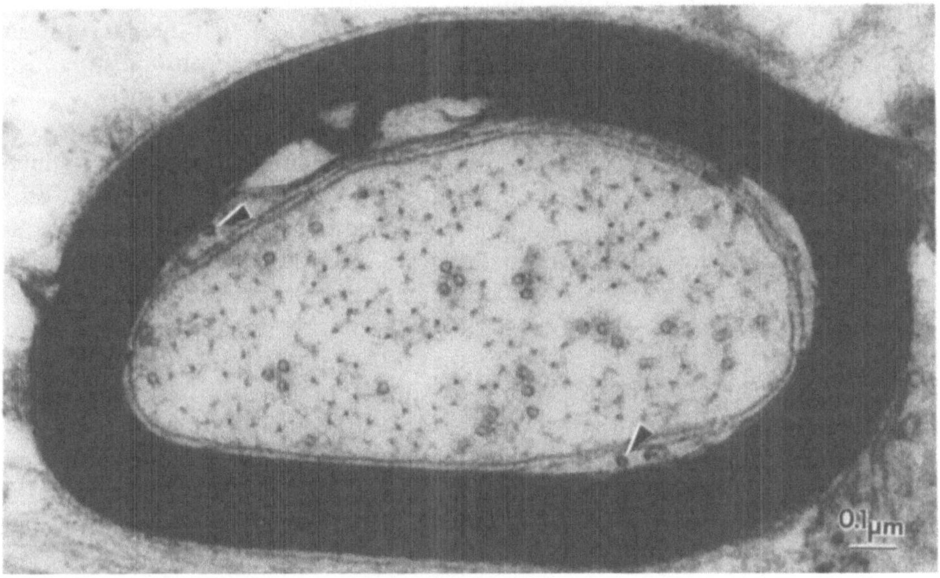

Fig. 1. Transverse section of a myelinated axon from rabbit optic nerve. Note that most microtubule profiles are grouped together, defining microtubule domains. When a single microtubule is present, it is clearly in a region from which neurofilaments are excluded. Many of the neurofilaments appear to be interconnected by cross bridges. These regions of axoplasm are neurofilament domains. Distances between neighboring microtubules in the microtubule domains are smaller than distances between neighboring neurofilaments or between microtubules and neurofilaments. Microtubules run parallel to the axon in the oligodendroglial adaxonal cytoplasm (arrows). Magnification ×65,000

2.1.1c. Microtubule-Associated Domains. The axoplasmic domains defined by the presence of microtubules are also the region where most membranous structures of the axoplasm are to be found.[109,203] Schnapp and Reese[203] suggested that "The microtubule-associated domains comprise the longitudinal channels through which the organelles are transported." They noted that organelles of pathologically swollen axons tend to accumulate in longitudinal islands, which are presumably the microtubule-associated domains. Recent studies on the mechanism of IDPN neuropathy support the notion that the microtubule domains are channels in which fast axoplasmic transport is carried out.[96,174,175]

Following treatment of rats with β,β-iminodipropionitrile (IDPN), many of their motor neuron axons become enlarged and filled with neurofilaments.[36–38] Immunocytochemical and electron microscopic studies demonstrate that following IDPN treatment the axoplasm becomes reorganized such that the several smaller microtubule-associated domains are now fused into a single axial domain surrounded by an annular, neurofilament-associated domain.[96,174,175,270] Virtually all membranous organelles are found in the microtubule domain. Fast axoplasmic transport is maintained in nerves from IDPN-treated animals; radiolabeled, fast-transported material passes through the central, microtubule-associated domains.[96,175] Taken together, the studies of axoplasmic transport using the IDPN model support the idea that the microtubule domains are the locus of fast axoplasmic transport. How microtubules function in this system is not known.

2.1.1d. Microtubule Length in Axons. How long are the axial microtubules of neurites? Reconstruction of serial sections through nematode nerve,[34] cultured rat dorsal root axons,[24] and mouse saphenous nerve[109,235] all demonstrate that at least some microtubules do not extend for the total length of axons. Calculations of average microtubule length based on the percent of microtubule ends detected in serial sections through 10-μm-long segments[235] and 12.8-μm-long segments[24] of axons and on the assumption that microtubules are all about equal in length gave values ranging between 100 and 700 μm for the average length of an axonal microtubule. An alternative that cannot be excluded is that there are more than one population of microtubules; some are very long, and others are short. Indeed, in the study of dorsal root ganglia, three microtubules were seen to both start and end within the 12.8-μm segments of axons that were examined.[24] No microtubules that short were seen in the segments of saphenous nerve.[235]

2.1.1e. Polarity of Axonal Microtubules. Studies of cat renal nerve,[102] frog olfactory nerve,[32] and chicken sciatic nerve[77] all indicate that the fast-assembling end of axonal microtubules is oriented away from the perikaryon (see discussions of microtubule polarity and assembly in Section 4.2). The similarity in orientation of microtubules both within a given axon and among all the axons in a nerve may reflect underlying features of the differentiation and maintenance of axoplasm. The identical polarity of most, if not all, axonal microtubules places limits on models that implicate a role for microtubules in such vectorial processes as axoplasmic transport.[219]

2.1.1f. Dendritic Microtubules. In favorable EM sections, microtubules may be seen passing from perikaryon into the basal regions of dendrites as well as into the axon hillock.[180] As many as 800 microtubules have been counted extending from the Purkinje cell soma into a single dendritic tree.[106] Many of these microtubules arise from microtubule-organizing centers (MTOCs) in pericentriolar regions of the apical soma of the Purkinje cell. Other dendritic microtubules arise from additional electron-dense, granular masses of material (similar to the pericentriolar MTOCs) within the ring formed in the Purkinje cell soma by the Golgi complex.[106]

2.1.2. Microtubules in Pre- And Postsynaptic Cytoplasm

Neuronal microtubules appear to be excluded from axon terminals. In general, they are seen in the preterminal axoplasm but not in the terminal areas themselves. However, when pieces of neural tissue are teased in solutions of albumin and are allowed to soak in such solutions prior to fixation, structures indistinguishable from microtubules are seen to course among the synaptic vesicles and to form close associations with the presynaptic membrane.[95] In dendrites, microtubules approach the postsynaptic membrane in the region of the postsynaptic densities (PSDs). Indeed, both tubulin and microtubule-associated proteins (MAPs) have been reported to be constituents of the PSDs.[121]

2.2. Distribution in Glia

Microtubules are obvious constituents of all types of glial cells. Those glia able to undergo mitosis have spindle microtubules. Ependymal cells have microtubules as elements of their cilia. Astrocytic processes often include arrays of axially oriented microtubules. There have been numerous descriptions of microtubules in the oligodendroglial cytoplasm of the paranodal loops of myelin at nodes of Ranvier. These glial microtubules form a complex annulus around the axon, which defines the lateral extent of each myelin leaflet.[180] Microtubules are often found in other parts of the oligodendroglial cytoplasm as well (Fig. 1). Since glia comprise about 80% of the total cell number and 50% of the cell mass of the brain, their complement of microtubules obviously contributes greatly to the total of microtubule proteins that may be isolated from pieces of brain.

3. THE COMPOSITION AND STRUCTURE OF MICROTUBULES

3.1. Proteins

3.1.1. Tubulins

Microtubules are formed of structural proteins, the tubulins, as well as a series of microtubule-associated proteins (MAPs). (Many properties of tubulin and MAPs have been reviewed by Margaretta Wallin and Johanna Deinum in Volume 5 of this series.) There are two groups of tubulins, termed α and β tubulins. Most evidence indicates that the prominent form of tubulin in cells is as a heterodimer of α and β tubulins. Heterodimeric tubulin is the structural subunit of tubulin oligomers and of assembled microtubules.

The heterodimeric nature of the dimer is assumed from cross-linking studies,[144] from optical diffraction studies,[3,50] and from measurements of α : β ratios under various conditions. Most current models of microtubule structure assume a heterodimeric subunit. The possibility of homodimers must be considered, however, especially since there are skewed distributions of α and β tubulins among axonal microtubule proteins (which may or may not be assembled into microtubules).

3.1.2. Microtubule-Associated Proteins

In the following discussion, two classes of MAPs are considered. One class, with apparent molecular weights between 270,000 and 350,000 on SDS gels, is designated HMW. The second class includes a group of polypeptides

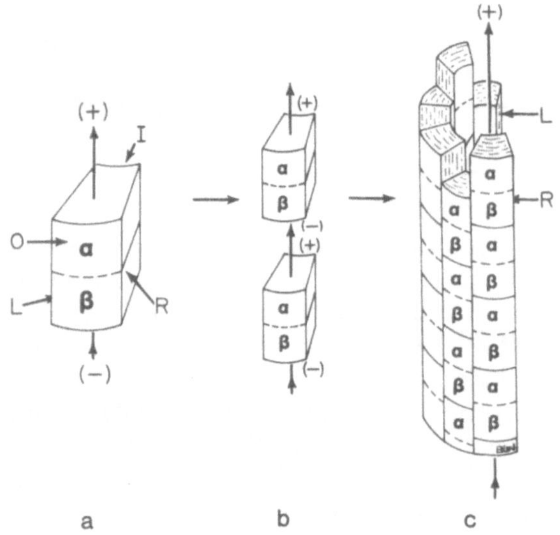

Fig. 2. Schematic drawing showing some aspects of polymerization of tubulin. Positions of α and β tubulins are arbitrarily assigned in this drawing. a: The tubulin dimer. The (+)/(−), Outside (O)/Inside (I), and Right (R)/Left (L) polarities of the dimer are indicated. b: Relationship between dimers in a single protofilament formed by head-to-tail polymerization of dimers. c: Vectorial relationships between dimers within a single protofilament of an assembled microtubule are preserved. Vectorial relationships between dimers on adjacent protofilaments also may be visualized. The surface lattice in this drawing is arbitrary and is designed to show the stagger between adjacent protofilaments.

with apparent molecular weights between 52,000 and 70,000; these are referred to as tau. Subgroups of HMW and tau are referred to by numbers. There are a variety of other MAPs that have been described. Those relevant to the nervous system are designated in terms related to the context in which they are discussed.

3.2. Structural Elements

3.2.1. Protofilaments

Microtubules are formed of parallel arrays of protofilaments grouped together to give a circular cross section. In mammalian nerve; indeed, in all vertebrate tissue described until now; each microtubule is formed of 13 protofilaments. These protofilaments are formed of head-to-tail strings of heterodimeric tubulin. The head-to-tail alignment of the heterodimers confers a polarity to the entire protofilament. Each of the protofilaments of a microtubule has the same polarity as all of the others.[3] The spatial relationships of heterodimer, protofilament, and microtubule are summarized in Fig. 2.

A critical aspect of research on microtubules has been the assumption that microtubules assembled *in vitro* are the same as intracellular microtubules. A combination of experimental results have supported that assumption. When antibodies are raised against tubulin and various MAPs, purified by cell-free assembly procedures, the antibodies decorate microtubules, *in situ*, in immunocytochemical reactions.[46,47,110,246] But the microtubules of vertebrate nervous tissue are invariably formed of 13 protofilaments,[181] whereas microtubules formed *in vitro* vary in protofilament number, the most frequent being not 13, but 14.[157,181,199]

The protofilaments that form microtubules are staggered slightly with respect to each other. As a result of this stagger, the protein units of the 13-

protofilament microtubules describe a left-handed three-start helix.[3,4,65,66,68] The surface lattices of the microtubules with other than 13 protofilaments present various packing problems, which have recently been analyzed in detail.[157]

3.2.2. Structural Polarity

The staggered array of protofilaments in the microtubules of vertebrate cells indicates another polar axis to the tubulin heterodimer. Not only is there a head-to-tail polarity and an inside-to-outside polarity, but right and left are distinguishable. The bonds that hold the 13 protofilaments together must include interaction between heterologous regions of the tubulin heterodimers. If the heterodimeric subunits are aligned such that the α subunit is at one end and the β subunit at the other, it is easy to describe this left–right polarity. The left side of an α subunit interacts with the right side of a β subunit, and the left side of a β subunit with the right side of an α subunit. The tubulin dimer thus is a three-dimensional vector (see Fig. 2).

3.2.3. Side Arms and MAPs

Examination of electron micrographs of microtubules in favorable sections often gives the impression of the presence of side arms. These may vary in size and regularity. They may extend between adjacent microtubules, may connect microtubules to other cytoplasmic constituents, or may stand out from the microtubule walls into surrounding space. Microtubules assembled *in vitro* from tubulin alone are smooth walled; they have no side-arms. Microtubules assembled from mixtures of tubulin and MAPs have side arms.[2,56,122,158,161,212,213,274] The HMW MAPs themselves extend from the microtubule surface.[2,3,201,210,239] Antibodies against HMW MAPs and tau generally exhibit the same staining patterns as do antibodies to tubulins,[46,47,140,208–210] although recent studies with monoclonal antibodies suggest a heterogeneous disposition of HMW MAPs with respect to types of cells in the CNS[110,156] and to subcellular distribution within neurons.[156,176] In sum, studies of microtubule coassembly with MAPs and the coincident immunocytochemical demonstration of tubulin and MAPs have led to the conclusion that the MAPs form the side arms on microtubules. On the basis of optical diffraction analysis of electron micrographs of microtubules assembled with HMW MAPs, a model has been proposed that places the MAPs at regular intervals along and around the microtubules. In this model, a part of the MAP is anchored between two protofilaments, and another part of the MAP extends away from the surface.[2,3]

The current view of the basic structure of microtubules is a hollow cylinder formed of 13 protofilaments staggered with respect to each other. Each protofilament is a head-to-tail linear polymer of tubulin dimers. The MAPs are intercalated between dimers and form side arms but do not contribute to the basic tubular form of the microtubule.

4. MICROTUBULE ASSEMBLY

A critical feature of the biology of microtubules is their dynamic ability to form or reform on demand. When a cell requires microtubules, it assembles them from preexisting pools of tubulin and other proteins. The process is reversible, and as the cell's requirements change, microtubules can be disassembled. The subunits thus made available serve as a pool for the assembly of other microtubules in the same cell. The dynamic plasticity of microtubules has been described in many cell activities. Striking among these have been studies of microtubules of the mitotic spindle. Immunocytochemical localization of tubulin before, during, and after mitosis in cultured vertebrate cells demonstrates the disassembly of cytoplasmic microtubules and assembly of spindle microtubules from the MTOCs associated with the reorienting centrioles during early prophase. Throughout mitosis, the organization, orientation, and length of the spindle microtubules all vary. By the end of telophase, the spindle microtubules have depolymerized, and cytoplasmic microtubules have reformed. The understanding of the assembly and disassembly of microtubules as a dynamic process in both dividing and nondividing cells has been a major focus of research on microtubules (see refs. 45, 187 for recent reviews).

It is important to recognize that *in vitro* assembly studies of microtubule proteins from brain homogenates introduce the possibility of artifactual copolymerization of microtubule proteins that are not associated with each other *in vivo*. For example, certain MAPs are found in neurons but not in glia;[110] within neurons, they are found primarily in dendrites.[156] When microtubules are disassembled in cold homogenates and then reassembled, a recombination of the tubulins and MAPs would be expected. Similar recombination might mix tubulin isomorphs into chimeric polymers.

4.1. Preparation of Assembly-Competent Tubulin

Once it became possible to assemble and disassemble microtubules *in vitro*, the mechanism of assembly became accessible to experimental manipulation. Two recent monographs include multiple approaches to cell-free microtubule assembly in numerous systems.[82,264] In general, cell-free assembly (and disassembly) procedures are based on the principle that an equilibrium between free tubulin dimer and polymerized microtubules may be regulated by variation of one or more of several factors including temperature, GTP concentration, GTP/GDP, and Ca^{2+} and Mg^{2+} concentrations. Assembled microtubules may be stabilized with glycerol,[205] sucrose,[119,205] and DMSO.[78,80,107,108]

4.1.1. In-Vitro Assembly/Disassembly Methods

For the study of microtubule assembly, microtubule proteins are prepared using sequential disassembly and assembly procedures. Brain is homogenized in an ice-cold buffer (approx. 1 ml/g wet tissue). Various buffers have been used, most commonly MES or PIPES.[6,19,119,160,162,205,218,225,226,249,263] The cold homogenate is centrifuged sufficiently to sediment all but the smallest cytoplasmic structures. The cold-disassembled microtubule proteins are recovered from the high-speed supernatant by first promoting their assembly (by addition of GTP, warming to 25–37°C, and chelation of free Ca^{2+}) and then collecting them by centrifugation at 25–35°C. The reassembled microtubules obtained in the pellet are washed repeatedly by alternating temperature-dependent disassembly and assembly steps with centrifugations between successive cold and warm incubations. After two or three cycles of disassembly and assembly, the microtubules are stored either frozen as droplets directly in liquid nitrogen, as pellets of assembled microtubules frozen to liquid nitrogen temperatures or to −80°C, or in 4 M glycerol in assembly buffer at −20°C.

The protein compositions of the reassembled, washed microtubule preparations vary. All include tubulin as the major component. All include the HMW MAPs and tau, but in varying amounts and ratios. In general, the ratio of MAPs to tubulin is reduced when 4 M glycerol is included in the assembly buffer. Neurofilament proteins are often present, but the relative amount of neurofilament protein seems to depend on the species and age of the animal from which the brain was taken and also on the procedure chosen for purification. For example, compare Fig. 2 of Murphy[160] with Fig. 1 of Williams and Lee.[263] The porcine brain microtubule protein prepared without glycerol includes tau and the HMW MAPs, but neurofilament proteins are not evident.[160] Bovine brain microtubule protein fractions include significant amounts of neurofilament protein, especially when glycerol is used in the assembly buffer.[190,263] (Indeed, glycerol is used to assemble neurofilaments from brain extracts.[54,86])

When the expensive buffers, PIPES or MES, generally used in these assembly procedures are replaced with phosphate–glutamate buffer, similar results are obtained.[6] We have compared bovine microtubule preparations obtained from brain extracts subjected to parallel fractionations, one[162] using PIPES at $pH^{23°}$ 6.62 and the other[6] employing phosphate–glutamate buffer at $pH^{23°}$ 6.75. We find similar yields and similar protein compositions for both preparations (D. Soifer and K. Mack, unpublished observations; see also ref. 226). It must be emphasized that optimum conditions for preparing microtubule proteins by sequential disassembly and assembly vary considerably from one species to another.

4.1.2. Separation of Tubulin from MAPs

Microtubule protein obtained after two or three cycles of polymerization is normally the material used for studies of the assembly process. Tubulin may be separated from MAPs on phosphocellulose,[40,212,249,262] DEAE,[164,254] or gel filtration columns[101] and allowed to grow onto seeds or to self-assemble in the presence or absence of MAPs and/or various other substances. Purified HMW and tau fractions may be obtained from the ion-exchange columns[164,212,239,249] or directly from microtubule preparations exposed to elevated temperature.[74,103,122]

4.2. Microtubule Elongation and Assembly

The analysis of the elongation process has dealt with a number of questions: (1) Where on the microtubule do new subunits add onto the existing structure? (2) In what form are subunits added on to microtubules? (3) What role(s) do MAPs play in assembly? (4) How does the cell regulate microtubule assembly? Along with these questions have come kinetic and thermodynamic analyses of the assembly process.

4.2.1. *Biased Polar Growth*

From the discussion of microtubule structure above, it is evident that there are only two ways for microtubules to grow. Either the tubulin is inserted into the microtubule, forcing its way into a protofilament between subunits already there, or growth is by subunit addition at the ends of the protofilaments. Morphological, biochemical, and theoretical considerations all point to the elongation of microtubules being a reaction of the type called condensation polymerization,[172] involving addition of subunits to the ends of linear polymers.

Not only do microtubules elongate by end addition but, as the microtubules are formed of parallel polar protofilaments, the microtubules themselves are polar and show different assembly and disassembly kinetics at each end. The next few pages include a selective review of the evidence for biased polar growth of microtubules and a discussion of some aspects of microtubule assembly, especially with respect to possible roles for microtubules in the function of cells of the nervous system.

Once it was possible to assemble microtubules *in vitro*, several attempts were made to add tubulin onto preexisting seeds. These studies all used microtubule fragments that could be identified in electron micrographs. Tubulin was added to the identifiable seeds under conditions favoring microtubule assembly. Following addition of the tubulin, the microtubule seeds appeared longer than before. Under most conditions, the identifiable seeds were at one end of the new structures, indicating that the newly assembled tubulin had been added preferentially to one end of each seed.[1,57,149,171,189,267] The kinetics of addition of brain tubulin subunits to flagellar axonemes was measured by Bergen and Borisy.[10] From length measurements on electron micrographs of microtubules at various times after addition of subunits to the flagellar seeds, rate constants were derived for association and dissociation of tubulin at both the proximal (−) and distal (+) ends of the flagellar axonemes. They found that both the assembly and disassembly rates on the (+) end were significantly greater than those rates for the (−) end. Their observations confirmed and extended the earlier electron microscopic (EM) studies of microtubule elongation and provided quantitative data that the condensation polymerization of tubulin is biased to one end of a microtubule.

Polar growth of microtubules was also observed by Summers and Kirschner,[231] who monitored the assembly of individual microtubules by dark-field microscopy. The rates of elongation of individual microtubules could be followed by recording the images on motion picture film and measuring microtubule lengths on succeeding frames. Under the assembly conditions used in these studies, microtubules were seen to elongate with the growth rate at one end about three times that at the other end. When ciliary axonemes were used as seeds, the fast-growing end was the distal end. The two morphological approaches—dark-field microscopic observations of individual microtubules and electron microscopy of microtubules grown onto flagellar axonemes—produced similar kinds of information. They suggested that microtubule elongation occurs by biased polar growth and that (where polarity may be determined) the fast-growing end is usually the distal end of the microtubule. The use of organized seeds such as ciliary or flagellar axonemes has the added advantage that the microtubules that elongate from these seeds generally are formed of 13 subunits,[199] the same protofilament number as found *in vivo*.[81,84]

4.2.2. Kinetics of Microtubule Assembly

4.2.2a. Measurement of Assembly Rates. In addition to direct visualization of elongating microtubules, as described above, several other approaches have been used to determine kinetic and thermodynamic parameters of polymerization and depolymerization. Among these, viscometry[131,173] and light scattering[85] have been the most extensively used. The principle behind viscometry is that the viscosity of a preparation of microtubule protein will increase as a function of the degree of polymerization of subunits. Elongating microtubules scatter light; the more tubulin assembled, the more light scattered. This may be seen as an increase in apparent optical density (turbidity). The turbidity is approximately proportional to the total concentration of polymerized microtubule protein and independent of microtubule length (as long as the microtubule length exceeds the wavelength of the light).[13,85,116]

The availability of viscometry and turbidimetry have made it possible to follow the rate of assembly of microtubules from the initiation of assembly to the establishment of steady state. The effect of alteration of steady-state conditions (by adding subunits or drugs, by changing temperature or ionic environment, etc.) on the amount of tubulin assembled could also be monitored by turbidimetry.

4.2.2b. Nucleation of Assembly. From self-assembly experiments, it has become clear that an initial nucleation step is necessary for assembly to take place in brain extracts or in preparations of washed (cycled) microtubule proteins. At this time, the nucleation step is not understood. It is manifested, in viscometry or turbidimetry studies, as a lag between the activation of assembly and the detection of assembly. When seeds, formed of sheared or sonicated microtubule fragments or of ciliary or flagellar axonemes, are present, assembly is detectable without a lag. To block spontaneous nucleation in elongation experiments, either polyanions[25,116,227] or MAP-free tubulin can be used.

4.2.2c. Critical Concentration. The assembly of microtubules onto seeds will not occur unless there is an adequate concentration of tubulin. Under experimental conditions in which the number of microtubule ends available for assembly remains constant (seeds are present; spontaneous nucleation is inhibited), assembly was recorded using varying amounts of tubulin dimer.[116,227] Analysis by plotting the initial rate (dn/dt)[116] or the steady-state optical density (OD)[227] as a function of subunit concentration demonstrates that there is a critical concentration (C_c) of subunits that must be present for assembly to take place. At steady state, the difference between the amount of subunits

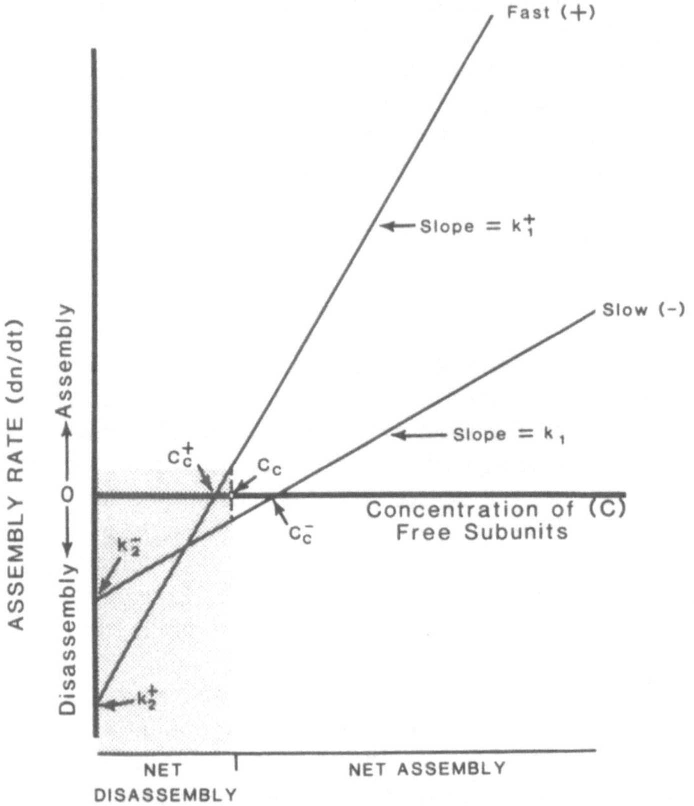

Fig. 3. Assembly rates at the fast- (+) and slow- (−) assembling microtubule ends as a function of the concentration of free subunits of tubulin dimer. The two curves represent typical plots of equations 1a and 1b in Section 4.2.2h. The dissociation constants (k_2^+ and k_2^-) are the *y* intercepts; the association constants (k_1^+ and k_1^-) are the slopes of the curves. The separate critical concentrations for each end are indicated by C_c^+ and C_c^-. The critical concentration for net assembly on the microtubule is indicated by C_c. At dimer concentrations $C < C_c$, there is net disassembly of a microtubule with two free ends (shaded area).

assembled onto seeds and the total amount of active subunits is equal to the critical concentration. The initial rate of assembly is a function of the number of ends available for assembly. The number of subunits that assemble is essentially independent of the seed concentration; there can be many short microtubules or a small number of long microtubules. Various factors may increase or decrease the critical concentration.

4.2.2d. Assembly at Each End of a Microtubule. The rates of addition of bovine brain tubulin to the (+) and (−) ends of flagellar seeds were determined by Bergen and Borisy[10] in their elongation experiments. When they plotted the individual rates as a function of dimer concentration, they observed that there are dimer concentrations at which the polymerization rates for each end are zero (Fig. 3). These are the critical concentrations (C_c^+ and C_c^-) for polymerization at each end of the elongating microtubule. Below these critical dimer concentrations, there is depolymerization from the respective ends. When the dimer concentration is between C_c^+ and C_c^-, there is polymerization at the (+) end and depolymerization at the (−) end (but see Section 4.2.4a and Fig. 5). The dimer concentration when depolymerization at the (−) end equals polymerization at the (+) end is the critical concentration for assembly (C_c).

From this analysis, an approach to steady state may be predicted. At steady state, the net assembly at the (+) end and the net disassembly at the (−) end will be equal. With net assembly at one end and disassembly at the other, a flux of subunits through the microtubule is to be expected. Such a flux has been demonstrated by Margolis and Wilson[150,265] and by Cote and Borisy.[49]

4.2.2e. Subunit Flux at Steady State (Treadmilling). Margolis and Wilson[150,265] labeled dimeric tubulin with [^{3}H]GTP by taking advantage of the rapid equilibrium that is established between GTP in solution and GTP bound to the exchangeable site (E-site) of dimeric tubulin.[111] Since guanine nucleotide at the E-site becomes nonexchangeable when dimeric tubulin is incorporated into microtubules,[255] Margolis and Wilson could add [^{3}H]GTP to assembled microtubules at steady state without altering the concentration of free dimer in the polymer–dimer system. The label would exchange with bound GTP of tubulin dimers, which, now labeled themselves, could be followed as they were incorporated into assembled microtubules. The label could be chased with unlabeled GTP. Under appropriate conditions, incorporated label would remain associated with assembled microtubules for a period of time proportional to the average microtubule length. Alternatively, microtubules could be made to assemble and reach steady state in the presence of label. The labeled microtubules could then be chased at steady state with unlabeled GTP, and the rate of loss of labeled subunits from the microtubules could be followed. On the basis of these experiments and experiments in which they used drugs such as colchicine (which are presumed to act by blocking the assembly reaction at microtubule ends[149]), Margolis and Wilson[150,152,265,266] developed a model for subunit flux (which they called treadmilling). In subsequent studies, they and others have considered whether treadmilling microtubules can be made to do work *in vivo* (see Section 4.2.3).

4.2.2f. Nucleotide Hydrolysis and Subunit Flux. In his analysis of actin polymerization, which is also a biased polar process, Wegner showed that the hydrolysis of ATP is necessary for the generation of a steady-state flux of subunits through actin filaments.[247,248] There is strong evidence that the different reactions at the (+) and (−) ends of microtubules relate to GTP hydrolysis. This was predicted in the analyses of Bergen and Borisy[10] and of Kirschner[123] and is now supported by experimental evidence.[49]

Cote and Borisy used tritiated tubulin subunits to repeat and extend the experiments of Margolis and Wilson, in particular, to determine whether GTP hydrolysis is required for head-to-tail microtubule polymerization.[49] The presumption in their studies is that if the tubulin subunits that assemble are equivalent to those that disassemble, there will be a true equilibrium between subunits and free microtubule ends, and growth will be bidirectional rather than biased to one end. But if hydrolysis of bound GTP takes place during or right after the assembly process, disassembling subunits will have bound GDP, while assembling subunits will have bound GTP, and there will then be no equivalence between the subunits that are polymerizing and those that are depolymerizing. Microtubule elongation would indeed be biased to one end. By substituting a nonhydrolyzable GTP analogue (GMPPCP) for GTP,[5,118,178,182,232,255] Cote and Borisy could convert from biased assembly of microtubules to bidirectional assembly. In experiments in which tubulin subunits were directly labeled rather than labeled with radioactive bound nucleotide, Cote and Borisy found that although labeled tubulin was incorporated into microtubules at steady-state conditions in the presence of GMPPCP, the flux of tubulin through the microtubules was reduced to a value about two orders of magnitude below the flux of tubulin in the presence of GTP. Thus, blocking GTP hydrolysis greatly reduced treadmilling.

4.2.2g. Double-Labeled Microtubules. Farrell and Jordan have used GTP-labeled subunits to determine assembly–disassembly kinetics.[70] Microtubules were labeled throughout their lengths by assembly to steady state in the presence of [^{14}C]GTP. The labeled microtubules were washed free of unbound label under conditions in which the microtubules were stabilized. The [^{14}C]-labeled microtubules were then resuspended in tubulin dimer at the critical concentration and pulse-labeled with [^{3}H]GTP for up to 1 h. Since the microtubules were at steady state, a loss of [^{14}C]-labeled subunits is to be expected from one end with a gain of [^{3}H]-labeled subunits at the other end. The specific activity of the [^{3}H]GTP was maintained at a high enough level so that the effect of incorporation of ^{14}C label at the assembly end was negligible.

If double-labeled microtubules are rapidly diluted in buffer such that the free dimer concentration is below the critical concentration, the microtubules will disassemble. If the length distribution of the steady-state microtubules (prior to dilution) is known, and the initial free dimer concentration is determined, initial rates of microtubule depolymerization can be estimated.[70,117,130] Since the microtubule end that was the net assembly end at steady state was the source of [^{3}H]subunits under disassembly conditions, dissociation and association constants could be determined for the net assembly and disassembly ends of the microtubules. Farrell and Jordan have obtained association and dissociation constants for both bovine brain tubulin and sea urchin sperm tail outer doublet tubulin.[70] Comparison of the numbers obtained in various laboratories for the flux (Wegner's parameter, s[247]) indicates that many individual subunits must react with a microtubule end at steady state in order for one subunit to become incorporated into a microtubule (see Section 4.2.2i).

4.2.2h. Kinetic Parameters of Assembly. Despite the various experimental and theoretical approaches to the problem of microtubule assembly, most analyses are based on similar assumptions and use analogous terms. These derive from the treatment by Wegner[247] for actin filament assembly and have been variously formulated for the assembly of microtubules.[10,49,70,123] In these analytical treatments, only dimeric tubulin is assumed to participate in events

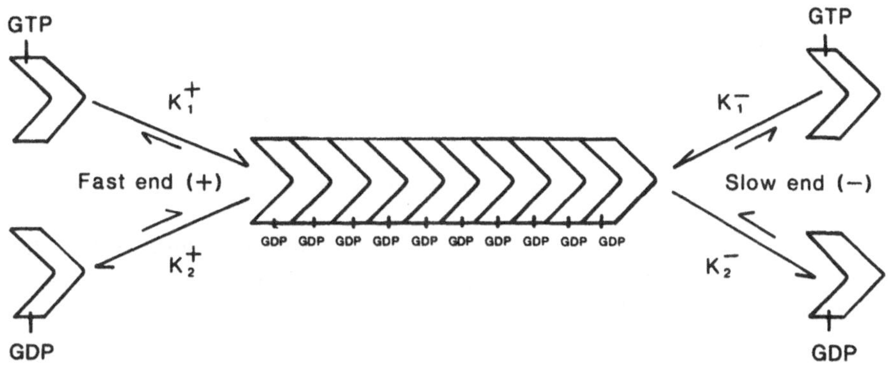

Fig. 4. Schematic representation of some of the reactions of microtubule assembly. For the simplified analysis included in Section 4.2.2h, only the four reactions represented by the large arrows, with their respective kinetic constants, are used. The reverse of these reactions (small arrows) are ignored. In this model, the assembling species at either end of the microtubule is presumed to have GTP at the E-site. The model does not distinguish whether GTP hydrolysis takes place during assembly or right after assembly.

at the ends of microtubules. Some features of the kinetic analyses of microtubule assembly are summarized here. For models including tubulin oligomers and MAPs, see Weisenberg.[252,253,271]

The primary reactions are assumed to include both the condensation of GTP-dimer onto each microtubule end and the dissociation of GDP-dimer from each end (Fig. 4, large arrows). For conditions in which there is effectively no nucleation, and the number of microtubule ends remains constant, the rate of polymerization of dimeric tubulin (dn/dt, where n = number of dimeric subunits polymerized) is a function of the concentration of dimer available for polymerization (C). Thus, the assembly rates at each end of a microtubule are given by

$$dn^+/dt = k_1^+ C - k_2^+, \text{ [for (+) end]} \tag{1a}$$

$$dn^-/dt = k_1^- C - k_2^-, \text{ [for (−) end]} \tag{1b}$$

Such curves are plotted in Fig. 3. The association constants, k_1^+ and k_1^-, are given by the slopes of each line. The dissociation constants, k_2^+ and k_2^-, are given by the y intercepts.

The critical concentrations for each end, C_c^+ and C_c^-, are those values of C for which dn^+/dt and $dn^-/dt = 0$. From equation 1a, when $dn^+/dt = 0$, $C = C_c^+$,

$$C_c^+ = k_2^+/k_1^+ \tag{2a}$$

and from equation 1b, when $dn^-/dt = 0$, $C = C_c^-$,

$$C_c^- = k_2^-/k_1^- \tag{2b}$$

Where net assembly at the (+) end is equal and opposite to net disassembly at the (−) end, the value of C equals the critical concentration, C_c, for microtubule assembly. At $C = C_c$, $(dn^+/dt) = -(dn^-/dt)$. Then, from equations 1a and 1b,

$$C_c = (k_2^+ + k_2^-)/(k_1^+ + k_1^-) \tag{3}$$

Wegner's[247] parameter, s, for the flux of subunits through a treadmilling polymer is given by the difference between the net assembly events at the (+) end and the net disassembly events at that end. Thus,

$$s = k_1^+/(k_1^+ + k_1^-) - k_2^+/(k_2^+ + k_2^-) \tag{4}$$

The value of s gives the fraction of all assembly events at the (+) end that result in net incorporation of subunit into the microtubule.

4.2.2i. The Efficiency of Treadmilling. The value s is essentially a measure of the efficiency of the flux of subunits through a microtubule. Not surprisingly, the efficiency of flux varies depending on the conditions of flux measurement. This variation appears to result, at least in part, from the great range of dissociation constants that have been measured. Indeed, the sum of the dissociation constants at each end (the denominator in the second term of equation 4) ranges from 4[70] to 232.[112] All values for s indicate that the treadmilling process is inefficient. Even the highest value, 0.26,[49] means that one subunit is incorporated into a microtubule at steady state for every four assembly and four disassembly events. More typical experiments have yielded flux efficiencies between 0.1[113] and near zero,[70] with typical values suggesting that one subunit is incorporated into a microtubule at steady state for every 15 or 20 associations and 15 or 20 dissociations. Maintaining a microtubule at steady state would appear to have a high energetic cost, since the net incorporation of a single subunit would require hydrolysis of 15 or 20 GTPs.

It is interesting to note that parallel studies of subunit flux in actin filaments suggest similar inefficiency and small levels of flux. Relatively high flux efficiencies were reported by Wegner[247] and by Pollard and Mooseker[181a] in seeded polymerization studies. About four or five assembly–disassembly events were required for net incorporation of a single subunit. However, under ionic conditions closer to those found *in vivo*, "subunit exchange" by actin filaments appears to be a more inefficient process.[177]

4.2.2j. Microtubule Polarity. It is easy to confuse terminology when discussing microtubule polarity and steady-state kinetics. In experiments in which microtubule assembly at steady state was followed using [^{3}H]GTP to label assembling subunits, the microtubule ends were defined as net assembly (A) or disassembly (D) ends.[70,150,152] Fast- (+) and slow- (−) growing ends were identified in studies in which microtubule elongation onto flagellar seeds was directly followed by electron microscopy.[10] The fast-growing ends were equivalent to the A ends if these elongation studies were analyzed in terms of the labeling experiments. But the converse was not necessarily true. As may be seen in Fig. 6, either the (+) end (6a,6c) or the (−) end (6b,6d) may be the end that is assembling at steady state. The A ends are those ends that are incorporating subunits at steady state, even if they are the slow-growing (−) ends (e.g., Fig. 6b).

To confuse matters more, studies of microtubule polarity by morphological methods presumably allow identification of (+) ends in electron micrographs of tissue sections.[102,158a] This is accomplished by adding tubulin to permeabilized cells under conditions in which tubulin will form small sheets or ribbons of protofilaments on the surface of existing microtubules. Viewed in cross section, the sheets appear as hooks extending from the circular profiles of microtubules. Repeated analyses of systems in which growth rates could be compared to the handedness of hooks (comparison with the results described in Section 4.2.1) led to the conclusion that when a microtubule is viewed from the (+) end toward the (−) end, clockwise hooks are to be expected.[158a] It is not at all clear that this is the case for other microtubules or whether it only holds for those conditions in which the (+) ends and the (A) ends are equivalent. Thus, although we can define microtubule polarity in terms of association constants (fast and slow ends), net assembly at steady state (A and D ends), or in terms of morphologic criteria (handedness of hooks), ends defined according to one set of criteria are not always equivalent to ends defined by another set of criteria.

4.2.3. Implications of Microtubule Treadmills

The phenomenon of treadmilling or subunit flux is prone to misinterpretation, especially by neuroscientists who are looking for a conveyor belt to account for the transport of substances along axons. A series of theoretical arguments has been developed describing how treadmilling between barriers might enable microtubules to transport associated material from (A) end to (D) end.[104,105] These are special case situations, such as where the (A) end is attached to a wall and the assembly process involves insertion of new subunits between the (A) end and the wall. For a microtubule in a test tube (or in a cell), the subunit does not move as it traverses the microtubule. Rather, there is net addition of subunits on the (A) side and loss of subunits on the (D) side of the subunit. The microtubule, in effect, grows past a given subunit. In Fig. 5, which is based on a drawing by Wilson and Margolis,[265] one can follow the flux of subunits through a microtubule. The representation given by most published drawings of the treadmilling phenomenon can give the misimpression that subunit flux can involve a useful translation of tubulin subunits through a fixed structure. Figure 5 is probably a more accurate description of the phenomenon.

If the phenomenon of treadmilling is to serve a work-related purpose in the cell, it must overcome a critical weakness: subunit flux only occurs at steady state, when the polymerization rate at the (A) end just equals the depolymerization rate at the (D) end. If the free dimer concentration is altered, or if C_c^+ or C_c^- is altered, there will be a net elongation or net decrease in length

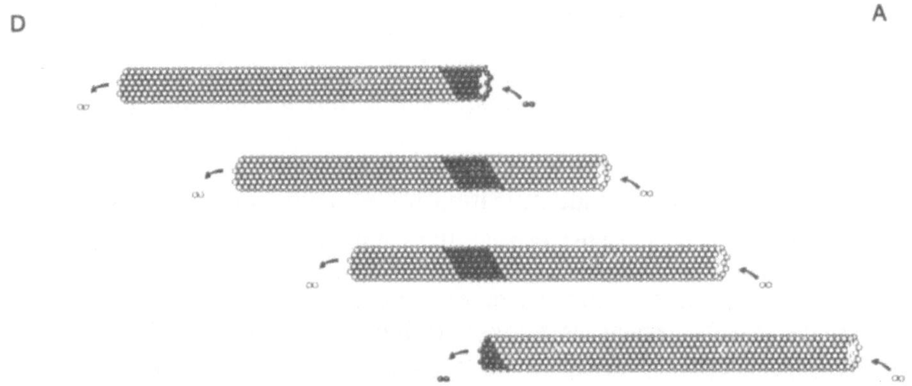

Fig. 5. The idea of steady-state treadmilling is that net dimer addition occurs at one end (A) of the microtubule and loss occurs at the opposite end (D). In a treadmilling microtubule, labeled subunits (shaded) assemble at the A end (upper diagram). They are not transported through the microtubule. Instead, the microtubule keeps assembling beyond the pulse of labeled subunits. Eventually, the region of disassembly reaches the labeled subunits, and they dissociate from the microtubule (bottom). (Redrawn from Wilson and Margolis.[265])

of the microtubule. Kirschner[123] has suggested that treadmilling might be an artifact of cell-free studies; in the cell, the (−) end of microtubules might be capped, and assembly may be regulated at the free (+) end by varying subunit concentrations. As long as subunit concentrations remain between C_c and C_c^+ (conditions as in Fig. 3), spontaneous nucleation will be inhibited, and (−)-end-capped microtubules would remain assembled.

There have been a number of other attempts to make models that would explain how treadmilling may be converted to work. One, proposed by Margolis and Wilson,[152,266] requires that a series of attachment–detachment reactions between dyneinlike molecules in the cytoplasm and assembled tubulin could push microtubules past these fixed dyneins. At the microtubule ends, the rate of assembly or disassembly would be somehow matched to this rowing mechanism, so that each new subunit would be pulled towards the dynein along with the microtubule as a whole.

4.2.4. Points for Regulation of Assembly

4.2.4a. Dissociation Constants. Analyses of assembly kinetics suggest various points at which parameters of microtubule function might be regulated. For example, net incorporation of subunits into a microtubule at steady state can happen at either end. In Fig. 6, we consider four sets of hypothetical rate curves for assembly. In the first (6a), which is like Fig. 4, the curves cross each other when the assembly rate is negative; the subunit concentration C is below C_c for both ends, and depolymerization is taking place at both ends. At steady-state conditions, when $C = C_c$, there is assembly on the (+) ends and disassembly (at an equal but opposite rate) on the (−) ends.

When the curves cross with $C > C_c$ (6b), the direction of steady-state flux is reversed. At steady state, the (−) end (with a smaller association constant)

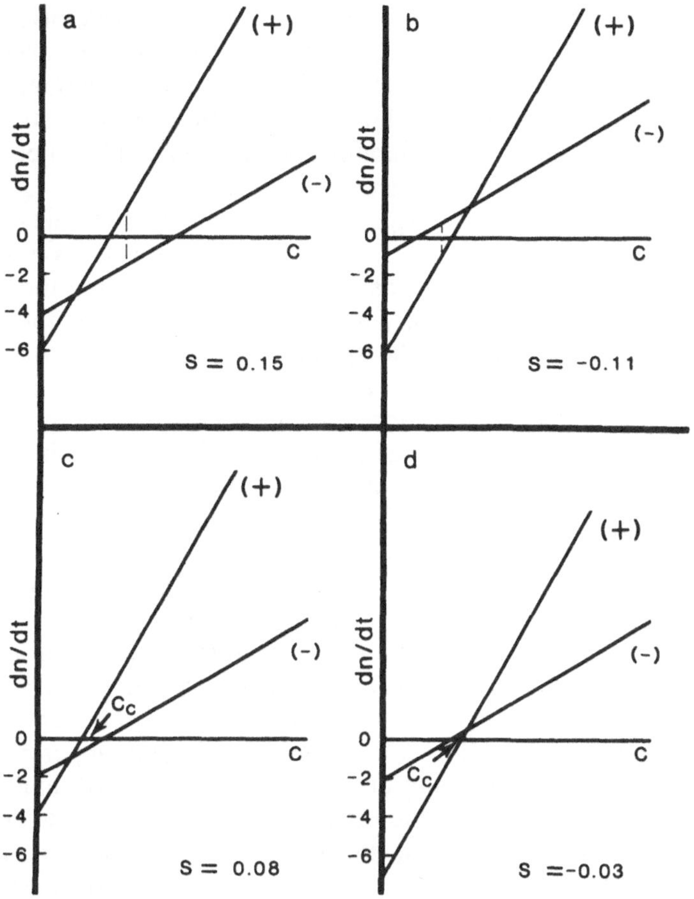

Fig. 6. Four possible solutions of equations 1a and 1b (see text, Section 4.2.2h). A small change in either k_2^- (panels a and b) or k_2^+ (panels c and d) is sufficient to reverse the direction of treadmilling. In panel b, at C_c, the (−) end is assembling, and the (+) end is disassembling. Thus, at steady state, the direction of subunit flux is reversed from the condition in a. Values of d*n*/d*t* are arbitrary. Values of *s* have been calculated from equation 4, using the same sets of arbitrary numbers for the four graphs. Dashed lines mark positions of C_c in curves a and b.

is assembling while the (+) end is disassembling. These arguments suggest a mechanism whereby the direction of flux may be regulated. In 6a and 6b, the curve for the (+) end is unchanged, whereas for the (−) end, the association constant k_1^- is unchanged, but the dissociation constant, k_2^- (*y* intercept), is increased. In Figs. 6c and 6d, the (−) curve is unchanged; the (+) curve is modified by varying k_2^+. Jameson and Caplow[112] and Berlin[12] have suggested that the dissociation rates may be modified by MAPs. Berlin has also suggested that modification of dissociation rate constants may reverse the direction of steady-state subunit flux.[12] Given the very low efficiency of subunit flux that has been obtained experimentally,[10,70] it is clear that small changes in k_2^+ or k_2^- can be sufficient to reverse the direction of treadmilling.

4.2.4b. Dimer Concentration. Variation of the concentration of free dimer will have a pronounced effect on steady-state microtubules whether or not one end is capped (Fig. 3). Raising C above C_c will promote assembly; lowering C below C_c will promote disassembly. If only one end is free to exchange subunits,[123] C_c^+ or C_c^- will be the critical value of C for regulation. The dimer concentration may be regulated at the level of synthesis (see Section 6.2.2) by sequestering of free dimer[256] or by modification of the equilibria between dimer and monomeric or oligomeric forms of tubulin.

Although there have been suggestions that tubulin oligomers participate in microtubule assembly,[68,125,253,271] the nature of the reactions governing the formation of oligomer from dimer as well as those governing the addition of oligomer to microtubules are poorly understood. Several attempts have been made to study the dissociation of the αβ tubulin dimer. With sufficiently dilute systems, Detrich and Williams have succeeded in demonstrating the reversible, concentration-dependent dissociation of tubulin dimer from bovine brain and have derived a dissociation constant of about 8×10^{-7} M for this reaction.[58,59] This means that at a concentration of about 0.1 mg/ml (0.8×10^{-7} M tubulin monomer + dimer), about half of the tubulin is dissociated. Since the K_D for αβ dissociation is only slightly less than the values of C_c determined for the assembly of bovine brain tubulin,[49,70,112] about half of the tubulin is dissociated at a concentration about the same as or just a bit less than C_c for tubulin assembly (1.0×10^{-7} M). Thus, the real concentration of dimer at C_c is only about half the measured C_c (monomer + dimer, calculated as if all tubulin is dimerized). Any factor that might perturb the monomer–dimer equilibrium will change the dimer concentration at C_c, thus changing the apparent C_c for assembly. A shift in the equilibrium of αβ dissociation toward dimer when microtubules are at steady state would increase the amount of dimer and promote microtubule assembly until a new steady state is achieved. An increase in dissociation of dimer would lower the dimer concentration in a system of steady-state microtubules, promoting disassembly. These studies of monomer–dimer equilibrium have been carried out in the ultracentrifuge. It has not been feasible to study monomeric tubulin or even to purify monomeric tubulin at a reasonable concentration under nondenaturing conditions. But it is evident that the dimerization of tubulin is one point at which microtubule assembly might be regulated.

Although the roles of MAPs in microtubule assembly and function are still poorly understood, regulation of MAP function or synthesis may also be a means for regulating the dynamics of specific subpopulations of microtubules (see Section 6.2.3).

5. *MICROTUBULE-ASSOCIATED PROTEINS*

5.1. *Distribution in the Nervous System*

5.1.1. *Axoplasmic Transport and Biochemical Studies*

It is important to realize that microtubules are heterogeneous structures. We know from the elegant experiments of Gozes and Sweadner[94] that single nerve cells are probably capable of expressing all of the tubulin genes and displaying all of the heterogeneous forms of tubulin found in the central nervous system. We do not know whether each microtubule is formed of only a single species of each of the α and β tubulins or even whether a given protofilament may include several tubulin isomorphs. It is even possible that the different isomorphs have slightly differing affinities, either for each other, to form dimers, or for other dimers or MAPs.

The distribution of MAPs in the nervous system is nonhomogeneous. The evidence for this is complicated and not without controversy, but the weight of the evidence suggests that only a subset of the tau proteins is axonally transported and that HMW2 is excluded from axons. Most studies of axoplasmic transport have failed to detect transport of HMW2.[17,22,23,90,134,237] Comparison of the proteins synthesized by retinal cells with the proteins transported along the optic nerve, which is formed of the axons of the retinal ganglion cells, indicates that the HMW2 and a full complement of tau proteins are synthesized in the retinal cells but only certain of the tau proteins enter the ganglion cell axons and are transported to the axon terminals.[134,237] There is some suggestion for the transport of very-high-molecular-weight proteins in the gel analyses published in several other reports. Most prominent is the polypeptide labeled 25,[7,138] which is transported with the proteins termed group II by Willard.[7,138,261] If HMW2 is associated with axonal microtubules or neurofilaments, one would expect to find it transported with tubulin and the neurofilament proteins, as are tau proteins,[134,237] and not at a rate nearly 50-fold faster. HMW2 has not been detected on one- and two-dimensional gels of homogenates of optic and sciatic nerves[223] but has been reported to be transported in rat peripheral nerve.[233]

5.1.2. Immunocytochemical Studies

Immunocytochemical studies have indicated that within the nervous system, HMW2 is found only in differentiated neurons and not in glia.[110,156] Within neurons, the binding of antitubulin antibodies could be visualized in all parts of the cell, whereas anti-HMW2 antibodies were demonstrable mostly in dendrites, with diffuse staining of perikarya.[156] Within dendrites, HMW2 was localized both with microtubules and with postsynaptic densities (PSDs).[156] Of a group of monoclonal antibodies raised against Chinese hamster brain microtubule proteins, those produced by five separate cloned hybridomas reacted specifically with HMW2 in various immunoassays.[16] One of these monoclonal antibodies reacted with axons to yield positive but weak staining;[176] however, most of the material that reacted with this antibody was located in the proximal regions of the axons.

5.1.3. HMW2 in Nerve Endings

In view of this evidence that HMW2 is excluded from axons, one must ask how HMW gets to nerve endings, where it has been reported to be included in synaptosomes.[28] One can only wonder how much perikaryal and dendritic cytoplasm is included in these synaptosome fractions.

5.2. Reactions with Tubulin

5.2.1. Polymerization of Tubulin with HMW2 and Tau

The role of various MAPs in microtubule function can only be surmised at this time, although there is evidence that tau and HMW2 act in different ways to influence the interactions of tubulin subunits. Microtubules have been reconstituted *in vitro* from phosphocellulose-purified tubulin (PC-tubulin) and from heat-stable MAPs, purified by ion-exchange chromatography, and separated into tau and HMW fractions by gel filtration.[274] In this study, there was minimal cross contamination of one MAP with the other. Microtubules formed from pure tubulin in the absence of MAPs have smooth surfaces.[56,108,161] Coassembly of the tubulin with HMW2 produces microtubules with side arms seen either as globular protrusions from the microtubule wall in negative-stained preparations or as 30-nm side arms in metal-shadowed or fixed and thin-sectioned preparations.[274] Tau proteins also coat the microtubules, giving them a rougher appearance than microtubules formed without MAPs.[274]

When tau or HMW2 polymerize with tubulin in the presence of the GTP analogue guanosine 5′-(α, β-methylene)triphosphate, they form different structures: the tau-tubulin polymers are microtubules, whereas the HMW–tubulin polymers are ribbons formed of about six protofilaments.[196] Such parameters of assembly as levels of nucleation, assembly rate, and dissociation constant all vary depending on which MAPs are used and at what concentration.[12,143,195]

Vinblastine has different effects on mixtures of tau and PC-tubulin and mixtures of HMW2 and PC-tubulin.[143] In the presence of tau, but not HMW2, vinblastine brings about the formation of loose helical polymers and paracrystalline arrays similar to those described for crude microtubule protein preparations.[67,84] In the presence of HMW2, without tau, vinblastine promotes microtubule depolymerization without the formation of coiled structures.[143]

The experiments using purified tau or HMW2 provide some hints about the possible roles of these proteins in the assembly of microtubules. On the basis of several models and discussions,[67,84,143,200,201,239] tau is viewed as stabilizing longitudinal associations between tubulin dimers, i.e., along the protofilament. HMW2 apparently interacts with tubulin subunits on adjacent protofilaments. The differential effects of vinblastine on HMW2–PC-tubulin mixtures and tau–PC-tubulin mixtures may be explained in these terms.[143]

5.2.2. Oligomers of Tubulin and MAPs

Another approach to the function of MAPs in the assembly process follows from studies of oligomeric forms of tubulin, the so-called "rings." Cold depolymerized microtubules, prepared from brains by *in vitro* polymerization from high-speed supernatants, may be fractionated to yield tubulin dimer (6 S) and two classes of oligomers of about 20 S and 30–36 S. The size, form, and peptide composition of these oligomers varies with the details of preparation of the microtubule proteins and the species from which they were obtained.[19,20,55,60,66,68,83,127,145,200,239,250,251] The 20 S oligomers appear to be single rings formed of tau and tubulin; the 30 S oligomers are double-layered rings formed of HMW and tubulin.[145,200, 201,239] The 36 S rings are double-walled, single-layered rings containing tau and some HMW.[126,127,201,250] In the model of Scheele and Borisy,[200,201] the oligomers of the 30 S rings formed of HMW2 and tubulin are arranged such that the tubulin subunits of the rings are arrayed in a two-turn helix of 14 subunits per turn. The adjacent tubulin dimers along the helix are aligned as the side-to-side (L–R) subunits of adjacent protofilaments are arranged in a microtubule. End-to-end interactions [(+) → (−)] occur between subunits on different turns of the helix. The HMW2 molecules lie in the grooves between the two helical turns.If this structure were open, linear oligomers would not be formed; only L–R oligomers would be present. Some forms of 36 S rings (those formed of tau and tubulin) appear to have the tubulins associated (+)→(−) (end to end). Accordingly, these rings may open up into oligomers roughly equivalent to microtubule protofilaments. Thus, HMW2 and tau also differ from each other in the nature of the oligomeric heteropolymers that each forms with tubulins.

5.2.3. MAP–Tubulin Oligomers and Microtubule Assembly

In the absence of seeds, MAPs have been reported to be necessary for the nucleation of microtubule assembly *in vitro*,[120,163,212] whereas under most experimental conditions MAPs are not essential for microtubule elongation. In contrast to these observations, tau has been reported to be required for microtubule elongation.[268] It should be noted, however, that MAPs are not absolutely required for microtubule assembly: MAP-free tubulin will polymerize at high concentration in standard assembly buffers or at low concentrations in the presence of glycerol or DMSO.[33a,107,134a,187a]

Most models of microtubule assembly consider only the kinetics of dimer addition.[10,49,123,152] There have been several models that consider oligomer addition.[66,125,252,253,271] The oligomers are generally considered to include MAPs. The possibility that oligomers play a significant role in the dynamics of microtubules in cells is suggested by recent studies of microtubule ends in outer segments of rabbit retinal rods.[257] At the distal ends of the doublet microtubules, the A tubules extend farther towards the pigment epithelium than the B tubules. Near the distal ends, the B tubules open at the point of intersection between protofilament 1 of the B tubule and protofilament 1 of the A tubule (terminology of Warner and Satir[244]). Profiles of the open B-tubule ends are seen that are composed of eight to 11 protofilaments. Since these are seen in sections that are 50–100 nm thick, the open tubules with differing numbers of subunits must differ from each other by relatively long stretches of the missing protofilaments. If only one or two subunits are missing, as might be expected for conditions of dimer addition at the microtubule end, the differences would not be detected at the microtubule ends within the thickness of the section. A statistical analysis of the lengths of B-tubule profiles suggests that protofilaments 4–11 are at least one, two, or three oligomers longer than protofilaments 3, 2, and 1, respectively.

5.3. Cold-Stable Microtubules

There is a subpopulation of microtubules in brain that are stable at low temperature but depolymerize in the presence of elevated Ca^{2+}.[99,245] Addition of ATP makes these microtubules cold labile.[147] A 64,000-dalton MAP has been described that is uniquely phosphorylated in the cold-labile microtubule population and has been termed switch protein.[147] The Ca^{2+} instability of the cold-stable microtubules requires substoichiometric concentrations of calmodulin; in the presence of Ca^{2+} alone, the cold-stable microtubules are insensitive to millimolar Ca^{2+} concentrations.[114] A blocking factor, consisting of a complex of several polypeptides, has been isolated from solubilized cold-stable microtubules.[148] Although some of the blocking factor polypeptides have molecular weights in the same range as tau, they differ from tau both in their effects on microtubule polymerization and in their physical properties.[148] The blocking factors appear to fit a model in which microtubules are rendered cold stable by the presence of calmodulin- and protein kinase-regulated substoichiometric blocks.[115]

6. EXPRESSION OF TUBULIN GENES

6.1. Multiple Forms of Tubulin

6.1.1. Tubulin Isomorphs

Brain tubulins have been shown to exhibit considerable microheterogeneity. Tubulins may be phosphorylated, detyrosylated, glycosylated, etc., but these sorts of modification do not account for all of the heterogeneity that has been described.

Isomorphic forms of tubulin have been resolved by isoelectric focusing, either alone or in two-dimensional electrophoretic systems.[73,87,92,153,154,168] Most of these tubulin isomorphs are present at the same time in a single neuron.[94] Analysis of tubulins synthesized in cell-free systems indicates that microheterogeneity is specified at the level of mRNA.[26,27,91,155,159,220,221,229] Separate mRNAs for several tubulin isomorphs have been identified, and the heterogeneity of tubulin transcripts has been shown to change during development.[91]

6.1.2. Complexity of Tubulin Genes

Cloned cDNA sequences complementary to tubulin mRNAs from chick brain[42] have been used to probe the genetic complexity of chicken tubulins. Such studies indicated that there are about four α-tubulin genes and four β-tubulin genes in the chicken genome. Four different chicken chromosomes contain sequences homologous to α-tubulin genes. The four β-tubulin gene sequences are on at least two different chromosomes. There is no evidence that α- and β-tubulin genes are linked.[39,43] Comparative studies using the chicken cDNAs to probe HindIII-restricted DNA fragments from various other species indicate that humans, rats, mice, and fish all have about ten α- and ten β-tubulin sequences.[42,43] At least some of the human α-tubulin genes differ from each other in sites,[259a] and two differ in the number and size[258] of intervening sequences. Two human β-tubulin pseudogenes have been identified with the chicken probes and have been sequenced.[258,259] One lacks intervening sequences and might have arisen by reverse transcription and subsequent insertion of the cDNA into the genomic DNA. The other pseudogene includes an intervening sequence but lacks the sequences encoding the 55 N-terminal amino acids of the β-tubulin polypeptide.

Three of the four β-tubulin genes in the chicken each contain three small intervening sequences clustered in the 5′ portion of the coding region.[141] Four distinct mRNAs carrying β-tubulin sequences were detected in chicken tissues. Three of these β-tubulin mRNAs were extraordinarily long (nearly three times the number of base pairs necessary to encode β-tubulin), but, when used to program a cell-free protein synthesis system, they were translated into authentic β-tubulin.[141]

6.2. *Regulation of Synthesis of Microtubule Proteins*

6.2.1. *Tubulin Synthesis in the Developing Nervous System*

The amount of tubulin synthesized in neuronal cells is clearly regulated in the developing nervous system.[8,35,62,75,135,202] When retinal photoreceptors of rats[188] or cats[51,52,179] are exposed to light during the course of development, the synthesis of tubulin is specifically enhanced in visual cortical neurons, as compared to motor cortical neurons of the same animals or to visual cortical neurons of dark-reared animals. To the extent that tubulin microheterogeneity reflects selective expression of individual tubulin genes, it would appear that at least some of the tubulin genes are independently regulated. The number of tubulin isomorphs detected in differentiating rat brain increases from five to six isomorphic forms in prenatal brain to nine forms during the postnatal period of brain maturation.[53,81,91]

Developmental variation in microheterogeneity and in relative synthesis rates for individual tubulin isomorphs has also been reported in studies of differentiating chick brain.[26,230,241] Thus, individual neuronal cells apparently have machinery to regulate tubulin synthesis in response to "private" signals, i.e., to fluctuation in the amount of tubulin dimer or monomer in the cell (see Section 6.2.2), or in response to "public" signals, i.e., to extracellular factors in differentiation.

6.2.2. *Control of Levels of Tubulin mRNA*

The availability of tubulin probes has allowed a reexamination of the interesting observation that tubulin synthesis is specifically inhibited when microtubules are depolymerized with colchicine, with a concomitant decrease in translatable tubulin mRNAs in treated cells.[9] *Vinca* alkaloids, which induce the formation of tubulin aggregates in cells, do not depress tubulin synthesis. These experiments have now been repeated using the chicken probes to identify α- and β-tubulin mRNAs. The transcription of the tubulin genes is clearly subject to negative feedback regulation based on the level of unpolymerized tubulin in the cell.[43,44] Raising the intracellular concentration of tubulin dimer by microinjection also is effective in shutting off tubulin synthesis.[44a] The regulatory step is apparently posttranscriptional but has not been identified.[44,44a]

6.2.3. Regulation of Microtubule Assembly by Regulation of Tau

A fetal form of tau protein has been described.[146,169] After 1 week, postnatal rats normally turn off production of this fetal tau and, by the end of the second week, produce the adult tau proteins.[169] This switch is apparently controlled by thyroid hormone.[169] Since the fetal form of tau is not effective in supporting microtubule assembly,[76,146,169] Nunez[169] has suggested that thyroid hormones control microtubule assembly during brain development by regulating the types of MAPs being made. It is not clear at this time whether the fetal and adult forms of tau represent different modifications of the same translation product or whether they are the products of different genes. Indeed, the fetal form of tau may simply be a fortuitous contaminant of MAP preparations that copurifies with the MAPs. Nevertheless, these studies indicate that the adult forms of tau do not appear in the rat brain until the second postnatal week and that the appearance of adult tau is blocked in hypothyroid animals but present both in thyroidectomized animals that receive thyroid hormone and in normal controls.

7. BIOSYNTHESIS AND PROCESSING OF MICROTUBULE PROTEINS

7.1. Free versus Membrane-Bound Polyribosomes

Although much of our knowledge of tubulin heterogeneity and processing, as well as of the heterogeneity of tubulin mRNAs, follows from studies of cell-free tubulin synthesis, there is no evidence for control of tubulin gene expression at the translational level. Where, in the neuronal cytoplasm, tubulin is synthesized and how nascent tubulin polypeptides are processed as they are released from the ribosomes on which they are synthesized are basic questions in understanding tubulin dynamics. In several laboratories that have applied "state-of-the-art" techniques to the separation of membrane-bound and free polyribosomes, a population of membrane-bound polyribosomes was prepared that synthesized both α and β tubulins.[41,79,197,198,220,221] In all of these studies, the amount of tubulin synthesized on membrane-bound polysomes represented only a small percentage of total tubulin synthesis. Thus, it is clear that most tubulin is synthesized on free polyribosomes in the nervous system, but a small portion of total brain tubulin is synthesized in the rough endoplasmic reticulum.[41]

7.2. Membrane-Associated Tubulin

Tubulin synthesized on membrane-bound polysomes is incorporated into the endoplasmic reticulum membrane by a mechanism closely associated with synthesis.[220,221] Several tubulin isomorphs are found as intrinsic constituents of smooth endoplasmic reticulum and plasma membrane fractions from brain.[228] More α tubulin than β tubulin is synthesized on membrane-bound polysomes, and this α tubulin differs somewhat from other α tubulins on one-dimensional peptide maps.[88] There is a preferential incorporation of α tubulin into the microsomal membranes: the membrane-associated α tubulin is protected from protease treatment[220,221] and is difficult to extract with detergents.[93,220,275] As would be expected for proteins synthesized on membrane-bound polyribosomes, inserted into the rough endoplasmic reticulum membrane as soon as they were synthesized, and immediately exported into the smooth endoplasmic reticulum, tubulins were not detectable among the Coomassie-blue-stained proteins of rough endoplasmic reticulum fractions. Tubulins are only demonstrable among membrane proteins of the rough ER when they are labeled by incorporation of radioactive amino acids during biosynthesis.[220,221]

Although there is considerable evidence for membrane-associated tubulins in nerve,[15,18,69,72,93,121,128,133,167,220,221,228,242,243,275] careful studies of axoplasmic transport have failed to demonstrate cotransport of tubulin with membranes[17,90,236] that usually move with one of the fast components of axoplasmic transport.[142] It is possible that at a given time, the amount of membrane tubulin being transported through any analytical window (i.e., piece of nerve being analyzed) is not sufficient to show up in these analyses. Since the membrane-associated tubulin represents only a small percentage of the total tubulin in nervous tissue, there may not be sufficient radioactive tubulin in the smooth endoplasmic reticulum and vesicles from pulse-labeled axons for detection on gels by fluorography. The insertion of membranes containing tubulin into the plasma membrane of the perikaryon must also be considered. This tubulin might then diffuse in the plasma membrane along with other protein complexes of the membrane.[215]

7.3. Microtubule-Associated Protein Kinases

There are several protein kinase activities associated with microtubules.[89,112,113,132,137,165,184–186,194,211,214,218,224,225,234,238,240] One of these, regulated by cyclic AMP, appears to be associated with HMW2 and uses HMW2 as a substrate.[112,113,214,234,240] A Ca^{2+}-and calmodulin-regulated protein kinase, quite distinct from the HMW2-phosphorylating enzyme, is a constituent of synaptic cytoplasm. This Ca^{2+}–calmodulin protein kinase specifically phosphorylates both α and β tubulins in the nerve terminal, including tubulins associated with synaptic vesicles.[29–31]

The cyclic-AMP-regulated HMW2 kinase appears to be a subfraction of the total brain RII kinase activity. The catalytic activity is released from HMW2 by 3′,5′-AMP, but the regulatory protein remains tightly bound to HMW2.[234] Reaction of microtubule proteins with 8-azido-[^{32}P]cyclic AMP, a photoaffinity probe for cyclic-AMP-binding sites, labels a pair of polypeptides.[226] In one analytical system, these 8-N_3-[^{32}P]cyclic-AMP-labeled proteins appear to behave like RII regulatory proteins.[234] However, analysis of the photoaffinity-labeled proteins on two-dimensional gels indicates that the microtubule-associated cyclic-AMP-binding protein is distinct from the RII protein of brain and other tissues.[226]

Phosphorylation of HMW2 may be a mechanism for regulation of the kinetics of microtubule assembly. Addition of ATP to steady-state microtubules increases the rate of treadmilling.[151,152] In microtubules prepared from PC-tubulin and phosphorylated MAPs, the subunit flux is greater than when nonphosphorylated MAPs are used.[112] In the nervous system, the machinery clearly exists for regulation of microtubules by transients in cyclic AMP or Ca^{2+} levels. But the precise mechanism of regulation, as well as the exact steps being regulated, remain to be discovered.

8. *RELATIONSHIPS BETWEEN MICROTUBULE PROTEINS AND OTHER COMPONENTS OF THE NEURONAL CYTOSKELETON*

8.1. *Interactions with Neurofilaments*

Cell-free studies have suggested associations between various microtubule proteins and other cytoskeletal components. Neurofilament proteins copurify with microtubules.[11] Neurofilaments and microtubules will form gels when mixed with ATP.[192] Both HMW and tau have been reported to bind to neurofilament proteins.[139,206,207] The protein kinase associated with MAP preparations will phosphorylate neurofilament proteins.[206]

Bridges between neurofilaments and microtubules have been described in high-voltage electron micrographs,[63] and periodic links between parallel neurofilaments and microtubules have been seen in thin longitudinal sections of olfactory nerve[32] and nerve explants.[166] When these facts are taken together with the evidence for separate axonal domains of microtubules and neurofilaments (see Section 2.1.1c), one must ask whether both pictures reflect different states of the axonal cytoskeleton or whether one is artifactual. The bridges between microtubules and neurofilaments seen in electron micrographs suggest orderly spacing of these axial structures, and the cotransport of neurofilaments with microtubules in SCa of axoplasmic transport[17] suggests active interaction between axonal microtubules and neurofilaments. But the discontinuity of most neurofilaments across nodes of Ranvier[235] as well as both the separation of tubulin transport from neurofilament transport and the isolation of the locus of fast axoplasmic transport from neurofilament-filled regions of axoplasm by IDPN[96,174,175] lead one to question the significance of interaction between neurofilaments and microtubules in axons. Perhaps most interactions between these structures happen at the interfaces between domains and serve to maintain the mosaic organization of axoplasm.

8.2. *Interactions with Actin*

Another type of interaction is to be predicted from cell-free studies with actin and microtubule proteins. HMW2 and tau will both form gels with actin.[97,98,204] The ability of MAPs to form gels with actin is enhanced by cyclic-AMP-stimulated phosphorylation of the MAPs.[204] The demonstration of phosphorylation-stimulated MAP : actin interaction suggests interesting possibilities for microtubule–microfilament interaction in cells, but there is as yet no good evidence that such interactions take place *in vivo*.

9. *CLOSING REMARKS*

A complex system such as that described in this chapter provides not only many points for regulation but many points at which something can go wrong. Without functional microtubules an organism cannot develop. But subtle lesions of the microtubule system are possible and appear to underly a variety of pathological disorders.[61,266a] Understanding of the control of the expression of genes for microtubule proteins and of the factors that control microtubule dynamics should provide a key not only to the study of cell function but to understanding fundamental changes underlying a variety of degenerative and other pathological processes, particularly in the nervous system.

The study of microtubules requires a multifaceted approach. In this chapter we have tried to limit our discussion to the nervous system and to specific features of microtubules. The emphasis has been on the organization of microtubules, the regulation of the pool(s) of tubulin available to assemble microtubules, and the process of assembly. We have also tried to relate various aspects of microtubule organization to the organization of neuronal cytoplasm. Very little is known about what microtubules do or how they do whatever it may be that they do. Yet it is clear that the structure and function of the cells of the nervous system depend on the disposition and function of microtubules. Understanding either requires an understanding of both.

ACKNOWLEDGMENTS. We thank Dr. Julia Currie for many valuable comments and suggestions during the preparation of this chapter; Dr. Henryk M. Wisniewski for his continuing interest and support and for his comments on the manuscript; Richard Weed, Lucille Donadio, and Dan Klitnick for their assistance with the figures; and Pat Casiano for her untiring patience in seeing this manuscript from drafts to final copy.

REFERENCES

1. Allen, C., and Borisy, G. G., 1974, *J. Mol. Biol.* **90**:381–402.
2. Amos, L. A., 1977, *J. Cell Biol.* **72**:642–654.

3. Amos, L. A., 1979, *Microtubules* (K. Roberts and J. S. Hyams, eds.), Academic Press, New York, London, pp. 1–64.
4. Amos, L. A., and Klug, A., 1974, *J. Cell Sci.* **14**:523–549.
5. Arai, T., and Kaziro, Y., 1976, *Biochem. Biophys. Res. Commun.* **69**:369–376.
6. Asnes, C. F., and Wilson, L., 1979, *Anal. Biochem.* **98**:64–73.
7. Baitinger, C., Levine, J., Lorenz, T., Simon, C., Skene, P., and Willard, M., 1982 *Axoplasmic Transport* (D. Weiss, ed.), Springer-Verlag, Berlin, Heidelberg, New York, pp. 110–120.
8. Bamburg, J. R., Shooter, E. M., and Wilson, L., 1973, *Biochemistry* **12**:1476–1482.
9. Ben-Ze'ev, A., Farmer, S. R., and Penman, S., 1979, *Cell* **17**:319–325.
10. Bergen, L. G., and Borisy, G. G., 1980, *J. Cell Biol.* **84**:141–150.
11. Berkowitz, S. A., Katagiri, J., Binder, H. K., and Williams, R. C., Jr., 1977, *Biochemistry* **16**:5610–5617.
12. Berlin, R. D., Pfeiffer, J. R., and Regula, C. S., 1982, *J. Cell Biol.* **95**:354a.
13. Berne, B. J., 1974, *J. Mol. Biol.* **89**:755–758.
14. Berthold, C., 1982, *Axoplasmic Transport* (D. Weiss, ed.), Springer-Verlag, Berlin, Heidelberg, New York, pp. 40–54.
15. Bhattacharyya, B., and Wolff, J., 1975, *J. Biol. Chem.* **250**:7639–7646.
16. Binder, L. I., Payne, M. R., Kim, H., Sheridan, V. R., Schroeder, D. K., Walker, C. C., and Rebhun, L. I., 1982, *J. Cell Biol.* **95**:349a.
17. Black, M. M., and Lasek, R. J., 1980, *J. Cell Biol.* **86**:616–623.
18. Blitz, A. L., and Fine, R. E., 1974, *Proc. Natl. Acad. Sci. U.S.A.* **71**:4472–4476.
19. Borisy, G. G., Marcum, J. M., Olmsted, J. B., Murphy, D. B., and Johnson, K. A., 1975, *Ann. N.Y. Acad. Sci.* **253**:107–132.
20. Borisy, G. G., and Olmsted, J. B., 1972, *Science* **177**:1196–1197.
21. Borisy, G. G., and Taylor, E. W., 1967, *J. Cell Biol.* **34**:525–533.
22. Brady, S. T., and Lasek, R. J., 1982, *Axoplasmic Transport* (D. Weiss, ed.), Springer-Verlag, Berlin, Heidelberg, New York, pp. 206–217.
23. Brady, S. T., and Lasek, R. J., 1982, *Methods Cell Biol.* **25**:365–398.
24. Bray, D., and Bunge, M. B., 1981, *J. Neurocytol.* **10**:589–605.
25. Bryan, J., 1976, *J. Cell Biol.* **71**:749–767.
26. Bryan, R. N., Bossinger, J., and Hayashi, M., 1981, *Dev. Biol.* **81**:349–355.
27. Bryan, R. N., Cutter, G. A., and Hayashi, M., 1978, *Nature* **272**:81–83.
28. Burgoyne, R. D., and Cumming, R., 1982, *FEBS Lett.* **146**:273–277.
29. Burke, B. E., and DeLorenzo, R. J., 1981, *Proc. Natl. Acad. Sci. U.S.A.* **78**:991–995.
30. Burke, B. E., and DeLorenzo, R. J., 1982, *J. Neurochem.* **38**:1205–1218.
31. Burke, B. E., and DeLorenzo, R. J., 1982, *Brain Res.* **236**:393–415.
32. Burton, P. R., and Paige, J. L., 1981, *Proc. Natl. Acad. Sci. U.S.A.* **78**:3269–3273.
33. Caplow, M., and Zeeberg, B., 1981, *J. Biol. Chem.* **256**:5608–5611.
33a. Carlier, M. F., and Pantaloni, D., 1978, *Biochemistry* **17**:1908–1915.
34. Chalfie, M., and Thomson, J. N., 1979, *J. Cell Biol.* **82**:278–289.
35. Chaudhury, S., Chaudhury, L., and Sarkar, P. K., 1982, *Dev. Brain Res.* **4**:241–243.
36. Chou, S. M., and Hartman, H. A., 1964, *Acta Neuropathol.* **3**:428–450.
37. Chou, S. M., and Hartman, H. A., 1965, *Acta Neuropathol.* **4**:590–603.
38. Clark, W. A., Griffin, J. W., and Price, D. L., 1980, *J. Neuropathol. Exp. Neurol.* **39**:42–55.
38a. Cleveland, D. W., 1982, *Cell* **28**:689–691.
39. Cleveland, D. W., Hughes, S. H., Stubblefield, E., Kirschner, M. W., and Varmus, H. E., 1981, *J. Biol. Chem.* **256**:3130–3134.
40. Cleveland, D. W., Hwo, S.-Y., and Kirschner, M. W., 1977, *J. Mol. Biol.* **116**:207–225.
41. Cleveland, D. W., Kirschner, M. W., and Cowan, N. J., 1978, *Cell* **15**:1021–1031.
42. Cleveland, D. W., Lopata, M. A., MacDonald, R. J., Cowan, N. J., Rutter, W. J., and Kirschner, M. W., 1980, *Cell* **20**:95–105.
43. Cleveland, D. W., and Kirschner, M. W., 1982, *Cold Spring Harbor Symp. Quant. Biol.* **46**:171–183.
44. Cleveland, D. W., Lopata, M. A., Sherline, P., and Kirschner, M. W., 1981, *Cell* **25**:537–546.

44a. Cleveland, D. W., Pittenger, M., and Feramisco, J., 1982, *J. Cell Biol.* **95**:335a.
45. Cold Spring Harbor Laboratory, 1982, *Cold Spring Harbor Symp. Quant. Biol.* **46**:.
46. Connolly, J. A., Kalnins, V. I., Cleveland, D. W., and Kirschner, M. W., 1977, *Proc. Natl. Acad. Sci. U.S.A.* **74**:2437–2440.
47. Connolly, J. A., Kalnins, V. I., Cleveland, D. W., and Kirschner, M. W., 1978, *J. Cell Biol.* **76**:781–786.
48. Cote, R. H,, Bergen, L. G., and Borisy, G. G., 1980, *Microtubules and Microtubule Inhibitors, 1980* (M. DeBrabander, and J. DeMey, eds.), Elsevier/North-Holland, Amsterdam, pp. 325–338.
49. Cote, R. H., and Borisy G. G., 1981, *J. Mol. Biol.* **150**:577–602.
50. Crepeau, R. H., McEwen, B., and Edelstein, S. J., 1978, *Proc. Natl. Acad. Sci. U.S.A.* **75**:5006–5010.
51. Cronly-Dillon, J. R., and Perry, G. W., 1976, *Nature* **261**:5561–5563.
52. Cronly-Dillon, J. R., and Perry, G. W., 1979, *J. Physiol. (Lond.)* **287**:26p–27p.
53. Dahl, J. L., and Weibel, V. J., 1979, *Biochem. Biophys. Res. Commun.* **86**:822–828.
54. Delacourte, A., Filliatreau, G., Boutteau, F., Biserte, G., and Schrevel, J., 1980, *Biochem. J.* **191**:543–546.
55. Delacourte, A., Plancot, M. T., Boutteau, F., Han, K.-K., Hildebrand, H. F., and Biserte, G., 1977, *Biochimie* **59**:479–486.
56. Dentler, W. L., Granett, S., and Rosenbaum, J. L., 1975, *J. Cell Biol.* **65**:237–241.
57. Dentler, W. L., Granett, S., Whitman, G. B., and Rosenbaum, J. L., 1974, *Proc. Natl. Acad. Sci. U.S.A.* **71**:1710–1714.
58. Detrich, H. W. III, and Williams, R. C., Jr., 1978, *Biochemistry* **17**:3900–3907.
59. Detrich, H. W. III, Williams, R. C., Jr., and Wilson, L., 1982, *Biochemistry* **21**:2392–2400.
60. Doenges, K. H., Biedert, S., and Paweletz, N., 1976, *Biochemistry* **15**:2995–2999.
61. Dustin, P., 1978, *Microtubules*, Springer-Verlag, Berlin, Heidelberg, New York.
62. Dutton, G. R., and Barondes, S., 1969, *Science* **166**:1637–1638.
63. Elisman, M. H., and Porter, K. R., 1980, *J. Cell Biol.* **87**:464–479.
64. Engelborghs, Y., DeMaeyer, L. C. M., and Overbergh, N., 1977, *FEBS Lett.* **80**:81–85.
65. Erickson, H. P., 1974, *J. Cell Biol.* **60**:153–167.
66. Erickson, H. P., 1974, *J. Supramol. Struct.* **2**:393–411.
67. Erickson, H. P., 1975, *Ann. N.Y. Acad. Sci.* **253**:51–52.
68. Erickson, H. P., 1975, *Ann. N.Y. Acad. Sci.* **253**:60–77.
69. Estridge, M., 1977, *Nature* **268**:60–63.
70. Farrell, K. W., and Jordan, M. A., 1982, *J. Biol. Chem.* **257**:3131–3138.
71. Farrell, K. W., Kassis, J. A., and Wilson, L., 1979, *Biochemistry* **18**:2642–2647.
72. Feit, H., and Barondes, S., 1970, *J. Neurochem.* **17**:1355–1364.
73. Feit, H., Neudeck, U., and Baskin, F., 1977, *J. Neurochem.* **28**:697–706.
74. Fellous, A., Francon, J., Lennon, A. M., and Nunez, J. 1977, *Eur. J. Biochem.* **78**:167–174.
75. Fellous, A., Francon, J., Virion, A., and Nunez, J., 1975, *FEBS Lett.* **57**:5–8.
76. Fellous, A., Lennon, A. M., Francon, J., and Nunez, J., 1979, *Eur. J. Biochem.* **101**:365–376.
77. Filliatreau, G., and DiGiamberardino, L., 1981, *Biol. Cell* **42**:69–72.
78. Filner, P., and Behnke, O., 1973, *J. Cell Biol.* **59**:99a.
79. Floor, E. R., Gilbert, J. M., and Nowack, T. S., Jr., 1976, *Biochim. Biophys. Acta* **442**:285–297.
80. Forer, A., and Zimmerman, A. M., 1975, *Ann. N.Y. Acad. Sci.* **253**:378–382.
81. Forgue, S. T., and Dahl, J. L., 1979, *J. Neurochem.* **32**:1015–1025.
82. Frederiksen, D. W., and Cunningham, L. W., 1982, *Methods in Enzymology*, Volume 85, Academic Press, New York, London.
83. Frigon, R. P., and Timasheff, S. N., 1975, *Biochemistry* **14**:4559–4566.
84. Fujiwara, K., and Tilney, L. G., 1975, *Ann. N.Y. Acad. Sci.* **253**:27–50.
85. Gaskin, F., Cantor, C. R., and Shelanski, M. L., 1974, *J. Mol. Biol.* **89**:737–755.
86. Geisler, N., and Weber, K., 1981, *J. Mol. Biol.* **151**:565–571.
87. George, H. J., Misra, L., Field, D. J., and Lee, J. C., 1981, *Biochemistry* **20**:2402–2409.
88. Gilbert, J. M., Strocchi, P., Brown, B. A., and Marotta, C. A., 1981, *J. Neurochem.* **36**:839–846.

89. Goodman, D. B. P., Rasmussen, H., DiBella, F., and Guthrow, C. E., 1970, *Proc. Natl. Acad. Sci. U.S.A.* **67**:652–659.
90. Goodrum, J. F., and Morell, P., 1982, *J. Neurochem.* **39**:443–451.
91. Gozes, I., deBaetselier, A., and Littauer, U.Z., 1980, *Eur. J. Biochem.* **103**:13–20.
92. Gozes, I., and Littauer, U. Z., 1978, *Nature* **276**:411–413.
93. Gozes, I., and Littauer, U. Z., 1979, *FEBS Lett.* **99**:86–90.
94. Gozes, I., and Sweadner, K. J., 1981, *Nature* **294**:477–480.
95. Gray, E. G., 1975, *Proc. R. Soc. (Lond.)* [*Biol.*] **190**:369–372.
96. Griffin, J. W., Hoffman, P. N., and Price, D. L., 1982, *Axoplasmic Transport in Physiology and Pathology* (D. G. Weiss and A. Gorio, eds.), Springer-Verlag, Berlin, Heidelberg, New York, pp. 109–118.
97. Griffith, L. M., and Pollard, T. D., 1978, *J. Cell Biol.* **78**:958–965.
98. Griffith, L. M., and Pollard, T. D., 1982, *J. Biol. Chem.* **257**:9143–9151.
99. Grisham, L., 1976, Ph.D. Thesis, Stanford University, Stanford, California.
100. Gross, G. W., and Weiss, D. G., 1982, *Axoplasmic Transport* (D. G. Weiss, ed.), Springer-Verlag, Berlin, Heidelberg, New York, pp. 330–341.
101. Haga, T., and Kurokawa, M., 1975, *Biochim. Biophys. Acta* **392**:335–345.
102. Heidemann, S. R., 1980, *Microtubules and Microtubule Inhibitors, 1980* (M. DeBrabander and J. DeMey, eds.), Elsevier/North-Holland, Amsterdam, pp. 341–355.
103. Herzog, W., and Weber, K., 1978, *Eur. J. Biochem.* **92**:1–8.
104. Hill, T., 1981, *Proc. Natl. Acad. Sci. U.S.A.* **78**:5614–5617.
105. Hill, T., and Kirschner, M. W., 1982, *Proc. Natl. Acad. Sci. U.S.A.* **79**:490–494.
106. Hillman, D. E., and Chen, S., 1982, *Proc. Soc. Neurosci.* **8**:787.
107. Himes, R. H., Burton, P. R., and Gaito, J. M., 1977, *J. Biol. Chem.* **252**:6222–6228.
108. Himes, R. H., Burton, P. R., Kersey, R. N., and Pierson, G. B., 1976, *Proc. Natl. Acad. Sci. U.S.A.* **73**:4397–4399.
109. Ishikawa, H., and Tsukita, S., 1982, *Axoplasmic Transport* (D. G. Weiss, ed.), Springer-Verlag, Berlin, Heidelberg, New York, pp. 251–259.
110. Izant, J. G., and McIntosh, J. R., 1980, *Proc. Natl. Acad. Sci. U.S.A.* **77**:4741–4745.
111. Jacobs, M., Smith, H., and Taylor, E. W., 1974, *J. Mol. Biol.* **89**:455–468.
112. Jameson, L., and Caplow, M., 1981, *Proc. Natl. Acad. Sci.* **78**:3413–3417.
113. Jameson, L., Frey, T., Zeeberg, B., Dalldorf, F., and Caplow, M., 1980, *Biochemistry* **19**:2472–2479.
114. Job, D., Fischer, E. H., and Margolis, R. L., 1981, *Proc. Natl. Acad. Sci. U.S.A.* **78**:4679–4682.
115. Job, D., Rauch, C. T., Fischer, E. H., and Margolis, R. L., 1982, *Biochemistry* **21**:509–515.
116. Johnson, K. A., and Borisy, G. G., 1977, *J. Mol. Biol.* **117**:1–31.
117. Karr, T. L., Kristofferson, D., and Purich, D. L., 1980, *J. Biol. Chem.* **255**:8560–8566.
118. Karr, T. L., Podrasky, A. E., and Purich, D. L., 1979, *Proc. Natl. Acad. Sci. U.S.A.* **76**:5475–5479.
119. Karr, T. L., White, H. D., Coughlin, B. A., and Purich, D. L., 1982, *Methods Cell Biol.* **24**:51–60.
120. Keates, R. A. B., and Hall, R. H., 1975, *Nature* **257**:419.
121. Kelly, P. T., and Cotman, C. W., 1978, *J. Cell Biol.* **79**:173–183.
122. Kim, H., Binder, L. I., and Rosenbaum, J. L., 1979, *J. Cell Biol.* **80**:266–276.
123. Kirschner, M. W., 1980, *J. Cell Biol.* **86**:330–334.
124. Kirschner, M. W., Honig, L. S., and Williams, R. C., 1975, *J. Mol. Biol.* **99**:263–276.
125. Kirschner, M. W., Suter, M., Weingarten, M., and Littman, D., 1975, *Ann. N.Y. Acad. Sci.* **253**:90–106.
126. Kirschner, M. W., and Williams, R. C., 1974, *J. Supramol. Struct.* **2**:412–428.
127. Kirschner, M. W., Williams, R. C., Weingarten, M., and Gerhart, J. C., 1974, *Proc. Natl. Acad. Sci. U.S.A.* **71**:1159–1163.
128. Kornguth, S. E., and Sunderland, F., 1975, *Biochim. Biophys. Acta* **393**:100–114.
129. Kreutzberg, G. W., and Gross, G. W., 1977, *Cell Tissue Res.* **181**:443–457.
130. Kristofferson, D., Karr, T. L., and Purich, D. L., 1980, *J. Biol. Chem.* **255**:8567–8572.
131. Kuriyama, R., and Sakai, H., 1974, *J. Biochem.* **75**:463–471.

132. Lagnado, J. R., Lyons, C. A., Weller, M., and Phillipson, O., 1972, *Biochem. J.* **128**:95P.
133. Lagnado, J. R., Lyons, C., and Wickremasinghe, G., 1971, *FEBS Lett.* **15**:254–258.
134. Lasek, R. J., and Brady, S. T., 1982, *Cold Spring Harbor Symp. Quant. Biol.* **46**:113–124.
134a. Lee, J. C., and Timasheff, S. N., 1977, *Biochemistry* **16**:1754–1764.
135. Lennon, A. M., Francon, J., Fellous, A., and Nunez, J., 1980, *J. Neurochem.* **35**:804–813.
136. Leterrier, J.-F., Liem, R. K. H., and Shelanski, M. L., 1981, *J. Cell Biol.* **90**:755–760.
137. Letterier, J.-F., Rappaport, L., and Nunez, J., 1974, *Mol. Cell. Endocrinol.* **1**:65–75.
138. Levine, J., and Willard, M., 1980, *Brain Res.* **194**:137–154.
139. Liem, R. K. H., Keith, C. H., Leterrier, J.-F., Trenkner, E., and Shelanski, M. L., 1982, *Cold Spring Harbor Symp. Quant. Biol.* **46**:341–351.
140. Lockwood, A. H., 1978, *Cell* **13**:613–627.
141. Lopata, M. A., Chow, L. T., and Cleveland, D. W., 1982, *J. Cell Biol.* **95**:335a.
142. Lorenz, T., and Willard, M., 1978, *Proc. Natl. Acad. Sci. U.S.A.* **75**:505–509.
143. Luduena, R. F., Fellous, A., Francon, J., Nunez, J., and McManus, L., 1981, *J. Cell Biol.* **89**:680–683.
144. Luduena, R. F., Shooter, E. M., and Wilson, L., 1977, *J. Biol. Chem.* **252**:7006–7014.
145. Marcum, J. M., and Borisy, G. G., 1978, *J. Biol. Chem.* **253**:2825–2833.
146. Mareck, A., Fellous, A., Francon, J., and Nunez, J., 1980, *Nature* **284**:353–355.
147. Margolis, R. L., and Rauch, C. T., 1981, *Biochemistry* **20**:4451–4458.
148. Margolis, R. L., and Rauch, C. T., 1982, *J. Cell Biol.* **95**:347a.
149. Margolis, R. L., and Wilson, L., 1977, *Proc. Natl. Acad. Sci. U.S.A.* **74**:3466–3470.
150. Margolis, R. L., and Wilson, L., 1978, *Cell* **13**:1–8.
151. Margolis, R. L., and Wilson, L., 1979, *Cell* **18**:673–679.
152. Margolis, R. L., and Wilson, L., 1981, *Nature* **293**:705–711.
153. Marotta, C. A., Harris, J. L., and Gilbert, J. M., 1978, *J. Neurochem.* **30**:1431–1440.
154. Marotta, C. A., Strocchi, P., and Gilbert, J. M., 1979, *Brain Res.* **167**:93–106.
155. Marotta, C. A., Strocchi, P., and Gilbert, J. M., 1979, *J. Neurochem.* **33**:231–246.
156. Matus, A., Bernhardt, R., and Hugh-Jones, T., 1981, *Proc. Natl. Acad. Sci. U.S.A.* **78**:3010–3014.
157. McEwen, B., and Edelstein, S. J., 1980, *J. Mol. Biol.* **139**:123–145.
158. McIntosh, J. R., 1974, *J. Cell Biol.* **61**:166–187.
158a. McIntosh, J. R., Euteneuer, U., and Neighbors, B., 1980, *Microtubules and Microtubule Inhibitors, 1980* (M. De Brabander, and J. De Mey, eds.), Elsevier/North Holland, Amsterdam, pp. 357–371.
159. Morrison, M. R., Pardue, S., and Griffin, W. S. T., 1981, *J. Biol. Chem.* **256**:3550–3556.
160. Murphy, D. B., 1982, *Methods Cell Biol.* **24**:31–49.
161. Murphy, D. B., and Borisy, G. G., 1975, *Proc. Natl. Acad. Sci. U.S.A.* **72**:2696–2700.
162. Murphy, D. B., and Hiebsch, R. R., 1979, *Anal. Biochem.* **96**:225–235.
163. Murphy, D. B., Johnson, K. A., and Borisy, G. G., 1977, *J. Mol. Biol.* **117**:33–52.
164. Murphy, D. B., Vallee, R. B., and Borisy, G. G., 1977, *Biochemistry* **16**:2598–2605.
165. Murray, A. W., and Froscio, M., 1971, *Biochem. Biophys. Res. Commun.* **44**:1089–1095.
166. Nagele, R., and Roisen, F., 1982, *Brain Res.* **253**:31–37.
167. Nath, J., and Flavin, M., 1978, *FEBS Lett.* **95**:335–338.
168. Nelles, L. P., and Bamburg, J. R., 1979, *J. Neurochem.* **32**:477–489.
169. Nunez, J., Francon, J., Lennon, A., Fellous, A., and Mareck, A., 1980, *Microtubules and Microtubule Inhibitors, 1980* (M. DeBrabander and J. DeMey, eds.), Elsevier/North Holland, Amsterdam, pp. 213–225.
170. Olmsted, J. B., and Borisy, G. G., 1973, *Biochemistry* **12**:4282–4289.
171. Olmsted, J. B., Marcum, J. M., Johnson, K. A., Allen, C., and Borisy, G. G., 1974, *J. Supramol. Struct.* **2**:429–450.
172. Oosawa, F., and Kasai, M., 1962, *J. Mol. Biol.* **4**:10–21.
173. Osborn, M., Webster, R. E., and Weber, K., 1978, *J. Cell Biol.* **77**:R27–R34.
174. Papasozomenos, S. C., Autilio-Gambetti, L., and Gambetti, P., 1981, *J. Cell Biol.* **91**:866–871.
175. Papasozomenos, S. C., Autilio-Gambetti, L., and Gambetti, P., 1982, *Axoplasmic Transport* (D. Weiss, ed.), Springer-Verlag, Berlin, Heidelberg, New York, pp. 241–250.

176. Papasozomenos, S. C., Binder, L. I., Bender, P., and Payne, M. R., 1982, *J. Cell Biol.* **95:**341a.
177. Pardee, J. D., Simpson, P. A., Stryer, L., and Spudich, J. A., 1982, *J. Cell Biol.* **94:**316–324.
178. Penningroth, S. M., and Kirschner, M. W., 1978, *Biochemistry* **17:**734–740.
179. Perry, G. W., and Cronly-Dillon, J. R., 1978, *Brain Res.* **142:**374–378.
180. Peters, A., Palay, S. L., and Webster, H. de F., 1976, *The Fine Structure Of The Nervous System*, W. B. Saunders, Philadelphia.
181. Pierson, G. B., Burton, P. R., and Himes, R. H., 1978, *J. Cell Biol.* **76:**223–228.
181a. Pollard, T. D., and Mooseker, M., 1981, *J. Cell Biol.* **88:**654–659.
182. Purich, D. L., and MacNeal, R. K., 1978, *FEBS Lett.* **96:**83–86.
183. Raine, C. S., Ghetti, B., and Shelanski, M., 1971, *Brain Res.* **34:**389–393.
184. Rappaport, L., Letterier, J.-F., and Nunez, J., 1972, *FEBS Lett.* **26:**349–352.
185. Rappaport, L., Letterier, J.-F., Virion, A., and Nunez, J., 1976, *Eur. J. Biochem.* **62:**539–549.
186. Reddington, M., and Lagnado, J. R., 1973, *FEBS Lett.* **30:**188–194.
187. Roberts, K., and Hyams, J. S. (eds.), 1979, *Microtubules*, Academic Press, London, New York.
187a. Robinson, J., and Engelborghs, Y., 1982, *J. Biol. Chem.* **257:**5367–5371.
188. Rose, S. P. R., Sinha, A. K., and Jones-Lecointe, A., 1976, *FEBS Lett.* **65:**135–139.
189. Rosenbaum, J. L., Binder, L. I., Granett, S., Dentler, W. L., Snell, W., Sloboda, R., and Haimo, L., 1975, *Ann. N.Y. Acad. Sci.* **253:**147–177.
190. Runge, M. S., Detrich, H. W. III, and Williams, R. C. Jr., 1979, *Biochemistry* **18:**1689–1698.
191. Runge, M. S., El-Maghrabi, M. R., Claus, T. H., Pilkis, S. J., and Williams, R. C. Jr., 1981, *Biochemistry* **20:**175–180.
192. Runge, M. S., and Williams, R. C., Jr., 1982, *Cold Spring Harbor Symp. Quant. Biol.* **46:**483–493.
193. Saborio, J. L., Palmer, E., and Meza, I., 1978, *Exp. Cell. Res.* **114:**365–373.
194. Sandoval, I. V., and Cuatrecasas, P., 1976, *Biochemistry* **15:**3424–3432.
195. Sandoval, I. V., and Vandekerckhove, J. S., 1981, *J. Biol. Chem.* **256:**8795–8800.
196. Sandoval, I. V., and Weber, K., 1980, *J. Biol. Chem.* **255:**8952–8954.
197. Sato, M., and Takahashi, Y., 1977, *Proc. Jpn. Acad.* **53:**99–102.
198. Sato, M., Yoshida, Y., and Takahashi, Y., 1978, *J. Neurochem.* **31:**1361–1370.
199. Scheele, R. B., Bergen, L. G., and Borisy, G. G., 1982, *J. Mol. Biol.* **154:**485–500.
200. Scheele, R. B., and Borisy, G. G., 1978, *J. Biol. Chem.* **253:**2846–2851.
201. Scheele, R. B., and Borisy, G. G., 1979, *Microtubules* (K. Roberts and J. S. Hyams, eds.), Academic Press, New York, London, pp. 175–254.
202. Schmitt, H., Gozes, I., and Littauer, U. Z., 1977, *Brain Res.* **121:**327–342.
203. Schnapp, B. J., and Reese, T. S., 1982, *J. Cell Biol.* **94:**667–679.
204. Selden, S. C., and Pollard, T. D., 1982, *J. Cell Biol.* **95:**348a.
205. Shelanski, M. L., Gaskin, F., and Cantor, C. R., 1973, *Proc. Natl. Acad. Sci. U.S.A.* **70:**765–768.
206. Shelanski, M. L., Letterier, J.-F., and Liem, R. K. H., 1981, *Neurosci. Res. Prog. Bull.* **19:**32–43.
207. Shelanski, M. L., Liem, R. K. H., Letterier, J.-F., and Kieth, C. H., 1981, *International Cell Biology, 1980–1981* (H. G. Schweiger, ed.), Springer-Verlag, Berlin, Heidelberg, New York, pp. 428–439.
208. Sherline, P., and Schiavone, K., 1977, *Science* **198:**1038–1040.
209. Sherline, P., and Schiavone, K., 1978, *J. Cell Biol.* **77:**R9–R12.
210. Sheterline, P., 1978, *Exp. Cell Res.* **115:**460–464.
211. Sheterline, P., and Schofield, J. G., 1975, *FEBS Lett.* **56:**297–302.
212. Sloboda, R. D., Dentler, W. L., and Rosenbaum, J. L., 1976, *Biochemistry* **15:**4497–4504.
213. Sloboda, R. D., and Rosenbaum, J. L., 1979, *Biochemistry* **18:**48–55.
214. Sloboda, R. D., Rudolph, S. A., Rosenbaum, J. L., and Greengard, P., 1975, *Proc. Natl. Acad. Sci. U.S.A.* **72:**177–181.
215. Small, R., Blank, M., and Pfenninger, K. H., 1982, *J. Cell Biol.* **95:**249a.
216. Smith, D. S., Jarlfors, U., and Cameron, B. F., 1975, *Ann. N.Y. Acad. Sci.* **253:**472–506.
217. Soifer, D. (ed.), 1975, The Biology of Cytoplasmic Microtubules, New York Academy of Sciences, New York.

218. Soifer, D., 1975, *J. Neurochem.* **24**:21–33.
219. Soifer, D., 1982, *Axoplasmic Transport* (D. G. Weiss, ed.), Springer-Verlag, Berlin, Heidelberg, New York, pp. 81–90.
220. Soifer, D., and Czosnek, H. H., 1980, *J. Neurochem.* **35**:1128–1136.
221. Soifer, D., and Czosnek, H. H., 1980, *Microtubules and Microtubule Inhibitors, 1980* (M. DeBrabander and J. DeMey, eds.), Elsevier/North Holland, Amsterdam, pp. 429–447.
222. Soifer, D., Czosnek, H. H., Mack, K., and Wisniewski, H. M., 1982, *Axoplasmic Transport*, (D. G. Weiss, ed.), Springer-Verlag, Berlin, Heidelberg, New York, pp. 64–72.
223. Soifer, D., Iqbal, K., Czosnek, H. H., DeMartini, J., Sturman, J., and Wisniewski, H. M., 1981, *J. Neurosci.* **1**:461–470.
224. Soifer, D., Laszlo, A. H., and Scotto, J. M., 1972, *Biochim. Biophys. Acta* **271**:182–192.
225. Soifer, D., Laszlo, A., Mack, K., Scotto, J., and Siconolfi, L., 1975, *Ann. N.Y. Acad. Sci.* **253**:598–610.
226. Soifer, D., Mack, K., and Chambers, D. A., 1982, *Arch. Biochem. Biophys.* **219**:388–393.
227. Sternlicht, H., and Ringel, I., 1979, *J. Biol. Chem.* **254**:10540–10550.
228. Strocchi, P., Brown, B. A., Young, J. D., Bonventre, J. A., and Gilbert, J. M., 1981, *J. Neurochem.* **37**:1295–1307.
229. Strocchi, P., Marotta, C. A., Bonventre, J., and Gilbert, J. M., 1981, *Brain Res.* **121**:206–210.
230. Sullivan, K. F., Farrell, K. W., and Wilson, K., 1979, *J. Cell Biol.* **83**:351a.
231. Summers, K., and Kirschner, M. W., 1979, *J. Cell Biol.* **83**:205–217.
232. Sutherland, J. W. H., 1976, *Biochem. Biophys. Res. Commun.* **72**:933–938.
233. Takenaka, T., and Inomata, K., 1981, *J. Neurobiol.* **12**:479–486.
234. Theurkauf, W. E., and Vallee, R. B., 1982, *J. Biol. Chem.* **257**:3284–3290.
235. Tsukita, S., and Ishikawa, H., 1981, *Biomed. Res.* **2**:424–437.
236. Tytell, M., Black, M. M., Garner, J. A., and Lasek, R. J., 1981, *Science* **214**:179–181.
237. Tytell, M., Brady, S. T., and Lasek, R. J., 1984, *Proc. Natl. Acad. Sci. U.S.A.* (in press).
238. Vallee, R. B., 1980, *Proc. Natl. Acad. Sci. U.S.A.* **77**:3206–3210.
239. Vallee, R. B., and Borisy, G. G., 1978, *J. Biol. Chem.* **253**:2834–2845.
240. Vallee, R. B., DiBartolomeis, M. J., and Theurkauf, W. E., 1981, *J. Cell Biol.* **90**:568–576.
241. Von Hungen, K., Chin, R. C., and Baxter, C. F., 1981, *J. Neurochem.* **37**:511–514.
242. Walters, B. B., and Matus, A. I., 1975, *Biochem. Soc. Trans.* **3**:109–112.
243. Wang, Y. J., and Mahler, H. R., 1976, *J. Cell Biol.* **71**:639–658.
244. Warner, F. D., and Satir, P., 1974, *J. Cell Biol.* **63**:35–63.
245. Webb, B. C., and Wilson, L., 1980, *Biochemistry* **19**:1993–2001.
246. Weber, K., and Osborn, M., 1979, *Microtubules* (K. Roberts and J. S. Hyams, eds.), Academic Press, London, New York, pp. 279–313.
247. Wegner, A., 1976, *J. Mol. Biol.* **108**:139–150.
248. Wegner, A., 1977, *Biophys. Chem.* **7**:51–58.
249. Weingarten, M. D., Lockwood, A. H., Hwo, S.-Y., and Kirschner, M. W., 1975, *Proc. Natl. Acad. Sci. U.S.A.* **72**:1858–1862.
250. Weingarten, M. D., Suter, M. M., Littman, D. R., and Kirschner, M. W., 1974, *Biochemistry* **13**:5529–5537.
251. Weisenberg, R. C., 1974, *J. Supramol. Struct.* **2**:451–465.
252. Weisenberg, R. C., 1980, *J. Mol. Biol.* **139**:660–677.
253. Weisenberg, R. C., 1980, *Microtubules and Microtubule Inhibitors, 1980* (M. DeBrabander and J. DeMey, eds.), Elsevier/North Holland, Amsterdam, pp. 161–172.
254. Weisenberg, R. C., Borisy, G. G., and Taylor, E. W., 1968, *Biochemistry* **7**:4466–4479.
255. Weisenberg, R. C., Deery, W. J., and Dickenson, P. J., 1976, *Biochemistry* **15**:4248–4254.
256. Weisenberg, R. C., and Rosenfield, A., 1975, *Ann. N.Y. Acad. Sci.* **253**:78–89.
257. Wen, G. Y., Soifer, D., and Wisniewski, H. M., 1982, *Anat. Embryol.* **165**:315–328.
258. Wilde, C. D., Chow, L. T., Wefald, E. C., and Cowan, N. J., 1982, *Proc. Natl. Acad. Sci. U.S.A.* **79**:96–100.
259. Wilde, C. D., Crowther, C. E., and Cowan, N. J., 1982, *Science* **217**:549–552.
259a. Wilde, C. D., Crowther, C. E., and Cowan, N. J., 1982, *J. Mol. Biol.* **155**:533–538.
260. Wilde, C. D., Crowther, C. E., Cripe, T. D., Lee, M. G.-S., and Cowan, N. J., 1982, *Nature* **297**:83–84.

261. Willard, M., Cowan, W. M., and Vagelos, P. R., 1974, *Proc. Natl. Acad. Sci. U.S.A.* **71**:2183–2187.
262. Williams, R. C., Jr., and Detrich, H. W. III, 1979, *Biochemistry* **18**:2499–2503.
263. Williams, R. C., Jr., and Lee, J. C., 1982, *Methods Enzymol.* **85**:376–385.
264. Wilson, L. (ed.), 1982, *Methods in Cell Biology*, Volume 24, *The Cytoskeleton*, Academic Press, New York, London.
265. Wilson, L., and Margolis, R. L., 1978, *ICN–UCLA Conference on Cell Reproduction* (E. R. Dirksen, D. M. Prescott, and C. F. Fox, eds.), Academic Press, New York, p. 241.
266. Wilson, L., and Margolis, R. L., 1982, *Cold Spring Harbor Symp. Quant. Biol.* **46**:199–205.
266a. Wisniewski, H. M., and Soifer, D., 1979, *Mech. Ageing Dev.* **9**:119–142.
267. Witman, G. B., 1975, *Ann. N.Y. Acad. Sci.* **253**:178–191.
268. Witman, G. B., Cleveland, D. W., Weingarten, M. D., and Kirschner, M. W., 1976, *Proc. Natl. Acad. Sci. U.S.A.* **73**:4070–4074.
269. Wuerker, R. B., and Kirkpatrick, J. B., 1972, *Int. Rev. Cytol.* **33**:45–75.
270. Yokoyama, K., Tsukita, S., Ishikawa, H., and Kurokawa, M., 1980, *Biomed. Res.* **1**:537–547.
271. Zackroff, R. V., Deery, W. J., and Weisenberg, R. C., 1980, *J. Mol. Biol.* **139**:641–659.
272. Zeeberg, B., and Caplow, M., 1981, *J. Biol. Chem.* **256**:12051–12057.
273. Zeeberg, B., Reid, R., and Caplow, M., 1980, *J. Biol. Chem.* **255**:9891–9899.
274. Zingsheim, H.-P., Herzog, W., and Weber, K., 1979, *Eur. J. Cell. Biol.* **19**:175–183.
275. Zisapel, N., Levi, M., and Gozes, I., 1980, *J. Neurochem.* **34**:26–32.

11

Neuronal Intermediate Filaments

Charles A. Marotta

1. INTRODUCTION

Neuronal intermediate filaments, or neurofilaments, are among the major fibrous organelles of neurons. Elaborated within the perikaryon, assembled, and extruded to the axon, neurofilaments, together with microtubules, are essential to the neuronal cytoskeletal architecture. Neurofilaments were first recognized by early 19th century light microscopists who used the term neurofibrils to describe the parallel arrays of these elongated structures. Initial investigations in the field of neurofibrillary research have been elegantly discussed by Parker.[1] The pioneering electron microscopic observations of Schmitt and his associates[2–6] in the middle of this century led to the first descriptions of neurofilament fine structure, which studies have served as a basis for a host of later investigations. Although these ubiquitous neuronal organelles have been recognized and measured for over a century, it was not until recently that biochemical and physicochemical methods were applied. Progress was slow in these areas because of obstacles to obtaining undegraded neurofilament protein sufficiently free of other cellular constituents to make studies at the molecular level meaningful. The major portion of the present chapter focuses on this experimentation, with particular attention given to mammalian neurofilaments. Certain aspects of neurofilament research have been evaluated in other recent reviews.[7,8]

2. NEUROFILAMENTS COMPARED WITH OTHER INTERMEDIATE FILAMENTS

Neurofilaments (NFs) are intracellular, unbranching fibrous organelles of 8–10 nm[5,9–13] that are a subclass of intermediate filaments (IFs)[14]; the latter are fibrous elements of similar morphology occurring in different eucaryotic

Charles A. Marotta • Molecular Neurobiology Laboratory of the Laboratories for Psychiatric Research, Mailman Research Center, McLean Hospital, Belmont, Massachusetts 02178; and Department of Psychiatry, Massachusetts General Hospital and Harvard Medical School, Boston, Massachusetts 02115.

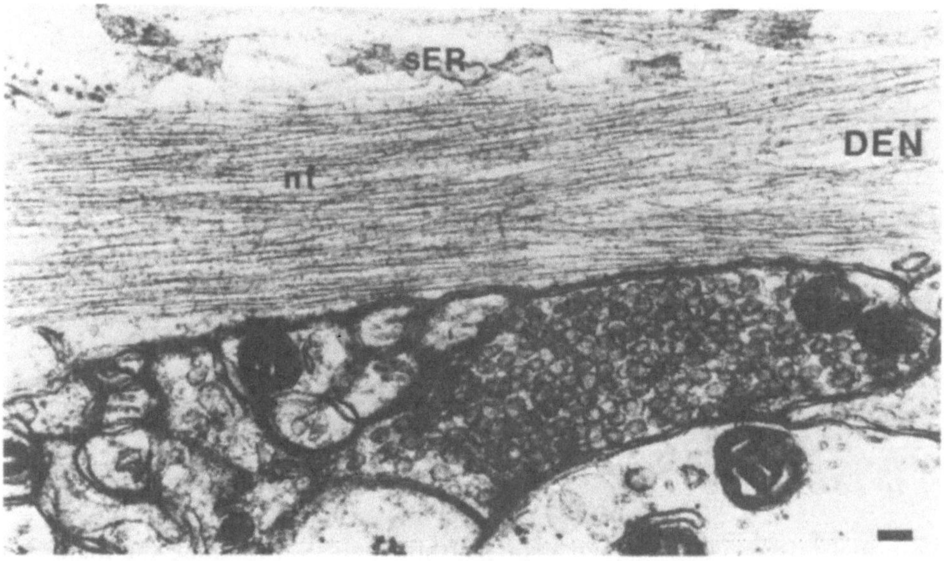

Fig. 1. An electron micrograph from the neuropil of the rat substantia nigra, pars reticulata showing a portion of a longitudinally sectioned dendrite (DEN) containing an extensive array of neurofilaments (nf). The characteristically smooth and regular outlines of these structures can be contrasted with the more varicose appearance of the smooth endoplasmic reticulum (sER). Bar, 0.1 μm. (Courtesy of Peter A. Paskevich.)

cell types. The diameter of the fibers is "intermediate" in size between that of the smaller actin filaments (7 nm) and the larger myosin filaments (15 nm) and microtubules (25 nm).[14,15,17] Biochemical and immunologic studies have established the existence of five classes of IFs that are associated predominantly, although in some cases not exclusively, with unique cell types. The classes are (1) keratin (tono) filaments of epithelial cells; (2) desmin filaments of muscle cells; (3) vimentin filaments of cells of mesenchymal origin; (4) glial filaments of glial cells; and (5) neurofilaments of neurons.[14]

Neurofilaments can be distinguished from other IFs by morphology, distribution, composition, and antigenicity. Unlike other IFs, neurofilaments possess side arms, which may play a role in maintaining the organization of the neuronal cytoskeleton by interacting with microtubules and other axonal elements.[18–20] In some cases, one type of IF is found in more than one cell type. Vimentin has the widest distribution, being found in muscle, glial, epithelial, and neuronal cells.[14,21,23,77] By contrast, neurofilaments have been identified in CNS (Fig. 1) and PNS neurons only and have been immunologically detected early in neuronal development.[24]

A further distinction between neurofilaments and other IFs concerns the subunit protein composition. Vimentin, desmin, and glial filaments are each composed of a single polypeptide of 50,000–52,000 mol. wt., and keratin is composed of a family of polypeptides ranging in size from 40,000 to 65,000 mol. wt.[14] By contrast, the neurofilament proteins (NFPs) are three species of approximately 68,000–70,000, 140,000–160,000, and 200,000–225,000 mol. wt.; the experimental evidence is described below. Although IFs of each class are immunologically distinct,[14] at least one antigenic site appears to be common to all classes.[25]

In subsequent sections the NFPs will be referred to as the triplet proteins after Hoffman and Lasek,[26] and the individual subunits will be abbreviated as 68K, 145K, and 200K. Variations in the assigned molecular weight values reflect the data of the individual reports under discussion.

3. *DISTRIBUTION, MORPHOLOGY, AND FINE STRUCTURE*

By electron microscopy or by immunohistological methods, NFs are easily demonstrated in the myelinated axons of cerebrum, cerebellum, brainstem, spinal cord, and peripheral nerve and to a lesser extent in axons of gray matter of cerebral cortex, cerebellar cortex, spinal cord, and dorsal root ganglia; although less readily detected, NFs are present in neuronal perikarya and dendrites.[8,10,27–34] Neurofilaments have not been found in nonneuronal cells, including glia, fibroblasts, Schwann cells, and connective tissue.[27–29,33]

Neurofilaments are oriented parallel to the long axis of the axon, in which they tend to be organized in clusters, typically in central zones rather than at the periphery.[10,13,35–38] At the level of the electron microscope, the NF appears tubular in cross section, with walls 3 nm thick surrounding an electron-lucent core.[5,10,11,15,39] Compared with other intermediate filaments, neurofilaments appear less tightly organized in the cytoplasm, where they remain 30–40 nm apart; this may in part be because of the spikelike sidearms (2.5–7 nm in length) perpendicular to the filament axis and occurring at 20- to 40-nm intervals.[7,10,15,40] Neurofilaments appear to interconnect with one another, microtubules, and possibly other axonal constituents to form a three-dimensional cytoplasmic network that is considered to provide cytoskeletal support for the neuron. Birefringence studies on squid giant axons and extruded axoplasm[41,42] support the presence of an ordered neuroaxoplasmic lattice. Subsequent electron microscopic examination of squid axonal fibers and of the brain system of the nudibranch *Hermissenda crassicornis* has demonstrated a cytoskeletal network of longitudinally oriented neurofilaments and microfilaments.[40] Together with microtubules, when present, the cytoskeletal elements appeared linked together by a system of thin transverse filamentous bridges (see also Section 5.5). Evidence derived from high-voltage electron microscopy supports the concept of a cytoplasmic ground substance of interconnecting strands of 5–10 nm, which are designated the microtrabecular lattice.[18,43,44] According to this view, the neurofilament–microtubular cytoskeletal network is enmeshed in the finer microtrabecular filaments.[44a]

The organization of the triplet proteins within the neurofilament has been examined by antibody decoration procedures. The interaction between neurofilaments and anti-NFP antibodies can be detected in electron micrographs.[45] As discussed below (Section 6), the three NFPs contain both shared and unique antigenic sites. Using antibodies enriched for unique determinants, Willard and

Simon[46] determined that although the three NFPs were shown to be physically associated with the same NF, the three proteins were not uniformly distributed. The 73K protein was associated with a "central core" of the filament, whereas the 195K species was more peripherally attached to the core and periodically arranged along its axis. The latter complex appeared in some cases to form cross bridges connecting two filaments and in other cases appeared as a helix wrapping the central core. Filaments decorated with anti-73K suggested that the 73K NFP may also be involved in a helical structure. In other experiments, Schlaepfer *et al.*[27] observed that isolated neurofilaments became decorated with a uniform coat of antibodies when exposed to specific antisera for each of the NFPs. Debus *et al.*[30] used a monoclonal antibody to the 200K NFP to examine the distribution of the antigen in cultured dorsal root ganglion cells. Immunoelectron microscopy of neuronal cytoskeletons treated with the antibody showed discontinuous decoration of NFs. By contrast, antibodies to the 145K and 68K proteins gave continuous decoration.

X-ray diffraction analyses of neurofilaments of the marine worm *Myxicola infundibulum* revealed patterns consistent with considerable coiled-coil α-helical configuration.[16] In other studies, α-helix was estimated to account for 40–80% of the amino acid sequence of subunits of nonneuronal IFs.[47] Based on chemical and physical analyses, Steinert and co-workers[47,48] suggested a common structural unit for IFs: three intertwined polypeptide subunits with two discrete coiled-coil α-helical segments, each 18 nm long, interspersed with nonhelical domains of 4 nm. The intermediate filament then would consist of many of these units associated both end to end and side by side. The relevance of the proposed model to NFs, which contain three subunits of different sizes, antigenic determinants, and amino acid composition (see below) has not as yet been tested directly. However, the existence of a monoclonal antibody that reacts with all classes of IFs including neurofilaments[25] suggests the presence of a conserved domain common to all classes of filaments. It was recently demonstrated that isolated 68K protein can, under the appropriate conditions,[49] undergo self-assembly into morphologically normal-appearing intermediate-size filaments (see Section 5.4). When examined by the electron microscope after high-resolution metal shadowing, the filaments exhibited a longitudinal periodicity of about 21 nm, similar to self-assembled desmin and keratin. In a preliminary communication, a periodicity of 21 nm was reported for isolated neurofilaments.[50] The measured distance of 21 nm is consistent with the intermediate filament model of Steinert.[47,48]

4. PURIFICATION

Initial studies on the purification of neuronal intermediate filaments from the CNS employed the axon flotation procedure of Norton and associates[51,52] in which partially purified axons rich in NFs were obtained. Homogenization of CNS white matter in isotonic buffer allowed myelin sheaths to be preserved; on centrifugation in 0.85 M sucrose, the myelinated axons remained as a floating pad, separated from denser cellular constituents. Myelin was removed from axons by homogenization in hypotonic solutions followed by centrifugation. The pellet, containing myelin-free axons, was seen by electron microscopy to contain a mixture of tightly and loosely packed bundles of 8–10 nm, the former resembling glial filaments, and the latter neurofilaments. Subsequent modification of the axon flotation procedure using longer myelin-stripping steps in hypotonic solution yielded filament pellets of uniform morphology containing primarily loose bundles. When the protein composition was examined by SDS–PAGE, a major band of 50,000 mol. wt. was observed and incorrectly identified as neurofilament protein. It is now known that this species was predominantly glial fibrillary acidic protein (GFAP) rather than NFP (for reviews see references 7 and 8).

Studies on the solubilization of peripheral nerve NFs led to the discovery that prolonged exposure of axoplasm to media of low ionic strength allowed the solubilization and subsequent loss of NFs.[7,8,31,54,54a] Modified versions of the axon flotation procedure reduce the length of time that axons are exposed to hypotonic buffer or eliminate this procedure entirely; in the latter case, demyelination can be carried out by a nonionic detergent.[31,53] Norton and Goldman[7] and Shelanski and Liem[8] have reviewed the axon flotation procedure and the various modified versions. Methods for the isolation of CNS neurofilaments often include a calcium-chelating agent in preparative buffers. As discussed in a subsequent section, the use of reagents such as EDTA or EGTA was based on the assumption that calcium-activated proteases known to exist in peripheral nerves[55,56,56a] were also active in the CNS. However, it was not until more recently that this proteolytic activity was demonstrated in mammalian CNS tissue.[57–60]

Modified versions of the axon flotation procedure remain among the simplest and most useful methods to prepare apparently intact neurofilaments containing subunit proteins of approximately 200,000, 145,000, and 70,000 mol. wt., as shown in Fig. 2. The NFPs thus obtained are in reasonably high yield (0.2–0.5 mg of protein per gram of white matter) and purity (23–38% of protein in NF pellets is NFP as determined by densitometric scans of stained gels following SDS–PAGE).[22,53] In modified axon flotation procedures, neurofilaments may be contaminated with lipid, myelin proteins, actin, tubulin, GFAP, and other unidentified polypeptides (see Fig. 2).[22,53,61–64] The more elaborate purification procedures described below can significantly increase the purity of NFPs.

Preparation of neurofilaments by direct extraction of mammalian peripheral nerve was first described by Schlaepfer.[65,66] The absence of GFAP in peripheral nerve avoided confusion of this protein with the neurofilament triplet proteins.[31,66–68] Axoplasmic proteins from peripheral nerves, containing abundant amounts of NFPs, were prepared from desheathed nerves after osmotic shock in hypotonic solutions. After centrifugation of tissue extracts, membranes and particulate constituents were separated from a 68K component.[45] Antibodies raised to this protein reacted with isolated NF structures in rat and human CNS and PNS.[56a] Later versions of the direct extraction procedure applied to peripheral nerve ascertained that the NFPs were composed of 200K, 150K, and 69K species.[66] The similarity between PNS and CNS neurofilament

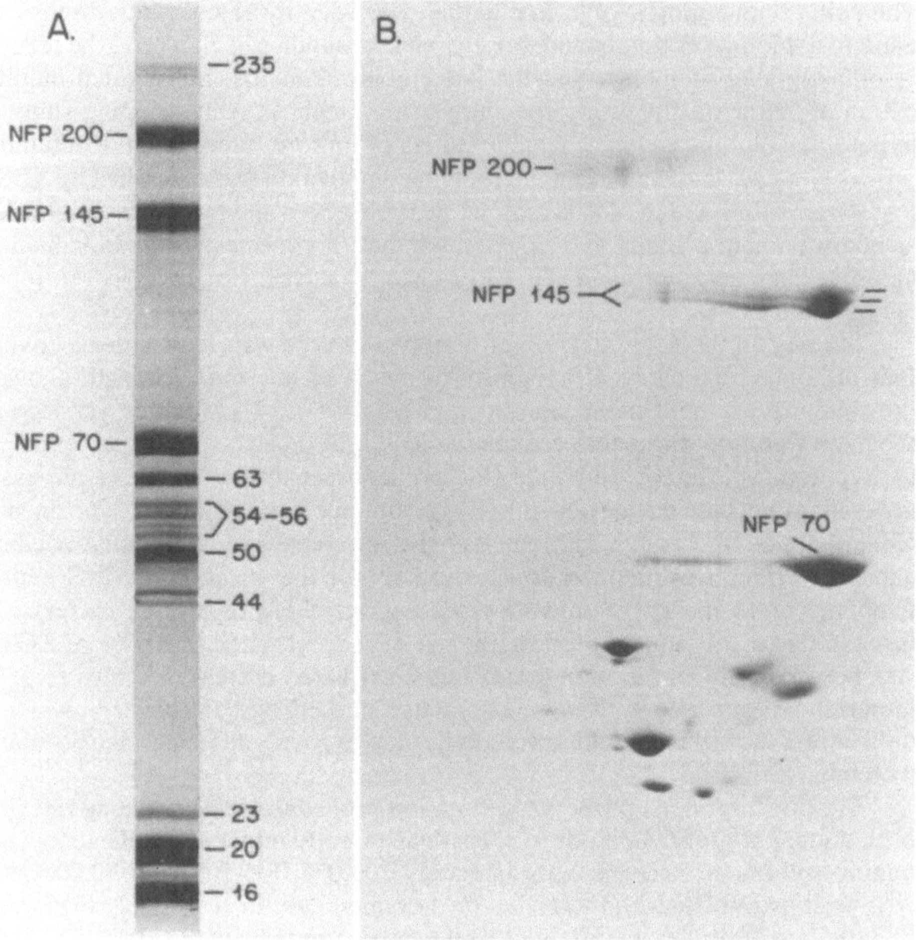

Fig. 2. Polypeptides in NF pellets. A: Polyacrylamide gel electrophoresis (PAGE) of proteins in a NF pellet prepared by a modified axon flotation procedure[53] from frozen rat CNS white matter. Electrophoresis was carried out in a SDS gel[185a] containing a 5–15% acrylamide gradient. The NFPs are indicated at 200K, 145K, and 70K. Tubulin (56K–54K), GFAP (50K), actin (44K), and myelin proteins (23K–16K) are also present. B: Two-dimensional gel[185b] of rat NFPs. Molecular weight markers[53] were electrophoresed in parallel with the NFPs (not shown). The basic end of the gel is on the left.

triplet proteins has now been established by several different investigations.[31,67–69] Biochemical and immunologic studies of peripheral nerve NFs were significant for their contribution to our understanding of the composition of NFPs: these experiments defined some conditions under which NFPs disassemble, in particular, their instability in hypotonic media[54a]; they helped establish that NFP subunits of peripheral nerve could not be represented by a 50K protein; and they provided initial support for the hypothesis that NFPs consist of a triad of proteins as suggested by Hoffman and Lasek.[26] Studies concerning peripheral nerve neurofilaments have recently been reviewed.[70]

Numerous variations of the axon flotation procedure for CNS tissue and the direct extraction procedure for peripheral nerves have been developed.[22,31,53,64,67,68,71] Since the methods often yield a mixture of NFPs and polypeptide contaminants, the partially purified NFPs may not be suitable for certain investigations (e.g., the detailed macromolecular events of filament assembly and disassembly). Recently Liem[72] described a procedure whereby bovine brain NFs can be prepared in high yield and to a high degree of purity. Protein obtained by axon flotation was solubilized in 8 M urea and chromatographed on hydroxylapatite. The eluted proteins were competent to reassemble into filaments, and reassembly served as the final step in the purification. A modification of the direct extraction procedure that is applicable to both peripheral nerve and spinal cord was described by Schecket and Lasek.[73] Nervous tissue extracts were fractionated by ammonium sulfate precipitation and Sepharose 4B chromatography and then precipitated with ethanol. The advantage of this method is that a relatively high yield of quite pure NFs can be obtained from both CNS and PNS using a single procedure.

When only small amounts of tissue are available, or when cultured cells are used, it may be useful to obtain a cytoskeleton preparation using a nonionic detergent.[74–76b] By use of Triton X-100 or NP-40, cell membranes can be solubilized and cytoplasmic contents released. The insoluble portion is composed primarily of filaments and microtubules.[77–80] This method was successfully applied to spinal cord for the rapid preparation of CNS intermediate filaments in high yield.[81] Cytoskeletal preparations from cultured sensory or sympathetic neurons were reported to be rich in NFs.[23]

Microtubules isolated by the *in vitro* assembly–disassembly procedure[83] copurify with other proteins referred to as microtubule-associated proteins or MAPs.[83–86] Neurofilament proteins are among the copurifying species, and this procedure can be used to prepare the triplet proteins in high yield (see Section 5.5).[64,87,88] During the second cycle of assembly–disassembly, MAPs remained as aggregates separable from the disassembled tubulin from which they were separated; the aggregate formed 10-nm fibers that subsequently were identified as neurofilaments.[89,90] Both brain and spinal cord have been used for this preparative procedure. Although the NFPs are obtained in high yield (1 mg protein per gram spinal cord[64]), they remain contaminated with other polypeptides, most notably tubulin.

The giant axons of squid (*Dosidicus gigas* or *Loligo pealii*) and marine worm (*Myxicola infundibulum*) are frequently used to investigate invertebrate neurofilaments since their axons are highly enriched in NF protein[4] (more than 70% of *Myxicola* axoplasmic protein is NFP[91]) and since the axoplasm can be obtained easily and quickly by mechanical extrusion.[16,92] The rapidity of preparation is an important consideration since both squid[93] and *Myxicola*[54] contain calcium-activated protease activity that degrades axoplasmic proteins including NFPs (see Section 7.3).

Invertebrate NFs were purified by sedimentation through a discontinuous sucrose gradient in which they settled at the 1.0–2.5 M interface.[92] Invertebrate NFs were also separated from other axoplasmic constituents by gel filtration using Sephadex G-200 or Sepharose 4B, after which they can be recovered in the void volume.[91,92] Alternative methods for purification of *Myxicola* filaments include precipitation by cytochrome *c* or by repeated cycles of polymerization–depolymerization[91] (see Section 5.4).

5. PROPERTIES OF ISOLATED NEUROFILAMENT PROTEINS

5.1. Evidence That the Triplet Proteins are Constituents of Neurofilaments

Before we consider biochemical and physiological studies on NFPs, we can summarize the evidence that the triplet proteins are constituents of neurofilaments as follows. (1) Hoffman and Lasek[26] first observed the synchronous movement of 200K, 145K, and 68K proteins along with tubulin in the slowest component of axoplasmic transport (0.2–1 mm per day) in rat motor neurons and spinal sensory neurons; the triplet proteins coelectrophoresed with major proteins of sciatic nerve and were deduced to be components of neuronal intermediate filaments. These observations were supported by subsequent reports from other workers.[94,95] (2) The major proteins, in addition to tubulin, that can be extracted from CNS and PNS axons are species with molecular weights similar to those of the triplet proteins of Hoffman and Lasek[26] (see Section 4). (3) Antibodies to each of the purified triplet proteins specifically bind to CNS and PNS neuronal filaments *in situ* (see Section 6). (4) Antibodies to each of the purified triplet proteins bind to individual isolated neurofilaments.[27,46] (5) The presence of neurofilaments as visualized by electron microscopy was correlated with the presence of the triplet proteins.[31,96] (6) The triplet proteins were selectively lost at the same time that neurofilaments underwent granular disintegration in transected nerve.[97,98] (7) The granular disintegration of peripheral nerve neurofilaments by calcium (Section 7.3) was accompanied by the disappearance of the triplet proteins.[55] (8) Isolated triplet proteins undergo self-assembly into filamentous structures *in vitro*.[72,99,100] (9) As neurofilaments accumulate in nerves treated with aluminum[101] or β,β′-iminodiproprionitrile,[102] the triplet proteins also increase.

As a whole, these findings provide evidence that three neuronal proteins of approximately 200,000, 145,000, and 70,000 mol. wt. are major components of neurofilaments. However, these data do not rule out the possibility that other polypeptides, in addition to the triplet, are also intimately associated with NFs; nor do they imply that all neurofilaments of all nerve cells are composed of all three proteins. This point appears a germane consideration with respect to the developing neuron.[108]

5.2. Characteristics of the Isolated Proteins

5.2.1. Stoichiometry

When prepared by a modified axon flotation procedure,[53] the NFP triplet proteins appear on SDS polyacrylamide gels as shown in Fig. 1; the contaminating proteins are identified in the legend to Fig. 1. Stoichiometric determinations were made by densitometric scanning of stained gels; values of 9:2:1[103] to 6:2:1[73] were reported for the ratio of 70K:160K:210K purified by different methods. However, the stoichiometry may vary with the species.[22]

5.2.2. *Molecular Weight*

The most often reported variation concerns molecular weight. This may represent to some extent true interspecies differences. However, the reported variability may be contributed to by the use of different preparative procedures (which may yield fractions containing varying amounts of proteases) and by the method used for molecular weight estimation. Table I lists reported molecular weights for NFPs prepared using the various methods discussed previously. In Table I, the three NFPs are referred to as components I, II, and III, listed in decreasing order of size. It is interesting to note that in at least one cytoskeleton preparation (the method that exposes intracellular structures to the least amount of experimental manipulation), NFPs were obtained with the highest reported molecular weight for component I (footnote *g*, Table I).

Neurofilaments of the commonly studied invertebrates squid and *Myxicola* are composed of subunit proteins that differ in size from one another and from mammalian NFPs. In the early studies of Huneeus and Davison,[6] freeze-dried squid axoplasm was extracted initially with low-ionic-strength buffer followed by 6 M guanidine HCl to yield a major polypeptide of 70,000–74,000 mol. wt. referred to as filarin. The isolated filarin was competent to form long filaments under conditions of reassembly. In later studies, buffers of high ionic strength containing EGTA were used to prepare squid and *Myxicola* NFs.[92] Squid NFs were purified by sucrose gradient centrifugation and gel filtration; in these preparations NFs are composed of 200,000- and 60,000-mol.-wt. species.[92] The same proteins were also noted in NFs polymerized from a 100,000 × *g* supernatant prepared from squid axoplasm.[104] The 200,000- and 60,000-mol.-wt. proteins contained a minor 70,000-mol.-wt. species, which may correspond to filarin. With the same extraction procedure as used for squid, it was determined that the major NFPs of *Myxicola* are 160,000- and 150,000-mol.-wt. proteins.[92] Similar or slightly higher values were obtained by Gilbert and co-workers,[54,91] who prepared *Myxicola* NFs by repeated cycles of polymerization and depolymerization, precipitation with cytochrome *c*, or by chromatography on Sepharose 4B. Values of 172,000 and 155,000 mol. wt. were reported. The relationship of these two polypeptides to each other and to other axoplasmic polypeptides of *Myxicola* is discussed in Section 5.3.

5.2.3. *Heterogeneity*

Neurofilaments can be dissolved in 4–6 M guanidine hydrochloride[6,104a] and dissociated into subunits in high concentration of urea, as in the two-dimensional PAGE procedure.[53,64,67] As shown in Fig. 1B, on two-dimensional gels the NFPs appear as heterogeneous proteins.[53,64,67] The isoelectric points were reported as 5.2, 5.2–5.5, and 5.8–5.9[53] for the 70K, 145K, and 200K

Table 1
Molecular Weights of Purified Vertebrate Neurofilament Proteins

Species	Component I ($\times 10^{-3}$)	Component II ($\times 10^{-3}$)	Component III ($\times 10^{-3}$)	References
Central nervous system				
Bird	205	156	69	190[a]
Bovine	210	155	70	64[b]
	210	160	70	22[b]
	210	160	70	71[c]
	210	160	69	89[d]
	210	160	68	31[b]
	200	160	70	72[e]
Dog	200	160	70	110[a]
Guinea pig	215	145	68	31[b]
	200	145	68	73[f]
Human	200	160	70	63[b]
	210	160	70	22[b]
	210	160	68	21[b]
Mouse	200	140–145	70	53[b]
Rabbit	200	145	68	31[b]
	210	155	70	64[b]
	200	150	68	67[b]
	200	150	68	189[a]
	200	150	70	68[b]
Rat	210	155	70	64[b]
	200	140–145	70	53[b]
	210	160	70	22[b]
	200	145	68	89[d]
	200	150	70	68[b]
	200	150	69	66[a]
	225	160	68	77[g]
	210	160	70	81[g]
Peripheral nervous system				
Guinea Pig	200[h]	145	68	73[f]
Human	210	160	69	67[b]
Rabbit	200	150	68	189[a]
	200	150	70	68[b]
Rat	200	150	70	68[b]
	200	150	69	66[a]
	200	150	69	67[b]

[a] Prepared by a direct extraction method.
[b] Prepared by a modified axon flotation procedure.
[c] Obtained during the assembly–disassembly of microtubules.
[d] Prepared by differential centrifugation followed by gel filtration.
[e] Prepared by a modified axon flotation procedure and hydroxylapatite chromatography followed by reassembly into filaments.
[f] Prepared by ammonium sulfate fractionation followed by gel filtration and ethanol precipitation.
[g] Obtained in a cytoskeletal preparation.
[h] Exists also as a genetic variant in which component I is composed of two polypeptides of 200,000 and 192,000 mol. wt.

species, respectively; a second report[64] estimated the pI values as 5.0–5.1, 5.1–5.2, and 5.4–5.7. In either case, the acidic nature of the proteins was verified. The 200K, the least acidic, often appears as a broad streak[53]; the 145K appears as heterogeneous forms in terms of both molecular weight and isoelectric point[53,59,105,106]; the 70K, the most acidic species, exhibits the least heterogeneity but often appears as a doublet.[53,64,67] The heterogeneous appearance on two-dimensional gels observed for both CNS and PNS neurofilament proteins may be contributed to by incomplete solubility of NFs in the buffers used for two-dimensional electrophoresis. However, other contributing factors are the specific *in vivo* posttranslational proteolytic modification of the 145K species,[59,105,106] the *in vivo* posttranslational phosphorylation of all three proteins,[59,106,107] as well as nonspecific proteolysis that may occur during the extraction procedures. In addition, rabbit 200K can exist in allelic forms in certain strains.[69,108,109]

5.3. Peptide Mapping and Amino Acid Composition

After the triplet protein composition of NFPs became known,[22,26,31,55] the relationship of the proteins to one another was examined by peptide mapping with attention to interspecies differences. In particular, the hypothesis tested was whether or not the smaller proteins were derived from the larger.[31] The NFPs were prepared from bovine, rat, and human CNS by the axon flotation procedure in which a short time was used for removal of myelin in 0.1 M phosphate buffer. The 210K, 160K, 70K, and 50K species were separated by SDS–PAGE and compared by limited digestion with *S. aureus* V8 protease or by cyanogen bromide cleavage.[22] Within the same molecular weight group, filament proteins from different species gave similar maps with both mapping techniques; however, the three size classes of NFPs of the same species gave different maps. These results were also obtained using *S. aureus* V8 protease to digest NFPs prepared by axon flotation followed by Triton X-100 treatment to remove myelin.[53,62] The latter study further showed that GFAP and the 63K polypeptide that copurifies with rodent CNS neurofilament proteins (Fig. 1) have peptide maps different from NFPs. Similar results have been reported for NFPs from a wide variety of animal sources.[8,64,110]

Data from the foregoing lead to the following conclusions: (1) among mammalian species, the NFPs in any one molecular weight group are similar (although not necessarily identical); (2) NFPs of different molecular weight groups are not related to one another by simple oligomerization; (3) the lower-molecular-weight species are not derived from the larger. Peptide mapping studies also established that the neurofilament triplet proteins and α and β tubulin have different maps.[53] Thus, the observed temporal relationship between the appearance or loss of microtubules and neurofilaments during neuronal development[11] and in cells treated with mitotic spindle inhibitors[111] cannot be accounted for by the simple redistribution of major subunit proteins.[53]

The amino acid compositions of the NFPs were first described by DeVries *et al.*,[112] who established their acidic nature (30% of residues were aspartic and glutamic acids). Improved preparative and analytic techniques have recently shown that rat CNS 70K and 160K proteins each contain 30% acidic residues, whereas the 210K protein contains 25%.[81] A similar acidic amino acid content was established for bovine PNS NFPs.[113] This is consistent with the more acidic isoelectric points of the 70K and 160K species compared with the 210K species. Certain features of the amino acid composition of NFPs from rat CNS[81] and bovine PNS[113] appear to be shared in common. The particularly high proline content of the largest NFP (8.5% in rat 210K and 7.8% in bovine 200K) relative to the smaller NFPs suggests that the 200–210K protein may have unique structural features. The tyrosine content of the 70K is greater than those of the other two proteins. The differences in amino acid composition among the NFPs are in accord with the earlier peptide mapping results; i.e., there is no clear indication for an oligomeric or precursor relationship among the three proteins.

Amino acid sequence analysis was carried out on a small polypeptide of the 68K NFP and compared with sequences from desmin and vimentin.[114] A fragment produced after double chemical cleavages at an internal tryptophan and cysteine, approximately 40 residues in length, was released from similar sites from the three intermediate filament proteins. With respect to vimentin and desmin sequences, 28 of 39 (72%) residues were identical. Sequence homology with the NF 68K was substantial but less extensive: 43% with vimentin and 41% with desmin. The results provide compelling evidence that nonneuronal intermediate filament proteins have their counterpart in the 68K member of the NF triplet proteins.

Myxicola infundibulum NFs are composed of two major polypeptides estimated by Eagles *et al.*[91] to be 172,000 and 155,000 mol. wt. These workers carried out extensive peptide mapping studies to examine the relationship of the polypeptides to each other and to the numerous polypeptides of 140,000–50,000 mol. wt. obtained from *Myxicola* axoplasm. The two larger polypeptides had nearly identical peptide maps, and certain characteristic fragments were shared with the smaller polypeptides. Based on these comparisons, it was concluded that all the *Myxicola* NFPs were derived from the 172,000-mol.-wt. species by proteolysis that occurred *in vivo* prior to extraction.

5.4. Reassembly into Filaments

Removing neurofilaments from media in which they remain disassembled (e.g., in low-ionic-strength buffers or in the presence of urea or guanidine HCl) has been used to examine the reassembly properties of the triplet proteins under *in vitro* conditions. As indicated in Section 4, Liem[72] used reassembly as the final step in the purification of NFPs. Moon *et al.*[100] solubilized bovine brain neurofilaments in a mixture containing 8 M urea, pyridine, formic acid, and 2-mercaptoethanol; the 200K protein was separated from a 160K–78K complex. After incubation at 37°C in an assembly buffer containing KCl, $MgCl_2$, and ATP, the 160K–78K complex elongated up to 500 nm; however, the formation of longer filaments required the presence of 200K. The results suggested that one or both of the smaller proteins formed a core structure that required 200K for further assembly.

Geisler and Weber[49] solubilized porcine spinal cord NFs in 8 M urea and purified the individual triplet proteins to 95% purity by anion-exchange chromatography and gel filtration. The 68K, denatured in 6 M guanidine HCl, self-assembled into smooth 8- to 10-nm filaments after dialysis at 37° against MES buffer containing EGTA, dithiothreitol, and NaCl. By contrast, addition of 210K to 68K during self-assembly led to the formation of rough-surfaced short filaments with whiskerlike protrusions. Intermediate filaments could not be reconstituted from the purified 160K and 210K triplet proteins either alone or together under the described conditions. Self-assembled 68K was further shown to have a periodicity of 21 nm when viewed in the electron microscope,[114] similar to purified neurofilaments.[50]

Similar results were recently obtained by Zackroff *et al.*[99,114a] following the *in vitro* reassembly of bovine spinal cord NFs from the triplet proteins. After disassembly in low-salt solution or 8 M urea, only the 68K component was capable of reassembly either alone or in combination with the other two proteins. The 150K and 200K components did not form filaments of 10 nm diameter. The 10-nm filaments reconstituted from the 68K protein were shorter and had smoother walls than freshly isolated NFs or NFs reassembled from the three NFPs.

The studies described lend experimental support to the suggestion of Willard and Simon[46] that the 68K NFP may be a core filamentous protein, since the 160K and 210K do not reassemble under conditions that favor 70K filament formation. This interpretation was also reached by Liem and Hutchison,[76b] who definitively demonstrated that the mammalian 70,000-mol.-wt. protein by itself is capable of forming an intermediate filament, whereas the other two are not. It is tempting to speculate that 160K or 210K serves a unique function; e.g., the whiskerlike projections of 210K–68K filaments[100] may indicate that the 210K forms the side arms observed in electron micrographs of neurofilaments *in situ*. However, it is also possible that 160K and 210K are capable of forming filaments under conditions not yet tested experimentally. The primary molecular information for intermediate filament formation resides within the polypeptides[49]; however, intracellularly this process may be modulated by local ionic conditions, availability of energy sources, and protein or nonprotein cofactors. Thus, further *in vitro* reconstitution experiments are required before one can draw firm conclusions concerning the self-assembly of neurofilaments from purified triplet proteins.

Several reports have briefly described the *in vitro* reassembly of invertebrate NFPs, from axoplasmic protein, into filaments.[6,54,104] A more detailed study by Zackroff and Goldman[115] examined the reassembly of squid brain filaments and their further purification by cycles of assembly and disassembly. Brain tissue homogenized in buffer containing 1.0 M salt yielded a clarified supernatant; on dilution with low-salt buffer, filaments were reassembled and then subjected to a second cycle. Sodium dodecyl sulfate PAGE of the pelleted filaments revealed predominantly a polypeptide of 60,000 mol. wt. and lesser amounts of 220,000-, 100,000-, and 74,000-mol.-wt. polypeptides. Some of these may have been contributed by nonneuronal cells. Based on turbidity measurements, it was suggested that the initial formation of short protofilaments serve as nucleation centers for a slower phase of polymer elongation.

5.5. Interaction with Microtubule Protein

Cytoskeletal elements of nerve cells consist of neurofilaments and microtubules arranged parallel to the long axis of the axon and interconnected to form a three-dimensional filamentous network.[4,10,17,18,37,39,116,116a] Cross bridges are known to exist between neurofilaments,[7,15,37] and recent reports concerning cross bridges between neurofilaments and microtubules[40,117] confirmed earlier observations on the occurrence of sidearms in both structures.[10] Electron microscopic examination of neurons of the teleost fish *Tinca tinca*[118] revealed a highly ordered array of nine or ten neurofilaments surrounding a microtubule, suggesting one-to-one cross bridging between adjacent subunits in filaments and tubules. Electron microscopic observations demonstrating a direct association between neurofilaments and microtubules in the neuronal cytoskeleton have been complemented to some extent by *in vitro* studies.

It has been observed that microtubules prepared by cycles of assembly–disassembly contain cosedimenting neurofilaments.[64,87,119] Runge *et al.*[120] used both viscosity and sedimentation velocity measurements to show that purified neurofilaments form complexes with microtubule protein or purified tubulin when the components are incubated together at 37°C in the presence of ATP. These authors speculated that the cyclic-AMP-independent neurofilament-associated protein kinase they described previously[89] acts on one or more proteins in the system, triggering formation of a complex. The phosphorylation of NFPs by the kinase was stimulated by the high-molecular-weight microtubule-associated protein MAP-2, which was also phosphorylated. It should be noted that microtubule protein prepared by assembly–disassembly contains a phosphokinase that phosphorylates the 150K component of NFPs in the presence of cyclic AMP[121] (See Section 7.2).

Neurofilaments prepared from intradural roots that are free of tubulin were useful for examining the interaction of NF with added microtubule proteins and tubulin. Shelanski and co-workers[20] noted that NFs decrease the rate of assembly of microtubules, and, in more detailed studies using purified microtubule components, they found that both high-molecular-weight MAP and tau proteins bind to the neurofilaments. Even excess amounts of pure 6 S tubulin did not bind to filaments in the absence of the accessory proteins. Proteins that bind to both filaments and tubules may be related to the cross bridges seen in electron micrographs.

The cited works are initial reports concerning a direct interaction between NFPs and microtubule components. Although more detailed investigations are necessary to define the detailed mechanism of interaction, it appears to be dependent on phosphorylation of one or more components. Both NF and microtubule preparations contain kinases with distinct properties that may be involved in the interaction of the fibrous elements; moreover, NF preparations have been reported to contain phosphodiesterase and ATPase activities. Known properties of these enzymes are described in Section 7.2; however, the specific influence of the individual enzymatic activities on microtubule–neurofilament interactions is currently unknown.

6. IMMUNOLOGIC STUDIES

Neurofilament proteins are highly antigenic and have been used in a variety of studies designed to examine the following: (1) the existence of shared and unique antigenic determinants among the triplet proteins; (2) the structural organization of the triplet proteins within NFs; (3) the distribution of triplet protein antigenic sites in different neural tissues; and (4) the immunologic distinction between NFPs and other intermediate filament proteins, particularly GFAP. By use of immunohistofluorescent procedures with antibodies to purified triplet proteins, neurofilaments are readily demonstrated in myelinated axons. As shown in Fig. 3, NFs have a characteristic morphology that easily distinguishes them from glial filaments (see legend to Fig. 3). Neurofilaments prepared by the previously described methods usually contain other proteins, which may be antigenic and give spurious results when the entire preparation is used for the production of antibodies (see refs. 7 and 8 for a discussion). This section considers investigations that made use primarily of purified NFPs as immunogens.

In early studies by Schlaepfer,[45,65] antibodies raised in rabbits to a 68,000-mol.-wt. protein directly extracted from neurofilament-rich rat peripheral nerve reacted with surface components on isolated rat neurofilaments and localized to NF structures in rat and human CNS and PNS. Using a different approach, Liem *et al.*[31] avoided contamination by glial protein and collagen by preparing neurofilaments from excised rabbit intradural spinal roots subjected to a modified axon flotation procedure. Antibodies were raised in guinea pigs to NFPs separated by SDS–PAGE and subsequently eluted. By radioimmunoassay, it was established that anti-68K and anti-160K were both reactive towards 68K, 160K, and 210K. In immunohistological studies, the anti-NFP antibodies were shown to stain neural structures in brain and to strongly stain sciatic nerve. The pattern of staining was distinctly different from a 51K contaminant that arose during long low-ionic-strength extraction of axon preparations. Thus, it was established that the 51K protein was predominantly of glial origin. The studies of Schlaepfer[45,65] and those of Liem *et al.*[31] provided early confirmatory data that the triplet proteins of Hoffman and Lasek[26] were components of neuronal fibrous structures.

In other investigations some inconsistent results have been obtained concerning the immunologic response to NFPs and the extent to which anti-NFP antibodies are cross reactive with heterologous polypeptides. Rabbit antibodies were prepared to bovine brain triplet proteins and GFAP that had been purified by PAGE.[122] Anti-200K and anti-150K antisera preabsorbed with either 200K or 150K proteins failed to detect rat brain neurofilaments by immunofluorescent staining; however, preabsorption of anti-70K serum with 70K protein did not diminish staining. Anderton *et al.*[32] also showed immunofluorescent staining of neuronal filaments in tissue using anti-210K and anti-155K proteins but not with anti-70K. Yen and Fields[21] used antibody to human 210K NFP, obtained by the aforementioned procedure of Liem *et al.*,[31] that had been purified by PAGE prior to immunization of rabbits. The antibody was extensively char-

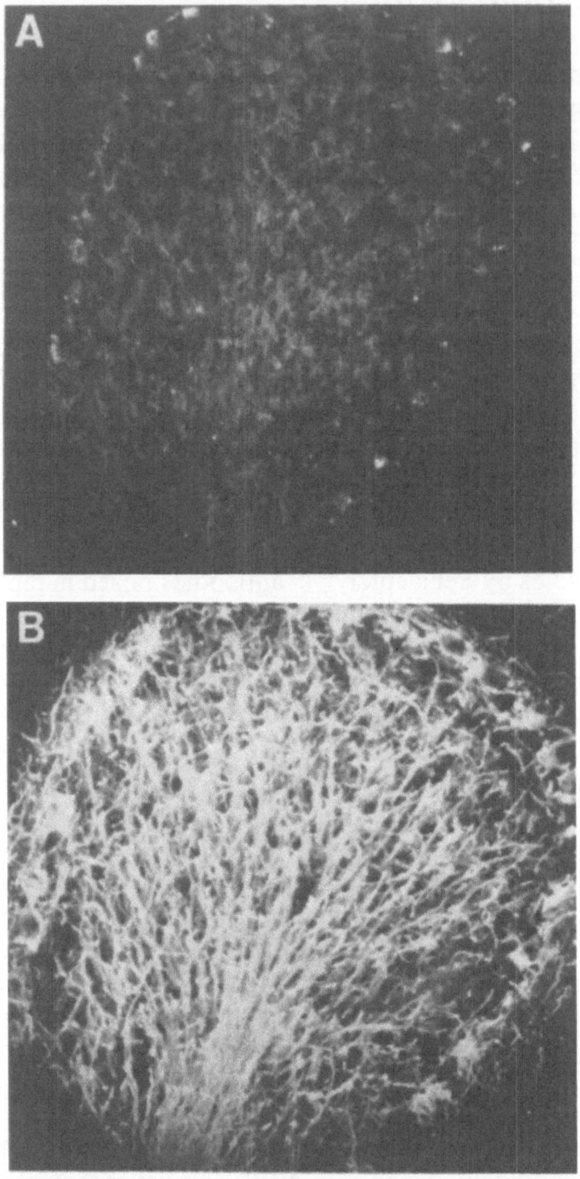

Fig. 3. Immunohistofluorescent staining of rat cerebellar folia with anti-NFP and anti-GFAP. A: Preimmune serum (1:20 dilution), ×200. B: Anti-200K (1:50 dilution), ×200. Note the intense staining of the myelinated fiber tract at the center of each folium. C: Anti-140K (1:50 dilution), ×200. Staining was similar to B. D: Anti-140K staining seen at higher magnification showed basket cell neuritic processes surrounding a Purkinje cell. ×800. E: Anti-GFAP allowed visualization of radial glial fibers, presumably Bergmann glia, ×200. F: Individual astrocytes from panel E, ×700. (Courtesy of Ronald E. Majocha.)

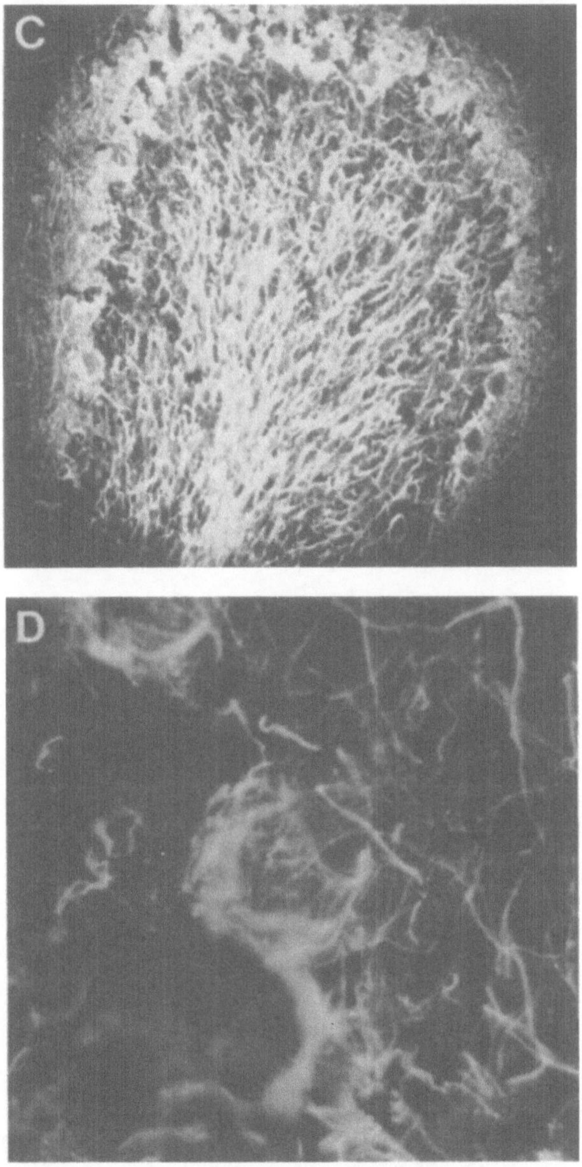

Fig. 3. (*Continued*)

acterized by immunohistological procedures and shown to bind exclusively to filamentous structures within axons. In rocket immunoelectrophoresis studies anti-210K formed precipitin lines with 210K, 160K, and 68K NFPs but not with GFAP.

Studies on cross specificity were also performed using gel-purified rat NFPs to prepare antibodies that were characterized by a peroxidase–antiperoxidase (PAP) procedure[124] directed against NFPs in neural tissues or SDS

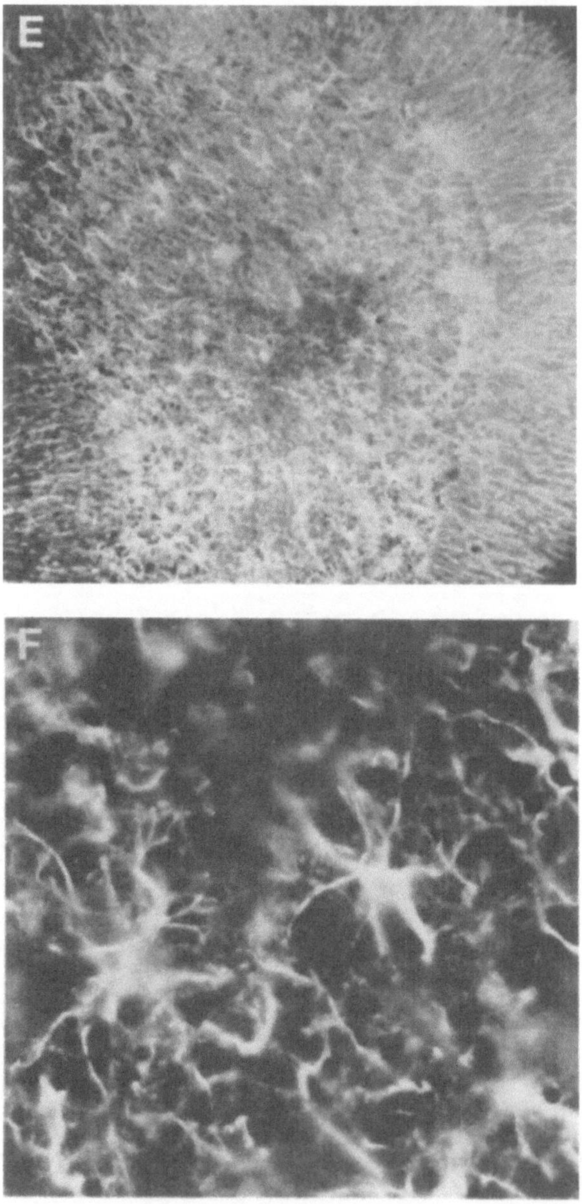

Fig. 3. (*Continued*)

gels.[123,124] Both CNS and PNS axonal structures were reported to stain similarly with the three antisera. Each antiserum reacted intensely with the corresponding NFP and more weakly with the other two proteins. Anti-145K serum also bound to chicken and goldfish NFPs. It was concluded that although the molecular weights of the NF subunits vary widely among species (chicken NFPs were reported as 180K, 145K, 130K, and 69K, and fish NFPs as 145K,

115k, and 85K), they probably share a common region, which has been phylogenetically conserved. Willard and Simon,[46] using rabbit spinal cord NFPs as immunogens in goats, obtained sera that were further purified on affinity columns containing each of the NFPs. By this procedure it was shown that a substantial fraction of the antibodies against each NFP bound to affinity columns containing either of the other two. Since each antibody bound to its corresponding antigen after preabsorption with the heterologous proteins, it was concluded that each protein also contained antigenic determinants that are unique. The specificity of the column-purified antibodies was established by an enzyme-linked immunoassay (ELISA) procedure.

Thus, although it appears that NFPs contain shared and unshared antigenic sites, there is nonuniformity in the specificity of different antisera. For example, Brown *et al.*[28,124a] demonstrated, using standard procedures, that antisera can be prepared that react nearly exclusively with unique determinants of each NFP. Long one-dimensional gels were used to purify rat 200K and 140–145K NFPs, and two-dimensional gels were used to separate the 70K NFP from other constituents. The three antisera were prepared in rabbits, and each was shown to strongly stain similar axonal filaments in cerebellar neurons by immunohistofluorescence. With a sensitive horseradish peroxidase (HRP)-conjugated indirect antibody procedure, however, differences were detected in the specificity of the sera when directed towards rat NFPs separated by SDS–PAGE and electrophoretically transferred to nitrocellulose membranes. Each antiserum reacted nearly exclusively with its homologous antigen. This study also demonstrated that rabbit preimmune serum may, in some cases, contain autoantibodies to NFPs as well as GFAP.[28]

The variable specificities obtained with antineurofilament sera may be caused by one or more of the following factors: the purity of the NFP used as an antigen; the physical state of the antigen, i.e., whether or not it was deatured with a detergent prior to immunization; the animal species used to prepare NFPs and for immunization; the time period prior to collection of serum; the presence of autoantibodies to intermediate filament proteins in preimmune serum; and the sensitivity of the immunoassay.

Two of these factors, assay sensitivity and interspecies differences, have been specifically addressed. In the work of Brown *et al.*,[28] antisera were compared by immunohistofluorescence and by the HRP-conjugated indirect antibody method applied to proteins attached to nitrocellulose membranes. With the same antibodies it was shown that the latter procedure was the more sensitive by two to three orders of magnitude. Interspecies differences in the immune response to NFPs were examined by Schlaepfer *et al.*,[27] who prepared both rabbit and guinea pig antibodies to gel-purified NFPs from bovine spinal cord. By ELISA, it was shown that the same antigenic inoculum elicited antibodies exhibiting more extensive cross reactivity when prepared from guinea pig than from rabbit. For example, rabbit anti-200K exhibited the strongest reaction with 200K, a weaker reaction with 150K, and no reaction with 68K; guinea pig 200K reacted equally with the three NFPs. Whereas rabbit anti-68K was specific for the 68K NFP, the guinea pig anti-68K detected the three NFPs. It was concluded that differences in cross reactivity of antisera towards heterologous NFPs may reflect the presence of numerous antigenic sites and variability in the extent of their antigenicity. The immune response to NFPs appeared dependent on both the antigenicity of the protein and the species used to prepare the antibody.

Depending on the experimental goal, certain difficulties inherent in the use of polyclonal antibodies can be overcome with the use of monoclonal antibodies to neurofilament proteins. With this technique, it was demonstrated that a monoclonal antibody can be obtained with specificity for 200K only[30] or 200K and 150K.[125] Thus, there is convincing evidence for the existence of unique and shared antigenic determinants among the NF proteins. Recently, Pruss *et al.*[25] produced a monoclonal antibody to human GFAP that reacted with all classes of intermediate filament proteins including those from mammalian brain, squid, and marine worm. This finding demonstrates the existence of a common antigenic site shared by vertebrate and invertebrate intermediate filament proteins.

An interesting application of the monoclonal antibody technique was reported by Weber and colleagues.[30] These authors showed that certain neurons of adult rat brain that are rich in NFs, as determined by electron microscopy and immunofluorescence microscopy using polyclonal antibodies against 145K and 68K NFPs, were not decorated by a monoclonal antibody with exclusive specificity for the 200K protein. The authors proposed the existence of subpopulations of neurons (e.g., certain pyramidal cells of hippocampus) that lack a specific 200K NFP antigen. It was hypothesized that this may be attributable to the absence of the 200K NFP, to the expression of an immunologically distinct form of 200K, or to the action of a specific protease.

Few immunologic studies are available on invertebrate neurofilaments. The early experiments of Huneeus and Davison[6,125a] defined certain characteristics of antibodies prepared to squid axoplasmic proteins enriched in NF subunits. The antisera failed to cross react with proteins of nonneuronal squid tissues but did detect protein of optic nerve and brain. Although the antibodies cross reacted with neuronal extracts from other invertebrates, they did not cross react with mammalian brain extracts. In the more extensive investigation of Eng *et al.*,[126] immunocytologic analyses were carried out on neurofilaments of the *Myxicola* giant axon. At the light microscopic level, anti-NFP staining was confined to neuronal perikarya, neurites, and the giant axon. The staining patterns at the electron microscopic level corresponded to the NFs within axons and neurons. Similarly to mammalian studies, there was no detectable staining of glial filaments or cytoskeletal elements of striated muscle by antibody to neurofilaments.

7. METABOLISM

7.1. Synthesis

Axoplasmic transport studies have shown that the three neurofilament proteins are readily labeled by radioactive amino acid precursors.[26,94,95] The labeled NFPs are transported along the axon in the slowest component of axoplasmic flow at a rate of 0.2–1.0 mm/day. Recent studies on the synthesis of neurofilament proteins in the mouse retinal ganglion cell (RGC) axon closely examined the newly synthesized middle-sized neurofilament protein(s)[105,106,127] that in the mouse are microheterogeneous forms of 145,000, 143,000, and 140,000 mol. wt.[53] After injection of radioactive proline into the RGC, and following the onset of axoplasmic transport of labeled protein, a time-dependent increase in 143K and 140K species occurred concomitantly with a decrease in the 145K NFP. The enzyme responsible was identified as a calcium-activated neutral proteinase active at endogenous concentrations of calcium.[59,105] The 200K and 70K NFPs appeared unaltered under the same experimental conditions. This enzyme, as well as others that degrade NFPs, are considered in more detail in Section 7.3.

In contrast to the numerous studies on the *in vitro* synthesis of neuronal tubulin and actin,[128–132a] only limited information is available concerning the *in vitro* synthesis of the neurofilament proteins. Although an initial study failed to synthesize the NFPs using rabbit spinal cord polysomes translated in a homologous cell-free protein synthesizing system,[133] a later investigation established that a homologous system primed with free polysomes from spinal cord synthesized three polypeptides of the same size and isoelectric point as the NFP triplet.[134] An attempt was made to synthesize NFPs using the spinal cord polysomes in a reticulocyte lysate translation system. Under the stated conditions, numerous products were synthesized and analyzed by one-dimensional SDS–PAGE, among which were species of 200K and 150K in trace amounts and a band at 68K. A preliminary attempt was made to synthesize NFPs by purified mRNA from rat spinal cord free polysomes in a wheat germ homogenate.[135] However, neither the 200K nor the 150K was synthesized, and a degradation product (at 139,000 mol. wt.) was obtained. Degradation may have been caused by the high levels of endogenous protease present in the wheat germ homogenate.[136] Although it is not unreasonable to speculate that the triplet proteins are products of separate genes, rigorous proof is not yet available.

7.2. Phosphorylation

Studies on enzymes of phosphate metabolism have focused on the posttranslational phosphorylation of neurofilament proteins and whether kinase activity is associated with the NFPs themselves. In addition, phosphorylation studies have provided direct biochemical evidence for an *in vitro* interaction between the triplet proteins and components of microtubule preparations.

Runge *et al.*[137] purified bovine brain NFPs that contained a neurofilament-associated protein kinase with distinct characteristics: the enzyme was not stimulated by cyclic AMP to a significant degree, and activity was not altered in the presence of the protein kinase inhibitor of Walsh *et al.*[138] or by preincubation with ADP. The kinase activity remained associated with the NFPs even after repeated centrifugation through 0.5 M ionic strength buffer, demonstrating a tight association between the measured activity and the filament proteins under the experimental conditions. The neurofilament-associated kinase was active over a broad pH range with a maximum at pH 6.5. When it was incubated with divalent cations, activity was highest with magnesium and lowest with calcium; calmodulin was without effect.

The particularly interesting aspect of this kinase was its stimulation twofold by high-molecular-weight microtubule-associated protein, MAP-2. MAP-2 stimulated the phosphorylation of the three NFPs and was itself a substrate for the kinase, as was tubulin. In twice-cycled microtubule preparations containing the NFP triplet proteins, the cyclic-AMP-independent NF-associated kinase activity was separable from cyclic-AMP-dependent microtubule-associated protein kinase. The bovine brain NF-associated kinase was similar in some respects to the guinea pig peripheral nerve enzyme of Shecket and Lasek.[139] Their enzyme too was a cyclic-AMP-independent NF-associated kinase that phosphorylated the three NFPs. The peripheral nerve enzyme, with a pH optimum of 8, was maximally stimulated by manganese, whereas calcium was inhibitory. A second enzyme of phosphate metabolism associated with NFs was identified by Runge *et al.*[120] as a calcium/magnesium-dependent cyclic nucleotide phosphodiesterase present in bovine brain neurofilament preparations. The enzymatic activity was also observed in microtubule preparations; however, the activity was associated with the neurofilaments present in twice-cycled microtubules.[119,120] The calmodulin-activated phosphodiesterase activity was not separated from NFs after washing with 0.5 M salt, suggesting that the enzyme is a component of bovine brain neurofilaments.

The cyclic-AMP-dependent phosphorylation of at least one component of the NFP triplet proteins has been observed. The NFPs prepared from rabbit intradural nerve roots and bovine brain essentially lacked cyclic-AMP-dependent phosphokinase activity under conditions in which high activity was found in microtubule protein prepared by cycles of assembly–disassembly.[121] Addition of microtubule protein to NFP stimulated phosphate incorporation into protein in the presence and absence of cyclic AMP. Without cyclic AMP, MAP-2 and tau were phosphorylated; in the presence of cyclic AMP, phosphorylation of these same proteins occurred in addition to the 150K NFP. The NFP phosphorylation was almost totally cyclic AMP dependent, and activity was abolished in the presence of the kinase inhibitor of Walsh *et al.*[138] In related preliminary studies, it was shown that NF preparations contain an EDTA-inhibited, magnesium/calcium-stimulated ATPase activity.[20] The effects appeared similar to those previously obtained with microtubule-associated ATPase, raising the possibility that they are the same enzyme. More recently, it was shown that the ATPase activity of a neurofilament preparation was neither inhibited nor precipitated by antibody to NFP[139a]; thus, ATPase does not appear to be an integral component of filaments.

Julien and Mushynski[107] compared both *in vitro* and *in vivo* phosphorylated NFPs in order to assess the physiological significance of the protein kinase activity that copurifies with neurofilaments. The NFPs were prepared from rat CNS after injection with radioactive phosphate; SDS–PAGE of the triplet proteins demonstrated the high incorporation of radioactive label into the 200K component relative to the 145K and 68K proteins. By contrast, NFPs incubated *in vitro* with radioactive phosphate contained more highly labeled 145K and

68K components relative to 200K. Peptide maps of the phosphorylated residues showed similarities between the *in vitro* and *in vivo* labeled species in the three size classes. However, 145K had an additional phosphorylated site *in vivo*, whereas the reverse was true for 200K; 68K had a single phosphorylated residue. In these comparative studies, it was not possible to control for certain factors that may affect the results, e.g., the potential action of phosphatases active *in vivo* or during tissue disruption and the possibility that *in vivo* NFPs may be subject to phosphorylation by a cyclic-AMP-dependent kinase in addition to a cyclic-AMP-independent enzyme.

In other studies it was shown by Nixon *et al.*[59,105,106,127] that NFPs phosphorylated *in vivo* in retinal ganglion cells undergo axoplasmic transport. After mice were injected intravitreally with radioactive phosphate, phosphorylated forms of the neurofilament triplet proteins migrated into the optic nerve. The posttranslational proteolytic processing of the mouse 145K protein, described in detail in the subsequent section, also occurred with the phosphorylated protein. The 145K phosphoprotein was present in the most proximal segments of the optic pathway, and 143K and 140K phosphoproteins appeared in more distal segments. Thus, some of the phosphorylated NFPs resemble their unmodified counterparts with respect to axoplasmic transport and to posttranslational processing.

Pant *et al.*[140,142] examined the *in vitro* and *in situ* phosphorylation of squid axoplasmic proteins. The 200,000-mol.-wt. NFP and another protein at 400,000 mol. wt., which copurifies with NFs, became radiolabeled when intact axons or extruded axoplasm were incubated with [^{32}P]ATP. The 60,000-mol.-wt. NFP was not labeled. Phosphorylation was stimulated by magnesium, unaffected by cyclic nucleotides, and inhibited by calcium. The latter effect may have represented calcium-activated proteolysis. This point was examined more explicitly by Eagles *et al.*[141] using phosphorylated *Myxicola* NFPs. Incubation of extruded axoplasm with [^{32}P]ATP resulted in the phosphorylation of two major NFPs of 172,000 and 155,000 mol. wt. Addition of calcium led to the production of poorly phosphorylated polypeptides (70,000–47,000 mol. wt.) and soluble degradation products containing the bulk of the radioactivity. Although a similar mechanism may operate *in vivo*, the physiological function of NFP phosphorylation in invertebrates and mammals has not yet been determined.

7.3. Degradation

At the present time, considerably more extensive information is available on the degradation of neurofilament proteins than on their synthesis. In a series of reports, Schlaepfer and associates characterized the calcium-mediated loss of neurofilaments in mammalian peripheral nerve. Disruption of NFs occurred during incubation in isotonic saline containing 1 mM calcium.[143] Electron microscopy verified the loss of NFs and revealed a granular disintegration of axonal filaments and microtubules. The changes were minimal in the absence of calcium or in the presence of a calcium chelator such as EDTA present in concentrations equivalent with calcium. Transected sciatic nerves left to degenerate *in vivo* exhibited granular disintegration of neurofilaments and microtubules similar to excised nerve incubated in the presence of calcium.[144] The calcium-induced loss of neurofilaments was accelerated under conditions that favor an influx of calcium into the axon: freeze-thawing, addition of detergents, and addition of the calcium ionophore A23187. In excised peripheral nerves[97] or tissue homogenates,[55] the calcium-induced loss of NFs occurred under the same conditions that led to loss of NFs in transected nerves. The rate of disappearance of the triplet proteins was in the order 68K > 150K > 200K in initial stages; by 72 h, the NFPs were no longer detectable in transected nerve.[55,57]

Exposure of a crude preparation of NFs to calcium also led to disintegration of NFs and loss of NFPs.[58] The factor responsible for the changes was identified as a calcium-activated protease active at pH 6.0–8.0, inhibited by sulfhydryl reagents, and only partly sensitive to PMSF. The properties of the CANP of rat peripheral nerve were previously listed in some detail.[146]

With respect to central nervous system proteases, Nixon and associates have identified two enzymes active towards NFPs in the mouse retinal ganglion cell optic pathway.[59,60,105,106,127,147] By introducing radioactive amino acids into the RGC, a radiolabeled axon-specific protein population was transported into the optic pathway; the optic pathway protein was examined by SDS–PAGE and shown to contain prominent radioactive triplet proteins. By incubating the optic pathway under a variety of different conditions (varying pH, ionic strength, and the presence or absence of inhibitor), proteases with affinity for NFPs were identified and shown to be two types of calcium-activated neutral proteinases[59,60] designated CANP A and CANP B.[147]

When studied in excised optic pathway segments, CANP A was proximally active at millimolar calcium concentrations (>1.0 mM) with a pH range of 7.0–8.0. The enzyme was extremely sensitive to leupeptin, antipain, and sulfhydryl reagents and exhibited only slight, if any, sensitivity to PMSF. Thus, CANP A of RGC neurons[59,60] shares certain properties with the CANP of peripheral nerve.[145] CANP A represented at least 90% of the total calcium-dependent proteolytic activity in RGC axons and utilized as substrate other proteins of large size in addition to the NFP triplet.

In contrast to the broad substrate specificity of CANP A, RGC calcium-activated neutral protease B was active towards the middle-sized member of the triplet proteins only.[59,60,105] At endogenous levels of calcium, CANP B mediates the conversion of the 145K NFP to two smaller forms at 143K and 140K. The three forms correspond in their two-dimensional electrophoretic patterns and peptide maps to the 145K, 143K, and 140K NFPs present in fresh mouse optic pathway or in NFP-enriched preparations from whole mouse brain. Heterogeneous forms of the NFPs were discussed above and shown in Fig. 2B. These forms also undergo phosphorylation in the RGC. The 145K NFP conversion occurred in the absence of exogenous calcium and was inhibited by EGTA, indicating that this calcium-dependent process is active at submicromolar calcium levels that may normally be present within axons. CANP B was partially inhibited by the serine proteinase inhibitors PMSF and DFP but not by leupeptin and antipain; CANP B is thiol dependent and optimally active in the neutral pH range. The enzymatic processing of 145K to smaller polypeptides occurred during axoplasmic transport nonuniformly along the length of the axon. Proteolytic modification of the 145K NFP was much less in the most proximal region of the optic nerve than more distally, indicating regional specialization of the axon with respect to this enzymatic activity and to certain forms of the 145K NFP.[105] Thus, CANP A and CANP B were distinguished from one another by their inhibitor profile, requirement for calcium ions, substrate specificity, and, possibly, location in the RGC axon. The proteolytic processing of the 145K NFP represents posttranslational modification of an NFP other than phosphorylation. Other types of NFP modifications may exist, similar to those for tubulin; α and β tubulin undergo a wide variety of processing events including the modification of heterogeneous forms of α tubulin during axoplasmic transport.[148,148a]

Cathepsin D, one of several acidic brain proteinases active at pH 3.0–5.0,[149,150] was recently shown to be active toward NFPs.[147,151] Purified cathepsin D from postmortem human brain, when incubated with purified NFPs from human CNS white matter, caused the partial degradation of 200K, 160K, and 70K proteins within 1 h. Pepstatin, a specific inhibitor of cathepsin D, completely blocked the loss of neurofilament protein.

The proteolytic activity causing the calcium-mediated disruption of neurofilaments previously noted to occur in rat spinal cord[58] was partially purified and identified as a calcium-activated neutral thiol protease.[152] Proteolysis of the triplet proteins followed simple Michaelis–Menton kinetics with decreasing reactivity in the order 68K, 150K, 200K.

Calcium-stimulated proteolysis of neurofilaments has been demonstrated both in squid and *Myxicola*. Addition of exogenous calcium causes the liquefaction of squid axoplasm.[153–155] The 200K NFP appeared to undergo a stepwise degradation, proceeding through a 100,000-mol.-wt. stage to smaller polypeptides and finally to amino acids.[93] In *Myxicola*, the larger NFPs with reported molecular weights of 160,000 and 152,000 underwent degradation on incubation with calcium at millimolar concentrations concurrent with the dispersal of axoplasm.[54] Incubation of radioiodinated *Myxicola* filament proteins with axoplasm in the presence of calcium caused degradation to lower-molecular-weight forms.[92] Based on the observation that the calcium-induced cleavage of *Myxicola* NFPs was inhibited by sulfhydryl reagents, TPCK, TLCK, and EGTA, Gilbert *et al.*[54] first suggested the presence of a calcium-activated proteinase in axoplasm. More recently, Pant and Gainer[56,93] distinguished between a calcium-activated neutral proteinase in squid that is a relatively nonspecific and one with specificity for the 200 NFP. Activity of both was maximal at millimolar calcium. The specific CANP appeared to be similar to a protease purified from human platelets.[156]

8. *NEUROFIBRILLARY TANGLES*

The possibility that the neurofibrillary lesion observed in some neuronal perikarya of the brain in Alzheimer's disease (AD) or senile dementia of the Alzheimer type and several other diseases may be associated with NFPs has been the subject of intense experimentation and critical debate. Although additional data are needed before one can draw definitive conclusions, it appears worthwhile to review the several recent reports that deal with this question. Various experimental approaches to elucidating the pathogenesis of AD have been extensively reviewed[157–162a]; the present review focuses almost exclusively on investigations that implicate a role for neurofilaments in the formation of neurofibrillary tangles. In the following discussion a distinction is not made between Alzheimer's disease and senile dementia of the Alzheimer type with respect to histopathological features.

The neurofibrillary tangle (NFT) of Alzheimer's disease is prominent in the neuronal perikaryon, particularly in hippocampal and neocortical neurons.[163] In the light microscope, NFTs appear as argentophilic tangles of coarse fibers.[164] Ultrastructurally, the bundles of abnormal twisted elements appear as clusters of paired helical filaments (PHFs),[165,166] each member of which is a fiber 10 nm in width. The pair of filaments together have a maximum width of 20 nm to 25 nm and a minimum width of 10 nm with cross-over points at approximately 80 nm.[167] Each member of the pair appears to resemble the normal straight neurofilament; however, in PHFs, the presence of side arms has been questioned,[168] and they were reported to have unusual solubility characteristics.[169] Although PHFs are present in aged humans free of dementia, they are far more prominent and widespread in AD. They are not, however, found in the Alzheimer brain exclusively, since PHFs identical to those in AD have been described in postencephalitic Parkinson's disease,[170] in Guam–Parkinson–dementia complex,[171] in patients with Down's syndrome who survive beyond age 30,[172] and in certain other conditions.[168,173] The distinctly human paired helical filaments have not been found in other species, and at present there are no suitable animal models for producing this type of lesion. In addition to the prominent neuronal loss that accompanies NFTs in AD, neuritic or senile plaques are also typically observed. The plaques are composed of enlarged neuronal processes, primarily axonal boutons, containing PHFs as well as degenerating lysosomes and mitochondria.[174]

Evidence that implicates NFPs in the neurofibrillary lesion of AD has been obtained primarily from immunocytochemical investigations. However, at present there is no compelling evidence to suggest that the neurofilament triplet proteins are the sole or even the primary components of NFTs, and, as discussed below, there is limited evidence for the direct involvement of neurofilaments in the pathogenesis of this lesion.

With respect to the immunologic staining of NFTs, some success has been reported using antisera to purified NFPs and to a microtubule protein fraction. Human postmortem 210K was purified by SDS–PAGE and used for the production of antiserum that reacted with all members of the triplet.[21] Brain sections of both normal and AD specimens treated with anti-NFP serum followed by the PAP method revealed a strong reaction product in white-matter axons but little reaction with cell bodies or their processes.[176] There was no preferential binding to tanglelike structures in AD tissue sections; however, when they were applied to isolated perikarya and examined by immunofluorescent staining, approximately 5% of tangle-bearing cells bound the antibodies.[176] By contrast, antiserum to twice-cycled human microtubule fractions stained structures resembling neurofibrillary tangles and neurites of neuritic plaques in the AD cortex and in approximately 40% of isolated tangle-bearing neurons. However, the reactive polypeptide of the NFT was antigenically different from the major proteins found in microtubule fractions.

It was suggested that heterogeneity exists with respect to the types of tangles within different neurons. This possibility, as well as differences in the clinical material and experimental procedures used in different laboratories, may partially account for the varying results obtained from immunologic studies. In another report, gel-purified 200K from rat spinal cord was used as immunogen, and the serum obtained reacted with the corresponding NFP only.[177] Alzheimer's disease NFTs as well as material in the periphery of the senile plaque were stained with the antiserum by the PAP method. Antiserum was also prepared to tubulin purified by the assembly–dissassembly procedure followed by phosphocellulose chromatography; the antiserum to the pure tubulin was not reactive with NFTs.

Staining of NFTs by antibody to an antigen in a microtubule fraction other than tubulin has now been found in several, although not all, laboratories that have addressed this question. The approach taken by Iqbal and coworkers[178,179] was to carry out biochemical and immunologic studies on a protein fraction purified from PHF-enriched neurons from the demented brain. The major PHF protein appeared to be a 50,000-mol.-wt. species distinct from α and β tubulin as well as GFAP. Antisera raised against the gel-purified PHF protein[175] and normal human microtubule protein purified by the assembly–disassembly procedure, when tested by the Ochterlony double-diffusion method, both formed precipitin lines with microtubule protein and with PHF protein but not with purified neurofilament proteins from normal human brain.[178] Both sera bound to NFT in isolated neurons and in sections from autopsied brains.[178,179] It was concluded that the paired helical filament protein, although distinct from α and β tubulin, neurofilament protein, and GFAP, is present in *in vitro* assembled microtubule preparations. These results can be contrasted with other studies on the immunocytological localization of microtubule protein in the AD brain.[180] Antisera were prepared to reassembled microtubule protein (containing tubulin and microtubule-associated protein) and to 55,000-mol.-wt. tubulin purified by SDS–PAGE. Although these sera stained neurons in normal and AD brain, preferential staining of neurons containing NFTs was not observed.

As with antisera to proteins in microtubule fractions, immunocytochemical observations based on the use of antisera to only partially purified neurofilament proteins have not led to straightforward interpretations. Antiserum was prepared to urea-extracted chicken brain protein containing a major 54,000-mol.-wt. species.[181] In the AD brain, the antiserum produced intense staining of NFTs, and when applied to proteins on gels, the anti-chicken-brain serum stained the triplet NFPs and tubulin.[29,182] Based on these data, it was concluded that the main component of AD neurofibrillary tangles appeared to be a protein antigenically related to neurofilament protein.[182] The chicken brain serum was contrasted with antisera to gel-purified mammalian brain NF triplet proteins.[29]

Although the anti-NFP sera stained NFPs in fresh animal tissue, there was no staining of routine human brain autopsy specimens; structures highly suggestive of NFT were occasionally stained in brain from patients with Alzheimer's disease but only when they were obtained shortly after death.

These results were in contrast to those obtained with anti-chicken-brain serum, which consistently stained NFT irrespective of the postmortem interval.[182] The conflicting results were ascribed to different methods of tissue preparation. In experiments alluded to, it appeared that the use of acetone-fixed cryostat sections rather than formalin-fixed, paraffin-embedded material allowed antisera to gel-purified NFP triplet proteins to react with tangles.[182] A further contributory factor to the differing results was suggested, i.e., that the anti-chicken-brain serum was raised against a breakdown product to neurofilament protein.[29] This interpretation may apply to another series of reported experiments in which antibodies were prepared to a calf brain subcellular fraction containing GFAP and an apparent NF degradation product, both of which migrated at 50,000 mol. wt. on a SDS–urea gel.[82] The antiserum stained axons, glia, and NFT in the AD brain. Nevertheless, the data were interpreted as indicating that NF protein is the structural component of AD neurofibrillary tangles.

The divergent results from different laboratories may reflect differences in the activity and specificity of the various antisera, the state of preservation of postmortem neuronal structures, differences in fixation procedures, and differences in the details of the various immunologic procedures. Indeed, all the cautions previously listed (Section 6) with respect to antineurofilament protein antibodies apply to the immunologic studies under consideration. Attempts to identify a neurofilament triplet protein in the NFT by immunocytochemical methods depend in part on the premise that specific antisera can be routinely obtained to NFP antigenic sites that are not also shared by a number of other neuronal polypeptides. However, it was recently shown that NFP antisera that appear to react almost exclusively with the corresponding triplet protein when tested at high dilutions also react with a variety of lower-molecular-weight neuronal polypeptides when examined at low dilutions.[28] The use of highly specific monoclonal antibodies to NFPs would be useful in helping to clarify the extent to which NFPs are directly involved in the neurofibrillary tangle of Alzheimer's disease. A report by Anderton *et al.*[182a] used this methodology. Monoclonal antibodies were prepared to 210K from rat brain and 155K from the hippocampi of two cases of senile dementia of the Alzheimer type. Immunoperoxidase staining of affected brain tissues revealed that both antibodies reacted with NFTs as well as senile plaques. Since immunoblotting analysis revealed that each antibody reacted exclusively with its homologous NFP, it was concluded that NFTs share at least two antigenic determinants with neurofilaments.

One of the few reports that directly implicates neurofilaments in the pathogenesis of AD neurofibrillary tangles used a vastly different approach from immunocytochemistry. Cultured human fetal cerebral cortical neurons were exposed to an extract prepared from an Alzheimer's brain.[183] Fourteen days after exposure, occasional paired helical filaments were observed by electron microscopy; by 35 days, 3% of all neuronal processes were affected, and the PHFs resembled those found in the human disease. It was further observed that many of the PHFs seemed to be assembled from separate, morphologically normal-appearing neurofilaments.[183] Replication and extension of such experiments would be necessary to elucidate the basis for the observed transformation in cultured cells. The contention that the PHFs induced in cultured neurons resemble the PHF in the AD cortex has been questioned on morphological grounds.[184]

The NFP triplet proteins from human brain are 200K, 160K, and 70K species.[63,185] When examined in homogenates prepared from affected regions of the AD cortex and examined by PAGE, the electrophoretic migration of proteins that enter into conventionally prepared SDS gels[185a,b] is similar to normal control NFPs.[187] In some AD neuronal fractions and neurofilament preparations, a relative increase in a 20,000-mol.-wt. protein was observed and believed to be related to the occurrence of PHFs.[186] The protein was subsequently identified as myelin basic protein of intracortical origin, which copurifies with neurofilaments during the axon flotation procedure[53,62,187] (see Fig. 1). A recent report suggested that PHFs are not susceptible to analysis by standard PAGE techniques because of the insolubility of abnormal filament protein in buffers normally used for electrophoresis.[169]

At the present, there is no adequate experimental model for inducing PHFs morphologically identical to those of the AD brain. A number of pharmacological agents have been applied to various neural tissues to induce neurofibrillary changes. Although tanglelike filaments accumulate in cells treated with colchicine, vinblastine, podophyllotoxin, maytansine, and, most notably, aluminum salts, ultrastructural studies demonstrated that they are composed of 10-nm neurofilaments distinctly different from PHFs found in Alzheimer's disease. These investigations are reviewed elsewhere.[184,188,188a]

ACKNOWLEDGMENTS. I am grateful to Leta Sinclair and Beatrice Bradley for typing this manuscript and to R. A. Nixon, F.-C. Chiu, J. E. Goldman, and W. T. Norton for critical reading prior to publication. This chapter was completed while the author was the recipient of NIH Research Career Development Award AG00084 and a MacArthur Foundation Award.

REFERENCES

1. Parker, G. H., 1929, *Q. Rev. Biol.* **4**:155–178.
2. Schmitt, F. O., 1950, *J. Exp. Zool.* **113**:499–515.
3. Schmitt, F. O., and Geren, B. B., 1950, *J. Exp. Med.* **91**:499–504.
4. Schmitt, F. O., 1968, *Proc. Natl. Acad. Sci. U.S.A.* **60**:1092–1101.
5. Schmitt, F. O., 1968, *Neurosci. Res. Prog. Bull.* **6**:119–144.
6. Huneeus, F. C., and Davison, P. F., 1970, *J. Mol. Biol.* **52**:415–428.
7. Norton, W. T., and Goldman, J. E., 1980, *Proteins of the Nervous System*, 2nd ed. (R. A. Bradshaw and D. M. Schneider, eds.), Raven Press, New York, pp. 301–329.
8. Shelanski, M. L., and Liem, R. K. H., 1979, *J. Neurochem.* **33**:5–13.
9. Weiss, P. A., and Mayr, R., 1971, *Proc. Natl. Acad. Sci. U.S.A.* **68**:846–850.

10. Peters, A., Palay, S. L., and Webster, H. deF., 1976, *The Fine Structure of the Nervous System: The Neurons and Supporting Cells*, W. B. Saunders, Philadelphia.
11. Peters, A., and Vaughn, J. E., 1967, *J. Cell Biol.* **32**:113–119.
12. Friede, R. L., and Samorajski, T., 1970, *Anat. Rec.* **167**:379–388.
13. Metuzals, J., 1969, *J. Cell Biol.* **43**:480–498.
14. Lazarides, E., 1980, *Nature* **283**:249–256.
15. Wuerker, R. B., 1970, *Tissue Cell* **2**:1–9.
16. Gilbert, D. S., 1972, *Nature* (*New Biol.*) **237**:195–224.
17. Wuerker, R. B., and Kirkpatrick, J. B., 1972, *Int. Rev. Cytol.* **33**:45–75.
18. Ellisman, M. H., 1981, *Neurosci. Res. Prog. Bull.* **19**:43–58.
19. Tsukita, S., and Ishikawa, H., 1980, *J. Cell Biol.* **84**:513–530.
20. Shelanski, M. L., Leterrier, J.-F., and Liem, R. K. H., 1981, *Neurosci. Res. Prog. Bull.* **19**:32–43.
21. Yen, S.-H., and Fields, K. L., 1981, *J. Cell Biol.* **88**:115–126.
22. Chiu, F.-C., Korey, B., and Norton, W. T., 1980, *J. Neurochem.* **34**:1149–1159.
23. Jacobs, M., Choo, Q. L., and Thomas, C., 1982, *J. Neurochem.* **38**:969–977.
24. Tapscott, S. J., Bennett, G. S., and Holtzer, H., 1981, *Nature* **292**:836–838.
25. Pruss, R. M., Mirsky, R., Raff, M. C., Thorpe, R., Dowding, A. J., and Anderton, B. H., 1981, *Cell* **27**:419–428.
26. Hoffman, P. N., and Lasek, R. J., 1975, *J. Cell Biol.* **66**:351–366.
27. Schlaepfer, W. W., Lee, V., and Wu, H.-L., 1981, *Brain Res.* **226**:259–272.
28. Brown, B. A., Majocha, R. E., Staton, D. M., and Marotta, C. A., 1983, *J. Neurochem.* **40**:299–308.
29. Gambetti, P., Velasco, M. E., Dahl, D., Bignami, A., Roessmann, U., and Sindely, S. D., 1980, *Aging of the Brain and Dementia* (L. Amaducci, A. N. Dawson, and P. Antuono, eds.), Raven Press, New York, pp. 39–48.
30. Debus, E., Flugge, G., Weber, K., and Osborn, M., 1982, *EMBO J.* **1**:41–45.
31. Liem, R. K. H., Yen, S.-H., Salomon, G. D., and Shelanski, M. L., 1978, *J. Cell. Biol.* **79**:637–645.
32. Anderton, B. H., Thorpe, R., Cohen, J., Selvendran, S., and Woodhams, P., 1980, *J. Neurocytol.* **9**:835–844.
33. Matus, A. I., Meelian, N., and Jones, H. D., 1979, *J. Neurocytol.* **8**:513–525.
34. Shaw, G., Osborn, M., and Weber, K., 1981, *Eur. J. Cell Biol.* **24**:20–27.
35. Cravioto, H., 1966, *J. Comp. Neurol.* **126**:453–462.
36. Krishnan, N., Kaiserman-Abramof, I., and Lasek, R. J., 1979, *J. Cell Biol.* **82**:323–335.
37. Yamada, K. M., Spooner, B. S., and Wessells, N. K., 1971, *J. Cell Biol.* **49**:614–635.
38. Lentz, T. L., 1972, *J. Cell Biol.* **52**:719–732.
39. Wuerker, R. B., and Palay, S. L., 1969, *Tissue Cell* **1**:387–402.
40. Hodge, A. J., and Adelman, W. J., Jr., 1980, *J. Ultrastruct. Res.* **70**:220–241.
41. Bear, R. S., Schmitt, F. O., and Young, J. Z., 1937, *Proc. R. Soc.* (*Lond.*) [*Biol.*] **123B**:505–519.
42. Metuzals, J., and Izzard, C. S., 1969, *J. Cell Biol.* **43**:456–479.
43. Schliwa, M., van Blerkom, J., and Porter, K. R., 1981, *Proc. Natl. Acad. Sci. U.S.A.* **78**:4329–4333.
44. Ellisman, M. H., and Porter, K. R., 1980, *J. Cell Biol.* **87**:464–479.
44a. Ellisman, M. H., and Porter, K. R., 1983, *Neurofilaments* (C. A. Marotta, ed.), University of Minnesota Press, Minneapolis, pp. 3–26.
45. Schlaepfer, W. W., 1977, *J. Cell Biol.* **74**:226–240.
46. Willard, M., and Simon, C., 1981, *J. Cell Biol.* **89**:198–205.
47. Steinert, P. M., Idler, W. W., and Goldman, R. D., 1980, *Proc. Natl. Acad. Sci. U.S.A.* **77**:4534–4538.
48. Steinert, P. M., 1978, *J. Mol. Biol.* **123**:49–70.
49. Geisler, N., and Weber, K., 1981, *J. Mol. Biol.* **151**:565–571.
50. Milam, L., and Erickson, H. P., 1981, *J. Cell Biol.* **91**:235a.
51. DeVries, G. H., Norton, W. T., and Raine, C. S., 1972, *Science* **175**:1370–1372.
52. Shelanski, M. L., Albert, S., DeVries, G. H., and Norton, W. T., 1971, *Science* **174**:1242–1245.

53. Brown, B. A., Nixon, R. A., Strocchi, P., and Marotta, C. A., 1981, *J. Neurochem.* **36**:143–153.
54. Gilbert, D. S., Newby, B. J., and Anderton, B. H., 1975, *Nature* **256**:586–589.
54a. Schlaepfer, W. W., 1978, *J. Cell Biol.* **76**:50–56.
55. Schlaepfer, W. W., and Micko, S., 1979, *J. Neurochem.* **32**:211–219.
56. Pant, H. C., and Gainer, H., 1980, *J. Neurobiol.* **11**:1–12.
56a. Schlaepfer, W. W., and Lynch, R. G., 1977, *J. Cell Biol.* **74**:241–250.
57. Schlaepfer, W. W., and Zimmerman, U.-J. P., 1981, *Neurochem. Res.* **6**:243–255.
58. Schlaepfer, W. W., and Freeman, L. A., 1980, *Neuroscience* **5**:2305–2314.
59. Nixon, R. A., Brown, B. A., and Marotta, C. A., 1982, *Trans. Am. Soc. Neurochem.* **13**:237.
60. Nixon, R. A., and Froimowitz, 1982, *Trans. Am. Soc. Neurochem.* **13**:238.
61. Shook, W. J., and Norton, W. T., 1976, *Brain Res.* **118**:517–522.
62. Brown, B. A., Strocchi, P., Nixon, R. A., and Marotta, C. A., 1980, *Fed. Proc.* **39**:2049.
63. Marotta, C. A., Strocchi, P., Brown, B. A., Bonventre, J. A., and Gilbert, J. M., 1981, *Genetic Research Strategies for Psychobiology and Psychiatry* (E. S. Gershon, S. Matthysse, X. O. Breakefield, and R. D. Ciaranello, eds.), The Boxwood Press, Pacific Grove, CA, pp. 39–57.
64. Thorpe, R., Delacourte, A., Ayers, M., Bullock, C., and Anderton, B. H., 1979, *Biochem. J.* **181**:275–284.
65. Schlaepfer, W. W., 1977, *J. Ultrastruct. Res.* **61**:149–157.
66. Schlaepfer, W. W., and Freeman, L. A., 1978, *J. Cell Biol.* **78**:653–662.
67. Czosnek, H., and Soifer, D., 1980, *FEBS Lett.* **117**:175–178.
68. Anderton, B. H., Ayers, M., and Thorpe, R., 1978, *FEBS Lett.* **96**:159–163.
69. Czosnek, H., Soifer, D., Mack, K., and Wisniewski, H. M., 1981, *Brain Res.* **216**:387–398.
70. Schlaepfer, W. W., 1983, *Neurofilaments* (C. A. Marotta, ed.), University of Minnesota Press, Minneapolis, pp. 57–85.
71. Plancke, Y., Delacourte, A., and Biserte, G., 1981, *Biochimie* **63**:365–367.
72. Liem, R., 1982, *J. Neurochem.* **38**:142–150.
73. Schecket, G., and Lasek, R. J., 1980, *J. Neurochem.* **35**:1335–1344.
74. Small, J. V., and Sobieszek, A., 1977, *J. Cell Sci.* **23**:243–268.
75. Starger, J. M., Brown, W. E., Goldman, A. E., and Goldman, R. D., 1978, *J. Cell Biol.* **78**:93–109.
76. Hynes, R. O., and Destree, A. T., 1978, *Cell* **13**:151–163.
76a. Osborn, M., and Weber, K., 1977, *Exp. Cell Res.* **106**:339–349.
76b. Liem, R. K. H., and Hutchison, S. B., 1982, *Biochemistry* **21**:3221–3226.
77. Chiu, F.-C., Norton, W. T., and Fields, K. L., 1981, *J. Neurochem.* **37**:147–155.
78. Heuser, J. E., and Kirschner, M. W., 1980, *J. Cell Biol.* **86**:212–234.
79. Trotter, J. A., Foerder, B. A., and Keller, J. M., 1978, *J. Cell Sci.* **31**:369–393.
80. Webster, R. E., Henderson, D., Osborn, M., and Weber, K., 1978, *Proc. Natl. Acad. Sci. U.S.A.* **75**:5511–5515.
81. Chiu, F.-C., and Norton, W. T., 1982, *J. Neurochem.* **39**:1252–1260.
82. Ishii, T., Haga, S., and Tokutake, S., 1979, *Acta Neuropathol.* (*Berl.*) **48**:105–112.
83. Shelanski, M. L., Gaskin, F., and Cantor, C. R., 1973, *Proc. Natl. Acad. Sci. U.S.A.* **70**:765–768.
84. Stephens, R. E., and Edds, K. T., 1976, *Physiol. Rev.* **56**:709–777.
85. Snyder, J. A., and McIntosh, J. R., 1976, *Annu. Rev. Biochem.* **45**:699–720.
86. Olmsted, J. B., and Borisy, G. G., 1973, *Annu. Rev. Biochem.* **42**:507–534.
87. Berkowitz, S. A., Katagiri, J., Binder, H. K., and Williams, R. C., Jr., 1977, *Biochemistry* **16**:5610–5617.
88. Delacourte, A., Plancot, M.-T., Han, K.-K., Hildebrande, H., and Biserte, G., 1977, *FEBS Lett.* **77**:41–46.
89. Runge, M. S., Schlaepfer, W. W., and Williams, R. C., Jr., 1981, *Biochemistry* **20**:170–175.
90. Delacourte, A., Filliatreau, G., Boutteau, F., Biserte, G., and Schrevel, J., 1980, *Biochem. J.* **191**:543–546.
91. Eagles, P. A. M., Gilbert, D. S., and Maggs, A., 1981, *Biochem. J.* **199**:89–100.
92. Lasek, R. J., Krishnan, N., and Kaiserman-Abramof, I., 1979, *J. Cell Biol.* **82**:336–346.

93. Pant, H. C., Terakawa, S., and Gainer, H., 1979, *J. Neurochem.* **32:**99–102.
94. Willard, M. B., and Huleback, K. L., 1977, *Brain Res.* **136:**289–306.
95. Mori, H., Komiya, Y., and Kurokawa, M., 1979, *J. Cell Biol.* **82:**174–184.
96. Mori, H., and Kurokawa, M., 1980, *Biomed. Res.* **1:**24–31.
97. Schlaepfer, W. W., and Micko, S., 1978, *J. Cell Biol.* **78:**369–378.
98. Soifer, D., Iqbal, K., Czosnek, J., DeMartini, J., Sturman, J., and Wisniewski, H. M., 1981, *J. Neurosci.* **1:**461–470.
99. Zackroff, R. V., Idler, W. W., Steinert, P. M., and Goldman, R. D., 1981, *J. Cell Biol.* **91:**236a.
100. Moon, H. M., Wisniewski, T., Merz, P., DeMartini, J., and Wisniewski, H. M., 1981, *J. Cell Biol.* **89:**560–567.
101. Selkoe, D. J., Liem, R. K. H., Yen, S.-H., and Shelanski, M. L., 1979, *Brain Res.* **163:**235–252.
102. Griffin, J. W., Hoffman, P. N., Clark, A. W., Carroll, P. T., and Price, D. L., 1978, *Science* **202:**633–635.
103. Mori, J., and Kurokawa, M., 1979, *Cell Struct. Funct.* **4:**163–167.
104. Lasek, R. J., and Kaiserman-Abramof, I., 1977, *J. Cell Biol.* **75:**266a.
104a. Davison, P. F., and Winslow, B., 1974, *Neurobiology* **5:**119–133.
105. Nixon, R. A., Brown, B. A., and Marotta, C. A., 1982, *J. Cell Biol.* **94:**150–158.
106. Nixon, R. A., Brown, B. A., and Marotta, C. A., 1981, *Trans. Am. Soc. Neurochem.* **12:**172.
107. Julien, J.-P., and Mushynski, W. E., 1981, *J. Neurochem.* **37:**1579–1585.
108. Willard, M., 1983, *Neurofilaments* (C. A. Marotta, ed.), University of Minnesota Press, Minneapolis, pp. 86–116.
109. Willard, M. B., 1976, *Proc. Natl. Acad. Sci. U.S.A.* **73:**3641–3645.
110. Davison, P. F., and Jones, R. N., 1980, *Brain Res.* **182:**470–473.
111. Wisniewski, H., Shelanski, M. L., and Terry, R. D., 1968, *J. Cell Biol.* **38:**224–229.
112. DeVries, G. H., Eng, L. F., Lewis, D. L., and Hadfield, M. G., 1976, *Biochim. Biophys. Acta* **439:**133–145.
113. Hogue-Angeletti, R. A., Wu, H.-L., and Schlaepfer, W. W., 1982, *J. Neurochem.* **38:**116–120.
114. Geisler, N., Plessmann, U., and Weber, K., 1982, *Nature* **296:**448–450.
114a. Zackroff, R. V., Idler, W. W., Steinert, P. M., and Goldman, R. D., 1982, *Proc. Natl. Acad. Sci. U.S.A.* **79:**754–757.
115. Zackroff, R. V., and Goldman, R. D., 1980, *Science* **208:**1152–1155.
116. Metuzals, J., and Tasaki, I., 1978, *J. Cell Biol.* **78:**597–620.
116a. Metuzals, J., and Mushynski, W. E., 1974, *J. Cell Biol.* **61:**701–722.
117. Rice, R. V., Roslansky, P. F., Pascoe, N., and Houghton, S. M., 1980, *J. Ultrastruct. Res.* **71:**303–310.
118. Bertolini, B., Monaco, G., and Rossi, A., 1970, *J. Ultrastruct. Res.* **33:**173–186.
119. Runge, M. S., Detrich, H. W., and Williams, R. C., Jr., 1979, *Biochemistry* **18:**1689–1698.
120. Runge, M. S., Hewgley, P. B., Puett, D., and Williams, R. C., Jr., 1979, *Proc. Natl. Acad. Sci. U.S.A.* **76:**2561–2565.
121. Leterrier, J.-F., Liem, R. K. H., and Shelanski, M. L., 1981, *J. Cell Biol.* **90:**755–760.
122. Dahl, D., 1980, *FEBS Lett.* **111:**152–156.
123. Burridge, K., 1978, *Methods in Enzymology*, Volume 50 (S. P. Colowick and N. O. Kaplan, eds.), Academic Press, New York, pp. 119–133.
124. Autilio-Gambetti, L., Velasco, M. E., Sipple, J., and Gambetti, P., 1981, *J. Neurochem.* **37:**1260–1265.
124a. Brown, B. A., Majocha, R. E., Staton, D. M., and Marotta, C. A., 1982, *Trans. Am. Soc. Neurochem.* **13:**241.
125. Eng, L. F., Sternberger, N., and Sternberger, L., 1982, *Trans. Am. Soc. Neurochem.* **13:**108.
125a. Davison, P. F., 1968, *Neurosci. Res. Prog. Bull.* **6:**176–179.
126. Eng, L. F., Lasek, R. J., Bigbee, J. W., and Eng, D. L., 1980, *J. Histochem. Cytochem.* **28:**1312–1318.
127. Nixon, R. A., Brown, B. A., and Marotta, C. A., 1981, *Int. Soc. Neurochem.* (*Abstr.*) **8:**82.
128. Marotta, C. A., Strocchi, P., and Gilbert, J. M., 1979, *J. Neurochem.* **33:**231–246.
129. Bryan, R. N., Cutter, G. A., and Hayashi, M., 1978, *Nature* **272:**81–83.

130. Cleveland, D. W., Kirscher, M. W., and Cowan, N. J., 1979, *Cell* **15**:1021–1031.
131. Portier, M. M., Jeantet, C., and Gros, F., 1980, *Eur. J. Biochem.* **112**:601–609.
132. Hunter, T., and Garrels, J. I., 1977, *Cell* **12**:767–781.
132a. Bryan, R. N., Cutter, G. A., and Hayashi, M., 1978, *Nature* **272**:81–83.
133. Czosnek, H. H., Soifer, D., and Wisniewski, H. M., 1979, *Int. Soc. Neurochem. (Abstr.)* **7**:284.
134. Czosnek, H., Soifer, D., and Wisniewski, H. M., 1980, *J. Cell Biol.* **85**:726–734.
135. Strocchi, P., Dahl, D., and Gilbert, J. M., 1982, *Trans. Am. Soc. Neurochem.* **13**:242.
136. Mumford, R. A., Pickett, C. B., Zimmerman, M., and Strauss, A. W., 1981, *Biochem. Biophys. Res. Commun.* **103**:565–572.
137. Runge, M. S., El-Magharabi, M. R., Claus, T. H., Pilkis, S. J., and Williams, R. C., Jr., 1981, *Biochemistry* **20**:175–180.
138. Walsh, D. A., Ashby, C. D., Gonzales, C., Calkins, D., Fischer, E. W., and Krebs, E. G., 1971, *J. Biol. Chem.* **246**:1977–1985.
139. Schecket, G., and Lasek, R. J., 1979, *Trans. Am. Soc. Neurochem.* **10**:140.
139a. Glicksman, M. A., and Willard, M., 1982, *J. Neurochem.* **38**:1774–1776.
140. Pant, H. C., Yoshioka, T., Tasaki, I., and Gainer, H., 1979, *Brain Res.* **162**:303–313.
141. Eagles, P. A. M., Gilbert, D. S., and Maggs, A., 1981, *Biochem. J.* **199**:101–111.
142. Pant, H. C., Schecket, G., Gainer, H., and Lasek, R. J., 1978, *J. Cell Biol.* **78**:R23–R27.
143. Schlaepfer, W. W., 1971, *Exp. Cell Res.* **67**:73–80.
144. Schlaepfer, W. W., 1974, *Brain Res.* **69**:203–215.
145. Schlaepfer, W. W., and Zimmerman, U.-J., 1981, *J. Neuropathol. Exp. Neurol.* **40**:315.
146. Schlaepfer, W. W., Zimmerman, U.-J., and Micko, S., 1981, *Cell Calcium* **2**:235–250.
147. Nixon, R. A., 1983, *Neurofilaments* (C. A. Marotta, ed.), University of Minnesota Press, Minneapolis, pp. 117–154.
148. Brown, B. A., Nixon, R. A., and Marotta, C. A., 1982, *J. Cell Biol.* **94**:159–164.
148a. Brown, B. A., Nixon, R. A., and Marotta, C. A., 1981, *Trans. Am. Soc. Neurochem.* **12**:205.
149. Marks, N., and Lajtha, A., 1965, *Biochem. J.* **97**:74–83.
150. Marks, N., and Lajtha, A., 1971, *Handbook of Neurochemistry*, Volume 5 (A. Lajtha, ed.), Plenum Press, New York, pp. 49–139.
151. Nixon, R. A., and Marotta, C. A., 1984, *J. Neurochem.* (in press).
152. Schlaepfer, W. W., and Zimmerman, U.-J. P., 1981, *Int. Soc. Neurochem. (Abstr.)* **8**:20.
153. Hodgkin, A. L., and Katz, B., 1949, *J. Exp. Biol.* **26**:292–294.
154. Hodgkin, A. L., and Keynes, R. D., 1956, *J. Physiol. (Lond.)* **131**:592–616.
155. Orrego, F., 1971, *J. Neurochem.* **18**:2249–2254.
156. Truglia, J. A., and Stracher, A., 1981, *Biochem. Biophys. Res. Commun.* **100**:814–822.
157. Terry, R. D., and Davies, P., 1980, *Annu. Rev. Neurosci.* **3**:77–95.
158. DeBoni, U., and Crapper McLachlin, D. R. C., 1980, *Life Sci.* **27**:1–14.
159. Katzman, R., Terry, R. D., and Bick, K. L. (eds.), 1978, *Alzheimer's Disease: Senile Dementia and Related Disorders*, Raven Press, New York.
160. Wisniewski, H. M., and Iqbal, K., 1980, *Trends Neurosci.* **3**:226–228.
161. Wisniewski, H. M., and Terry, R. D., 1976, *Neurobiology of Aging* (R. D. Terry and S. Gershon, eds.), Raven Press, New York, pp. 265–280.
162. Amaducci, L., Davison, A. N., and Antuono, P. (eds.), 1980, *Aging of the Brain and Dementia*, Raven Press, New York.
162a. Wisniewski, H. M., Merz, G. S., Merz, P. A., Wen, G. Y., and Iqbal, K., 1983, *Neurofilaments* (C. A. Marotta, ed.), University of Minnesota Press, Minneapolis, pp. 196–221.
163. Hirano, A., and Zimmerman, 1962, *Arch. Neurol.* **7**:227–242.
164. Terry, R. D., 1963, *J. Neuropathol. Exp. Neurol.* **22**:629–642.
165. Kidd, M., 1963, *Nature* **197**:192–193.
166. Wisniewski, H. M., Narang, H. K., and Terry, R. D., 1976, *J. Neurol. Sci.* **27**:173–181.
167. Terry, R. D., Gonatas, N. K., and Weiss, M., 1964, *Am. J. Pathol.* **44**:269–297.
168. Terry, R. D., 1980, *Aging of the Brain and Dementia* (L. Amaducci, A. N. Davison, and P. Antuono, eds.), Raven Press, New York, pp. 49–54.
169. Selkoe, D. J., Ihara, Y., and Salazer, F. J., 1982, *Science* **215**:1243–1246.
170. Wisniewski, H. M., Terry, R. D., and Hirano, A., 1970, *J. Neuropathol. Exp. Neurol.* **29**:163–176.

171. Hirano, A., Dembitzer, H. M., and Kurland, L. T., 1968, *J. Neuropathol. Exp. Neurol.* **27**:167–182.
172. Burger, P. G., and Vogel, F. S., 1973, *Am. J. Pathol.* **73**:457–476.
173. Roth, M., 1980, *Aging of the Brain and Dementia* (L. Amaducci, A. N. Davison, and P. Antuono, eds.), Raven Press, New York, pp. 1–21.
174. Gonatas, N. K., Anderson, A., and Evangelista, I., 1967, *J. Neuropathol. Exp. Neurol.* **26**:25–39.
175. Grundke-Iqbal, I., Johnson, A. B., Terry, R. D., Wisniewski, H. M., and Iqbal, K., 1979, *Ann. Neurol.* **6**:532–537.
176. Yen. S.-H. C., Gaskin, F., and Terry, R. D., 1981, *Am. J. Pathol.* **104**:77–89.
177. Ihara, Y., Nukina, N., Sugita, H., and Toyokura, Y., 1981, *Proc. Jpn. Acad.* **57B**:152–156.
178. Iabal, K., Grundke-Iqbal, I., Johnson, A. B., and Wisniewski, H. M., 1980, *Aging of the Brain and Dementia* (L. Amaducci, A. N. Davison, and P. Antuono, eds.), Raven Press, New York, pp. 39–48.
179. Grundke-Iqbal, I., Johnson, A. B., Wisniewski, H. M., Terry, R. D., and Iqbal, K., 1979, *Lancet* **1**:578–580.
180. Eng, L. F., Forno, L. S., Bigbee, J. W., and Forno, K. I., 1980, *Aging of the Brain and Dementia* (L. Amaducci, A. N. Davison, and P. Antuono, eds.), Raven Press, New York, pp. 49–54.
181. Dahl, D., and Bignami, A., 1977, *J. Comp. Neurol.* **176**:645–658.
182. Dahl, D., Selkoe, D. J., Pero, R. T., and Bignami, A., 1982, *J. Neurosci.* **2**:113–119.
182a. Anderton, B. H., Breinburg, D., Downes, M. J., Green, P. J., Tomlinson, B. E., Ulrich, J., Wood, J. N., and Kahn, J., 1982, *Nature* **298**:84–86.
183. DeBoni, U., and Crapper, D. R., 1978, *Nature* **271**:566–568.
184. Wisniewski, H. M., Sinatra, R. S., Iqbal, K., and Grundke-Iqbal, I., 1981, *Aging and Cell Structure*, Volume 1 (J. E. Johnson, Jr., ed.), Plenum Press, New York, pp. 105–142.
185. Marotta, C. A., Strocchi, P., Brown, B. A., and Gilbert, J. M., 1981, *Psychiatry and the Biology of the Human Brain* (S. Matthysse, ed.), Elsevier/North Holland, New York, pp. 71–87.
185a. Laemmli, U. K., 1970, *Nature* **227**:680–685.
185b. O'Farrell, P. H., 1975, *J. Biol. Chem.* **250**:4007–4021.
186. Selkoe, D. J., 1980, *Ann. Neurol.* **8**:468–478.
187. Selkoe, D. J., Brown, B. A., Salazar, F. J., and Marotta, C. A., 1981, *Ann. Neurol.* **10**:429–436.
188. Selkoe, D. J., and Shelanski, M. L., 1977, *The Aging Brain and Senile Dementia* (K. Nandy and I. Sherwin, eds.), Plenum Press, New York, pp. 247–263.
188a. Kosik, K. S., and Selkoe, D. J., 1983, *Neurofilaments* (C. A. Marotta, ed.), University of Minnesota Press, Minneapolis, pp. 155–195.
189. Czosnek, H., Soifer, D., and Wisniewski, H. M., 1980, *Neurochem. Res.* **5**:777–793.
190. Filliatreau, G., Giamberadino, I. D., Delacourte, A., Boutteau, F., and Biserte, 1981, *Biochimie* **63**:369–371.

12

Local Synthesis of Axonal Protein

Edward Koenig

1. BACKGROUND

1.1 Introduction

Given that the axon is an extended cellular process remote from the neuron's metabolic center, the supply of newly synthesized proteins to the axon is a question of fundamental importance. It is well established that one mode by which this is accomplished is by synthesis of axonal proteins in the cell body, whereupon products reach the axon by intracellular transport. There are, nonetheless, two additional modes that need to be considered. One involves local synthesis in cells of ensheathment (e.g., glial cells) and the supply of products to the axon by intercellular transfer. This appears to be a means for locally supplementing some axonal proteins in unmyelinated invertebrate axons (see Section 4). Another mode is that of local synthesis in the axon. The prevailing view, however, is that the axon is capable of little or no endogenous protein synthesis other than that perhaps associated with mitochondria.

The latter view is predicated by two principal assumptions that need to be carefully scrutinized. One assumption is that intracellular transport from a central site of synthesis is sufficient to supply all local macromolecular needs of the axon. The second assumption is based on the paucity or apparent lack of axoplasmic ribosomes, which is taken as indicative of little or no endogenous protein-synthesizing activity.

The first assumption, particularly as it applies to polypeptides transported at slow rates of <4 mm/day, implies that the biological half-life of such proteins in long axons must be unusually extended when compared to the average half-life of these same proteins in whole brain; otherwise, they would not survive the time required for transit. The validity of this assumption is questionable, as there is now evidence that degradation occurs during slow transport (see Section 1.3). The significance of the "negative" evidence on which the second assumption is based has now also been brought into question by at least two

Edward Koenig • Division of Neurobiology/Department of Physiology, State University of New York at Buffalo, Buffalo, New York 14214.

recent analytical studies of axoplasm in which ribosomal RNA (rRNA) species have been demonstrated, notwithstanding apparent lack of visible ribosomes (see Section 2.2).

This chapter focuses on the general problem of local synthesis and reviews published and unpublished evidence that there is measurable endogenous protein-synthesizing activity in the axon that is of biological significance. Although attention centers primarily on the vertebrate myelinated axon, evidence related to the sheath cell as a local source of axonal proteins in invertebrate unmyelinated axons is also reviewed. Selected aspects of the biology and pathobiology of the axon are also included in order to provide a context for interpreting the possible significance of local synthesis. Before proceeding, however, it would be well to consider briefly what contributes to the uniqueness of the axon and why a specialized approach is needed to study endogenous axonal protein synthesis.

The myelinated fiber has certain structural features that set it apart from the neuronal perikaryon, dendrites, and other cells as well. Specifically, the myelin sheath, which is intimately associated with the axon, probably plays a significant role in isolating the axon as a metabolic compartment. Thus, direct access to endoneural space is available to the axon only at a node of Ranvier, a rather narrow gap in the myelin sheath of 1–1.5 μm. Moreover, the nodal axolemma is specialized and densely packed with Na^+,K^+-ATPase and sodium channels. This type of structural arrangement and restricted specialized gap membrane are likely to impede the free exchange of amino acids across the axonal membrane. On the other hand, there may be little need for a dependence on an exogenous supply of amino acids since endogenous protein catabolism (see Section 1.3) in the confined axonal compartment may be sufficient to maintain saturation of the free amino acid pool in the axon.

Studies of axonal protein synthesis are technically difficult. The principal problem is, again, related to the presence of cells of ensheathment (i.e., Schwann cells, oligodendrocytes). Such cells are metabolically active and, therefore, would mask the potentially low level of endogenous axonal protein synthesis. This makes it essential to analyze axoplasmic samples uncontaminated by myelin, an imperative that cannot be realized by standard biochemical approaches. Although histoautoradiographic techniques can yield information regarding relative rates of incorporation and localization of labeled product, the technique is indirect and is also subject to limitations in spatial resolution and sensitivity. A specialized approach has been developed in the author's laboratory in which microscopic samples of myelin-free axons can be isolated for purposes of direct quantitative and compositional analyses of radioactive proteins on a microscale.

1.2. Major Axonal Proteins

The average protein mass of an axon in the hypoglossal nerve of the rabbit is about 4.3 ng/cm.[1] It is a reasonable assumption that the protein mass of a cell body in the hypoglossal nucleus is not likely to be much more. This leads to the conclusion that axons in the hypoglossal nerve exceed the mass of their cell bodies severalfold. No doubt, the axon-to-cell-body protein mass ratio can approach a value of 100 and much more, given the rather extended length of many axons in large vertebrate species. The preponderance of axonal mass is attributable to proteins comprising the three principal cytoskeletal systems of the neuron, namely, 10-nm neurofilaments, 25-nm microtubules, and 5-nm actin microfilaments.

Polypeptides comprising these systems, in addition to soluble proteins, are supplied to the axon by slow axoplasmic transport, i.e., group IV or slow component b (SCb) (2–4 mm/day) and group V or slow component a (SCa) (0.2–1 mm/day)[2–6] (see ref. 7 for comprehensive review). The slowest-moving group of polypeptides contains a triplet[2] that corresponds to proteins derived from purified neurofilaments, having molecular weights of 68–72 K, 140–170 K, and 200–220 K.[8–12] In addition, α-(57 K) and β-(53 K) tubulins, subunits of microtubules[6,13,14] as well as minor microtubule-associated tau proteins (62 K, 64 K[6]) are also transported in this group. The cotransport of polypeptides of these two cytoskeletal systems has prompted Black and Lasek[6] to propose that neurofilaments and microtubules move coherently as assembled systems in the axon. Although the principal role of neurofilaments appears to be cytoskeletal in a purely structural sense, microtubules and actin microfilaments are believed to play, among other things, a role in intracellular transport (see ref. 7).

The third cytoskeletal system is composed of actin microfilaments. Actin (43 K) is transported in the faster-moving group IV or SCb.[5,15] A high-molecular-weight doublet (240 K, 250 K) recently characterized in neural tissue and called fodrin[16] comigrates with actin[5] and has been shown to bind actin and calmodulin.[16,17] In many respects, fodrin bears some homology to red cell α spectrin with regard to both certain immunologic determinants[17,18] and its cortical distribution in the cell and apparent association with the cytoplasmic surface of the plasmalemma.[16–18] Although actin is associated with the plasmalemma, perhaps through its complex with fodrin (α spectrin), it is also distributed in the cytoplasm as well.[19]

Neurofilaments confer to the axon mechanical properties of an elastic solid,[20] which makes it possible to translate axons physically out of their myelin sheaths with microtweezers under conditions that preserve integrity of neurofilaments.[21,22] Unlike microtubules and microfilaments, which undergo assembly and disassembly reversibly in the cell, neurofilaments are dissassembled through a process of degradation catalyzed by a calcium-activated neutral protease.[23–25] The microtweezer technique of isolating axons, therefore, can only be applied under conditions that inhibit the enzyme.[22]

The compositional profile of major polypeptides in axons isolated from spinal roots of the rabbit with the microtweezer technique is shown in Fig. 1. From inspection, it is clear that identified proteins of the major cytoskeletal systems dominate, and it is likely that some presently unidentified components may also belong to the complement of "structural" axonal proteins.

1.3. Metabolic Stability of Slowly Transported Polypeptides

Implicit in the idea that there is no local axonal protein synthesis is that proteins transported at a slow rate (<4 mm/day) are metabolically stable. Until

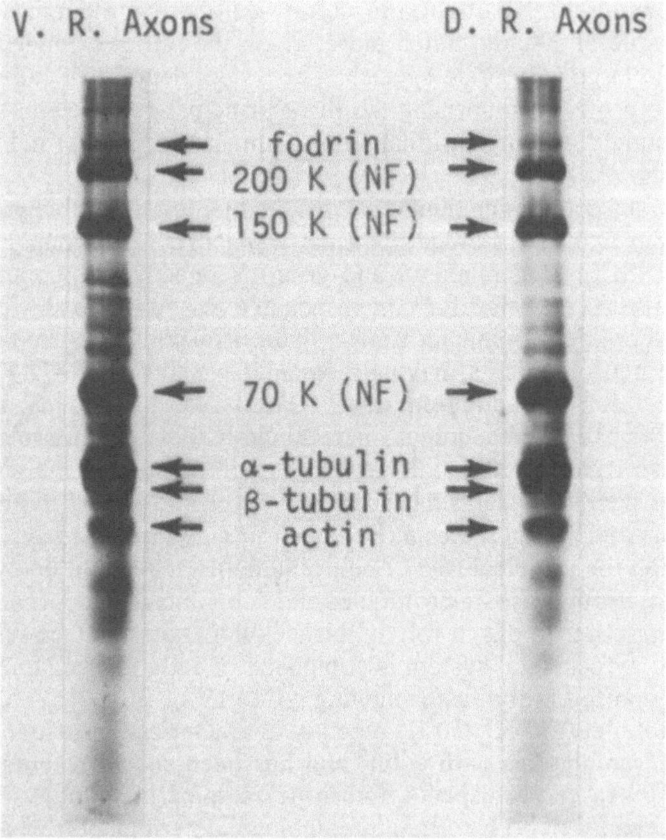

Fig. 1. Identifiable polypeptides of the three major cytoskeletal systems in axons of rabbit ventral root (V.R.) and dorsal root (D.R.). Original microelectrophoretogram, 1.4 cm (NF, neurofilament).

recently there were no studies that addressed the issue of metabolic stability of proteins undergoing slow transport while rigorously eliminating reutilization as a complicating factor. Nixon[26] has now published an analysis of endogenous proteolysis of slowly transported retinal ganglion cell proteins in the primary optic pathway of the mouse under conditions that exclude reutilization of labeled amino acids. He determined the rate of endogenous degradation of labeled axonal proteins by *in vitro* incubation in the presence or absence of cycloheximide after dissection of optic nerve and tract. Nixon observed both acid- (pH 3.8) and calcium-stimulated neutral (pH 7.4) proteolytic systems present. The rate of proteolysis was comparable to that of retina and to that reported by others for brain. In addition, Nixon observed that in the absence of cycloheximide during *in vitro* incubation, the estimated average half-life was increased threefold, indicating local reutilization of labeled amino acids.

Although the foregoing finding does not indicate that reutilization necessarily occurred intraaxonally, there are observations related to the temporal dispersion of labeled proteins during slow transport that are consistent with such an interpretation. For example, radioactive peaks comprising proteins of

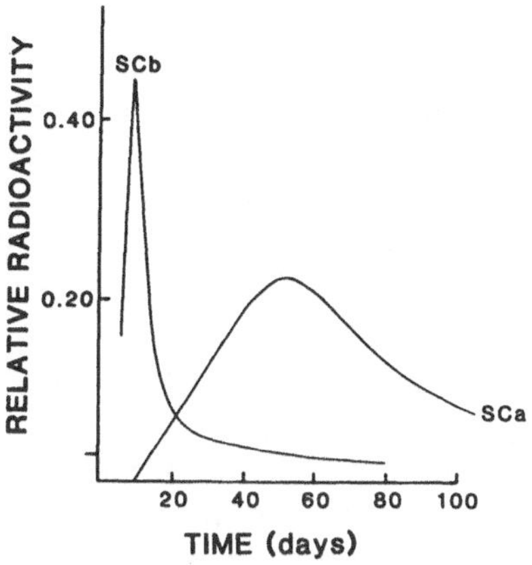

Fig. 2. Radioactive polypeptides in transport groups SCb (2–4 mm/day) and SCa (0.5–1 mm/day) "moving" with time through a 3-mm segment of the optic nerve of guinea pig. The fact that radioactive profiles do not decay to zero with time after the peak has passed may indicate local reutilization (see text for details). (Redrawn from Fig. 4 in ref. 6 with permission from the publisher.)

SCb and SCa (i.e., slow transport groups IV and V) leave a persistent residuum of labeled proteins long after the bulk of the proteins have "moved" through a given region of the axon[27] (Fig. 2). The apparent stasis of labeled proteins has been interpreted by Black and Lasek[27] as a reflection of "deposition" of a small fraction of proteins during transit. However, in view of Nixon's findings,[26] referred to above, an equally likely interpretation is that labeled amino acids released by endogenous proteolysis are reincorporated by local axonal protein synthesis.

A more extensive dispersion of slowly transported labeled proteins than that suggested by Fig. 2 appears to be indicated by a very recent study of Stromska and Ochs.[28] Indeed, these investigators challenge the validity of the concept of slow transport. Based on an analysis of pattern changes of longitudinal labeling profiles occurring with time along sensory axons in the sciatic nerve of the cat and rat and in axons in dorsal columns of the cat, Stromska and Ochs concluded that there were no discrete waves that systematically "moved" along the nerve at rates corresponding to those of SCb and SCa as characterized by Hoffman and Lasek[2] in motor axons of the rat. Rather, the longitudinal labeling profiles showed extended elevated plateaulike distributions with poorly defined low-amplitude "wavelets" appearing at irregular intervals. Obviously, the disparity between the findings of Hoffman and Lasek[2] and those of Stromska and Ochs[28] cannot be attributed to fundamental differences between motor and sensory axons. It seems likely that it will be necessary to give much more cognizance than has been the case in the past to the importance of local protein metabolic activity, both synthetic and catabolic, in order to explain some of the experimental findings related to slow transport.

Most recently, Nixon and co-workers have extended their observations by studying subtle endogenous proteolytic modifications of slowly transported proteins during transit in the primary optic pathway of the mouse. Specifically, the intermediate-size neurofilament polypeptide (i.e., "140 K") exhibited regional microheterogeneity, with two major species (145 K and 140 K) present along the whole pathway, whereas a third polypeptide (143 K) showed increasing prominance in the distal portion of the pathway.[29] The posttranslational proteolytic conversion showed calcium dependence with a sensitivity in the micromolar range. Consistent with this proximodistal difference in microheterogeneity was that Nixon *et al.* could demonstrate conversion to the 143 K *in vitro* in the optic tract but not in the more proximal optic nerve. Similarly, compositional makeup of α-tubulin isoforms was also modified during transport, whereas that of β-tubulin isoforms was not.[30] These observations indicate that there are regional axonal differences in posttranslational processing of selected polypeptides and caution in a compelling way against underestimating the potential complexity of metabolic processes that characterize the biology of the axon.

As a final point, it should be noted that the computed half-lives of the tubulins and actin in rat brain are 3.9 days for the former[31] and 4.4 days for the latter.[32] If these estimates can be extrapolated to corresponding polypeptides in transport groups V and IV, respectively, in the axon, then it is apparent that both actin and the tubulins would be reduced to fractions of their initial amounts after transport over a distance of several centimeters.

2. *AXONAL RNA*

2.1. *Morphological Evidence of Ribosomes*

A feature of axoplasm, noted repeatedly when it is examined at the electron microscopic level, is the paucity or apparent lack of ribosomal particles. In reports that have described the presence of ribosomes, they tend to be found in proximal portions of the axon, such as the initial segment,[33,34] or in proximal myelinated segments.[35–41] Ribosomes have also been reported in growing axons of embryos[42–44] and in neurites of cultured sympathetic ganglion cells.[45]

A curious cellular inclusion, called a nematosome,[46] has been described in a variety of neuronal cell types and has been characterized morphologically as being nucleoluslike[47] because it is composed of dense filamentous and granular components. Cytochemically, nematosomes exhibit a basophilia that can be eliminated by ribonuclease digestion,[47,48] which has led to the inference that they contain RNA. LeBeux[46] and others (for references, see ref. 47) have noted that nematosomes are occasionally identified in dendrites and axons and are relatively common in certain types of neurons. In addition, LeBeux[46] has noted that they are frequently linked to ribosomes through microfilaments. In a recent study by Hamóri and Lakos,[49] namatosomes, which are not ordinarily seen in normal Purkinje cells, were regularly observed in the initial segment of the axon and in recurrent axonal collateral terminals of Purkinje cells of the rat 15 and 30 days following transection of Purkinje cell axons in cerebellar white matter. The hypertrophic state of recurrent collateral axons of Purkinje cells and the incidence of nematosomes suggested to these workers that nematosomes may have contributed to local protein-synthesizing activity. Unfortunately, because these interesting cellular inclusions have not been characterized biochemically, it is difficult to draw any definitive conclusions regarding their biological significance.

2.2. *Analysis of Axonal RNA*

Questions about the endogenous capacity of axons to synthesize proteins locally prompted the first analysis of axonal RNA by Edström, Eichner, and Edström.[50] This study and subsequent ones probing questions related to axonal RNA and protein synthesis by the Göteborg group in the 1960s were conducted on a large-diameter myelinated axon in goldfish and carp. The model axon was that of the Mauthner neuron, which ranges approximately from 40 to 80 μm in diameter[51] and extends from the pons to the caudal spinal cord, mediating the tail-flip reflex, an escape response of fish and larval forms of amphibia (see ref. 52). The Mauthner axon studies have been reviewed recently in detail.[22]

Ribonuclease digests of microdissected segments of Mauthner axon (M-axon) revealed the presence of RNA.[50] One centimeter of M-axon contained about 1.5 ng of RNA, which ranged in concentration from 0.03% to 0.07% (w/v),[53] although in another study RNA content averaged about 3 ng/cm.[54] Although the concentration is about one-fifth of that of the M-cell body,[50] the total RNA content of a length of M-axon can exceed that of the cell body by severalfold.

Similar microanalytical methods were applied to axons isolated from the spinal accessory nerve root of the cat[21] and rabbit[55] using the microtweezer technique (see Section 1.2). Values of RNA concentration reported at that time, however, of 0.003%[21] and 0.006%[55] were underestimated by a factor of 10, which would now place the average concentration of cat and rabbit axons in the same range as that reported for the M-axon[53] and for invertebrate axons, including the axon of the stretch receptor sensory neuron (0.06%)[56] and the giant axon of the crayfish (0.02%)[57] as well as the giant axon of the squid.[58] However, RNA concentration in dystrophic regions of proximal motor axons of the rat (i.e., axonal balloons) produced by β,β′-iminodiproprionitrile intoxication is 0.25%.[59] The adenine/guanine base ratio of RNA extracted from those vertebrate axons analyzed is quite similar and indicative of a ribosomal–tRNA-like composition.

Earlier analytical studies of axoplasm of giant axons of the squid and the marine worm *Myxicola* led to the conclusion that there was no detectable ribosomal RNA (rRNA), only a preponderance of 4 S RNA.[60] Very recently, however, Giuditta *et al.*[61] have demonstrated that axoplasm of the squid giant axon does contain small amounts of rRNA and confirmed the prevalence of 4 S RNA. In addition, they occasionally observed a minor high-molecular-weight RNA component that comigrated in the gel with a similar minor component present in RNA extracts from the giant fiber lobe of the squid. Both Giuditta *et al.*[62] and Black and Lasek[63] observed aminoacyl-tRNA synthetase activity in axoplasm of squid giant fiber.

Analysis of the goldfish M-axon, isolated by the microtweezer technique, showed that rRNA was present in significant proportions in both the axon and the myelin sheath, although 4 S RNA still represented a major fraction in the axon (40%) (Fig. 3) and in the sheath as well.[64] In addition, a minor nominal 15 S component was consistently seen in all microelectrophoretic profiles of M-axonal RNA extracts, whereas the component was not detectable in extracts

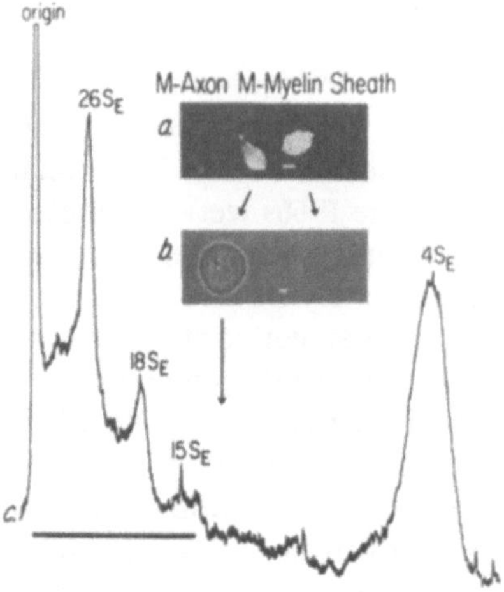

Fig. 3. A composite showing (a) dried microscopic samples of Mauthner axon and Mauthner fiber myelin sheath of similar size, (b) relative yield of RNA recovered by extraction of samples shown in a, and (c) the microdensitometric profile of Mauthner axonal RNA shown in b separated by microelectrophoresis. Note the difference in RNA yield from the axon and sheath samples in b; also note the 15 S_E (not detected in microsomal fraction from facial lobe or in myelin sheath) and the large proportion of 4 S_E. Calibration bars are (a) 0.2 mm, (b) 20 μm, and (c) 1 mm. (From Ref. 64 with permission of the publisher.)

from myelin sheath or from facial lobe of the brain. On a dry weight basis, RNA was at least tenfold more concentrated in the axon than in the sheath (Fig. 3); however, on a unit-length basis, RNA content was similar.[53] The disparity is explained by the likely distribution of RNA being limited to cytoplasmic spirals making up the Schmidt–Lantermann incisures of the myelin sheath, constituting a low volume-to-mass ratio.

It is important to emphasize in the context of the M-axon study that ribosomes have been observed neither in axoplasm[65] nor in cytoplasm of the Schmidt–Lantermann incisures of the myelin sheath.[65,66] It is clear, therefore, that the lack of electron microscopic evidence of ribosomes is not of itself sufficiently rigorous to rule out the presence of ribosomes.

The final point that should be made is that 4 S RNA in axoplasm is disproportionately large when compared to a perikaryonal extract or a typical microsomal fraction. The latter cases represent examples in which ribosomes are highly concentrated. The large proportion of 4 S RNA relative to the small proportions of ribosomal RNA in the axon can be explained if it is assumed that axoplasmic ribosomes are sparsely distributed, requiring disproportionately more tRNA in order to satisfy local requirements for translational activity. A disproportionately large 4 S RNA is also seen in studies of axoplasmic RNA transport (see Section 2.3.2 below)

2.3. Sources of Axonal RNA

There are three potential sources of axonal RNA. Two sources are local and include the ensheathing cell (oligodendroglial cell, Schwann cell) and endogenous axonal mitochondria. The third source, of course, is the cell body.

2.3.1. Local Synthesis of Axonal RNA

The first suggestive evidence that there may be a local synthesis of axonal RNA emerged when *in vitro* incubation of segments of rabbit spinal accessory nerve yielded actinomycin-D-sensitive incorporation of labeled RNA precursors into axons of the nerve.[55] Axons were isolated by the microtweezer technique. However, a large fraction of actinomycin-D-sensitive radioactivity was not solubilized by ribonuclease digestion of the axoplasmic protein residue even after exhaustive digestion.[67] In retrospect, this was most likely because of the inaccessibility of the labeled RNA associated with the highly insoluble axonal residue rather than an RNA that was resistant to ribonuclease, as was considered probable at the time.

A major study conducted on the Mauthner axon by Edström *et al.*[68] yielded strong evidence for a local DNA-dependent axonal RNA synthesis. Analysis of radioactive axonal RNA was carried out (1) after incubation of microdissected M-fibers, composed of the axon and its associated myelin sheath lacking glial cell bodies (i.e., M-fiber preparation), and (2) after incubation of spinal cord segments with radioactive RNA precursors. Both axon and cell-free myelin sheath components of the isolated fiber showed actinomycin-D-sensitive incorporation into RNA. The amount incorporated, calculated on the basis of length, was about the same or more for the axon as for the sheath. Sedimentational analysis of labeled RNA after *in vitro* incubation of the M-fiber preparation showed a single peak in the 4 S zone of the gradient. However, a similar analysis of labeled RNA extracted from the fiber microdissected after incubation of spinal cord segments also revealed radioactive peaks in the 16 S and 28–30 S as well as in the 4 S portions of the gradient.

The investigators commented on the likelihood that the labeled high-molecular-weight RNA species were probably synthesized in oligodendroglial cell bodies and supplied to the fiber in the *in situ* preparation and that mitochondria indigenous to the sheath and axon were the probable sources of the labeled 4 S in the *in vitro* fiber preparation. Unfortunately, the distribution of the high-molecular-weight RNAs in the M-fiber preparation was not determined, and it is possible that they may have been exclusively associated with the myelin sheath (it is perhaps worth noting here that the myelin sheath of the Mauthner axon is continuous,therefore, there are no nodal gaps.[22,66] The Schmidt–Lantermann incisures, containing microtubules and intermediate filaments, are quite complex from a fine-structural standpoint.[66,68] Finally, an additional conclusion that emerged from this study was that an estimate of RNA turnover in M-fiber components indicated a very high rate when compared to spinal cord RNA as a whole.

Other experiments that suggest the possibility of a close local metabolic coupling between the sheath and the axon of the M-neuron are the parallel changes in RNA base composition of the sheath and the axon induced by spinal cord transection.[53] Cordotomy in goldfish resulted in an increase in adenine/guanine ratio within 12 h in both the sheath and the axon of the proximal M-fiber stump; the ratio of purines then gradually declined in both components over the ensuing 2 to 3 weeks. This interesting effect merits reexamination from the standpoint of analyzing changes in compositional makeup of undegraded RNA species in the sheath and axonal components of the M-fiber. Pulse-chase experiments, in which rabbit spinal accessory nerves were incubated with [^{3}H]RNA precursors and chased with unlabeled precursors and actinomycin D, also suggested the possibility of an extraaxonal source of labeled axonal RNA.[67] The latter experiments should be repeated using newer techniques to isolate and characterize axonal RNA.

2.3.2. Axoplasmic Transport of RNA

Early studies[69–77] showed that after labeling of neuronal cell bodies, radioactive RNA appeared along the course of their axons. Unlike fast axoplasmic transport of labeled proteins, which shows up as a discrete wave, labeled RNA in the nerve is distributed in a highly dispersed fashion. The distributional profile appears as a steep proximodistal gradient that simply diminishes in amplitude with time while still maintaining a proximodistal gradient.[70,76,79]

Another important distinction that emerged from the foregoing studies was that radioactive RNA precursors also appeared in the axon in advance of labeled RNA and was also distributed along a steep proximodistal gradient, the amplitude of which decayed with time.[69–71,76,77] Autoradiographic studies indicated that radioactive RNA was distributed in periaxonal structures as well as in axons,[79,80] with the largest fraction of grains located over periaxonal glial cell nuclei, cytoplasm, and myelin sheath.[80] It was apparent from these studies that labeled RNA precursors were transported in advance of the appearance of labeled RNA in the nerve and were then transferred from axon to periaxonal sheath cells, where they underwent local incorporation into glial cell RNA.

At present, there appears to be some general agreement among those working in the field that RNA is transported to the axon, particularly in regenerating axons and in axons of neonatal animals, but there is disagreement as to what RNA species are being transported The importance of this problem merits discussion of some of the experimental work.

Ingoglia and Tuliszewski[81] concluded from their study that 4 S RNA is transported in regenerating axons of the optic nerve in goldfish. The design of the principal experiment entailed injecting [^{3}H]uridine intraocularly, and, after time was allowed for "transport" in the optic nerve, [^{14}C]uridine was injected intracranially. The rationale for the double-labeling procedure was that ^{3}H-labeled tectal RNA would comprise both locally synthesized and axonally transported RNA, whereas ^{14}C-labeled RNA would represent only locally synthesized products. Regenerating/control ratios of radioactivities associated with each of the 28 S, 18 S, and 4 S RNA species were compared on the basis of the route the precursor had been given. The ratios for 28 S, 18 S, and 4 S RNAs derived from the intraocular injection were 3.0, 3.4, and 8.0, respectively; the corresponding ratios derived from the intracranial injection were 1.2, 1.5, and 1.5. The disproportionate ratio of 4 S RNA in the former case indicated to the investigators that 4 S was transported in the newly regenerated optic nerve axons. The elevated ratios of the ribosomal RNAs, however, were not deemed sufficiently large to warrant the conclusion that they were also transported, particularly in view of autoradiographic evidence of extraaxonal labeling[80,82] and the finding by Lasek *et al*[60] that only 4 S RNA was present in squid giant axoplasm (see Section 2.2). Subsequent studies[83-85] confirmed and extended these initial observations.

Recently, the Australian group arrived at a similar conclusion after studying the chick optic pathway and employing a somewhat similar paradigm of using two routes of administration to evaluate transported and locally synthesized RNA.[86] Either [^{3}H]uridine or [methyl-^{3}H]methionine was used to label ribosomal and 4 S RNA species. Comparison of specific radioactivities with either precursor resulted in preferential elevated ratios of 4 S:29 S/18 S after intraocular but not after intracranial injection. In a subsequent study,[87] a more sophisticated technique was used to evaluate local glial cell RNA synthesis. The specific radioactivity of a low-molecular-weight nuclear RNA species, designated DD′, was used to normalize the specific radioactivities of optic nerve and tectal ribosomal and 4 S RNA species. In all instances, the 29 S/DD′, 18 S/DD′, and 5 S/DD′ ratios were similar irrespective of the route of administration; however, the 4 S/DD′ ratio was two to three times greater only after intraocular injection. The latter results provide very compelling evidence that 4 S RNA is transported but that ribosomal RNA is not transported in the chick optic nerve (however, see below for an alternative interpretation).

Earlier experiments by Bondy and co-workers appeared to provide evidence for the transport of ribosomes in the chick optic nerve pathway[88] as well as to suggest the possibility that polyadenylated RNA (i.e., mRNA) may also be transported.[89] The principal approach in the former study was to label retinal ribosomes of the chick with radioactive amino acids and then compare radioactivity associated with purified ribosomes from the retina and optic tecta after appropriate time intervals following intraocular injection. Ribosomes isolated from the tectum innervated by fibers from the noninjected retina were not labeled, whereas ribosomes from the contralateral tectum were labeled. They reported that they could not " . . . detect a significant migration of free amino acids within [their] system." In addition, injection of 0.1 μg actinomycin D blocked the appearance of labeled RNA in the tectum by about 65%. Unfortunately, in the latter experiment, no information was provided about the effect of actinomycin D on the axonal transport of acid-soluble RNA precursors. In any case, although these results are suggestive, they are not readily reconciled with those of Gunning *et al.*[87]

The conclusion that only a 4 S RNA is transported is based on the large difference in the labeling of this species when an RNA precursor is administered intraocularly as opposed to when administration is intracranial. A tacit assumption in all such studies is that "local synthesis" of RNA takes place exclusively or predominantly in periaxonal glial cells and not in the axon to any significant extent. In view of the direct demonstration by Edström *et al.*[68] that 4 S RNA is synthesized in the isolated Mauthner fiber in the absence of glial cell bodies (see Section 2.3.1), the question might well be asked whether such an assumption is valid. Intracranial injections do not provide a reliable measure of local synthesis of RNA that may take place in the axon because it is unlikely that the specific radioactivity of labeled RNA precursors in axoplasm after intracranial injection would be as high as that achieved by intraocular injection. Reasons for such a differential would include (1) dilution of precursor through dispersion in brain tissue and uptake by systemic circulation, (2) preferential uptake by metabolically active cells, and (3) the relative isolation of the ensheathed axonal compartment. Thus, it is possible that what is interpreted as an exclusive transport of 4 S RNA may be largely a manifestation of 4 S RNA synthesized locally in the axon.

The results of one experiment reported by Ingoglia[90] appear to be consistent with this point of view but were interpreted differently by the investigator. Specifically, intraocular injection of cordycepin inhibited both transport of labeled RNA precursors and the appearance of 4 S RNA in the tectum. The lack of appearance of 4 S RNA in the tectum was interpreted as a block of transport; nonetheless, retinal synthesis of 4 S RNA was not affected by cordycepin. The lack of labeled tectal 4 S RNA could also have been attributable to the block in transport of labeled RNA precursors and, therefore, to a reduce local incorporation into axonal 4 S RNA. Although it is reasonable that there should be transport of RNA, the evidence at present is indirect and inconclusive.

3. LOCAL SYNTHESIS OF AXONAL PROTEINS

Potential local sources of axonal proteins are the axon and its ensheathing cell. For reasons already cited above (see Section 1.1), the axon has not been considered a likely site of local protein synthesis by some investigators. In addition, experiments on invertebrate axons have provided indirect evidence for a transfer of proteins from glial sheath cells to the subjacent axon; however, evidence for such an intercellular transfer is lacking in mammalian myelinated axons (see below).

The intimate association of metabolically active sheath cells with a much less potentially active axon poses a major challenge in eliminating or minimizing the ambiguity inherent in discerning autochthonous axonal protein synthesis. If it could be demonstrated (1) that the pattern of polypeptides labeled in the axon is uniquely different from that in the sheath and/or (2) that there are polypeptides labeled in the axon whose expression is specific to the neuron, then such results would provide strong indirect evidence for an axonal site of synthesis, providing, of course, that the incorporation would be sensitive to inhibitors of cytoribosomally dependent protein synthesis. More direct evidence could be obtained by utilizing growing axons free of glial cells in tissue culture to investigate protein synthesis. Recent studies and work in progress in the author's laboratory have employed these strategies to probe the problem with some success.

3.1. Early Evidence of Local Protein Synthesis

Early suggestive evidence of axonal protein synthesis was reported for the Mauthner axon,[65,91,92] myelinated axons of spinal accessory nerve root of rabbit,[55,67] and the giant axon of the squid.[93,94] In most of these studies, amino acids were shown to be incorporated into acid-insoluble axonal proteins in isolated nerves *in vitro*.

Inhibitors of protein synthesis, such as puromycin, cycloheximide, acetoxycycloheximide, or chloramphenicol, have been used to investigate the extent to which labeling of axonal proteins is dependent on nonribosomal, cytoribosomal, or mitoribosomal systems. In squid axon experiments,[93,94] puromycin, cycloheximide, and chloramphenicol inhibited incorporation into axonal and sheath cell proteins.

Incorporation into the M-axon in spinal cord segments and in the isolated M-fiber preparation exhibited a significant sensitivity to puromycin,[65,91] whereas acetoxycycloheximide in the isolated fiber had a highly variable effect such that the mean inhibition was not significant. In this context, it may be worthwhile noting that rRNA rapidly degrades after the M-axon is isolated from its *in situ* location by the microtweezer technique.[64] It is possible, therefore, that instability of ribosomal machinery in the isolated M-fiber preparation could have been a factor in reducing efficacy of inhibition and increasing variability. In rabbit axons, both puromycin[55] and cycloheximide[67] produced significant inhibition of incorporation. Chloramphenicol had no significant effect on either mammalian axons[55,67] or the M-axon.[91] The latter finding was consistent with subcellular fractionation of the labeled M-fiber preparation, which showed very low labeling of the mitochondrial fraction.[91]

Further research relating to the problem of axonal protein synthesis has not been very extensive. Indeed, it seems to have been limited largely to that conducted in the author's laboratory. The latter work has taken three basic directions: (1) effects of nerve injury on local protein synthesis, (2) use of explant culture to investigate protein synthesis in regenerating axons free of glial cells, and (3) identification of polypeptides labeled in axons.

3.2. *Effects of Nerve Injury*

An obvious question that arises in the context of local synthesis relates to whether nerve injury leads to changes in the rate and/or composition of protein synthesis. This was investigated in the hypoglossal nerve of the rabbit.[95,96] In these studies, axons were isolated by the microtweezer technique after *in vitro* incubation of nerves. Neurotomy of the hypoglossal produced a very significant transient increase in protein content and rate of incorporation of [^{3}H]leucine into axonal protein in the traumatized region of the nerve (i.e., the 3-mm segment proximal to lesion site). The effects on protein content and rate of incorporation were first apparent at 15 h but not at 12 h after neurotomy (Fig. 4). At 21 h protein content had doubled, representing a maximum, but the apparent maximum rate of incorporation peaked at 15 h, at which time it exceeded the control by a factor of 10 (see specific radioactivity profile, Fig. 4). However, when specific radioactivity was corrected for the increase in protein content resulting from damming, the peak rate of synthesis had actually occurred at 18 h, at which time it was 23-fold greater than control (see protein-synthesizing activity profile, Fig. 4). A similar effect, except of lower magnitude, was observed in an untraumatized 3-mm portion of the nerve about 8 mm proximal to the lesion site (i.e., 5 mm from traumatized segment analyzed).

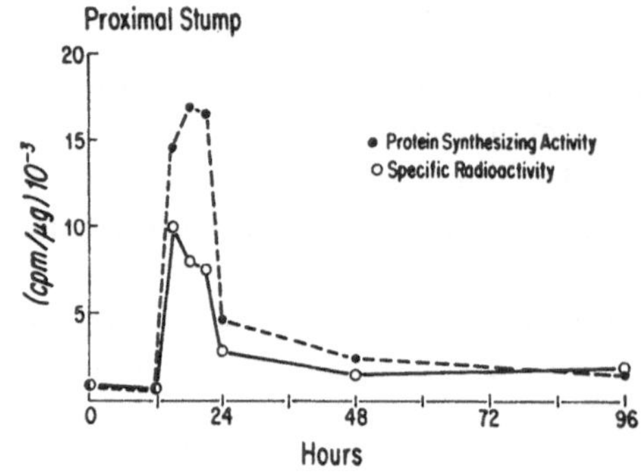

Fig. 4. The time course of change in the rate of incorporation of [^{3}H]leucine into protein of axons located in a 3-mm segment proximal to the site of neurotomy of the hypoglossal nerve of rabbit. The protein-synthesizing activity profile is corrected for the accumulation of protein mass as a result of damming. From ref. 95 with permission.

Thus, nerve trauma induced a gradient of elevated protein-synthesizing activity with the maximum closest to the site of injury. Incorporation was inhibited at least 95% by cycloheximide, and the effect was not observed in myelin microsamples taken from the traumatized region. The stimulation was shown to be a local response because it did not depend on continuity of the axon with cell bodies. Later, microelectrophoretic analysis of axons taken from the traumatized zone[97] indicated labeling of a component that was not seen in autoradiograms of axons from normal nerve. However, the resolving power of the gel microslab system was poor, and this question merits reexamination.

The local response to injury is an interesting phenomenon. The experiments indicate that local protein synthesis can be affected quantitatively, and perhaps qualitatively, in a dramatic way. The evidence indicating an intraaxonal site of synthesis can be summarized as follows: (1) amino acid incorporation into axonal protein occurs without a perceptible delay[96]; (2) autoradiograms after microelectrophoresis showed that labeled axonal proteins are uniquely different from those released into the incubation medium of the traumatized nerve, the latter of which would have been available for uptake by naked axon terminals[97]; (3) labeled proteins recovered from the incubation medium are neither taken up by unlabeled axons in traumatized nerve segments *in vitro* nor when focally microinjected into the traumatized region of the nerve *in vivo*.[97]

The mechanism underlying the striking and transient elevation of protein synthesis in traumatized axons and its significance are not known. An alternative to possible mechanisms operating either at transcription (seemingly unlikely) or translation is simply a concentration of protein-synthesizing machinery in the axon through the damming process. An argument against the mechanical explanation is that the rate of incorporation drops precipitously between 21 and 24 h (see Fig. 4), whereas axonal protein content is still abnormally high and remains so through 48 h.[95] Whether the phenomenon has a significance for intraaxonal processes associated with the outgrowth phase of regeneration is unknown. The answer may be forthcoming when polypeptides synthesized under such circumstances have been identified.

3.3. Local Synthesis in Axons Regenerating in Tissue Culture

A persistant problem in studying local synthesis of axonal proteins is that periaxonal cells of the myelin sheath cannot be ruled out as a potential source, especially since there is evidence for intercellular transfer in invertebrate axons (see Section 3.5). One solution to the problem is to study axons free of glial cells growing in tissue culture. One such system is the sympathetic ganglion cell explant culture system developed by Estridge and Bunge[98]; another explant culture system is that of the goldfish retina developed by Landreth and Agranoff.[99] The former system was used by Estridge and Bunge[98] to study polypeptide transport and composition of growing axons. The latter system has been used by us to study endogenous protein synthesis of axons regenerating *in vitro*.[100]

The approach taken in this study was to incubate decentralized axonal fields (i.e., regenerating retinal ganglion cell axons severed from the explant) with ^{3}H-labeled amino acids. Each isolated field was analyzed separately utilizing available quantitative microanalytical methods. Amino acids were incorporated into axonal protein in a manner that was inhibited significantly by cycloheximide but not by chloramphenicol.[100] Preliminary fluorographic patterns of labeled axonal proteins showed that primarily tubulin and actin were labeled.

Recent analysis using a gel microslab system in which α- and β-tubulins are resolved show that α-tubulin is not labeled significantly. The principal polypeptides labeled are β-tubulin and actin; in addition, there are two minor components having apparent molecular weights of 145 K and 105 K (Fig. 5) (E. Koenig, unpublished data). The 145 K component comigrates with a 145 K putative neurofilament polypeptide in goldfish axons. Inhibition of labeling of these polypeptides by cycloheximide provides highly suggestive evidence that they are synthesized *de novo* in these axons. Conclusive proof must await peptide-mapping experiments.

3.4. Polypeptides Labeled in Mammalian Axons

Early attempts to identify polypeptides synthesized locally in mammalian axons had limited success for several reasons, the principal one of which was that the gel microslab system did not yield good resolution. Nevertheless, significant labeling of three polypeptides was observed in extracts of normal axons of the rabbit hypoglossal nerve after incubation with [^{3}H]leucine.[101] An additional novel polypeptide appeared in autoradiograms of axonal extracts from the traumatized region of the hypoglossal nerve 18 h after lesioning (i.e., at the time of peak rate of synthesis; see Section 3.2).[97]

Recently, a gel microslab system that yields resolution comparable to standard macro polyacrylamide gel electrophoresis has been developed and has been applied to studies of rat spinal roots. These experiments have been undertaken to compare polypeptides labeled in axons and in Schwann cell–myelin sheath complex following *in vitro* incubation with [^{35}S]methionine. Labeling patterns of axonal polypeptides of dorsal and ventral roots were similar, generally showing multiple polypeptides labeled (Fig. 6). The labeling was inhibited by cycloheximide. In addition, the labeling pattern of major polypeptides of the sheath was qualitatively different from that of the axon, signifying that labeled axonal polypeptides could not have been derived from the sheath either by intercellular transfer or by contamination during the isolation procedure.

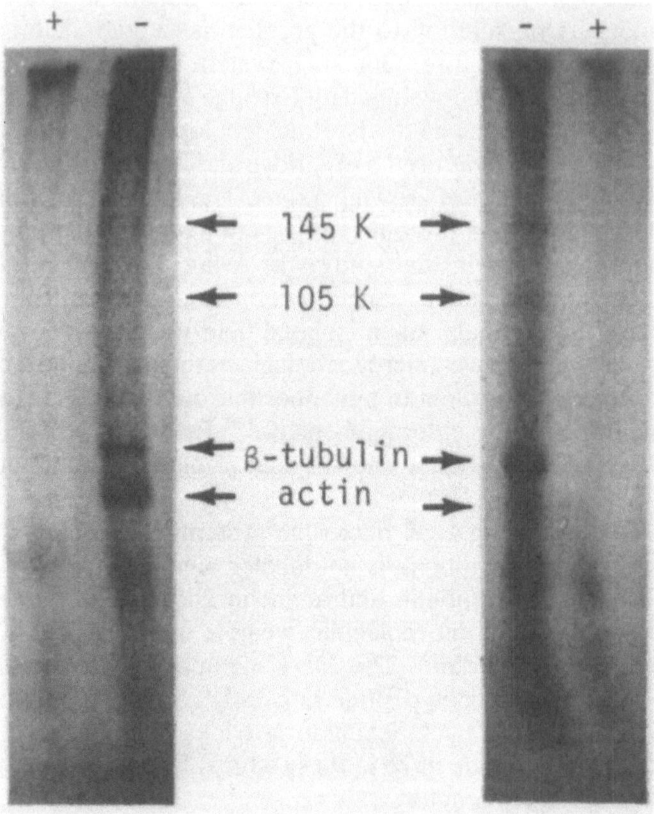

Fig. 5. Autoradiograms of two experiments showing polypeptides of retinal ganglion cell axons labeled with [^{35}S]methionine in the presence (+) or absence (−) of 1 mM cycloheximide (E. Koenig, unpublished data). Incubation was carried out on isolated glial-cell-free axons regenerating in tissue culture (see ref. 100). Labeled polypeptides exhibiting sensitivity to cycloheximide include a 145 K and a 105 K polypeptide, β-tubulin, and actin. Original microelectrophoretograms, 1.3 cm.

Radioactive axonal polypeptides after 4-h incubation of the roots included the following identifiable components in one-dimensional SDS gel microslab electrophoresis: actin, α- and β-tubulins, 150 K neurofilament protein, 200 K neurofilament protein, and fodrin (spectrin) (Fig. 6) (E. Koenig, unpublished data).

In summary, the experimental results indicate that (1) axonal proteins exhibit cycloheximide-sensitive labeling, (2) there are dissimilarities between labeling patterns of the axon and its myelin sheath, and (3) there is labeling of neuron-specific polypeptides (i.e., 200 K and 150 K neurofilament proteins). These findings lead to the conclusion that there is measurable endogenous synthesis of axonal proteins locally in nonregenerating mammalian axons. Further comments regarding the possible significance of these findings in the context of the normo- and pathobiology of the axon are given below (Section 6) after discussion of the sheath cell as a potential source of axonal proteins.

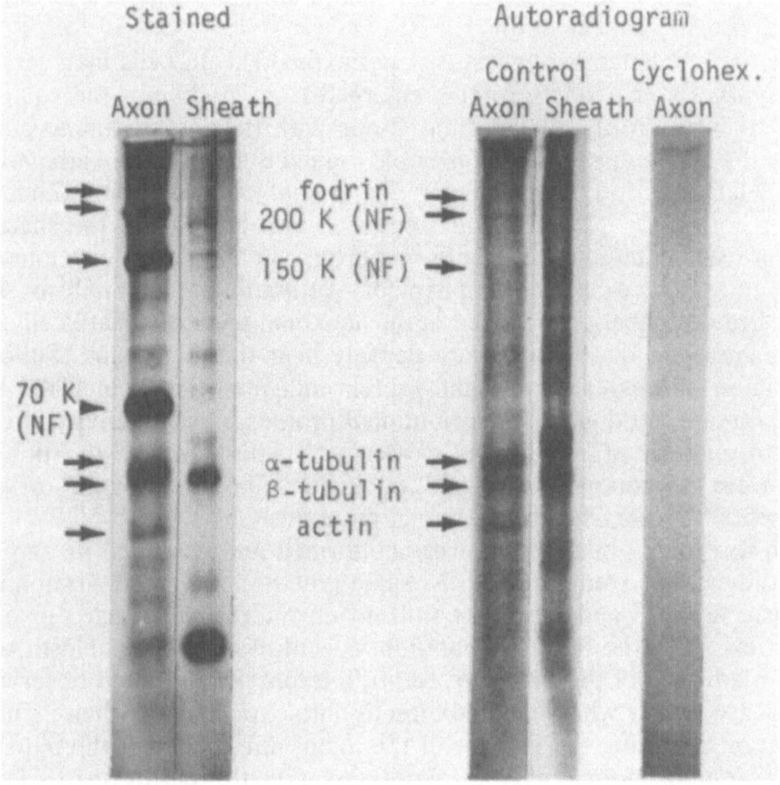

Fig. 6. Comparison of *in vitro* labeling of axonal and myelin sheath polypeptides of the dorsal root of the rat after incubation (4 h) with [^{35}S]methionine in the presence or absence of 1 mM cycloheximide. Most identifiable cytoskeletal proteins are labeled in the absence of cycloheximide; however, the 70 K neurofilament polypeptide shows little discernible labeling. Original microelectrophoretograms, 1.5 cm.

4. *SHEATH CELLS AS A SOURCE OF AXONAL PROTEINS*

An alternative or an adjunct to an intraaxonal site of local synthesis of axonal proteins is the periaxonal sheath cell (e.g., Schwann cell or glial cell). A number of studies conducted on crustacea in recent years provide indirect evidence that sheath cells may serve as an accessory metabolic center for the subjacent axon because axons severed from their cell bodies survive structurally and functionally for up to 200–300 days, although there is a progressive involution.[102–108] Nevertheless, it appears that the long survival time is selective; e.g., certain peripheral sensory axons degenerate rapidly in crayfish, whereas motor axons survive for long periods.[109] Periaxonal sheath cells associated with decentralized axons that survive undergo hypertrophy and hyperplasia, and they show cytologic signs of increased protein synthetic activity. Such morphological changes strongly suggest that the sheath serves as an essential trophic center for the decentralized axon and may simply reflect an enhancement of metabolic support the sheath normally supplies to the intact axon (for review, see ref. 110).

The first evidence to suggest that periaxonal sheath cells may serve as a site of synthesis of axonal proteins emerged from studies of the squid giant axon.[111] Incubation of isolated giant axons with radioactive amino acids resulted in the appearance of acid-insoluble radioactivity in the sheath and in the axon; the latent period, determined in a later study, [112] was approximately the same for both (i.e., ~25 min). Incorporation into proteins of the sheath and axoplasm was inhibited significantly by puromycin and acetoxycycloheximide but was not affected by chloramphenicol. Autoradiographic analysis showed that the heaviest labeling occurred in the adaxonal layer of sheath cells; in the axon, grains were distributed more densely near the membrane. Subcellular fractionation of labeled axoplasmic protein indicated that about 80% was in the supernatant fraction. The distributional profiles of radioactivity after SDS gel electrophoresis of soluble and particulate fractions of the sheath and axoplasm were too complex to permit conclusions about homologies of labeled components between sheath and axonal fractions.

The foregoing observations were confirmed and extended in two subsequent studies. One study utilized the same type of *in vitro* preparation as that previously used,[112] and the other utilized an internally perfused giant axon preparation.[113] In the latter preparation, a central core of axoplasm was removed in advance of perfusion. In addition to previously noted observations, ribonuclease, either when injected directly into axoplasm or when included in the solution perfusing the interior of the axon, did not affect the appearance of acid-insoluble radioactivity in axoplasm or in the perfusate. In perfused preparations, acid-insoluble radioactivity appeared in the perfusate after a latency of 45–60 min and continued to be released into the perfusate for up to 8 h. Under these conditions, puromycin in the incubation medium blocked the appearance of acid-insoluble radioactivity in the perfusate.

There were, however, significant quantitative discrepancies between experiments involving the nonperfused preparation and those involving the internally perfused preparation. For example, the latencies of appearance of axoplasmic acid-insoluble radioactivity in the two preparations were quite disparate, i.e., 25 min in the nonperfused and 45–60 min in the perfused preparation. In addition, there was a marked reduction in the amount of acid-insoluble radioactivity appearing in the perfusate and residual axoplasm of the perfused preparation compared to the axoplasm of the nonperfused preparation for a 3-h incubation interval.

Some experiments[112] were also carried out in an attempt to characterize the mode of presumed sheath cell–axonal transfer of labeled proteins. The labeling profiles after electrophoresis of products released by the preparation into medium indicated that labeled proteins were not taken up secondarily by the axon despite the fact that exogenously labeled albumin is taken up by the axon in a transient fashion.[112,114] This suggested a direct intercellular transfer mechanism, and a potential role for extracellular calcium was adduced from ionic substitution experiments.

In the crayfish, where decentralized axons have been shown to survive for long periods (references cited above), studies were undertaken to investigate labeling of axons and glial sheaths in intact[115,116] and decentralized axons.[116] Unlike frog dorsal root ganglia and dorsal roots, whose rates of incorporation into protein were similar, incorporation into protein of peripheral crayfish nerves was greater than that of the ganglion or the muscle.[115,116] When a comparison was made between normal nerves and distal stumps of nerves that had been decentralized for 1 month or more, grain densities over adaxonal sheath cells associated with decentralized axons were twice those of controls, whereas grain density over decentralized axons was about fourfold greater.[116] Unlike the squid giant axon, however, chloramphenicol significantly inhibited incorporation into nerves.[110,115] The observation that a calcium-free incubation medium reduced incorporation into both sheath cells and axoplasm of crayfish nerves[117] rather than axoplasm alone signifies another distinction between crayfish and squid axons (see above).

It is apparent from the work on invertebrates that the evidence for a sheath cell–axonal transfer of locally synthesized proteins is compelling. Whether local synthesis is exclusive of any endogenous axonal protein synthesis is not clear at present, however, particularly in view of the recent demonstration by Giuditta, Cupello, and Lazzarini[61] that small proportions of rRNA are present in the squid giant axon. The experiments reviewed above certainly do not exclude some axoplasmic protein synthesis.

On the other hand, experiments on unmyelinated vertebrate axons have not been performed. It is quite possible that, as in invertebrate unmyelinated axons, local synthesis of axonal proteins in vertebrate unmyelinated axons could be shared with or take place primarily in periaxonal sheath cells. If this were the case, then a shift to an exclusively endogenous mode of local synthesis in myelinated axons may simply reflect an imperative imposed by virtue of the nature of the ensheathment (i.e., myelin). This could place a greater metabolic burden on the neuron having a myelinated axon and, at the same time, could confer on the myelinated axon a greater vulnerability to toxic agents that might inactivate local protein-synthesizing machinery in an irreversible manner (see Section 5, below).

Attempts to demonstrate transfer of locally synthesized axonal protein from Schwann cells to mammalian myelinated axons *in vitro* have not been successful. In one set of experiments, a pulse-chase paradigm was used in segments of rabbit spinal accessory nerve.[67] After a 3-h incubation with [^{3}H]leucine and a 17-h chase with unlabeled leucine and cycloheximide, the specific radioactivity of axonal protein was the same as that after a 3-h incubation alone; axons in a control nerve continued to incorporate during the chase period. An increase in specific radioactivity after the chase period would have indicated an exogenous source of labeled protein. Subsequent experiments involving neurotomy of hypoglossal nerve in rabbit were similarly negative (see Section 3.2). For example, incorporation of [^{3}H]leucine into protein of myelin-denuded portions of axons proximal to the lesion site occurred without a measurable lag.[96] A lag in incorporation would have signified an extra-axonal source of labeled protein. In addition, labeled proteins released from traumatized nerve into the incubation medium were uniquely different from those labeled in myelin-denuded axons proximal to the lesion site.[97] Moreover, labeled proteins in the medium were not taken up by unlabeled axons in the traumatized zone during incubation *in vitro* or after focal microinjection of concentrated labeled medium proteins *in vivo*[97] (see Section 3.2). Although these experiments were not conclusive, they indicated that there did not appear to be measurable Schwann cell–axonal transfer of proteins in myelinated axons.

5. POSSIBLE SIGNIFICANCE OF LOCAL SYNTHESIS FOR TOXIC AXONOPATHIES

If the assumption is correct that local synthesis fulfills an essential biological need for the axon, then impairment of local synthesis could lead to a loss of structural and/or functional integrity, i.e., axonopathy. Primary axonopathies can be produced after a latent period by several classes of toxic organic compounds, including industrial hexacarbon solvents, acrylamide (for review, see ref. 118), and certain organophosphorous compounds in susceptible species.[118] Pathomorphological signs occur after a lag as a multifocal derangement in distribution and in structural organization of the cytoskeleton and organelles initially involving the distal preterminal portion of myelinated axons in the central and peripheral nervous system. For this reason, it is referred to as central–peripheral distal axonopathy in preference to the earlier designation of "dying back" neuropathy.[120]

The basis for this vulnerability of the distal axon has long puzzled investigators. One favored explanation is that axoplasmic transport may be blocked as a result of intoxication.[121] Although there is evidence that fast axoplasmic transport is impaired by acrylamide,[122,123] by hexacarbons,[122] and by organophosphorous intoxication,[124] this impairment is partial, and it is difficult, therefore, to evaluate the significance of a fractional reduction. Sabri *et al.*[125] have reported that neurotoxic hexacarbon solvents inhibit glycerophosphate dehydrogenerase *in vitro* and could, thereby, provide a mechanism for blocking energy-dependent fast axoplasmic transport. Unfortunately, energy production and energy stores have not been studied experimentally in nerves during intoxication. Moreover, fast transport appears to be only partially blocked (see above), and much of what is transported at a rapid rate is targeted for terminals; degenerative events do not begin at the nerve terminal.[126–128] Slow transport, on the other hand, does not appear to be affected to a significant extent.[123,129,130]

A hypothesis that could take into account the differential vulnerability of the distal axon is based on the premise that local axonal protein synthesis is irreversibly impaired through primary or secondary effects of intoxication. Indeed, this was one of several possible explanations put forward by Prineas.[131] It follows, then, that if local supplementation of polypeptides vital to maintaining structural and/or functional integrity of the axon is compromised, then there would be a progressive deterioration of the distal axon because essential polypeptides supplied to the distal axon from the cell body would have become depleted through biological decay (i.e., degradation) during transit. The hypothesis is quite compatible with the viewpoint that toxic actions are exerted locally at the level of the axon, perhaps entailing inactivation of essential metabolic cofactors and/or enzymes.[132]

The effects of acrylamide intoxication on protein synthesis have been investigated in the rat by Schotman *et al.*[133] Protein-synthesizing activity was assayed *in vivo* in spinal cord, brainstem, and heart of rats (1) during the latent period before neurological signs appeared, (2) after neurological signs appeared, and (3) during recovery from intoxication. They observed a small but significant generalized reduction of protein synthesis in both neural tissue and heart during the development of intoxication and a larger inhibition after neuropathy had developed. During the recovery phase, a small but significantly elevated rate of protein synthesis was observed only in neural tissue, which was attributed to regeneration. Acrylamide had no effect on incorporation into proteins of spinal cord slices *in vitro*, indicating that if intoxication entailed an inhibitory action, it was likely caused by a metabolically converted product.

The studies by Schotman *et al*[133] were designed to assess the effects of acrylamide intoxication on protein synthesis in whole neural tissue. An underlying assumption of their work was that all axonal proteins were synthesized in cell bodies and that a severely curtailed supply to the axon could be a cause for axonopathy. In any case, it would not have been possible in their experiments to discern a local direct effect on axonal protein synthesis. The axon, as already noted, constitutes a restricted metabolic compartment surrounded by metabolically active sheath cells with very limited access to endoneural space and in which most of the free amino acids are probably supplied by endogenous protein breakdown.

Tests of the hypothesis that toxic axonopathies may be a consequence of irreversible impairment of local axonal protein synthesis are now being undertaken in the author's laboratory. Experimental results, although qualitative and preliminary at this juncture, indicate a marked reduction in the *in vitro* labeling with [^{35}S]methionine of axonal polypeptides of rat dorsal and ventral roots 24 h after a single moderately large dose of acrylamide (75 mg/kg, i.p.) had been administered (Fig. 7). Although such findings are suggestive of a local impairment of axonal protein synthesis during acrylamide intoxication, a systematic quantitative investigation of the question is needed. If a complete and persevering impairment of local axonal protein synthesis can be demonstrated to be an antecedent common to toxin-induced distal axonopathies, it would provide strong support for the hypothesis. Finally, a causal linkage would provide, in addition, a strong argument for the notion that autochthonous protein synthesis is an endogenous activity of vital importance to the axon.

6. *POSSIBLE SIGNIFICANCE OF LOCAL SYNTHESIS OF AXONAL PROTEINS*

As the only line of communication between macroneurons and their target cells, the axon constitutes an extended cellular appendage that undergoes continuous renewal while still retaining a potential for further growth (e.g., regeneration, sprouting). A balance must be maintained between degradation and

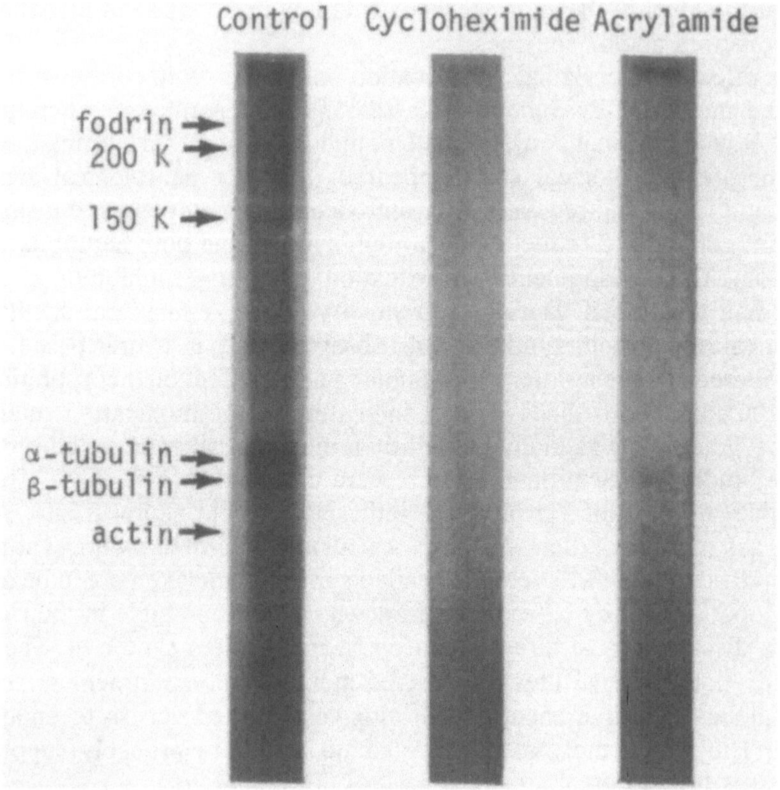

Fig. 7. Comparison of *in vitro* labeling of polypeptides with [^{35}S]methionine in axons of the ventral root of the rat in the presence or absence of cycloheximide and 24 h after one i.p. injection of acrylamide (75 mg/kg). Original microelectrophoretograms, 1.5 cm.

renewal processes in the distal axon; otherwise, functional and/or structural integrity would be compromised. Renewal of membranes and their constituents in the axon pose no serious strategic problem to the cell because of rapid intracellular transport mechanisms that shuttle such components quickly from site of synthesis to remote sites of utilization. On the other hand, cytosolic constituents and, in particular, cytoskeletal components, which comprise the bulk of the axon's proteins (see Section 1.2), are transported at rates that may not be sufficient to offset the consequences of catabolism in transit without a local resupply of essential proteins.

The important question, therefore, centers on the relative metabolic stability of axoplasmic proteins transported in groups IV and V that play key roles in vital processes of the axon. Axoplasmic transport itself would certainly qualify as a vital process, and the tubulins and actin, which are constitutents of cytoskeletal systems that are doubtlessly involved in intracellular translocation, have been shown in brain to have half-lives of a few days (see Section 1.3). Although the half-lives of tubulin and actin undergoing slow transport have not been determined, Nixon[26] has shown that slowly transported proteins as a group undergo degradation in transit at a rate that is not significantly different from that of retina. The available data from work on vertebrate axons indicate that many of the cytoskeletal polypeptides are synthesized locally, including actin, tubulin, and certain of the neurofilament polypeptides (see Sections 3.3 and 3.4). Thus, it would seem that polypeptides supplied to the axon by slow transport are also supplemented by local synthesis. Nonetheless, the available data also indicate that local supplementation is highly selective. One example will suffice to illustrate this point.

Retinal ganglion cell axons *in vitro* synthesize β-tubulin but not α-tubulin to a significant extent (see Section 3.3). The significance of only one component of a heterodimer being synthesized is unclear at present. Nonetheless, it is worth noting in this context that tubulin is an integral membrane protein of microvesicles and synaptosomes and that α-tubulin predominates over β-tubulin.[134] This raises the possibility that there could be a disproportionate cytosolic β-tubulin requirement, especially in these particular growing axons. Retinal ganglion cell axons have varicosities that contain a large amount of a highly branched tubular smooth endoplasmic reticulum (SER) that is translocated at a slow rate distally.[100] The SER-filled varicosities may reflect a mode for bulk transport of precursor membrane destined to be inserted at the growth cone. In any case, in spite of a selective local β-tubulin synthesis, inspection of microelectrophoretograms of such axons suggest similar concentrations of α- and β-tubulins (E. Koenig, unpublished observations). Thus, a disproportionate amount of α-tubulin reaching the axon as an integral membrane protein, may be compensated for by a local selective synthesis of β-tubulin. It is also worth noting here that Eddé *et al.*[135] have reported that a particular β-tubulin isoform (i.e., isoform β-2) is specifically related to neurite outgrowth of mouse neuroblastoma cell clone C 1300.

In summary, the evidence strongly suggests that there is local endogenous synthesis of axonal proteins. The rate of synthesis appears to be low compared to that in nerve cells or cells of ensheathment, but local requirements of the axon must be very modest compared to those of a cell as a whole. The proteins labeled to the greatest extent are largely selected components included in transport groups IV and V. This suggests that local synthesis serves to supplement selected polypeptides supplied by slow transport. It will be of importance to determine whether there is a correlation between the relative half-lives of individual slowly transported polypeptides and the extent to which the same polypeptides are synthesized locally. If a strong correlation exists, then it would provide a simple explanation of the need for local synthesis.

REFERENCES

1. Tobias, G. S., and Koenig, E., 1975, *Exp. Nuerol.* **49**:221–234.
2. Hoffman, P. N., and Lasek, R. J., 1975, *J. Cell Biol.* **66**:351–366.
3. Lasek, R. J., and Hoffman, P. N., 1976, *Cell Motility*, Volume 3 (R. Goldman, T. Pollard, and J. Rosenbaum, eds.), Cold Spring Harbor Laboratory, Cold Spring Harbor, New York, pp. 1021–1049.
4. Lorenz, T., and Willard, M., 1978, *Proc. Natl. Acad. Sci. U.S.A.* **75**:505–509.

5. Willard, M., Wiseman, M., Levine, J., and Skene, P., 1979, *J. Cell Biol.* **81**:581–591.
6. Black, M. M., and Lasek, R. J., 1980, *J. Cell Biol.* **86**:616–623.
7. Grafstein, B., and Forman, D. S., 1980, *Physiol. Rev.* **60**:1167–1283.
8. Anderton, B. H., Ayers, M., and Thorpe, R., 1978, *FEBS Lett.* **96**:159–163.
9. Liem, R. K. H., Yen, S.-H., Salomon, G. D., and Shelanski, M. L., 1978, *J. Cell Biol.* **79**:637–645.
10. Shelanski, G., and Liem, R. K. H., 1979, *J. Neurochem.* **33**:5–13.
11. Chiu, F.-C., Korey, B., and Norton, W. T., 1980, *J. Neurochem.* **34**:1149–1159.
12. Brown, B. A., Nixon, R. A., Strocchi, P., and Marotta, C. A., 1981, *J. Neurochem.* **36**:143–153.
13. Mori, H., Komiya, Y., and Kurokawa, M., 1979, *J. Cell Biol.* **82**:174–184.
14. Brown, B. A., Nixon, R. A., and Marotta, C. A., 1982, *J. Cell Biol.* **94**:159–164.
15. Black, M. M., and Lasek, R. J., 1979, *Brain Res.* **171**:401–413.
16. Levine, J., and Willard, M., 1981, *J. Cell Biol.* **90**:631–643.
17. Glenney, J. R., Jr., Glenney, P., Osborn, M., and Weber, K., 1982, *Cell* **28**:843–854.
18. Repasky, E. A., Granger, B. L., and Lazarides, E., 1982, *Cell* **29**:821–833.
19. LeBeux, Y. J., and Willemot, J., 1975, *Cell Tissue Res.* **160**:1–36.
20. Gilbert, D. S., 1975, *J. Physiol. (Lond.)* **253**:257–301.
21. Koenig, E., 1965, *J. Neurochem.* **12**:357–361.
22. Koenig, E., 1978, *Neurobiology of the Mauthner Cell* (D. S. Faber and H. Korn, eds.), Raven Press, New York, pp. 167–182.
23. Gilbert, D. S., Newby, B. J., and Anderton, B. A., 1975, *Nature* **256**:586–589.
24. Schlaepfer, W. W., and Hasler, M. B., 1979, *Brain Res.* **168**:299–309.
25. Pant, H. C. and Gainer, H., 1980, *J. Neurobiol.* **11**:1–12.
26. Nixon, R. A., 1980, *Brain Res.* **200**:69–83.
27. Black, M. M., and Lasek, R. J., 1980, *J. Cell Biol.* **86**:616–623.
28. Stromska, D. P., and Ochs, S., 1981, *J. Neurobiol.* **12**:441–453.
29. Nixon, R. A., Brown, B. A., and Marotta, C. A., 1982, *J. Cell Biol.* **94**:150–158.
30. Brown, B. A., Nixon, R. A., and Marotta, C. A., 1982, *J. Cell Biol.* **94**:159–164.
31. Hemminki, K., 1973, *Biochim. Biophys. Acta* **310**:285–288.
32. Hemminki, K., 1973, *Brain Res.* **57**:259–260.
33. Palay, S. L., Sotelo, C., Peters, A., and Forkand, P. M., 1968, *J. Cell Biol.* **38**:193–201.
34. Peters, A., Proskauer, C. G., and Kaiser-Abramof, I. R., 1968, *J. Cell Biol.* **39**:604–619.
35. Conradi, S., 1969, *Acta Physiol. Scand.* [*Suppl.*] **332**:65–84.
36. Barron, K. D., and Doolin, P. F., 1968, *J. Neuropathol. Exp. Neurol.* **27**:401–420.
37. Zelená, J., 1970, *Brain Res.* **24**:359–363.
38. Barron, K. D., Chiang, T. Y., Daniels, A. C., and Doolin, P. F., 1971, *Progress in Neuropathology* (H. M. Zimmerman, ed.), Grune & Stratton, New York, pp. 255–280.
39. Zelená, J., 1972, *Z. Zellforsch.* **124**:217–229.
40. Dimova, R. V., and Maykov, D. V., 1976, *Acta Neuropathol. (Berl.)* **36**:235–242.
41. Barron, K. D., and Dentinger, M. P., 1978, *Acta Neuropathol. (Berl.)* **44**:1–8.
42. Caley, D. W., and Maxwell, D. S., 1968, *J. Comp. Neurol.* **133**:17–43.
43. Tennyson, V. M., 1970, *J. Cell Biol.* **44**:62–79.
44. Skoff, R. P., and Hambruger, V., 1974, *J. Comp. Neurol.* **153**:107–148.
45. Bunge, M. B., 1973, *J. Cell Biol.* **56**:713–735.
46. LeBeux, Y. J., 1972, *Z. Zellforsch.* **133**:289–325.
47. Santolaya, R. C., 1973, *Z. Zellforsch.* **146**:319–328.
48. Hindelang-Gertner, C., Stoeckel, M.-E., Porte, A., Dellmann, H.-D., and Madarasz, B., 1974, *Cell Tissue Res.* **155**:211–219.
49. Hámori, J., and Lakos, J., 1980, *Cell Tissue Res.* **212**:415–427.
50. Edström, J.-E., Eichner, D., and Edström, A., 1962, *Biochim. Biophys. Acta* **61**:178–184.
51. Funch, P., Kinsman, S. L., Faber, D. S., Koenig, E., and Zottoli, S. J., 1981, *Neurosci. Lett.* **27**:159–164.
52. Faber, D. S., and Korn, H., 1978, *Neurobiology of the Mauthner Cell* (D. S. Faber and H. Korn, eds.), Raven Press, New York, p. 290.
53. Edström, A., 1964, *J. Neurochem.* **1**:309–314.

54. Jakoubek, B., and Edström, J.-E., 1965, *J. Neurochem.* **12**:845–849.
55. Koenig, E., 1967, *J. Neurochem.* **14**:437–446.
56. Grampp, W., and Edström, J.-E., 1963, *J. Neurochem.* **10**:725–731.
57. Andersson, E., Edström, A., and Jarlstedt, J., 1970, *Acta Physiol. Scand.* **78**:491-502.
58. Lasek, R. J., 1970, *J. Neurochem.* **17**:103–109.
59. Slagel, D. H. Hartmann, H. A., and Edström, J.-E., 1966, *J. Neuropathol. Exp. Neurol.* **25**:244–253.
60. Lasek, R. J., Dabrowski, C., and Nordlander, R., 1973, *Nature [New Biol]* **244**:162–165.
61. Giuditta, A., Capello, A., and Lazzarini, G., 1980, *J. Neurochem.* **34**:1757–1760.
62. Giuditta, A., Metafora, S., Felsani, S., and Del Rio, A., 1977, *J. Neurochem.* **28**:1393–1395.
63. Black, M. M., and Lasek, R. J., 1977, *J. Neurobiol.* **8**:229–237.
64. Koenig, E., 1979, *Brain Res.* **174**:95–107.
65. Edström, A., 1966, *J. Neurochem.* **13**:315–321.
66. Celio, M. R., 1976, *Brain Res.* **108**:221–235.
67. Koenig, E., 1970, *Biochemical Psychopharmacology*, Volume 2 (E. Costa and E. Giacobini, eds.), Raven Press, New York, pp. 303–315.
68. Edström, A., Edström, J.-E, and Hökfelt, T., 1969, *J. Neurochem.* **16**:53–66.
69. Austin, L., Bray, J. J., and Young, R. J., 1966, *J. Neurochem.* **13**:1267–1269.
70. Bray, J. J., and Austin, L., 1968, *J. Neurochem.* **15**:731–740.
71. Casola, L., Davis, G. A., and Davis, R. E., 1969, *J. Neurochem.* **16**:1037–1041.
72. Bondy, S. C., 1971, *Exp. Brain Res.* **13**:135–139.
73. Bondy, S. C., 1972, *J. Neurochem.* **19**:1769–1776.
74. Rahmann, H., and Wolburg, H., 1971, *Experentia* **27**:903–904.
75. Wolburg, H., 1972, *Exp. Brain Res.* **15**:348–363.
76. Autilio-Gambetti, L., Gambetti, P., and Shafer, B., 1973, *Brain Res.* **53**:387–398.
77. Ingoglia, N. A., Grafstein, B., McEwen, B. S., and McQuarrie, I. G., 1973, *J. Neurochem.* **20**:1605–1615.
78. Bondy, S. C., and Marchisio, P. C., 1973, *Exp. Neurol.* **41**:29–37.
79. Peterson, J. A., Bray, J. J., and Austin, L., 1968, *J. Neurochem.* **15**:741–745.
80. Gambetti, P., Autilio-Gambetti, L., Shafer, B., and Pfaff, L., 1973, *J. Cell Biol.* **59**:677–684.
81. Ingoglia, N. A., and Tuliszewski, R., 1976, *Brain Res.* **112**:371–381.
82. Gambetti, P., Ingoglia, N. A., Autilio-Gambetti, L., and Weis, P., 1978, *Brain Res.* **154**:285–300.
83. Ingoglia, N. A., 1978, *J. Neurochem.* **30**:1029–1039.
84. Ingoglia, N. A., 1979, *Science* **206**:73–75.
85. Politis, M. J., and Ingoglia, N. A., 1979, *Brain Res.* **169**:343–356.
86. Por, S. B., Komiya, Y., McGregor, A., Jeffrey, P. L., Gunning, P. W., and Austin, L., 1978, *Neurosci. Lett.* **8**:165–169.
87. Gunning, P. W., Por, S. B., Langford, C. J., Scheffer, J., Austin, L., and Jeffrey, P. L., 1979, *J. Neurochem.* **32**:1737–1743.
88. Bondy, S. C., and Purdy, J. L., 1975, *Biochim. Biophys. Acta* **390**:332–341.
89. Bondy, S. C., Purdy, J. L., and Balitch, J. A., 1977, *Neurochem. Res.* **2**:407–415.
90. Ingoglia, N. A., 1978, *J. Neurochem.* **30**:1029–1039.
91. Edström, A., and Sjöstrand, J., 1969, *J. Neurochem.* **16**:67–81.
92. Alvarez, J., and Chen, W. Y., 1972, *Acta Physiol. Latin Am.* **22**:266–269.
93. Fischer, S., and Litvak, S., 1967, *J. Cell Physiol.* **70**:69–74.
94. Giuditta, A., Dettbarn, W.-D., and Brzin, M., 1968, *Proc. Natl, Acad. Sci. U.S.A.* **59**:1284–1287.
95. Tobias, G. S., and Koenig, E., 1975, *Exp. Neurol.* **49**:221–234.
96. Tobias, G. S., and Koenig, E., 1975, *Exp. Neurol.* **49**:235–245.
97. Frankel, R. D., and Koenig, E., 1978, *Brain Res.* **141**:67–76.
98. Estridge, M., and Bunge, R., 1978, *J. Cell Biol.* **79**:138–155.
99. Landreth, G. E., and Agranoff, B. W., 1976, *Brain Res.* **118**:299–303.
100. Koenig, E., and Adams, P., 1982, *J. Neurochem.* **39**:386–400.
101. Frankel, R. D., and Koenig, E., 1977, *Exp. Neurol.* **57**:282–295.
102. Hoy, R. R., Bittner, G. D., and Kennedy, D., 1967, *Science* **156**:251–252.

103. Nordlander, R. H., and Singer, M., 1972, *Z. Zellforsch.* **126**:157–158.
104. Atwood, H. L., Govind, C. K., and Bittner, G. D., 1973, *Z. Zellforsch.* **146**:155–165.
105. Bittner, G. D., 1973, *Am. Zool.* **13**:379–408.
106. Wine, J. J., 1973, *Exp. Neurol.* **38**:157–169.
107. Bittner, G. D., Ballinger, M. L., and Larimer, J. L., 1974, *J. Exp. Zool.* **180**:13–36.
108. Bittner, G. D., and Johnson, A. L., 1974, *J. Comp. Physiol.* **89**:1–21.
109. Bittner, G. D., and Mann, D. W., 1976, *Cell Tissue Res.* **169**:301–311.
110. Bittner, G. D., 1980, *Comp. Biochem. Physiol.* **68A**:299–306.
111. Lasek, R. J., Gainer, H., and Pryzbylski, R. J., 1974, *Proc. Natl. Acad. Sci. U.S.A.* **71**:1188–1192.
112. Lasek, R. J., Gainer, H., and Barker, J. L., 1977, *J. Cell Biol.* **74**:501–523.
113. Gainer, H., Tasaki, I., and Lasek, R. J., 1977, *J. Cell Biol.* **74**:524–530.
114. Giuditta, A., D'Udine, B., and Pepe, I. M., 1971, *Nature* **229**:29–30.
115. Sarne, Y., Neale, E. A., and Gainer, H., 1976, *Brain Res.* **110**:73–89.
116. Meyer, M. R., and Bittner, G. D., 1978, *Brain Res.* **143**:195–211.
117. Grossfeld, R. M., Bittner, G. D., Ballinger, M. L., and Viancour, R. A., 1980, *Neurosci. Abstr.* **6**:385.
118. Spencer, P. S., and Schaumburg, H. H., 1976, *Progress in Neurobiology* (H. M. Zimmerman, ed.), Grune & Stratton, New York, pp. 253–295.
119. Cavanagh, J. B., 1973, *CRC Rev. Toxicol.* **2**:365–417.
120. Spencer, P. S., and Schaumburg, H. H., 1977, *J. Neuropathol. Exp. Neurol.* **36**:300–320.
121. Prineas, J. B., 1969, *J. Neuropathol. Exp. Neurol.* **28**:598–621.
122. Sahenk, Z., and Mendell, J. R., 1981, *Brain Res.* **219**:397–405.
123. Souyri, F., Chretien, M., and Droz, B., 1981, *Brain Res.* **205**:1–13.
124. Reichert, B. L., and Abou-Donia, M. B., 1980, *Mol. Pharmacol.* **17**:56–60.
125. Sabri, M. I., Moore, C. L., and Spencer, P. S., 1979, *J. Neurochem.* **32**:683–689.
126. Schaumburg, H. H., Wisniewski, H., and Spencer, P. S., 1974, *J. Neuropathol. Exp. Neurol.* **33**:260–284.
127. Spencer, P. S., and Schaumburg, H. H., 1977, *J. Neuropathol. Exp. Neurol.* **36**:276–299.
128. Chretien, M., Patey, G., Souyri, F., and Droz, B., 1981, *Brain Res.* **205**:15–28.
129. Bradley, W. G., and Williams, M. H., 1973, *Brain* **96**:235–246.
130. Sumner, A., Pleasure, D., and Ciesielka, K., 1976, *J. Neuropathol. Exp. Neurol.* **35**:319.
131. Prineas, J. B., 1969, *J. Neuropathol. Exp. Neurol.* **28**:571–597.
132. Schoental, R., and Cavanagh, J. B., 1977, *Neuropathol. Appl. Neurobiol.* **3**:125–136.
133. Schotman, P., Gipon, L., Jennekins, F. G. I., and Gispen, W. H., 1977, *Neuropathol. Appl. Neurobiol.* **3**:125–136.
134. Zisapel, N., Levi, M., and Gozes, I., 1980, *J. Neurochem.* **34**:26–32.
135. Eddé, B., Jeantet, C., and Gros, F., 1981, *Biochem. Biophys. Res. Commun.* **103**:1035–1043.

13

The Axonal Plasma Membrane

George H. DeVries

1. INTRODUCTION

1.1. Definition of Axonal Plasma Membrane

The organelles and cytoplasm within the axonal process are enclosed by a plasma membrane known as the axonal plasma membrane or axolemma. This membrane may be defined as the surface membrane of the axonal process from the point at which it emerges from the axon hillock of the neuronal perikaryon to the axonal terminal or nerve ending, which forms a synapse on another cell.[1] The axonal plasma membrane may or may not be ensheathed by myelin synthesized by the oligodendroglia (CNS) or Schwann cells (PNS). The axonal plasma membrane is characterized by functional molecular heterogeneity. Therefore, when dealing with this membrane, one must define the state of myelination of the axonal plasma membrane in question and, if myelinated, state the localization of the membrane within the myelinated axon (internodal or nodal).

1.2. Scope of the Chapter

This chapter is intended to summarize our knowledge concerning the molecular properties of the axonal plasma membrane gleaned from a number of different axolemma-enriched preparations. The characteristic proteins, lipids, enzymes, and metabolic functions of the axolemma-enriched preparations obtained to date are summarized. A concluding statement points out the many crucial molecular properties of this membrane that remain to be elucidated based on our present state of knowledge concerning this membrane.

George H. DeVries • Department of Biochemistry, Medical College of Virginia, Richmond, Virginia 23298.

2. PREPARATION OF AXOLEMMA-ENRICHED FRACTIONS FROM NERVOUS TISSUE

2.1. Axolemma-Enriched Fractions Derived from Invertebrate Unmyelinated Axons

Initial attempts to isolate the axolemma utilized biological sources greatly enriched in axolemma such as the giant axons and retinal axons of squid.[2,3] Therefore, it is not surprising that the pioneer workers in the field of axolemma isolation were found in South America, close to the natural habitat of the squid *Dosidicus gigas*. A summary of the biological sources of starting material for axonal plasma membrane preparations, the preparative procedures used, the yield, and the ultrastructural appearance of the isolated preparation are described in Table I.

Most of the initial isolation procedures utilized large nerves of invertebrate poikilotherms as the starting material (preparations A through E). In these cases, the strategy for the isolation of the axolemma is the same: vigorous homogenization of the starting material to completely disrupt the axons, followed by isolation of a microsomal fraction by differential centrifugation and further purification of the microsomal fraction on either a continuous gradient (A, B, C, and G) or a discontinuous gradient (D, E, and F). Based on the density at which the axolemma fraction is obtained, the preparations fall into two categories. Axolemma preparations A, C, and E, all derived from invertebrate peripheral nerves, have a density approximately equivalent to that of 20% sucrose. Preparations derived from axons related to the central nervous system (preparations B and D) have a greater density, equivalent to approximately 30% sucrose. The lipid content of all axolemma-enriched fractions derived from unmyelinated axons is related to their origin; unmyelinated fibers in the CNS have a greater density (hence lower lipid content) than fibers originating in the peripheral nervous system. Analysis of the lipid content of these preparations supports this view, since the lipid content of preparations A, C, and E (derived from unmyelinated fibers of the PNS) is approximately 70% to 75%, whereas the lipid content of the CNS-derived preparations (B and D) is approximately 50% to 60% (see Table V).

The yield ranges from 0.4 mg to 20 mg of axolemma-enriched fraction per gram wet weight of starting material. The yield is related to the axonal plasma membrane content of the starting material. Garfish olfactory nerve contains fibers that are directly in contact with one another in groups of several hundred and Schwann cells that form a thin layer at the periphery of the fiber and make up a small proportion of the total cellular material.[11] Therefore, it is not surprising that the yield of axolemma from this source is the highest (20 mg per gram of wet weight). On the other hand, fibers in the walking legs of crabs and lobsters contain a greater proportion of connective tissue and Schwann cells relative to axonal membrane,[12] and, therefore, the yield is correspondingly lower. In spite of these differences in lipid/protein ratio, the invertebrate-derived preparations show some uniformity in their ultrastructural appearance:

Table I
Biological Source, Preparative Procedure, Yield, and Morphological Appearance of Axolemma-Enriched Preparations

	Preparative procedure						
Biological source	Method of disruption of axons	Method for isolation of axolemma	Isopynic density of axolemma fraction	Yield	Ultrastructural appearance	Preparation	Reference
First stellar nerve of *Dosidicus gigas* (giant squid)	Virtis homogenizer followed by Teflon® pestle homogenization	Crude microsomal fraction applied twice to a discontinuous gradient 0.33 M / 1.195 M; interface material separated on a 0.66 M to 1.195 M continuous sucrose gradient	16% to 25% sucrose	4 mg total weight per g protein in initial homogenate	Unilamellar membranes 95–110 Å thick, vesicles 0.1 μm to 0.5 μm in diameter	A	2
Retinal axons of *Dosidicus gigas* (giant squid)	Glass–glass homogenizer in presence of glass beads, Teflon® pestle, homogenization, sonication	Washed microsomal preparation separated on a 10% to 40% continuous sucrose gradient	27% sucrose	7.5 mg total weight per g wet weight of retinal axons	Unilamellar membranes, 70–80 Å thick, vesicles 0.03 μm to 1 μm in diameter	B	3
Nerve bundles of walking legs of *Homarus americanus*	Homogenization in ground glass Potter–Elvehjem homogenizer	Differential centrifugation to obtain microsomal fraction, isopycnic centrifugation, 10% to 40% linear sucrose gradient	20% sucrose	10.0 mg protein per g wet weight of walking leg axons	Entirely membranes, no organelles present, vesicles when fixed in the presence of lobster Ringers	C	4

(*Continued*)

Table I. (Continued)

Biological source	Preparative procedure: Method of disruption of axons	Preparative procedure: Method for isolation of axolemma	Preparative procedure: Isopynic density of axolemma fraction	Yield	Ultrastructural appearance	Preparation	Reference
Olfactory nerve of garfish *Lepisosteus osseus*	15 passes of a Teflon® glass Duall grinder	Separation twice on 0.25/1.195 M sucrose gradient, separation of interfacial material on 20%/30%/35%/40%/50%/density gradient	Interface of 20%/30% sucrose	20 mg protein per g wet weight of nerve	Rounded membrane vesicles, 0.05 μm to 0.5 μm in diameter, 75 Å thick	D	5
Sensory nerve bundles of crab, *Cancer pagurus*, lobster walking leg nerves, and nerves of spider crab, *Maia squinado*	Sorvall omnimixer followed by Teflon®–glass homogenizer	Hypotonically shocked microsomal fraction separated on 17.5%/19.5%/21.5% discontinuous sucrose gradient	Fraction I, 10%/17.5% sucrose interface; fraction II, 19.5% sucrose	0.07 mg protein fraction I and 0.35 mg protein fraction II per g wet weight of nerve	Vesicles of 0.15 μm to 0.30 μm diameter	E	6

Corpus callosum of bovine brain, human brain, and brainstem of rat	Hypotonic and mechanical shock of myelinated axons	Myelinated axons isolated by flotation in buffered-salt–sucrose; after disruption, shocked myelinated axons are applied to 0.8 M/1.0 M/1.2 M discontinuous sucrose gradient	Fraction I at interface of 25%/30% sucrose; fraction II at interface of 30%/36% sucrose	1 mg total protein in both fractions per g wet weight of starting white matter	Unilamellar membrane, some vesicles less than 1 μm in diameter, some linear pieces less than 1 μm in length. Freeze fracture morphology consistent with *in situ* axonal membrane	F	7
Rat sciatic nerve, rabbit sciatic nerve, bovine spinal accessory nerve, bovine intradural roots	30-s homogenization in Tekmar® Tissumizer®	Separation of microsomal fraction in 10%–40% sucrose gradient	26% to 29% sucrose	1.0 mg protein per 3.0 g wet weight peripheral nerve	Unilamellar vesicles less than 1 μm in diameter	G	8,9
Cultured nerve cells of dorsal root ganglion	35 strokes of a Dounce homogenizer	Supernatant from 300 *g* × 10 min centrifugation, centrifuged at 35,000 *g* for 1 h	Not determined	Not reported	Small empty vesicles less than 0.1 μm in diameter, floculent material outside and inside the vesicles, occasional mitochondria and lysosomes	H	10

they tend to form vesicles less than 1 μm in diameter and are 70 Å to 110 Å in thickness.

2.2. Axolemma-Enriched Fractions Derived from Vertebrate Myelinated Axons

The strategy to isolate the axolemma from CNS myelinated axons is to use the myelinated axon as a vehicle for the isolation. The detailed methodology used for these preparations has recently been reviewed.[7] A buffered salt–sucrose medium used in the initial homogenization of the CNS white matter maintains the integrity of the myelin around the axon. The lipid-rich myelin sheath acts as a kind of life preserver to allow only the myelinated axons (plus fragments of disrupted myelin) to float to the surface when centrifuged. The floating layer is subsequently disrupted by mechanical and osmotic shock. The major cellular constituents found in the shocked myelinated axons (axonal neurofilaments, axolemma, myelin, and myelin-related membranes) differ widely in lipid content, allowing an effective separation on a discontinuous gradient. More recently, this preparation procedure has been adapted for the zonal rotor, which allows a large-scale separation of the cellular elements of the myelinated axon on a linear sucrose gradient.[13] This preparative procedure has proven useful in studies of the binding of neurotoxins to axolemma-related membranes[14] as well as in studies to evaluate changes in the molecular composition of constituents derived from myelinated axons that may take place in normal-appearing white matter of multiple sclerosis brain.[15]

The presence of connective tissue in mammalian PNS fibers presented a dilemma for the isolation of axolemma. The vigorous homogenization required for effective disruption of these fibers destroyed, to a large degree, the integrity of the myelin sheath around the axon. On the other hand, conditions that favored the preservation of myelinated axons were ineffective in disrupting the tissue.[8] Therefore, the myelinated axon was abandoned as a vehicle by which to isolate the PNS axolemma. Instead, a crude microsomal fraction was isolated by differential centrifugation after complete disruption of the peripheral nerve in a Tissumizer® homogenizer. The microsomal fraction was then fractionated on a 10% to 40% linear sucrose gradient. In contrast to the invertebrate-derived axolemma-enriched preparations, both the CNS and PNS preparations derived from vertebrate sources have similar densities, equivalent to approximately 30% sucrose. The densities of the vertebrate-derived axolemma preparations (F and G) are similar to those of axolemma preparations derived from CNS-related invertebrate sources (preparations B and D).

2.3. Axolemma-Enriched Fractions Derived from Cultured Unmyelinated Axons

A membrane preparation derived from cultured axons has been described (Table I, preparation H). The dorsal root ganglion that is initially cultured contains both neurons and glial cells. Bromodeoxyridine is subsequently introduced into the culture and incorporated into the proliferating glial cell DNA.

Table II
Enzymatic Profile and Sodium Channel Density of Axonal Plasma Membrane Preparations Isolated from Various Species[a]

Species	Na^+, K^+-activated ATPase	Acetyl-cholinesterase	TTX (pmol/mg protein)	Reference
Rat				
CNS	42.0	11.7	1.0	14,16
PNS	—[b]	15.5	—	8
Bovine				
CNS	7.4	0.7	—	17
PNS	—	5.0	—	9
Human				
CNS	5.1	0.4	—	18
Lobster leg nerve	6.9	165.0	9.5	4,19
Garfish olfactory nerve	23.0	2.4	3.7	5,20,43
Crab leg nerve	59.4	150.0	11.5	6
Squid retinal nerve	126	8.9	—	20
Squid stellar nerve	1338	44.9	—	20

[a] Enzymatic activities represent μm substrate hydrolyzed/mg protein per h.
[b] Dashes indicate activity was not evaluated in the preparation.

Subsequent irradiation of the cultures selectively destroys the glial cells, whereas neurons appear to remain structurally normal. The neuronal cells are then maintained on a medium containing fluorodeoxyuridine to suppress glial growth. Within 1 month, the neurons produce a rich halo of processes termed "neurites," which are essentially myelin-competent unmyelinated axons.

When the cell bodies are surgically excised, the remaining pure population of neurites provides the source for the axolemma-enriched fraction. This axolemma-enriched preparation contains alkaline phosphodiesterase and is able to stimulate a quiescent population of Schwann cells to proliferate.[10] Further characterization will be contigent on the ability to obtain sufficient material by tissue culture so that after fractionation biochemical characterization can be carried out.

3. PROTEINS OF AXOLEMMA-ENRICHED PREPARATIONS

3.1. Enzymes

3.1.1. Na^+, K^+-ATPase

It was appreciated in the earliest axolemma-enriched preparations that since the axonal plasma membrane was concerned with ion translocation it should be enriched in this enzyme. As shown in Table II, the preparations derived from invertebrates have a substantially higher specific activity of Na^+, K^+-ATPase than the mammalian axolemma-enriched preparations derived from myelinated axons. These differences in specific activity may be explained by the following considerations. In the unmyelinated axons, which conduct in a continuous manner, Na^+, K^+-ATPase would not be expected to be restricted to the specialized regions of the axon. Rather, it may be continuously distributed along the entire axon. However, this enzyme may be restricted to the paranodal regions of the mammalian myelinated axon. Histochemical evidence for the localization of this enzyme at the nodal and paranodal regions has been presented.[21,22] The mammalian axolemma preparations are probably enriched in internodal axolemma since they are isolated via a purified preparation of myelinated axons, which, on the average, contain more internodal than nodal axolemma. It is anticipated that the internodal axolemma would contain only a small proportion of Na^+-, K^+-ATPase since ion translocation does not occur to any extent beneath the myelin sheath. Therefore, the net effect would be a dilution of the specific activity of nodal and paranodal activities such as Na^+, K^+-ATPase in preparations derived from myelinated axons.

Comparison of the specific activities of this enzyme in the preparations derived from frozen CNS sources (human and bovine CNS) with the specific activity of Na^+, K^+-ATPase in axolemma freshly prepared from rat CNS (Table II) demonstrates that the freshness of tissue and the dilution of internodal axolemma both contribute to the lowered specific activity of this enzyme found in the mammalian preparations. Sweadner has found that the Na^+, K^+-ATPase in the rat CNS axolemma-enriched preparations (termed the $\alpha+$ form of the enzyme) is unique in several aspects: it shows increased affinity for the inhibitor strophanthidin and has a molecular weight 2000 greater (because of an extra 20-amino-acid sequence) than another form of Na^+, K^+-ATPase (the α form).[23] She has postulated that the extra 2000-molecular-weight piece of the $\alpha+$ form of the enzyme may serve to anchor the enzyme to another structure at the node. The partially purified Na^+, K^+-ATPase of garfish axonal membrane contains two major subunits of 42,000 and 110,000 molecular weight.[24] The 110,000-molecular-weight subunit was identified as the catalytic subunit by phosphorylation with ATP and had a lower molecular weight than the analogous enzyme isolated from kidney. The smaller subunit of the garfish olfactory nerve enzyme had a molecular weight of 42,000 and did not appear to contain any carbohydrate in contrast to other small subunits of ATPase that have been analyzed. This preliminary analysis also demonstrates the unique nature of this enzyme in the axonal plasma membrane. The kinetic properties of the Na^+, K^+-ATPase in crab axonal membranes have been extensively studied by the laboratory of Lazdunski to reveal the nature of the interaction among the binding sites for sodium, potassium, and ATP with the enzyme in its native membrane environment.[25,26]

3.1.2. Acetylcholinesterase

Table II demonstrates that the specific activity of acetylcholinesterase in the axolemma preparations that have been characterized to date shows wide variation. Part of this variation may reflect the relative freshness of the tissue used for fractionation. For example, both the human and bovine preparations are derived from tissues that had been frozen after some degree of autolysis; the specific activity of the acetylcholinesterase in such preparations is therefore approximately tenfold lower than the corresponding fractions isolated from fresh tissue. Chacko *et al.* have postulated that axonal–glial contact dictates the level of acetylcholinesterase in the axonal membrane.[20] In lobster walking leg axon, the enzyme can be demonstrated histochemically at the interface between the axon and the Schwann cell plasma membranes. However, histochemical activity was virtually absent in the garfish olfactory nerve in accordance with the biochemical data. In the myelinated axons of the mammalian CNS, Stanley *et al.* could not find any evidence for selective acetylcholinesterase staining at the paranodal regions of the axon.[27] In addition, it has been demonstrated through the use of specific inhibitors that the acetylcholinesterase present in the mammalian and crab axolemma preparations is a true rather than pseudocholinesterase.[6,7] By using labeled disopropylfluorophosphate, Balerna *et al.* were able to determine that the crab and lobster axonal plasma membrane contained about 70 pmol of acetylcholinesterase per milligram of membrane protein[6]; similar studies have not been done on other preparations. The axolemma-enriched preparations derived from lobster or crab leg nerve are approximately tenfold higher in acetylcholinesterase activity than the mammalian preparations. Nachmansohn and Newman have suggested that this enzyme is related to the mechanism of nerve conduction.[28] However, others have suggested that the wide variation in the level of this enzyme in nerve membrane preparations (Table II) does not support this view.[20] It is probable that some of the activity associated with the axolemma is being transported to the nerve ending where it will have a role to play in synaptic transmission. In addition, as pointed out by Chacko *et al.*, the axonal acetylcholinesterase may mediate a cholinergic system associated with axon–Schwann cell interaction.[20]

3.2. Other Metabolic Functions Associated with the Axonal Plasma Membrane

A number of metabolic activities have been associated with the axolemma via histochemical, enzymatic, or metabolic studies, as shown in Table III. Estridge and Bunge found that cultured axons from dorsal root ganglion contained an intrinsic set of enzymatic activities that allowed the incorporation of glucosamine into macromolecules irrespective of whether the neurites were attached to the cell body.[32] In contrast, they reported that radioactively labeled protein and macromolecules labeled with fucose were found in the axon only when the cell bodies were attached. In this case, the labeled precursor was incorporated into macromolecules in the cell body followed by transport to the axon.

Metabolic experiments that utilize cultured axons are more informative and less equivocal than experiments in which metabolic activities are evaluated in isolated cell fractions. The finding of a metabolic activity in an isolated membrane preparation is always limited by the possibility that a small percentage of contaminating membrane may be responsible for the observed activity. For this reason, it is imperative to characterize the metabolic activity under consideration in membrane fractions (usually a microsomal fraction) that are likely to contaminate the axolemmal preparations. Striking differences in

Table III
Metabolic Activities Associated with Axonal Plasma Membrane Preparations

Source of preparation	Metabolic activity	Evidence	Reference
Stellar nerves of squid, *Doryteuthis plei*	ATPase	Histochemical staining	29
Retinal axons of *Dosidicus gigas*	Synthesis of ATP, GTP, plus several enzymes of glyolysis	Enzymatic activities in washed membrane preparations	30
Rat trigeminal nerve	Acyltransferase	Histochemical staining	31
Rat dorsal root ganglion neurites in culture	Incorporation of glucosamine into macromolecules	Ability of attached/detached neurites to incorporate labeled glucosamine into macromolecules	32
Bovine corpus callosum	Sugar nucleotide pyrophosphatase; synthesis of mannosylphosphoryldolichol, synthesis of N-acetylglucosaminyl pyrophosphoryl dolichol	Metabolic labeling using axolemma-enriched preparations	33
Rat brainstem	UDP-galactose: ceramide galactosyltransferase	Characterization of enzymatic activity in axolemma-enriched preparations	34
Rat brainstem	Phospholipase D	Similarity in the distribution and enrichment of specific activity of phospholipase D and acetylcholinesterase	35

the cofactor requirements of an enzymatic activity or in the developmental profile of that activity would distinguish between contaminating and true activity. Experiments along these lines have been conducted in the case of the enzymes of polyisoprenoid metabolism[33] and galactosyltransferase,[34] which are associated with axolemma-enriched preparations.

However, contamination with organelles closely associated with the axolemma, such as the smooth endoplasmic reticulum beneath the axonal plasma membrane, cannot be assessed because of a lack of unequivocal markers for these organelles. The majority of enzymatic activities that have been associated with the axolemma are concerned with lipid or glycolipid metabolism. The axonal membrane may be required to synthesize its own lipid from appropriate precursors to provide the correct molecular architecture at its surface for glial cell interaction. For example, Costantino-Ceccarini *et al.* have reported that the activity of UDP-galactose: ceramide galactosyltransferase in axolemma-enriched preparations is greatest during the initial stages of myelination; there is a sharp decrease in this activity after myelination is complete.[34] If phospholipase D, which is associated with the axolemma, can utilize its own endogenous lipid as a substrate, this lipolytic activity may modulate enzymatic activities closely associated with membrane lipid such as Na^+, K^+-ATPase and possibly sodium channel function.[37]

3.3. Molecular Weight Profile of Proteins

As expected for any membrane preparation, the axolemma-enriched preparations contain proteins that range in molecular weight from 20,000 to over 200,000. A summary of the molecular weight distribution of polypeptides found in the various preparations is shown in Table IV. Visual inspection of the polyacrylamide gel patterns of the various preparations shows the following similarities: the 50,000-molecular-weight region of the profile always contains the most quantitatively prominent polypeptides; there are a series of prominent polypeptides in the 30,000–40,000 molecular-weight range of the profile; there is usually a prominent polypeptide in the 100,000 region of the polyacrylamide gel profile; and there is a paucity of polypeptides in the 150,000- to 200,000-molecular-weight range.

It is difficult to assess to what degree there may be common polypeptides in all of these preparations, since direct comparisons of a wide variety of preparations have not been carried out in the same laboratory. However, Chacko *et al.* found that eight out of the 11 polypeptides found in the garfish olfactory nerve axolemma were also found in the axolemma preparation derived from lobster leg nerve, although the relative proportions of the polypeptides were different.[20] Yoshino and DeVries reported that the prominent polypeptides found at 190,000, 77,000, 50,000, and 39,000 molecular weight in the rabbit PNS preparation were also found in the rat CNS axolemma-enriched preparation.[9] However, the polypeptide pattern of the CNS axolemma-enriched preparation appears to be much more complex than that of the comparable fraction isolated from the PNS.

Some element of similarity in the protein content of all axolemma-enriched preparations is not unexpected since the unity of axonal plasma membrane function in all these preparations should be reflected by some unity of protein composition. For example, since the majority of the preparations contain Na^+, K^+-ATPase, acetylcholinesterase, and the sodium channel, it may be expected that the polypeptides that constitute these functional activities should be present in all preparations. In addition, since tubulin[32] and contractile proteins[41] are associated with the axonal membrane, they may be expected to be prominent in the polypeptide profile of most axolemma-enriched preparations. Balerna *et al.* demonstrated that tropomyosin, actin, tubulin, and myosin are present in the crab axonal membrane preparation.[6] To date, a protein that is uniquely present in the axonal membrane has not been described. However, Goodrum and Morell have provided indirect evidence from studies of axoplasmic transport that a 49,000-molecular-weight protein containing fucose may be uniquely present in the axonal plasma membrane.[42]

Table IV
Molecular Weight Ranges of Polypeptides Found in Axolemma-Enriched Preparations

	Source of axolemma-enriched preparation							
Molecular weight range[a]	Garfish olfactory nerve	Lobster leg nerve	Crab leg nerve	Neurites of dorsal root ganglion	Rat CNS	Bovine CNS	Bovine PNS[b]	Rabbit PNS[b]
Reference	20,38	20	6	32	7	39	9	9
200–250K			235K		250K	233K		
195–150K			155K	180K, 170K	190K			190K
145–100K	130K, 110K	130K, 110K	125K		130K, 105K	127K		
99–80K			95K	96K	87K	95K, 86K, 82K		
79–70K	78K	78K		70K		75K		77K
69–60K			63K	62K		66K, 62K	60K	
59–55K	57K		57K	58K	58K	58K		
54–50K	53K	53K		53K	54K	53K		50K
49–45K	47K		45K		48K		46K	
44–40K	42K	42K						
39–35K	38K	38K	37K		38K	39K	39K	39K
34–30K	32K	32K	32K, 30K	30K	34K	34K, 32K		
29–25K	26K			25K	27K			
20–24K	22K	22K				20K		

[a] Molecular masses are given in kilodaltons (K).
[b] Prominent polypeptides only; 16 and 40 proteins of different molecular weights were observed in the bovine and rabbit PNS, respectively.

4. SODIUM CHANNELS IN AXOLEMMA-ENRICHED PREPARATIONS

The sodium channels in axolemma-enriched preparations have been investigated using sodium-channel-specific neurotoxins such as tetrodotoxin or saxitoxin. The maximum binding of tetrodotoxin is related on a one-to-one basis to the density of sodium channels in an axolemma-enriched preparation. Based on this evidence, the sodium channel density in axolemma-enriched preparations derived from lobster and crab is about 10 pmol/mg membrane protein[6,19]; the comparable figure for preparations obtained from garfish olfactory nerve is about 4 pmol/mg membrane protein,[43] whereas mammalian derived axolemma-enriched preparations contain only 1 pmol/mg membrane protein[14] (see Table II).

The density of sodium channels in a preparation will be related to the mode of conduction in the fiber from which the preparation is derived. The poorly myelinated invertebrate fibers conduct in a continuous fashion and should contain uniformly distributed sodium channels along the entire length of the axonal membrane. On the other hand, in the mammalian myelinated systems, the sodium channels are thought to be clustered in the nodal and paranodal regions of the axon.[44] Therefore, if the mammalian preparation contains a greater proportion of internodal than nodal or paranodal membrane (as argued previously), it will contain on the average a smaller proportion of sodium channels because of the dilution effect of the internodal axolemma, which may contain very few sodium channels. On the other hand, recent evidence suggests that the internodal axolemma is rich in potassium channels.[45] On this basis, the potassium channel would appear to be a good marker for the internodal axolemma. However, the low affinity of presently known drugs that interact specifically with this channel make the detection of this marker difficult.

If the true density of sodium channels in the mammalian preparations is equal to or greater than that observed in the unmyelinated preparations, the axolemma-enriched preparations must contain only 10% or less of the paranodal and nodal axolemma.This proportion would appear to be reasonable based on the fact that the myelinated axons from which the fractions are derived would not be expected to contain a high proportion of nodal and paranodal axolemma.

Sodium channels contain three classes of receptor sites: (1) a site to which tetrodotoxin or saxitoxin binds and inhibits ion transport, (2) a site to which the polypeptide neurotoxins (sea anemone and scorpion toxin) bind and inhibit the inactivation of sodium current, and (3) a binding site to which the alkaloid neurotoxins veratridine, batrachotoxin, and aconitine bind and cause persistent activation.[46] The first of these three functional sites of the sodium channel has been most extensively investigated in axonal plasma membrane preparations.

In addition to the determination of the dissociation constant and maximal binding capacity of axonal plasma membrane preparations for these neurotoxins, the inhibitory effect of purified phospholipases and unsaturated fatty acids on the binding of saxitoxin or tetrodotoxin to axon plasma membranes has been described.[37,47,48] A cooperative interaction between the sodium ion and saxitoxin binding has been reported in mammalian CNS axolemma-enriched preparations.[49] Balerna *et al.* reported the selective binding of veratridine to crab axonal membranes and, more specifically, to the phospholipids in those membranes.[6] We have recently found that sea anemone toxin and scorpion toxin bind to axolemma derived from the rat CNS.[14] The maximal binding capacity and dissociation content of sea anemone toxin and the dissociation constant for scorpion toxin were similar to what has been previously reported for synaptosomes derived from rat CNS.[50] However, the maximal binding capacity for scorpion neurotoxin was fourfold less than that of synaptotomes. Since scorpion toxin binding is dependent on a polarized state of the membranes,[51] the less effective polarization in the axolemma preparation may be responsible for the lowered binding ability.

5. *ACETYLCHOLINE-BINDING MACROMOLECULE IN AXONAL MEMBRANE PREPARATIONS*

Denberg *et al.* first reported that axonal plasma membrane preparations from lobster legs contain a macromolecule that specifically binds nicotine.[52] The characteristics of this macromolecule suggest that it is a classical acetylcholine receptor. Balerna *et al.* confirmed the presence of this molecule in lobster axonal membranes to the extent of 310 pmol/mg membrane protein.[6] However, no nicotine binding could be detected with crab axonal membranes. Chester *et al.* demonstrated the specific binding of a peroxidase-labeled α-bungarotoxin to an acetylcholine-binding macromolecule in both intact axons and an isolated axonal membrane preparation derived from lobster nerves.[53] Jumblatt *et al.* have shown that the α-bungarotoxin binds specifically to the acetylcholine receptor in axolemma-enriched fractions derived from lobster leg nerve.[53,54] The possible function of a nicotinic acetylcholine receptor in the axonal plasma membranes remains obscure.

6. *LIPIDS OF AXOLEMMA-ENRICHED PREPARATIONS*

6.1. *Phospholipids*

As shown in Table V, the bulk of the lipid in axolemma-enriched preparations is phospholipid, which constitutes from half to three-quarters of the total lipid in the preparations. The invertebrate-derived axolemma-enriched preparations tend to have a higher phospholipid content than the mammalian CNS preparations, which appear to contain galactolipid in lieu of some phospholipid. The major phospholipids in all axolemma-enriched preparations that have been characterized to date are the ethanolamine and choline phosphatides in approximately equal molar amounts. The fatty acid composition of the axonal plasma membrane phospholipids has been reported for preparations from squid retinal axons,[60] garfish olfactory nerve,[5,57] and human CNS.[55] A high degree

Table V
Lipid Composition of Axolemma-Enriched Preparations[a]

	Bovine	Human	Rat	Lobster leg nerve	Garfish olfactory nerve	Crab leg nerve	Squid retinal nerve	Squid stellar nerve
Reference	39	55	56	57	5,20	6	3,58	2
Lipid (% of dry weight)	42.7	47.3	51.5	76.0	66	67.8	45.4	70.5
Total phospholipid	49.7	48.9	67.4	77.3	74.0	71.2	66.7	58.5
Ethanolamine phosphatides	16.6	19.8	24.2	33.5	25.0	25.9	24.9	20.0
Choline phosphatides	16.5	18.7	27.8	19.6	31.3	21.8	26.4	26.7
Sphingomyelin	4.3	3.0	2.0	9.9	5.7	13.0	2.6	5.8
Serine phosphatides	4.8	3.0	5.3	9.3	8.9	9.6	6.5	6.0
Inositol phosphatides	3.9	4.4	2.8	2.4	3.1	0.7	0[c]	—[d]
Cholesterol	26.1	25.3	21.3	22.7	26.0	28.8	22.0	28.1
Phospholipid/cholesterol[b]	0.95	0.90	1.58	1.70	1.41	1.23	1.51	1.04
Total galactolipid	23.8	25.8	11.3	—	—	0	0	—
Cerebroside	19.0	21.3	6.3	—	—	0	0	—
Sulfatide	4.8	4.8	5.0	—	—	0	0	—
Ganglioside (μg/mg protein)	15.0	13.9	24.0	—	—	0	0	—

[a] All lipids are reported as weight percent.
[b] Molar ratio.
[c] Zero indicates that the constituent was assayed but was not detected.
[d] Dash (—) indicates no assay was carried out for that constituent.

of fatty acid unsaturation was found in the major phospholipids of all these preparations; the content of 22:6 reached one-third of the total fatty acids in some cases.[5,57,58] This high degree of unsaturation and hence fluidity in the membrane is similar to what has been described for the synaptic plasma membrane[59] and was considered to be consistent with the active metabolic role of the axonal plasma membrane in ion translocation. However, comparison of the fatty acid compositions of phospholipids of an axonal plasma membrane preparation and a simultaneously isolated nonexcitable periaxonal cell plasma membrane preparation revealed no differences in the fatty acid compositions.[57] The phospholipids from either source had a high percentage of polyunsaturated fatty acids. These results suggest that although polyunsaturated fatty acids may contribute to the functional properties of the axonal plasma membrane, they are not characteristic of the excitable membrane only but may be found in nonexcitable membranes as well.

6.2. Cholesterol

Plasma membranes in general are characterized by a greater content of cholesterol than other cellular membranes.[60] It is not surprising, therefore, that cholesterol comprises approximately one-quarter of the lipid weight in all axonal plasma membrane preparations. The molar ratio of phospholipid to cholesterol is usually greater than 1, a situation that favors the permeability of the membrane, since decreased cholesterol content will allow increased phospholipid fatty acid chain mobility and hence permeability.[61] The phospholipid-to-cholesterol ratio is lowest in the mammalian axolemma-enriched preparations prepared from frozen tissue. The ratio of phospholipid to galactolipid may be higher in the nodal and paranodal axonal region but lower in the internodal axonal plasma membrane, which is closely apposed to the galactolipid-rich myelin sheath. In support of this view, a number of recent studies of axoplasmic transport in myelinated fibers have suggested that there may be an exchange of lipid between the axon and myelin.[62] This could mean that there may be regional specialization of lipid composition at the site of interaction of the myelin with the axolemma.

6.3. Glycolipids

Glycolipids such as cerebroside and sulfatide are usually considered to be characteristic of glial-derived membranes such as myelin.[63] However, it can be calculated that not all the galactolipid found in whole brain can be accounted for by myelin.[64] Galactolipids have been described in myelin-free axonal preparations at a level that cannot be accounted for by myelin contamination.[65] In addition, the fatty acid composition of the axonal-related galactolipid was distinct from that of the comparable lipids in a myelin fraction.[66] Although the lipids in these myelin-free axonal preparations have been described in association with neurofilaments,[67] recent evidence suggests that some of the lipid originates from axolemmal membrane found in such preparations.[68] The similarity in the lipid composition of the bovine axolemma-enriched preparation and the bovine myelin-free axons also supports this view.

Glycolipids have been described only in axolemma preparations that originate from myelinated axons, raising the possibility that this glycolipid may originate from myelin-related membranes that contaminate the preparation. However, the level of myelin-related markers, such as myelin basic protein and 2′,3′-cyclic nucleotide 3′-phosphohydrolase, that are found in these preparations is not sufficient to account for all of the galactolipid present.[39,55,56] In addition, the high ganglioside content of axolemma preparations derived from myelinated axons is consistent with the ganglioside content reported for other neuronal membranes.[69] These axonal glycolipids may originate via some metabolic interplay between the axonal plasma membrane and the myelin by which it is ensheathed. Recent evidence suggests that such metabolic exchange between the axon and its myelin sheath is possible.[70] The presence of sulfatide in an excitable membrane is not expected, since this sphingolipid has been found in other membranes involved in ionic flux.[71] The distribution of gangliosides in the axonal plasma membrane is not unique; all the major brain type gangliosides are present.[65] Ganser has recently demonstrated a ganglioside specifically associated with the axolemma at the node of Ranvier.[72]

7. CONCLUSIONS

7.1. Summary of Common Properties of Axonal Plasma Membrane Preparations

As shown in Table VI, there are a number of properties that can be gleaned from the axonal plasma membrane preparations that have been characterized to date. This partial listing makes it obvious that there are still many interesting properties of the axonal plasma membrane yet to be elucidated.

7.2. Future Directions for Defining the Molecular Properties of the Axonal Plasma Membrane

The neurochemistry of the axonal plasma membrane has lagged behind the many interesting morphological and physiological observations that have been made concerning this vital membrane. It can reorganize itself after demyelination,[73] it can form nodal regions without any direct contact from the myelin-generating glial cell,[74] and it can dictate to an ensheathing glial cell whether or not to myelinate.[75] It is of interest to note that the axonal plasma membrane isolated from myelinated axons[75,76] or cultured axons[10] contains a membrane-bound mitogenic signal that will cause a quiescent population of cultured Schwann cells to divide. A major task that remains is to define the neurochemical basis for these physiological changes. As shown in Table VI, the molecular characteristics of the axonal plasma membrane fraction have been defined only in the broadest of brush strokes. The finer details of molecular events that involve specific and specialized regions of the axonal plasma mem-

Table VI
General Characteristics of Axonal Plasma Membrane Preparations

1. Lipids comprise at least 50% of the dry weight of all preparations; those derived from unmyelinated peripheral axons of invertebrates have the highest lipid content.
2. After fractionation, the axonal plasma membrane is isolated as unilamellar vesicles less than 1 μm in diameter with a unit thickness of 70 Å to 110 Å.
3. Most preparations contain an active Na^+, K^+-ATPase and "true" acetylcholinesterase.
4. The axonal plasma membrane appears to contain a number of metabolic activities associated with lipid, glycoplipid, and glycoprotein metabolism.
5. The polypeptide profiles of all preparations characterized to date have a series of quantitatively prominent polypeptides in the 50,000 to 60,000 molecular weight range of the polyacrylamide gel.
6. Other polypeptides common to most axonal plasma membrane preparations can be identified; these polypeptides may serve similar functions, such as the large subunit of Na^+, K^+-ATPase.
7. Phospholipid comprises from one-half to three-quarters of the lipid; the ethanolamine and choline phosphatides are present in the highest amount and at approximately equal levels.
8. The fatty acid composition of phospholipids associated with the axonal plasma membrane contains a high degree of polyunsaturated fatty acids.
9. Glycolipids such as cerebroside, sulfatides, and gangliosides are found only in axolemma preparations derived from myelinated axons.
10. As revealed by neurotoxin-binding studies, axolemma-enriched preparations derived from unmyelinated, continuously conducting axons have a tenfold higher sodium channel content than axolemma-enriched preparations derived from myelinated axons with saltatory conduction.

brane remain to be elucidated. Future investigations will require more sophisticated separation techniques, which will allow the isolation of nodal and internodal axolemma to allow the characterization of the molecular properties of these specialized regions of the axonal plasma membrane. In addition, developmental changes in the molecular architecture of this membrane have not been adequately explored. For example, it would be of interest to know how the molecular composition of the axonal membrane changes during myelination. There are now sensitive micromethods for the separation and analysis of membrane molecular components. This development should allow studies of the axonal plasma membrane in tissue culture systems in which the normal axonal development takes place *in vitro*, including the interaction of the neuronal membrane with its ensheathing and myelinating glial cell. The production of axolemma-specific antisera in conjunction with membrane isolation techniques that utilize these specific antibodies should also allow great progress to be made in the elucidation of the molecular properties of this membrane.

ACKNOWLEDGMENTS This chapter is dedicated to the memory of C.C., a faithful friend and a constant inspiration to the author for the past 17 years. The work in the author's laboratory was made possible by support from the National Institute of Health (NS 10821 and NS 15408), a grant from the Kroc Foundation, and the National Multiple Sclerosis Society. The cheerful and expert secretarial assistance of Ms. Judy Watts is gratefully acknowledged. The author thanks H. Thomson McKean for a critical reading of the manuscript.

REFERENCES

1. Peters, A., Palay, S. L., and Webster, H. de F., 1976, *The Fine Structure of the Nervous System*, W. B. Saunders, Philadelphia, p. 106.
2. Camejo, G., Villegas, G., Barnola, F., and Villegas, R., 1969, *Biochim. Biophys. Acta* **193**:247–259.
3. Fischer, S., Cellino, M., Zambrano, F., Zampighi, G., Tellez-Nagel, M., Marcus, D., and Canessa-Fischer, M., 1970, *Arch. Biochem. Biophys.* **138**:1–15.
4. Denburg, J. L., 1972, *Biochim. Biophys. Acta* **282**:453–458.
5. Chacko, G. K., Goldman, D. E., Malhotra, H. C., and Dewey, M. M., 1974, *J. Cell Biol.* **62**:831–843.
6. Balerna, M., Fosset, M., Chicheportiche, R., Romey, G., and Lazdunski, M., 1975, *Biochemistry* **14**:5500–5511.
7. DeVries, G. H., 1981, *Research Methods in Neurochemistry*, Volume 5 (W. Marks and R. Rodnight, eds.), Plenum Press, New York, pp. 3–38.
8. Yoshino, J. E., Griffin, J. W., and DeVries, G. H., 1983, *J. Neurochem.* **41**:1126–1130.
9. Yoshino, J. E., and DeVries, G. H., 1983, *Trans. Amer. Soc. Neurochem.* **14**:111.
10. Salzer, J. L., Williams, A. K., Glaser, L., and Bunge, R. P., 1980, *J. Cell Biol.* **84**:753–766.
11. Easton, D. M., 1971, *Science* **172**:952–955.
12. Steinbrecht, R. A., 1969, *J. Cell Sci.* **4**:39–53.
13. DeVries, G. H., Anderson, M. G., and Johnson, D. L., 1983, *J. Neurochem.* **40**:1709–1717.
14. DeVries, G. H., and Lazdunski, M., 1982, *J. Biol. Chem.* **257**:11684–11688.
15. Rulfs, J., and DeVries, G. H., 1982, *Trans. Amer. Soc. Neurochem.* **13**:173.
16. DeVries, G. H., Matthieu, J.-M., Beny, M., Chicheportiche, R., Lazdunski, M., and Dolivo, M., 1978, *Brain Res.* **147**:339–352.
17. DeVries, G. H., 1976, *Neurosci. Lett.* **3**:117–122.
18. Zetusky, W., Calabrese, V. P., Zetusky, A., Anderson, M. G., Cullen, M., and DeVries, G. H., 1979, *J. Neurochem.* **32**:1103–1109.
19. Barnola, F., Villegas, R., and Camejo, G., 1973, *Biochim. Biophys. Acta* **298**:84–94.
20. Chacko, G., Villegas, G., Barnola, F., Villegas, and R., Goldman, D., 1976, *Biochim. Biophys. Acta* **443**:19–32.
21. Wood, J. G., Jean, D. H., Whitaker, J. N., McLaughlin, B. J., and Albers, R. W., 1977, *J. Neurocytol.* **6**:571–581.
22. Schwartz, M., Ernst, S. A., Siegel, G. A., and Agranoff, B. W., 1981, *J. Neurochem.* **36**:107–115.
23. Sweadner, K., 1979, *J. Biol. Chem.* **254**:6060–6067.
24. Kracke, G. R., and Chacko, G. K., 1979, *Life Sci.* **25**:2125–2129.
25. Gache, C., Rossi, B., and Lazdunski, M., 1976, *Eur. J. Biochem.* **65**:293–306.
26. Rossi, B., Gache, C., and Lazdunski, M., 1978, *Eur. J. Biochem.* **85**:561–570.
27. Stanley, J., Saul, R. G., Hadfield, M. G., and DeVries, G. H., 1979, *Neuroscience* **4**:155–167.
28. Nachmansohn, D., and Newman, E., 1975, *Chemical and Electrical Basis of Nerve Activity*, Academic Press, New York, pp. 253–259.
29. Sabatini, M., Dipolo, R., and Villegas, R., 1968, *J. Cell Biol.* **38**:176–183.
30. Cecchi, X., Canessa-Fischer, M., Maturana, A., and Fischer, S., 1971, *Arch. Biochem. Biophys.* **145**:240–247.
31. Benes, F., Higgins, J. A., and Barnett, R., 1973, *J. Cell Biol.* **57**:613–629.
32. Estridge, M., and Bunge, R., 1978, *J. Cell Biol.* **79**:138–155.
33. Harford, J. B., Waechter, C. J., Saul, R., and DeVries, G. H., 1979, *J. Neurochem.* **32**:91–98.
34. Costantino-Ceccarini, E., Cestelli, A., and DeVries, G. H., 1979, *J. Neurochem.* **32**:1175–1182.
35. DeVries, G. H., Chalifour, R. J., and Kanfer, J. N., 1983, *J. Neurochem.* **40**:1189–1191.
36. Stahl, W. L., 1973, *Arch. Biochem. Biophys.* **154**:56–65.
37. Chacko, G., 1979, *J. Membr. Biol.* **47**:285–301.
38. Grefrath, S., and Reynolds, J. A., 1973, *J. Biol. Chem.* **248**:6091–6094.

39. DeVries, G. H., Payne, W., and Saul, R. G., 1981, *Neurochem. Res.* **6**:521–537.
40. Estridge, M., 1977, *Nature* **268**:60–63.
41. Chang, C.-M., and Goldman, R. D., 1973, *J. Cell Biol.* **57**:867–874.
42. Goodrum, J. F., and Morell, P., 1982, *J. Neurochem.* **38**:696–704.
43. Chacko, G. K., Barnola, F. V., Villegas, R., and Goldman, D. E., 1974, *Biochim. Biophys. Acta* **373**:308–312.
44. Ritchie, J. M., and Rogart, R. B., 1977, *Proc. Natl. Acad. Sci. U.S.A.* **74**:211–215.
45. Chiu, S. Y., and Ritchie, J. M., 1980, *Nature* **284**:170–171.
46. Catterall, W. A., 1980, *Annu. Rev. Pharmacol. Toxicol.* **20**:15–43.
47. Denburg, J. L., 1976, *Life Sci.* **18**:751–758.
48. Baumgold, J., 1980, *J. Neurochem.* **34**:327–334.
49. Rhoden, V. A., and Goldin, S. M., 1979, *J. Biol. Chem.* **254**:11199–11201.
50. Vincent, J. P., Balerna, M., Barhanin, J., Fosset, M., and Lazdunski, M., 1980, *Proc. Natl. Acad. Sci. U.S.A.* **77**:1646–1650.
51. Catterall, W. A., Morrow, C. S., and Hartshorne, R. P., 1979, *J. Biol. Chem.* **254**:11379–11387.
52. Denburg, J. L., Eldefrawi, M. E., and O'Brien, R. D., 1972, *Proc. Natl. Acad. Sci. U.S.A.* **69**:177–181.
53. Chester, J., Lentz, T. L., Marguis, J. K., and Mautner, H., 1979, *Proc. Natl. Acad. Sci. U.S.A.* **76**:3542–3546.
54. Jumblatt, J. E., Marguis, J. K., and Mautner, H., 1981, *J. Neurochem.* **37**:392–400.
55. DeVries, G. H., Zetusky, W. J., Zmachinski, C. J., and Calabrese, V. P., 1981, *J. Lipid Res.* **22**:208–216.
56. DeVries, G. H., and Zmachinski, C. J., 1980, *J. Neurochem.* **34**:424–430.
57. Chacko, G. K., Barnola, F. V., and Villegas, R., 1977, *J. Neurochem.* **28**:445–447.
58. Zambrano, F., Cellino, M., and Canessa-Fischer, M., 1971, *J. Membr. Biol.* **6**:289–303.
59. Cotman, C., Blank, L. M., Moehl, A., and Snyder, F., 1969, *Biochemistry* **11**:4606–4612.
60. Houslay, M. D., and Stanley, K. K., 1982, *Dynamics of Biological Membranes*, John Wiley & Sons, New York, p. 20.
61. De Gier, J., Mandershool, J. G., and Van Deenen, L. L. M., 1968, *Biochim. Biophys. Acta* **150**:666–685.
62. Haley, J. E., and Ledeen, R. W., 1979, *J. Neurochem.* **32**:735–742.
63. Norton, W. T., Abe, T., Poduslo, S., and DeVries, G. H., 1975, *J. Neurosci. Res.* **1**:57–75.
64. Norton, W. T., and Autilio, L. A., 1966, *J. Neurochem.* **13**:213–222.
65. DeVries, G. H., and Norton, W. T., 1974, *J. Neurochem.* **22**:259–264.
66. DeVries, G. H., and Norton, W. T., 1974, *J. Neurochem.* **22**:251–257.
67. Shook, W., and Norton, W. T., 1976, *Brain Res.* **118**:517–522.
68. Henderson, T. J., Bigbee, J. W., and DeVries, G. H., 1984, *Brain Res.* (in press).
69. Breckenridge, W. C., Gombos, G., and Morgan, I., 1972, *Biochim. Biophys. Acta* **266**:695–707.
70. Kriegler, J. S., Krishman, N., and Singer, M., 1981, *Adv. Neurol.* **31**:479–504.
71. Zambrano, F., Morales, M., Fuentes, N., and Rojas, M., 1981, *J. Membr. Biol.* **63**:71–75.
72. Ganser, A., Kirchner, D., and Willinger, M., 1983, *J. Neurocytol.* **12**:921–938.
73. Foster, R. E., Whalen, C. C., and Waxman, S. G., 1980, *Science* **210**:661–663.
74. Smith, K. J., Bostock, H., and Hall, S. M., 1982, *J. Neurol. Sci.* **54**:13–31.
75. DeVries, G. H., Salzer, J. L., and Bunge, R. P., 1982, *Dev. Brain Res.* **3**:295–299.
76. DeVries, G. H., Minier, L., and Lewis, B. L., 1983, *Dev. Brain Res.* **9**:87–93.

14

Central Nervous System Myelin

Joyce A. Benjamins, Pierre Morell, Boyd K. Hartman, and Harish C. Agrawal

1. INTRODUCTION: STRUCTURE AND FUNCTION OF MYELIN

In the first edition of the *Handbook of Neurochemistry*, Mokrasch[1] reviewed the biochemical literature on myelin through 1967. In the intervening 15 years, hundreds of research articles and numerous reviews about myelin have been published, encouraged both by the growth of neuroscience and the development of several methods for detailed analysis of membrane proteins and lipids. Although much new biochemical information is available, our basic concepts of the morphological structure[2,3] and principal function[4] of myelin have changed little since 1967. The concept that the multilamellar myelin membrane is produced by the oligodendroglial cell is well accepted, although the molecular events involved in the proliferation of oligodendroglial cells and the mechanics of wrapping of a single oligodendroglial plasma membrane around multiple axons are an enigma to neurobiologists. The role of myelin in the establishment of saltatory conduction is still thought to be its primary function. However, the details of how this conduction is established and how it is reestablished in remyelinating systems continue to be active areas of investigation,[5,6] particularly with regard to the relationship between axonal ion channels and glial membrane. This chapter emphasizes advances in the biochemistry of myelin over the past 15 years. Review articles are cited extensively where appropriate to minimize redundancy.

2. ISOLATION OF MYELIN

Myelin is characterized by a low content of water and protein and a high content of lipid relative to other membranes. In mammalian species, about 70%

Joyce A. Benjamins • Department of Neurology, Wayne State University School of Medicine, Detroit, Michigan 48201. *Pierre Morell* • Department of Biochemistry, University of North Carolina, Chapel Hill, North Carolina 27514. *Boyd K. Hartman* • Department of Psychiatry, Washington University School of Medicine, St. Louis, Missouri 63178. *Harish C. Agrawal* • Department of Pediatrics, Washington University School of Medicine, St. Louis, Missouri 63178. Work in the authors' laboratories was supported by grants from NIH.

of the dry weight of myelin is lipid, compared to 60% for white matter and 45% for gray matter.[7] Because of its high lipid content, myelin can be readily isolated from homogenates by virtue of its low density. A variety of methods for isolation of myelin have been devised. The one in most common use is that of Norton and Poduslo.[8,9] In this method, homogenate in 0.32 M sucrose is layered over 0.85 M sucrose, and the myelin is collected from the interface after centrifugation. Variations of this method with different numbers of gradients and washes have been used by many investigators. Other methods of isolation involve centrifuging the myelin up rather than down,[10,11] isolating myelin from crude mitochondrial fractions,[12] adding buffer, Na^+,[13] or Cs^+,[14] salts to the homogenization medium, using continuous gradients or zonal centrifugation (see ref. 15), and scaling down the procedures for small sample sizes.[16,30] Since myelin is quite different in density and average particle size from most other membranes, it is relatively easy to isolate in good yield with high purity. All of these aspects of myelin isolation have been discussed in detail in a recent review.[17] The most common sources of other membranes in the myelin fraction are thought to be axolemma,[18] plasma membranes derived from oligodendrocytes and astrocytes, endoplasmic reticulum,[17] and Golgi membranes.[19]

3. PROTEINS OF CNS MYELIN

Properties of specific proteins of CNS myelin have been described in numerous reviews.[20–24] Most of our information about specific proteins in isolated myelin has come from studies on a few mammalian species. Small differences in protein composition between species and between various CNS regions in the same species have been noted, as discussed below.

3.1. Composition

3.1.1. Identification on Gels

The proteins of myelin are readily soluble in sodium dodecyl sulfate (SDS) and can be separated by disk or slab gel electrophoresis in buffers containing SDS. The composition of proteins seen in a given study depends on a variety of methodological factors, including procedures for myelin isolation,[14] gel electrophoresis,[25–27] sample preparation,[28,29] and staining of gels. Molecular weight determinations on gels present well-known problems.[25,26] The use of "apparent molecular weight" with regard to known molecular-weight markers in a given system appears to provide the most reproducible means of describing a particular protein.

The protein profiles of CNS myelin from four species are shown in Fig. 1. Examination of myelin proteins by this method reveals three major protein bands in most mammalian species, namely, proteolipid protein (PLP), large basic protein (LBP), and a high-molecular-weight doublet called Wolfgram proteins (WP). Rat and mouse brain myelin have an additional protein, called small basic protein (SBP). Basic proteins and proteolipid protein account for over

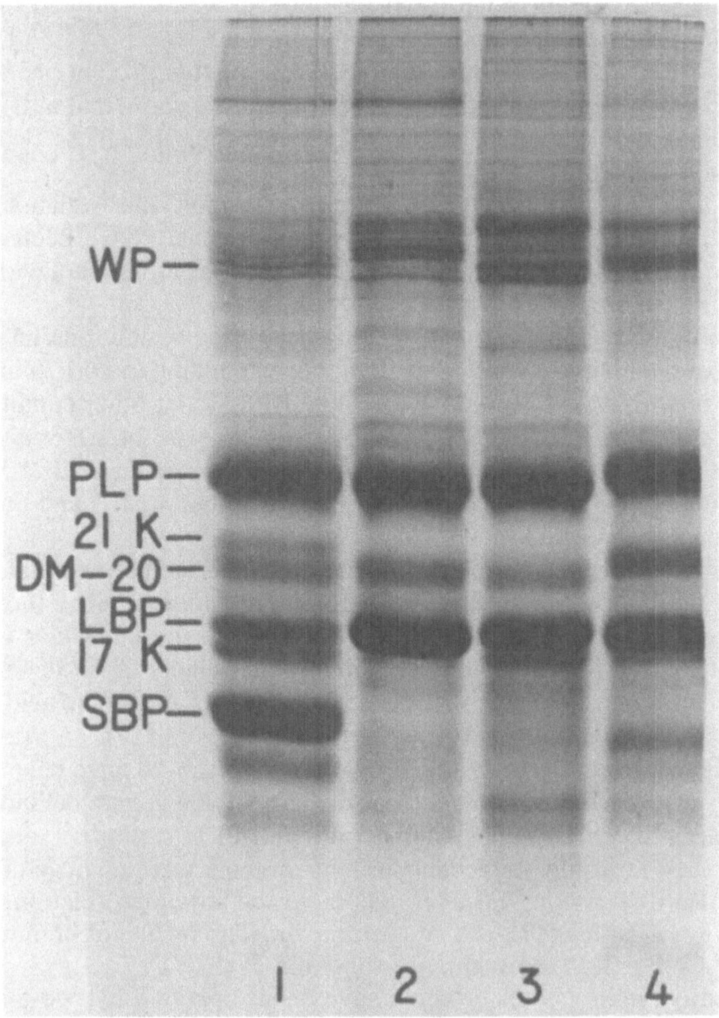

Fig. 1 SDS-slab gel electrophoresis of proteins of myelin from four species; 75 μg of protein were applied to each lane. Lanes 1–4 show protein profiles of mouse, guinea pig, human, and bovine brain myelin, respectively. A Tris–glycine buffer system was used. (Figure from H. C. Agrawal, unpublished observations.)

80% of the membrane protein as determined by dye-binding capacity[14] or by radioimmunoassay.[31,32] In addition, a large number of high-molecular-weight proteins (30,000 daltons and above) are also seen; the origin and relationship of most of these proteins to the myelin membrane *in situ* remain to be elucidated. However, two classes of high-molecular-weight proteins, namely, WP[33] and myelin-associated glycoprotein,[34] have been shown to be intrinsic components of oligodendrocytes and myelin.[35,36] The Wolfgram proteins of 43–48 K probably contain the enzyme 2′,3′-cyclic nucleotide 3′-phosphohydrolase (CNP'ase), which is highly enriched in myelin (see Section 3.3.3).

Another proteolipid protein, designated intermediate protein or DM-20, has also been shown to be a component of the CNS myelin.[14,24,37] The amino acid composition[38,39] and N-terminal sequence of this protein are similar to those of PLP.[40] In addition, immunochemical studies show that antisera to rat brain myelin PLP cross react with DM-20, suggesting that these two proteins share similar antigenic sites.[41]

Recently, two subsidiary basic proteins with molecular weights of 21,500 and 17,500 (originally designated prelarge and presmall) have been described in both mouse[42] and rat[43] brain myelin. These proteins cross react with antisera to BP.[42,43]

It should be pointed out that in some gel systems, only one intermediate band is observed between PLP and LBP, corresponding to DM-20 or possibly a mixture of DM-20 and 21.5 K BP (prelarge BP). Under other conditions, two bands are clearly visible between PLP and LBP (see refs. 24, 29 for discussion). Thus, DM-20 corresponds to the small intermediate protein, and 21.5 K BP to the large intermediate protein observed by Greenfield *et al.*[14] and by Cammer and Norton[29] (see Fig. 2).

The molecular weights of the major myelin proteins, as determined by SDS gel electrophoresis, are listed in Table I. In addition, since the complete amino acid sequences of LBP and SBP are known, the molecular weights of these proteins can be calculated. The molecular weights of BP determined by SDS gel electrophoresis agree with those determined from amino acid sequence data.

Myelin BP from the CNS and PNS, PLP, and DM-20 have been identified after transferring myelin proteins from the acrylamide gels to cellulose nitrate sheets followed by immunostaining with anti-PLP serum[41,51] and anti-BP serum.[52–54] Several high-molecular-weight proteins were also found to cross react with anti-BP serum, but their metabolic and structural relationship to the 21 K, LBP, 17 K, and SBP are not yet known. The results of immunostaining of BPs of CNS and PNS myelin are shown in Fig. 2.

Recently, white matter proteins have been separated by two-dimensional gel electrophoresis.[55] Because of the complex pattern seen on two-dimensional gels, the data are difficult to interpret. Purified BP will not enter from the basic side of isoelectric focusing gels (i.e., a pI of 10 or greater). It can be run into the acid side of the nonequilibrium pH-focusing gels described by O'Farrell.[49] However, PLP does not appear as a spot under these conditions (J. Goodrum and P. Morell, unpublished observations). Althaus, *et al.*[144] have reported that PLP can be analyzed by isoelectric focusing if myelin proteins are first solubilized in a mixture of tetramethyl urea and dimethylene urea instead of SDS.

3.1.2. Changes during Development

The developmental pattern of the major myelin proteins has recently been reviewed with regard to changes seen in whole brain, in myelin *per se*,[56,57] and in cultures containing oligodendroglia.[56,58] The levels of the major myelin proteins increase markedly in various CNS regions as myelination begins.[59,60] Increases in 2′,3′-CNP′ase activity in brain precede increases in BP and are

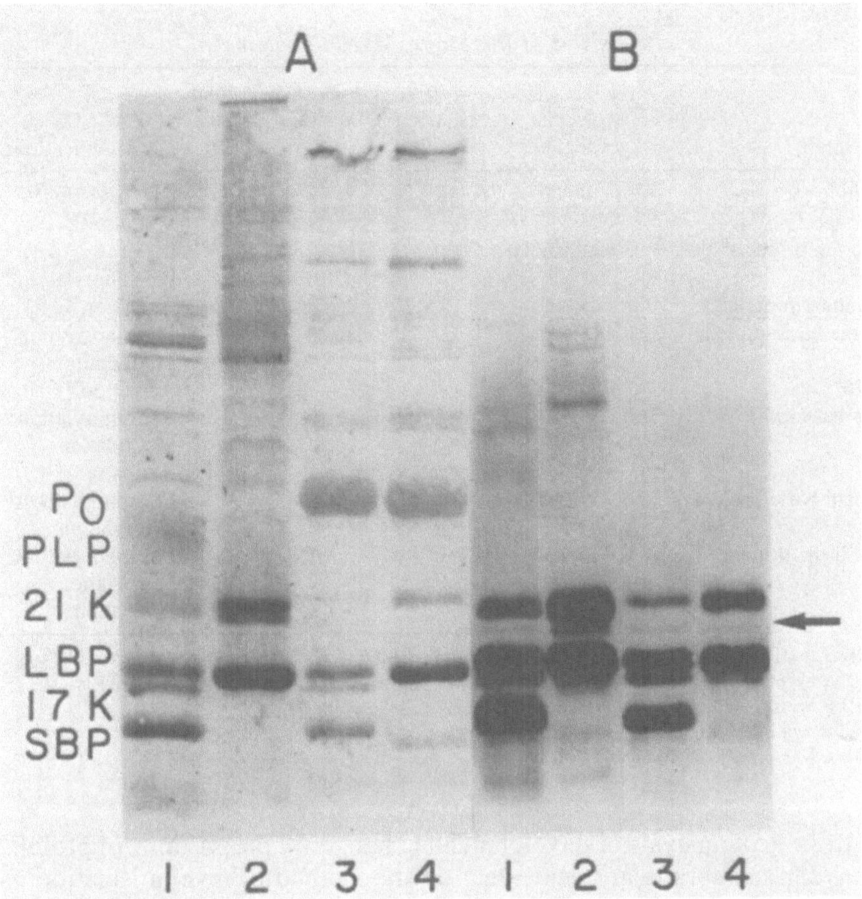

Fig. 2. Immunoblot identification of myelin basic proteins. Panel A: Cellulose nitrate sheets stained with amido black. Panel B: Cellulose nitrate sheets immunostained with antiserum to chick basic protein; 100 μg of protein were applied to each lane. Lanes 1–2: Proteins of rat and rabbit CNS myelin, respectively. Lanes 3–4: Proteins of rat and rabbit PNS myelin, respectively. Arrow indicates the presence of a BP migrating between 21 K and LBP. (Figure from Gilbert *et al.*,[54] with permission.)

followed by increases in PLP. Similar conclusions have been reached in studies on optic nerve.[16,61,62] In several culture systems, activity of CNP'ase increases more rapidly than BP levels, as *in vivo*.[58,64] A recent study with dissociated cell cultures showed that increases in levels of BP are delayed, with the result that PLP levels increase before BP levels.[63] The availability of immunoassays for the major myelin proteins should allow more precise determination of developmental changes and allow analysis of small regions of brain or cord. With the development of immunoblot methodology, the question of whether oligodendroglia contain forms of the proteins other than those that appear in myelin can be investigated. Changes in rates of synthesis of myelin proteins during development are discussed in Section 3.4.1.

Table I
Properties of the Major Myelin Proteins[a]

Protein	Molecular weight	Isoelectric point	Basic residues (%)	Apolar residues (%)	Modifications
Myelin-associated glycoprotein[44]	110,000	acidic (3–4)[d]			Glycosylation[171] Sulfation[172]
Wolfgram proteins[b]	43,000–62,000				
(CNP'ase)[45]	(43,000–48,000)	9.7[45]	16	44	Phosphorylation[46,135]
Proteolipid protein[24,47]	25,000	9.2[47]	10	66	Acylation[48,162]
Prelarge basic protein[42]	21,500				Phosphorylation[43] Methylation
DM-20[37]	20,500				Acylation[48]
Large basic protein[21,23]	18,500	>10.6[50]	24	41	Phosphorylation[21,23] Methylation Amidation
Presmall basic protein[42c]	17,500				Phosphorylation[43] Methylation
Small basic protein[21c]	14,000				Phosphorylation[21,23] Methylation Amidation

[a] References are not necessarily to original articles but to recent reviews or articles that summarize the properties of a given protein; see text for fuller discussion.
[b] Heterogeneous group of proteins; see Section 3.3.3.
[c] Found in some rodents.
[d] Quarles, R. H., personal communicaion.

Results of analysis of developmental changes in proportions of proteins in the myelin membrane are dependent on the method of myelin isolation used. Thus, myelin isolated from younger brain may be more contaminated with other membranes than that isolated from older animals. A given method of isolation may select membranes more enriched in one component than another. However, most studies show an increase in basic protein and PLP with an accompanying decrease in high-molecular-weight proteins. This change has been demonstrated in CNS myelin from a number of species (see ref. 20), although one study in mouse[65] and another in pig[66] found no changes in the proportions of the various proteins in myelin during development.

In both mouse and rat, LBP decreases and SBP increases during maturation.[67–69] Similarly, Fishman *et al.*[70] also observed an increase in LBP in myelin isolated from developing human brain, which is essentially similar to the earlier findings of Eng *et al.*[71] Barbarese *et al.*[72] showed increases in 21.5 K BP, LBP, 17.5 K BP, and SBP in myelin isolated from 15- to 60-day old mouse brain. Between 30 and 60 days, the proportion of SBP protein in myelin increases, whereas that of LBP remains constant, and those of 21.5 K and 17.5 K proteins decrease. Thus, SBP accumulates relative to the other proteins.

The developmental changes in PLP are not as readily apparent. A number of studies have analyzed the ratio of basic proteins to proteolipid protein in myelin isolated at various stages of development, with conflicting results.[16,68,70,73,74] Aggregation or loss of PLP during sample preparation, method of myelin isolation, and the brain region or species used may all contribute to the difficulty in resolving this issue. For example, Morell *et al.*[68] found an increase in the concentration of PLP in mouse brain myelin, whereas Zgorzalewicz *et al.*[69] observed little change in the concentration of PLP in myelin isolated from different regions of developing rat brain.

The levels of DM-20 relative to PLP appear to decrease during development.[67,69] The WPs decrease relative to BP and PLP; MAG has been reported to remain constant[75] or decrease[76] relative to other proteins. However, nonspecific methods, i.e., PAS staining and Con-A binding, were used rather than the immunoassay now available.

3.1.3. Regional Variation

Regions of the CNS differ with regard to time of onset of myelination, extent of myelination, and composition of the myelin. Myelination proceeds along a rostral–caudal gradient; thus, differences found in composition of myelin from various brain regions isolated during the period of active myelination may reflect developmental differences as well as any intrinsic regional differences. From adult CNS, the yield of myelin decreases in the order spinal cord > brainstem > cerebrum = cerebellum.[77] Spinal cord myelin has a lower content of protein than that from cerebrum, with brainstem again intermediate (see ref. 78 for a summary of values). Typical values are 17% and 22% protein for myelin from spinal cord and cerebrum, respectively.[77] In a number of species, the ratio of BP to PLP is highest in spinal cord, followed by brainstem, cerebrum, and cerebellum.[74,77,79–81] Regional differences have also been found in the stability of myelin to dissociation during incubation of brain slices in Krebs–Ringer buffer.[77]

Regional differences in immunohistochemical localization of basic protein and proteolipid protein are described in Section 5.2. Some regional differences have also been noted in the distribution of myelin into subfractions (see ref. 82).

3.1.4. Species Variation

When myelin from other vertebrates is examined, the composition varies much more widely than among mammals. Phylogenetically, multilamellar myelin is first seen in sharks,[83,84] with lower chordates lacking both identifiable oligodendroglia and myelin. A recent survey[85] reviewed previous studies and presented an analysis of the SDS–PAGE patterns of myelin from a variety of species. They concluded that mammalian, avian, and reptilian myelin showed similar protein patterns, apparently including basic proteins, proteolipid protein, DM-20, and Wolfgram protein, whereas amphibians lacked DM-20. The myelin proteins of carp and trout were markedly different from those of other species, with the pattern dominated by a protein of lower molecular weight than SBP and another in the Wolfgram region. In the absence of immunochemical identification of BP, PLP, and WP, it is doubtful if myelin from fish and reptiles indeed contains BP and PLP. However, myelin from bony fish showed a somewhat more "typical" pattern but included a protein migrating like the P_0 protein from mammalian species. Besides differences in distribution of proteins, differences in Con-A binding among various species were noted as well. It is of interest that the protein composition of myelin from various species differs so widely, yet the ultrastructural organization appears remarkably similar.

3.2. Extraction and Purification

A variety of methods have been developed for bulk isolation of the major myelin proteins from either white matter or myelin (see refs. 23,24,47,86,87 for discussion). Extraction of whole brain or white matter with 2:1 chloroform–methanol results in a fraction highly enriched in myelin PLP. Several other hydrophobic proteins including DM-20 are present in this fraction, depending on the starting material and conditions used. Extraction of delipidated whole brain or white matter or of isolated myelin with dilute salt or acid yields a fraction enriched in the myelin basic proteins.[87] If isolated myelin is extracted with 2:1 chloroform–methanol, BP as well as PLP is soluble in the organic solvent, as noted below. However, if white matter or whole-brain homogenate is used, much less BP protein is extracted relative to PLP. This difference in solubility properties of BP depending on starting material has been attributed to differences in ionic content[88] under the two conditions.

Proteins in isolated myelin can be extracted with salts and detergents[71] or with chloroform–methanol (2:1, v:v) as described by Gonzales-Sastre.[89] In the latter procedure, both BP and PLP are soluble in chloroform–methanol; BP can be precipitated out of solution by the addition of 1 M KCl. The BP can be further extracted with dilute HCl. The PLP and DM-20 remain in the organic phase. The chloroform–methanol fraction can be reduced to one-fifth its original volume, and PLP and DM-20 can be precipitated from this concentrated solution by the addition of diethyl ether. The chloroform–methanol-insoluble fraction contains predominantly high-molecular-weight proteins including WP and MAG.

It must be emphasized that these bulk procedures for preparation of myelin proteins from whole tissue or isolated myelin provide fractions enriched in BP and PLP that have to be further purified for chemical or immunologic studies: PLP has been purified by solvent extraction followed by LH-Sephadex chromatography[90,91] or SDS-gel electrophoresis,[48] whereas basic proteins have been purified by Sephadex gel filtration, ion-exchange chromatography,[87,94] HPLC,[95] SDS-gel electrophoresis,[92] and by affinity chromatography.[97]

3.3. Properties of Myelin Proteins

3.3.1. Basic Proteins

3.3.1a. Structural and Chemical Characterization. As noted above, myelin from all mammalian species contains the large basic proteins (18.5 K), but mouse, rat, and other rodents have in addition a basic protein of 14 K designated small basic protein. The complete amino acid sequence of LBP from human,[98] bovine,[99] and rabbit[100] brain myelin is known. The amino acid sequence of SBP is similar to that of LBP except for an internal deletion of 40 amino acid residues between 117 and 157.[101] Barbarese *et al.*[42] examined the structure of 21.5 K (prelarge) and 17.5 K (presmall) proteins in mouse and found that the sequences of these two proteins are similar to those of LBP and SBP, respectively, except for the N-terminus addition of 25–35 amino acid residues in the 21.5 K and 17.5 K proteins.

Much interest has centered on these proteins because of their ability to induce EAE in a variety of species (see refs. 102,103). The disease-inducing determinants appear to differ from one species to another; a variety of peptides and modified peptides have been used to map the regions of the molecule responsible for disease. Similar studies to determine the structural requirements for suppression of disease are under way as well.

Myelin BPs are characterized by a high content of basic amino acids, with most of the glutamic acid amidated, giving it an isoelectric point greater than 10.6.[50] The proteins contain no cysteine (except for BP from frog brain[104]), a single tryptophan, monomethyl- and dimethylarginines, and several prolines. A triproline sequence has been proposed to give rise to a 180° hairpin bend in the molecule (see ref. 21), but recent NMR studies did not detect such a structure (see ref. 23). The protein appears to have an open, highly unfolded structure in aqueous solution; addition of lipid causes a change toward more α-helical structure (see ref. 23). Boggs and Moscarello have proposed that several hydrophobic sequences in the amino-terminal half of the molecule may interact with lipids in the membrane.[105] Physical measurements and chemical probe studies indicate that the protein is not deeply buried within the membrane, however. Exposure of intact myelin lamellae to surface-labeling reagents gives no labeling of BP, indicating it is on the inner cytoplasmic surface of the membrane.[106] With a few exceptions, most recent immunohistochemical studies are in agreement with this result, showing BP at the major dense line (see ref. 107 for discussion).

Cross-linking studies suggest that interaction may occur between the carboxyl and amino terminal regions of neighboring basic proteins; these interactions may occur between BPs on the apposed inner surfaces of the myelin lamellae, thus playing a role in myelin compaction.[108] The cross-linking studies further indicated interaction between BP and PLP in the membrane, but the nature and function of this interaction remain to be characterized. The recent observation that CNS myelin of the mouse mutant shiverer lacks MBP, frequently shows splitting of the major dense line, and has uncompacted myelin lamellae[96,109,110] indicates that BP plays a major role in normal CNS myelination. However, shiverer PNS myelin, which also lacks BP, appears normal, indicating that the presence of the major dense line is not strictly correlated with the presence or absence of BP and that other factors besides BP regulate compaction of the the lamellae in the PNS.[111]

3.3.1b. Posttranslational Modification. Myelin BP has been shown to be acetylated, methylated, glycosylated, and phosphorylated. Details of these reactions have been discussed by Braun and Brostoff,[21] Martenson,[112] and Carnegie and Moore.[23]

Acetylation. The N-terminal amino acid of bovine, human, rabbit, rat, and monkey LBP, as well as rat SBP and mouse 21.5 K protein and 17.5 K protein, has been shown to be alanine; this amino acid has been shown to be acetylated in all cases.[101]

Methylation. Arginine at position 107 has been shown to be methylated in human and bovine BP. In fact, this one arginine is present as arginine, monomethylarginine, or dimethylarginine.[23,113] The ratio of dimethyl- to monomethylarginine is highest in the BP of human, monkey, and rabbit BP, lowest in rodent BP, and intermediate in bovine and turtle BP.[112,114] Incubation of isolated BP with cytoplasmic arginine methyltransferase *in vitro* results in the methylation of the same arginine residue as *in vivo*,[115] suggesting that BP is methylated in the cytosol of oligodendrocytes before its incorporation into the myelin membrane.[112] This conclusion is also supported by the recent finding that *in vivo* methylation of BP is proportional to the rate of synthesis of BP; that is, it occurs more rapidly during active myelination than in adulthood.[116] Recently, Small and Carnegie[117] demonstrated *in vivo* methylation of arginine (residue 107) of chicken BP after injection of [methyl-^{3}H]methionine in 2-day-old chick. The half-life of chicken BP, as determined by the decay rate of methylated arginine, was about 40 days.[117]

The biological significance of methylation of BP at Arg-107 remains to be determined. However, Brostoff and Eylar[113] postulated that the methylated base might play a role in stabilizing the conformation of protein by cross-chain interactions. Baldwin and Carnegie[118] and London and Vossenberg[119] suggested that methylation of BP facilitates specific interactions between BP and myelin lipids because methylation of arginine renders it more hydrophobic. Martenson[112] postulated that the primary function of methylation of Arg-107 may be to specifically prevent phosphorylation of Ser-110, because Carnegie *et al.*[120,121] reported cyclic-AMP-dependent *in vitro* phosphorylation of Ser-110 primarily in the unmethylated BP molecules.

Phosphorylation. Myelin BP has been shown to be an excellent substrate for phosphorylation *in vitro* by both endogenous and exogeneous protein kinases.[120–125] Miyamoto and Kakiuchi[124] and Steck and Appel[125] also demonstrated that BPs of rat brain myelin were phosphorylated after intracerebral injection of [^{32}P]orthophosphoric acid. These investigators found that BP phosphorylated by the endogeneous protein kinase contained 0.2 mol phosphate/mol BP.[124–126] Several investigators have subsequently shown the presence of protein kinases associated with CNS myelin fractions which phosphorylate SBP and LBP *in vitro*.[127–134] Turner *et al.*[135] reported that Ca^{2+}-dependent phosphorylation of BP and WP *in vitro* by the endogeneous protein kinase required phosphatidylserine and not calmodulin.

Recently, it was demonstrated that both 21 K BP and LBP in rabbit myelin are phosphorylated *in vivo* in both CNS and PNS.[43,136] Similarly, four basic proteins of rat brain myelin have also been shown to be phosphorylated *in vivo*[43] (Fig. 3). In a comparison of CNS and PNS myelin, Gilbert *et al.*[54] used immunoblot techniques to demonstrate the *in vivo* phosphorylation of four basic

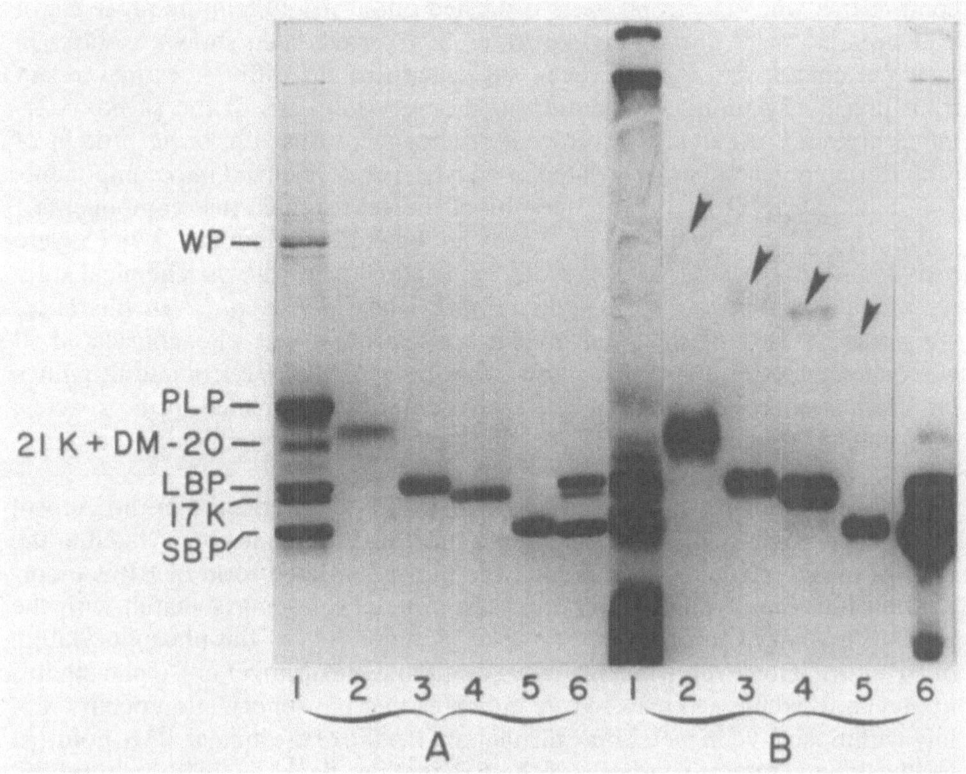

Fig. 3. SDS-slab gel electrophoresis of four purified phosphorylated basic proteins of rat brain myelin. Panel A: Stained proteins. Panel B: Fluorogram of panel A. Lane 1, myelin proteins; lane 2, 21 K BP; lane 3, LBP; lane 4, 17 K BP; lane 5, SBP; lane 6, basic-protein-enriched fraction. Arrows indicate the phosphorylated high-molecular-weight polymer. (Figure from Agrawal *et al.*,[43] with permission.)

proteins in rat PNS myelin with molecular weights of 21 K, 18 K (LBP), 17 K, and 14 K (SBP) and two basic proteins (21 K and LBP) in rabbit PNS myelin.

The sites of phosphorylation of BP vary depending on the species and the conditions used. For example, Carnegie and co-workers[120,121] demonstrated *in vitro* phosphorylation of Ser-55 by endogeneous myelin protein kinase, but incubation of myelin with rabbit muscle protein kinase *in vitro* resulted in the phosphorylation of Ser-110 as well as Ser-55.[121] Chou *et al.*[137] found that Thr-98 and Ser-165 in bovine LBP are phosphorylated *in vivo*, and Martenson *et al.*[138] showed that LBP from adult rabbit brain was phosphorylated *in vivo* on five different sites, Ser-7, Ser-56, Thr-96, Ser-113, and Ser-163, with the total amount of phosphate at these sites about 0.2 g·atom/mol of BP. The observations that different sites of BP are phosphorylated *in vitro* and *in vivo* may indicate that the same protein kinase is not responsible for phosphorylation under the two conditions. At present, the role of phosphorylation in metabolism or function of the basic proteins is still unknown.

When LBP is examined electrophoretically on an acid–urea or SDS-polyacrylamide gel, it migrates as a single band. However, it is resolved into five components when electrophoresis is carried out at pH 10.6 in the presence of 8 M urea.[126,139,140] These multiple forms of BP have been shown to differ in their net charges.[126,137] It has been suggested that the differences may result from phosphorylation, deamidation, and/or possibly loss of the carboxy-terminal arginine residues.[126,137] Recent studies show that the basic protein of rabbit brain myelin phosphorylated *in vivo* is clearly resolved into components 1, 3, and 5 at pH 10.6.[141] Fluorography of the gel revealed that components 1 and 2 were completely devoid of phosphate, whereas components 3 and 5 were heavily phosphorylated. These results are in agreement with the chemical studies previously reported by Deibler *et al.*[126] and Chou *et al.*[137] In this latter study, about 50% of the total myelin basic protein was phosphorylated,[141] whereas Deibler *et al.*[126] have shown that only 20% of BP from adult guinea pig brain is phosphorylated. Whether this difference is related to the species, age, method of isolation of BP, or some other variable remains to be determined.

It is not known whether the phosphorylation of BP occurs in the cytosol of oligodendroglial cells or in the myelin sheath. If BP is phosphorylated in the cytosol of the oligodendrocytes, and the phosphorylated form of BP is incorporated into the myelin membrane, then protein kinases associated with the isolated myelin membrane may not play a major role in the phosphorylation of BP *in situ*. However, a recent study comparing the turnover of BP phosphates in myelin in young and mature rats indicates that phosphorylation occurs rapidly within the myelin membrane throughout the life of the animal.[116] In contrast to the slowing rate of synthesis of the BP peptides backbone with maturation, incorporation of [^{33}P]phosphoric acid into myelin BP was low at 18 days and high at 60 days. The rate of incorporation was directly proportional to the amount of myelin present, not the rate of synthesis. Further, the MBP phosphates turned over much more rapidly than the peptide backbone in both young and mature rats. If BP is phosphorylated and dephosphorylated within the myelin membrane *in situ* as these results indicate, there must be some mechanism to provide a continuous supply of ATP from the oligodendroglial cells to the myelin membrane.

3.3.2. Proteolipid Proteins

3.3.2a. Structural and Chemical Characterization. Folch and Lees[142] first described a protein fraction extracted from bovine white matter by chloroform–methanol, which they called proteolipid protein (PLP). When the Folch–Lees PLP fraction from bovine white matter is subjected to SDS gel electrophoresis, the predominant species is myelin PLP, with traces of DM-20 and components of higher and lower molecular weights.[143] The amino acid composition of Folch–Lees PLP shows a high content of nonpolar, aromatic, and sulfur-containing amino acids (Table I). Its moderate content of basic amino acids along with amidation of glutamic acid gives the protein a highly basic PI of around 9.2 as determined by isoelectric focusing.[47] Under some conditions, multiple bands are found in the Folch–Lees PLP fraction, ranging in molecular weight from 10,000 to 60,000, with major protein bands migrating to the position of myelin PLP, DM-20, and a low-molecular-weight species.[143] Analysis of electrophoretic mobilities[143] and amino acid composition[39] of the bands suggests that they may represent an oligomeric series, but this point is still under investigation (see ref. 47).

When whole brain is extracted with chloroform–methanol, an even more complex pattern is observed.[38,145–147] Since these extracts contain proteins from both gray and white matter and from many subcellular sites, they cannot be equated with proteolipid protein derived from white matter or from isolated myelin. In contrast, when isolated myelin is extracted with chloroform–methanol, and a proteolipid protein-enriched fraction is prepared by solvent extraction[89] or LH-Sephadex chromatography,[90] only two bands of the Folch–Lees preparation are observed, myelin PLP and DM-20. The current status of research on Folch–Lees PLP and myelin PLP has been recently reviewed by Lees and co-workers,[47,147] Boggs and Moscarello,[105] Agrawal and Hartman,[24] and Schlesinger.[148]

Myelin PLP is an intrinsic protein of CNS myelin and accounts for 40–60% of the total membrane protein as determined by extraction[80,149] or by radioimmunoassay.[32] Somewhat lower values are obtained from gel scans.[150] Until methods for isolating a homogeneous preparation of PLP from CNS myelin were developed, its immunologic properties and amino acid sequence could not be determined. Techniques have now been developed in several laboratories to isolate pure preparations of myelin PLP.

Gagnon *et al.*[90] first isolated a protein from human brain myelin by chromatography on Sephadex LH-20 resin and designated this protein N2 or, more recently, lipophilin. The amino acid composition and solubility properties of N2 were found to be similar to those of the Folch–Lees proteolipid protein fraction from bovine white matter, and glycine was the only N-terminal amino acid found in N2.[90] Subsequently, Nicot and co-workers[38,40] isolated two protein bands of molecular weights of 25,000 (PLP) and 20,000. Nussbaum *et al.*[151] isolated PLP from rat brain myelin by preparative PAGE in SDS and determined its molecular weight, amino acid composition, C-terminal amino acid, and N-terminal sequence. Subsequently, the N-terminal sequence of human and bovine brain proteolipid protein has been determined.[40,105,151,152] There is a sequence homology among different species. Jolles *et al.*[153] reported the sequence of 11 tryptic peptides of rat brain myelin PLP. The sequencing of these tryptic peptides showed the presence of six arginine residues, which has also calculated from amino acid analysis of purified myelin PLP.[151] This agreement confirmed the molecular weight of 25,000 for myelin PLP as previously described.[37,151]

Jolles *et al.*[154] isolated three cyanogen bromide (CNBR) peptides after treatment of the 25 K component purified from bovine white matter by gel filtration. The largest fragment separated into two species after performic acid oxidation of disulfide bonds. The fragment CNBR 1, with a molecular weight of 10,000, corresponded to the N-terminal sequence of the intact protein on the basis of amino-terminal sequence data. CNBR 1 is disulfide linked to CNBR 2. CNBR 3 has 13 residues and is adjacent to a 6-residue CNBR piece that contains the carboxyl-terminal phenylalanine residue. From the sequence described thus far, the N-terminal is very hydrophobic, whereas the C-terminal is highly positively charged.[154]

Lees and co-workers have also determined the sequence of seven tryptic peptides[144] of Folch–Lees PLP and a hydrophobic, chloroform-soluble tryptic peptide of molecular weight approximately 4000.[156] This peptide spans the three CNBR peptides located by Jolles *et al.*[154] and Stoffel *et al.*[156a] at the C-terminal end of PLP. Nussbaum *et al.*[157] cleaved PLP with BNPS-skatole and generated four long peptides and determined their sequences. This resulted in the sequencing of 50 additional amino acids from rat brain myelin PLP. Recently, the completed sequence of bovine PLP has been reported by Lees, *et al.*[165] and Stoffe *et al.*[394]

Myelin PLP is very hydrophobic, with two-thirds of the amino acids nonpolar. The hydrophobicity of PLP may also be caused in part by the presence of covalently linked fatty acids. In addition, myelin PLP has been shown to be extremely susceptible to aggregation if delipidated myelin proteins or isolated preparations of PLP are heated in the presence of high concentrations of reducing reagents such as 2-mercaptoethanol.[28,38,158] Irreversible aggregation also occurs if organic solvent is removed by evaporation or if water is added too rapidly during conversion to a water-soluble form.[159] The aggregation of myelin PLP may result from hydrophobic bonding,[28,38,158] formation of inter- or intradisulfide linkages, or the presence of covalently linked fatty acids.[158]

3.3.2b. Posttranslational Modification. The work of Folch and collaborators[160–162] revealed that the proteolipid apoprotein contains 2 to 3% fatty acid, which is covalently linked to the apoprotein by ester linkage. Apoprotein was shown to contain palmitic, stearic, and oleic acids (60%, 10%, and 25% respectively). It also contained other fatty acids (5%), which were not identified.[162] These results were subsequently confirmed in isolated N2 from human myelin by Gagnon *et al.*[90] Jolles *et al.*[154] digested rat brain myelin PLP with trypsin, and two of the tryptic peptides were suggested to contain fatty acids. The contribution of the fatty acids to the structure and function of PLP in the myelin membrane is not known. Presumably, the two fatty acids are oriented so that they are intercalated into the lipid bilayer. However, we have no evidence whether the fatty acids interact with the inner or outer leaflet of the membrane or whether, in fact, the fatty acids might interact with the bilayer apposed to the bilayer containing the peptide backbone. Completion of the sequence and determination of the position of the fatty acids will allow the formulation of more specific models.

Acylation of proteolipid protein and DM-20 with [^{3}H]palmitic acid has been demonstrated both *in vivo* (Fig. 4)[48] and in slices.[163] The distribution of labeled fatty acids is similar to the composition, with palmitic the predominant fatty acid. The subcellular site at which acylation occurs has not been identified. In slices, cycloheximide did not prevent the acylation, indicating that the fatty acids were added after translation.[163] The time course of addition of fatty acid to proteolipid protein in myelin was linear, in contrast to the delay seen for entry of the peptide backbone of proteolipid protein into myelin. These results suggest that in the slices acylation of proteolipid protein is occurring late in

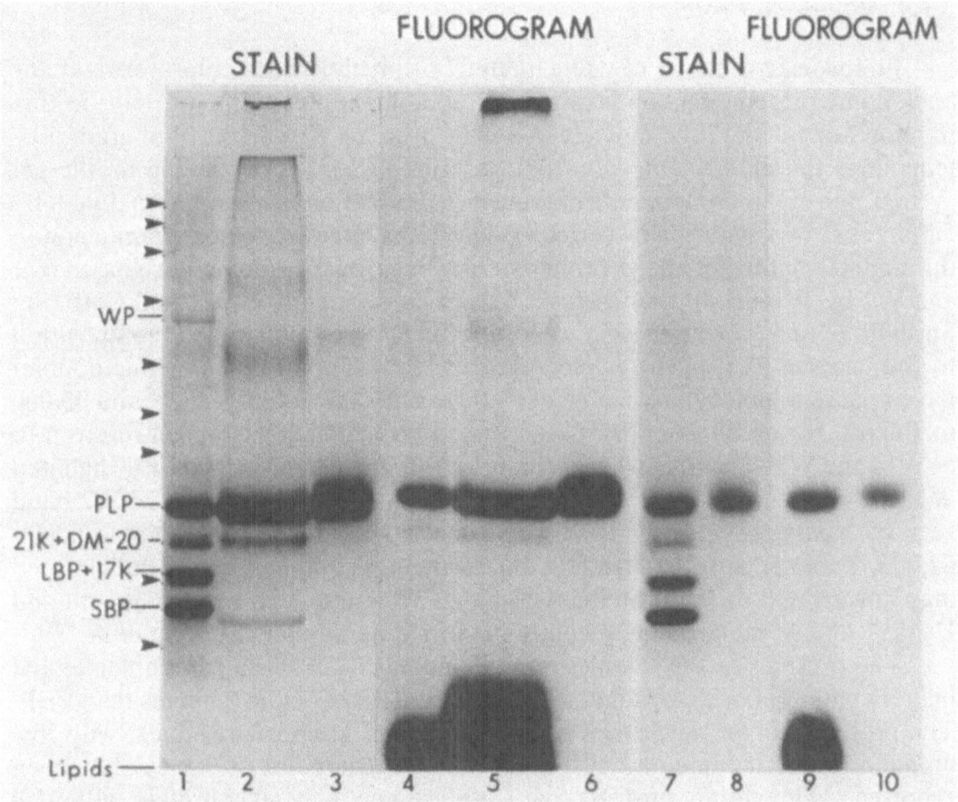

Fig. 4. SDS-slab gel electrophoresis of proteins of whole myelin, PLP-enriched fraction, and purified PLP. Electrophoresis followed by fluorography clearly shows that both PLP and DM-20 are acylated 24 h after the intracerebral injection of [^{3}H]palmitic acid. The gel was stained with Coomassie blue. Lane 1, whole myelin proteins (100 μg); lane 2, proteins of PLP-enriched fraction (100 μg); lane 3, purified PLP (73 μg); lanes 4, 5, and 6, fluorographs of lanes 1, 2, and 3, respectively; lane 7, whole myelin protein (92 μg); lane 8, purified PLP (59 μg); lanes 9 and 10, fluorographs of lanes 7 and 8, respectively. WP, Wolfgram protein; LBP, large basic protein; SBP, small basic protein. (Figure from Agrawal, et al.,[48] with permission.)

processing, either by exchange or *de novo* addition, possibly within myelin itself. When isolated myelin was incubated directly with [^{3}H]palmitic acid, PLP was not acylated, but various conditions that might be conducive to acylation were not investigated. Myelin PLP and DM-20 have been shown to be acylated after intracerebral injection of [^{3}H]palmitic acid in 9-, 15-, and 30-day-old rats.[41] However, no differences were observed in the relative specific activity of acylated PLP in myelin between 9 and 30 days. This may be related to the developmental changes in the activity of the enzymes responsible for acylation of PLP. Since PLP is acylated in 9-day-old rat brain, which has only a minimal amount of compact myelin, it is tempting to suggest that acylation of PLP might be a prerequisite for the compaction and structural organization of the myelin membrane.[41]

3.3.3. Wolfgram Proteins

Following extraction of white matter[33] or myelin[89] with chloroform–methanol, some proteins remain insoluble. Predominant among the proteins in this fraction are several high-molecular-weight proteins designated Wolfgram proteins after the investigator who first described them.[33] Depending on the gel system, the major components migrate as a doublet or as a set of two doublets. Differences in separation in various gel systems have led to some confusion in the nonmenclature of these proteins. However, two proteins designated W1 and W2 by Waehneldt and Malotka[164] and subsequently CNP1 and CNP2 by Sprinkle *et al.*[45] correspond in molecular weight and amino acid composition to the enzyme 2′,3′-cyclic nucleotidase.[45,166,167] The proteins in this doublet have apparent molecular weights of 43,000 to 48,000 on SDS gels[164] and 96,000 to 100,000 by gel filtration.[167] These same proteins have also been referred to as W1a and W1b[166] and appear to comprise the single band originally designated W1 by Nussbaum *et al.*[168] and W2 by Wiggins *et al.*[169] The original W2 band seen by Nussbaum *et al.*[168] has a somewhat higher molecular weight (52,000–62,000; Table I). In other studies, this protein also appears as a doublet and may correspond to the proteins designated W2a and W2b by Drummond and Dean[166] and to the proteins recently identified as tubulins by Reig *et al.*[170]

The proteins in the doublet corresponding to CNP'ase have a somewhat higher content of basic residues and lower content of acidic residues than PLP. The proteins contain over 40% hydrophobic and nonpolar residues, with low amounts of methionine and cystine. An isoelectric point of 9.5–9.7 has been reported.[45] One report suggests that CNP'ase may be a glycoprotein, although this has not been confirmed. The conclusion was based on indirect evidence from PAS staining and binding to lectin columns.[167] Binding to wheat germ agglutinin could be reversed by the appropriate sugar, whereas the binding to Con-A and ricinus could not, suggesting nonspecific binding.

3.3.4. Glycoproteins

Central nervous system myelin has a relatively low content of glycoproteins. The most prominent glycoprotein in the membrane is the "myelin-associated glycoprotein" or MAG, a protein of about 100,000 to 120,000 daltons, which comprises less than 1% of the total myelin protein. The protein was first detected by Quarles *et al.*[171] using radioactive sugars *in vivo*. The protein contains about 33% carbohydrate in contrast to the P_0 glycoprotein of PNS, which contains about 5% carbohydrate. The sugars detected include N-acetylglucosamine, mannose, galactose, fucose, and sialic acid.[44] In addition the protein is also sulfated.[172] Acidic and hydrophobic residues predominate over basic ones.

Because of the low content of MAG in isolated myelin, the protein was originally designated "myelin-associated" since the possibility existed that it was part of the axonal or oligodendroglial plasma membrane associated with myelin during isolation. Immunocytochemical staining has shown that MAG is a myelin constituent.[173] There is evidence that MAG has a periaxonal location and is not present in compact myelin lamellae[173], although these conclusions have been questioned.[391] This periaxonal location suggests that MAG may play a role in glia–axon interactions during the formation and maintenance of myelin.

In the CNS, the content of MAG increases as myelination proceeds. The MAG in immature myelin has a higher apparent molecular weight than that in mature myelin. Available evidence suggests the difference between the two forms resides in the peptide chain.[231]

By radioimmunoassay, MAG appears to be specific to the nervous system.[174] Interestingly, it is found in both PNS and CNS. Although the concentration in adult sciatic nerve is only one-sixth that in whole rat brain,[175] the MAG in PNS is found in Schwann cells and has a periaxonal location as in the CNS.

In addition to MAG, other glycoproteins have been detected in isolated myelin fractions by PAS staining of proteins on gels,[171] by binding of lectins to myelin,[176–178] and by [^{3}H]fucose labeling.[44] Four additional major glycoproteins (mol. wt. 95,000, 88,000, 43,000, and 38,000) and many minor glycoproteins were revealed only after the treatment of isolated myelin membrane with neurominidase before oxidation with galactose oxidase and subsequent reduction with sodium borotritide.[179] Since all the major and minor glycoproteins with the exception of the 120,000-dalton glycoprotein were detected only after heating of myelin with neuraminidase, N-acetylneuraminic acid is a terminal sugar residue in these glycoproteins. Which of these are truly myelin constituents and which are constituents of axolemma or other membranes is not yet known with certainty. Recently, Poduslo[180] examined 16 lectin-binding constituents in myelin fractions for developmental changes. Although both increases and decreases in binding of specific lectins were observed, in general the oligosacchride branches appeared more complex with maturation.

3.3.5. *Myelin-Associated Enzymes*

This topic has been reviewed in an earlier volume of this *Handbook* by Lees and Sapirstein.[181] We previously noted that the enzyme 2′,3′-CNP'ase is contained in the WPs of 43,000–48,000 daltons. Lees and Sapirstein have classified the other myelin-associated enzymes into three groups, including (1) enzymes involved in lipid metabolism, (2) enzymes involved in protein modification, and (3) enzymes that influence transport processes.

3.4. *Metabolism of Myelin Proteins*

Extensive reviews of this topic have appeared within the last 5 years.[20,22,182,183] The following sections summarize recent findings about the synthesis and turnover of each of the major myelin proteins.

3.4.1. *Synthesis*

3.4.1a. Changes in Rates of Synthesis during Development. As expected from the dramatic increases in myelin and its proteins during development, the rates of synthesis of myelin proteins increase rapidly during development.

Available evidence indicates that the rate of synthesis decreases as the rate of myelination slows, suggesting that decreased synthesis rather than increased degradation is primarily responsible for the slower rate of accumulation of myelin proteins seen in adult compared to developing CNS.

In a comprehensive study of developing CNS in rat, Smith[184] investigated the capacity of brain slices to synthesize proteins isolated in the myelin fraction. A survey of five CNS regions showed the highest rate of synthesis at 20 days (the earliest age examined), with a sharp decrease in all regions between 20 and 60 days. Over the same period, incorporation of amino acids into proteins in the nonmyelin fraction showed little change. Comparison of brain regions at 20 days demonstrated that the rate of synthesis of myelin proteins was greatest in those regions with highest myelin content and earliest onset of myelination, that is, spinal cord > brainstem > cerebral cortex.

Relatively few studies have examined rates of synthesis of myelin proteins in whole brain during development. Analysis of the rates of synthesis of the 14 K and 18.5 K basic proteins in mouse brain between 14 and 39 days showed a peak of synthesis for both proteins at 18 days of age, coincident with the period of maximal myelin deposition.[185] Throughout development, the smaller basic protein was synthesized at a greater rate than the larger basic protein. This is consistent with the relative enrichment of the 14 K protein relative to the 18.5 K protein in myelin and indicates that the rate of synthesis is a major factor in this enrichment. In agreement with these findings, a later study in rat brain showed that relative incorporation of amino acids into myelin SBP remained stable between 17 and 40 days, while incorporation into LBP declined markedly.[187] Recent studies have demonstrated that synthesis of PLP in mouse brain also increases during myelination.[186] Compared to basic protein, proteolipid protein begins to be synthesized somewhat later and does not reach its maximal rate until 22 days, 4 days later than the peak for basic proteins, again in agreement with changes in levels of these proteins in brain.

Biochemical studies on RNA-directed synthesis of basic proteins have so far utilized tissue from either mouse or rat brain between 8 and 25 days after birth, spanning the period of onset and increase in rate of myelination. With the exception of one early study,[188] all of the studies indicate that free ribosomes have the capacity to synthesize the two major basic proteins.[185,189–193] However, in some of these investigations, evidence for synthesis of basic proteins on membrane-bound ribosomes has been found as well.[190–192]

Whether these differences reflect differences in preparation of ribosomes, conditions for translation, or ages of the animals used is not yet clear. In addition to finding synthesis of MBP primarily on free ribosomes, Colman *et al.*[193] report evidence for a class of ribosomes associated with myelin that are enriched in message for basic protein (see Benjamins[194] for more extensive comparison of these studies).

Results from studies with mRNA extracted from whole tissue[195] or from polysome fractions[155,193,196] indicate that the two major basic proteins are synthesized in their mature form. Yu and Campagnoni[155] and Carson *et al.*,[196] using mRNA from 15-day-old mouse brain, have demonstrated *in vitro* synthesis of the four proteins commonly identified in myelin (21 K, 18.5 K, 17.5

Table II
Metabolic Half-Lives of Proteins in Myelin and Other Subcellular Fractions of Adult Rat Brain[a]

Proteins	Fast components	Slow components (days)
Total homogenate	3 days	13, 52
Microsomal	3, 20 h	8, 25
Supernatant	14 h	8, 20–22
Total myelin	5 days	110
Wolfgram	5 days	20, 90–100
Proteolipid	7 days	100–120
Basic	5 days	80–110

[a] Apparent half-lives of proteins were determined from the semi-logarithmic plots of the decay of [^{14}C]leucine radioactivity incorporated into proteins. Data from Singh and Jungalwala, with permission.[199]

K, and 14 K). These results support one conclusion made from earlier structural studies of these proteins, showing that differences between the 18.5 K and 14 K proteins arose from internal deletion,[42] thus suggesting that these proteins should arise by synthesis from independent messages rather than by posttranslational cleavage of one parent species. In addition to the four major proteins, Carson *et al.*[196] reported the synthesis of a number of high-molecular-weight proteins cross reactive with antisera to the 14 K basic protein. These species predominate early in development, and their synthesis decreases as myelination proceeds.

As predicted from its location as an integral protein in myelin, proteolipid protein has been shown to be synthesized predominantly on membrane-bound polysomes.[193] As with the basic proteins, PLP appears to be synthesized in its mature form.

3.4.2. Turnover

The rate of removal of proteins from myelin has been investigated by following long-term isotopic decay of these proteins after injection of radioactive amino acids. Comparison of results from a number of studies shows that a wide range of values has been obtained for half-lives of myelin proteins, with results depending on a number of variables including route of injection, reutilization of isotope, methods of myelin isolation, methods of protein separation, age of animals, and intervals examined (see ref. 22). When both short and long intervals after injection of isotope were examined, the major myelin proteins exhibited a biphasic rate of decay.[197,199] For example, Singh and Jungalwala found that the major proteins of myelin exhibit at least two rates of isotopic decay, one around 5–7 days, the other around 100 days or greater (Table II). Since the animals in this study were 65 days old at injection, almost all oligodendroglia should be relatively mature, so the results probably reflect what is happening in any given oligodendroglia rather than two populations of oligodendroglia, one less mature than the other, with more rapid turnover.

When turnover rates of myelin proteins were compared in brains of rapidly myelinating versus slowly myelinating rats, the major myelin proteins labeled in the younger animals showed virtually no turnover, whereas half-lives of 100 days were found for proteins labeled in the older animals.[200] Since only long intervals after injection were examined in this study, the more rapid phase of decay was not compared at the two ages.

As indicated above, most of the studies on turnover of myelin components have utilized short-term pulse labeling. An alternative approach is that used by Lajtha *et al.*[201] In those studies, a pellet of labeled tyrosine was implanted in adult mice to keep the specific radioactivity of precursor constant over several days. In this way, several problems of labeling with a single injection may be circumvented. First, a pulse may not label all proteins in a uniform manner, since those most metabolically active will be most highly labeled. Secondly, reutilization of label may occur, especially in myelin. With the constant-precursor method, the investigators found that the rate of protein turnover was about 40% of that of whole brain proteins, and about one-fifth of the myelin proteins were replaced within 10 days. Each of the major myelin proteins showed some turnover, in the order Wolfgram > DM-20 > basic = proteolipid protein. Thus, by 10 days, about 29% of the Wolfgram proteins had been replaced compared to 12–13% of the basic and proteolipid proteins. The similarity in half-lives of these two proteins with this method compared to the slightly shorter half-life for basic protein found in most studies using a pulse of precursor may be attributable to the longer labeling period.

One study of fucose incorporation as a function of age suggests that myelin glycoproteins are relatively stable.[202] However, the high-molecular-weight proteins as a group appear to have a more rapid turnover than the major myelin proteins.[197,198,200]

3.4.3. Catabolism

The mechanisms involved in breakdown of myelin proteins under normal and demyelinating conditions have been reviewed by Smith.[93] Several endogenous proteases have been described, including a neutral protease that is found in highly purified myelin. A recent study[203] indicates that this protein preferentially degrades BP and the myelin-associated glycoprotein. The role of this enzyme compared to lysosomal enzymes in the normal turnover of myelin proteins is not known.

4. LIPIDS OF CNS MYELIN

4.1. Chemical Studies

4.1.1. Isolation of Myelin Lipids and Their Analysis

The first step in lipid analysis involves the solubilization of myelin lipids using organic solvents, usually a combination of chloroform and methanol.

Most laboratories utilize some modification of the classical methodology of Folch, Lees, and Sloane-Stanley[204] (for review, see ref. 205), which involves homogenization with 19 volumes of chloroform: methanol (2:1, v/v). An alternative procedure using a less toxic solvent system containing hexane and isopropanol is also available.[206] The insoluble residue is removed by filtration or centrifugation. Certain minor lipid constituents may not be recovered quantitatively by such a procedure unless the insoluble material is further extracted with an acidified chloroform: methanol mixture (to recover polyphosphoinositides) or with chloroform: methanol (1:2, v/v) containing 5% water (to recover gangliosides). Addition of water to the chloroform: methanol extract results in phase separation; gangliosides are present in the upper aqueous phase, and less polar lipids are retained in a lower organic phase. As mentioned earlier, much of the total protein of CNS myelin is also soluble in organic solvents and thus is still present in the organic phase. If necessary, and for many routine purposes it is not, these proteins can be removed by repeated evaporation of the extract from a chloroform: methanol: water mixture; this eventually results in the denaturation of protein, which can then be centrifuged away from the lipid. The lipids in the organic phase can then be concentrated by evaporation of the solvent.

The exact analytical scheme utilized then depends on the goal of the study. Thin-layer chromatographic procedures are relatively rapid and reproducible and are generally used to separate different classes of myelin lipids from each other. Single-dimensional procedures are sufficient to separate major classes of neutral lipids and phospholipids (for discussion of relevant solvent systems, see ref. 207 or 208). Two-dimensional methods offer higher resolution. One such system[209] also allows for separation of plasmalogens from their diacyl and alkylacyl analogues; this is accomplished by exposure of the lipids on the TLC plate to acid vapors to cleave the vinyl ether linkage between the two chromatography steps. Lipids separated on silica gel plates can be identified by comparison with standards and quantitated by charring the TLC plate and quantitating the resultant spots by densitometry, by using more specific spray reagents, or by chemical analyses of material eluted from the chromatography plate; reviews of the methodology involved are presented in various books such as those of Lowenstein,[210] Marinetti,[211] and Kates[212]. Even after two-dimensional chromatography, the resultant spots on the TLC plates will contain lipids of a given class that will be somewhat heterogeneous with respect to their fatty acid composition. The fatty acid composition can be established by gas chromatographic techniques. This and many specialized procedures useful in elucidating the structure of individual components of the various lipid classes are beyond the scope of this chapter. However, it should be noted that the considerable structural heterogeneity of lipid classes related to such variables as fatty acid length or minor variations in the species of long-chain base in single-lipid components presumably does have biological significance.

4.1.2 Composition

Some 70% or more (depending on the species) of the myelin of the central nervous system consists of lipid.[213,214] Among the lipids present (Table III) are

Table III
Rat Brain Myelin Composition during Development[a]

	Age (days)							Adult bovine[b]
	15	20	30	60	144	190	425	
Cholesterol	100	100	100	100	100	100	100	100
Total galactolipid (GL)	39	42	47	54	55	50	53	48
Cerebroside[c]	31	33	41	43	39	40	39	39
Sulfatide	7	8	9	11	10	11	12	6
Total phospholipid[d]	102	96	93	85	78	78	82	76
Ethanolamine phosphatides (EP)	40	37	36	33	29	32	34	33
Choline phosphatides (CP)[e]	32	29	25	22	19	18	20	18
Serine phosphatides	13	13	14	12	11	12	12	11
Inositol phosphatides	3	2	2	2	2	2	2	2
Sphingomyelin	7	7	6	7	6	5	6	12
Plasmalogens	29	27	28	30	27	26	28	27
Mole ratio EP/CP	1.25	1.28	1.44	1.50	1.53	1.78	1.70	1.83
Mole ratio GL/CP	1.2	1.4	1.9	2.5	2.9	2.8	2.7	2.7

[a] Data taken from Norton and Poduslo,[8] except for the last column. The figures are expressed as mol/100 mol of cholesterol. This base was chosen because the cholesterol content of myelin does not change significantly during development.
[b] Data recalculated from Norton and Autilio.[390]
[c] Cerebroside increases from 16.6 to 23.3% of lipid weight.
[d] Total phospholipids decrease from 51.4 to 44.0% of lipid weight.
[e] Choline phosphatides decrease from 16.8 to 11.0% lipid weight.

those common to most plasma membranes such as phosphatidylcholine, phosphatidylethanolamine, sphingomyelin, phosphatidylserine, and, of course, cholesterol. A distinguishing feature of myelin is its high concentration of galactolipids. These include cerebroside (although not exclusively localized in myelin, it is greatly enriched in this structure and is often referred to as the most "myelin-typical" lipid) and its sulfated derivative, sulfatide. Plasmalogen, primarily ethanolamine plasmalogen, is also present in a high concentration in myelin (it is found in lower concentrations in other brain membranes and in low concentrations in many nonneural membranes). Phosphatidylinositol is present at low concentrations, as are the polyphosphoinositides di- and triphosphatidylinositide. These last two lipids are not shown in the table since in these assays the insoluble residue was not extracted with acidified solvents (see refs. 215, 216).

Several other minor lipid constituents intrinsic to myelin have also been identified. They are of interest because of the possibility that they may have some dynamic function or recognition properties as opposed to the primarily structural significance of the more quantitatively significant lipids. Some 0.2% of myelin lipid is composed of the major monosialoganglioside G_{M1}[217]; some species contain, in addition, monosialosylgalactosyl ceramide, or ganglioside G_{M4}.[218] Myelin also contains small amounts of different fatty acid esters of cerebroside.[219] Also of interest are the mono- and digalactosyldiglycerides,[220] related alkylacyl analogues,[219,221,222] and some sulfated derivatives.[223]

4.1.3. Changes during Development

Changes in the lipid composition of myelin during development are at least as marked as those observed with myelin proteins. Such developmental studies were initially conducted by Horrocks *et al.*[224]; results have subsequently been confirmed and extended in many laboratories.[225–229] Results of a study of development in rats are shown in Table III.[227] Myelin isolated from rats early in development has only about half the galactolipid content (as a percentage of myelin lipid) of material isolated from mature animals. Much of this percentage increase in galactolipid is at the expense of a decrease in lecithin. The monosialogangliosides also increase in prominence during development, although sialic-acid-containing lipids always account for only a tiny fraction of total myelin lipid. The small amount of desmosterol present initially also declines. A most unusual observation is that the concentration of monogalactosyldiglyceride peaks during the time period of rapid myelin accumulation; thus, these lipids appear to be a marker for the process of myelination itself rather than a marker for accumulation of myelin.[220,230]

Few studies of developmental changes in the fatty acid composition of individual lipid classes isolated from myelin have been conducted; one such study is that of Fishman *et al.*,[70] which showed that, in developing rat, the phosphatidylcholine and phosphatidylethanolamine of myelin became somewhat more unsaturated, whereas sulfatides and cerebroside tended toward longer fatty acids in the N-acyl position with increasing age. In rats, this changing composition involves a time span from between 12–13 days of age (when myelin can first be isolated) to about 8 weeks of age. In summary, the work is consistent with a hypothesis that early myelin is an extension of (or is compositionally related to) the oligodendroglial cell membrane and is increasingly differentiated, that is, assumes the composition of "mature" myelin, with later development.

4.1.4. Species Variation

There are some differences with respect to quantitative distribution of the major lipids among samples of myelin isolated from different species. For example, Table III shows that the sphingomyelin concentration of rat myelin is lower than it is in ox myelin. However, the details of lipid composition are known for only relatively few species; no particular pattern of species differences has emerged from these studies. This is in contrast to the situation with respect to proteins, where, for example, many rodents have basic protein components that are not present in the primates. A unifying scheme with respect to lipid composition is the observation[214] that all CNS myelin preparations studied contain cholesterol, phospholipid, and galactolipid in a molar ratio varying between 4:3:2 and 4:4:2. There may be more dramatic compositional differences with respect to some of the minor lipid components, especially ganglioside. In mature rat, mouse, ox, cat, and rabbit, the major whole-brain monosialoganglioside, G_{M1}, is relatively enriched in myelin, where it accounts for 70 mol % of the total ganglioside.[217,232] In contrast, myelin from human and other primates and from avian CNS also contains a prominent sialosylgalactosyl ceramide (G_{M4} or G_7). This ganglioside has a fatty acid pattern typical of cerebroside and may be relatively specific to myelin and oligodendroglial cells. A later study extends these results and points out differences among mammals, amphibians, and fish; a unique ganglioside pattern (predominantly di- and trisialo gangliosides) occurs in alligator myelin.[233]

4.2. Metabolism of Myelin Lipids

It is assumed that many aspects of the pathways for synthesis and catabolism of myelin lipids are similar to those taking place in other organs. Thus, in the case of lipids that are generally distributed throughout the body, only a general outline is given of the metabolic pathways involved. More details and references are given with respect to the lipids that are enriched in myelin and sparsely distributed in other structures (e.g., cerebroside).

4.2.1. Synthesis

4.2.1a. Diacylglycerophospholipids. Dihydroxyacetone phosphate, which is formed in the brain primarily as a product of glycolysis, is the precursor for the glycerol backbone of all phospholipids. Much of the dihydroxyacetone phosphate is reduced to glycerol phosphate by NADH, which is in turn acylated by fatty acyl-CoAs to form phosphatidic acid (a reaction first demonstrated in liver by Kornberg and Pricer.[234,235] Some of the dihydroxyacetone phosphate is also acylated directly at the 1 position followed by an NADPH-mediated reduction of the keto group in the 2 position and a subsequent acylation, reactions first demonstrated by Hajra and co-workers in liver mitochondria[236–238] and brain microsomes.[239]

If the resultant phosphatidic acid is to be utilized for synthesis of phosphatidylcholine or phosphatidylethanolamine, the phosphate group must first be removed. The resultant diacylglycerol acts as the acceptor for a phosphorylated base moiety, which is transferred from CDP-ethanolamine or CDP-choline to form phosphatidylethanolamine or phosphatidylcholine, respectively. Synthesis of the CDP-base moiety begins with the formation of phosphocholine or phosphoethanolamine (at the expense of ATP); this is followed by reaction with CTP to give CDP-choline or CDP-ethanolamine. The involvement of CDP in this synthesis was elucidated by Kennedy and his co-workers (see ref. 240) and first demonstrated in brain by McMurray *et al.*[241] Brain (and most other tissues) contains a glycerol kinase so that radioactive glycerol can be injected and used as a tracer for studying the metabolism of phospholipids. Most of the synthesis of these phospholipids takes place in the microsomal fraction.[242–246] Presumably, that lipid destined for myelin is synthesized on the endoplasmic reticulum of oligodendroglial cells.

Since myelin contains much of the brain phospholipid, it would be expected that the activity of relevant synthetic enzymes increases during myelinogenesis. This has been shown for phosphocholine transferase and phosphoethanolamine transferase.[247–249] After the rise in enzyme activity, which parallels to a certain extent accumulation of myelin, enzyme activity remains high, suggesting that there is a continued need to replace phospholipids that are undergoing metabolic turnover.

In addition to the pathways utilizing preformed choline or ethanolamine, there is also a base-exchange pathway (energy-independent, Ca^{2+}-dependent exchange of serine, ethanolamine, or choline for the polar moieties of preexisting phospholipids), which is operative in the brain tissue.[250–252] It is of greatest significance as the major (probably the only) pathway for the synthesis of phosphatidylserine. The presence of exchange enzymes suggests the possibility of a *de novo* pathway for phospholipid synthesis in brain (such a pathway does exist in liver; see ref. 253). This would initially involve the formation of phosphatidylserine by displacement of choline or ethanolamine from a preexisting phospholipid. There could then be a decarboxylation to form phosphatidylethanolamine. Phosphatidylserine decarboxylase in brain has recently been characterized by Butler and Morell.[254] The base initially displaced by serine can enter the salvage pathway and end up in another molecule of phospholipid, in which case the end result is the net synthesis of phosphatidylethanolamine. Although a significant amount of whole-brain phosphatidylethanolamine of brain is synthesized from serine (about 7%; ref. 254), it is not known how much of this goes to myelin.

Most of the enzymes involved in glycerophospholipid synthesis are thought to be localized in the endoplasmic reticulum, presumably in oligodendroglial cells if the resulting lipids are to be used for myelin formation. At least one brain enzyme, phosphatidylserine decarboxylase, is clearly mitochondrial in location.[254,255] However, CDP-ethanolamine:diacylglyceroethanolamine phosphotransferase is in part associated with highly purified myelin, as is the equivalent enzyme for phosphatidylcholine biosynthesis.[256] CDP:phosphoethanolamine cytidylyltransferase, the enzyme that makes the phosphorylated base for synthesis of the nucleotide base, has also been found in purified myelin.[257] Thus, it is possible that some of the myelin phospholipid is synthesized *in situ*, although the physiological significance of such reactions is not known.

Cytoplasmic lipoproteins that facilitate exchange of phospholipids between membranes have been described in brain as well as other tissues.[257a] One study demonstrated exchange of phosphatidyl choline between microsomes and myelin[257b] that was mediated by a soluble exchange protein from brain. Whether this mechanism plays a role in transport of lipids of myelin *in situ* is not clear.[257c,d]

4.2.1b. Plasmalogens. Synthesis of plasmalogens (in myelin, primarily ethanolamine plasmalogen) also involves an initial acylation of dihydroxyacetone phosphate.[258,259] In an unusual reaction, the fatty acid on 1-acyldihydroxyacetone phosphate exchanges with a long-chain alcohol is given 1-alkyldihydroxyacetone phosphate. A series of steps analogous to the synthesis of diacylphospholipids then follows: the keto group in the 2 position is reduced, and an acyl transferase reaction then yields 1-alkyl, 2-acylglycerol-3-phosphate. This analogue of phosphatidic acid then has its phosphate group hydrolyzed, and there is a reaction with CDP-ethanolamine or CDP-choline to give the 1-alkyl analogue of phosphatidylcholine or phosphatidylethanolamine. A mixed-function oxidase then converts the alkyl linkage to a vinyl ether bond. The specificity of this reaction has been studied *in vivo* during myelinogenesis.[260] Some of the extensive literature regarding plasmalogen biosynthesis is reviewed by Snyder[261] and Horrocks and Sharma.[262]

4.2.1c. Glyceroglycolipids Although they are quantitatively only minor components of the myelin sheath, phosphatidylinositol (which, of course, is also classified as a glycerophospholipid) and its di- and triphosphorylated analogues are of considerable interest because of their metabolic properties. The formation of phosphatidylinositol, as does the synthesis of the other phospholipids, also involves the dephosphorylation of phosphatidic acid. However, the pathway diverges from other phospholipids at this point, since it is the released diglyceride moiety that is activated by a reaction with CTP to yield CDP-diglyceride. This activated moiety then reacts with free inositol to give phosphatidylinositol.[263,264] The phosphatidylinositol (which is synthesized on the endoplasmic reticulum) can then be phosphorylated on the inositol moiety at the 4 position (by an enzyme localized primarily in the plasma membrane) to give diphosphoinositide; this lipid can be further phosphorylated at the 5 position by a cytosolic enzyme[264] to give triphosphoinositide. However, both of these enzymes, which require Mg^{2+} and are inhibited by Ca^{2+}, are also present in myelin, and polyphosphoinositides are enriched relative to whole brain in highly purified myelin fractions.[215,265] Of interest are the facts that the phosphate moieties on the 4 and 5 positions of inositol phospholipids turn over with a half-life of the order of minutes or less and that these lipids are enriched in the denser myelin subfractions.[265,266]

There are also several glycosylated glycerolipids that do not contain a phosphate moiety. Most prominent are the mono- and digalactosyldiglycerides, which are synthesized by a reaction involving diglyceride and UDP-galactose; in the case of a digalactosyldiglyceride, there is a second UDP-galactose step involved.[267,268] In equal concentration with the diacyl form of the lipid, there are also alkylacyl forms; detailed studies of the synthesis of these lipids have not been published. Finally, sulfated derivatives of the galactosyldiglyceride are also associated with myelin and are formed by a phosphoadenosine phosphosulfate-mediated reaction.[223,269]

The pattern of developmental activity seen with respect to the synthesis of the galactosyldiglycerides[230,267,268] differs somewhat from that seen for phospholipid synthetic enzymes. The biosynthetic activity increases as a function of the rate of myelin accumulation; however, the concentration of these lipids in myelin declines along with a decrease in enzyme activity after the peak of myelin accumulation. The activity of the enzyme responsible for sulfogalactosyldiglyceride synthesis also rises during myelination and declines later. This lipid (and its alkylacyl analogue) remain at constant levels on a per-brain basis, but their concentration in myelin is diluted as myelin accumulates.

4.2.1d. Sphingolipids. Myelin is unusually rich in lipids containing the long-chain base sphingosine. In brain, as in other organs,[270] the formation of the 18-carbon long-chain bases involves condensation of palmitic acid and serine. The resultant product, 3-ketodihydrosphingosine, is reduced by NADPH to dihydrosphingosine. The desaturation to form sphingosine can occur at the level of palmitic acid or, more likely, at the level of ceramide.[271,272] The next step in the synthesis of sphingolipids of myelin involves condensation of acyl-CoA and the long-chain base (this has been studied *in vitro* in several laboratories, e.g., refs. 273,274). The specificity of this reaction *in vitro* is in line with that expected of a system in large part dedicated to cerebroside production. That is, the 24-carbon acyl-CoA is a better substrate than, for example, the 16-carbon acyl-CoA[274]; CoA-derivatives of hydroxy fatty acids are also good substrates.[275]

In connection with this, it should be noted that the α-hydroxy fatty acids are formed from their corresponding nonhydroxy fatty acids in brain.[276–278] Another pathway for synthesis of ceramide, the reverse reaction of ceramidase to bring together a free fatty acid and sphingosine, can be demonstrated *in vitro*.[279] However, its lysosomal location and its activity primarily with short-chain fatty acids are not compatible with its playing a major role in the synthesis of myelin sphingolipids. Finally, an unusual reaction involving the synthesis of ceramide from lignoceric acid and sphingosine has been reported.[280] In contrast to the lysosomal system, this enzyme system requires both soluble and particulate components and is pyridine nucleotide dependent. However, the possibility that this pyridine nucleotide dependence is secondary to an ATP requirement has not been eliminated.

Much of the ceramide formed in oligodendroglial cells is presumed to act as a substrate for UDP-galactose: ceramide galactosyltransferase, an enzyme activity studied by Morell and Radin.[281] A reaction that can be demonstrated *in vitro*, acylation of psychosine,[281–283] may be catalyzed by the same enzyme. However, it is probable that acylation of sphingosine followed by galactosylation is the predominant *in vivo* route; the alternative, galactosylation of sphingosine followed by acylation, does not appear to be a quantitatively significant pathway.[284,285]

The developmental pattern for the enzyme involved in the committed step in cerebroside biosynthesis UDP-galactose: ceramide galactosyltransferase, which transfers galactose from UDP-galactose to ceramide,[286,287] is different from that for the phospholipid synthetic enzyme. Here, the enzyme activity follows a function of the rate of accumulation of myelin; i.e., after the period of peak myelin accumulation is over, enzyme activity declines markedly. This, of course, correlates with the considerable metabolic stability of cerebroside (see Section 3.4.2), since, once synthesized, a relatively low level of enzymatic activity is sufficient to maintain cerebroside levels.

Some of the cerebroside is converted to sulfatide by a phosphoadenosine phosphosulfate-dependent reaction.[288–290] It is of interest to note that there appears to be a metabolic compartmentalization; it is not the most recently synthesized cerebroside that is selected for sulfation.[291–293] The enzymes that catalyze the final steps in the synthesis of cerebroside and sulfatide are largely in the microsomal fraction. Some synthesis probably also occurs in the Golgi apparatus,[294] and the cerebroside sulfotransferase is enriched in this fraction.[295]

Cerebroside sulfotransferase[296] and presumably ceramide galactosyltransferase[296a] are enriched in oligodendroglia relative to other brain cells. There is also good evidence that some of the galactosyltransferase activity is associated with myelin itself.[297–300] Cytoplasmic proteins that specifically bind cerebroside[300a,b] and sulfatide[300c] have been identified. In the case of sulfatide, one study concludes that lipoproteins might transport this galactolipid to myelin.[300c]

Another fate for cerebroside (at least in some species; see Section 4.1.4) is to react with CMP-sialic acid to form ganglioside G_{M4} (G_7), since the fatty acid composition of this myelin-specific ganglioside resembles that of cerebroside.

Sphingomyelin, which contains the long-chain base sphingosine, is also found in myelin, although it is not enriched in that structure as are the galactolipids. It is generally assumed that spingomyelin is formed by the condensation between ceramide and the choline moiety delivered from CDP-choline.[301–303] However, the specificity of this enzyme *in vitro* raises complications, since *threo*-sphingosine is a better substrate than is the natural *erythro*-isomer, and only synthetic ceramide with very-short-chain fatty acids is of significant activity as a substrate. An interesting reaction involving the transfer of the phosphocholine moiety of phosphatidylcholine to ceramide has been described.[304,305] Although the presence of this reaction in several organs has been verified by Ullman and Radin,[306] it was not found in brain. Still another possible pathway involves the acylation of sphingosylphosphorylcholine.[307] Thus, despite the quantitative significance of this phospholipid in brain, its synthetic pathway is not well understood.

4.2.1e. Cholesterol. Most of the synthetic steps involving formation of cholesterol in brain are similar to those in other organs.[308] It is a common observation that, following injection of labeled acetate into the brain, much of the label is incorporated into cholesterol. However, the specialized environment of the brain accounts for some differences in the supply of substrate and in the control of cholesterol synthesis. Although some cholesterol may be obtained by a transfer from the plasma,[309] there is clearly no possibility of direct deposition of cholesterol into myelin from circulating lipoproteins. Unlike liver, the brain efficiently utilizes circulating ketone bodies for cholesterol synthesis (see ref. 310 for review). This preferential use of β-hydroxybutyrate and acetoacetate derives from the ability of brain to convert the substrate to acetoacetyl-CoA (acetoacetyl-CoA synthetase has a lower activity in liver). In contrast, circulating mevalonate is a relatively poor precursor for brain cholesterol.[311] This is presumably because, in most organs, a shunt mechanism exists whereby mevalonate is cycled back through hydroxymethylglutaryl-CoA and then to acetoacetate, which may serve as the quantitatively more significant substrate for cholesterol formation. However, the shunt is not complete in brain since hydroxymethylglutaryl-CoA acetoacetate lyase is not present in brain.[311]

A major difference in control of cholesterol synthesis in peripheral tissue versus brain is that inhibition of HMG Co-A reductase by low-density lipoprotein (a major step in control of cholesterol synthesis in the periphery) is probably not of great significance in brain. Other possible control steps in synthesis of brain cholesterol are discussed by Wykle.[308] A recent paper by Shah[312] discusses recent findings on brain HMG Co-A-reductase with respect to phosphorylated and dephosphorylated forms of the enzyme and implications of this with respect to control of enzyme activity.

4.2.2. Turnover

It has long been known that if myelin lipids are biosynthetically labeled, they are relatively more stable than lipids in other brain subcellular fractions (for review, see ref. 229). In part, this is because half of the total lipid of myelin consists of cholesterol and cerebroside, and these components indeed are very stable metabolically with a half-life of the order of many months. Thus, when the animals are labeled with a general precursor (such as radioactive glucose or acetate) during the period of rapid myelin accumulation, much of the radioactivity is incorporated into these very metabolically stable compounds. However, it has also been known for many years that most phospholipids of the myelin sheath do have reasonable turnover rates, which are different for the different lipids.[314–316] Although many relevant studies have been published, the quantitative details relating to turnover of myelin lipid components are difficult to arrange into a consistent picture. A number of factors are relevant to an understanding of the apparent discrepancies in the literature.

A major complicating factor is that the various moieties that comprise lipids are differentially reutilized during metabolism. For example, phosphatidylcholine consists of a glycerol backbone, two fatty acyl residues, a phosphate group, and choline. Turnover studies of myelin usually involve intracranial injection of a radioactive precursor into a series of animals followed by measurement of the decrease in radioactivity in the lipid of interest as a function of time. Depending on the precursor used, radioactivity may wind up primarily in a particular portion of the molecule; for example, glycerol will be found mostly in the glycerol backbone, fatty acid will be incorporated largely into the long-chain acyl residues, choline will be incorporated intact as a base, and, of course, radioactive phosphate can be inserted into only a single position. [^{14}C]Glucose will be incorporated into most of the carbon positions. When the molecule is degraded, various portions of it may be reutilized intact as part of a new molecule. This problem was illustrated with respect to phosphatidylcholine from brain microsomes in a study by Abdel-Latif and Smith,[317] which showed that the turnover of this lipid, as measured by decay of radioactivity following intracranial injection of a radioactive precursor, varied greatly depending on the precursor utilized. Data from the laboratory of one of the authors (Table IV) illustrate differences in the apparent turnover rate of phosphatidylcholine in myelin and microsomes depending on the precursor utilized. The best estimates of absolute turnover rates are probably those based on work with [2-^{3}H]glycerol, which is least subject to reutilization.[318]

Two other factors that are closely interrelated are the age of the animals being studied and the period of time following injection that is being used to establish the turnover kinetics. If animals are injected at a young age with

Table IV
Metabolic Half-Lives of Phospatidylcholine in Myelin and Microsomes of Rat Brain[a]

Precursor	Phosphatidylcholine	
	MY	MIC
A. Animals injected at 17 days of age		
[^{3}H]Glycerol	25(10)	13(4)
[^{14}C[Choline	39	26
[^{14}C[Ethanolamine	—	—
[^{32}P[Phosphate	30	17
[^{14}C[Glucose	56	35
[^{14}C]Acetate	54	28
B. Animals injected at 60 days of age		
[^{3}H]Glycerol	11	6.0
[^{14}C[Choline	22	11
[^{14}C]Glucose	27	14

[a] Data in part A from Miller *et al.*[319]; those in part B from Miller and Morell.[320] All times are in days, and, except for the numbers in parentheses in line 1, represent half-lives of the slow phase. The numbers in parentheses are half-lives of the rapid phase and are calculated for the first 15 days following injection.

[2-^{3}H]glycerol, there appear to be at least two phases of turnover. For example, phosphatidylcholine of myelin has a turnover rate of abour 10 days when followed for the first 2 weeks after injection, but after that time a slower turnover is measured. In contrast, when older animals are injected, no prominent slow phase of metabolic turnover is detected for phosphatidylcholine.

These matters are discussed in some detail in a paper by Miller *et al.*[319]; see also Miller and Morell[320] and Horrocks *et al.*[321] These data on lipid turnover, along with those for protein turnover, suggest a model of the cellular and molecular mechanisms involved in assembly and turnover of the myelin bilayer. This working hypothesis is described in Section 5.3.

As mentioned previously, the turnover of one class of myelin phospholipids, the polyphosphoinositides, is extremely rapid. It is also possible that a small amount of the "major" phospholipids also turnover rapidly; it is very difficult to study rapid turnover of a small pool of a lipid against the background of a large pool of metabolically stable lipid of the same class. Thus, there is a distinct possibility that some of the lipid turnover may be related to a dynamic function of myelin (e.g., osmotic regulation) rather than being related only to slow turnover of structural lipids.

4.2.3. Catabolism

Degradative pathways for the phospholipids are not understood as well as the synthetic pathways. There is relatively little information as to how degradation of phosphatidylcholine and phosphatidylethanolamine of myelin may differ from that in other brain membranes. One unique aspect of myelin lipid catabolism is the mechanism by which lipids in the compacted myelin sheath get to the lysosomes (hypotheses relevant to this process are discussed in a subsequent section). Many phospholipases are found in brain, and enzymes specific for fatty acids in the 1 position (phospholipase A_1) and for fatty acids in the 2 position (phospholipase A_2) are well known. Phospholipase C, which cleaves between the phosphate and the glycerol moiety, and the recently characterized (for review, see ref. 322) phospholipase D, which cleaves between the base moiety and the phosphate, are even less well understood. Other enzymes that are lysophospholipases exist; these catalyze the hydrolysis of acyl groups from 1-acyl- or 2-acyllysophosphoglycerides but not on the diacyl substrates.[323] This and many other aspects of brain phospholipid metabolism are discussed by Wykle.[308]

Activity for the catabolism of plasmalogens (even when studied in whole-brain preparations) is presumed to be directly relevant to myelin metabolism since this lipid is concentrated in myelin. The enzyme activities involved in cleavage of the 1-alkyl-1′-enyl moiety of plasmalogens to release of fatty aldehyde has been studied in a number of laboratories.[324,325] A lysophospholipase D specific for removal of the ethanolamine moiety of ethanolamine plasmalogen has been studied.[326] The idea that plasmalogenase activity is involved in different demyelinating lesions has been proposed by Horrocks *et al.*[327]

Perhaps most specific of all with respect to degradation of myelin is the enzyme cerebroside galactosidase, the enzyme responsible for catabolism of cerebroside. In contrast to synthesis of cerebroside, which is concentrated during a narrow developmental time window, turnover of cerebroside takes place at a relatively slow rate over the whole age span. As noted by Bowen and Radin,[328] hydrolytic activity for catabolism of cerebroside corresponds to the total amount of myelin present in brain. The literature on this subject is large and is reviewed by Radin in a chapter in this series. Activity of aryl sulfatase, presumably related to sulfatide catabolism, also increases in line with myelin accumulation.[328]

Little is known about how cholesterol turnover is handled in myelin. It is important to note that unlike other lipids, which are catabolized in the brain, cholesterol is metabolically stable. No degradative pathway is available as a control step to regulate the total amount of cholesterol present. Cholesterol in brain, as in other parts of the body, is removed only when it can be transported from brain to the liver, where it can equilibrate with the enterohepatic shunt. In this manner, cholesterol—some of which may be converted to bile salt—is lost in the feces. Thus, the only way cholesterol is removed from the brain is by transfer to plasma lipoproteins, a rather slow process,[329] which is not understood. The only known metabolic transformation involving myelin cholesterol that takes place *in situ* is hydrolysis of cholesterol esters (some cholesterol esters accumulate during the period of myelination). A cholesterol ester hydrolase is found in myelin.[330,331]

5. *INTRACELLULAR PROCESSING OF MYELIN COMPONENTS*

5.1. *Current Models*

From a variety of morphological and autoradiographic studies, oligodendroglia appear to originate from a pool of dividing precursors and migrate toward axons (for review, see ref. 332). Interaction of the immature oligodendroglia with the axons is thought to trigger the subsequent onset of myelination. At some point in maturation, the glial processes overlap themselves, and actual formation of multilamellar myelin begins. This process involves the addition of large amounts of lipid and protein to the oligodendroglial plasma membrane. The three mechanisms most commonly proposed for addition of molecules to newly forming membrane include (1) fusion of vesicles, which in the case of myelin might occur either in the oligo plasma membrane around the cell body or at an initiation site in the glial process for enrichment of myelin components, (2) membrane flow from preexisting plasma membrane, and (3) addition of components transported by cytoplasmic carriers, usually proteins. These mechanisms apply to addition of either lipids or proteins, and all probably contribute to formation of myelin. Conversely, these mechanisms, along with lateral diffusion of molecules within the bilayers of myelin lamellae, probably serve to remove molecules from the myelin membrane, as slow turnover occurs in the mature animals. These concepts have led investigators to postulate the existence of precursor membranes and vesicles, presumably in the oligodendroglial process. In addition to these mechanisms, which are thought to be common to membrane biogenesis in a variety of cell types, the special geometry of the axon–myelin unit has suggested that the cytoplasmic loop regions of the myelin sheath may play an important role in myelin metabolism, both in periaxonal and intralamellar regions.

As these mechanisms are considered in more detail, an understanding of the structural organization of the membranes involved in the process becomes essential to building models to be tested. The sidedness of lipids and proteins in the bilayer, the interactions between lipids and proteins, and the organization and function of the cytoplasmic loop regions are thought to be important factors in the sequence of events leading to compact myelin lamellae.

A variety of experimental approaches have provided information that contributes to our understanding of how oligodendroglia assemble and maintain myelin. Results from three of these experimental approaches are summarized below. These approaches include (1) immunohistochemical studies; (2) kinetics of assembly and turnover, and (3) analysis of myelin subfractions and other subcellular fractions.

5.2 Immunohistochemical Localization of Myelin Proteins and Lipids

Immunohistochemistry offers a means of obtaining precise information regarding the localization of protein within the central and peripheral nervous system. This method has an advantage over conventional biochemical techniques by permitting localization of proteins and enzymes in tissue with preserved morphological characteristics. Thus, localization of proteins within specific cell types (oligodendrocytes and Schwann cells) and membranes (myelin) can be accomplished more precisely than in isolated neurons, oligodendrocytes, astrocytes, and subcellular structures, where cross contamination is a potentially serious problem.

It must be emphasized that the purity of the antigen as well as the specificity of antibodies generated must be ascertained chemically (e.g., homogeneity demonstrated by gel electrophoresis, amino acid composition, N-terminus sequence) and/or immunochemically (e.g., double immunodiffusion analysis, cross immunoadsorption) prior to the localization of proteins by immunohistochemistry. Immunohistochemical localization to structures known to contain the antigen is not a sufficient test for antibody specificity for a particular protein, since other antigens with similar cellular distribution will also give a similar localization pattern. Thus, appropriate localization, although supporting the immunochemical tests for specificity, cannot be used as a substitute for rigorous biochemical and immunochemical characterization.

Four major proteins of CNS myelin, namely, PLP, BP, WP, and MAG, and a major lipid (galactocerebroside) have been localized (for recent reviews, see refs. 333, 334).

5.2.1. Proteolipid Protein and Basic Protein

Proteolipid protein was the first CNS myelin-associated protein shown to be specifically and exclusively localized to the myelin sheath and oligodendrocytes.[335] The localization of the protein has been examined in detail using immunohistochemistry and shown not to be present in astrocytes, neuronal cell bodies, or their dendrites. The absence of any staining at the nodes of Ranvier indicated that PLP was not associated with the axonal membrane. During the active process of myelination, PLP was demonstrated in oligodendrocytes and their processes.[335]

Myelin basic protein was subsequently localized by Sternberger *et al.*[336] in the rat brain and by Hartman *et al.*[333,337] in the rat and chicken brain. Basic protein was also found to be restricted to the oligodendrocytes and myelin sheath in the CNS of rat and chick. The same immunohistochemical criteria previously applied to PLP were used to demonstrate that this protein was also exclusively localized to myelin and oligodendrocytes. The localization was similar except that BP was present both in the CNS and PNS, whereas PLP was completely absent in the PNS.[337]

Some degree of differentiation of oligodendrocytes occurs before BP or PLP is observed because these proteins appear only after the precursor cells have migrated into the preformed axonal pathways. Proteolipid protein and BP were always observed in oligodendrocytes in close association with axons, and isolated immunostained oligodendrocytes were never seen; thus, it was concluded that the onset of the synthesis of these proteins occurs only after the glial processes have made contact with axon to be myelinated and some degree of ensheathment has occurred. If these proteins were produced prior to making contact with the axons to be myelinated, isolated immunostained oligodendrocytes ought to be seen. These results indicate that the signal for proliferation of oligodendrocytes appears to be different from the signal for the synthesis of BP and PLP.[338]

The specific immunofluorescence of PLP within the oligodendroglial cells was not present diffusely within the cytoplasm but appeared to be concentrated

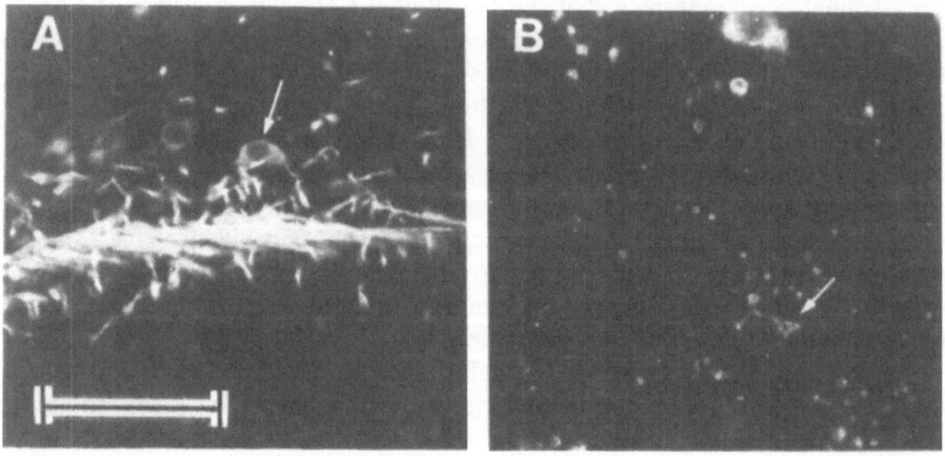

Fig. 5. Comparison of the immunocytochemical localization of BP (A) and PLP (B) in oligodendrocytes. Specific BP immunofluorescence appears homogeneous in the cytoplasm, whereas PLP exhibits a granular appearance (brain of 10-day-old rat; fields from cerebellum and lateral pons, respectively). Arrows indicate oligodendrocytes. Fibers showing varying degrees of myelination are also visible in both photomicrographs. Scale bar, 50 μm. (Figure from Hartman *et al.*[338] with permission.)

in "clumps" of immunoreactive material. In contrast, BP is uniformly and homogeneously distributed throughout the cytoplasm.[335,337,338] The coarse or clumped appearance of PLP in oligodendrocytes may indicate storage of this protein within the Golgi complex prior to its transport from the cell body to the myelin sheath (Fig. 5).

A comparative study of the localization of PLP and BP in adjacent sections showed that oligodendroglial cells do not appear to stain intensely with both PLP and BP simultaneously. This observation suggests that individual oligodendrocytes do not produce both proteins in large quantities at the same time.[337,338] Furthermore, a comparison of localization of BP and PLP also showed that in general BP appears to be synthesized by oligodendrocytes and incorporated into the developing myelin membrane before PLP. Therefore, it was concluded that there is a shift in priority from the synthesis of BP to the synthesis of PLP during the active period of myelination, in agreement with biochemical studies (see Section 3.4.1a).

There is a dramatic loss of immunofluorescence of BP from oligodendroglial cells in both rat and chicken brain (Fig. 6) with a concomitant increase in immunostaining in myelin.[337] Similarly, a loss of PLP immunofluorescence (Fig. 7) from oligodendrocytes in the medulla of developing rat brain has been observed.[338] As myelination progressed, staining for both proteins eventually became undetectable. Since BP and PLP turn over throughout life (see Section 3.4), some PLP and BP are continuously synthesized. Thus, the inability to visualize those proteins by immunohistochemistry in oligodendroglial cells in the adult brain reflects insufficient sensitivity of the method rather than cessation of synthesis.

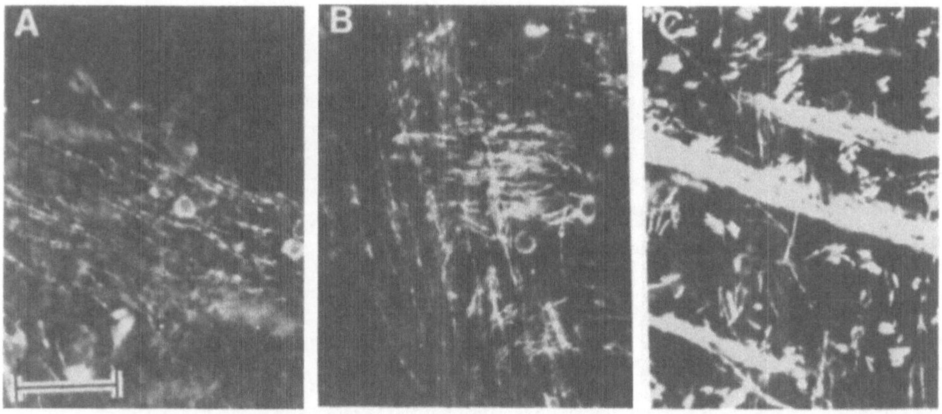

Fig. 6. Maturation of myelin. Sequence demonstrating various stages in the process of maturation of the myelin sheath as seen by immunohistochemical localization of BP in rat brain. A: Early stage. Oligodendrocytes are immunoreactive. Prominent varicosities are clearly visible along fibers, with BP concentrated at the outer edge of the dilation. The intervaricose segments are faintly immunoreactive. B: Middle stage. Oligodendrocytes are faintly fluorescent or absent. Varicosities are somewhat smaller in size but are still prominent, and the intervaricose segment is more brightly fluorescent. C: Late stage. Fibers are thickened and brightly immunoreactive. Varicosities are no longer prominent, though a slight hint of their presence can still be seen. Scale bar, 50 μm. (Figure from Hartman *et al.*,[337] with permission.)

One of the characteristics of fibers during the early stages of myelination is the presence of regularly spaced periodic enlargements and constrictions, which produce a distinctly varicosed appearance.[337] Individual varicosities vary from 3 to 10 μm in diameter. As myelination progresses, the fibers thicken, the intensity of immunoreaction increases, and the varicosities become less prominent. Eventually, the fibers take on the smooth appearance of mature myelinated fibers. This varicosed appearance of immature myelinated fibers has been shown with BP localization in chick and rat and PLP localization in rat[337,338] and therefore appears to represent a general structural characteristic of myelin during formation (Fig. 6).

The varicosities have been postulated to represent local regions of increased oligodendrocytic cytoplasmic volume along the newly ensheathed axons. These regions may serve as channels of transport or areas of transient storage of myelin proteins to be incorporated into the myelin sheath during development. Alternatively, they may serve as reservoirs of cytoplasmic contents displaced during compaction of myelin membrane. A major difference observed between the localization with BP and PLP is that the number of fibers exhibiting large and prominent varicosities is substantially less when PLP rather than BP is used as a marker. In contrast, when BP is localized at an early stage of maturation, almost all fibers exhibit distinct varicosed appearances. Therefore, the nature and the appearance of varicosities can be used as an index of the relative stage of maturation of myelin membrane. It is of interest that serial-section electron micrographic studies of developing myelin[339,340] have demonstrated that different numbers of turns around an individual axon within the

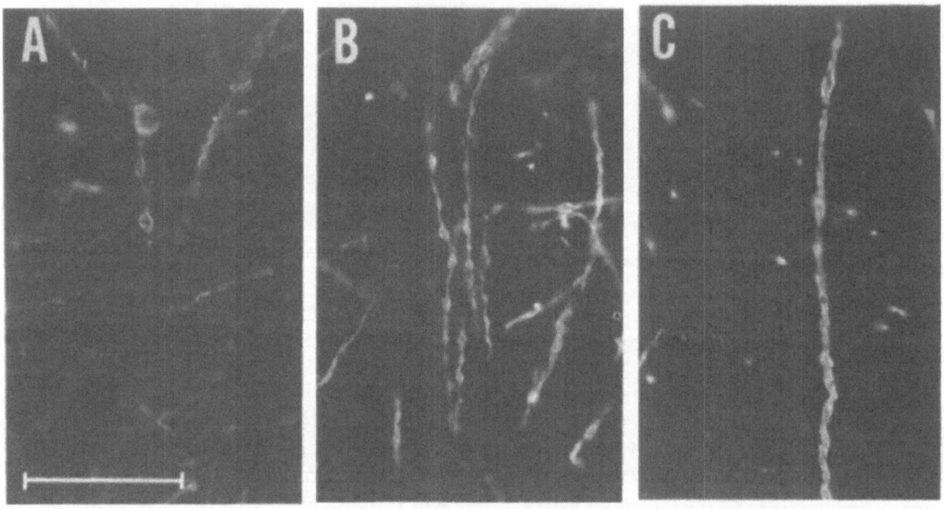

Fig. 7. Proteolipid proteins in oligodendrocytes during myelination. Progressive decrease in the immunoreactivity of PLP in oligodendrocytes with a concomitant increase in the intensity of fluorescence in myelinated fibers is observed during the period of active myelination. A, B, and C show the gradual loss of PLP immunofluorescence from oligodendrocytes in the medulla of rat brains at 1, 3, and 10 days after birth. In this series, because the examples are taken from the same anatomic area of the brain, the chronological age correlates well with the stage of maturation of myelin. However, because different brain regions or fiber pathways are myelinated at different times, a similar sequence could be shown with sections taken from different regions of the same brain, which are in different stages of myelination. Scale bar, 50 μm. (Figure from Hartman *et al.*,[338] with permission.)

same internode occur at different levels along the nerve fiber. Furthermore, compaction occurs at different rates within the same internode. These two phenomena would produce variations in the thickness of the sheath along the fiber. A more complete understanding of the function of these varicosities will be helpful in elucidating mechanics of myelin formation and maturation.

A substantial variation in the relative intensity of staining of BP and PLP in individual myelinated fibers was observed in the bovine, rat, and rabbit spinal cord. The myelin sheaths of very large fibers are heavily stained with BP but appear essentially negative with PLP. Conversely, the myelin of small-diameter fibers stains more intensely for PLP than for BP (Fig. 8).

Thus, the relative concentration of these proteins in myelin appears not to be constant but may vary as a function of the size of the myelinated fibers even in the mature myelin membrane. However, it must be emphasized that not all large fibers contain high BP and low PLP concentrations, because large fibers of optic nerve show equally intense immunofluorescence with these two myelin proteins.[338]

5.2.2. *Wolfgram Protein(s)*

As discussed previously (Section 3.3.3), the molecular weights of proteins called Wolfgram proteins include species from 43,000 to 62,000, and identifi-

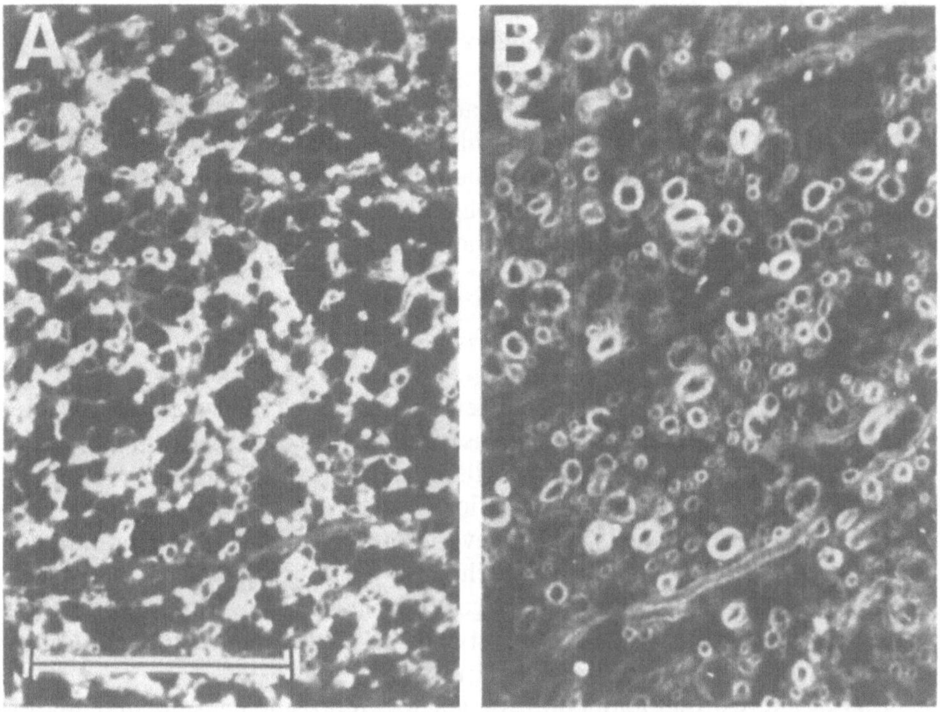

Fig. 8. Differences between localization of PLP and BP in bovine spinal cord. PLP immunofluorescence (A) is intense in the small myelinated fibers but is not visible in the very large fibers. In contrast, BP immunofluorescence (B) is intense in some large myelinated fibers and only faintly observed in many of the small-diameter fibers. Photomicrographs are from adjacent sections of lateral cervical spinal cord from adult bovine tissue. Scale bar, 100 μm (Figure from Hartman *et al.*,[338] with permission.)

cation of Wolfgram proteins is more confusing than that of PLP. Nevertheless, Nussbaum *et al.*[168] isolated two proteins (designated W1 and W2) with molecular weights of 54,000 and 62,000 (see Section 3.3.3) and generated antisera to these proteins. These antisera were used to show by immunofluorescence that W1 and W2 are restricted to the oligodendroglial cell and myelin sheath. Recently, Roussel and Nussbaum[341] compared the localization of BP to W1 and W2 and presented evidence that W1 and W2 appeared in the oligodendrocytes earlier than BP. The major dense line of myelin shows staining for their W1 and W2 proteins, indicating that its antigenic site is located on the cytoplasmic surface[342] in analogy to BP.[107] Antisera raised specifically to CNPase (W1) have now been shown to stain oligodendroglia and myelin. In the glia, the antigen has a primarily intracellular location.[343]

5.2.3. Myelin-Associated Glycoprotein

Myelin-associated glycoprotein (MAG) has been purified to homogeneity, and specific precipitating antibodies have been produced.[344] The specific staining for MAG within oligodendroglial cells is not homogeneous within the cytoplasm, unlike BP, but has a coarse granular appearance like that for PLP.[36] Antisera to MAG stain the periaxonal regions of oligodendrocytes with sparse staining of compact myelin[173] (see p. 376). The MAG has been shown by immunohistochemical technique[173,345] and radioimmunoassay[175] to be a component of both CNS and PNS myelin like BP but unlike PLP and WP. Sternberger *et al.*[173] have also shown by immunohistochemical technique that MAG and BP are not detected in the same oligodendrocytes at the same time. This observation suggests a similar shift in priority of synthesis for these two proteins as was hypothesized for BP and PLP.

5.2.4. Galactolipids

Antibodies to cerebroside have been characterized in several laboratories in the past (see ref. 346) and have been used recently to demonstrate localization of cerebroside on oligodendroglial cells in culture.[347–349] In tissue sections, antibodies to cerebroside react specifically with oligodengroglia and myelin rather than other neural cells.[350] However, cerebroside has also been found in choroid plexus, both by immunohistochemistry and by direct analysis.[350] Antibodies to sulfatide stain not only oligodendroglia and myelin but also ependymal cells and Bergmann glia of cerebellum,[351] indicating that sulfatide may not be as highly localized to oligodendroglia as cerebroside.

It is of interest that antibodies to galactocerebroside cause demyelination and swelling of myelin lamellae in myelinating organ cultures,[352,392] suggesting that galactocerebroside may be a major target in antibody-mediated demyelination. In contrast, antibodies to BP[352,353], PLP[354], or MAG[393] have little effect on myelin in culture.

5.3. Kinetics of Assembly and Turnover

Examination of the time course of appearance of newly synthesized lipids and proteins in myelin has provided information about the ways oligodendroglia process these molecules. Studies that examined entry of newly synthesized proteins into myelin either *in vivo* or in brain slices showed that in rats at 6 weeks of age, the high-molecular-weight proteins showed the most rapid initial rate of appearance in myelin, followed by BPs and then PLP.[355–357] In younger animals, a similar pattern was seen at 1 h after injection of radioactive amino acids; that is, the specific radioactivity of myelin proteins was highest in the high-molecular-weight proteins, followed by BPs, and lowest in PLP.[357] However, by 6 h the specific activity of PLP was two- to threefold higher than that of BP and WP, indicating a greater rate of deposition of this protein relative to the other proteins in rapidly myelinating animals.

In addition to demonstrating a shift in the relative rates of deposition of individual proteins into myelin during maturation, this latter study indicated that newly synthesized PLP showed a delay in appearance in myelin relative to other proteins. The kinetics of this phenomenon were investigated in more detail in brain slices.[358] With a constant level of radioactive amino acids present in the incubation medium, the newly synthesized BPs and WP appeared in myelin in a nearly linear fashion, indicating little delay between their synthesis and appearance in myelin. Proteolipid protein showed a delay in appearance in myelin, entering at a slow rate for 45 min and then at a faster rate similar to the other two proteins. Inhibition of protein synthesis and chase studies with nonradioactive amino acids revealed an extramyelin pool of PLP, which continued to provide proteolipid protein for myelin for 30 min after synthesis of newly labeled protein had been stopped. Conversely, entry of basic proteins and Wolfgram proteins stopped within several minutes, indicating that only small pools of these proteins were available within the cell for subsequent myelin assembly.

Similar conclusions have been reached from short-term labeling studies *in vivo*. Colman *et al.*[193] demonstrated that newly synthesized BP appears in myelin within 5 min after intracerebral injection of radioactive amino acids. By contrast, labeled PLP was detected in a microsomal fraction within 5 min after labeling. By 30 min, labeled PLP could no longer be detected in the microsomal fraction but was detected in myelin, indicating transport from endoplasmic reticulum to myelin during this interval. Pereyra *et al.*[359] have calculated an interval of 10–14 min for transport of PLP from site of synthesis to myelin; they suggest that this may be too rapid for transport through Golgi *per se* and propose that other mechanisms may be involved.

The kinetics of posttranslational modification of basic proteins or the specific myelin glycoproteins MAG and Wolfgram proteins have not been investigated *in vivo* or in slices. Processing of total [^{3}H]fucose-labeled glycoproteins in CNS myelin has been examined in slices.[358] These studies showed a delay between addition of [^{3}H]fucose to protein and the appearance of newly synthesized glycoproteins in myelin. An extramyelin pool was present, which continued to provide glycoproteins to myelin for about 30 min after inhibition of protein synthesis.

One possible explanation for the differences in kinetics of entry of PLP and MAG relative to BP and WP is that basic and Wolfgram proteins may be synthesized near the site of myelin assembly and travel only short distances through cytoplasm or on vesicles before insertion into the myelin membrane; PLP may be synthesized primarily in the cell body proper and undergo transport over longer distances than basic protein. The distance from cell body to the myelinating process has been estimated to be 10–40 μm[336]; however, even the longer distance is too short to allow detection of a transport process such as fast axoplasmic transport at a rate of 400 mm/day. Thus, the delay seen for entry of proteolipid protein into myelin is likely to result primarily from processing within endoplasmic reticulum, Golgi, or other intracellular membranes rather than transport down the oligodendroglial process *per se*.

The kinetics of appearance of newly synthesized lipids in myelin have also been studied *in vivo* and in slices (see ref. 22 for review). The choline and ethanolamine phospholipids and cerebroside show a nearly linear rate of appearance in myelin, whereas sulfatide shows a delay.[291] In analogy to PLP, radioactive sulfatide continues to enter myelin for up to 30 min after a chase, whereas radioactivity of the other three lipids in myelin no longer continues to increase. These results suggest that sulfatide undergoes longer intracellular processing than the other lipids studied. Following inhibition of protein synthesis for 1 h *in vivo*, phosphatidylcholine continues to enter myelin at a normal rate, whereas sulfatide shows a 30% decrease in entry.[361] Thus, its assembly into myelin may be more dependent on entry of newly synthesized protein than is the case for phosphatidylcholine.

In addition to inhibitors of protein synthesis, several other agents have been investigated for their ability to block entry of proteins and lipids into myelin. Triethyl lead[362] blocks entry of proteins into myelin, but the effect appears secondary to a depression in synthesis of myelin proteins relative to other proteins. Antiserum to basic protein also blocks entry of protein into myelin; in this case, synthesis of the myelin proteins was not directly investigated.[363] Further, the antiserum used was not heated to removed complement activity, so the effect might be caused by serum components other than antibody.

The sodium ionophore monensin has been shown to block secretion and assembly of plasma membrane proteins in several systems, presumably by disrupting budding of vesicles from the Golgi. When applied to CNS slices, monensin causes disappearance of normal Golgi structures and appearance of dilated vesicles in the oligodendroglial cytoplasm.[364] Entry of proteolipid protein[365] and sulfatide[364] into myelin is preferentially inhibited in the presence of monensin, whereas entry of basic protein and cerebroside are not affected. Colchicine has similar effects[360,364], suggesting a role for microtubules as well.

These results indicate that proteolipid protein and sulfatide are processed through the Golgi or related vesicular structures and that basic protein and cerebroside are processed by other routes. In the presence of monensin, basic protein can enter myelin independently of proteolipid protein, at least for short periods. This is the converse of the previous observation that proteolipid protein can enter myelin without basic protein when protein synthesis is inhibited. Similarly, cerebroside enters at a nearly normal rate under conditions in which sulfatide entry is decreased.

Results from long-term turnover studies of myelin proteins and lipids (Sections 3.4.2 and 4.2.2) have suggested several mechanisms by which multilamellar myelin exchanges components with oligodendroglial cytoplasm. The primary conclusions from these turnover studies are that (1) after labeling, both lipids and proteins show an initial rapid phase of decay, followed by a more stable phase; (2) on the average, molecules inserted into myelin early in development have longer half-lives than those inserted later; and (3) there is heterogeneity of turnover; that is, lipids generally turn over more rapidly than proteins, and some lipids and proteins turn over more rapidly than others. One interpretation of the first two results is that for several days after injection and synthesis of lipids and proteins, the molecules are not yet buried deep within the myelin sheath and are more accessible to the systems involved in membrane turnover. Thus, when older animals (only slowly accumulating myelin) are injected with a radioactive precursor for myelin, it is primarily the more rapid turnover phase that is observed.[200,320] Since myelin is being formed less rapidly in older than in younger animals, the period of time during which myelin components are accessible to turnover systems prior to being deeply buried under successive layers of myelin is greatly extended.

Another hypothesis proposed for lipid turnover[319,321,366] is that the pool rapidly turning over represents the lipids at the major dense line (cytoplasmic face) that are more likely to come into contact with cytoplasm. In contrast, the more metabolically stable pool would consist of lipid at the intraperiod line (extracellular face) separated from the cytoplasmic face by the energy barrier and able to "flip–flop" across the lipid bilayer. In young animals, both metabolic pools would be labeled, as myelin is rapidly accumulating. In older animals, the cytoplasmic face lipids would be preferentially labeled but would also decay rapidly. A further refinement of this hypothesis is based on a study showing that in myelin of mature rats, sphingomyelin containing short-chain fatty acids (C18 sphingomyelin) turned over at a significant rate (of the order of two months) whereas the long-chain lipid (C24 sphingomyelin) was much more stable metabolically[366] (see also ref. 367 for a contradictory report). It has been suggested that the more rapidly turning over sphingomyelin may be on the cytoplasmic face, and the longer-chain-containing lipid may be on the extracellular face.[366,368]

This hypothesis might also explain some of the heterogeneity in turnover of various protein components. For example, the observation that myelin BP has a slightly shorter half-life than PLP suggests that BP is more susceptible to exchange and removal than PLP. If true, this may arise by several mechanisms. By virtue of its localization on the cytoplasmic surface of the membrane, BP may be more accessible for degradation than PLP, which is more deeply buried in the membrane. These results may also mean that BP could be more enriched in cytoplasmic loop regions, with PLP more enriched in compact lamellar regions. Although staining patterns of myelin sheaths with MBP and PLP differ from each other, immunohistochemical studies to date have not shown a clear difference in distribution. In contrast, MAG appears to be more enriched in cytoplasmic regions.

It would be of interest to compare the turnover of MAG with that of MBP. In general, the high-molecular-weight proteins have a faster turnover than MBP, but the turnover of MAG *per se* is not known. Perhaps the rapid decay phase of MBP turnover would correspond to that of MAG turnover. If MBP were more enriched than PLP in the membrane at the cytoplasmic loops, subcellular fractions thought to be enriched in these membranes should contain more MBP, with the less dense membranes more enriched in PLP. However, many of the published data show just the opposite (see Section 5.4). Thus, the most likely explanation for the slightly more rapid turnover of basic protein appears to be its cytoplasmic peripheral location in the membrane. Also, although the overall distribution of BP and PLP in compact myelin versus cytoplasmic loops may be relatively even, BP may be able to move more rapidly by lateral diffusion into the cytoplasmic loops. On the other hand, PLP may be anchored in the membrane via intermembranous interactions on the external surface which slow its diffusion in the membrane.

Similar considerations may apply to the more rapid turnover of phospholipids compared to galactolipids and cholesterol. However, in addition to the possible roles of membrane sidedness, cytoplasmic loops, and rates of lateral diffusion, another factor to be considered in lipid turnover is the availability of enzymes that degrade lipids. For example, phospholipases are known to be present and active in white matter, but enzymes that degrade cholesterol are absent. Thus, the slow turnover of cholesterol may reflect this lack of degradative pathway in brain.

5.4. *Myelin Subfractions and Precursor Membranes*

Several reviews have discussed in detail the results of recent studies on characterization and metabolism of myelin subfractions and of extra-myelin membranes which might serve as precursors to myelin.[22,82,369–371] The objective of these studies has been to determine the sequence of events that occur between synthesis of myelin components and their assembly into myelin. Particular emphasis has been placed on the basic proteins, proteolipid protein, and galactolipids in these fractions, since they are highly specific for oligodendroglia and enriched in myelin. These studies can be placed in three categories depending on the strategy used for subcellular fractionation: (1) isolation of myelinlike or myelin-associated membranes from a crude myelin fraction by osmotic shock; (2) fractionation of an isolated myelin fraction into subfractions, either by continuous or discontinuous gradients; and (3) fractionation of homogenates to survey all subcellular fractions. All of these have the inherent problems of any subcellular fractionation method with identity and "purity" of a given membrane fraction. However, these studies have led to the development of several hypotheses about the mechanism of myelin assembly, which can now be tested by a variety of experimental approaches.

Because of the physical continuity between the oligodendroglial plasma membrane and the myelin sheath, numerous investigators have attempted to isolate membrane fractions that might represent a "transition membrane" stage between the undifferentiated oligodendroglial plasma and mature myelin. The assumption is that this "immature myelin" and possibly some oligodendroglial plasma membrane fragments may be attached to, or associated with, mature myelin during the initial isolation steps. One fraction that appears to be enriched in such membranes is the "myelinlike" fraction derived from myelin isolated from developing brain, as first described by Davison and co-workers.[372,373] The myelinlike fraction was shown to contain predominantly single vesicles; it is undoubtedly contaminated with the plasma membranes and subcellular membranes from a variety of cell types. However, this fraction contains substantially higher specific activity of CNPase than other subcellular fractions and is enriched in the higher-molecular-weight protein and impoverished in BP and PLP.[374] Subsequently, the myelinlike fraction prepared by various laboratories using different fractionation protocols has been shown to be highly enriched in CNPase and the high-molecular-weight proteins[200,375] and impoverished in the low-molecular-weight proteins.

It has also been shown, particularly in developing brain, that the specific activity of the myelin marker enzyme CNPase is higher in a myelinlike fraction (designated SN4) than in myelin.[376] Recent work from two laboratories[377,378] has shown that the specific activity of CNPase in the myelinlike fraction is two to three times higher than that in myelin isolated from young and adult rat brain. This "myelinlike" fraction has a composition intermediate between microsomes and mature myelin and consists of vesicles larger than microsomes but smaller than myelin vesicles, suggesting that it may consist of oligodendroglial plasma membrane, cytoplasmic loops, and/or immature uncompacted myelin.[200,373] In general, the components of the myelinlike fraction show more rapid incorporation and turnover than those of mature myelin. Although the results of current biochemical and metabolic studies suggest that the myelinlike fraction is a portion of the glial plasma membrane en route to being incorporated into the myelin sheath, it is not known if this fraction actually represents myelin precursor membranes, a mixture of myelin and microsomes, or possibly even a completely unrelated fraction.

A refinement of the above approach has been to separate a crude myelin or purified myelin into artibrary subfractions on the basis of differences in size and density of the myelin vesicles (for references, see refs. 67,370,379–381). The densest subfraction has a composition similar to that of the myelinlike fraction, with increasing amounts of myelin-enriched lipids and proteins present in progressively lighter subfractions.[375,379,380,382] Isolated myelin can be further separated on discontinuous sucrose gradients into myelin subfractions. The myelin subfractions have been designated "A," "B," "C," and "D" by Benjamins *et al.*,[379] medium, light, and heavy myelin subfractions by Matthieu *et al.*,[375] and light myelin, heavy myelin, and membrane subfractions by Agrawal *et al.*[380] The light and heavy myelin subfractions exhibit multilamellar structures on examination by electron microscopy.

It is notable that the light myelin subfraction contains more basic protein than PLP[67,383–385] and *vice versa* in the heavy myelin (Fig. 9). In fact, Fujimoto *et al.*[67] demonstrated that BPs account for 50–60% of the total myelin proteins in the light myelin subfraction isolated from both developing and mature brain. Matthieu *et al.*[375] did not find enrichment of either BP or PLP in their light, medium, and heavy myelin subfractions prepared by a different method. However, the heavy myelin subfraction was enriched in MAG. How the differences in protein composition in the myelin subfractions reflect the *in situ* situation remains to be established. One study using immunohistochemical techniques suggests that there are, in fact, myelinated fibers in the CNS that exhibit a preponderance of BP as opposed to PLP or *vice versa* (Fig. 8).[338] The direct interrelationship of myelin subfractions either during development or in the adult brain remains to be determined.

A variety of metabolic studies have provided some circumstantial evidence that the subfractions may represent stages of myelin assembly or maturation. At short times following radioactive precursor injection, the specific activity of sulfatide,[379] phospholipids,[382] and proteins[380,383] is higher in the denser subfractions than in the lighter subfractions, suggesting that the denser subfractions may be precursors to the less dense subfractions.

In addition, at longer times after precursor injection, the least dense fractions have the highest specific activities,[379,381,383,386] suggesting a possible transfer of components from the dense to the lighter subfractions. All of these

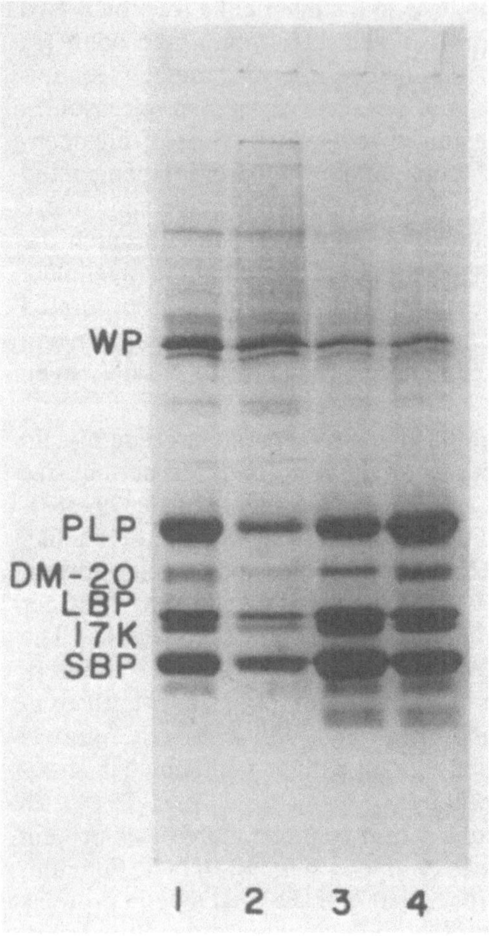

Fig. 9. SDS-slab gel electrophoresis of proteins of myelin subfractions; 70 μg of protein were applied to each lane. Lanes 1–4 show protein profiles of whole myelin, myelinlike, light myelin, and heavy myelin subfractions, respectively. The light myelin subfraction shows the enrichment of BP (lane 3). (Figure from H. C. Agrawal, unpublished observations.)

results suggest that these arbitrarily defined subfractions may actually have some functional significance with regard to myelin assembly. However, as noted recently,[359] the situation is quite complex, and it should be emphasized that these data are only consistent with, but do not prove, precursor–product relationships, since other kinetic explanations are also possible.

Time-staggered injections of precursors labeled with different isotopes (^3H and ^{14}C) have been utilized in an attempt to better define possible precursor–product relationships among subfractions. The idea is to determine if the various components of myelin are added simultaneously to all subfractions or if they enter one subfraction and then move sequentially through the others.[382,383] If a given lipid or protein was added simultaneously, the $^3H/^{14}C$ ratio for that substance at short times following a time-staggered injection sequence should be the same in all fractions. However, if newly synthesized molecules are added to the most dense subfraction and then proceed through increasingly lighter subfractions, the isotope ratio should decrease as one examines progressively lighter subfractions.

In a typical series of experiments, [^{14}C]precursor was injected intracranially; then, 30 min later, [^{3}H]precursor was injected. Rats were killed 15 min after the second injection, and myelin subfractions were isolated. Thus, lipids or proteins were labeled for 45 min with [^{14}C]precursor and 15 min with [^{3}H]precursor.[382,383] Isotope ratios for choline phosphoglycerides were relatively similar in the subfractions, suggesting simultaneous or nonordered addition of these molecules. However, isotope ratios for ethanolamine phosphoglycerides (both the diacyl and plasmalogen species) and proteolipid protein decreased from the bottom to the top of the gradient, consistent with a sequential addition. These results suggest that ethanolamine phosphoglycerides and proteolipid protein are inserted in the maturing membrane at a specific stage (and perhaps at a specific location along the myelinating cell process), whereas most other major myelin lipids and basic protein can be inserted in a less specific manner at any stage. The "simultaneous" entry of cerebrosides, sulfatides, and choline phosphoglycerides may be accomplished by the soluble lipid-exchange proteins mentioned earlier. It is also possible that some myelin lipid arises from its *in situ* synthesis in myelin, since some enzymes of lipid synthesis have been reported to be components of purified myelin (see Section 4.2.1).

Using this approach, Peregra *et al.*[359] determined isotope ratios in BP and PLP in a series of membranes prepared from mouse brain. They showed that turnover of the proteins is a factor in the ratios observed and that BP and PLP are found in a variety of endomembrane compartments.

6. *SUMMARY*

This chapter has reviewed current studies on myelin composition, metabolism, and assembly. Future studies should provide new information on genetic regulation of myelin formation, intracellular processing of myelin components, and interactions between oligodendroglia and axons that control the deposition and maintenance of myelin. Models of abnormal myelination and demyelination will continue to be important tools in these future studies as they have in the past. That topic is beyond the scope of this chapter but has been discussed in several recent reviews (see refs. 387–389).

REFERENCES

1. Mokrasch, L. C., 1968, *Handbook of Neurochemistry*, Volume 1 (A. Lajtha, ed.), Plenum Press, New York, pp. 171–193.
2. Raine, C. S., 1977, *Myelin* (P. Morell, ed.), Plenum Press, New York, pp. 1–49.
3. Kirschner, D. A., and Caspar, D. L. D., 1977, *Myelin* (P. Morell, ed.), Plenum Press, New York, pp. 51–89.
4. Bogart, R. B., and Ritchie, J. M., 1977, *Myelin* (P. Morell, ed.), Plenum Press, New York, pp. 117–159.
5. Ritchie, J. M., 1982, *Proc. R. Soc.* (*Lond.*) [*Biol.*] **215**:273–287.
6. Waxman, S. G., 1982, *N. Engl. J. Med.* **306**:1529–1532.

7. Norton, W. T., 1981, *Basic Neurochemistry*, 3rd ed. (G. J. Siegel, R. W. Albers, B. W. Agranoff, and R. Katzman, eds.), Little, Brown, Boston, pp. 63–92.
8. Norton, W. T., and Poduslo, S. E., 1973, *J. Neurochem.* **21**:749–757.
9. Norton, W. T., 1974, *Methods in Enzymology*, Volume 31 (S. Fleischer and L. Packer, eds.), Academic Press, New York, pp. 435–444.
10. Laatsch, R. H., Kies, M. W., Gordon, S., and Alvord, E. C., 1962, *J. Exp. Med.* **115**:777–787.
11. Waehneldt, T. V., and Mandel, P., 1972, *Brain Res.* **40**:419–436.
12. Cuzner, M. L., Davison, A. N., and Gregson, N. A., 1965, *J. Neurochem.* **12**:469–481.
13. DeVries, G. H., Norton, W. T., and Raine, C. S., 1972, *Science* **175**:1370–1372.
14. Greenfield, S., Norton, W. T., and Morell, P., 1971, *J. Neurochem.* **18**:2119–2128.
15. Bourre, J. M., Jacque, C., Delassalle, A., Nguyen-Legros, J., Dumont, O., Lachapelle, F., Raoul, M., Alvarez, C., and Baumann, N. 1980, *J. Neurochem.* **35**:458–464.
16. Detering, N. K., and Wells, M. A., 1976, *J. Neurochem.* **26**:253–257.
17. Norton, W. T., 1977, *Myelin* (P. Morell, ed.), Plenum Press, New York, pp. 161–199.
18. Haley, J. E., Samuels, F. G., and Ledeen, R. W., 1981, *Cell. Mol. Neurobiol.* **1**:175–187.
19. Benjamins, J. A., Hadden, T., and Skoff, R. P., 1982, *J. Neurochem.* **38**:233–241.
20. Benjamins, J. A., and Morell, P., 1978, *Neurochem. Res.* **3**:137–174.
21. Braun, P. E., and Brostoff, S. W., 1977, *Myelin* (P. Morell, ed.), Plenum Press, New York, pp. 201–231.
22. Benjamins, J. A., and Smith, M. E., 1977, *Myelin* (P. Morell, ed.), Plenum Press, New York, pp. 233–270.
23. Carnegie, P. R., and Moore, W. C., 1980, *Proteins of the Nervous System* (R. A. Bradshaw and D. A. Schneider, eds.), Raven Press, New York, pp. 119–143.
24. Agrawal, H. C., and Hartman, B. K., 1980, *Proteins of the Nervous System* (R. A. Bradshaw and D. A. Schneider, eds.), Raven Press, New York, pp. 145–169.
25. Magno-Sumbilla, C., and Campagnoni, A. T., 1977, *Brain Res.* **126**:131–148.
26. Poduslo, J. F., and Rodbard, D., 1980, *Anal. Biochem.* **101**:394–406.
27. Allison, J. H., Agrawal, H. C., and Moore, B. W., 1974, *Anal. Biochem.* **68**:592–601.
28. Morell, P., Wiggins, R. C., and Gray, M. J., 1975, *Anal. Biochem.* **68**:148–154.
29. Cammer, W., and Norton, W. T., 1976, *Brain Res.* **109**:643–648.
30. Keenan, R. W., and Jones, M., 1980, *J. Neurochem.* **34**:231–235.
31. Cohen, S. R., McKhann, G. M., and Guarnieri, M., 1975, *J. Neurochem.* **25**:371–376.
32. Trotter, J. L., Lieberman, L., Margolis, F. L., and Agrawal, H. C., 1981, *J. Neurochem.* **36**:1256–1262.
33. Wolfgram, F. A., 1966, *J. Neurochem.* **13**:461–470.
34. Quarles, R. H., Everly, J. L., and Brady, R. O., 1972, *Biochem. Biophys. Res. Commun.* **47**:491–497.
35. Roussel, G., Delaunoy, J. P., Nussbaum, J. L., and Mandel, P., 1977, *Neuroscience* **2**:307–313.
36. Sternberger, N. H., Quarles, R. H., Itoyama, Y., and Webster, H. deF., 1979, *Proc. Natl. Acad. Sci. USA* **76**:1510–1514.
37. Agrawal, H. C., Burton, R. M., Fishman, M. A., Mitchell, R. F., and Prensky, A. L., 1972, *J. Neurochem.* **19**:2083–2089.
38. Nicot, C., Nguyen Le T., Lepretre, M., and Alfsen, A., 1973, *Biochim. Biophys. Acta* **322**:109–123.
39. Chan, D. S., and Lees, M. B., 1978, *J. Neurochem.* **30**:983–990.
40. Vacher-Lepreter, M., Nicot, C., Alfsen, A., Jolles, J., and Jolles, P., 1976, *Biochim. Biophys. Acta* **420**:323–331.
41. Garwood, M. M., Gilbert, W. R., and Agrawal, H. C., 1983, *Neurochem. Res.* **8**:649–659.
42. Barbarese, E., Braun, P. E., and Carson, J. H., 1977, *Proc. Natl. Acad. Sci U.S.A.* **74**:3360–3364.
43. Agrawal, H. C., O'Connell, K., Randle, C. L., and Agrawal, D., 1982, *Biochem. J.* **201**:39–47.
44. Quarles, R. H., 1980, *Myelin: Chemistry and Biology* (G. A. Hashim, ed.), Alan R. Liss, New York, pp. 55–77.

45. Sprinkle, T. J., Wells, M. R., Garver, F. A., and Smith, D. B., 1980, *J. Neurochem.* **35**:1200–1208.
46. Sprinkle, T. J., 1982, *Trans. Am. Soc. Neurochem.* **13**:215.
47. Lees, M. B., 1982, *Scand. J. Immunol.* **15**:147–166.
48. Agrawal, H. C., Randle, C. L., and Agrawal, D., 1982, *J. Biol. Chem.* **257**:4588–4592.
49. O' Farrell, P. Z., Goodman, H. M., and O' Farrell, P. H., 1977, *Cell* **12**:1133–1142.
50. Martenson, R. E., and Gaitonde, M. K., 1969, *J. Neurochem.* **16**:333–347.
51. Macklin, W. B., Braun, P. E., and Lees, M. B., 1982, *J. Neurosci. Res.* **7**:1–10.
52. Carson, J. H., 1981, *Trans. Am. Soc. Neurochem.* **12**:102.
53. Greenfield, S., Weise, M. J., Gantt, G., Hogan, E. L., and Brostoff, S. W., 1982, *J. Neurochem.* **39**:1278–1282.
54. Gilbert, W. R., Garwood, M. M., Agrawal, D., Schmidt, R. E., and Agrawal, H. C., 1982, *Neurochem. Res.* **7**:1495–1506.
55. Comings, D. E., and Pekkula-Flagan, A., 1982, *Clin. Chem.* **28**:813–818.
56. Norton, W. T., 1982, *Advances in Cellular Neurobiology*, Volume 4 (S. Federoff and L. Hertz, eds.), Academic Press, New York, pp. 3–55.
57. Benjamins, J. A., 1983, *Advances in Neurochemistry*, Volume 5 (W. T. Norton, ed.), Plenum Press, New York, pp. 87–123.
58. Barbarese, E., and Pfeiffer, S. E., 1981, *Proc. Natl. Acad. Sci. U.S.A.*, **78**:1953–1957.
59. Banik, N. L., and Smith, M. E., 1977, *Biochem. J.* **162**:247–255.
60. Cohen, S. R., and Guarnieri, M., 1976, *Dev. Biol.* **49**:294–299.
61. Tennekoon, G. I., Cohen, S. R., Price, D. L., and McKhann, G. M., 1977, *J. Cell. Biol.* **72**:604–616.
62. Sprinkle, T. J., Zoruba, M. E., and McKhann, G. M., 1978, *J. Neurochem.* **30**:309–314.
63. Macklin, W. B., and Pfeiffer, S. E., 1983, *Trans. Am. Soc. Neurochem.* **14**:212.
64. Trapp, B. D., Webster, H. De F., Johnson, D., Quarles, R. H., Cohen, S. R., and Murray, M. R., 1982, *J. Neurosci.* **2**:986–993.
65. Kelly, P. T., and Luttges, M. W., 1976, *J. Neurochem.* **27**:1163–1172.
66. Foulkes, J. A., and Patterson, D. S. P., 1974, *Brain Res.* **82**:139–149.
67. Fujimoto, K., Roots, B. I., Burton, R. M., and Agrawal, H. C., 1976, *Biochim. Biophys. Acta* **426**:659–668.
68. Morell, P., Greenfield, S., Costantino-Ceccarini, E., and Wisniewski, H., 1972, *J. Neurochem.* **18**:2545–2554.
69. Zgorzalewicz, B., Neuhoff, V., and Waehneldt, T. V., 1974, *Neurobiology* **4**:265–276.
70. Fishman, M. A., Agrawal, H. C., Alexander, A., Golterman, J., Martenson, R. E., and Mitchell, R. F., 1975, *J. Neurochem.* **24**:689–694.
71. Eng, L. F., Chao, F. C., Gerstl, B., Pratt, D., and Tavaststjerna, M. G., 1968, *Biochemistry* **7**:4455–4465.
72. Barbarese, E., Carson, J. H., and Braun, P. E., 1978, *J. Neurochem.* **31**:779–782.
73. Banik, N. L., Davison, A. N., Ramsey, R. B., and Scott, T., 1974, *Dev. Psychobiol.* **7**:539–546
74. Einstein, E. R., Dalal, K. B., and Csejtey, J., 1970, *Brain Res.* **18**:35–49.
75. Druse, M. J., Brady, R. O., and Quarles, R. H., 1974, *Brain Res.* **76**:423–434.
76. Lane, J. D., and Fagg, G. E., 1980, *J. Neurochem.* **34**:163–171.
77. Smith, M. E., and Sedgewick, L. M., 1975, *J. Neurochem.* **24**:763–770.
78. Wolfgram, F., and Kotorii, K., 1968, *J. Neurochem.* **15**:1291–1295.
79. Morell, P., Lipkind, R., and Greenfield, S., 1973, *Brain Res.* **58**:510–514.
80. Lees, M. B., and Waxman, S. A., 1974, *J. Neurochem.* **23**:825–831.
81. Fagg, G. E., Schipper, H., and Neuhoff, V., 1979, *Brain Res.* **167**:251–258.
82. Quarles, R. H., 1979, *Biochemistry of Brain* (S. Kumar, ed.), Pergamon Press, Oxford, pp. 81–102.
83. Bakey, L., and Lee, J. C., 1966, *Arch. Neurol.* **14**:644–660.
84. Agrawal, H. C., Banik, N. L., Bone, A. H., Cuzner, M. L., Davison, A. N., and Mitchell, R. F., 1971, *Biochem. J.* **124**:70.
85. Franz, T., Waehneldt, T. V., Neuhoff, V., and Wachtler, K., 1981, *Brain Res.* **226**:245–258.
86. Lees, M. B., and Sakura, J. D., 1978, *Research Methods in Neurochemistry*, Volume 4 (N. Marks and R. Rodnight, eds.), Plenum Press, New York, pp. 354–370.

87. Deibler, G. E., Martenson, R. E., and Kies, M. W., 1972, *Prep. Biochem.* **2:**139–165.
88. Lees, M. B., 1968, *J. Neurochem.* **15:**153–160.
89. Gonzales-Sastre, F., 1970, *J. Neurochem.* **17:**1049–1056.
90. Gagnon, J., Finch, P. R., Wood, D. D., and Moscarello, M. A., 1971, *Biochemistry* **10:**4756–4763.
91. Bizzozero, O., Besio-Moreno, M., Pasquini, J. M., Soto, E. F., and Gomez, C. J., 1982, *J. Chromatogr.* **227:**33–34.
92. Waehneldt, T. V., *Anal. Biochem.* **43:**306–312.
93. Smith, M. E., 1977, *Neurochem. Res.* **2:**233–257.
94. Dunkley, P. R., and Carnegie, P. R., 1974, *Research Methods in Neurochemistry*, Volume 2 (N. Marks and R. Rodnight, eds.), Plenum Press, New York, pp. 219–245.
95. Hinman, C. L., Rauch, H. C., and Pfeifer, R. F., 1982, *Life Sci.* **30:**989–993.
96. Inoue, Y., Nakamura, R., Mikoshiba, K., and Tsukada, Y., 1981, *Brain Res.* **219:**85–94.
97. Reidl, L. S., Campagnoni, C. W., and Campagnoni, A. T., 1981, *J. Neurochem.* **37:**373–380.
98. Carnegie, P. R., 1971, *Biochem. J.* **123:**57–67.
99. Eylar, E. H., Brostoff, S., Hashim, G., Caccom, J., and Burnett, P., 1971, *J. Biol. Chem.* **246:**5770–5784.
100. Martenson, R. E., Luthy, V., and Deibler, G. E., 1981, *J. Neurochem.* **36:**58–68.
101. Dunkley, P. R., and Carnegie, P. R., 1974, *Biochem. J.* **141:**243–255.
102. Day, E. D., 1981, *Contemporary Topics in Molecular Immunology*, Volume 8 (F. P. Inman and W. J. Mandy, eds.), Plenum Press, New York, pp. 1–39.
103. Hashim, G. A., 1980, *Myelin: Chemistry and Biology* (G. A. Hashim, ed.), Alan R. Liss, New York, pp. 79–122.
104. Macklin, W. B., Braun, P. E., and Lees, M. B., 1982, *J. Neurosci. Res.* **7:**1–10.
105. Boggs, J. M., and Moscarello, M. A., 1978, *Biochim. Biophys. Acta* **515:**1–21.
106. Poduslo, J. F., and Braun, P. E., 1975, *J. Biol. Chem.* **250:**1099–1105.
107. Omlin, F. X., Webster, H. deF., Palkovitz, C. G., and Cohen, S. R., 1982, *J. Cell Biol.* **95:**242–248.
108. Golds, E. E., and Braun, P. E., 1978, *J. Biol. Chem.* **253:**8162–8170.
109. Privat, A., Jacque, C., Bourre, J. M., Dupouey, P., and Baumann N., 1979, *Neurosci. Lett.* **12:**107–112.
110. Rosenbluth, J., 1980, *J. Comp. Neurol.* **194:**639–648.
111. Kirschner, D. A., and Ganser, A. L., 1980, *Nature* **283:**207–210.
112. Martenson, R. E., 1980, *Biochemistry of Brain* (S. Kumar, ed.), Pergamon Press, Oxford, pp. 49–79.
113. Brostoff, S., and Eylar, E. H., 1971, *Proc. Natl. Acad. Sci. U.S.A.* **68:**765–769.
114. Deibler, G. E., and Martenson, R. E., 1973, *J. Biol. Chem.* **248:**2392–2396.
115. Miyake, M., 1975, *J. Neurochem.* **24:**909–915.
116. DesJardins, K. C., and Morell, P., 1983, *J. Cell Biol.* **97:**438–446.
117. Small, D. H., and Carnegie, P. R., 1982, *J. Neurochem.* **38:**184–190.
118. Baldwin, G. S., and Carnegie, P. R., 1971, *Biochem. J.* **123:**69–74.
119. London, Y., and Vossenberg, F. G. A., 1973, *Biochim. Biophys. Acta* **307:**478–490.
120. Carnegie, P. R., Kemp, B. E., Dunkley, P. R., and Murray, A. W., 1973, *Biochem. J.* **135:**569–572.
121. Carnegie, P. R., Dunkley, P. R., Kemp, B. E., and Murray, A. W., 1974, *Nature* **249:**147–150.
122. Dunkley, P. R., and Carnegie, P. R., 1974, *Biochem. J.* **141:**243–255.
123. Miyamoto, E., Kakiuchi, S., and Kakimoto, Y., 1974, *Nature* **249:**150–151.
124. Miyamoto, E., and Kakiuchi, S., 1974, *J. Biol. Chem.* **249:**2769–2777.
125. Steck, A. J., and Appel, S. H., 1974, *J. Biol. Chem.* **249:**5416–5420.
126. Deibler, G. E., Martenson, R. E., Kramer, A. J., and Kies, M. W., 1975, *J. Biol. Chem.* **250:**7931–7938.
127. Miyamoto, E., 1975, *J. Neurochem.* **24:**503–512.
128. Miyamoto, E., and Kakiuchi, S., 1975, *Biochim. Biophys. Acta* **384:**458–465.
129. Miyamoto, E., 1976, *J. Neurochem.* **26:**573–577.
130. McNamara, J. O., and Appel, S. H., 1977, *J. Neurochem.* **29:**27–35.

131. Petrali, E. H., Theissen, B. J., and Sulakhe, P. V., 1980, *Arch. Biochem. Biophys.* **105:**520–535.
132. Sulakhe, P. V., Petrali, E. H., Theissen, B. J., and Davis, E. R., 1980, *Biochem. J.* **186:**469–473.
133. Sulakhe, P. V., Petrali, E. H., Davis, E. V., and Theissen, B. J., 1980, *Biochemistry* **19:**5363–5371.
134. Endo, T., and Hidaka, H., 1980, *Biochem. Biophys. Res. Commun.* **97:**553–558.
135. Turner, S. R., Jean Chou, C.-H., Kibler, R. F., and Kuo, J. F., 1982, *J. Neurochem.* **39:**1397–1404.
136. Agrawal, H. C., Randle, C. L., and Agrawal, D., 1981, *J. Biol. Chem.* **256:**12243–12246.
137. Chou, F. C.-H., Chou, C.-H. J., Shapira, R., and Kibler, R. F., 1976, *J. Biol. Chem.* **251:**2671–2679.
138. Martenson, R. E., Law, M. J., and Deibler, G. E., 1983, *J. Biol. Chem.* **258:**930–937.
139. Martenson, R. E., Kramer, A. J., and Deibler, G. E., 1976, *J. Neurochem.* **27:**1529–1531.
140. Shult, C. W., Whitaker, J. N., and Wood, J. G., 1978, *J. Neurochem.* **30:**1543–1551.
141. Agrawal, H. C., Martenson, R. E., and Agrawal, D., 1982, *J. Neurochem.* **39:**1755–1758.
142. Folch, J., and Lees, M., 1951, *J. Biol. Chem.* **191:**807–817.
143. Chan, D. S., and Lees, M. B., 1974, *Biochemistry* **13:**2704–2712.
144. Althaus, H., Klöppner, S., Poehling, H., and Neuhoff, V., 1983, *Electrophoresis* **4:**347–353.
145. Lerner, P., Campagnoni, A. T., and Sampugna, J., 1974, *J. Neurochem.* **22:**163–170.
146. Nussbaum, J. L., and Mandel, P., 1973, *Brain Res.* **61:**295–310.
147. Lees, M. B., Sakura, J. D., Sapirstein, S., and Curatolo, W., 1979, *Biochim. Biophys. Acta* **559:**209–230.
148. Schlesinger, M. J., 1981, *Annu. Rev. Biochem.* **50:**193–206.
149. Eng., L. F., Chao, F. C., Gerstl, B., Pratt, D., and Tavastsjerna, M. G., 1968, *Biochemistry* **4:**4455–4465.
150. Morell, P., Greenfield, S., Costantino-Ceccarini, E., and Wisniewski, H., 1972, *J. Neurochem.* **19:**2545–2554.
151. Nussbaum, J. L., Rouayrenc, J. F., Jolles, J., Jolles, P., and Mandel, P., 1974, *FEBS Lett.* **45:**295–298.
152. Lees, M. B., Chang, D. S., and Foster, J. A., 1976, *Trans. Am. Soc. Neurochem.* **7:**222.
153. Jolles, J., Nussbaum, J.-L., Schoentgen, F., Mandel, P., and Jolles, P., 1976, *FEBS Lett.* **74:**190–194.
154. Jolles, J., Schoentgen, F., Jolles, P., Vacher, M., Nicot, C., and Alfsen, A., 1979, *Biochem. Biophys. Res. Commun.* **87:**619–626.
155. Yu, Y. T., and Campagnoni, A. T., 1982, *J. Neurochem.* **39:**1559–1568.
156. Lees, M. B., Chao, B. H., Laursen, R. A., and L'Italien, J. J., 1982, *Biochim. Biophys. Acta* **702:**117–124.
156a. Stoffel, W., Hillen, H., Schröder, W., and Deutzmann, R., 1982, *Hoppe Seylers Z. Physiol. Chem.* **363:**855–864.
157. Nussbaum, J.-L., Jolles, J., and Jolles, P., 1982, *Biochimie* **64:**405–410.
158. Magno-Sumbilla, C., and Campagnoni, A. T., 1977, *Brain Res.* **126:**131–148.
159. Sherman, G., and Folch-Pi, J., 1970, *J. Neurochem.* **17:**597–605.
160. Folch-Pi, J., 1973, *Proteins of the Nervous System* (D. J. Schneider, ed.), Raven Press, New York, pp. 45–66.
161. Stoffyn, P. J., and Folch-Pi, J., 1971, *Biochem. Biophys. Res. Commun.* **44:**157.
162. Folch-Pi, J., and Stoffyn, P. J., 1972, *Ann. N.Y. Acad. Sci.* **195:**86–107.
163. Townsend, L. E., and Agrawal, D., Benjamins, J. A., and Agrawal, H. C., 1982, *J. Biol. Chem.* **257:**9745–9750.
164. Waehneldt, T. V., and Malokta, J., 1980, *Brain Res.* **189:**582–587.
165. Lees, M., Chao, B. H., Lin, L. H., Samiullah, N., and Laursen, R. A., 1983, *Arch. Biochem. Biophys.* **226:**643–656.
166. Drummond, R. J., and Dean, G., 1980, *J. Neurochem.* **35:**1155–1165.
167. Muller, H. W., Clapshaw, P. A., and Seifert, W., 1981, *J. Neurochem.* **36:**2004–2013.
168. Nussbaum, J. L., DeLaunoy, J. P., and Mandel, P., 1977, *J. Neurochem.* **28:**183–191.
169. Wiggins, R. C., Joffe, S., Davidson, D., and Del Valle, U., 1974, *J. Neurochem.* **22:**171–175.

170. Reig, J. A., Ramos, J. M., Cozar, M., Aguilar, J. S., Craido, M., and Monreal, J., 1982, *J. Neurochem.* **39**:507–511.
171. Quarles, R. H., Everly, J. H., and Brady, R. P., 1972, *Biochem. Biophys. Res. Commun.* **47**:491–497.
172. Matthieu, J. M., Quarles, R. H., Poduslo, J. F., and Brady, R. O., 1975, *Biochim. Biophys. Acta* **392**:159–166.
173. Sternberger, N. H., Quarles, R. H., Itoyama, Y., and Webster, H. D., 1979, *Proc. Natl. Acad. Sci. U.S.A.* **76**:1510–1514.
174. Johnson, D., Quarles, R. H., and Brady, R. O., 1980, *Fed. Proc.* **39**:1831.
175. Figlewicz, D. A., Quarles, R. H., Sternberger, N. H., and Barbarash, G. R., 1981, *J. Neurochem.* **37**:749–758.
176. Mc Intyre, L. J., Quarles, R. H., and Brady, R. O., 1979, *Biochem. J.* **183**:205–212.
177. Quarles, R. H., McIntyre, L. J., and Pasnak, C. F., 1979, *Biochem. J.* **183**:213–221.
178. Poduslo, J. F., Harman, J. L., and McFarlin, D. E., 1980, *J. Neurochem.* **34**:1733–1744.
179. Mena, E. E., Moore, B. W., Hogen, S., and Agrawal, H. C., 1981, *Biochem. J.* **195**:525–528.
180. Poduslo, J. F., 1981, *J. Neurochem.* **36**:1924–1931.
181. Lees, M. B., and Sapirstein, V., 1982, *Handbook of Neurochemistry,* Volume 4 (A. Lajtha, ed.), Plenum Press, New York, pp. 435–457.
182. Smith, M. E., and Benjamins, J. A., 1984, *Myelin* 2nd ed. (P. Morell, ed.), Plenum Press, New York, pp. 441–487.
183. Benjamins, J. A., and Smith, M. E., 1984, *Myelin* 2nd ed. (P. Morell, ed.), Plenum Press, New York, pp. 225–258.
184. Smith, M. E., 1973, *J. Lipid Res.* **14**:541–551.
185. Campagnoni, C. W., Carey, G. D., and Campagnoni, A. T., 1978, *Arch. Biochem. Biophys.* **190**:118–125.
186. Campagnoni, A. T., and Hunkeler, M. H., 1980, *J. Neurobiol.* **11**:355–364.
187. Walters, S. N., and Morell, P., 1981, *J. Neurochem.* **36**:1792–1801.
188. Lim. L., White, J. O., Hall, C., Berthold, W., and Davison, A. N., 1974, *Biochim. Biophys. Acta* **318**:313–325.
189. Campagnoni, A. T., Carey, G. D., and Yu, Y.-T., 1980, *J. Neurochem.* **34**:677–686.
190. Townsend, L. E., and Benjamins, J. A., 1979, *Trans. Am. Soc. Neurochem.* **11**:157.
191. Hall, C., and Lim, L., 1981, *Biochem. J.* **196**:327–336.
192. Hall, C., Mahadevan, L., Whatley, S., Ling, Tit-Soong, and Lim, L., 1982, *Biochem. J.* **202**:407–417.
193. Colman, D. R., Kreibich, G., Frey, A. B., and Sabatini, D. B., 1982, *J. Cell Biol.* **95**:598–608.
194. Benjamins, J. A., 1984, *Oligodendroglia* (W. T. Norton, ed.), Plenum Press, New York, pp. 87–123.
195. Matthees, J., and Campagnoni, A. T., 1980, *J. Neurochem.* **35**:867–872.
196. Carson, J. H., Nielson, M. L., and Barbarese, E., 1983, *Dev. Biol.* **96**:485–492.
197. Smith, M. E., 1968, *Biochim. Biophys. Acta* **164**:285–293.
198. Sabri, M. I., Bone, A. H., and Davison, A. N., 1974, *Biochem. J.* **142**:499–507.
199. Singh, H., and Jungalwala, F. B., 1979, *Int. J. Neurosci.* **9**:123–131.
200. Fischer, C. A., and Morell, P., 1974, *Brain Res.* **74**:51–65.
201. Lajtha, A., Toth, J., Fujimoto, K., and Agrawal, H. C., 1977, *Biochem. J.* **164**:323–329.
202. Glasgow, M. S., Quarles, R. H., and Grollman, S., 1972, *Brain Res.* **42**:129–137.
203. Sato, S., Quarles, R. H., and Brady, R. O., 1982, *J. Neurochem.* **39**:97–105.
204. Folch, J., Lees, M., and Sloane-Stanley, G. H., 1957, *J. Biol. Chem.* **226**:497–509.
205. Radin, N. S., 1969, *Methods in Enzymology*, Volume XIV (S. P. Colowick and N. O. Kaplan, eds.), Academic Press, New York, pp. 245–254.
206. Hajra, A., and Radin, N. S., 1978, *Anal. Biochem.* **90**:420–426.
207. Skipski, V. P., and Barclay, M. 1969, *Enzymology* **14**:530–598.
208. Renkonen, O., and Varo, P., 1967. *Lipid Chromatographic Analysis* (G. Y. Marinetti, ed.), Marcel Dekker, New York, pp. 41–98.
209. Horrocks, L. A., and Sun, G. Y., 1972, *Research Methods in Neurochemistry*, Volume 1 (N. Marks and R. Rodnight, eds.), Plenum Press, New York, pp. 223–232.

210. Lowenstein, J. M. (ed.), 1969, *Methods in Enzymology*, Vol. XIV, *Lipids* (S. P. Colowick and N. O. Kaplan, eds.), Academic Press, New York.
211. Marinetti, G. V. (ed.), 1976, *Lipid Chromatographic Analysis*, Marcel Dekker, New York.
212. Kates, M., 1972, *Techniques of Lipidology: Isolation, Analysis and Identification of Lipids*, North-Holland, Amsterdam.
213. Norton, W. T., and Autilio, L. A., 1966, *J. Neurochem.* **13**:213–222.
214. Norton, W. T., 1981, *Basic Neurochemistry* (G. J. Siegel, R. W. Albers, R. Katzman, and B. W. Agranoff, eds.), Little Brown, Boston, pp. 63–92.
215. Eichberg, J., and Dawson, R. M. C., 1965, *Biochem. J.* **96**:644–650.
216. Hauser, G., and Eichberg, J., 1973, *Biochim. Biophys. Acta* **326**:201–209.
217. Ueno, K., Ando, S., and Yu, R. K., 1978, *J. Lipid Res.* **19**:863–871.
218. Ledeen, R. W., Yu, R. K., and Eng, L. F., 1973, *J. Neurochem.* **21**:829–840.
219. Norton, W. T., and Brotz, M. 1963, *Biochem. Biophys. Res. Commun.* **12**:198–203.
220. Pieringer, R. A., Deshmukh, D. S., and Flynn, T. J., 1973, *Prog. Brain Res.* **40**:397–405.
221. Norton, W. T., and Brotz, M., 1967, *Fed. Proc.* **26**:675.
222. Rumsby, M. G., and Rossiter, R. J., 1968, *J. Neurochem.* **15**:1473–1476.
223. Pieringer, G., Subba Rao, G., Mandel, P., and Pieringer, A., 1977, *Biochem. J.* **166**:421–428.
224. Horrocks, L. A., Mechkler, R. J., and Collins, R. L., 1966, *Variations in the Chemical Composition of the Nervous System as Determined by Development and Genetic Factors* (G. B. Ansell,ed.), Pergamon Press, Oxford, p. 46.
225. Eng, L. F., and Noble, E. P., 1968, *Lipid* **3**:157–161.
226. Horrocks, L. A., 1968, *J. Neurochem.* **15**:483–488.
227. Norton, W. T., and Poduslo, S. E., 1973, *J. Neurochem.* **21**:759–773.
228. Horrocks, L. A., 1973, *Prog. Brain Res.* **40**:383–395.
229. Davison, A. N., and Peters, A., 1970, *Myelination* Charles C Thomas, Springfield, Illinois.
230. Inoue, T., Deshmukh, D. S., and Pieringer, R. A., 1971, *J. Biol. Chem.* **246**:5688–5694.
231. McIntyre, L. J., Quarles, R. H., and Brady, R. O., 1979, *Trans. Am. Soc Neurochem.* **9**:106.
232. Suzuki, K., Poduslo, J. F., and Poduslo, S. E., 1968, *Biochim. Biophys. Acta* **152**:576–586.
233. Cochran, F. B., Yu, R. K., and Ledeen, R. W., 1982, *J. Neurochem.* **39**:773–780.
234. Kornberg, A., and Pricer, W. E., Jr., 1953, *J. Biol. Chem.* **204**:329–343.
235. Kornberg, A., and Pricer, W. E., 1953, *J. Biol. Chem.* **204**:345–357.
236. Hajra, A. K., 1968, *J. Biol. Chem.* **243**:3458–3465.
237. Hajra, A. K., and Agranoff, B. W., 1967, *J. Biol. Chem.* **242**:1074–1075.
238. Hajra, A. K., and Agranoff, B. W., 1968, *J. Biol. Chem.* **243**:3542–3543.
239. Hajra, A. K., and Burke, C., 1978, *J. Neurochem.* **31**:125–134.
240. Kennedy, E. P., 1961, *Fed. Proc.* **20**:934–940.
241. McMurray, W. C., Strickland, K. P., Berry, J. F., and Rossiter, R. J., 1957, *Biochem. J.* **66**:634–644.
242. Porcellati, G., Biasion, M. G., and Pirotta, M., 1970, *Lipids* **5**:734–742.
243. Miller, E. K., and Dawson, R. M. C., 1972, *Biochem. J.* **126**:805–821.
244. Possmayer, F., Meiners, B., and Mudd, J. B., 1973, *Biochem. J* **132**:381–394.
245. Jungalwala, F. B., 1974, *Brain Res.* **78**:99–108.
246. Toews, A. D., Horrocks, L. A., and King, J. S., 1976, *J. Neurochem.* **27**:25–31.
247. McCaman, R. E., and Cook, K., 1966, *J. Biol. Chem.* **241**:3390–3394.
248. Ansell, G. B., and Metcalfe, R. F., 1971, *J. Neurochem.* **18**:647–665.
249. Freysz, L., Horrocks, L. A., and Mandel, P., 1980, *J. Neurochem.* **34**:963–969.
250. Porcellati, G., Arienti, G., Pirotta, M., and Giorgini, D., 1971, *J. Neurochem.* **18**:1395–1402.
251. Kanfer, J. N., 1972, *J. Lipid Res.* **13**:468–476.
252. Arienti, G., Brunetti, M., Gaiti, A., Orlando, P., and Porcellati, G., 1976, *Function and Metabolism of Phospholipids in the Central and Peripheral Nervous Systems* (G. Porcellati, L. Amaducci, and C. Galli, eds.), Plenum Press, New York, pp. 63–78.
253. Bremer, J., Figard, P. H., and Greenberg, D. M., 1960, *Biochim. Biophys. Acta* **43**:477–488.
254. Butler, M., and Morell, P., 1983, *J. Neurochem.* **41**:1445–1454.
255. Percy, A. K., Moore, J. F., Carson, M. A., and Waechter, C. J., 1983, *Arch. Biochem. Biophys.* **223**:484–494.
256. Wu, Po-Shun, and Ledeen, R. W., 1980, *J. Neurochem.* **35**:659–666.

257. Kunishita, T., and Ledeen, R. W., 1984, *J. Neurochem.* **42**:326–333.
257a. Harvey, M. S., Wirtz, K. W. A., Kamp, H. H., Zegers, B. J. M., and Van Deenen, L. L. M., 1973, *Biochim. Biophys. Acta* **323**:234–239.
257b. Brammer, M. J., 1978, *J. Neurochem.* **31**:1435–1440.
257c. Benjamins, J. A., and McKhann, G. M., 1973, *J. Neurochem.* **20**:1121–1129.
257d. Pasquini, J. M., Gomez, C. J., Najle, R., and Soto, E. F., 1975, *J. Neurochem.* **24**:439–434.
258. Hajra, A., 1969, *Biochem. Biophys. Res. Commun.* **37**:486–492.
259. Wykle, R. L., and Snyder, F., 1969, *Biochem. Biophys. Res. Commun.* **37**:658–662.
260. Tjiong, H. B., Gunawan, J., and Debuch, H., 1976, *Hoppe-Seylers Z. Physiol. Chem.* **357**:707–712.
261. Snyder, F., 1972, *Ether Lipids: Chemistry and Biology* (F. Snyder, ed.), Academic Press, New York, pp. 121–156.
262. Horrocks, L. A., and Sharma, M., 1982, *Phospholipids* (J. N. Hawthorne and G. B. Ansell, eds.), Elsevier Biomedical Press, Amsterdam, pp. 51–93.
263. Paulus, H., and Kennedy, E. P., 1960, *J. Biol. Chem.* **235**:1303–1311.
264. Hawthorne, J. N., and Kai, M., 1970, *Handbook of Neurochemistry,* Volume 3 (A. Lajtha, ed.), Plenum Press, New York, pp. 491–508.
265. Deshmukh, D. S., Bear, W. D., and Brockerhoff, H., 1978, *J. Neurochem.* **30**:1191–1193.
266. Deshmukh, D. S., Kuizon, S., Bear, W. D., and Brockerhoff, H., 1981, *J. Neurochem.* **36**:594–601.
267. Wenger, D. A., Petitpas, J. W., and Pieringer, R. A., 1968, *Biochemistry* **7**:3700–3707.
268. Wenger, D. A., Subba Rao, K., and Pieringer, R. A., 1970, *J. Biol. Chem.* **245**:2513–2519.
269. Subba Rao, G., Norcia, N., Pieringer, J., and Pieringer, A., 1977, *Biochem. J.* **166**:429–435.
270. Braun, P. E., Morell, P., and Radin, N. S., 1970, *J. Biol. Chem.* **245**:335–341.
271. Stoffel, W., and Bister, K., 1974, *Hoppe-Seylers Z. Physiol. Chem.* **355**:911–923.
272. Ong, D. E., and Brady, R. N., 1973, *J. Biol. Chem.* **248**:3884–3888.
273. Sribney, M., 1966, *Biochim. Biophys. Acta* **125**:542–547.
274. Morell, P., and Radin, N. S., 1970, *J. Biol. Chem.* **245**:342–350.
275. Ullman, M. D., and Radin, N. S., 1972, *Arch. Biochem. Biophys.* **152**:767–777.
276. Hajra, A. K., and Radin, N. S., 1963, *J. Lipid Res.* **4**:448–453.
277. Hoshi, M., and Kishimoto, Y., 1973, *J. Biol. Chem.* **248**:4123–4130.
278. Murad, S., Chen, R. H. K., and Kishimoto, Y., 1977, *J. Biol. Chem.* **252**:5206–5210.
279. Yavin, E., and Gatt, S., 1969, *Biochemistry* **8**:1692–1698.
280. Singh, I., and Kishimoto, Y., 1980, *Arch. Biochem. Biophys.* **202**:93–100.
281. Morell, P., and Radin, N. S., 1969, *Biochemistry* **8**:506–512.
282. Cleland, W. W., and Kennedy, E. P., 1960, *J. Biol. Chem.* **235**:45–51.
283. Neskovic, N. M., Nussbaum, J. L., and Mandel, P., 1969, *FEBS Lett.* **3**:199–201.
284. Hammarstrom, S., 1972, *Biochem. Biophys. Res. Commun.* **45**:468–475.
285. Hammarstrom, S., and Samuelsson, B., 1972, *J. Biol. Chem.* **247**:1001–1011.
286. Costantino-Ceccarini, E., and Morell, P., 1972, *Lipids* **7**:656–659.
287. Brenkert, A., and Radin, N. S., 1972, *Brain Res.* **36**:183–193.
288. McKhann, G. M., Levy, R., and Ho., W., 1965, *Biochem. Biophys. Res. Commun.* **20**:109–113.
289. McKhann, G. M., and Ho, W., 1967, *J. Neurochem.* **14**:717–724.
290. Farrell, D. F., and McKhann, G. M., 1971, *J. Biol. Chem.* **246**:4694–4702.
291. Benjamins, J. A., and Iwata, R., 1979, *J. Neurochem.* **32**:921–926.
292. Hayes, L., and Jungalwala, F. B., 1976, *Biochem. J.* **160**:195–204.
293. Shoyama, Y., and Kishimoto, Y., 1978, *J. Neurochem.* **30**:377–382.
294. Siegrist, H. P., Burkhart, T., Wiesman, U. N., Herschkowitz, N. N., and Spycher, M. A., 1979, *J. Neurochem.* **33**:397–404.
295. Benjamins, J. A., Hadden, T., and Skoff, R. P., 1982, *J. Neurochem.* **38**:233–241.
296. Benjamins, J. A., Guarnieri, M., Sonneborn, M., and McKhann, G. M., 1974, *J. Neurochem.* **23**:751–757.
296a. Deshmukh, D. S., and Lee, P. K., 1983, *Trans. Am. Soc. Neurochem.* **14**:249.
297. Neskovic, N. M., Sarlieve, L. L., and Mandel, P., 1973, *J. Neurochem.* **20**:1419–1430.

298. Jungalwala, F. B., 1974, *J. Lipid Res.* **15**:114–123.
299. Koul, O., Chou, K.-H., and Jungalwala, F. B., 1980, *Biochem. J.* **186**:959–969.
300. Costantino-Ceccarini, E., and Suzuki, K., 1975, *Brain Res.* **93**:358–362.
300a. Yahara, S., Singh, I., and Kishimoto, Y., 1980, *Biochim. Biophys. Acta* **619**:177–185.
300b. Mertz, R. J., and Radin, N. S., 1982, *J. Biol. Chem.* **257**:12901–12907.
300c. Herschkowitz, N., McKhann, G. M., Saxena, S., and Shooter, E. M., 1968, *J. Neurochem.* **15**:1181–1189.
301. Sribney, M., and Kennedy, E. P., 1958, *J. Biol. Chem.* **233**:1315–1322.
302. Kopaczyk, K. C., and Radin, N. S., 1965, *J. Lipid Res.* **6**:140–145.
303. Fujino, Y., Nakano, M., Negishi, T., and Ito, S., 1968, *J. Biol. Chem.* **243**:4650–4651.
304. Diringer, H., Marggraf, W. D., Koch, M. A., and Anderen, F. A., 1972, *Biochem. Biophys. Res. Commun.* **47**:1345–1352.
305. Marggraf, W. D., and Anderer, F. A., 1974, *Hoppe-Seylers Z. Physiol. Chem.* **335**:803–810.
306. Ullman, M. D., and Radin, N. S., 1974, *J. Biol. Chem.* **249**:1506–1612.
307. Brady, R. O., Bradley, R. M., Young, O. M., and Kaller, H., 1965, *J. Biol. Chem.* **240**:3693–3694.
308. Wykle, R. L., 1977, *Lipid Metabolism in Mammals,* Volume 1 (F. Snyder, ed.), Plenum Press, New York, pp. 317–366.
309. Dobbing, J., 1963, *J. Neurochem.* **10**:739–742.
310. Koper, J. W., Lopes-Cardogo, M., and Van Golde, L. M. G., 1981, *Biochim. Biophys. Acta* **666**:411–417.
311. Edmond, J., 1974, *J. Biol. Chem.* **249**:72–80.
312. Shah, S. N., 1981, *Arch. Biochem. Biophys.* **211**:439–446.
313. Sato, S., Quarles, R. H., and Brady, R. O., 1982, *J. Neurochem.* **39**:97–105.
314. Smith, M. E., and Eng, L. F., 1965, *J. Am. Oil Chem. Soc.* **42**:1013–1018.
315. Smith, M. E., 1967, *Advances in Lipid Research,* Volume 5 (R. Paoletti and D. Kritchevsky, eds.), Academic Press, New York, pp. 241–278.
316. Smith, M. E., 1968, *Biochim. Biophys. Acta* **164**:283–293.
317. Abdel-Latif, A. A., and Smith, J. P., 1970, *Biochim. Biophys. Acta* **218**:134–140.
318. Benjamins, J. A., and McKhann, G. M., 1973, *J. Neurochem.* **20**:1111–1120.
319. Miller, S. L., Benjamins, J. A., and Morell, P., 1977, *J. Biol. Chem.* **252**:4025–4037.
320. Miller, S. L., and Morell, P., 1978, *J. Neurochem.* **31**:771–777.
321. Horrocks, L. A., Toews, A. D., Thompson, D. K., and Chin, J. Y., 1976, *Function and Metabolism of Phospholipids in the Central and Peripheral Nervous Systems* (G. Porcellati, L. Amaducci, and C. Galli, eds.), Plenum Press, New York, pp. 37–54.
322. Kanfer, J. N., 1980, *Can. J. Biochem.* **58**:1370–1380.
323. Leibovitz-Ben Gershon, Z., Kobiler, I., and Gatt, S., 1972, *J. Biol. Chem.* **247**:6840–6847.
324. Ansell, G. B., and Spanner, S., 1965, *Biochem. J.* **94**:252–258.
325. D'Amato, R. A., Horrocks, L. A., and Richardson, K. E., 1975, *J. Neurochem.* **24**:1251–1255.
326. Wykle, R. L., and Schremmer, J. M., 1974, *J. Biol. Chem.* **249**:1742–1746.
327. Horrocks, L. A., Spanner, S., Mozzi, R., Fu, S. C., D'Amato, R., and Krakowka, S., 1978, *Myelination and Demyelination* (J. Palo, ed.), Plenum Press, New York, pp. 423–437.
328. Bowen, D. M., and Radin, N. S., 1969, *J. Neurochem.* **16**:457–460.
329. Serougne, C., Lefeure, C., and Chevallier, F., 1976, *Exp. Neurol.* **51**:229–240.
330. Eto, Y., and Suzuki, K., 1973, *J. Biol. Chem.* **248**:1986–1991.
331. Igarashi, M., and Suzuki, K., 1977, *J. Neurochem.* **28**:729–738.
332. Skoff, R. P., 1980, *Pathol. Res. Pract.* **168**:279–300.
333. Agrawal, H. C., and Hartman, B. K., 1980, *Biochemistry of Brain* (S. Kumar, ed.), Pergamon Press, Oxford, pp. 583–613.
334. Sternberger, N., 1983, *Advances in Neurochemistry*, Volume 5 (W. T. Norton, ed.), Plenum Press, New York, pp. 125–173.
335. Agrawal, H. C., Hartman, B. K., Shearer, W. T., Kalmbach, S., and Margolis, F., 1977, *J. Neurochem.* **28**:495–508.
336. Sternberger, N. H., Itoyama, Y., Kies, M. W., and Webster, H. deF., 1978, *J. Neurocytol.* **7**:251–263.

337. Hartman, B. K., Agrawal, H. C., Kalmbach, S., and Shearer, W. T., 1979, *J. Comp. Neurol.* **188**:273–290.
338. Hartman, B. K., Agrawal, H. C., Agrawal, D., and Kalmbach, S., 1982, *Proc. Natl. Acad. Sci. U.S.A.* **79**:4217–4220.
339. Knobler, R. L., and Stempak, J. G., 1973, *Prog. Brain Res.* **40**:407–423.
340. Knobler, R. L., Stempak, J. G., and Saurencin, M., 1974, *J. Ultrastruct. Res.* **49**:34–39.
341. Roussel, G., and Nussbaum, J. L., 1981, *Histochem. J.* **13**:1029–1048.
342. Roussel, G., Delaunoy, J. P., Mandel, P., and Nussbaum, 1978, *J. Neurocytol.* **7**:155–163.
343. McMorris, F. A., Kim, S. U., and Sprinkle, T. J., 1984, *Brain Res.* **179**:123–131.
344. Quarles, R. H., Johnson, H. D., Brady, R. O., and Sternberger, N. H., 1981, *Neurochem. Res.* **6**:1115–1128.
345. Trapp, B. D., and Quarles, R. H., 1982, *J. Cell. Biol.* **92**:877–882.
346. Rapport, M. M., and Graf, L., 1969, *Prog. Allergy* **13**:273–331.
347. Raff, M. C., Mirsky, R., Fields, K. L., Lisak, R. P., Dorfman, S., Silberberg, D., Gregson, N., Leibowitz, S., and Kennedy, M., 1978, *Nature* **274**:813–816.
348. Raff, M. C., Fields, K. L., Hakomori, S. I., Mirsky, R., Pruss, R. M., and Winter, J., 1979, *Brain Res.* **174**:283–308.
349. Mirsky, R., Winter, J., Abney, E. R., Pruss, R. M., Gavrilovic, J., and Raff, M. C., 1980, *J. Cell. Biol.* **84**:483–494.
350. Zalc, B., Monge, M., Dupouey, P., Hauw, J. J., and Baumann, N. A., 1981, *Brain Res.* **211**:341–354.
351. Uchida, T., Takahashi, K., Yamaguchi, H., and Nagai, Y., 1981, *Jpn. J. Exp. Med.* **51**:29–35.
352. Raine, C. S., Johnson, A. B., Marcus, D. M. Suzuki, A., and Bornstein, M. B., 1981, *J. Neurol. Sci.* **52**:117–131.
353. Kies, M. W., Driscoll, B. F., Seil, F. J., and Alvord, E. C., 1973, *Science* **179**:689–690.
354. Seil, F. J., and Agrawal, H. C., 1980, *Brain Res.* **194**:273–277.
355. D'Monte, B., Mela, P., and Marks, N., 1971, *Eur. J. Biochem.* **23**:355–365.
356. Smith, M. E., and Hasinoff, C. M., 1971, *J. Neurochem.* **18**:739–747.
357. Benjamins, J. A., Jones, M., and Morell, P., 1975, *J. Neurochem.* **24**:1117–1122.
358. Benjamins, J. A., Iwata, R., and Hazlett, J., 1978, *J. Neurochem.* **31**:1077–1085.
359. Pereyra, P. M., Braun, P. E., Greenfield, S., and Hogan, E. L., 1983, *J. Neurochem.* **41**:974–988.
360. Bizzozero, O. A., Pasquin, J. M., and Sato, E. F., 1983, *Neurochem. Res.* **7**:1415–1425.
361. Benjamins, J. A., Herschkowitz, N., Robinson, J., and Mc Khann, G. M., 1971, *J. Neurochem.* **18**:729–728.
362. Konat, G., and Clausen, J., 1978, *J. Neurochem.* **30**:907–909.
363. Pellkofer, R., and Jatzkewitz, H., 1976, *J. Neurochem.* **27**:351–364.
364. Townsend, L. E., and Benjamins, J. A., 1983, *J. Neurochem.* (in press).
365. Townsend, L. E., and Benjamins, J. A., 1983, *J. Neurochem.* **40**:1333–1339.
366. Freysz, L., and Mandel, P., 1980, *J. Neurochem.* **34**:305–308.
367. Le Baron, F. N., Sanyal, S., and Jungalwala, F. B., 1981, *Neurochem. Res.* **6**:1081–1089.
368. Freysz, L., and Horrocks, L. A., 1980, *Neurological Mutations Affecting Myelination* (N. A. Baumann, ed.), Elsevier/North Holland Biomedical Press, Amsterdam, pp. 223–230.
369. Danks, D. M., and Matthieu, J. M., 1979, *Life Sci.* **24**:1425–40.
370. Braun, P. E., Pereyra, P. M., and Greenfield, S., 1980, *Prog. Clin. Biol. Res.* **49**:1–17.
371. Waehneldt, T. V., and Linington, C., 1980, *Neurological Mutations Affecting Myelination* (N. A. Baumann, ed.), North-Holland Biomedical Press, Amsterdam, pp. 389–412.
372. Banik, N. L., and Davison, A. N., 1969, *Biochem. J.* **115**:1051.
373. Agrawal, H. C., Banik, N. L., Bone, A. H., Davison, A. N., Mitchell, R. F., and Spohn, M., 1970, *Biochem. J.* **120**:635–642.
374. Agrawal, H. C., Trotter, J. L., Mitchell, R. F., and Burton, R. M., 1973, *Biochem. J.* **136**:1117–1119.
375. Matthieu, J. M., Quarles, R. H., Poduslo, J. F., Brady, R. O., and Webster, H. DeF., 1973, *Biochim. Biophys. Acta* **329**:305–317.
376. Waehneldt, T. V., Matthieu, J. M., and Neuhoff, V., 1977, *Brain Res.* **138**:29–43.

377. Waehneldt, T. V., 1975, *Biochem. J.* **151**:435–437.
378. Mc Intyre, R. J., Quarles, R. H., Webster, H. DeF., and Brady, R. O., *J. Neurochem.* **30**:991–1002.
379. Benjamins, J. A., Miller, K., and Mc Khann, G. M., 1973, *J. Neurochem.* **20**:1589–1603.
380. Agrawal, H. C., Trotter, J. L., Burton, R. M., and Mitchell, R. F., 1974, *Biochem. J.* **140**:99–109.
381. Eng, L. F., and Bignami, A., 1972, *Trans. Am. Soc. Neurochem.* **3**:75.
382. Benjamins, J. A., Miller, S. L., and Morell, P., 1976, *J. Neurochem.* **27**:565–570.
383. Benjamins, J. A., Gray, M., and Morell, P., 1976, *J. Neurochem.* **27**:571–575.
384. Schults, C. W., Whittaker, J. N., and Wood, J. G., 1978, *J. Neurochem.* **30**:1543–1551.
385. Shapira, R., Mobley, W. C., Thiele, S. B., Wilhelmi, M. R., Wallace, A., and Kibler, R. F., 1978, *J. Neurochem.* **30**:735–744.
386. Daniel, A., Day, E. D., and Kaufman, B., 1972, *Fed. Proc.* **31**:490.
387. Smith, M. E., and Benjamins, J. A., 1984, *Myelin*, 2nd ed. (P. Morell, ed.), Plenum Press, New York, pp. 441–487.
388. Palo, J. (ed.), *Myelination and Demyelination, Advances in Experimental Biology and Medicine*, Volume 100, Plenum Press, New York.
389. Cammer, W. E., *Experimental and Clinical Neurotoxicology* (P. S. Spencer, and H. H. Schaumburg, eds.), Williams & Wilkins, New York, pp. 239–256.
390. Norton, W. T., and Autilio, L. A., 1966, *J. Neurochem.* **13**:213–222.
391. Webster, H. deF., Palkovits, C. G., Stoner, G. L., Favilla, J. T., Frail, D. E., and Braun, P. E., 1983, *J. Neurochem.* **41**:1469–1479.
392. Fry, J. M., Weissbarth, S., Lehrer, G. M., Bornstein, M. B., 1974, *Science* **183**:540–542.
393. Seil, F. J., Quarles, R. H., Johnson, D., and Brady, R. O., 1981, *Brain Res.* **209**:470–475.
394. Stoffel, W., Hillen, H., Schröder, W., and Dentzmann, R., 1983, *Hoppe-Seyler's Z. Physiol, Chem.* **364**:1455.

15

Autonomic Ganglia and Their Biochemistry

Jack D. Klingman

1. INTRODUCTION

The systemic involuntary control of animal visceral function is supported and controlled by the autonomic nervous system. The biochemical subtlety and the pathologies of this system are not often appreciated (*re*: pineal, iris, spleen T-cells, heart, and gastrointestinal tract, to mention a few). An understanding, then, of this system's biology, particularly its ontogeny, developmental control, and biochemical functions, allows not only an understanding of this system but also of the more complex systems of the CNS.

Historically, an understanding of physiology as well as the autonomic system's morphology may be credited to two investigators: anatomically to Gaskell,[1] who described the ganglionic system and called it involuntary, and to Langley,[2] who not only called this the autonomic system but subdivided it into the sympathetic, parasympathetic, and enteric ganglionic systems. Langley also described the original nicotinic effects and was one of the first to consider that there were pre- and postganglionic fibers. From his work on the functionality of the sympathetic nervous system, Langley may be considered the father of receptor theory. From these earlier anatomical and physiological studies, the sympathetic ganglia became one of the most used tissues for testing the physiological and pharmacological actions of nicotinic versus muscarinic receptors.

This chapter attempts to make current our present understanding of the autonomic ganglia and their biochemistry from the embryo to the adult. The reader will find more developed discussions of trophic and nerve growth factors and specific enzymes dealing with the adrenergic and cholinergic systems elsewhere in this *Handbook of Neurochemistry*. In order to present a coherent picture, we shall lightly touch on some of these, but our purpose will be to explore the ganglion's functional biochemistry. Several reviews have appeared

Jack D. Klingman • Department of Biochemistry, School of Medicine, State University of New York at Buffalo, Buffalo, New York 14214.

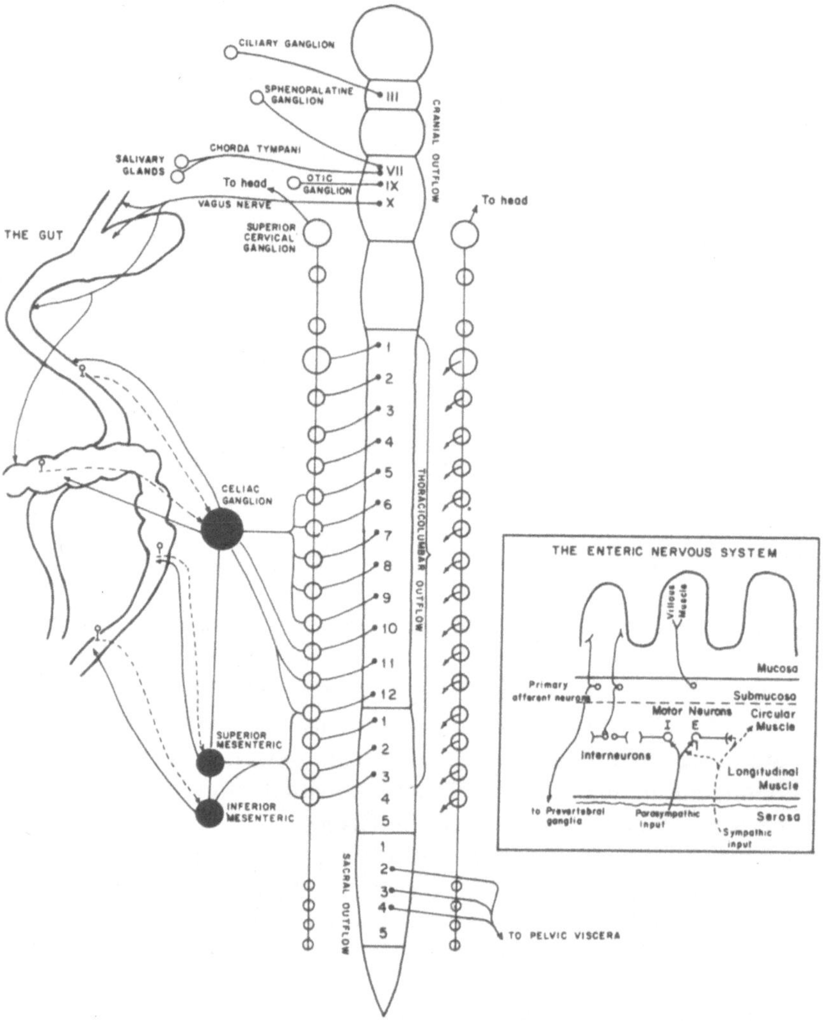

Fig. 1. Schematic representation of the autonomic nervous system and the enteric nervous system. (Reproduced by permission of M. D. Gershon and the MIT Press.[289])

recently on the ontogeny,[3–5] trophic factors,[6–12] SIF cells,[13] and pharmacology[14] of the autonomic ganglia.

2. EMBRYOLOGY

The primal sympathetic neurons develop embryonically from the neural tube margin of neural crest cells. These cells migrate ventrally to further derive into the autonomic sensory and glial, neuroendocrine, and mesenchymal derivatives of the cephalic region and calcitonin-producing cells[5] (see Fig. 1 and Table I). Their timed migration appears to be dependent on a matrix rich in

Table I
Derivatives of the Neural Crest[a]

Skeletal and connvective tissue (cranial crest derivatives)
- Cartilages
  - visceral arch cartilage
  - chondrocranial cartilage
- Bones
  - upper and lower jaw odontoblasts
  - palate
  - cranial vault floor
- Mesenchyme
  - corneal endothelium and stromal fibroblasts
  - contribution to adenohypophysis, lingual gland, parathyroid, thymus, and thyroid
  - contribution to dermis and subcutaneous adipose of face, jaw, and upper neck
  - leptomeninx of diencephalon and telencephalon
- Muscles
  - ciliary muscles (striated)
  - cranial vasculature and dermal smooth muscle

Neural derivatives
- Sensory neurons
  - spinal (dorsal root) ganglia
  - trigeminal (V) ganglion
  - facial (VII) root (geniculate)[b]
  - glossopharyngeal (IX) root (superior)[b]
  - vagal (X) root (jugular)[b]
- Sympathetic (adrenergic) neurons
  - superior cervical
  - paravertebral chain
  - prevertebral complexes (celiac, mesenteric, adrenal, and retroaortic)
- Parasympathetic (cholinergic) neurons
  - ciliary
  - submandibular, ethmoid, otic, lingual, and sphenopalatine
  - Remak's ganglion
  - Meissner's and Auerbach's plexi
  - pelvic plexus
  - visceral intrinsic ganglia
- Neurosecretory cells
  - carotid body type 1 cells
  - calcitonin-producing C-cells of the thyroid
  - possible contribution to adrenocorticotropic hormone/melanocyte-stimulating hormone-producing tissue
  - adrenal medulla
- Supportive cells of the peripheral nervous system[b]
  - glia, Schwann sheath cells, and satellite cells
- Pigment cells (melanocytes) of skin, hair, and irises

[a] Reference 5. Reproduced by permission of J. A. Weston and Raven Press.
[b] Other neural and supportive components of cranial ganglia and nerves derived from placodal ectoderm.

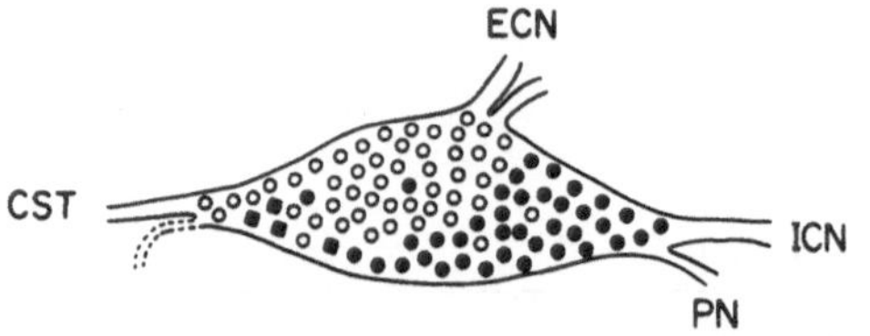

Fig. 2. Schematic diagram of a rat superior cervical ganglion summarizing the location of neurons that project into the internal carotid nerve (ICN) (filled circles), the external carotid nerve (ECN) (open circles), and the cervical sympathetic trunk (CST) (squares). The dotted lines indicate the location of a nerve trunk that is often observed leaving the caudal portion of the SCG. Pterygopalatine nerve (PN). (Reproduced from Bowers and Zigmond[290] with permission of the authors and *Journal of Comparative Neurology*.)

collagen, glycoproteins, and glycosaminoglycans.[3,5] One of the glycoproteins found within this migratory space is fibronectin, known to be produced by a number of embryonic cells. These neural crest cells lack this glycoprotein, but they do contain a large amount of plasminogen activator, a protease for fibronectin. This dependence on a matrix of collagen and glycoproteins is apparent in tissue cultures of neurons and their myelination by Schwann cells.[15,16] Excellent discussions of crest cell migration and development in embryos may be found in several reviews.[3,5]

The neural crest in the trunk region gives rise to the adrenergic sympathetic neurons and the adrenal medulla,[11] and that from somites 1–7 gives rise to the cholinergic enteric ganglionic cells. This division of cholinergic and adrenergic types is one of differentiation as the system matures and is strongly influenced by the innervated organ. The immature ganglia are first visible in chick and rat embryos at days 6–8.

The migration of the primordial sympathetics is along the neuronal axis of their preganglionic fibers. From work on the parasympathetic ciliary ganglia (CG) of the chick, there are initially more ganglionic cells than later, when their final postganglionic synapses are made. This is not necessarily true for all ganglia.[3,17] Whether there is an excess of cells that becomes reduced in number after the sympathetic–target tissue contact or the original cell population is maintained at maturity is still a controversial point for each ganglion and/or across species. (Fig. 2).

3. *ANATOMIC PHYSIOLOGY*

As in the CNS, in addition to the neuronal cells there are accompanying supporting cells, often called satellite cells, and, in some sympathetic ganglia (SG), small intensely fluorescent (SIF) cells. The supporting cells are the glial elements. Little biochemistry has been done on these cells except recently in culture, undoubtedly because of the paucity of material. A procedure has been worked out for the separation of neuronal and glial cells from embryonic chick and rat ganglia.[18] I am unaware of anyone having attempted to isolate cells from other species or successfully from adult ganglia for biochemical studies. The number of supporting cells per neuronal cell varies for each ganglion.[17] Their physiological role is not primarily that of the Schwann cells, since few of the SG neurons are myelinated. Most probably their function is more akin to that of the CNS astrocyte.

The number of preganglionic fibers per ganglionic neuron also varies for each ganglion and each species. Ratios from a low of two ganglionic cells per fiber in the cat CG[19] to as high as approximately 196 in the human superior cervical ganglia (SCG) have been demonstrated.[20] In the intestinal intramural ganglia, upwards of 6000 preganglionic fibers per cell have been found, but this is primarily because of the presence of both afferent and efferent components[21] as well as internuncial neurons.[22,23]

Each neuron has inputs from various numbers of preganglionic fibers: convergence can vary from as low as ten in the guinea pig pelvic ganglia to 240 cells receiving a single fiber input in the rabbit SCG.[24] Whereas the firing of one of these singular inputs causes a small excitatory postsynaptic potential, it requires synchrony of the various preganglionic fibers impinging on a single neuron to cause excitation. (This will be brought up again in discussing one of the possible roles of prostaglandins.) Only in the chick and amphibian CG is there one fiber input per cell.[25,26]

Both myelinated and nonmyelinated preganglionic fibers exist. Their conduction velocities vary from 0.5 to 55 m/s in rat SCG[27] to 6.5 to 31 m/s in other species' ganglia.[28,29] The smallest preganglionic fibers are of peripheral origin with conduction velocities of 0.2–0.3 m/s.[30]

Unlike amphibian neurons, which are unipolar, all mammalian sympathetic neurons are multipolar, having two to ten dendrites.[31,32] Most of the contacts of the preganglionic fibers are found on these dendrites, although some make neuronal soma contacts.

One of the more interesting morphological features of autonomic ganglia is the paucity of myelinated postganglionic fibers. This is useful for metabolic studies since little concern has to be given to possible Schwann cell metabolism. The postganglionic fibers have conduction velocities from 0.4[27] to 9 m/s.[33]

The presynaptic autonomic terminals may contain vesicles of two sizes. Those that contain acetylcholine are lucent and 500 Å, and the larger dense adrenergic granules are 1000 Å. The synaptic space is between 150 and 300 Å with a dense postganglionic area.

It would appear that there are few, if any, internuncial neurons within the autonomic ganglia, although they are present in the intestinal intramural ganglia.[11,34] Depending on the species and ganglia, the only other nonneuronal cell is the SIF cell. Whether the SIF cells make direct synaptic contact with neurons is still a debated issue. It has been ascertained that the SIF cells of rat and human contain norepinephrine,[35] whereas SIF cells of cat, rabbit, and pig contain dopamine.[13,36] In human SCG,[37,39] two types of SIF cells are present in the fetus. Type I cells are true interneurons of efferent or afferent projections onto the neurons; these SIF cells appear to become nonfunctional in the adult. Type II cells are next to capillary epithelium. Eränkö and Soinila[40,41] demonstrated that, in the axotomized rat SCG, a loss of all but $\frac{1}{20}$ of the neurons occurs, whereas there are still 200–600 SIF cells per ganglion on day 23 post-axotomy. An intriguing and more complete discussion of SIF cells and their development and function may be found in a recent review.[13]

4. TROPHIC FACTORS

4.1. Ontogeny and Phenotypic Expression; Nerve Growth Factor

The last three decades have seen an explosive development in efforts to understand the ontogeny and phenotypic expression of sympathetic neurons. The developmental embryologist has long expressed the belief that juxtaposition of cell types was important for the inductive differentiation of a given cell. With the observation that transplanted sarcomas in mice acquired an increased sensory fiber input[42,43] came the development of two great advances in neurobiology. Although a few embryologists, such as Ruth Geiger, were attempting to develop free neuron cultures, the isolation of nerve growth factor (NGF) made it even more imperative that defined culture systems be developed. Trying to understand the mode of action of NGF and the phenotypic expression in the autonomic nervous system has also led to the characterization of other trophic substances as well as an explanation of the interaction between the supporting glial and neuronal cells.[6,10]

The early observation that NGF induced neurite sprouting in dorsal root ganglia (DRG) and the SG[7,8] obviously raised the questions of its biochemical mechanism and specificity. Since NGF is a protein, an antiserum was developed,[43,44] which produces an immunosympathectomy in neonatal rats and mice or in the offspring when administered to the pregnant female. Two important findings were that not only was this sympathectomy dose dependent but the extent or ablation of the particular ganglion depended on the gestational time of antiserum treatment.[44–48] An unexpected observation was that when ablation of the ganglia appeared to have been completed, there still remained inappropriately high acetylcholinesterase activity.[47] These immunoganglionectomies are irreversible.

Purified mouse submaxillary NGF is composed of three protein subunits,[49,50] α, β, and γ, with a total molecular weight of 131,500. Respectively, these subunits are in a ratio of 2:1:2, with two Zn^{2+} ions producing stability. The α subunit is composed of two peptide chains, a heavy chain with a molecular weight of 17,200 and a light chain with a molecular weight of 9500, linked by disulfide bridges.[51] No function has been ascertained for the α protein, although it does increase the yields of neuronal and nonneuronal dissociated DRG cells in culture.[52,53] The γ subunit is an arginine esteropeptidase having a molecular weight of 26,000, which cleaves the pro-β protein to the active peptide.[50] The two polypeptides of the active β subunit have a molecular weight of 26,518.

Dorsal root ganglion neurons contain two membrane binding sites for β-NGF: type I, a high-affinity receptor, $K_D = 3.3 \times 10^{-11}$ M, and type II, a lower-affinity site, $K_D = 1.7 \times 10^{-9}$ M, with association rate constant at 20° C of $1 \times 10^{7}\ M^{-1}\ s^{-1}$ (type I) and rates of dissociation of $6.5 \times 10^{-4}\ s^{-1}$ (type I) and $3.2 \times 10^{-2}\ s^{-1}$ (type II).[49,54] The type I site is involved in the sequestration necessary for initiation of the biological responses. *In vitro*, a maximum outgrowth of neurites is seen at β-NGF concentrations at which less than 20% of the type I receptor sites are occupied.[49] These binding sites are seen in the presumptive chick ganglia at 6 days. The reader is directed to more thorough reviews on the history,[7,44] chemistry,[7] and relation to other trophic materials[9,10] in addition to the chapters in these volumes.

The NGF-induced increased neurite sprouting 12 h after addition to embryonic SCG or DRG in culture posed questions about whether this was related to increased protein synthesis and of the specificity of these effects. Increased incorporation of thymidine into DNA,[55] uridine into RNA,[56,57] and leucine into protein [58] and an increased activity of tyrosine hydroxylase (TH)[59] were reported. In culture, NGF was necessary for survival of neonatal but not the corresponding embryonic ganglia.[60] Cocultures of SG with target tissues would also cause neurite outgrowth. It is necessary to ascertain if this was attributable to NGF or to other trophic agents. Further study found that there were increased amounts of cyclic AMP,[61] increased ornithine decarboxylase activity, and polyamine biosynthesis.[62,63] It appeared as if the classical protein synthesis story was unfolding: a specific membrane receptor, functioning through a cyclic-AMP-stimulated system, to yield specific DNA, RNA, and protein syntheses for neurite growth and possible neurotransmitter enzyme systems.

Many of these initial findings arose from studies of intact ganglia. In studies of neurite outgrowth and the possible increased protein and RNA syntheses, it was learned that puromycin-treated ganglia were still able, on addition of NGF, to produce neurites.[55] In later studies with isolated ganglionic neurons versus ganglionic glial cells,[18] the incorporation of thymidine was found to occur in the glial elements and be related to an action on the neurons.[64] (Is this a demonstration of a neuronal trophism on glial cells as the glial cells, in turn, induce a neuronal effect?) Further, neurite extension *per se* does not require new protein synthesis.[65] Explanted embryonic mouse SCG without NGF still produce outgrowth when their protein synthesis is inhibited.[66] The increased cyclic AMP seen in neonatal NGF-treated SCG[61] is not seen in the adult SCG[67]; however, it does occur in embryonic chick DRG but not SG.[68] Neither does the cyclic AMP response appear to be related to neurite extension or to transmitter synthesis.[58,69,70] The increased ornithine decarboxylase activity may be linked to those tissues showing a responsive NGF-mediated cyclic AMP.[62] This ornithine decarboxylase induction by NGF does not appear to be similar to that evoked in pheochromocytoma clones (PC-12) by epidermal growth factor or to the increase produced by glucocorticoids.[63]

Hori and Varon[71] observed that NGF enhances entry of RNA precursors, glucose, and amino acids into the cells within minutes of NGF addition. Partlow and Larrabee[55] noted an immediate increase in uridine incorporation into the ganglionic RNA. The percentage incorporated increased with embryonic age, with peaks of incorporation at 10 and 12.5 days, suggesting a bimodal RNA function. One sees in many biological systems an initial subtle action and a delay followed by a later discernable expression. This has been most succinctly expressed by Varon and his co-workers, who have been trying to explain this confusing array of events over the past few years.[72–74] They have discerned other neurotrophic survival factors, which have led them to separate their effects on the biological events into those of "short latency" and others of "cosequential actions"—the traditional later responses—which are caused by the indirect effects of the trophic factor(s) and may involve cell–cell interactions.[72]

The "short-latency" effects with which Varon and co-workers are concerned[75] are best exemplified by dissociated cultures of DRG or SG neurons, which will not survive more than 6 h unless NFG is added. One of their prime findings is that the neuron is unable to maintain the high intracellular K^+ and low Na^+ concentrations in the absence of NGF. If NGF is added at 6 h to deprived neurons, an active extrusion of the acquired Na^+ occurs with a K^+ gain. The required dose of NGF is related to the type I receptor and appears to involve activation of the ionic pumps. This action, in turn, could be the linkage to the increased nutrient transport seen earlier.[76]

General statements concerning the permissive action of NGF can best be summarized as follows:[77]

1. Regulation of peripheral sympathetic and sensory neurons' survival at defined periods in development.
2. Growth effects on both types of neurons, which are independent of their developmental stage.
3. Maintenance of specific functions of differentiated sympathetic neurons and an enhancement of differentiation of both sensory and sympathetic neurons during development.
4. Possible chemotactic action on these neurons and/or possible mediation by target organs of other trophic influences.

4.2. Other Survival, Maintenance, and Phenotypic Agents

It has long been known, although not understood, that cells in culture require sera for survival. In the case of SG, fetal calf serum (FCS) has long been used. The fact that agents within these sera are important for survival and possibly aid in differentiation is seen in the reported effect on growing neural crest mesenchymal cells with either FCS or horse serum (HS).[78] When such cells are grown in FCS, they develop normally into their melanocytic phenotypes, which contain little choline acetyltransferase activity (ChAT). If, however, they are grown in HS, they differentiate into neuronlike cells with little pigmentation but containing a large amount of ChAT. (One immediately suspects the varied steroid components contained in HS.) This transformation can be blocked by pretreatment of the cells with DMSO.

A minimal medium, independent of serum, was developed by Bottenstein and Sato,[79] which promoted survival of PC-12 cells. The required constituents are insulin, transferrin, putrescine, progesterone, and selinite. An adaptation of this medium[80] has been shown to foster survival of SG and DRG neurons and, in many cases, to decrease the survival of nonneuronal cells. Such neuronal survival is independent of NGF. However, the omission of any medium constituent but selinite results in the loss of DRG-derived neurons, whereas omission of insulin or selinite alone affects the survival of SG neurons.

One of the early papers in a series of studies by Patterson and Chun[81,82] demonstrated among other results that 11-day chick embryo ganglionic cultures produced outgrowth when cocultured with heart muscle. This led Ebendal *et al.*[83,84] to study heart tissue effects on CG, DRG, SCG, and Remak ganglia, and they found that in coculture there was a marked ontogenic dependence. Heart extracts yielded a 40,000-dalton material that produced both neurite survival and outgrowth of DRG that were not blocked by the addition of NGF antiserum. Heart-conditioned medium has also been shown to be a survival agent for 11-day chick embryo SG and neonatal mouse DRG. They survived with the medium extract about as well as if they had been treated with NGF or a combination of NGF and conditioned heart medium. The possibility that the same receptors or group of neurons might be responding was ruled out by the finding that NGF antiserum blocked only the NGF activity but not that of conditioned heart medium.[30,85]

Chick CG neurons survive better in culture if macromolecule(s) derived from chick intraocular smooth and striated muscle extracts are added.[72] This material[86] has also been shown to support the survival of 11-day chick embryo SG and DRG. Besides the already-defined CG neurotrophic factor, there appears to be another factor(s) present in the intraocular muscle extracts. The material coelutes from DE52 cellulose by salt gradients, has similar isoelectric points, is heat and acid labile but base stable, and is inactivated by tryspin. Neither of these agents supports the survival of 8-day embryonic SG or DRG, although NGF does. Recall that the NGF receptor is known to be present on the cells of 6-day-old chick embryonic ganglia. It has yet to be demonstrated when the receptors for the other possible trophic agents appear. (One may ask whether this is a function of NGF.)

Numerous other possible materials, their specificity, and their survival/neurite outgrowth effects are listed and discussed in a review by Varon and Adler.[72] Few of these possible trophic agents have been tested on SG other than chick ganglia.

4.3. To Be or Not to Be: Phenotypic Transmitter Expression

As mentioned, a presumptive melanocyte can be transformed into what appears to be a neuronal cell type simply by replacing FCS with HS. More recently, Levi-Montalcini has shown that NGF can have a chameleon effect on the adrenal chromaffin cells by transforming them into neurite-producing cells.[7] One of the earliest biochemical effects of NGF is the induced increase in ganglionic TH activity.[87,88] The time course with which the SCG acquires the enzymes necessary for the biosynthesis of the catecholamines has been determined in the mouse and chick.[46,89] The 1-day-old chick embryo can convert tyrosine to DOPA prior to migration of the neural crest cells. The activity of TH, DOPA decarboxylase (DDC), dopamine β-hydroxylase (DBH), and phenylethanolamine N-methyltransferase (PNMT) are first seen, respectively, at 1, 2, 4, and 6 days.

Black and his coworkers[10,11,90] have long been concerned with the phenotypic adrenergic expressions of intestinal intramural ganglia and the SCG of mice and rats. The autonomic neuron begins as an adrenergic neuron. How then does it become a cholinergic neuron when it innervates certain organs such as the heart? This question became answerable when Patterson and Chun[84,91,92] cocultured normal adrenergic ganglionic cells with heart muscle and found an induction of the cholinergic system. (A more complete description on cholinergic development is found in Chapter 21 of Volume 5 in this series.) It has recently been demonstrated that the target organ apparently elaborates a macromolecular material that is possibly retrogradedly transported to the perikaryon and causes the induction of the cholinergic enzymes involved in acetylcholine formation and vesicular storage. This has been most clearly shown in the recent experiment by Weber.[93] Such transformed cells still retain the ability to take up and store norepinephrine.[94–98]

Other nonneuronal cells or nonneuronal-cell-conditioned media have been shown to induce acetylcholine synthesis[91,92] with decreased conversion of tyrosine to catecholamines. Superior cervical ganglion neurons treated with the material isolated by Weber[93] acquire 100- to 1000-fold higher activity of ChAT than do untreated controls. Such transformed cells can form functional synapses with target cells or with themselves.[97,99] It should be noted that adrenergic neurons grown on formaldehyde-fixed heart cells also become transformed into cholinergic cells.[100] It would appear that either such fixation does not remove all of the diffusible heart material, there is a membrane-bound factor, or a collagen matrix effect. Weber's diffusible heart cell material is a macromolecular material with an apparent molecular weight on Sephadex of 40–45,000. (Ebendall's material[83] had roughly the same molecular weight.) This macromolecule is insensitive to treatment with mercaptoethanol but is inactivated by periodate. Neuraminidase treatment does not affect its activity, but the activity is lost when it is exposed to pronase. One of its curious properties is its inhibitory action on the protease activity of trypsin and chymotryspin. (There is a series of such protease inhibitors present in normal serum.) Addition of 1 μg of protein/ml of culture medium increases the acetylcholine (ACh) production 50-fold in the cultured sympathetic neurons after 18 days with a concomitant threefold decrease in catecholamine production. Several studies have shown that once the transformation has taken place, NGF can aid in maintaining this induction.[7,9,11,12]

4.3.1. *Acetylcholinesterase*

The reported release of acetylcholinesterase (AChE) into the medium from cultured SCG[56,57] has prompted two areas of work: what types of AChE exist in ganglia at various developmental periods, and where are they located in the system?

Electron microscopy has given an assessment of the intracellular localization of AChE and butyrylcholinesterase (BChE) during maturation of neonatal rat ganglia and adult cat ganglia.[101,102] Acetylcholinesterase appears postsynaptically prior to appearing in the presynaptic or axolemma regions; BChE follows a similar postsynaptic appearance but is not found in the presynaptic or axolemma regions. Acetylcholinesterase is first found in 1-day SG endoplasmic reticulum cisternae and, by 4 days, also appears in the postsynaptic junction. By the end of 2 weeks, a further increase of activity in the endoplasmic reticulum as well as moderate activity on the axolemma of the nerve endings are seen. In chick embryo CG and SG,[103] the ACh and choline (Ch) levels were followed from 5–7 days to 1 year of age: 1–10 pmol ACh were found in 5-day iris and in 7-day CG and SG; these rapidly increased ninefold by 14 days of incubation. The ACh levels increased in the iris and the CG during the first year, whereas maximal levels in the SG were attained by 3 months. The ACh increase appears to follow the functional needs of the tissue as reflected in the increased number of ganglionic presynaptic terminals. (The possible importance of neuronal cell death during the maturation was not discussed.) The Ch concentration followed a similar time course in these tissues as that of the ACh concentration. In a study on aging rats,[104] the AChE and ChAT activities exhibited a decline with age. This was found to be the result of a decrease in the V_{max} of the enzymes rather than a change in the K_m.

Sedimentation centrifugation has shown four forms of AChE in neonate cultured SCG: 16, 10, 6.6, and 4 S components.[105] The 16 S component is located in the neuron and is maintained independently of the original preganglionic fiber synaptic elements. This cellular pool is not depleted by tetanus toxin. Taylor *et al.*,[106] like Gisiger, reported that the 4 S and 6.5 S components are found in the culture medium of SCG. The 4 S component is cytoplasmic, and the 10 S component is on the external cell membrane. The preganglionic fibers contained less than $\frac{1}{50}$ of the total ganglionic AChE. The 16 S component remained after decentralization and is partly postsynaptic. Without added NGF, the 16 S is lost from cultured cells, and if no postganglionic contacts are made, there is also no 16 S component. Thus, most of the 16 S component is postsynaptic. The total AChE in a non-NGF-supplemented culture is only 45% of the total activity in NGF-supported cultures.

In a somewhat similar study[107] on chick CG, sedimentation centrifugation showed 5, 7.5, 11.5, and 20 S components of AChE. The 7.5 S component was found to be confined to postganglionic regions, possibly on ganglionic perikarya and in the cell body cytoplasm. The 11.5 S had the same amount of activity in pre- and postganglionic axons and recovered slowly after denervation. The 20 S component was preferentially located presynaptically; it was shown to be the one found in fast axonal flow and, thus, is made in the preganglionic neurons. The stability of all forms was dependent on pre- and postganglionic fiber integrity. The 20 S component, which almost completely disappeared after denervation or after denervation and axotomy, responded by a partial loss of activity on the unoperated contralateral side. The contralateral activity recovered in 5 days, whereas it took 10 days on the operated side. In these studies, 44% of the total AChE activity was accounted for by the 7.5 S, 40% by the 11.5 S, 10% by the 20 S, and 6% by the 5 S component. The 5 S component is thought to be formed by a protease during isolation.

Coculturing neonatal rat SCG with pineal gland caused this adrenergic ganglion to produce a large increase in ChAT activity, a decrease in TH activity,[108] and increased amounts of 4.5 S, 6.5 S, and 10 S AChE, with the 16 S component only occurring in neurons. Cocultures increased the SCG AChE activity with age, whereas the pineal AChE activity, when pineal was cultured alone, decreased. Therefore, interaction between neurons and target cells results in increased 4 S and 10 S but not 16 S component. The 16 S component occurs in neurons and is transported down to the presynaptic region. This form is only seen, as stated above, in NGF-treated SCG.

The choline kinase activity of embryonic to 120-day-old rat SCG has been determined.[109,110] The activity is highest at 5 days before birth, decreases some 50% at birth, and then remains at this level. Any increase in activity parallels the increase in total ganglionic protein. After axotomy the activity per ganglion increases twofold between 1 and 5 days postoperative and then decreases over the next 30 days to a basal level. The peak activity is associated with the peak in chromatolysis, in response to the need of phosphorylcholine for regeneration. Previously, axotomy has been shown to cause an increased ornithine decarboxylase activity after 7 h, which stabilizes after 10 h. On day 6 postaxotomy, it begins to drop to basal levels. Addition of polyamines, however, does not protect the choline kinase activity.

A series of elegant papers by Vigny and co-workers concerning the actual compositional forms of rat SCG AChE has enlightened us on a number of the polymeric forms of this enzyme.[111] Two of the AChEs exist as both globular and collagen-tailed forms. These collagen forms are saline extractable—denoted A_{12} and A_8—representing about 15% of the total AChE. The remaining activity, 85%, consists of the globular forms, G_4 and G_1 being the predominant ones, with smaller amounts of G_2. These hydrophobic forms are best extracted with Triton X-100. The authors most conclusively reinforce the finding that the 4.5 S form results from protease activity. The activities of the various forms are: G_1, 40%; G_2, 3%; G_4, 41%; A_8, 4%; and A_{12}, 11.5%. When these are extracted with Triton X-100, the sedimentation coefficients are 3.5 S (G_1), 5.5 S (G_2), 9.7 S (G_4^H), and 1.5 S (G_4^S). When the AChEs are saline extracted, these coefficients change to 4.5 S, 7.1 S, 11.6 S, and 10.5 S, respectively. The transformations to hydrated and collagen forms are clearly demonstrated in this series of publications.

4.3.2. α-Bungarotoxin–Nicotinic Receptors

The nicotinic receptors of SG and their agonists and antagonists have been concisely reviewed (Table III).[14] Since synaptic transmission develops between 5–6 days and 8 days, respectively, in chick embryo CG and SCG,[112] which coincides with the appearance of AChE activity,[113] it is pertinent to determine when the nicotinic and muscarinic receptors first appear. By means of α-bungarotoxin (αBTX), a high level of ACh receptor has been demonstrated in 7-day chick embryo CG, with a maximum binding by day 12, which remains relatively constant from day 15 through 8 days post-hatching.[114] A secondary level of activity is seen from day 8 to day 50. Thus, unlike the embryonic choroid muscle neurons, which have adult patterns, the neurons of the CG develop adult receptor patterns after hatching.

Chick embryo iris and CG[115] have αBTX binding constants of K_D = 2.5 nM and 2.7 nM, respectively, which saturate at a level of 10 nM and are inhibited by *d*-tubocurarine. In the iris, binding is seen at 12 days and increases over the next 4 months. The αBTX binding in the CG increases fourfold from 7 to 11 days in the embryo and then remains unchanged up to 4 months. At incubation day 16, the CG binds 3×10^6 nM αBTX per neuron.

Similar studies have been carried out on the SG of rats and chicks.[116] When cultured chick SG are blocked with αBTX at 12 days, a rapid reappearance of new receptors occurs—some 40% of the total receptors. There is a precursor pool of receptors—about 18% of the total receptor pool—which rapidly supplies the neurons with new αBTX binding sites within 1–2 h without new protein synthesis by an energy-independent process. The reported binding is 0.75 ± 0.07 fmol αBTX bound per explant or 4.5×10^8 receptors per explant.

In the adult rat SCG,[117] αBTX has binding constants of $K_{on} = 1.587 \times 10^5\ M^{-1}\ s^{-1}$ and a $K_{off} = 9.24 \times 10^{-6}\ M^{-1}\ s^{-1}$ at 24°C. This binding is not affected by decentralization, axotomy, reserpine, or administration of 6-hydroxydopamine. After decentralization of the ganglion, a 10% binding loss is seen by 7 days, whereas axotomy results in a 90% loss. The authors concluded that the binding sites in the rat SCG are to a greater extent controlled by postganglionic rather than preganglionic signals. Drugs used for postganglionic fiber destruction do not mimic the effects of axotomy. (Is this related to the trophic factors from the target organs, which still maintain the receptors in drug-treated animals but cannot be transported to the neurons of axotomized SCG?)

It would appear that there are two types of nicotinic receptor channels in sympathetic neurons[118]: a slow and a fast channel, which do not differ in their single-channel conductance. The slow channel determines the kinetics of the extra postsynaptic current decay, whereas the fast channel is seen only when pharmacological doses of ACh are given. The latter channel is not the one that is involved in neuronally stimulated ACh release.

4.3.3. Acetylcholine

The origin of ACh in the sympathetic nervous system has been of biochemical and physiological interest since its release and metabolism were first demonstrated in the SCG.[119,120] The SG may be a useful tool in answering some of the perplexing questions of the ACh metabolic origin as well as the control mechanisms imposed on its formation, reuptake, and possible reformation in the various compartments. One of the problems that may not be present in the SG and which is a major concern in the CNS is that of the tightly coupled junctions of the capillary epithelium. I am not aware of studies that have specifically ascertained whether tightly coupled epithelium is or is not present in the ganglia. From the types of compounds that have been perfused and taken up by the SG, we can at least state that the capillary epithelium is less tightly coupled than that in the CNS.

Recent experiments on ACh formation in SG include those by Collier and co-workers, who have been studying SG ACh formation and release.[121–124] In Ch transport mechanisms into synaptosomes, a carrier-mediated mechanism is most plausible.[121] It was shown that homocholine can be transported by the same choline carrier into brain and SCG synaptosomes. The specificity of this carrier has more clearly been defined by the demonstration that 4-hydroxybutyryltrimethylammonium cannot be transported by this carrier mechanism into either synaptosomes or SCG. When the cat is infused with increasing concentrations of Ch into the right femoral vein, an increased amount of ACh appears in the SCG.[122] However, direct perfusion of the SCG with increasing Ch concentrations in a Krebs solution, even over physiological levels of 10^{-5} M, increased the tissue Ch levels but not the level of ACh. In similarly Krebs-solution-perfused ganglia, there also is no increased release of ACh on stimulation. If, however, the Ch was perfused in cat plasma, there was not only increased tissue Ch but also increased ACh. When these perfused ganglia were stimulated, a transient release of increased ACh occurred, although the major portion of the newly acquired ACh was not released. The authors concluded that increased Ch made available to the SCG does not have a sustained effect on the turnover of releasable ACh. We are left with the phenomenon that in the SCG as in the CNS, Ch uptake is perhaps facilitated by plasma factor(s), but the Ch converted to ACh does not become incorporated into the functional pool. (Could part of this choline uptake in Krebs-perfused SCG be an osmotic effect?)

Studies dealing with the origin of the acetate moiety of ACh[123,124] in cat SCG, as in the CNS, left another series of unanswered questions. Acetate perfused in the SCG does enter the ACh pool and is found to be released in a Ca^{2+}-dependent manner on preganglionic stimulation. There is a reduced amount of acetate incorporated during stimulation, but when SCG is perfused with glucose and stimulated, there is an enhanced labeling of ACh. The only real increased incorporation of acetate is seen in ganglia previously stimulated at 20 Hz for 60 min and rested for 15 min. The authors concluded on the basis of sound arguments that during stimulation there is an increased delivery of acetyl-CoA to ChAT, but the source of the acetyl moiety is not from acetate (see Section 7).

4.3.4. Adrenergic Receptors

The autonomic neurons are, as previously stated, all adrenergic, at least originally. That NGF also induces an increased amount of TH activity in cultured and *in vivo* rat SCG[11,58,59] has made this tissue the one of choice for studying the kinetics and compartmentation of the catecholamine system relative to function and drug effects. The rates of dopamine and DOPAC formation in the SCG have been measured[125] (see Table IV). With [^{3}H]dihydroergocriptine as the receptor binder,[126] it has been demonstrated that decentralization of the rat SCG lowers the number of binding sites. Approximately half of the sites are located on the cholinergic preganglionic axon terminals, and the remainder on the neuronal cell membranes. The dissociation constant (K_D) is 14.6 nM for this single class of receptor binding sites, with a density of 1074 fmol/mg protein and a half-life of dissociation of 3.7 min (Table III). The dihydroergocriptine binding is inhibited by epinephrine (E), which is four times as effective as norepinephrine (NE) or clondine with K_Is of 4×10^{-7}, 1.6×10^{-7}, and 1.5×10^{-7} M, respectively. The NE released from the dendritic ends acts as a negative feedback control on ACh release from the preganglionic fibers.[127]

4.3.5. *Tyrosine Hydroxylase*

Stimulation of the preganglionic fibers has been shown to lead to an increase in TH activity, but this occurs 3 days after stimulation without a detectable increase in protein synthesis.[128] This increase is frequency and duration dependent and is blocked by hexamethonium. Orthodromic stimulation of the SCG internal or external carotid nerve produces no change in activity.

The regulation of the rate of tyrosine hydroxylation is both a long- and short-term process.[129] The effects of various stimulants in various tissues (cells) make the short-term increase in TH activity appear to be mediated by an induced covalent event. Such increases either occur through kinetic changes in the enzyme or are accounted for by an increase affinity for the cofactor, not the substrate,[130] which does not reflect increased inhibition by catecholamines. The finding that dibutyryl cyclic AMP (dbcAMP) also increases tyrosine hydroxylation[131] suggests that there may be a regulatory phosphorylation of this enzyme mediated via a cyclic AMP- protein kinase system in response to Ca^{2+}–K^{+} fluxes during electrical stimulation.[132]

To measure these parameters,[129] rat SCG were incubated with carbachol, and the rapid increase in TH activity *in situ* was studied. Such increases are found in both neonatal and adult rat SCG or in animals pretreated with reserpine. The increase is Ca^{2+} dependent and is additive to the stimulation produced by dbcAMP. It is not dependent on the rate of substrate transport. This receptor-mediated response does not increase DBH activity. The latter enzyme is mainly concerned with the vesicular storage process.[133] The carbachol-induced stimulation is inhibited by atropine or by low concentrations of phenoxybenzamine or haloperidol. The authors concluded that the receptor-mediated response is evoked by the muscarinic receptors on interneurons rather than the muscarinic receptors on ganglion neurons. Readers should be aware that the existence of ganglionic interneurons is still an unclarified issue. Arguments for the presence of interneurons can be found in several papers[134–137] and involve an explanation for the generation of slow inhibitory postsynaptic potential (sIPSP). Counterarguments based on the variability of SIF cells—possible interneurons—in different ganglia and species variance of SIF cells and the types of catecholamines they contain are presented by a number of investigators.[13,138,139]

In the rat SCG[140] decentralization leads to an inhibition in TH and DBH activities. In target organs, such as iris, there also is a reduction in both TH activity and the number of nerve terminals. The DBH activity of the iris is unaffected, but its storage capacity is increased. When the irises from rats that had their SCG decentralized 4 weeks previously are incubated in 100 mM K^{+}, an increased release of NE occurs. When pretreated with reserpine 24 h before incubation, similarly treated rat irises take up the same amount of NE but cannot put it into vesicular stores. Treatment with pargyline, MAO blocking agent, showed that the NE taken up was cytoplasmic, not vesiculated. It would appear that decentralization allows a greater vesicular uptake of NE, and this excess is released on potassium depolarization.

5. OTHER TRANSMITTERS

5.1. γ-Aminobutyric Acid

In our studies[141] on amino acids and amines of the rat SCG, we were unable to find credible amounts of γ-aminobutyric acid (GABA). However, Bowery *et al.*[142] later showed that it is present in the mammalian SCG and in frog SG.[143] The GABA in the SCG is found in the glial cells. Its release is nonexocytotic since the release is neither stimulation dependent nor calcium augmented or controlled and is induced by potassium only at concentrations above 70mM.[144] Bowery *et al.* critically analyzed their data and concluded that autonomic glial cells remove excess GABA from the interstitial space. The release of this GABA is an obligatory leakage as a consequence of the high intraglial concentration created by a carrier-mediated uptake mechanism.

5.2. Substance P

Substance P is found and is a putative neurotransmitter in the autonomic nervous system.[145–147] If the carotid sinus nerve is decentralized from the submaxillary gland or the gland deafferentiated, no effect is seen on substance P levels in the rat SCG.[147] If the SCG preganglionic fibers are sectioned, an increased amount of substance P-like material is seen in the ganglion.[149,150] Treatment of the rat with chlorisondamine (nicotinic blocker) causes an increase, phenoxybenzamine (α-receptor blocker) results in a decrease, and axotomy produces no change. Treatment of neonatal rats with 6-hydroxydopamine (6-OH-DA) results in decreased levels of substance P in the adult ganglion. Phentolamine and reserpine have no effect on substance P levels. Stimulation produces a reduction of substance P, which the authors concluded occurs by a transsynaptic process.

5.3. Enkephalins

Immunologic assessments of the enkephalins in rat and guinea pig ganglia have been made.[151] The guinea pig intramural ganglia show the greatest density, followed by a moderate amount in the celiac and superior mesenteric ganglia, with very little in the SCG. Similar immunoreactive densities are seen in rat ganglia. A quantitative assessment between the celiac ganglion and SCG of spontaneously hypertensive (SHR) Kyoto and Wistar rats at 4 and 12 weeks of age has been reported[152] (Table II). It would appear that the enkephalins are not made in the ganglia but are transported into the ganglia.

5.4. LHRH-like Peptides

One cannot leave this area of neurotransmitters and receptors without mentioning the recent work on the frog sympathetic ganglion[153] and the demonstration that a peptide found in the preganglionic nerve terminals similar to lutenizing hormone-releasing hormone (LHRH) may be the mediator of the late

Table II
Constituents of the Rat Superior Cervical Ganglia

Constituent	Quantity	Reference
Water space	809.8 μg/mg wet	223
Urea	0.83 ± 0.001 ml/g dry	259
Mannitol	0.39 ± 0.003 ml/g dry	259
Inulin	0.308 ± 0.004 ml/g dry	259
Volume	0.6 mm^3	259, 260
Synapses	8.78 × $10^6/mm^3$	261
K^+	0.489 μmol/mg dry	262, 263
Protein	154.6 μg/mg wet	223
TCA-insoluble	91.7 μg/mg wet	184
RNA	7.1 μg/mg wet	223
DNA	0.6398 μg/mg wet	223
Glycogen	21.9 pg/mg wet	223
Glucose	6.03 nmol/mg wet	182
Glucose-6-phosphate	1.057 nmol/mg wet	223
6-P-Gluconate	36 pmol/mg wet	223
Fructose	59 pmol/mg wet	223
Pyruvate	0.845 ± 0.170 nmol/mg wet	179
Lactate	2.070 nmol/mg wet	223
α-Ketoglutarate	0.192 ± 0.028 nmol/mg wet	179
NAD^+	200 pmol/mg wet	223
NADH	15 pmol/mg wet	223
NAD^+/NADH	1.249 ± 0.233 nmol/mg wet	179
ATP	1.48 nmol/mg wet	182
Phosphocreatine	1.90 nmol/mg wet	182
Cyclic AMP	4.67 ± 0.45 pmol/ganglion (rat)	250
	45.1 ± 0.6 pmol/ganglion (rabbit)	250
	42.7 ± 2.5 pmol/ganglion (guinea pig)	250
	33.2 ± 3.5 pmol/ganglion (calf)	250
Cyclic AMP (neuronal)	4 pmol/mg wet	264
Cyclic AMP (glial)	14 pmol/mg wet	264
Cyclic GMP	0.9 ± 0.01 pmol/mg wet	264
PGE	23.3 ± 4.5 pg/ganglion	250
Acetylcholine	0.0214 nmol/mg wet	223
Epinephrine	1.6 nmol/mg wet	46
Norepinephrine	0.267 nmol/mg wet	265
Dopamine	0.023 nmol/mg wet	265
DOPAC	0.030678 nmol/mg wet	265
GABA	47.77 nmol/mg wet	265
Putrescine	1.52 ± 0.16 nmoles/ng protein neonate	165
Spermine	11.21 ± 0.96 nmoles/ng protein neonate	165
Spermidine	8.69 ± 0.93 nmoles/ng protein neonate	165
Substance P values		
SCG	70 ± 8 fmol/mg protein	267
Middle inferior	84 ± 14 fmol/mg protein	267
Thoracic	269 ± 38 fmol/mg protein	267
Celiac–superior mesenteric	508 ± 22 fmol/mg protein	267
P_1 Protein	1,040 fmol/mg protein	259
Enkephalin (4 wk; 12 wk)		
Wistar Celiac	3.4 ± 0.37; 1.7 ± 0.14 ng/mg protein	152
ME SCG	1.9 ± 0.12; 1.4 ± 0.081 ng/mg protein	152
SHR Celiac	2.0 ± 0.16; 0.93 ± 0.18 ng/mg protein	152
SCG	119 ± 0.059; 0.93 ± 0.18 ng/mg protein	152
Wistar Celiac	1.8 ± 0.097 ng/mg protein	152
LE		
SHR Celiac	1.1 ± 0.048 ng/mg protein	152

slow excitatory postsynaptic potentials (EPSP). The frog ganglion has two cell types, B and C, each having common and specific afferents, making direct one-to-one dendritic synapses. Pressure ejection of LHRH near the membrane of either cell type elicits only in that cell the late slow EPSP in a calcium-dependent manner and with an increase in membrane resistance. Analysis has shown this membrane depolarization not to be related to the K^+ equilibrium potential. The authors further demonstrated that this LHRH effect is not similar to that of substance P. By appropriate afferent sectioning and immunohistochemical analysis, the LHRH-like peptide was found only to be present in the preganglionic terminals on the C cells. They postulate that the peptide is released from the presynaptics onto the C cells, eliciting the late slow EPSP, and then migrates to the B cells to produce a similar but later response. Besides offering an explanation for the generation of the late slow EPSP, this finding raises the possibility that many circulating peptides, not heretofore considered, may act on neurons to modify their function and metabolism.[154,155]

Neither glutamate, aspartate, taurine, cysteic acid, nor glycine functions as a putative transmitter in the rat SCG.

6. DESTRUCTION OF SYMPATHETIC GANGLIA

Aside from surgical axotomy and deafferentation, chemical procedures can ablate the sympathetics. As mentioned, either total or selective destruction of the SG and postganglionic fibers can be achieved by using NGF antiserum. This destruction is ontogenic and dose dependent.[44,47,156] Administration of 6-OH-DA can also cause destruction of the sympathetic neuron.[157,159] Again, a dose-dependent relationship exists, and the procedure is best performed on neonates. (See the αBTX binding study in ref. 117, which compares ACh binding following these various ablations.)

Another interesting antisympathetic agent is guanethidine.[160–163] Guanethidine has an inhibition constant of 25 μM for neonatal rat SCG S-adenosylmethionine decarboxylase but does not affect methyltransferases or ornithine decarboxylase.[164,165] It has no influence on the level of polyamines in the neonatal rat SCG. (Table II) Therefore, its toxicity does not derive from inhibition of polyamine synthesis. It is also interesting to note that the sympathetic toxicity of guanethidine can be blocked by the administration of NGF[166] and by immunosuppressive agents.[167] These cytotoxic agents, besides being of research interest, have considerable therapeutic interest for humans if more selective agents could be developed that would produce effects at lower doses.

7. METABOLISM

The metabolism of the autonomic ganglia has many similarities to that of the CNS. The major differences are associated with, as discussed in the preceding sections, the homogeneity of cell types, the limited number of known functional transmitters, and the possible absence of the tightly coupled capillary

Table III
Receptors in Sympathetic Ganglia

Receptor	K_D	Association rate	Dissociation rate	Binding sites
Nerve growth factor[54]				
Type I	$3.3 \pm 1.9 \times 10^{-11}$ M	$1.0 \pm 0.3 \times 10^{7}$ $M^{-1}s^{-1}$	6.5×10^{-4} s	$4.0 \pm 0.5 \times 10^{3}$/cell
Type II	$1.7 \pm 0.4 \times 10^{-9}$ M	1.0×10^{7} $M^{-1}s^{-1}$	3.2×10^{-2} s	$4.7 \pm 0.5 \times 10^{4}$/cell
Nicotinic (α-bungarotoxin)				
Ciliary ganglia[115]	2.7 nM			3×10^{6} molecules/cell
Iris[115]	2.5 nM			
SCG[117]		1.587×10^{5} $M^{-1}s^{-1}$ (K_{on})	9.24×10^{-6} $M^{-1}s^{-1}$ (K_{off})	$3.43 \pm 0.68 \times 10^{10}$/SCG 0.166×10^{-6} AChR/synapse
Chick embryo[116]				0.75 ± 0.7 fmol/explant 4.51×10^{8} AChR/explant
Muscarinic (quinuclidinyl benzilate)[277]				
SCG	0.83 ± 0.13 nM			64 ± 11 fmol/ganglion 580 pmol/g protein
α-Adrenergic[126]				
Dihydroergocriptine	14.6 nM		3.7 min ($t_{1/2}$)	1074 fmol/mg protein
Epinephrine	4×10^{-7} M (K_I)			
Norepinephrine	1.6×10^{-6} M			
Clonidine	1.5×10^{-6} M			
Glucocorticoid[278]				1750 fmol/mg protein

Table IV
Activity of Enzymes in Rat Superior Cervical Ganglion

Enzyme	Activity (nmol/mg protein per min)	Reference
Acyl-CoA synthetase	1.33	268
Acyl-CoA hydrolase	27.	268
Acyl-CoA lysophosphatidylcholine acyltransferase	3.09	268
Dihydropteridine reductase	208 ± 10	269
Transketolase	4.9 ± 0.3	269
Glucose-6-phosphate dehydrogenase	16.7 ± 0.2	270
Hexokinase	37.7 ± 3.3	270
Isocitrate dehydrogenase	25.1 ± 1.5	270
Malic enzyme	4.8 ± 0.4	270
NADPH–Cytochrome *c* reductase	5.8	270
Phosphofructokinase	27.6 ± 2.4	270
6-Phosphogluconate dehydrogenase	9.9 ± 0.4	270
Pyruvate kinase	66.8 ± 6.0	270
Prostaglandin E synthetase	0.154	249
Ribose-5-phosphate reductase	21.5 ± 10	270
	(nmol/mg Fresh tissue per min)	
Glutamic–oxaloacetic transaminase	20.5 ± 2.40	271
Lactate dehydrogenase	51.5 ± 8.0	271
RNAase polymerase I	$0.6 \rightarrow 1.18 \times 10^{-3}$	222
RNAase polymerase II	$0.57 \rightarrow 1.16 \times 10^{-3}$	222
Succinic dehydrogenase	10.2 ± 2.51	271
	(μmol Substrate hydrolyzed/g wet tissue per min)	
Cholineesterase	28.8 ± 3.3	45
Acetylcholinesterase	18.5 ± 2.1	45
Pseudocholinesterase	9.6 ± 1.2	45
	(mol Product formed/ganglion per hr)	
Choline kinase (newborn)	0.009	109
Choline kinase (adult)	0.038	109
Choline acetyltransferase	250	110
Catechol-O-methyltransferase	3.71 ± 0.63	272
DOPA decarboxylase	0.175 ± 0.015	46
Monamine oxidase	5.6 ± 0.19	273
Ornithine decarboxylase	$0.63 \pm 0.25 \times 10^{-3}$	276
Phenylethanolamine-N-methyltransferase	2.8×10^{-3} (newborn)	274
	0.29×10^{-3} (adult)	274
Protein carboxymethylase	$3.3 \pm 0.3 \times 10^{-6}$ (newborn)	275
	$28.3 \pm 3.0 \times 10^{-6}$ (adult)	275
Tyrosine hydroxylase	14.4 ± 0.7 (stimulated)	128
	10.0 ± 0.8 (contralateral)	128

epithelium in autonomic ganglia. The neuronal mitochondria, as in the CNS, are distributed mainly in the dendritic and presynaptic endings, and there are similar metabolic interactions between the neurons and glial cells. What makes the metabolism of any neuronal system interesting are those metabolic events that can be associated with neuronal function, the action potential. Autonomic neurons display the same critical need as those in the CNS for glucose and oxygen. Most metabolic studies have been performed with the adult rat SCG because of its ease of surgical removal to *in vitro* electrode systems and its long physiological survival in such systems. Only recently, as an outgrowth of tissue culture work, are basic metabolic questions being asked of embryonic or neonatal ganglia.

7.1. Oxygen

The first functional metabolism studies were concerned with the critical oxygen and glucose needs of stimulated SCG.[169] This paper and the ones reviewed in it contain a wealth of information concerning stimulation frequencies and oxygen consumption[170] and some of the basic work on how anesthetics relate to these variables.[171,172] Dependence on oxygen of cat and rabbit SCG was also shown, like the rat's, to be related to the mitochondrial cytochrome system, with the rate of oxygen consumption related to frequency of preganglionic stimulation. The SCG oxygen consumption is an increasing function of frequency and reaches a maximum value, in the rat, at 15 Hz. However, the amount of oxygen consumed for each species' SG per preganglionic stimulus declines as the frequency of stimulation increases. A similar relationship is seen in myelinated fibers.[173] The oxygen uptake *in vitro* of desheathed rat SCG can be measured in stirred bathing solution because its small cross-sectional diameter is sufficient for oxygen diffusion to its core[169,170]; cat, dog, and rabbit SCG, being of larger cross section, must be perfused or, like calf ganglia, sliced, as they easily become anoxic on *in vitro* incubation. In rat SCG there is a delay of about 280 ms[174] between the rise of a single action potential and the activation of the mitochondrial cytochromes. Undoubtedly, this delay and the diminished oxygen consumption at frequencies greater than 15 Hz is coupled to the delays in ionic pumps and intracellular metabolic programming as the ganglion switches in and out of physiological work.

Neuronal tissue has a critical need for oxygen, and its mitochondria are more easily damaged by ischemia than are the mitochondria of other tissues.[175–179] The SCG is similarly affected by ischemia. When SCG are perfused with a deoxygenated medium or when the oxygenated fluid flow is stopped, the resting action potential begins to fail after 30 min. However, if oxygenated perfusion is immediately reinstituted, the action potential loss can be reversed.[169,171]

As with the CNS sensitivity to anoxia, this critical biochemical need by ganglia for oxygen has been a focus of investigation. One of the earlier observations dealt with the need to keep the *in vitro* oxygenated media flowing past the SCG[170,178] to prevent formation of any anoxic cores. Recently,[179] an assessment of the NAD/NADH ratio was made in whole rat SCG versus sliced SCG in static cultures. It was found that within 3 h in intact SCG this ratio had dropped to half its initial value and did not recover over the next 21 h. The SCG slices were able to maintain their initial normal ratio. This again shows that one cannot depend on simple diffusion of oxygen to maintain the required redox state. Another possible explanation is that movement of the medium around the whole ganglia, besides increasing oxygen diffusion, prevents materials from accumulating in the interstitial space of the intact ganglia.

7.2. Glucose

Concomitant with the increased oxygen consumption of stimulated SCG are increased glucose consumption and lactate and CO_2 production.[180] With increased work—ionic pumps—a greater energy expenditure is required. The questions to ask are which biochemical event might be rate limiting: oxygen consumption, glucose uptake or utilization, NAD/NADH ratio, ATP, phosphocreatine (PCr) levels, or some critical secondary metabolite (cofactor) not directly considered to participate in glycolysis or oxidative decarboxylation. The glucose concentration of normal freshly excised desheathed SCG is 30 nmol/mg dry weight[181] or 6 nmol/mg wet weight.[182] With [U-^{14}C]D-glucose, a resting steady-state *in vitro* SCG glucose concentration is achieved in 30 min.[183] In resting ganglia, the glucose carbons are partitioned into CO_2 (64%), lactate (24%), lipids, and amino acids (approximately 12%), with less than 1% as volatile components. Within this initial 30-min period, measurable amounts of amino acids, amines, nucleotides, and lactate efflux from the ganglia[141,168,184] (P. Cancalon and J. D. Klingman, unpublished data). To achieve a steady-state level of labeled carbons from glucose in the internal pools requires 45 min for CO_2, 30 min for lactate, and some 10 h for the total lipid pool.[185] Such partitioning of glucose carbons must also take into account the external pool of materials that have effluxed from the ganglion and determine the extent to which such materials may be reutilized, especially during periods of stimulation or in the critical few minutes of recovery following a period of stimulation.

7.3. Glycogen

The SCG does contain a store of glycogen.[186] When the SCG is incubated in increasing concentrations of glucose, there is an increased synthesis of glycogen.[187] In resting ganglia *in vitro*, the maximum glycogen store is achieved after 16 h of incubation. Although cortisol and glucagon have no *in vitro* effect on increasing glycogen synthesis, insulin, vasopressin, and oxytocin accelerate the rate of synthesis. Neither glucagon nor epinephrine affects the rate of glycogen utilization. When made anoxic, the total SCG glycogen pool is seen to diminish, and no new glycogen is synthesized. If deprived of glucose, the SCG glycogen store is quickly depleted. However, restoration of glucose to the medium causes a redepositing of glycogen in the neurons. It is interesting that in rested SCG the glycogen pool can be best maintained by fructose, among the other hexoses, but at a 50-fold higher concentration than the requisite glucose. Similarly, lactate can support the glycogen pool, but only when supplemented with glucose. The glucose supplement needed, however, is only one-tenth of the required basal concentration.

The basic need of these highly ordered systems which require both a continuous supply of glucose and oxygen is the critical need for energy production and/or the maintenance of a redox potential. Freshly excised and frozen ganglia contain 1.5 and 1.9 nmol/mg wet weight of ATP and PCr, respectively.[182,188] In an oxygenated glucose Krebs–Ringer's solution, a resting SCG that is intermittently stimulated physiologically maintains its transsynaptic action potential for 36 h. The ATP levels are found to decrease in 12 h to about one-third the initial values[189]; glycogen falls as well, but at a much slower rate. If a glucose-deficient medium is added, the glycogen store is depleted in less than 2 h with a rapid loss of the internal glucose such that by 2 h glycogen is barely detectable, and the action potential has begun to fail.[188,189] The ATP and PCr levels in glucose-deprived SCG at this 2-h period are still at more than 50% of their initial values. Anoxia along with glucose deprivation produces a loss of ATP and PCr to barely detectable levels in 15 min.[188] The failure of the action potentials at 30 min in anoxic resting SCG, whether perfused or *in vitro*,[169,190] is reversible if the ganglia are immediately presented with an oxygenated medium.

When SCG are incubated *in vitro* with lithium-containing Krebs–Ringer's solution formed by substituting lithium for equivalent sodium ions (acute SCG), only 10% and 30% of the initial ATP and PCr remain after 20 min.[182] If, however, weanling rats are raised on a lithium-containing diet to achieve blood lithium levels of 0.5 mEq/liter (chronic SCG), their freshly excised SCG have only 42% and 75% of their normal ATP and PCr levels, respectively. When chronic SCG are incubated in normal-glucose Krebs–Ringer's, regardless of whether the chronic SCG is stimulated or rested, the levels tend to increase to almost normal values but are not maintained and, by 30 min, decline below those of control SCG. The ATP and PCr levels in the SCG of chronic lithium-fed rats, however, do not achieve the very low levels found in the acute lithium-treated SCG. These altered energy states seen in the lithium-treated ganglia may partially explain the altered transsynaptic action potentials.[191,192]

8. *STIMULATION; METABOLISM*

8.1. *Glucose*

In the *in vitro* SCG subjected to continuous preganglionic stimulation, there are an increased uptake of glucose and oxygen and release of CO_2, an accompanying increased concentration of lactate, both internally and externally,[170,193,194] and a loss of glycogen.[187] Deviation from the normal partitioning of glucose carbons occurs for both the internal and external metabolic pools.[182–184,194] Even though there is an increased distribution of glucose carbons, the increased molar glucose uptake during stimulation can account for all of the increased molar CO_2 that is released. As the sole carbon source, glucose can maintain the SCG's physiological response for several hours.[189,196]

It has been ascertained that if the ganglion's initial pool of glucose[181] is maintained over a 3-h period of stimulation, some 30% of the increased CO_2 and about 40% of the lactate carbons are not derived directly from the medium's [U-^{14}C]glucose. A portion of the glucose carbons are also effluxing into the bathing solution at an increased rate during stimultion. Thus, some other pool(s) of oxidizable substrates must be used during stimulation to account for these increased steady-state levels of CO_2 and lactate. The extent to which the hexose monophosphate shunt (HMP) may be of importance is discussed in Section 8.4. It should also be emphasized that much still needs to be done beyond this initial work relative to the reutilization of possible substrates elaborated into the medium.[184,194,197]

8.2. Glucose Lack

Before we discuss several of the above events, it is interesting to consider what happens in SCG incubated in media devoid of glucose. The effects of stimulation in a glucose-free medium were first described by Larrabee and Bronk.[169] Within 30 min of glucose withdrawal, the transsynaptic action potential fails. If the SCG is not immediately returned to a glucose medium, complete irreversible failure of the transsynaptic action potential occurs in 2–3 h.[194] The pre- and postganglionic fibers are still functional when stimulated for 10 and 30 h, respectively.

Marked morphological alterations in the glucose-deprived SCG are seen histologically or by electron microscopy.[198] The ganglionic neurons have dispersed ribosomes, dissociated polysomes, elongated mitochondria, and enlarged Golgi. The presynaptic area is most severely damaged, swollen and with a loss of fibrillar material, and the mitochondria are few, and those seen are vesicated and swollen.[187] Likewise, the presynaptic area does not regain its glycogen after glucose lack. In the presynaptic area, vesicles, cytolysosomes, and amorphous masses are typically seen as in myelin degeneration.[199] Few synaptic vesicles or synapses are present in the most extensively damaged area. If normal SCG are incubated for 18 h at 6°C without glucose and then returned to a glucose-containing Krebs–Ringer's medium and electrically stimulated, the action potential can be elicited. The lowered temperature maintains the presynaptic integrity.

The glucose-lack presynaptic damage is not solely an osmotic damage, since SCG bathed in electrolytic or polyvinylpyrrolidone osmotic equivalent media still show the presynaptic lesions.[199] When glucose-deprived SCG are tested with physostigmine, the addition of acetylcholine still produces a postaxonal potential. The inability of a transsynaptic action potential to be elicited by glucose-deprived ganglia must result from irreversible damage in the presynaptic area. The only chemical alteration noted at the time of failure is a slight increase in nitrogenous material, but there is no change in oxygen consumption.[180] One can speculate about why the presynaptic mitochondria are more sensitive to glucose lack than those of the dendritic areas. This cannot merely be caused by a difference in the initial amount of glycogen stores but more likely reflects either limited alternate metabolic pathways available to sustain the redox potential or the extreme sensitivity of the coupling in presynaptic mitochondria.

8.3. Alternate Substrates

Ever since Kahlson and MacIntosh[120] perfused the cat SCG *in situ* with non-glucose-containing media, producing a loss of the nictitating membrane response, possible alternate metabolites have been sought. Although complete recovery was never attained, only mannose, lactate, or pyruvate maintained the nictitating membrane response and the continued synthesis of acetylcholine. *In vitro* rested rat SCG maintain their action potentials if lactate and glutamine, but not glutamate or PCr, replaced glucose.[180] A more complete investigation[196] has determined that as sole substrates only mannose, fructose, lactate, or pyruvate would allow an 80% or greater survival time of rat SCG when compared to glucose. Oxaloacetate alone among the TCA intermediates gives the next best survival time. Amino acids by themselves are very poor substrates, but some can be combined with α-ketoglutarate to give survival times of from 50 to 72% of normal. Our experience has shown that acetate is also a very poor substrate (G. D. DeVincenzo and J. D. Klingman, unpublished data, 1968; J. D. Klingman, unpublished data, 1965; L. Opler and J. D. Klingman, unpublished data, 1967).

These alternate substrate studies have looked at the survival of the SCG physiological response—the action potential. One may ask what alterations in metabolic pathways and compartmentations may be taking place, i.e., what critical intermediates must be biochemically sustained. Such questions have not been thoroughly investigated. We have found in our work[199,200] that pyruvate can sustain the rat SCG at 6 Hz for 2 h. The partitioning of the pyruvate carbons is completely different from that of glucose. Glucosamine[141,201] (J. D. Klingman, unpublished data, 1965), when substituted for glucose, produces an immediate loss of the transsynaptic potential in less than 1 min of stimulation.

8.4. CO_2 and Hexose Monophosphate Shunt

As stated, it would appear that in the adult SCG some 30% of the elaborated CO_2 is derived from glucose carbons.[185] While studying the effects of NGF on chick embryo SG, Larrabee[202,203] reported that in culture the [1-^{14}C]glucose-derived CO_2 reached a maximum labeling at 10–12 days of embryonic age whereas that from [6-^{14}C]glucose showed only a slight increase, contributing but 20% to the total CO_2 output. The specific activity of lactate was 10- to 15-fold higher at day 7 in similar labeled glucose experiments, falling rapidly by day 12 for [1-^{14}C]glucose and by day 14 for [6-^{14}C]glucose. He reasoned that this demonstrated a turning on of the HMP. The intriguing aspect of these findings is the differential labeling seen in the lactate and CO_2. In the embryonic chick DRG,[204,205] a similar delayed labeling of released CO_2 is seen when the specifically labeled glucoses are used. When [2-^{14}C]glucose is used, the final maximum labeling of CO_2 is delayed similarly to that for [6-^{14}C]glucose.

From these data and the finding that there is sufficient fructose-1,6-diphosphate to allow for recycling of the HMP-generated intermediates,[206] a two-compartment model system has been developed.[205] This two-compartment model system is dependent on one of the compartments having an associated pool of HMP intermediates. It has been determined that pyruvate carboxylase does not act as an important regulator between these pools.[204] Further experiments[206] with DRG have attempted to clarify the various internal and external pools that are generated by [1-^{14}C]glucose and [6-^{14}C]glucose. The model generated presumed there are two cell types.

In one cell type (P), all of the metabolized glucose-6-P is diverted to the HMP, a negligible amount of glucose carbon is incorporated into tissue constituents with slow turnover rates, and, other than CO_2, negligible amounts of material are released into the medium. The P cell is, more or less, a classical cell with glycolysis, HMP, and pyruvate formation; the latter enters a tightly coupled TCA system, with the bulk of CO_2 generated from the 1-C of glucose. The other cell (D) has no HMP, and the glucose is directly glycolyzed to pyruvate, which forms and releases lactate, while the derived acetyl-CoA enters both fatty acids and the TCA cycle. In the D cell, a release of glutamate and alanine from the TCA cycle occurs. The delay in CO_2 output from the 6-C of glucose in P cells is explained by the presence of a pool of intermediates within the TCA cycle or of products that can exchange with it, whereas in D cells this pool is located prior to the TCA cycle to yield mainly acetyl-CoA reactions. It is assumed that in the D cells there are significant rates of incorporation into the slowly equilbrating constituents, such as lipids, amino acids, and nucleotides. This is insignificant in the P cells, which apparently have a very low steady-state level of HMP intermediates. A second assumption is that there is an irreversible loss of materials from the D cells to the medium, which is not true for the P cells.

These models predict that elaborated amino acids are not readily reused *in vitro* and that only some 20% of any generated glyceraldehyde-3-phosphate produced by the HMP cycle is converted into fructose-6-phosphate with a suggested 60% recycling. One would, of course, be interested in knowing which of these model cells (P or D) best fits or describes the glucose metabolism of the ganglionic neuronal and nonneuronal cells. In studies of intact 15-day chick embryo SCG and those dissociated into neuronal and nonneuronal cells, several findings were made.[197] All three systems put out less CO_2 from the 6-C than from the 1-C of glucose. Thus, both cell types as well as the intact ganglion have active HMP systems. The 1-C/6-C ratio is greater for intact ganglion and neurons than for nonneuronal cells, indicating that perhaps the neurons have less HMP activity than the nonneurons or that the neurons' HMP intermediates recycle faster. Thus, it would appear that no one cell is devoid of the HMP system. These thoughtful papers raise several issues, one of which is the demonstrated importance of tissue-to-culture volume ratio as having an effect on the data obtained. A second is the obvious synergism seen in the whole ganglia between the two cell types, which is lost or uncontrolled in isolated neuronal and nonneuronal cultures. An important experiment will be to determine what controls the redox potentials of the two SG cell types and their metabolic synergism. For those interested in neuronal cell culture and/or metabolic model systems, this is a prime series of papers to read.

8.5. Amino Acids

The free amino acid pool in the rat SCG is equivalent to about 15 nmol/mg wet weight, some twofold greater than the free glucose pool.[184] The distribution of these amino acids between the cell types or between the pre- and postsynaptic regions is not known. Interest in amino acids and their utilization in the resting and excited states of ganglionic metabolism has importance for understanding a number of issues. Aside from their involvement with transmitter synthesis, the amino acids—as in other tissues—interact with carbohydrate metabolism in a unique manner. Historically, neuronal amino acid metabolism has been centered around glutamate–glutamine systems and the compartmentation occurring in the CNS. Little has been done, however, in determining how amino acids may be used, or their use curtailed, during or after stimulation of a synaptic system such as the SG.

Our interest in the amino acids began with the initial glucose carbon partitioning studies on rat SCG,[178] when labeled amino acids were found to be present in the bathing medium. Subsequently, the intraganglionic free amino acids of freshly excised sterile SCG and the bathing solutions from incubated SCG were analyzed for amino acids and amines.[184,207–210] As one would expect, the fresh SCG contains predominantly glutamate(amine), glycine, aspartate(amine), serine, and alanine, in that order, followed by smaller amounts of the remaining amino acids (Table V). Under sterile conditions, even in rested ganglia, there is leakage of amino acids[184,208] (P. Cancalon and J. D. Klingman, unpublished data) into the medium such that by 1 h of incubation the extraganglionic pool of amino acids is almost equivalent to the free intraganglionic pool.[141] The free intraganglionic amino acid pool concentrations increase with time. By 30 min, significant increases in alanine and serine have occurred along with increases in several of the essential amino acids.

In attempts to inhibit the elaboration of amino acids into the medium as well as to inhibit the intraganglionic pool increase, amino acids at concentrations present in rat serum have been added to the bathing medium, but without any noticeable effect in preventing these changes. The only significant increases in the intraganglionic pool have occurred for glutamate, aspartate, and lysine. This suggests that there is proteolysis caused by the bathing medium imbalance and/or the effect of axotomy. Some of the more pertinent questions concerning amino acid metabolism include the rates at which amino acids are produced during or after stimulation, their entry into or release from proteins, their ionic dependence for uptake, their compartmentalization, and any unique metabolic uses in support of the action potential. It must be kept in mind that, as previously mentioned, many of the desired *in vitro* amino acid studies cannot be easily performed since glucose must always be present in the bathing media.

8.5.1. Glutamate

Stimulation has been shown to increase the rate of entry of glucose carbons into alanine, glutamate, and aspartate.[184,210] McBride and Klingman[184] reported

Table V
Free Intraganglionic Amino Acids in Rat Superior Cervical Ganglia[a]

Amino acid		nmol/Wet weight of ganglia	
		Fresh tissue[b]	0.5 h[c] incubation
Alanine	(*R*)	1.69 ± 0.004	1.9 ± 0.1
	(*S*)		2.9 ± 0.3
Serine	(*R*)	1.90 ± 0.023	2.0 ± 0.3
	(*S*)		1.2 ± 0.3
Glycine	(*R*)	2.30 ± 0.018	3.1 ± 0.3
	(*S*)		3.9 ± 0.5
Aspartate (asparagine)	(*R*)	2.06 ± 0.046	2.3 ± 0.2
	(*S*)		3.3 ± 0.7
Glutamate (glutamine)	(*R*)	4.09 ± 0.044	4.2 ± 0.4
	(*S*)		3.4 ± 0.7
Proline	(*R*)	0.50 ± 0.012	0.56 ± 0.05
	(*S*)		0.60 ± 0.09
Ornithine	(*R*)	0.28 ± 0.048	0.20 ± 0.08
	(*S*)		0.38 ± 0.02
Threonine	(*R*)	0.52 ± 0.011	0.81 ± 0.07
	(*S*)		0.82 ± 0.11
Valine	(*R*)	0.13 ± 0.032	0.50 ± 0.06
	(*S*)		0.41 ± 0.07
Leucine	(*R*)	0.21 ± 0.008	0.60 ± 0.09
	(*S*)		0.44 ± 0.06
Isoleucine	(*R*)	0.10 ± 0.009	0.42 ± 0.08
	(*S*)		0.43 ± 0.23
Lysine	(*R*)	0.24 ± 0.021	0.11 ± 0.03
	(*S*)		0.29 ± 0.10
Phenylalanine	(*R*)	0.08 ± 0.004	0.23 ± 0.02
	(*S*)		0.31 ± 0.04
Tyrosine	(*R*)	0.03 ± 0.002	
Methionine	(*R*)	0.05 ± 0.002	

[a] The data represent the mean ± S.E.M. of four to six determinations for freshly dissected desheathed ganglia, which were then quick frozen.
[b] P. Cancalon and J. D. Klingman, unpublished data.
[c] Reference 184.

that the specific activity of aspartate and glutamate were significantly increased by stimulation for short incubation times, whereas the incorporation into glycine and serine was not significantly increased by stimulation during the first hours, and by 4 h only serine's specific activity was increased. These four amino acids were the only ones to show a significantly increased specific activity related to stimulation. However, I would caution against overinterpretation of data from *in vitro* rat SCG incubations for longer than 2 h. Beyond this time period, an entirely different labeling pattern is apparent, which may represent the more profound and acute effects of axotomy.

Only glutamate, aspartate, and serine uptake and metabolism have been studied to any extent[168,184,210,211] (P. Cancalon and J. D. Klingman, unpublished data). The accumulation of glutamate is an energy-dependent process.[184,210] The uptake of aspartate and serine is not an active process (P. Cancalon and

J. D. Klingman, unpublished data). It is interesting, however, that more serine is taken up by the SCG than either aspartate or glutamate.

In stimulated rat SCG, the initial 30-min [U-^{14}C]L-glutamate uptake is almost the same as that of rested SCG.[141,184] Yet, an additional 30 min of stimulation produces a significant increase over rested SCG. The amount of label appearing in aspartate is only about one-fifth of that which can be derived from glucose carbons. Addition of amino acids in concentrations equivalent to those in rat serum did not alter the entry of glutamate into stimulated SCG but did increase that in rested SCG such that no difference between the two states existed. The entry of glutamate carbons into aspartate under these conditions shows a more linear rate, but one still well below that of aspartate formation from glucose carbons. It would appear that glutamate usage is increased by stimulation but that very little of its carbon enters the TCA cycle or is transaminated to aspartate. One reason may be the use of other amino acids or lactate present in the medium. After incubation in [U-^{14}C]L-glutamate-containing glucose Krebs–Ringer's medium for 30 min, the specific activities of glutamate, aspartate, and alanine are 1569, 436, and 19 dpm/nmol per mg wet weight, respectively. The conversion of glutamate to aspartate is equivalent to that seen from labeled glucose, whereas that entering alanine is more than 20-fold smaller. Similarly, label from glucose is also found in serine and glycine, but none in these amino acids is derived from glutamate, nor is label found in any other amino acids even over a 2-h period of incubation. Thus, the glutamate taken up can be transaminated to aspartate, but little appears to be converted to cytoplasmic oxaloacetate, from which serine is derived. This observation is a good indicator of why glutamate may not sustain the energy needs of the SCG when used as a glucose replacement.

8.5.2. Aspartate and Serine

The partitioning of aspartic acid carbons[168] (P. Cancalon and J. D. Klingman, unpublished data) is in sharp contrast to that of glutamate or glucose carbons. Like glucose, aspartate supplies carbon for all nonessential amino acids. Significant labeling is found after 5 min of incubation. When the total labeled pools of both extra- and intraganglionic amino acids, into which the aspartate carbons have flowed, are summed, no difference is seen between rested and stimulated SCG. However, if only the free intraganglionic amino acid pool is considered, significantly higher incorporation occurs in the stimulated SCG. The rested-to-stimulated intraganglionic amino acid pool ratio of specific activities is 0.35 compared to 0.73 for the ratio of both pools.

The specific activity of glutamate derived from the [U-^{14}C]L-aspartate is approximately one-half that of the intraganglionic aspartate. The specific activity of alanine is only appreciably higher than that derived from glucose but some 30-fold higher than the specific activity derived from glutamate. Likewise, the specific activity of serine and glycine are five and 40 times, respectiviely, greater than that produced from glucose (remember, none is obtained from glutamate).

The only effect of stimulation on these free intraganglionic amino acids is to curtail the flow of aspartate carbons into glutamate by more than 50%. It would appear that aspartate enters an entirely different pool than does glutamate, in which aspartate is readily transaminated to oxaloacetate and from which serine and glycine can be derived. That portion going to glutamate is also in a pool convertible to proline but not to ornithine. On stimulation, the proline label is readily diluted by glucose carbons, or conversion from glutamate is curtailed. This partitioning is somewhat similar for aspartate utilization by brain slices.[212] The amount and rate of labeling of the extraganglionic amino acids directly reflect both the concentration and the rate of labeling of the intraganglionic amino acids.

Of the three amino acids studied to date, [U-^{14}C]L-serine[168] (P. Cancalon and J. D. Klingman, unpublished data) enters the rat SCG faster than either glutamate or aspartate, the specific activity of the intraganglionic serine being 40-fold higher than that of aspartate. Within 30 min, significant labeling is seen in alanine, glycine, aspartate, glutamate, proline, and ornithine. On stimulation, no labeling of ornithine is seen. The specific activity of glycine is, as expected, almost identical to that of serine, which decreases with stimulation. In the resting state, aspartate has one-tenth the activity of the serine, but when the SCG is stimulated, this falls to about 20% of the value of rested ganglia. The results for glutamate are just the opposite of those of aspartate. In stimulated SCG, the glutamate specific activity is fivefold higher than that in rested ganglia, being about one-fourth of the aspartate activity. Alanine produced from serine carbons in rested SCG has an activity similar to that of glutamate, but on stimulation alanine loses more than 50% of its activity, acting similar to aspartate.

The serine data are an excellent example of the extent to which transamination is taking place, coupled with the utilization of the major metabolite, glucose. The loss of label in the alanine and aspartate on stimulation could be to a large measure a dilution by glucose carbons if it were not for the almost equivalent increase in the specific activity of glutamate in the stimulated SCG. Simplistically, there appears to be a diversion of the carbons into the TCA cycle with glutamate as the nitrogen sink rather than aspartate, as is seen in the rested ganglia. With respect to the utilization of these three amino acids—glutamate, aspartate, and serine—glutamate appears to be taken up into a compartment that has little ability to transaminate and to retain the derived carbons for the subsequent generation of other amino acids. Serine, on the other hand, appears to enter all compartments, because even ornithine derives labeled carbons from it. Again, this ornithine pool is not one into which aspartate carbons are entering. Thus, the aspartate must mainly enter a compartment similar to that of glutamate, which gives no carbons to serine, glycine, ornithine, or proline, and a smaller compartment, which contains a gluconeogenic pathway. A series of experiments with specific carbon-labeled amino acids and shorter time periods of incubation and stimulation would allow us to more clearly define the kinetics in these systems.

8.5.3. *Ionic Effects on Amino Acids*

Earlier,[178,213] we had noted that calcium had a profound effect on the partitioning of glucose carbons into certain metabolites. Having developed the capability to quantitate the individual amino acids from a single SCG,[207–209] we began a series of studies to determine how the deletion of particular cations from the glucose Krebs–Ringer's medium would affect the partitioning of [U-^{14}C]D-glucose into the free intra- and extraganglionic amino acids.[184] The effects of cation deletion are quite interesting. There is no effect on the intraganglionic amino acid concentrations except when potassium is deleted. This results in higher concentrations of alanine, aspartate, serine, valine, and threonine. The proline concentration declines in SCG bathed in 50% Na^+–sucrose medium. Stimulation, as stated previously, tends to increase the free intraganglionic amino acid concentrations in standard medium, but this is not the case in the ion-deleted media. The concentrations are similar to those of rested SCG controls.

The effect of reducing medium Na^+ alone is to decrease the efflux of almost all amino acids. The amino acids of rested ganglia have higher specific activities than those of controls in standard medium. When these ganglia are stimulated, only alanine and glycine have significantly increased specific activities. The entry of labeled glucose carbons into ornithine in both rested and stimulated SCG is blocked.

Potassium deletion from the medium results in a specific increased efflux of serine into the medium. This ion loss effects a significantly decreased specific activity in all amino acids but alanine in both stimulated and rested ganglia. A plausible explanation for this increase in alanine labeling might be that the pyruvate derived from the labeled glucose cannot be utilized by the potassium-depleted mitochondria and piles up as alanine. Alternatively, an increase in transamination may occur with K^+ lack, and the effluxed alanine is now the end product of this unwanted nitrogen. Again, lack of this ion, like Na^+, prevents entry of label into ornithine.

The SCG action potential is unaffected by either the K^+-deleted or 50% Na^+–sucrose incubation medium. This is not true when Li^+ replaces 50% of the Na^+ in the standard medium.[190,192,214] The effect lithium has on metabolism in the SCG is also quite profound[182,184,200,213] (P. Cancalon and J. D. Klingman, unpublished data, 1971; G. D. DeVincenzo and J. D. Klingman, unpublished data, 1968). With respect to amino acids, lithium causes a significant increase in the efflux of alanine and serine and a decrease of proline and ornithine. The rested SCG has increased specific activities in alanine, aspartate, and glutamate, whereas glycine is below control values. When the SCG is stimulated, only glutamate has significantly increased activity. The overall effects of lithium appear at first to be a summation of the effects seen with the deletion of both sodium and potassium. Lithium is the only ion that causes an increased amino acid specific activity in rested SCG that increases further on stimulation. The other fascinating observation, besides the lowered activity of serine, is the decreased conversion of serine to glycine.

8.6. Proteins

The proteins of SG have not been subjected to as rigorous examination as have those of the CNS.[215] Studies that have been performed have been related either to the induction of a specific enzyme such as TH by NGF,[216] the effects of glucocorticoids,[217–220] measurements of a specific amino acid entry into the proteins,[221] or our measurements of the effects of stimulation on the entry of glucose, glutamate, aspartate, and serine carbons into the gross protein pool.[184] (P. Cancalon and J. D. Klingman, unpublished data.)

That the protein pools are being metabolized is seen in the increased total amino acid concentrations—both intra- and extraganglionic—during *in vitro* incubation. Not all of this is caused by axotomy, although it certainly accounts for a portion of the increase. This can easily be surmised by the increase in the essential amino acids after 2 h of incubation in the glucose-containing Krebs–Ringer's medium. Prior to this time period, stimulation causes a decrease in total ganglionic protein. This is borne out by studies with [^{14}C]valine,[221] [^{14}C]glucose, and [^{14}C]glutamate,[184] which show a decreased entry of label into protein of stimulated SCG compared to resting controls. In light of this, an experimental protocol for the ^{14}C-labeled aspartate and serine experiments was used to attempt to answer a portion of the protein utilization question[168] (P. Cancalon and J. D. Klingman, unpublished data). Ganglia that are kept at rest while incubated with either labeled amino acid have higher protein specific activities than do stimulated SCG. When similarly stimulated labeled SCG are immediately transferred to normal, nonradioactive medium and rested, the proteins acquire higher specific activities than controls within 5 min and this increase continues for at least 30 min at rest. These limited experiments imply that stimulation shuts off a major portion of the amino acid incorporation into proteins, but the ganglia seem to be primed with the necessary biochemicals, i.e., mRNA, etc., so that when a period of rest occurs, a rapid incorporation can take place. We hope, in the near future, to be able to define whether this increased incorporation in rested SCG following stimulation also occurs for essential amino acids and whether this is unique to a particular group of proteins.

8.7. RNA and Protein

If a resynthesis of protein does occur in the SCG following a period of stimulation, one should expect to find increased RNA metabolism. Only one study has been reported on the incorporation of bases into RNA related to stimulation.[56,57] With [^{14}C]uridine it was found that increased incorporation results from either electrical or ACh stimulation. This increase was prevented if mecamine or *d*-tubocurarine was added to the *in vitro* SCG. The data do not indicate whether this increased incorporation reflects breakdown or synthesis of RNA.

In a study of chick embryonic SG,[55] the time courses of incorporation of leucine and uridine were followed in 8- to 17-day embryos. Both protein and RNA show elevated incorporations at day 10, and both fall to lower values by day 12. The RNA shows only a slow increase over the next 5 days, whereas the incorporation into protein increases rapidly and by day 17 exceeds that seen at day 10. When such SG are treated with NGF, the 10-day maximums are increased: the RNA more than twofold; the protein about 1.8-fold. Both decline, but now a clearly defined increased incorporation into RNA is seen at day 13. The incorporation into protein begins earlier, between 11 and 12 days, and the maximum is seen at day 13 rather than day 17 as in the untreated controls.

The RNA polymerases I and II or rat neonatal SCG have been measured with and without NGF treatment.[222] The NGF increases the activity of both polymerases, with II peaking 8 h after injection, whereas the activity of I is maximal approximately 12 h after NGF injection, but the rise and fall of this enzyme's activity extends over a period of time twice that of polymerase II.

The concentration of RNA has been found to decrease following axotomy of the SCG.[223] The types of RNA and their synthesis following *in vivo* vagotomy of the nonautonomic nodose ganglion have been described.[224,225]

Others have noted the inductive effects of NGF on RNA and/or protein synthesis[70,185,217,225] or the effects of glucocorticoids.[219,226,227]Most of the remaining work deals with measuring RNA concentrations in various ganglia.[228–230]

8.8. Lipids

8.8.1. Phospholipids

The original impetus for the quantitative separation and measurement of the ganglionic lipids was to discern their possible function as secondary oxidative substrates during periods of extended excitation.[178] A combination of silicic acid-impregnated paper, our development of a unique micro-silicic acid column,[231] and our later combining of the microcolumn with a micro-TLC and GLC procedures[195,200,201] (G. D. DeVincenzo and J. D. Klingman, unpublished data) allowed for the complete lipid analysis of the SCG as well as assessment of the turnover of the various components within each lipid. In analysis of the embryonic lipids, it has been found that the ganglion can be directly crushed on a TLC plate and separation achieved.[232,233] The lipid composition of the ganglion is not unlike that of brain (Table VI).[235] As would be expected, the smallest fraction of SCG lipid is the gangliosides, only 0.3 nmol/mg wet weight. The SCG gangliosides have been found to consist of G_T, G_Q, G_{D2b}, G_{D1a}, and G_{M2} with no difference in these types between DRG and SG, but this does differ from the distribution in brain.[235] The largest constituents are phosphatidylcholine (PC) and phosphatidylethanolamine (PE), representing 22.5% and 14.31%, respectively, of the total lipid pool or, combined, approximately 75% of the glycerol phospholipids. In a single fresh-frozen ganglion, the amount of free fatty acids is below the amount detectable by GLC, although a faint area is seen on TLC.

The composition of the lipid pool is not altered dramatically by extended *in vitro* incubation. Over a 4-h period, there is actually a 20% increase in phos-

Table VI
Rat Superior Cervical Ganglionic Lipids[a]

Lipid	Quantity (nmol P/mg dry weight)
Phosphatidylethanolamine (PE)	30.1 ± 4.7
Phosphatidic acid (PA)	7.2 ± 1.7
Phosphatidylserine (PS)	9.4 ± 1.0
Phosphatidylcholine (PC)	46.4 ± 6.6
Phosphatidylinositol (PI)	8.4 ± 1.0
Sphingomyelin	24.2 ± 3.4
Cerebroside[b]	10.2
Gangliosides[d]	0.03
Sulfatide[b]	11.8
Cholesterol[c]	55.5 ± 0.9
Cholesterol esters[c]	6.9 ± 0.9

[a] Results are expressed as nanomoles P/mg tissue dry weight ± S.D. (n = 3) for ganglia phospholipids incubated 20 min in Krebs–Ringer buffer. All others are nmol/mg dry weight.
[b] Estimated as free sphingosine by the method of Siakotos.
[c] Determination by Liebermann–Burchard assay.
[d] Reference 201; all others ref. 195.

pholipid phosphorus[213] (G. D. DeVincenzo and J. D. Klingman, unpublished data). Turnover of lipids is seen when alterations are made in the ionic composition of the Krebs–Ringer's medium; the greatest effect on their quantity and their turnover is produced by lithium[195,200,213] (G. D. DeVincenzo and J. D. Klingman, unpublished data). Although only 8% of glucose carbons normally enter the lipid pool, pyruvate and serine carbons enter the lipids at a much faster rate, mainly into the fatty acids, ceramides, and cholesterols[200] (J. D. Klingman, unpublished data). Glucosamine carbons, although producing an initial burst of lipid labeling[201] (J. D. Klingman, unpublished data), fail after 2 h, and acetate carbons are barely used at all (J. D. Klingman, unpublished data; L. Opler and J. D. Klingman, unpublished data).

An early observation that has since been verified is that there are different turnover times for the various lipids and their individual components, i.e., phosphate, fatty acids, glycerol, bases, and carbohydrates[178,195,213,231,233,237] (G. D. DeVincenzo and J. D. Klingman, unpublished data). At tracer levels, 94% of glucosamine's carbons enter the glycerol moiety of the phospholipids, whereas 70% of glucose carbons enter the phospholipids' glycerol, and 30% are found in fatty acids; 75% of pyruvate carbons enter the fatty acids, and only about 25% are in the glycerol base portion of the phospholipids.[168] Although serine carbons partition in some ways like glucose carbons, labeling the glycerol base moiety of the phospholipids more than the fatty acids, its labeling of complex and neutral lipids is much more discrete[200] (J. D. Klingman, unpublished data).

8.8.1a. Calcium. Besides these basic partitioning studies, two broad areas of investigation have been concerned with the effects of stimulation. Is there an alteration that is related to the action potential, and what are the effects of the incubation medium's cations on ganglionic lipids? Obviously, the answers to these two questions are intertwined, since the action potential is an ionic event. Again, the substrates used will determine how the experiments are to be interpreted. The initial observations[178,236] with labeled [^{32}P]phosphate and [U-^{14}C]glucose on the rat SCG phospholipids, aside from the differential labeling rates, reveal that of all the phospholipids only the ^{32}P labeling of phosphatidylinositol increases with increasing frequencies of stimulation, as does that from [^{14}C]glucose. As the temperature is increased from 20 to 37°C, there is, for both substrates, the expected increase in temperature coefficient and incorporation into phospholipids; the carbon labeling of PC and PE has temperature coefficients of 7.0 and 7.2, respectively.[237] When the height of the action potential is measured against the ^{32}P incorporated into PI, the ^{32}P incorporation is found to be directly reduced as the action potential is reduced. Thus, as with oxygen consumption, there is a temperature independence[177] in which the extra work of producing an action potential places a fixed demand on particular molecular events independent of temperature.

This initial series of experiments shows that when Ca^{2+} in Krebs–Ringer's medium is lowered, the ^{32}P labeling of all phospholipids, except PI, decreases, whereas the reverse is true for ^{14}C labeling from glucose. Some of these findings are similar to those previously reported[238] in which the bisected cat stellate ganglia was stimulated with pharmacological doses of acetylcholine in the presence of sarin, which not only increases ^{32}P labeling of PI but also that of phosphatidic acid. To determine whether the PI effect is pre- or postganglionic, a [^{32}P]phosphate study of various neuronal tissues was performed including a graded dosage effect with *d*-tubocurarine on the rat SCG. As the dose of *d*-tubocurarine is increased, the action potential declines, as does the specific stimulus-increased labeling of PI. It is thus thought that the increased labeling of PI with increasing frequencies of stimulation results from postsynaptic effects of the neurotransmitter[237].

Further study[238] revealed that, unlike phrenic and vagus nerves, the SCG places most of its acquired ^{32}P into PI rather than DPI and TPI and that the reduction in the PI labeling by added delthexane or gammexane is a result of the lowered action potential and not metabolic blockade. Centrifugal gradient separation of the rat SCG[233] subcellular components has suggested that this stimulus-related effect on PI occurs in the synaptosomes and the mitochondria, although some caution with this interpretation must be exercised because possible cross-gradient contamination of the various fractions. It has been shown that this PI-labeling effect by [^{32}P]phosphate in the SCG may represent a muscarinic effect.[240,241] A good discussion of the various issues surrounding this phenomenon has been presented.[242] That Ca^{2+} is involved has recently been demonstrated.[243] One should use caution, however, in interpreting data from various ganglia or other neuronal tissues relative to nicotinic versus muscarinic events, as these events are not easily distinguished.[192]

The calcium efflux from various compartments in the rat SCG has been measured.[244] Assuming that there are varying efflux rates and uptakes, one anticipates that there might well be differential turnover rates of the various lipids in compartments that might be associated with function.

8.8.1b. Magnesium. The effect of magnesium reduction in the Krebs–Ringer's medium is to decrease by 50% the entry of glucose carbons into PC while increasing the incorporation into PE and PA. [^{32}P]Phosphate is incorporated less into PC and PA but is increased in PE. Although the latter increase is inhibited by stimulation, little effect has been seen in any of the other phospholipids studied[213] (G. D. DeVincenzo and J. D. Klingman, unpublished data).

8.8.1c. Potassium. Lowered potassium in the glucose Krebs–Ringer's medium reduces the ^{32}Pi incorporation into PC, PE, and PA, and the entry of glucose carbons is reduced by 40%. The effect of reduced potassium is more profound on the phospholipid labeling in the rested ganglia than in stimulated ganglia. Subsequent studies have determined that these lowered incorporations of glucose carbons result from lowered entry into the fatty acid moieties of the phospholipid[213] (G. D. DeVincenzo and J. D. Klingman, unpublished data). Addition of potassium to 80mM in the medium produces a rapid increase in ^{32}Pi uptake into all of the phospholipids[245] along with enhanced glucose utilization and lactate production. It should be remembered that high potassium levels provoke a general membrane depolarization throughout all ganglionic elements.

8.8.1d. Sodium. When an equiosmolar amount of sucrose replaces 50% of the Krebs–Ringer's Na^+, no increase in phospholipid concentrations occurs, but there is a 100% increase in the [^{14}C]glucose carbons incorporated[213] (G. D. DeVincenzo and J. D. Klingman, unpublished data). Whereas in control SCG the fatty acid to glycerol base ^{14}C ratio is 7:3, in these Na^+–sucrose SCG the ratio is 1:1. This may reflect a disinhibition of hexokinase and pyruvate kinase, resulting in a greater utilization of glucose.

8.8.1e. Lithium. More striking effects are seen when lithium is partially substituted for equivalent amounts of sodium in the Krebs–Ringer's medium (acute lithium experiments on normal SCG) and these results are compared to SCG from rats raised on a lithium-containing diet (chronic lithium experiments) and to normal SCG.[182,195,200] With labeled glucose[215] (G. D. DeVincenzo and J. D. Klingman, unpublished data), the acute lithium SCG again have increased incorporation of ^{14}C into their phospholipids, as seen in the above sodium–sucrose experiments, but the incorporation is even greater. However, the fatty acid to glycerol base ratio remains at 7:3 as in the normal SCG phospholipid labeling pattern. In normal SCG (Fig. 3), when [U-^{14}C]pyruvate is the substrate,[167,230] the phospholipid fatty acids acquire more label than does the glycerol base, which is sustained at a slightly higher level in acute lithium-incubated SCG.

In chronic lithium SCG incubated with [^{14}C]pyruvate, there is an early increased labeling, which decreases by 80 min. This observed change in the labeling of the fatty acids takes place in both chronic and acute lithium-treated SCG. Acute lithium SCG phospholipid fatty acids almost maintain the entry of carbon into palmitic acid compared to the normal entry rate but have much lower specific radioactivity in stearic and oleic acids. Stimulation further lowers

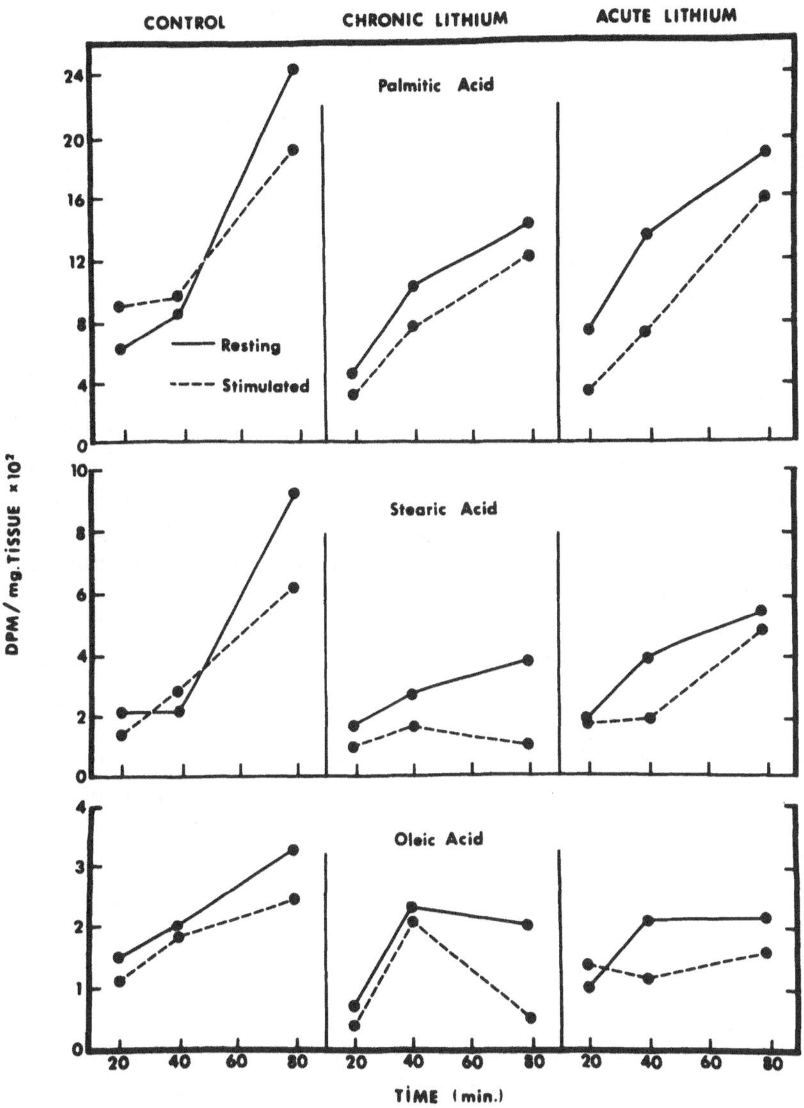

Fig. 3. Specific activity of total glycerolphospholipid fatty acids from normal, acute, and chronic lithium-fed rat SCG incubated *in vitro* in Krebs–Ringer's media aerated with 95% O_2 and 5% CO_2, pH 7.4, and [U-^{14}C]pyruvate. (Reproduced from Klingman *et al.*[168] by permission of Springer-Verlag.)

the activity in palmitate, with a constant low entry of label into oleic acid over the 80-min incubation period. The labeling of these three fatty acids in chronic lithium SCG is more depressed than in acute lithium ganglia. Upon stimulation of the chronic lithium ganglia, only a slight increase in labeling of palmitate, below the rested values, is seen during the course of the experiment. For oleic acid, the activity increases for the first 40 min and then abruptly falls to the initial 20-min value at 80 min. It would appear that lithium impairs not only entry into palmitic acid, which is depressed by the energy demands of stimulation, but also the elongation and desaturation pathways. It is unclear at this time whether this inhibition of fatty acid metabolism is a site-specific impairment in either the endoplasmic reticulum or mitochondria or is a result of the lowered high-energy phosphate or altered NAD^+/NADH seen in lithium-treated ganglia.[179,182] It has been found recently[246] in male rats receiving s.c. injections of lithium that lithium acts as a noncompetitive inhibitor of D- and L-myoinositol-1-phosphatase.

8.8.2. Sphingomyelin

The ceramide lipids in the SCG constitute 22.1% of the total lipids, with sphingomyelin being 11.5% of the total lipids and 52% of the ceramide pool.[168,195] The sulfatides and cerebrosides amount to 5% each of the total lipid pool (see Table VI). With [^{14}C]glucose (L. Opler and J. D. Klingman unpublished data), the sphingomyelin contains 10–20% of the total lipid radioactivity whether the ganglion is rested or stimulated. When [^{14}C]pyruvate is the substrate, by far the greatest recipient of radioactivity is the sphingomyelin, 74% of the total; 90% of this activity is in the sphingosine portion, and only 4% in its fatty acids. Acute lithium-incubated rested ganglia are similar to normal controls, but when these are stimulated, the sphingomyelin has only about 50% of the rested or normal control values. The loss in activity is accounted for by the loss of sphingosine labeling. This inhibition in sphingomyelin labeling is further exacerbated in the chronic lithium SCG. In this case, however, even the rested lithium ganglia fail to incorporate radioactivity into sphingomyelin: about 25% of the controls, with little effect when stimulated. Whether this is another example of shunting acetyl-COA into the TCA cycle in these chronic lithium-treated ganglia or an inhibition in sphingosine biosynthesis or another compartmentation process is not presently known.

8.8.3. Cholesterol

Normally we relegate neuronal cholesterol and its esters to a quiescent metabolic pool that, when metabolically active, is considered to be myelin related. In the SCG, slightly more pyruvate carbons enter the total cholesterol pool than the phospholipids. Stimulation increases only slightly the activity in cholesterol and/or its esters. In resting SCG, acute lithium treatment curtails cholesterol labeling, which on stimulation is further lowered to about 50% of the control value. The activity of cholesterol esters in these resting acute lithium ganglia is twice as high as that of controls, but on stimulation the activity is similar to that found in normal stimulated controls.

Chronic lithium treatment lowers cholesterol labeling from pyruvate in rested SCG, but the esters are labeled almost three times the control values. On stimulation, these chronic lithium-treated SCG have 50% less label in cholesterol than the controls, similar to acute lithium SCG, but the cholesterol esters have tripled in incorporated activity over acute lithium or normal control ester values. At first glance, one would assume that lithium again is decreasing the availability of acetyl-CoA, since a lowered activity is seen in the cholesterol of acute lithium stimulated and both chronic rested and stimulated lithium-treated SCG. This is partly true, but analysis of the cholesterol esters reveals that most of the radioactivity, almost 70%, is found in the fatty acids. These observations might lead one to speculate that biosynthesis of fatty acids is still going on albeit at a depressed rate in lithium-treated SCG, but the fatty acids cannot enter the sphingomyelin or the phospholipid pool. The fatty acids are removed from the cytoplasm by being fixed as cholesterol esters. This is a brief description of some of the more intriguing aspects of the metabolic effects produced by cations on lipid metabolism and substrate utilization.

8.8.4. Prostaglandins

The *raison d'être* for much of the interest in the ionic influences on neuronal tissue phospholipid metabolism is their involvement in eicosanoid biosynthesis (recently reviewed[247]). Stimulation of sympathetic neurons is known to cause the release of prostaglandin E, whereas addition of dopamine or norepinephrine inhibits this release. Thus, a feedback control on neurotransmitter release has been postulated.[248] The rat SCG has been shown to make PGE from exogenous arachidonic acid; this reaction is inhibited by prostaglandin synthetase inhibitors and stimulated by NE and dopamine.[249] The SCG *in vitro* releases PGE into the medium, which occurs in a time-dependent fashion.

Two recent reports[250,251] on the stimulated release of prostaglandins from whole guinea pig, rat, and rabbit and sliced calf SCG have attempted to correlate this to the increased cyclic AMP levels. Addition of PGE_1 and PGE_2 to these preparations increases cyclic AMP levels above control values. Added PGFs are also effective on calf SCG in this regard, ineffective on the guinea pig, and were not tested on rat and rabbit SCG. The guinea pig SCG, the only one tested, is somewhat sensitive to PGE but not PGB in increasing cyclic AMP levels. Supramaximal electrical stimulation of rat, guinea pig, and calf SCG produces about the same increases in cyclic AMP as the addition of 10^{-5} M PGE. This increase in cyclic AMP is reduced to almost control levels in the presence of aminophylline.

When the Ca^{2+} and Mg^{2+} concentrations in the bathing medium are changed from 2.50 mM and 1.15 mM, respectively, to 0.50 mM and 12.00 mM, the stimulated ganglia show only slight increases in both PGE and cyclic AMP; these are conditions of impaired transmitter release. When the guinea pig SCG is stimulated and treated with propranolol, a β-adrenergic blocker, no effects on the increases in PGE and cyclic AMP are seen; atropine, a muscarinic blocker, inhibits the rise of PGB and produces only a 50% increase in cyclic AMP; phentolamine, an α-adrenergic blocker, produces a response similar to atropine. These observations lend support to the proposal that the prostaglandins cause presynaptic formation of cyclic AMP, regulating negatively the release of acetylcholine. Microelectrode impalements of guinea pig SCG nerve fibers show that addition of PGE at 10^{-7} M does not affect the conduction velocities or peak amplitudes of the compound action potential, the resting potential, or the total membrane resistance.[251] Addition of NE to phentolamine-superfused ganglia in the presence of PGE_1 also does not impair synaptic transmission. Measurements of the excitatory postsynaptic potentials (EPSP) of stimulated ganglia in the presence of PGE show a quantal loss of EPSP as well as an increased number of evoked potential failures. There is an impairment of transmission by PGE via decrease presynaptic release of ACh resulting from a reduction in the elementary quanta of ACh released at each impulse by a single fiber impinging on a neuron, although the quantal size—the sensitivity of the postsynaptic membrane—remains unchanged. These data suggest that this release of PGE is a mechanism of presynaptic inhibition for weak signals and conserves the quantal ACh output.

8.9. Cyclic AMP

The interest in this area arose from the initial findings that preganglionic nerve stimulation,[252] catecholamines,[253] and muscarinic cholinomimetics[254] all cause an increase in cyclic AMP. It was proposed[255] that the catecholamines function in ganglia as inhibitory transmitters by increasing presynaptic cyclic AMP present through activation of muscarinic receptors. Rather than produce here all the arguments—anatomic, biochemical, pharmacological, and physiological—that have developed in this area, the reader is directed to pertinent publications.[137,256,257] The interest in cyclic AMP is further heightened by the finding of phosphoprotein 1 in the vesicles of autonomic ganglia[258,259] as well as in other presynaptic neuron areas. This protein is a substrate for both cyclic-AMP-dependent and Ca^{2+}–calmodulin-dependent protein kinases. The amount of this protein is increased by stimulation, serotonin, dopamine, and depolarizing agents.

8.10. Cyclic GMP

Similarly, there have been the reports of increased amounts of cyclic GMP in stimulated frog SG,[260] which are not blocked by nicotinic ACh receptor blockade but are blocked by atropine. High-magnesium-low-calcium Ringer's medium, which lowers ACh release, also inhibits the cyclic GMP increase. In bovine SCG slices, addition of either ACh or bethanecol significantly increases the level of cyclic GMP.[261] It has been proposed that there is an activation of excitatory muscarinic ACh receptors resulting in a depolarization and excitation by means of an increased cyclic GMP formation through a membrane-bound guanylate cyclase system.[262] An increase of cyclic GMP in sliced bovine SCG is also produced by addition to 100 mM K^+.[263] In rat SCG, elevated potassium causes a similar increase in the cyclic GMP level.[264] The absence of calcium does not allow for the potassium-induced increase. Denervation does not lower the cyclic GMP content of ganglia 7 days later, but, again, increasing the potassium level does not elevate cyclic GMP levels. Neither physostigmine, atropine, nor hexamethonium blocks the potassium activation. The effect of increasing the potassium, which prevents any increase in cyclic GMP, on denervated SCG indicates that depolarization of the neurons and glial cells is not the initial stimulus for activation of the guanylate cyclase system. Along with this is the calcium requirement, suggesting the involvement of a transmitter, which does not seem to be ACh. It is presently unknown whether the presynaptic terminals are the site of the increased cyclic GMP levels seen on stimulation.

9. *IMMUNOLOGY*

We cannot close this chapter on the autonomic nervous system without a short discussion of the role the sympathetic nervous system may play in viral diseases and their effects on the sympathetic system as well as its possible participation in lymphocytic differentiation.

A number of years ago,[265] it was shown that pseudorabies virus injected into the vitreous humor of the rat eye produced systemic expressions of the infection in 48–72 h. Aside from the fact that this study may have been the first demonstration of retrograde particle flow in neurons, physiological effects of the virus on the SCG were also determined. It was found that as the virus entered the ganglionic neurons, spontaneous discharges of the postganglionic fibers occurred. This was followed a short time later by spontaneous discharges from the preganglionic fibers. The virus had traveled into the ganglionic neurons, produced lesions as documented, multiplied, and then crossed the synaptic cleft into the preganglionic neurons. Undoubtedly, the virus could now travel into the CNS. More recently,[266] studies have ascertained how the virus produced the spontaneous discharges and defined the biochemical transformations that could account for them. The apparent increased sensitivity of the postganglionic neurons is accompanied by an increased release and SCG content of ACh and by an increase in ChAT.

After mice are infected with herpes simplex virus, there is an acute growth phase of virus in the SCG, which is followed by a latent phase.[267] Administration of 6-OH-DA prior to the infection increases the acute growth phase but does not affect the latent phase. The effect of 6-OH-DA was found to be directly opposite to the effect of immunization. As 6-OH-DA has recently been shown to affect the adrenergic neurons by alteration of their SCG immunochemistry, it is likely that such treated ganglia should allow the virus to proliferate.

Although the two studies provide us with a rather limited perspective, they do suggest that the immunochemistry of the SCG may interact differentially depending on the viron presented to it.

In addition to the possibility that the autonomic neurons may be involved in virus propagation is the possibility that the autonomic system is also involved in lymphocyte differentiation. If the sympathetic fibers to the spleen are sectioned, there is little activation of lymphocytes from the spleen of challenged animals. Aside from the autonomic innervation of the spleen, the lymph nodes and the parenchyma of the thymus are also known to receive sympathetic innervation. Felten *et al.*[268] have more recently shown that noradrenergic neurons innervate the lymphoid region of the rabbit appendix. This raises the possibility that the noradrenergic fibers may well play an inductive role in the differentiation of the immune system in each of these lymphoid tissues.

These initial papers suggest several possibilities concerning the immune system and virus infections. In viral infections, the participation of the sympathetic nervous system may solely provide a route of entry, and, depending on the type of virus, it may move across the synaptic cleft and enter the preganglionic fibers. It may also remain as a latent virus in the ganglionic neurons with periodic symptoms or lesions expressed. Much of this is undoubtedly dependent on the immunochemical determinants of the cells comprising the sympathetic nervous system. Since all lymphoid tissues are sympathetically innervated, it is not unlikely that various aspects of the immune system's response to an infective agent may be dependent on the sympathetic system inducing, by the release of catecholamines or peptides, the development of immune competent lymphocytes.

REFERENCES

1. Gaskell, W. H., 1916, *The Involuntary Nervous System IX*, Longmans, Green, London.
2. Langley, J. N., 1921, *The Autonomic Nervous System*, W. Heffer and Sons, Ltd., Cambridge.
3. Le Douarin, N. M., 1980, *Curr. Top. Exp. Biol.* **16**:32–85.
4. Le Douarin, N. M., and Smith, J., 1981, *Annu. Rev. Physiol.* **43**:653–671.
5. Weston, J. A., 1981, *Adv. Neurol.* **29**:77–85.
6. Varon, S., and Adler, R., 1981, *Adv. Cell. Neurobiol.* **21**:115–163.
7. Levi-Montalcini, R., 1982, *Annu. Rev. Neurosci.* **5**:341–362.
8. Yankner, B. A., and Shooter, E. M., 1982, *Annu. Rev. Biochem.* **51**:845–868.
9. Harper, G. P., and Thoenen, H., 1980, *J. Neurochem.* **34**:5–16.
10. Black, I. B., Coughlin, M. D., and Cochard, P., 1979, *Soc. Neurosci. Symp.* **4**:184–207.
11. Black, I. B., 1982, *Science* **215**:1198–1204.
12. Varon, S., and Bunge, R., 1978, *Annu. Rev. Neurosci.* **1**:327–362.
13. Eränkö, O., Soinila, S., and Paivarenta, H., 1980, *Adv. Biochem. Psychopharmacol.* **25**:35–42.
14. Kharkevich, D. A. (ed.), 1980, *Pharmacology of Ganglionic Transmission, Handbook of Experimental Pharmacology,* Volume 53, Springer-Verlag, Berlin.
15. Salzer, J. L., and Bunge, R. P., 1980, *J. Cell Biol.* **84**:739–752.
16. Salzer, J. L., Williams, A. K., Glaser, J., and Bunge, R. P., 1980, *J. Cell Biol.* **84**:753–778.
17. Skok, V. I., 1980, *Handbook of Experimental Pharmacology,* Volume 53 (D. A. Kharkevich, ed.), Springer-Verlag, Berlin, pp. 9–39.
18. McCarthy, K. D., and Partlow, L. M., 1976, *Brain Res.* **114**:391–414.
19. Wolf, G. A., 1941, *J. Comp. Neurol.* **75**:235–243.
20. Ebbesson, S. O., 1963, *Anat. Rec.* **4**:353–356.
21. Kosterlitz, H. W., 1968, *Handbook of Physiology,* Section 6: *Alimentary Canal,* Part 4, Williams & Wilkins, Baltimore, pp. 2147–2172.
22. Hirst, G. D. S., Holman, M. E., Prosser, C. L., and Spence, I., 1972, *J. Physiol.* (*Lond.*) **225**:60–61.
23. Wood, J. D., 1975, *Physiol. Rev.* **55**:307–324.
24. Wallis, D. I., and North, R. A., 1978, *Pflügers Arch.* **347**:145–152.
25. Landmesser, L., and Pilar, G., 1974, *J. Physiol.* (*Lond.*) **241**:715–736.
26. Nishi, S., Soeda, H., and Koketsu, K., 1965, *J. Cell. Comp. Physiol.* **66**:19–32.
27. Dunant, Y., 1967, *J. Physiol.* (*Paris*) **59**:17–38.
28. De Castro, F., 1951, *Arch. Int. Physiol. Biochim.* **59**:479–511.
29. Skok, V. I., and Heeroog, S. S., 1975, *Brain Res.* **87**:343–353.
30. Brown, C. L., and Pascoe, J. E., 1952, *J. Physiol.* (*Lond.*) **118**:113–123.
31. De Castro, F., 1932, *Cytology and Cellular Pathology of the Nervous System 1*, (W. Penfield, ed.), Hoeber, New York, pp. 317–380.

32. Elfvin, L. G., 1971, *J. Ultrstruct, Res.* **37**:432–448.
33. Melnichenko, L. V., and Skok, V. I., 1969, *Neurofiziologia* **1**:101–108.
34. Roper, S., 1976, *J. Physiol. (Lond.)* **254**:427–454.
35. Hervonen, A., Pickel, V. M., Joh, T. H., Reis, D. J., Linnoila, I., and Miller, R. J., 1981, *Cell Tissue Res.* **214**:33–42.
36. Libet, B., and Owman, C., 1974, *J. Physiol. (Lond.)* **237**:635–662.
37. Williams, T. H., Black, A. C., Jr., Tanemichi, C., and Jew, J., 1977, *Adv. Biochem. Psychopharmacol.* **16**:505–511.
38. Williams, T. H., Chiba, T., Black, A. C., Jr., Bhalla, R. C., and Jew, J., 1976, *SIF Cells*, (E. Eränkö, ed.), Fogarty International Publications, No. (NIH) 76-949, U.S. Government Printing Office, Washington, pp. 143–162.
39. Korum, F., Speciale, S. G., Jr., and Wyatt, R. J., 1979, *J. Pharmacol. Exp. Ther.* **211**:706–710.
40. Eränkö, O., and Soinila, S., 1981, *J. Neurocytol.* **10**:1–18.
41. Eränkö, O., and Soinila, S., 1981, *J. Neurocytol.* **10**:45–55.
42. Levi-Montalcini, R., and Booker, B., 1960, *Proc. Natl. Acad. Sci. U.S.A.* **46**:373–383.
43. Levi-Montalcini, R., and Booker, B., 1960, *Proc. Natl. Acad. Sci. U.S.A.* **46**:384–391.
44. Levi-Montalcini, R., and Angelletti, P. U., 1968, *Physiol. Rev.* **48**:534–569.
45. Klingman, G. I., and Klingman, J. D., 1969, *Int. J. Neuropharmacol.* **6**:501–508.
46. Klingman, G. I., 1965, *J. Pharmacol. Exp. Ther.* **148**:14–21.
47. Klingman, G. I., and Klingman, J. D., 1972, *Immunosympathectomy* (G. Steiner and E. Schonbaum, eds.), Elsevier, Amsterdam, pp. 91–95.
48. Kessler, J. A., Cochard, P., and Black, I. B., 1981, *Adv. Neurol.* **29**:115–123.
49. Greene, L. A., and Shooter, E. M., 1980, *Annu. Rev. Neurosci.* **3**:353–402.
50. Stach, R. W., Pignatti, P. F., Baker, M. E., and Shooter, E. M., 1980, *J. Neurochem.* **34**:850–855.
51. Stach, R. W., and Shooter, E. M., 1980, *J. Neurochem.* **34**:1499–1505.
52. Varon, S., and Raiborn, C., 1972, *Neurobiology* **2**:183–196.
53. Angelletti, R. H., and Bradshaw, R. A., 1971, *Proc. Natl. Acad. Sci. U.S.A.* **68**:2417–2420.
54. Olender, E. J., Wagner, B. J., and Stach, R. W., 1981, *J. Neurochem.* **37**:436–442.
55. Partlow, L. M., and Larrabee, M. G., 1971, *J. Neurochem.* **18**:2101–2118.
56. Gisiger, V., 1971, *Brain Res.* **33**:139–146.
57. Gisiger, V., and Gaide-Hugunin, A. C., 1969, *Prog. Brain Res.* **31**:125–129.
58. Black, I. B., Hendry, I. A., and Iversen, L. L., 1971, *Brain Res.* **34**:229–240.
59. Hendry, I. A., 1971, *Rev. Neurosci.* **2**:149–194.
60. Varon, S., Raiborn, C., and Burnham, P. A., 1974, *Neurobiology* **4**:317–327.
61. Nikodijevic, B., Nikodijevic, O., Yu, M.-Y. W., Pollard, H., and Guroff, G., 1975, *Proc. Natl. Acad. Sci. U.S.A.* **72**:4769–4771.
62. MacDonnell, P. C., Nagaiah, K., Lakshmanan, J., and Guroff, G., 1977, *Proc. Natl. Acad. Sci. U.S.A.* **74**:4681–4684.
63. Guroff, G., Dickens, G., and End, D., 1981, *J. Neurochem.* **37**:342–349.
64. McCarthy, K. D., and Partlow, L. M., 1976, *Brain Res.* **114**415–426.
65. Larrabee, M. G., 1970, *Fed. Proc.* **29**:1919–1928.
66. Bloom, E. M., and Black, I. B., 1979, *Dev. Biol.* **68**:568–578.
67. Otten, U., Hatanaka, H., and Thoenen, H., 1978, *Brain Res.* **140**:385–389.
68. Skaper, S. D., Bottenstein, J. E., and Varon, S., 1979, *J. Neurochem.* **32**:1845–1851.
69. Frazier, W., Ohlendorf, C., Boyd, L., Aloe, L., Johnson, E., Ferrendellia, J., and Bradshaw, R., 1973, *Proc. Natl. Acad. Sci. U.S.A.* **70**:2448–2452.
70. Mizel, S. B., and Bamburg, J. R., 1976, *Dev. Biol.* **49**:20–28.
71. Hori, Z.-I., and Varon, S., 1977, *Brain Res.* **124**:121–133.
72. Varon, S., and Adler, R., 1981, *Adv. Cell Neurobiol.* **2**:115–163.
73. Varon, S., and Manthorpe, M., 1982, *Neurotransmitter Interaction and Compartmentation* (H. F. Bradford, ed.), Plenum Press, New York, pp. 473–496.
74. Skaper, S. D., and Varon, S., 1983, *Dev. Biol.* **98**:257–264.
75. Skaper, S. D., and Varon, S., 1981, *J. Neurochem.* **196**:1654–1660.
76. Skaper, S. D., and Varon, S., 1981, *J. Neurosci. Res.* **6**:133–141.

77. Harper, G. P., and Thoenen, H., 1980, *Neurochemistry* **34**:5–15.
78. Greenberg, J. H., 1981, *Adv. Neurol.* **29**:105–113.
79. Bottenstein, J. E., and Sato, G., 1979, *Proc. Natl. Acad. Sci. U.S.A.* **76**:514–517.
80. Varon, S., and Skaper, S. D., 1981, *Ciba Found. Symp.* **83**:151–176.
81. Patterson, P. H., and Chun, L. L. Y., 1977, *Dev. Biol.* **56**:263–280.
82. Patterson, P. H., 1978, *Annu. Rev. Neurosci.* **1**:1–19.
83. Ebendal, T., 1979, *Dev. Biol.* **72**:276–290.
84. Ebendal, T., Belew, M., Jacobson, C.-O., and Porath, J., 1979, *Neurosci. Lett.* **14**:91–95.
85. Helfand, S. L., Riopelle, R. J., and Wessels, N. K., 1978, *Exp. Cell Res.* **113**:39–45.
86. Manthorpe, M., Skaper, S. D., Barbin, G., and Varon, S., 1982, *J. Neurochem.* **38**:415–421.
87. Black I. B., Hendry, I. A., and Iversen, L. L., 1971, *Brain Res.* **34**:229–240.
88. Thoenen, T. H., and Barde, Y.-A., 1980, *Physiol. Rev.* **60**:1284–1335.
89. Ignarro, L. J., and Shideman, F. E., 1968, *J. Pharmacol. Exp. Ther.* **159**:38–48,49–58.
90. Cochard, P., Goldstein, M., and Black, I. B., 1979, *Dev. Biol.* **71**:100–114.
91. Patterson, P. H., and Chun, L. L. Y., 1974, *Proc. Natl. Acad. Sci. U.S.A.* **71**:3607–3610.
92. Patterson, P. H., and Chun, L. L. Y., 1977, *Dev. Biol.* **60**:473–481.
93. Weber, M. J., 1981, *J. Biol. Chem.* **256**:3447–3453.
94. O'Lague, P. H., Obata, K., Claude, P., Furshpan, E. J., and Potter, D. D., 1974, *Proc. Natl. Acad. Sci. U.S.A.* **71**:3602–3606.
95. Nurse, C. A., and O'Lague, P. H., 1975, *Proc. Natl. Acad. Sci. U.S.A.* **72**:1955–1959.
96. Furshpan, E. J., MacLeish, P. R., O'Lague, P. H., and Potter, D. D., 1976, *Proc. Natl. Acad. Sci. U.S.A.* **73**:4225–4229.
97. Ko, C. P., Burton, H., Johnson, M. I., and Bunge, R. P., 1976, *Brain Res.* **117**:461–485.
98. O'Lague, P. H., Potter, D. D., and Furshpan, E. J., 1978, *Dev. Biol.* **67**:384–403.
99. Higgins, D., and Burton, H., 1982, *Neuroscience* **7**:2241–2254.
100. Hawrot, E., 1980, *Dev. Biol.* **74**:136–151.
101. Klinar, B., and Brzin, M., 1978, *Neuroscience* **3**:1129–1134.
102. Davis, R., and Koelle, G. B., 1978, *J. Cell Biol.* **78**:785–809.
103. Chiappinelli, V. A., Giacobini, E., and Fairman, K., 1978, *Dev. Neurosci.* **1**:191–202.
104. Marchi, M., and Giacobini, E., 1980, *Dev. Neurosci.* **3**:39–48.
105. Verdiere, M., Derer, M., and Reiser, F., 1982, *Dev. Biol.* **89**:509–515.
106. Taylor, P., Rieger, F., and Greene, L. A., 1980, *Brain Res.* **182**:383–396.
107. Couraud, J. Y., Koenig, H. L., and DiGiamberardino, L., 1980, *J. Neurochem.* **34**:1209–1218.
108. Rowe, V., Fernandez, H., Duell, M., Parr, J., and Battie, C., 1981, *J. Neurochem.* **37**:861–866.
109. Gilad, B. M., and Gilad, V. H., 1981, *Brain Res.* **220**:420–426.
110. Marchi, M., Hoffman, D. W., Giacobini, E., and Fredrickson, T., 1980, *Dev. Neurosci.* **3**:235–247.
111. Grassi, J., Vigny, M., and Massoulié, J., 1982, *J. Neurochem.* **38**:457–469.
112. Brown, D. A., and Fumagalli, L., 1977, *Brain Res.* **129**:165–168.
113. Giacobini, E., 1976, *J. Physiol.* (*Lond.*) **257**:749–766.
114. Gangitano, C., Fumagalli, L., deRenzis, G., and Sangiacomo, C. D., 1978, *Neuroscience* **3**:1101–1108.
115. Chiappinelli, V. A., and Giacobini, E., 1978, *Neurochem. Res.* **3**:465–478.
116. Gangitano, C., Fumagalli, L., and Miani, N., 1978, *Brain Res.* **161**:131–141.
117. Fumagalli, L., and deRenzis, G., 1980, *Neuroscience* **5**:611–616.
118. Skok, V. J., Selyanho, A. A., and Derkach, V. A., 1982, *Brain Res.* **238**:480–483.
119. Birks, R. I., and Fitch, S. J. G., 1974, *J. Physiol.* (*Lond.*) **240**:125–134.
120. Kahlson, G., and MacIntosh, F. C., 1939, *J. Physiol.* (*Lond.*) **96**:277–292.
121. O'Reagan, S., and Collier, B., 1981, *J. Neurochem.* **36**:420–430.
122. Collier, B., 1981, *J. Neurochem.* **36**:1292–1294.
123. Kwok, Y. N., and Collier, B., 1982, *J. Neurochem.* **39**:16–26.
124. Boska, P., and Collier, B., 1980, *J. Neurochem.* **34**:1470–1482.
125. Westerink, B. H., and Van Dene, J. C., 1980, *Eur. J. Pharmacol.* **65**:71–79.
126. Kafka, M. S., and Thoa, N. B., 1979, *Biochem. Pharmacol.* **28**:2485–2489.

127. Martinez, A. E., and Adler-Graschinsky, E., 1980, *J. Pharmacol. Exp. Ther.* **212**:527–532,533–535.
128. Chalazonitis, N., and Zigmond, R. E., 1980, *J. Physiol. (Lond.)* **300**:525–538.
129. Ikeno, T., Dickens, G., Lloyd T., and Guroff, G., 1981, *J. Neurochem.* **36**:1632–1640.
130. Roth, R. H., Morgenroth, V. H. III, and Slazman, P. M., 1975, *Biochem. Pharmacol.* **23**:2779–2784.
131. Goldstein, M., Bronaugh, R. L., Ebstein, B., and Roberge, C., 1976, *Brain Res.* **109**:563–574.
132. Weiner, N., Lee, F.-L., Dreyer, E., and Barnes, E., 1978, *Life Sci.*, **22**:1197–1216.
133. Hartman, B. K., and Udenfriend, S., 1972, *Pharmacol. Rev.* **24**:311–330.
134. Libet, B., and Owman, C., 1974, *J. Physiol. (Lond.)* **237**:635–662.
135. Ashe, J. H., and Libet, B. 1982, *Brain Res.* **242**:345–349.
136. Dun, N. J., and Karczmar, A. G., 1978, *Proc. Natl. Acad. Sci. U.S.A.*, **75**:4029–4032.
137. Volle, R. L., 1980, *Handbook of Experimental Pharmacology*, Volume 53 (D. A. Kharkevich, ed.), Springer-Verlag, Berlin, pp. 393–402.
138. Skok, V. I., and Heeroog, S. S., 1975, *Brain Res.* **87**:343–353.
139. Roper, S., 1976, *J. Physiol. (Lond.)* **254**:427–454.
140. Mytilineou, C., and Papaconstantinou, M. C., 1979, *J. Neurochem.* **33**:347–349.
141. McBride, W. J., and Klingman, J. D., 1975, *Prog. Neurobiol.* **3**:251–278.
142. Bowery, N. G., Brown, D. A., Collins, G. G., Galvin, M., Marsh, S., and Yamini, G., 1976, *Br. J. Pharmacol.* **57**:73–91.
143. Kato, E., Morita, K., Kuba, K., Yamada, S., Kuhara, T., and Shinka, T., 1980, *Brain Res.* **195**:208–214.
144. Bowery, N. G., Brown, D. A., and Marsh, S., 1979, *J. Physiol. (Lond.)* **293**:75–101.
145. Hökfelt, T., Elfin, L.-G., Schultzberg, M., Goldstein, M., and Nilsson, G., 1977, *Brain Res.* **132**:29–41.
146. Mudge, A. W., Leeman, S. E., and Fischbach, G. D., 1979, *Proc. Natl. Acad. Sci. U.S.A.* **76**:526–530.
147. Jiang, Z.-G., Dun, N. J., and Karczmar, A. G., 1982, *Science* **217**:739–741.
148. Robinson, S. E., Schwartz, S. P., and Costa, E., 1980, *Brain Res.* **182**:11–17.
149. Kessler, J. A., Adler, J. E., Bohn, M. C., and Black, I. B., 1981, *Science* **214**:335–336.
150. Kessler, J. A., and Black, I. B., 1982, *Brain Res.* **234**:182–187.
151. Shultzberg, M., Hökfelt, T., Terenius, L., Elfin, L. G., and Lundberg, J. M., 1979, *Neuroscience* **4**:249–270.
152. Di Giulio, A. M., Yans, H. Y., Fratta, W., and Costa, E., 1979, *Nature* **278**:646–647.
153. Jan, L. Y., and Jan, Y. N., 1982, *J. Physiol. (Lond.)* **327**:219–246.
154. Kuffler, S. W., 1980, *J. Exp. Biol.* **89**:257–286.
155. Katyama, Y., and North, R. A., 1978, *Nature* **274**:387–388.
156. Kessler, J. A., and Black, I. B., 1980, *Brain Res.* **180**:157–158.
157. Goldman, H., and Jacobowitz, D., 1971, *J. Pharmacol. Exp. Ther.* **176**:119–123.
158. Thoenen, H., and Tranzor, J. P., 1973, *Annu. Rev. Pharmacol.* **13**:169–180.
159. Angeletti, P. U., and Levi-Montalcini, R., 1970, *Proc. Natl. Acad. Sci. U.S.A.* **65**:114–121.
160. Burnstock, G., Evans, B., Gannon, B. J., Heath, J. W., and James, V., 1971, *Br. J. Pharmacol.* **43**:295–301.
161. Eränkö, L., and Eränkö, O., 1971, *Acta Pharmacol. Toxicol. (Kbh)* **30**:403–416.
162. Angeletti, P. U., Levi-Montalcini, R., and Caramia, F., 1972, *Brain Res.* **43**:515–525.
163. Heath, J., Eränkö, O., and Eränkö, L., 1973, *Acta Pharmacol. Toxicol. (Kbh)* **33**:203–218.
164. Johnson, E. M., Jr., and Macia, R. H., 1979, *Circ. Res.* **45**:243–249.
165. Johnson, E. M., Jr., and Taylor, A. S., 1980, *Biochem. Pharmacol.* **29**:113–115.
166. Johnson, E. M., Jr., and Aloe, L., 1974, *Brain Res.* **81**:519–532.
167. Manning, P. T., Russell, J. H., and Johnson, E. M., Jr., 1982, *Brain Res.* **241**:131–143.
168. Klingman, J. D., Organisciak, D. T., and Klingman, G. I., 1980, *Handbook of Experimental Pharmacology*, Volume 53 (D. A. Kharkevich, ed.), Springer-Verlag, Berlin, pp. 41–61.
169. Larrabee, M. G., and Bronk, D. W., 1952, *Cold Spring Harbor Symp. Quant. Biol.* **17**:245–266.
170. Larrabee, M. G., 1958, *J. Neurochem.* **2**:81–101.

171. Larrabee, M. G., Ramos, J. G., and Bülbring, E., 1952, *J. Cell Comp. Physiol.* **40**:461–494.
172. Larrabee, M. G., and Posternak, J. M., 1952, *J. Neurophysiol.* **15**:91–114.
173. Brink, F., Bronk, D. W., Carlson, F. D., and Connelly, C. M., 1952, *Cold Spring Harbor Symp. Quant. Biol.* **17**:53–67.
174. Dolivo, M., and DeRibaupierre, F., 1973, *Experientia* **29**:5.
175. Rechncrona, S., Mela, L., and Chance, B., 1979, *Fed. Proc.* **38**:2489–2492.
176. Folbergrová, J., Ingvar, M., and Siesjö, B. K., 1981, *J. Neurochem.* **37**:1228–1238.
177. Dolivo, M., and Larrabee, M. G., 1958, *J. Neurochem.* **3**:72–88.
178. Larrabee, M. G., and Klingman, J. D., 1962, *Neurochemistry*, 2nd ed., (K. A. C. Elliott, I. H. Page, and J. H. Quatsel, eds.), Charles C. Thomas, Springfield, Illinois, pp. 150–176.
179. Sinicropi, D. V., Dombrowski, A., Montgomery, C. W., Evans, R. K., and Kauffman, F. C., 1980, *J. Neurochem.* **34**:1280–1287.
180. Larrabee, M. G., Horowicz, P., Stekiel, W., and Dolivo, M., 1956, *Metabolism of the Nervous System* (D. Richter, ed.), Pergamon Press, Oxford, pp. 208–220.
181. Edwards, C., and Larrabee, M. G., 1955, *J. Physiol.* (*Lond.*) **130**:456–466.
182. Organisciak, D. T., and Klingman, J. D., 1974, *J. Neurochem.* **22**:341–345.
183. Horowicz, P., and Larrabee, M. G., 1962, *J. Neurochem.* **9**:1–22,402–407.
184. McBride, W. J., and Klingman, J. D., 1972, *J. Neurochem.* **19**:865–880.
185. Larrabee, M. G., 1970, *Fed. Proc.* **29**:1919–1928.
186. DeRibaupierre, F., Siegrist, G., Dolivo, M., and Rouller, C., 1966, *Helv. Physiol. Pharmacol. Acta* **24**:C48–49.
187. DeRibaupierre, F., 1968, *Brain Res.* **11**:42–64.
188. Härkönen, M. H. A., Passonneau, J. V., and Lowry, O. H., 1969, *J. Neurochem.* **16**:1439–1458.
189. Larrabee, M. G., 1961, *Methods in Medical Research 9* (J. H. Quastel, ed.), Williams & Wilkins, Baltimore, pp. 241–247.
190. Larrabee, M. G., and Horowicz, P., 1956, *Molecular Structure and Functional Activity of Nerve Cells* (N. G. Grenell and L. J. Mullins, eds.), Waverly Press, Baltimore, pp. 84–102.
191. Klingman, J. D., 1966 *Life Sci.* **5**:365–373.
192. Volle, R. L., 1980, *Handbook of Experimental Pharmacology*, Volume 53 (D. A. Kharkevich, ed.), Springer-Verlag, Berlin, pp. 281–312.
193. Horowicz, P., and Larrabee, M. G., 1958, *J. Neurochem.* **2**:102–118.
194. Dolivo, M., 1974, *Fed. Proc.* **33**:1042–1048.
195. Organisciak, D. T., and Klingman, J. D., 1974, *Lipids* **9**:307–313.
196. Roch-Ramel, F., 1962, *Helv. Physiol. Acta* [*Suppl.*] **13**:1–64.
197. Larrabee, M. G., 1981, *J. Neurochem.* **38**:215–232.
198. Nicolescu, P., Dolivo, M., Roullier, C., and Foroglou-Kerameus, C., 1966, *J. Cell Biol.* **29**:267–285.
199. Dolivo, M., and Roullier, C., 1969, *Brain Res.* **31**:111–123.
200. Organisciak, D. T., and Klingman, J. D., 1976, *Brain Res.* **115**:467–478.
201. Harris, J. U., and Klingman, J. D., 1972, *J. Neurochem.* **19**:1267–1278.
202. Larrabee, M. G., 1978, *J. Neurochem.* **31**:492.
203. Larrabee, M. G., 1979, *J. Neurochem.* **32**:283.
204. Larrabee, M. G., 1979, *J. Neurochem.* **33**:123–131.
205. Larrabee, M. G., 1980, *J. Neurochem.* **35**:210–231.
206. Larrabee, M. G., and Kauffman, F. C., 1978, *Trans. Am. Soc. Neurochem.* **9**:118.
207. Cancalon, P., and Klingman, J. D., 1974, *J. Chromatogr. Sci.* **12**:349–355.
208. Cancalon, P., and Klingman, J. D., 1972, *J. Chromatogr. Sci.* **10**:253–256.
209. McBride, W. F., and Klingman, J. D., 1968, *Anal. Biochem.* **25**:109–122.
210. Nagata, Y., Yokoi, Y., and Tsiukada, Y., 1966, *J. Neurochem.* **13**:1421–1431.
211. Masi, I., Paggi, P., Pocchiari, F., and Toschi, G., 1969, *Brain Res.* **12**:467–470.
212. Sadsivudu, B., and Lajtha, A., 1970, *J. Neurochem.* **17**:1299–1312.
213. Klingman, J. D., 1966, *Life Sci.* **5**:1397–1407.
214. Pappano, A. J., and Volle, R. L., 1966, *J. Pharmacol. Exp. Ther.* **152**:171–180.
215. Hall, M. E., and Wilson, D. L., 1979, *Brain Res.* **168**:602–608.
216. Hendry, I. A., 1976, *Rev. Neurosci.* **2**:149–194.

217. Otten, U., and Thoenen, H., 1977, *J. Neurochem.* **29**:69–76.
218. Dibner, M. D., and Black, I. B., 1978, *J. Neurochem.* **30**:1479–1484.
219. Coughlin, M. D., Dibner, M. D., Boyer, D. M., and Black, I. A., 1978, *Dev. Biol.* **88**:513–528.
220. Bohn, M. C., Goldstein, M. M., and Black, I. B., 1981, *Dev. Biol.* **82**:1–10.
221. Banks, P., 1970, *Biochem. J.* **118**:813–818.
222. Huf, K., Lakshmanan, J., and Guroff, G., 1978, *J. Neurochem.* **31**:599–606.
223. Härkönen, M. H. A., and Kauffman, F. C., 1974, *Brain Res.* **65**:127–139.
224. Kaye, P. L., Gunning, P. W., and Austin, L., 1977, *J. Neurochem.* **9**:1–22.
225. Gunning, P. W., Kaye, P. L., and Austin, L., 1977 *J. Neurochem.* **28**:1237–1240,1245–1248.
226. Otten, U., Katanaka, H., and Thoenen, H., 1978, *Brain Res.* **140**:385–389.
227. Coughlin, M. D., and Black, I. B., 1978, *J. Neurochem.* **30**:1479–1484.
228. Gorelikov, P. L., and Em, V. S., 1974, *Arkh. Anat. Gistol. Embriol.* **66**:91–97.
229. Pevzner, L. Z., Nozdrachiev, A. D., Glushchenko, T. A., and Federova, L. D., 1973, *Dokl. Acad. Nauk. S.S.S.R.* **213**:1458–1460.
230. Levedev, D. B., 1977, *Eksp. Biol. Med.* **83**:748–758.
231. Larrabee, M. G., and Klingman, J. D., 1963, *Anal. Biochem.* **6**:111–124.
232. Burt, D. R., and Larrabee, M. G., 1973, *J. Neurochem.* **21**:255–272.
233. Burt, D. R., and Larrabee, M. G., 1976, *J. Neurochem.* **27**:753–763.
234. Wells, M. A., and Dittmer, J. C., 1967, *Biochemistry* **6**:3169–3174.
235. Cornbrooks, J., Bunge, R. P., and Gottlieb, D. I., 1980, *J. Neurochem.* **34**:800–874.
236. Larrabee, M. G., Klingman, J. D., and Leicht, W. S., 1963, *J. Neurochem.* **10**:549–570.
237. Larrabee, M. G., and Leicht, W. S., 1965, *J. Neurochem.* **12**:1–13.
238. Hokin, M. R., Hokin, L. E., and Shelp, W. D., 1960, *J. Gen. Physiol.* **44**:217–226.
239. White, G. L., and Larrabee, M. G., 1973, *J. Neurochem.* **20**:783–798.
240. Lapetina, E. G., Brown, W. E., and Michell, R., 1976, *J. Neurochem.* **26**:649–651.
241. Pickard, M. R., Hawthorne, J. N., Hayashi, E., and Yamada, S., 1977, *Biochem. Pharmacol.* **26**:448–450.
242. Hawthorne, J. N., and Pickard, M. R., 1979, *J. Neurochem.* **32**:5–14.
243. Fisher, S. K., and Agranoff, B. W., 1980, *J. Neurochem.* **34**:1231–1240.
244. Robinson, J. D., and Kabela, E., 1977, *J. Neurobiol.* **8**:511–522.
245. Nagata, Y., Mikoshiba, K., and Tsukada, Y., 1973, *Brain Res.* **56**:259–269.
246. Sherman, W. R., Leavitt, A. L., Honchar, M. P., Hallicher, L. M., and Phillips, B. E., 1981, *J. Neurochem.* **36**:1947–1951.
247. Wolfe, L. S., 1982, *J. Neurochem.* **38**:1–14.
248. Hedqvist, P., 1973, *The Prostaglandins*, Volume 1 (P. W. Ramwell, ed.), Plenum Press, New York, pp. 101–129.
249. Webb, J. G., Saelens, D. A., and Halushka, P. V., 1978, *J. Neurochem.* **31**:13–20.
250. Trevisani, A., Biondi, C., Belluzzi, O., Borasio, P. G., Capuzzo, A., Ferretti, M. E., and Perri, V., 1982, *Brain Res.* **236**:375–382.
251. Belluzzi, O., Biondi, C., Borasio, P. G., Cappuzzo, A., Ferretti, M. E., Trevisani, A., and Perri, V., 1982, *Brain Res.* **236**:383–392.
252. McAfee, D. A., Schorderet, M., and Greengard, P., 1971, *Science* **178**:310–312.
253. Cramer, H., Johnson, D. G., Hanbauer, I., Silberstein, S. D., and Kopin, I. J., 1973, *Brain Res.* **53**:97–104.
254. Kalix, P., McAfee, D. A., Schorderet, M., and Greengard, P., 1974, *J. Pharmacol. Exp. Ther.* **188**:676–687.
255. Greengard, P., McAfee, D. A., and Kebabain, J. W., 1972, *Adv. Cyclic Nucleotide Res.* **2**:337–355.
256. Busis, N. A., Weight, F. F., and Smith, P. A., 1978, *Science* **200**:1079–1981.
257. McAfee, D. A., Henon, B. K., Whiting, G. J., Horn, J. P., Yarowsky, P. J., and Turner, D. K., 1980, *Fed. Proc.* **39**:2997–3002.
258. De Camilli, P., Ueda, T., Bloom, F. E., Battenberg, E., and Greengard, P., 1979, *Proc. Natl. Acad. Sci. U.S.A.* **76**:5977–5981.
259. Fried, G., Nestler, E. J., De Camilli, P., Stjärne, L., Olson, L., Lundberg, J. M., Hökfelt, T., Ouimet, C. C., and Greengard, P., 1982, *Proc. Natl. Acad. Sci. U.S.A.* **79**:2717–2721.

260. Weight, F. F., Petzold, G., and Greengard, P., 1974, *Science* **186**:942–944.
261. Kebabian, J. W., Steiner, A. L., and Greengard, P., 1975, *J. Pharmacol. Exp. Ther.* **193**:474–488.
262. Casnelli, J. E., and Greengard, P., 1974, *Proc. Natl. Acad. Sci. U.S.A.* **71**:1891–1895.
263. Kalix, P., 1976, *J. Neurochem.* **27**:1563–1564.
264. Quenzer, T., Patterson, B. A., and Volle, R. L., 1980, *J. Neurochem.* **34**:1782–1784.
265. Dempsher, J., Larrabee, M. G., Bang, F. B., and Bodian, D., 1955, *Am. J. Physiol.* **182**:203–216.
266. George, C., and Dolivo, M., 1982, *Brain Res.* **242**:255–260.
267. Price, R. W., 1979, *Science* **205**:518–520.
268. Felten, D. L., Overhage, J. M., Felten, S. Y., and Schmedtje, J. F., 1981, *Brain Res. Bull.* **7**:595–612.
269. Brown, D. A., and Garthwaite, J., 1979, *J. Physiol.* (*Lond.*) **297**:597–620.
270. Garthwaite, J., 1979, *J. Neurosci. Methods* **1**:185–193.
271. Östberg, A.-J. C., Raisman, G., Field, P. M., Iverson, L. L., and Zigmond, R. E., 1976, *Brain Res.* **107**:445–470.
272. Brinley, F. J., Jr., 1967, *J. Neurophysiol.* **30**:1531–1560.
273. Galvin, M., Brussencate, G. T., and Senekowitsch, R., 1979, *Brain Res.* **160**:544–548.
274. Wallace, L. J., Partlow, L. M., and Ferrendelli, J. A., 1978, *J. Neurochem.* **31**:801–808.
275. Lutold, B. E., Karoum, F., and Neff, N. H., 1979, *Eur. J. Pharmacol.* **54**:21–26.
276. Bertilsson, L., and Costa, E., 1976, *J. Chromatogr.* **118**:395–402.
277. Gamse, R., Wax, A., Zigmond, R. E., and Leeman, S. E., 1981, *Neuroscience* **6**:437–441.
278. Doherty, F. C., and Rowe, C. E., 1979, *J. Neurochem.* **33**:819–822.
279. Nikodijevic, B., Yu, M. W., and Guroff, G., 1977, *J. Neurochem.* **28**:851–852.
280. Härkönen, M. H. A., and Kauffman, F. C., 1974, *Brain Res.* **65**:141–157.
281. Roch-Ramel, F., and Dolivo, M., 1967, *Helv. Physiol. Pharmacol. Acta* **25**:40–61.
282. Klingman, G. I., 1965, *Pharmacologist* **7**:157.
283. Klingman, G. I., 1966, *Biochem. Pharmacol.* **15**:1729–1736.
284. Luizzi, A., Poppen, P. H., and Kopin, I. J., 1977, *Brain Res.* **138**:309–315.
285. Gilad G. M., Gagon, C., and Kopin, I. J., 1980, *Brain Res.* **183**:393–402.
286. MacDonnell, P. C., Nagaiah, K., Lakshmanan, J., and Guroff, G., 1977, *Proc. Natl. Acad. Sci. U.S.A.* **74**:4681–4684.
287. Burt, D. R., 1978, *Brain Res.* **143**:573–579.
288. Towle, A. C., Sze, P. Y., and Lauder, J. M., 1979, *Trans. Am. Soc. Neurochem.* **10**:199.
289. Gershon, M. D., 1979, *Neurosci. Res. Prog. Bull.* **17**:384–388.
290. Bowers, C. W., and Zigmond, R. E., 1979, *J. Comp. Neurol.* **185**:381–391.

16

Brain Capillaries
Structure and Function

A. Lorris Betz and Gary W. Goldstein

1. INTRODUCTION

It is now well accepted that the brain capillary is the structure responsible for the formation of a blood–brain barrier (BBB). However, brain capillaries are also involved in a number of other metabolic and transport processes that may be equally important for brain function. The study of these biochemical properties was furthered by the development of methods to isolate metabolically active microvessels from brain.[1–6] This technical advance permitted investigation of capillary properties not accessible to study *in vivo*. As a result, many new ideas concerning brain capillary function were proposed during the past decade. In this chapter we review the structure and function of the brain capillary and attempt to integrate newer biochemical findings with the traditional concept of the BBB.

2. STRUCTURE

Capillaries are microvessels with diameters generally between 3 and 7 μm. The capillary wall does not contain smooth muscle cells as observed in arterioles and some venules. In addition to the absence of muscle and elastic fibers, capillaries can be distinguished from these other microvessels by their smaller size.[7] Surrounding the capillary endothelial cell is a basement membrane within which another cell, the pericyte, may be embedded (Fig. 1). In brain capillaries there is an additional close association with astrocytes and axon terminals. Each of these components is considered separately.

2.1. Endothelial Cell

The brain capillary endothelial cell forms a continuous, nonfenestrated endothelium that restricts the movement of many polar solutes between blood

A. Lorris Betz and Gary W. Goldstein • Departments of Pediatrics and Neurology, University of Michigan, Ann Arbor, Michigan 48109.

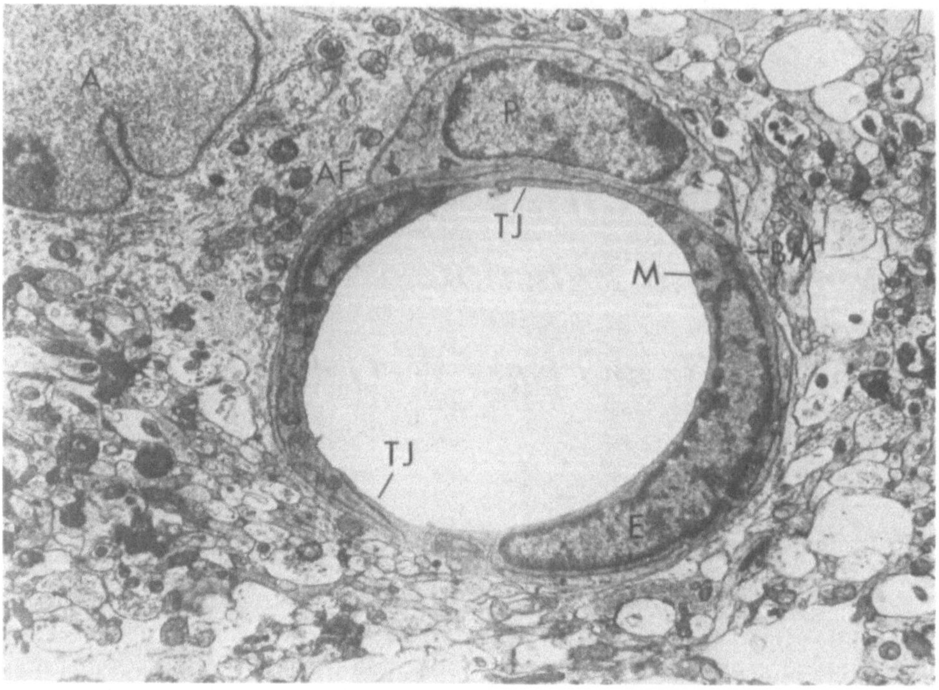

Fig. 1. Electron micrograph of brain capillary. A capillary in the cerebral cortex of rat brain is shown with the endothelial cell (E), pericyte (P), basement membrane (BM), astrocyte (A), and astrocytic foot processes (AF). Note the tight junction (TJ), numerous mitochondria (M), and few vesicles in the endothelial cell. (Courtesy of P. Cancilla.)

and brain. The classic study of Reese and Karnovsky[8] clearly demonstrated two unusual properties of endothelial cells in brain capillaries that explain the presence of a BBB. The first is the presence in brain capillaries of continuous tight junctions, which prevent the transcapillary movement of protein tracers. Subsequent studies using the freeze–fracture technique demonstrate the "very tight" nature of these junctions.[9,10] As shown in Fig. 2, the tight junctions of brain capillaries are composed of continuous, closely packed, anastomosing strands of membrane particles. This pattern is similar to the structure of tight junctions in relatively impermeable epithelia[11] but differs from the discontinuous particle arrays seen in the junctional complexes of nonbarrier capillaries such as those in the choroid plexus.[9] The complex nature of brain capillary tight junctions correlates well with the low permeability of the BBB to small organic molecules and ions.

Tight junctions are observed during the earliest stages of vasculogenesis in human brain, even before blood is flowing through the vessel lumen.[12,13] Since the BBB of the fetus appears more permeable to proteins and other polar solutes,[14] it is possible that these tight junctions are incomplete in the immature brain.[15] In the adult, the permeability of the BBB can be markedly enhanced by perfusion of the cerebrovascular bed with hypertonic agents.[16] Ultrastruc-

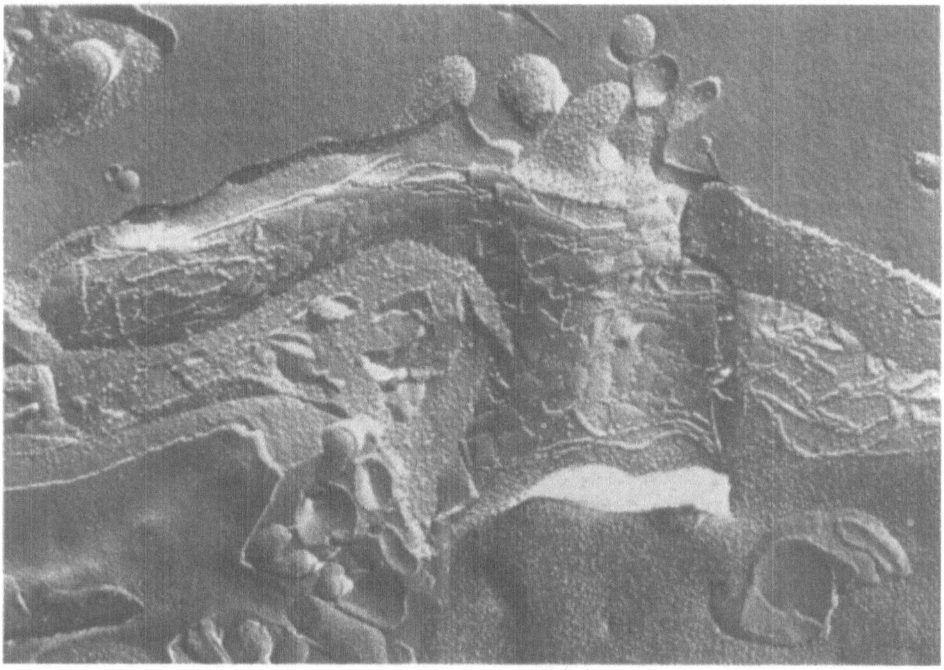

Fig. 2. Freeze–fracture replica of tight junction in isolated brain capillary. Microvessels were isolated from 30-day-old rats and then prepared for freeze–fracture analysis. Note the complex pattern of ridges and grooves formed by particles within the endothelial cell membrane. (Courtesy of R. Shivers.)

tural studies suggest that changes in tight junction permeability underlie this transitory increase in brain capillary permeability.[17] However, tight junction morphology during development or after osmotic injury has not been studied using freeze–fracture techniques.

A second major difference between brain and systemic capillaries is the smaller number of plasmalemma vesicles in brain capillary endothelium.[8] These vesicles have been proposed to play a role in the transendothelial movement of protein across muscle capillaries through the sequential steps of endocytosis from one side of the cell and exocytosis on the other side.[18] Alternatively, channels formed by chains of vesicles[19] or transitory fusion of fixed surface invaginations might be involved in protein movement across systemic capillaries.[20] In the brain capillary, the low density of these plasmalemma vesicles is consistent with the low permeability of the BBB to protein. After intravenous administration, protein tracers are occasionally found within cytoplasmic vesicles of brain endothelial cells even under normal conditions. In this situation, endocytotic vesicles appear to fuse with lysosomes and not result in transcapillary transport.[21] However, vesicular transport of proteins across the BBB may occur after a variety of insults such as hypertension, ischemia and reperfusion,[22] and administration of histamine.[23]

An additional morphological feature of brain capillary endothelial cells is their high density of mitochondria. Oldendorf *et al.*[24] calculated that the percentage of capillary endothelial cell volume occupied by mitochondria is 8–11% in brain regions that have a BBB and 2–5% in nonbarrier regions of brain or in systemic tissues. This increased mitochondrial density may reflect a higher metabolic activity related to the barrier function of these capillaries.

The biochemical composition of brain capillary endothelial cells has received less attention than their structure. Matheson *et al.* studied the types of phospholipids present in isolated brain microvessels and their response to dietary changes[25] and development.[26] Drewes and Lidinsky[27] found a difference in the labeling pattern of proteins on the luminal and antiluminal surfaces of brain capillaries. In two other studies, actin and myosin were identified in brain capillary endothelial cells and pericytes[28,29]; however, the role of these contractile proteins in capillary function is unknown.

2.2. Pericyte

As in virtually all other microvascular beds, pericytes are found within the basement membrane of brain capillaries. There are no specific structural features to indicate the function of these cells. However, because of their periendothelial location, proposed roles include synthesis of basement membrane, support and protection for endothelial cells, service as a precursor of smooth muscle cells,[30] phagocytosis of exogenous protein,[31,32] regulation of endothelial cell growth,[33] contraction of the capillary wall,[34] and mediation of information transfer to the endothelium.[30] The expression of these features may vary from organ to organ.[34]

In the brain, pericytes endocytose protein tracers administered into the cerebral ventricles.[32] Pericyte uptake of exogenous proteins from the blood is prominent after injury to the brain both adjacent to and remote from the lesion.[31] Thus, pericytes may function as phagocytic cells in the central nervous system. However, they probably are not the source of resting[35] or reactive[36] microglia.

Extensive interdigitation and peg-and-socket relationships are present between endothelial cells and pericytes of fetal human brain.[37,38] In addition, junctional complexes are sometimes seen between these cells.[39] Furthermore, pericytes contain abundant actin and myosinlike material,[28] and they appear to be innervated.[40] Thus, the pericyte might be involved in capillary contraction or dilatation. Junctions between endothelial cells and pericytes could also provide a pathway for electrochemical coupling leading to regulation of capillary permeability.[41]

In the retina, there is a temporal association between the appearance of pericytes during development and the cessation of endothelial cell growth.[42] There is also an association between the death of pericytes in disease and the proliferation of new blood vessels. Therefore, pericytes might have a role in the regulation of neural capillary growth.

2.3. Basement Membrane

The basement membrane of brain capillaries is a thin extracellular matrix located between the antiluminal membrane of the endothelial cell and the foot process of the astrocyte. As noted above, the pericyte is embedded within this structure. Early in brain development, distinct layers of filamentous material can be found in association with both the endothelial and the glial cells.[43] Later, these lamellae fuse, forming an apparently homogeneous structure, which thickens throughout postnatal development.[43,44] This apparent fusion of lamellae suggest that both glial and vascular (endothelial and pericyte) cells contribute to basement membrane formation. The functions most frequently ascribed to the basement membrane are to support cellular structures, to act as a semipermeable filter, and to provide a framework for cell migration and differentiation during development.[45]

Compositional studies of the basement membrane in brain microvessels were advanced by the development of a method for its purification. Beginning with freshly isolated brain microvessels, Carlson *et al.*[46] used detergents to selectively remove the cellular components, leaving behind ultrastructurally intact basement membrane. The amino acid composition of this material was similar to that of other basement membranes and indicated the presence of collagen. Further characterization of the types of collagen in brain microvessels was reported by Faris *et al.*[47]

Consistent with developmental thickening of the basement membrane, Betz and Goldstein[48] observed an increase in the hydroxyproline content of brain microvessels isolated from rats of different ages. This amino acid is found almost exclusively in collagen.[45] The enzyme that converts proline to hydroxyproline, prolylhydroxylase, is present in brain microvessels, and its activity increases in experimental hypertension and may be responsible for the basement membrane thickening found in this disease.[49] Basement membrane around capillaries also thickens in diabetes mellitus and may relate to the focal accumulation of collagen fibrils in brain microvessels that occurs in rats with experimental diabetes.[50] A functional significance for collagen in the basement membrane is suggested by the work of Robert *et al.*,[51] who produced an increase in BBB permeability by administration of collagenase *in vivo*.

2.4. Relationship to Astrocytes and Neurons

A special relationship exists between capillaries and astrocytes of the brain.[52] With few exceptions, cerebral capillaries are nearly completely ensheathed by astrocytic processes (Fig. 1). In fact, early investigators thought that the glial cells were the anatomic site of the BBB. However, studies with horseradish peroxidase demonstrated that proteins can freely permeate the space between astrocytic foot processes and enter the basement membrane up to but not past the tight junctions joining endothelial cells together.[53] The fact that astrocytic end feet are found attached to microvessels prepared by some methods[54] further emphasizes this association. This close contact between astrocytes and endothelial cells suggests a functional interaction. This possibility is supported by the fact that processes of the same astrocyte contact neurons or ependymal cells.[52] Potential interactions include transfer of substances, regulation of capillary activity, and induction of specialized endothelial properties.

Available evidence suggests that cerebral microvessels receive innervation from neurons originating within the brain.[39,55–58] If this is true, the potential exists for regulation by the brain of capillary blood flow and permeability.[41]

3. METABOLISM AND TRANSPORT

3.1. Energy Substrates

In contrast to other brain cells, cerebral capillaries are capable of using a variety of substrates for energy production. Isolated brain microvessels readily oxidize glucose,[1,3,48,59–61] fatty acids,[1,48,59,60] β-hydroxybutyrate,[48] and pyruvate[1] to carbon dioxide. Glucose is also converted to lactic acid[60] and lipids.[61] Biochemical[62] and histochemical[63] studies demonstrate the presence of enzymes for the glycolytic pathway, hexose monophosphate shunt, and citric acid cycle in brain microvessels. The conversion of glucose to carbon dioxide and lipids is increased by insulin[61]; however, the mechanism by which this occurs is unclear, since insulin does not directly stimulate the uptake of glucose analogues by isolated brain capillaries.[64,65]

Fatty acid metabolism may be particularly important to brain capillaries, since maximal ion transport into isolated cerebral microvessels can only be achieved if palmitate is present in addition to glucose.[59] In addition, brain microvessels from adult rats produce quantitatively similar amounts of carbon dioxide from glucose and palmitate.[48] Finally, brain microvessels contain lipoprotein lipase and acid lipase activity, which could mediate conversion of plasma lipoproteins to free fatty acids prior to their oxidation.[66]

3.2. Glucose Transport

There is considerable interest in correlating glucose transport into isolated brain capillaries with glucose transport at the BBB *in vivo*. Hexose uptake is usually studied using either of two glucose analogues; 3-0-methylglucose (3MG), which is transported but not metabolized, or 2-deoxyglucose (2DG), which is transported and then phosphorylated by hexokinase but not further metabolized. Using 3MG, Betz *et al.*[64] demonstrated that hexose transport into isolated rat brain microvessels is similar to hexose transport at the BBB *in vivo* (see ref. 67 for review). Thus, 3MG uptake is rapid, stereospecific, and equilibrative. Of the transport inhibitors tested, cytochalasin B is most potent, phloretin somewhat less effective, and phlorizin the least inhibitory. Ouabain, insulin, and 2,4-dinitrophenol are without effect. The stereospecificity is similar to that observed *in vivo* using the isolated perfused dog brain.[68] Kolber *et al.*[69] determined that the K_m for 3MG uptake into microvessels isolated from rat brains is 18 mM when assayed at 20°C. This compares to values of 6–9 mM determined *in vivo* in a number of different species.[67] Since the antiluminal membrane is exposed during incubations *in vitro*, whereas the luminal membrane is exposed *in vivo*, the similarities of sugar transport that are found using these two approaches suggest that glucose transport at the BBB is symmetrical, i.e., identical at the luminal and antiluminal membranes. Further, the rate of hexose uptake by microvessels is normally much more rapid than their rate of hexose metabolism.[64] This finding is consistent with the idea that glucose moves across the BBB passing through the endothelial cell cytoplasm. In this way, the brain capillary endothelial cells can facilitate exchange of glucose between blood and brain.

Spatz *et al.*[70] and Nell and Welch[71] found a decrease in 2DG uptake into brain microvessels isolated from gerbils subjected to bilateral carotid occlusion. Their observation correlates well with the decrease in blood-to-brain glucose flux observed during anoxia[72] or ischemia[73] in the isolated dog brain. However, use of 2DG to study hexose transport is complicated, since 2DG is phosphorylated by hexokinase. In fact, the K_m for 2DG uptake into isolated brain microvessels is between 0.09 and 0.23 mM,[4,65,71,74] which is similar to the 0.15 mM K_m of brain capillary hexokinase[75] and considerably less than the 6–9 mM K_m of glucose transport at the BBB *in vivo*.[67] Thus, when a decrease in 2DG uptake is observed in isolated capillaries after cerebral ischemia, it is not clear whether the change is in transport or in hexokinase activity. The problems in using 2DG to investigate hexose transport by isolated capillaries are discussed in more detail elsewhere.[64]

3.3. Amino Acid Transport

Amino acid transport from blood to brain *in vivo* has been well studied (see ref. 76 for review). Single-pass carotid injection techniques demonstrate saturable transport of large neutral, acidic, and basic amino acids from blood to brain. On the other hand, small neutral amino acids such as glycine and alanine show little uptake. This pattern favors uptake of essential over nonessential amino acids.

Isolated brain capillaries also exhibit saturable uptake of large neutral,[77–81] acidic,[80] and basic[78,80] amino acids. In contrast to the findings *in vivo*, small neutral amino acids are transported into isolated brain microvessels, and this transport occurs by a sodium-dependent process.[77] The discrepancy between the results obtained with small neutral amino acids *in vivo* and *in vitro* may be explained by the exposure of the antiluminal membranes of the endothelial cell to labeled amino acid *in vitro* whereas only the luminal membrane is exposed to isotopes following carotid injection *in vivo*. This suggests an asymmetry in distribution of small neutral amino acid carriers between the two sides of the brain capillary endothelial cell.[77] Such polarity would explain the active efflux of small amino acids from brain to blood[82–84] despite their limited entry from blood to brain. A similar situation may exist for cystine, since it is not transported across the BBB *in vivo*[84a] but is accumulated by microvessels isolated from rat brain.[6]

3.4. Potassium Transport

The distribution of potassium transport carriers at the BBB appears similar to that of the small neutral amino acid carriers discussed above. There is very little movement of potassium from blood to brain interstitial fluid,[85] yet potassium is readily pumped out of the brain by a saturable transport process.[86,87] This efflux occurs against a concentration gradient and is assumed to be mediated by the brain capillary. Despite the apparent lack of potassium uptake at the luminal membrane of the brain capillary endothelial cell, isolated brain microvessels accumulate potassium by a saturable and ouabain-inhibitable transport process.[59,88,89] This inhibition of potassium transport by ouabain correlates well with the presence of a substantial activity of Na^+,K^+-ATPase in isolated brain capillaries.[62,88] Taken together, these data suggest that Na^+,K^+-ATPase is involved in the active efflux of potassium from brain to blood and, further, that Na^+,K^+-ATPase is asymmetrically distributed between the two sides of the brain capillary endothelial cell. This hypothesis is supported by a cytochemical study showing Na^+,K^+-ATPase activity present in the antiluminal but not the luminal membrane of endothelial cells in rat brain capillaries.[90,91]

The polar distribution of Na^+,K^+-ATPase in brain capillary endothelial cells suggests that these cells play an important role in the regulation of the potassium concentration in the brain interstitial fluid. In the past, astrocytes were thought to act as the primary potassium buffer in brain.[92] However, these cells must eventually release potassium back into the interstitial fluid, and, therefore, acting alone, astrocytes could not long maintain the concentration of potassium in the interstitial fluid at 3 mM in the presence of plasma levels that normally range from 3.5 to 5.0 mM. The astrocytes could act in conjunction with the endothelial cells, and, after accumulating potassium from areas of intense neural activity, the astrocytes might release the potassium through their foot processes into the basement membrane surrounding the endothelial cells. The potassium could then be transported out of the brain through the endothelial cells. The active role of the brain capillary in pumping potassium from brain to blood suggests that the brain endothelial cell should be capable of greater energy production than systemic capillaries. This possibility is supported by their greater mitochondrial density.[24] The cooperation between astrocytes and endothelial cells in maintaining potassium homeostasis within the brain may be one reason for their close structural relationship (see Section 2.4).

The schematic shown in Fig. 3 summarizes our hypothesis for the distribution of the transport systems discussed in this chapter. Glucose and large neutral amino acids readily exchange between the blood and the brain, and, therefore, we propose that transport systems for these solutes are located on both the luminal and antiluminal membranes. In contrast, the transport activities mediated by Na^+,K^+-ATPase and the small neutral amino acid carrier are restricted to the antiluminal membrane. Although not shown in our figure, it is likely that other transport systems, such as those for iodide, organic acids, monoamines, and prostaglandins, may show a similar polarity (see refs. 93, 94 for reviews).

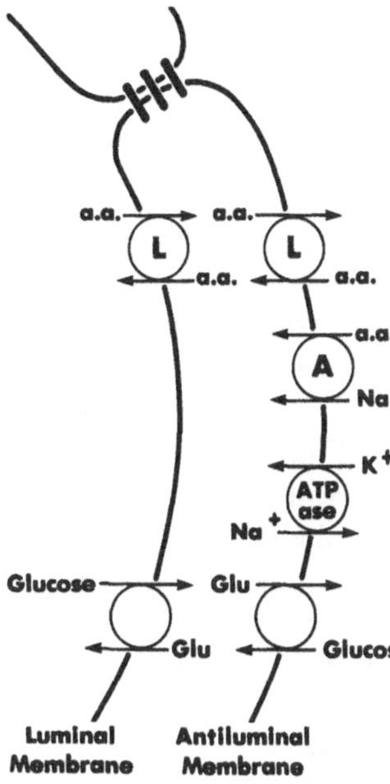

Fig. 3. Schematic diagram of the polar model of brain capillary endothelial cell. This proposed model is based on the results of transport studies *in vivo* and *in vitro*[77] and on the cytochemical localization of Na^+,K^+-ATPase.[91] The carriers for glucose and for large neutral amino acids (L) are shown on both the luminal and antiluminal membrane, whereas Na^+,K^+-ATPase and the sodium-dependent carrier for small neutral amino acids (A) are found on only the antiluminal membrane. This asymmetrical distribution of active ion and amino acid transport systems would provide the basis for potassium and amino acid transport from brain to blood against a concentration gradient.

3.5. Neurotransmitter Metabolism

Neurotransmitter metabolism by the BBB was first demonstrated by the histochemical studies of Bertler *et al.*[95] showing the presence of L-DOPA decarboxylase in brain capillaries. The localization of this enzyme in brain capillary endothelial cells and pericytes has led to the idea of an "enzymatic BBB," which limits entry into the brain of circulating neurotransmitters and their precursors. Studies with isolated brain capillaries have confirmed the original histochemical experiments. Isolated microvessels take up L-DOPA by a stereospecific and energy-dependent transport step.[96] L-DOPA is subsequently converted to dopamine within the capillary wall. Since capillaries also contain MAO,[48,96–100] the dopamine generated after L-DOPA uptake is eventually degraded to inactive products. Intracapillary MAO may also play a role in the inactivation of neurotransmitters released by neuronal activity, since monoamines are actively accumulated and metabolized by isolated brain capillaries.[101–103] It is likely that this monoamine uptake system is present only on the antiluminal membrane of the brain capillary endothelial cell, since monoamines show very little uptake from the luminal side.[104] Thus, MAO would degrade monoamines that are actively accumulated from brain interstitial fluid and those that are formed by L-DOPA decarboxylase as well as the small amount that might enter the endothelial cell from the blood.[105] Confirmation of this schema is provided by the study of Hardebo *et al.*[106] on monoamine uptake by the brain after disruption of the BBB.

Other enzymes of neurotransmitter metabolism such as tyrosine hydrosylase and dopamine β-hydroxylase, as well as norepinephrine itself, are also present in brain microvessels.[97,99] However, these enzymes are thought to be located within the axon terminals associated with isolated microvessel preparations.[97,105]

Butyryl- or nonspecific cholinesterase activity is found associated with capillaries in most regions of the brain,[107] whereas acetyl- or specific cholinesterase is found mainly in capillaries of those brain regions with high acetylcholinesterase activity in the neurons.[108] It is now clear that butyrylcholinesterase is synthesized by and located within the endothelial cells, whereas acetylcholinesterase is located in the basement membrane, perhaps after release from dendrites.[108] Capillary acetylcholinesterase may therefore assist in degradation of locally released acetylcholine. The function of butyrylcholinesterase is unknown.

Finally, Pardridge and Mietus[109] recently studied the interaction of enkephalin and isolated brain capillaries. Although no specific binding of leucine enkephalin was detected, the capillaries did contain an enkephalinase activity. This study provides further support to the notion that brain capillaries are involved in the degradation of a variety of potential neurotransmitter substances.

3.6. Receptors and Adenylate Cyclase

The presence or absence of receptors in brain capillaries for various agents has been an area of active investigation during the past 4 years. Early investigators studied increases in the cyclic AMP content of isolated brain microvessels after their exposure to potential agonists to show the presence of a receptor. More recent studies have used binding of radioactive ligands. Table I lists the receptors that are present in brain capillaries, and Table II lists those assayed for but not found. In general, most studies found that brain capillaries contain β-adrenergic receptors with the β_2 class in predominance. Receptors for dopamine were found by most but not all investigators. The presence of a histamine receptor is less clear, and in positive studies one group found only H_1 receptors,[118] whereas the other found H_2.[116] The α_2-adrenergic receptor and receptors for prostaglandins and adenosine are less well studied. Other enzymes involved in cyclic nucleotide metabolism including guanylate cyclase,[125,126] and phosphodiesterases for cyclic AMP and cyclic GMP are present in brain capillaries.[127]

Since isolated capillaries contain more than one cell type, the question arises as to where these receptors and enzymes are located. In a cytochemical study, Joó and Tóth demonstrated adenylate cyclase activity along the luminal and antiluminal surfaces of the endothelial cell as well as the astrocytic foot process.[128] The pericyte was not described. Membranes of astrocytic foot processes are sometimes found in isolated brain capillaries[54] and may therefore contribute to receptor binding. Axon terminals with presynaptic receptors might also contribute to measurable microvascular binding sites. Finally, histamine presumed to be within mast cells of the capillary wall is found in isolated brain capillaries,[129,130] and the presence of mast cells may also influence results. Because of the multiple possible binding sites present in isolated microvessels,

Table I
Receptors Demonstrated to Be Present in Brain Capillaries

Receptor	Assay[a]	Species	References
β-adrenergic	Cyclic AMP	Rat	110–112,115
	Cyclic AMP	Guinea pig	113
	Cyclic AMP	Cat	114,117
	Binding	Rat	120–122
	Binding	Cow	118,119
	Binding	Pig	120
$\beta_2 > \beta_1$	Cyclic AMP	Cat	117
	Binding	Rat	122
	Binding	Pig	120
α-Adrenergic	Binding	Cow	118
α_2	Binding	Pig	120
Dopamine	Cyclic AMP	Rat	110,111,115
	Cyclic AMP	Cat	114
Histamine	Cyclic AMP	Rat	110
	Cyclic AMP	Guinea pig	113
$H_2 > H_1$	Cyclic AMP	Guinea pig	116
$H_1 > H_2$	Binding	Cow	118
Adenosine	Cyclic AMP	Rat	112
	Cyclic AMP	Guinea pig	113
Prostaglandins			
PGE_1, PGE_2, prostacyclin	Cyclic AMP	Guinea pig	113
Insulin	Binding	Rat	123
	Binding	Cow	61,124

[a] The presence of neurotransmitter or hormonal receptors was assayed either by measuring increases in cellular cyclic AMP content or by quantitating the binding of radioactive ligands.

caution is advised in assuming that the receptors described above are located exclusively in endothelial cells.

Despite these limitations, the possibility that endothelial cells in brain capillaries are under hormonal regulation is appealing. For example, a functional significance for receptors in endothelial cells may be found in the observation of Raichle *et al.*,[41] that changes in cerebral blood flow and water permeability at the BBB follow adrenergic stimulation or in the enhanced pinocytosis within endothelial cells described by Joó[131] after administration of dibutyryl cyclic AMP.

3.7. Prostaglandins

Prostaglandins are another class of potential regulators of microvascular function. For this reason, the synthesis of prostaglandins by isolated brain microvessels has begun to receive attention. These early studies demonstrate the presence of various prostaglandins and their synthesizing enzymes.[132–137] The exact role of these substances in brain capillary function remains to be determined.

Table II
Receptors Demonstrated to Be Absent in Brain Capillaries

Receptor	Assay[a]	Species	References
α-Adrenergic	Cyclic AMP	Rat	110–112
α_1	Binding	Rat	120
	Binding	Pig	120
Dopamine	Cyclic AMP	Rat	112
Histamine	Cyclic AMP	Rat	111,112,115
H_1	Cyclic AMP	Guinea pig	116
Serotonin	Cyclic AMP	Rat	110,112
	Binding	Cow	118
Prostaglandins			
PGE_1, PGE_2, $PGF_{2\alpha}$	Cyclic AMP	Rat	111
Acetylcholine	Cyclic AMP	Rat	112
Muscarinic	Binding	Rat	120
	Binding	Cow	118
	Binding	Pig	120
Octopamine	Cyclic AMP	Rat	110
Apomorphine	Cyclic AMP	Rat	111
	Binding	Cow	118
Angiotensin II	Cyclic AMP	Rat	112
Substance P	Cyclic AMP	Rat	112
Vasopressin	Cyclic AMP	Rat	112
Benzodiazepine	Binding	Cow	118
Enkephalin	Binding	Cow	109

[a] The presence of neurotransmitter or hormonal receptors was assayed either by measuring increases in cellular cyclic AMP content or by quantitating the binding of radioactive ligands.

4. CULTURE OF ENDOTHELIAL CELLS

Isolated microvessels have been useful in characterizing the biochemical properties of the specialized endothelium found in brain capillaries. There are, however, several limitations to the use of freshly isolated brain microvessels. Studies are restricted to short incubations lasting minutes to several hours, so it is not possible to study the long-term response of microvessels to injury. In addition, investigations of transport by isolated capillaries are limited to studies of solute movement into the cells rather than across a layer of the cells. Furthermore, the microvessels contain pericytes and occasional smooth muscle cells as well as endothelial cells. This fact hinders interpretation of some of the metabolic responses of the isolated capillaries. Finally, although growth of new brain capillaries is a prominent reaction in several diseases, this proliferative response cannot be studied with isolated microvessels. Since these problems might be circumvented by growing purified capillary endothelial cells in culture, several laboratories have recently begun to develop methods for culturing endothelial cells from brain microvessels. However, the identification of endothelial cells in culture can be subject to misinterpretation. Therefore, we shall present in some detail the culture methods in current use and the criteria that should be used to establish the endothelial origin of brain capillary cells in culture.

4.1. Cell Identification

Most investigators acknowledge the difficulty of obtaining pure cultures of brain capillary endothelial cells. This is not surprising, since the isolated microvessels that serve as the source for endothelial cells also contain pericytes, smooth muscle cells, and, in some cases, also glial cells. It is therefore very important to characterize the cells present in the cultures. Ideally, cultured brain microvascular endothelial cells should express general endothelial cell markers as well as the more specialized properties of brain capillary endothelial cells.

4.1.1. General Markers for Capillary Endothelium

Factor VIII antigen is the most widely used marker for endothelial cells, and it is known to be present in brain capillaries *in vivo*.[138] Although the antibody used in the usual immunofluorescent test is raised against human factor VIII, it has been useful in detecting factor VIII antigen in cells from other species.[138–140]

Angiotensin-converting enzyme activity is present in most if not all endothelial cells including those of brain microvessels.[141–143] However, its specificity for endothelial cells has been questioned, since converting enzyme activity can be demonstrated under certain conditions in macrophages.[144] Nevertheless, presumptive pericytes[139] and smooth muscle cells[145] have a much lower activity than capillary endothelial cells.

Endothelial cells are known to have a luminal surface that does not bind platelets.[139,146] Although this fact has not often been used to characterize endothelial cell cultures, it is a simple and potentially useful test.

A cobblestone appearance has been proposed as characteristic of large-vessel endothelial cells in culture.[147] It is not known whether microvascular endothelial cells should have a similar morphology; however, such a criterion must be used with caution because of the known variations in cell morphology related to the phase of growth, the culture conditions, and the growth surface.[148] Well-differentiated brain capillary endothelial cells should grow in sheets of closely apposed cells since they will be joined by continuous tight junctions.

Capillary endothelial cells appear to require a growth surface coated with gelatin or collagen[149] for optimal attachment and proliferation. This need has also been noted for capillaries derived from brain[139]; however, it may be possible to overcome this requirement if sufficient quantities of attachment factors such as fibronectin are present in the serum added to the culture medium.

Weibel–Palade[150] bodies are sometimes used as markers for endothelial cells, although there is considerable species variation.[149] Although they are commonly observed in human large vessel endothelium, they are rarely found in capillaries of the human brain.[151,152]

4.1.2. Specific Markers for Brain Capillary Endothelium

Brain capillaries form a selective permeability barrier between plasma and the tissue interstitial fluid. To produce this barrier, brain capillary endothelial cells have certain features that are not present in systemic capillary endothelial cells, most notably the presence of continuous tight junctions and a low rate of pinocytosis. These properties should, therefore, be expressed in cultures of differentiated brain endothelial cells.

Brain capillaries contain a high activity of certain enzymes such as alkaline phosphatase,[3] γ-glutamyl transpeptidase,[2] and L-DOPA decarboxylase.[95] These enzymes may be useful markers for neural capillary endothelial cells; however, their specificity has not been proven. In fact, L-DOPA decarboxylase[153] and alkaline phosphatase[154] are clearly present in both brain capillary endothelial cells and pericytes. Since the pericyte is the most likely contaminant of endothelial cell cultures, these enzymes by themselves are not adequate markers to indicate the endothelial origin of capillary cells in culture. The distribution of γ-glutamyl transpeptidase between endothelial cells and pericytes has not yet been determined.

4.2. Culture Methods

In order to eliminate contamination by nonvascular cells, most investigators use isolated brain capillaries as the starting material for tissue culture. In some cases, the isolated capillaries are further treated with enzymes to release the various cell types from the basement membrane in which they are enclosed, and this step may be useful for minimizing pericyte contamination in the subsequent cultures. Other investigators use capillaries from young animals, where fewer pericytes are present,[44] in order to obtain an enriched endothelial population. Finally, a few investigators have been able to obtain pure endothelial cultures through serial passage of the cells, although most work has been done with primary cultures.

The first report of the successful culture of cells derived from brain microvessels was made in 1978 by Panula *et al.*,[155] who isolated microvessels from the brains of 3-day-old rats and plated these directly onto coverslips. They observed the growth of several different morphological cell types but found one particular cell, described as large and flat, that clearly arose from capillary fragments. Some but not all of these cells were histochemically positive for alkaline phosphatase, whereas L-DOPA was taken up by all cells.

DeBault *et al.*[156] described the growth of cells from microvessels isolated from weanling and young adult mouse brain. Capillaries were trapped on nylon meshes after homogenization of the brain and then transferred to tissue culture dishes or glass coverslips and incubated in modified Lewis medium containing 30% fetal bovine serum (FBS). Individual microvessel segments were then classified according to size and number of branching points, and it was determined that the smallest fragments contained endothelial cells but few pericytes. After several days, flat polygonal cells that were factor VII positive[138] emerged from the ends of these capillary segments and gave rise to plaquelike colonies. The cells in these colonies did not contain junctional complexes or Weibel–Palade bodies. Eventually, a cell line designated ME-2 was established and maintained for up to 30 passages.[138] These cells formed gap junctions when confluent but were factor VIII negative. When injured, however, some of the newly proliferating and/or migrating cells at the edge of the injury became transiently positive for factor VIII antigen.

Likewise, γ-glutamyl transpeptidase was not present in ME-2 cells but could be induced by contact with C-6 glioma cells.[157] Whole platelets and serum derived from platelet-rich plasma induce a mitogenic response in preconfluent ME-2 cells.[158] However, the failure of ME-2 cells to produce factor VIII antigen and the absence of tight junctions indicate that they no longer express important markers of differentiated function. The best evidence for the endothelial nature of the ME-2 cell line was provided by specific cross reactivity of an antibody raised against ME-2 cell plasma membranes with microvascular endothelium in frozen sections of brain, liver, and kidney.[138]

A different approach to capillary endothelial cell culture was taken by Phillips *et al.*[140] White matter was dissected from adult rat or cow brains, minced, and then treated briefly with tryspin. The resulting cells were plated onto tissue culture flasks in Medium 199 with 20% FBS. After several days, isolated flattened clumps of cells appeared and migrated out as loose sheets. By 2–3 weeks, the cultures became confluent with polygonal or fusiform cells that could be serially passaged with no apparent change in cell morphology. The endothelial origin of these cells was indicated by the presence of factor VIII antigen, alkaline phosphatase, and Weibel–Palade bodies.

Spatz *et al.*[159] described the culture of cells from microvessels that had been isolated from 2-day-old rat brain. The isolated microvessels were prepared for culture by brief treatment with trypsin and collagenase, and the dissociated cells were placed in tissue culture flasks in a modified Medium 199 containing 30% FBS. Cells grew as sheets or networks of elongated cells that became confluent in about 4 weeks. These primary cultures were subcultured twice and then examined histochemically. Alkaline phosphatase was present in most cells, and butyrylcholinesterase in all cells. γ-Glutamyl transpeptidase and L-DOPA uptake were present most prominently in multilayered groups of cells and some individual cells that were surrounded by nonreactive cells. Electron microscopy revealed gap junctions but no tight junctions, indicating loss of a specialized property of brain capillary endothelial cells. Pinocytotic vesicles were present in moderate numbers. The presence or absence of factor VIII antigen was not reported, and until this information is provided, the endothelial nature of these cells is uncertain. The presence in these cells of specific receptors linked to adenylate cyclase was recently reported.[160]

The most recent description of a method for culturing endothelial cells from brain microvessels is that by Bowman *et al.*[139] Microvessels were isolated from 30-day-old rat brain and then treated with collagenase and dispase. Clumps of endothelial cells were separated from pericytes by centrifugation on a Percoll gradient and then plated onto collagen-coated tissue culture dishes. After several days, the clumps spread into groups of closely associated elongated or polygonal cells that began to proliferate as discrete colonies. Confluence was reached after 7 to 10 days. The cells in these cultures were positive for factor VIII antigen, contained angiotensin-converting enzyme activity, and did not bind platelets. Electron microscopy revealed typical pentalaminar tight junctions and few pinocytotic vesicles.

5. CONCLUSIONS

The past decade has seen a great increase in information concerning brain capillary structure and function. Traditional histological and physiological approaches to the study of the BBB are now being expanded by the use of the latest techniques in cell isolation and biochemistry. The goal is to understand the regulation of brain capillary permeability at a cellular and molecular level. We anticipate continued rapid progress in this area and expect that this will lead to more effective treatment of disorders in brain capillary function.

ACKNOWLEDGMENTS. This work was supported in part by grants ES02380, EY03772, and HL26840 from the National Institutes of Health. Dr. Betz is an Established Investigator of the American Heart Association. We are indebted to Lillian Politser for continued assistance in the preparation of manuscripts.

REFERENCES

1. Brendel, K., Meezan, E., and Carlson, E. C., 1974, *Science* **185**:953–955.
2. Orlowski, M., Sessa, G., and Green, J. P., 1974, *Science* **184**:66–68.
3. Goldstein, G. W., Wolinsky, J. S., Csejtey, J., and Diamond, I., 1975, *J. Neurochem.* **25**:715–717.
4. Mršulja, B. B., Mršulja, B. J., Fujimoto, T., Klatzo, I., and Spatz, M., 1976, *Brain Res.* **110**:361–365.
5. Williams, S. K., Gillis, J. F., Matthews, M. A., Wagner, R. C., and Bitensky, M. W., 1980, *J. Neurochem.* **35**:374–381.
6. Hwang, S. M., Weiss, S., and Segal, S., 1980, *J. Neurochem.* **35**:417–424.
7. Baez, S., 1977, *Annu. Rev. Physiol.* **39**:391–415.
8. Reese, T. S., and Karnovsky, M. J., 1967, *J. Cell Biol.* **34**:207–217.
9. Dermietzel, R., 1975, *Cell Tissue Res.* **164**:45–62.
10. Shivers, R. R., 1979, *Brain Res.* **169**:221–230.
11. Claude, P., and Goodenough, D. A., 1973, *J. Cell Biol.* **58**:390–400.
12. Hauw, J.-J., Berger, B., and Escourolle, R., 1975, *Acta Neuropathol.* (*Berl.*) **31**:229–242.
13. Møllgåard, K., and Saunders, N. R., 1975, *J. Neurocytol.* **4**:453–468.
14. Saunders, N. R., 1977, *Exp. Eye Res.* **25**(Suppl):523–550.
15. Delorme, P., Gayet, J., and Grignon, G., 1970, *Brain Res.* **22**:269–283.
16. Rapoport, S. I., Fredericks, W. R., Ohno, K., and Pettigrew, K. D., 1980, *Am. J. Physiol.* **238**:R421–R431.
17. Brightman, M. W., Hori, M., Rapoport, S. I., Reese, T. S., and Westergaard, E., 1973, *J. Comp. Neurol.* **152**:317–326.
18. Simionescu, N., Simionescu, M., and Palade, G. E., 1978, *Microvasc. Res.* **15**:17–36.
19. Simionescu, N., Simionescu, M., and Palade, G. E., 1975, *J. Cell Biol.* **64**:586–607.
20. Bundgaard, M., Frøkjaer-Jensen, J., and Crone, C., 1979, *Proc. Natl. Acad. Sci. U.S.A.* **76**:6439–6442.
21. Broadwell, R. D., and Salcman, M., *Proc. Natl. Acad. Sci. U.S.A.* **78**:7820–7824.
22. Westergaard, E., 1977, *Acta Neuropathol.* (*Berl.*) **39**:181–187.
23. Joó, F., Dux, E., Karnushina, I. L., Halász, N., Gecse, A., Ottlecz, Á., and Mezei, Z., 1981, *Agents Actions* **11**:129–134.
24. Oldendorf, W. H., Cornford, M. E., and Brown, W. J., 1977, *Ann. Neurol.* **1**:409–417.
25. Matheson, D. F., Oei, R., and Roots, B. I., 1980, *Neurochem. Res.* **5**:43–59.
26. Matheson, D. F., Oei, R., and Roots, B. I., 1981, *Dev. Neurosci.* **4**:201–210.
27. Drewes, L. R., and Lidinsky, W. A., 1980, *Adv. Exp. Med. Biol.* **131**:17–27.

28. Owman, C., Edvinsson, L., Hardebo, J. E., Groschel-Stewart, U., Unsicker, K., and Walles, B., 1977, *Acta Neurol. Scand.* [*Suppl*] **64**:384–385.
29. Le Beaux, Y. J., and Willemot, J., 1978, *Exp. Neurol.* **58**:446–454.
30. Rhodin, J. A. G., 1968, *J. Ultrastruct. Res.* **25**:452–500.
31. Cancilla, P. A., Baker, R. N., Pollock, P. S., and Frommes, S. P., 1972, *Lab. Invest.* **26**:376–383.
32. Wagner, H.-J., Pilgrim, C., and Brandl, J., 1974, *Acta Neuropathol.* (*Berl.*) **27**:299–315.
33. Crocker, D. J., Murad, T. M., and Geer, J. C., 1970, *Exp. Mol. Pathol.* **13**:51–65.
34. Tilton, R. G., Kilo, C., Williamson, J. R., and Murch, D. W., 1979, *Microvasc. Res.* **18**:336–352.
35. Stensaas, L. J., 1975, *Cell Tissue Res.* **158**:517–541.
36. Wolinsky, J. S., Jubelt, B., Burke, S., and Narayan, O., 1982, *Ann. Neurol.* **11**:59–68.
37. Povlishock, J. T., Martinez, A. J., and Moossy, J., 1977, *Am. J. Anat.* **149**:439–452.
38. Allsopp, G., and Gamble, H. J., 1979, *J. Anat.* **128**:155–168.
39. Forbes, M. S., Rennels, M. L., and Nelson, E., 1977, *Am. J. Anat.* **149**:47–70.
40. Rennels, M. L., and Nelson, E., 1975, *Am. J. Anat.* **144**:233–241.
41. Raichle, M. E., Hartman, B. K., Eichling, J. O., and Sharpe, L. G., 1975, *Proc. Natl. Acad. Sci. U.S.A.* **72**:3726–3730.
42. Kuwabara, T., and Cogan, D. G., 1963, *Arch. Ophthalmol.* **69**:492–502.
43. Bar, T., and Wolff, J. R., 1972, *Z. Zellforsch.* **133**:231–248.
44. Donahue, S., and Pappas, G. D., 1961, *Am. J. Anat.* **108**:331–347.
45. Kefalides, N. A., 1980, *Advances in Microcirculation,* Volume 9 (B. M. Altura, E. Davis, and H. Harders, eds.), S. Karger, Basel, pp. 295–322.
46. Carlson, E. C., Brendel, K., Hjelle, J. T., and Meezan, E., 1978, *J. Ultrastruct. Res.* **62**:26–53.
47. Faris, B., Mozzicato, P., Ferrera, R., Glembourtt, M., Toselli, P., and Franzblau, C., 1982, *Microvasc. Res.* **23**:171–179.
48. Betz, A. L., and Goldstein, G. W., 1981, *J. Physiol.* (*Lond.*) **312**:365–376.
49. Ooshima, A., Fuller, G., Cardinale, G., Spector, S., and Udenfriend, S., 1975, *Science* **190**:898–900.
50. Mukai, N., Hori, S., and Pomeroy, M., 1980, *Acta Neuropathol.* (*Berl.*) **51**:79–84.
51. Robert, A. M., Godeau, G., Miskulin, M., and Moati, F., 1977, *Neurochem. Res.* **2**:449–455.
52. Wolff, J., 1963, *Z. Zellforsch.* **60**:409–431.
53. Brightman, M. W., and Reese, T. S., 1969, *J. Cell Biol.* **40**:648–677.
54. White, F. P., Dutton, G. R., and Norenberg, M. D., 1981, *J. Neurochem.* **36**:328–332.
55. McDonald, D. M., and Rasmussen, G. L., 1977, *J. Comp. Neurol.* **173**:475–496.
56. Itakura, T., Yamamoto, K., Tohyama, M., and Shimizu, N., 1977, *Stroke* **8**:360–365.
57. Swanson, L. W., Connelly, M. A., and Hartman, B. K., 1977, *Brain Res.* **136**:166–173.
58. Reinhard, J. F., Leibmann, J. E., Schlosberg, A. J., and Moskowitz, M. A., 1979, *Science* **206**:85–87.
59. Goldstein, G. W., 1979, *J. Physiol.* (*Lond.*) **286**:185–195.
60. Chan, C. T., Brecher, P., Haudenschild, C., and Chobanian, A. V., 1979, *Microvasc. Res.* **18**:353–369.
61. Pillion, D. J., Haskell, J. F., and Meezan, E., 1982, *Biochem. Biophys. Res. Commun.* **104**:686–692.
62. Djuričić, B. M., Rogač, L., Spatz, M., Rakić, L. M., and Mršulja, B. B., 1978, *Adv. Neurol.* **20**:197–205.
63. Cook, B. H., Granger, H. J., Granger, D. N., Taylor, A. E., and Smith, E. E., 1978, *Stroke* **9**:165–168.
64. Betz, A. L., Csejtey, J., and Goldstein, G. W., 1979, *Am. J. Physiol.* **236**:C96–C102.
65. Goldstein, G. W., Csejtey, J., and Diamond, I., 1977, *J. Neurochem.* **28**:725–728.
66. Brecher, P., and Kuan, H.-T., 1979, *J. Lipid Res.* **20**:464–471.
67. Lund-Andersen, H., 1979, *Physiol. Rev.* **59**:305–352.
68. Betz, A. L., Drewes, L. R., and Gilboe, D. D., 1975, *Biochim. Biophys. Acta* **406**:505–515.
69. Kolber, A. R., Bagnell, C. R., Krigman, M. R., Hayward, J., and Morell, P., 1979, *J. Neurochem.* **33**:419–431.

70. Spatz, M., Mršulja, B. B., Micic, D., Mršulja, B. J., and Klatzo, I., 1977, *Brain Res.* **120**:141–145.
71. Nell, J. H., and Welch, K. M. A., 1980, *Ann. Neurol.* **7**:457–461.
72. Betz, A. L., Gilboe, D. D., and Drewes, L. R., 1974, *Brain Res.* **67**:307–316.
73. Kintner, D., Costello, D. J., Levin, A. B., and Gilboe, D. D., 1980, *Am. J. Physiol.* **239**:E501–E509.
74. Stefanovich, V., and Gojowczyk, G., 1981, *Neurochem. Res.* **6**:431–440.
75. Djuričić, B. M., and Mršulja, B. B., 1979, *Experientia* **35**:169–171.
76. Pardridge, W. M., 1977, *Nutrition and the Brain,* Volume 1 (R. J. Wurtman and J. J. Wurtman, eds.), Raven Press, New York, pp. 141–203.
77. Betz, A. L., and Goldstein, G. W., 1978, *Science* **202**:225–227.
78. Sershen, H., and Lajtha, A., 1976, *Exp. Neurol.* **53**:465–474.
79. Hjelle, J. T., Baird-Lambert, J., Cardinale, G., Spector, S., and Udenfriend, S., 1978, *Proc. Natl. Acad. Sci. U.S.A.* **75**:4544–4548.
80. Cardelli-Cangiano, P., Cangiano, C., James, J. H., Jeppsson, B., Brenner, W., and Fischer, J. E., 1981, *J. Neurochem.* **36**:627–632.
81. Gozes, I., Cronin, B. L., and Moskowitz, M. A., 1981, *J. Neurochem.* **36**:1311–1315.
82. Lajtha, A., and Toth, J., 1961, *J. Neurochem.* **8**:216–225.
83. Murray, J. E., and Cutler, R. W. P., 1970, *Arch. Neurol.* **23**:23–31.
84. Lorenzo, A. V., and Snodgrass, S. R., 1972, *J. Neurochem.* **19**:1287–1298.
84a. Wade, L. A., and Brady, H. M., 1981, *J. Neurochem.* **37**:730–734.
85. Hansen, A. J., Lund-Andersen, H., and Crone, C., 1977, *Acta Physiol. Scand.* **101**:438–445.
86. Bradbury, M. W. B., Segal, M. B., and Wilson, J., 1972, *J. Physiol.* (*Lond.*) **221**:617–632.
87. Bradbury, M. W. B., and Štulcová, B., 1970, *Physiol.* (Lond.) **208**:415–430.
88. Eisenberg, H. M., and Suddith, R. L., 1979, *Science* **206**:1083–1085.
89. Chaplin, E. R., Free, R. G., and Goldstein, G. W., 1981, *Biochem. Pharmacol.* **30**:241–245.
90. Firth, J. A., 1977, *Experientia* **33**:1093–1094.
91. Betz, A. L., Firth, J. A., and Goldstein, G. W., 1980, *Brain Res.* **192**:17–28.
92. Katzman, R., and Pappius, H. M., 1973, *Brain Electrolytes and Fluid Metabolism,* Williams & Wilkins, Baltimore.
93. Davson, H., 1976, *J. Physiol.* (*Lond.*) **255**:1–28.
94. Betz, A. L., and Goldstein, G. W., 1980, *Adv. Exp. Med. Biol.* **131**:5–16.
95. Bertler, A., Falck, B., Owman, C., and Rosengren, E., 1966, *Pharmacol. Rev.* **18**:369–385.
96. Hardebo, J. E., Falck, B., and Owman, C., 1979, *Acta Physiol. Scand.* **107**:161–167.
97. Lai, F. M., Udenfriend, S., and Spector, S., 1975, *Proc. Natl. Acad. Sci. U.S.A.* **72**:4622–4625.
98. Lai, F. M., and Spector, S., 1978, *Arch. Int. Pharmacodyn.* **233**:227–234.
99. Hardebo, J. E., Emson, P. C., Falck, B., Owman, C., and Rosengren, E., 1980, *J. Neurochem.* **35**:1388–1393.
100. Haenick, D. H., Ladman, R. K., Weiss, J., Boehme, D. H., and Vogel, W. H., 1981, *Experientia* **37**:764–765.
101. Hardebo, J. E., and Owman, C., 1980, *Acta Physiol. Scand.* **108**:223–229.
102. Abe, T., Abe, K., Rausch, W. D., Klatzo, I., and Spatz, M., 1980, *Adv. Exp. Med. Biol.* **131**:45–55.
103. Spatz, M., Maruki, C., Abe, T., Rausch, W. D., Abe, K., and Merkel, N., 1981, *Brain Res.* **220**:214–219.
104. Oldendorf, W. H., 1971, *Am. J. Physiol.* **221**:1629–1639.
105. Hardebo, J. E., and Owman, C., 1980, *Ann. Neurol.* **8**:1–11.
106. Hardebo, J. E., Edvinsson, L., MacKenzie, E. T., and Owman, C., 1979, *Acta Neuropathol.* (*Berl.*) **47**:145–150.
107. Joó, F., Varkonyi, T., and Csillik, B., 1967, *Histochemie* **9**:140–148.
108. Kreutzberg, G. W., Kaiya, H., and Tóth, L., 1979, *Histochemistry* **61**:111–112.
109. Pardridge, W. M., and Mietus, L. J., 1981, *Endocrinology* **109**:1138–1143.
110. Baca, G. M., and Palmer, G. C., 1978, *Blood Vessels* **15**:286–298.
111. Palmer, G. C., and Palmer, S. J., 1978, *Life Sci.* **23**:207–216.
112. Herbst, T. J., Raichle, M. E., and Ferrendelli, J. A., 1979, *Science* **204**:330–332.

113. Huang, M., and Drummond, G. I., 1979, *Mol. Pharmacol.* **16**:462–472.
114. Nathanson, J. A., and Glaser, G. H., 1979, *Nature* **278**:567–569.
115. Palmer, G. C., 1979, *Biochem. Pharmacol.* **29**:2847–2849.
116. Karnushina, I. L., Palacios, J. M., Barbin, G., Dux, E., Joó, F., and Schwartz, J. C., 1980, *J. Neurochem.* **34**:1201–1208.
117. Nathanson, J. A., 1980, *Life Sci.* **26**:1793–1799.
118. Peroutka, S. J., Moskowitz, M. A., Reinhard, J. F., Jr., and Snyder, S. H., 1980, *Science* **208**:610–612.
119. Culvenor, A. J., and Jarrott, B., 1981, *Neuroscience* **6**:1643–1648.
120. Harik, S. I., Sharma, V. K., Wetherbee, J. R., Warren, R. H., and Banerjee, S. P., 1981, *J. Cereb. Blood Flow Metab.* **1**:329–338.
121. Kobayashi, H., Memo, M., Spano, P. F., and Trabucchi, M., 1981, *J. Neurochem.* **36**:1383–1388.
122. Kobayashi, H., Maoret, T., Ferrante, M., Spano, P., and Trabucchi, M., 1981, *Brain Res.* **220**:194–198.
123. van Houten, M., and Posner, B. I., 1979, *Nature* **282**:623–628.
124. Frank, H. J. L., and Pardridge, W. M., 1981, *Diabetes* **30**:757–761.
125. Karnushina, I., Tóth, I., Dux, E., and Joó, F., 1980, *Brain Res.* **189**:588–592.
126. Palmer, G. C., 1981, *Neuroscience* **6**:2547–2553.
127. Stefanovich, V., 1979, *Neurochem. Res.* **4**:681–687.
128. Joó, F., and Tóth, I., *Naturwissenschaften* **62**:397–398.
129. Jarrott, B., Hjelle, J. T., and Spector, S., 1979, *Brain Res.* **168**:323–330.
130. Head, R. J., Hjelle, J. T., Jarrott, B., Berkowitz, B., Cardinale, G., and Spector, S., 1980, *Blood Vessels* **17**:173–186.
131. Joó, F., 1972, *Experientia* **28**:1470–1471.
132. Gerritsen, M. E., Parks, T. P., and Printz, M. P., 1980, *Biochim. Biophys. Acta* **619**:196–206.
133. Maurer, P., Moskowitz, M. A., Levine, L., and Melamed, E., 1980, *Prostagland. Med.* **4**:153–161.
134. Birkle, D. L., Wright, K. F., Ellis, C. K., and Ellis, E. F., 1981, *Prostaglandins* **21**:865–877.
135. Gerritsen, M. E., and Printz, M. P., 1981, *Prostaglandins* **22**:553–566.
136. Goehlert, U. G., Ng Ying Kin, N. M. K., and Wolfe, L. S., 1981, *J. Neurochem.* **36**:1192–1201.
137. Gecse, A., Ottlecz, A., Mezei, Z., Telegdy, G., Joó, F., Dux, E., and Karnushina, I., 1982, *Prostaglandins* **23**:287–297.
138. DeBault, L. E., Henriquez, E., Hart, M. N., and Cancilla, P. A., 1981, *In Vitro* **17**:480–494.
139. Bowman, P. D., Betz, A. L., Ar, D., Wolinsky, J. S., Penney, J. B., Shivers, R. R., and Goldstein, G. W., 1981, *In Vitro* **17**:353–362.
140. Phillips, P., Kumar, P., Kumar, S., and Waghe, M., 1979, *J. Anat.* **129**:261–272.
141. Gimbrone, M. A., Jr., Majeau, G. R., Atkinson, W. J., Sadler, W., and Cruise, S. A., 1979, *Life Sci.* **25**:1075–1084.
142. Brecher, P., Hingorani, V., Reininga, K., and Chobanian, A. V., 1981, *Clin. Sci.* **61**:249s–251s.
143. Rix, E., Ganten, D., Schüll, B., Unger, T., and Taugner, R., 1981, *Neurosci. Lett.* **22**:125–130.
144. Friedland, J., Stetton, C., and Silverstein, E., 1977, *Science* **197**:64–65.
145. Hial, V., Gimbrone, M. A., Jr., Peyton, M. P., Wilcox, G. M., and Pisano, J. J., 1979, *Microvasc. Res.* **17**:314–329.
146. Wechezak, A. R., Holbrook, K. A., Way, S. A., and Mansfield, P. B., 1979, *Blood Vessels* **16**:35–42.
147. Ryan, U. S., Clements, E., Habliston, D., and Ryan, J. W., 1978, *Tissue Cell* **10**:535–554.
148. Gospodarowicz, D., Greenburg, G., and Birdwell, C. R., 1978, *Cancer Res.* **38**:4155–4171.
149. Folkman, J., and Haudenschild, C., 1980, *Nature* **288**:551–556.
150. Weibel, E. R., and Palade, G. E., 1964, *J. Cell Biol.* **23**:101–112.
151. Herrlinger, H., Anzil, A. P., Blinzinger, K., and Kronski, D., 1974, *J. Anat.* **118**:205–209.
152. Hirano, A., 1974, *Pathology of Microcirculation* (J. Cervos-Navarro, ed.), de Gruyter, Berlin, p. 203.

153. Wade, L. A., and Katzman, R., 1975, *Am. J. Physiol.* **228**:352–359.
154. Rowan, R. A., and Maxwell, D. S., 1981, *Am. J. Anat.* **160**:257–265.
155. Panula, P., Joó, F., and Rechardt, L., 1978, *Experientia* **34**:95–97.
156. DeBault, L. E., Kahn, L. E., Frommes, S. P., and Cancilla, P. A., 1979, *In Vitro* **15**:473–487.
157. DeBault, L. E., and Cancilla, 1979, *Science* **207**:653–655.
158. Fratkin, J. D., Cancilla, P. A., and DeBault, L. E., 1980, *Thromb. Res.* **19**:473–483.
159. Spatz, M., Bembry, J., Dodson, R. F., Hervonen, H., and Murray, M. R., 1980, *Brain Res.* **191**:577–582.
160. Karnushina, I. L., Spatz, M., and Bembry, J., 1982, *Life Sci.* **30**:849–858.

17

The Blood–Brain Barrier

William H. Oldendorf

1. INTRODUCTION

The concept of a barrier restricting exchange of certain tracer substances between blood and brain (the blood–barrier, BBB) originated with Paul Ehrlich's observation in the 1880's that intravenously injected aniline dyes failed to distribute to brain, whereas all other tissues were quickly colored. In general, it is the extracellular fluid (ECF) to which such dyes distribute to color the tissues. The blood plasma is a rapidly circulating subcompartment serving to disperse any local concentration of solute throughout the entire ECF. It can carry out this homogenizing function because of the ubiquity and permeability of capillaries. These capillary characteristics, together with the movement of blood plasma, bring the effective distances that solutes are required to diffuse between various body cells down to a few micrometers.

The capillary wall in brain, except for portions of the floor of the hypothalamus and the area postrema,[1] differs substantially in structure and function from capillaries in other tissues, and these characteristics are responsible for the unique permeability we refer to as the BBB.

Other than the few small periventricular regions noted above, the entire brain and spinal cord and, indeed, all tissues central to the arachnoid membrane have a BBB. The arachnoid membrane can usefully be considered the outer boundary of the CNS since the extracellular fluid of tissues external to it differs substantially from that of tissues central to it. Arachnoid membrane everywhere separates cerebrospinal fluid (CSF) from dura, which lies immediately external and in loose apposition to it.

Mammalian BBB develops variably in different species but appears in all at about the time of birth.[2] In mammals having a rather well-developed brain at birth (such as sheep and humans), it is quite well formed before delivery, and in those animals having less well-developed brains at birth (such as the rat), the BBB does not fully develop until 1–2 weeks after birth. In the newborn

---

William H. Oldendorf • VA Medical Center, Los Angeles, California 90073; and Departments of Neurology and Psychiatry, University of California, Los Angeles, School of Medicine, Los Angeles, California 90024.

rabbit some of the carrier transport mechanisms are much more fully developed and appear to diminish strikingly by about 1 month.[3]

The term BBB suggests that brain capillaries are impermeable, but it is obvious that, even though they are indeed impermeable to some plasma solutes, they must be freely permeable to others. From our fund of common experience, we know that some central nervous system (CNS) drugs (such as barbiturates) immediately alter brain function after intravenous injection, and so there cannot be a barrier against them. Additionally, the brain receives substantial amounts of metabolic substrates from blood. The "barrier" must be permeable to them. Rather than being impermeable, the BBB is more correctly considered selectively permeable in that there is a very wide range of permeability to various plasma solutes.

Based on many studies, there are fairly well-defined molecular criteria that determine this permeability. The following discussion considers the unique structural and functional characteristics of brain capillaries and concludes with some of the likely teleology of the BBB.

2. *NONNEURAL CAPILLARIES*

In the general capillary (in tissues other than brain), all blood plasma solutes of low molecular weight diffuse freely between plasma and the more stationary pericapillary ECF. This exchange is largely through narrow clefts between adjacent capillary endothelial cells and, in view of the very small fraction of capillary wall area that is cleft, is more effective than would be predicted intuitively because the diffusion distance is so short (of the order of 1 μm). This efficient exchange is indiscriminate of molecular characteristics other than size. Molecules greater than 20,000–40,000 mol. wt. are much less freely exchanged than smaller solutes.[4] This small residual exchange of even very large molecules probably occurs through pinocytotic transport. The role of the fenestrae present in some capillary cells is currently unsettled, although they seem to be numerous in capillaries with a large net flux of water through their walls such as in renal glomerulus and choroid plexus.

3. *NEURAL CAPILLARIES*

In brain capillaries the intercellular clefts are sealed shut,[5] and pinocytosis and fenestrae are virtually absent. Thus, exchange across the capillary wall must take place directly through the cell rather than between the cells as in the general capillary (Fig. 1).[6] To penetrate the BBB, a solute molecule must penetrate the two (inner and outer) plasma membranes and survive transit through the interposed endothelial cell cytoplasm. For most solutes, it is the membranes that constitute the barrier. This current view confirms Krogh's early postulate[7] that the BBB behaves as though there were a biological membrane separating blood from brain. In the light of subsequent observations, this "membrane" presumably is the combined effect of the two plasma membranes

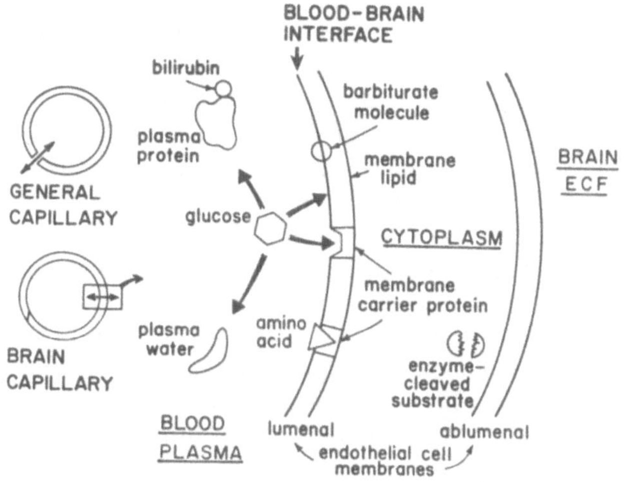

Fig. 1. On the left, the basic difference between the nonspecifically permeable capillary in other tissues is diagrammatically contrasted with the selectively permeable capillary in brain (the BBB). The open intercellular cleft of the general capillary is an unexpectedly efficient pathway for exchange because it is usually less than 1 μm in length, and diffusional exchange over such a short distance is very rapid. With the closure of the cleft pathway for plasma–brain exchange, substances must pass transcellularly to enter or leave brain. A section of brain capillary wall is enlarged on the right to clarify the mechanism by which BBB penetration takes place. The blood–brain interface is the contact surface between plasma and luminal membrane. A substance carried into brain by plasma flow and now at this interface must leave the plasma and enter the luminal membrane if it is to diffuse on into brain, toward the right in this diagram. The likelihood of its entering the membrane from plasma is established by its relative affinities for the four relevant species of molecules present at the interface. They are (1) plasma water, (2) membrane lipid, (3) plasma protein, and (4) membrane protein. Bilirubin, barbiturate, glucose, and an amino acid are shown at the interface. Bilirubin is firmly attached to plasma protein and is thus excluded from the membrane. The barbiturate molecule has an affinity for plasma water but also a significant affinity for membrane lipid, so it can make the transition into the lipid portion of the membrane. Glucose is very polar and so has a very high affinity for water relative to its affinity for either plasma protein or membrane lipid but still enters the membrane by virtue of a high affinity for a specific carrier protein embedded in the cell membrane. The amino acid has less affinity for water than does glucose but still requires a carrier to accelerate its partitioning from water into the membrane. The interplay between the plasma solutes carried into brain and the constituents of the plasma–brain interface explains the selective permeability of the BBB resulting from lipid mediation and carrier mediation, mechanisms requiring no specific work by the endothelial cell. This simple explanation does not explain energy-consuming processes such as the pumping of ions into and out of brain. The continuous pavement of endothelial cells making up the BBB creates an epithelial surface analogous to other layers of tight-junctioned cells (such as intestinal mucosa or renal tubule) separating tissue compartments differing substantially in chemical composition. The compartments separated by the BBB are blood plasma and brain extracellular fluid.

of the brain capillary endothelial cell. Thus, the BBB actually is a continuous double sheet of cell membrane. It probably is the most accessible of any living membrane system to experimental studies of permeability.

A wealth of information relevant to this topic is contained in two recent monographs,[8,9] and both are highly recommended. There is also an excellent compilation of individual papers published together in a supplement of *Experimental Eye Research*.[10]

The interface between blood and brain is at the inner capillary cell membrane. The BBB permeability to a particular solute in plasma is dependent on the likelihood that the solute molecule will leave the plasma and enter this membrane. This is established by the relative affinities of this molecule for the four major molecular species present at this interface (Fig. 1). These species are (1) blood plasma water, (2) endothelial cell membrane lipid, (3) blood plasma proteins, and (4) endothelial cell membrane proteins.

To have a possibility of entering the membrane, a molecule in plasma must have at least a small fraction that is unattached to plasma protein and is in free solution in plasma water, because the protein molecule will have virtually no chance of entering the membrane. Even a small fraction of unbound plasma solute may allow an unanticipatedly large membrane penetration and clearance by brain because the equilibrium between bound and unbound fraction can be very rapid and well within one capillary transit time, thus allowing a replenishment of unbound fraction.[11]

Such water-bound plasma solute must either enter the lipid portion of membrane or attach itself to membrane protein. Some membrane proteins exhibit specific affinites for certain classes of plasma solutes and are considered carrier proteins. In effect, these carrier proteins cause the membrane to exhibit a much higher effective solubility for their substrates than would be possible if the membrane were entirely lipid in nature.

The freedom of transition from water into the membrane lipid can be estimated *in vitro* by measuring partitioning in a two-phase oil: water system. Such a study of a number of drug partition coefficients correlates quite well with their observed freedom of BBB penetration.[12,13] This is an extension to the BBB of the classic work of Collander, who showed a similar relationship using plant protoplasts.[14] It is this relative affinity for lipid versus water that is relevant rather than the absolute solubility in either phase. To create a two-phase system most relevant to living membrane lipid, the lipid phase should be similar to membrane lipid. Which lipid or mixture of lipids is most physiologically relevant is unknown, since various lipids differ considerably in their polarity, and it is uncertain which of the commonly used lipid phases most closely represents the lipid of cell membranes. Quite likely there is no simple answer because the composition of membrane lipid is so complex. For convenience, octanol or refined olive oil is used, but coefficients for the former are of the order of five times greater than those for the latter.

The loss of consciousness within a few seconds after an intravenous injection of thiopental is a consequence of its lipid solubility, which allows the drug to partition into the brain capillary membranes from plasma water. It then proceeds into brain as it passes down its concentration gradient.

Lipid solubility of CNS drugs quite well explains their penetration of BBB, but highly polar solutes such as glucose may have a trivial lipid solubility and a very great affinity for water. Some other mechanism must be invoked to explain their BBB penetration. To attract glucose out of plasma water into the capillary cell membrane requires a protein in the membrane having a very high affinity for glucose, the hexose carrier. Each carrier protein has an affinity for a class of substances usually related structurally to the most commonly transported substrate. The glucose carrier is only one of many similar BBB transport systems that have been defined.

3.1. Methods of Study

The BBB permeability to a test substance is measured experimentally by establishing the amount of this substance that enters brain in a fixed length of time after a known concentration is generated in the blood entering brain. If a concentration is established abruptly and thereafter maintained constant,[15] the concentration in brain will rise at a high initial rate and accumulate at a rate expressing both BBB permeability and the distribution space of the test substance in brain.

The initial rate of entry is a function of permeability. If the substance is present in the blood for a very short time (<1 s), the amount taken up by brain is almost entirely a function of BBB permeability. This can be determined *in vivo* by a brief injection into the carotid artery, as pioneered by Chinard and Vosburgh[16] and independently by Crone[17] and later inverted and pursued in my laboratory.[18] After such an injection, the fraction of test substance lost to brain is a useful expression of BBB permeability and is quite independent of the substance's subsequent fate in brain. The translocation of a solute from blood into brain by virtue of a specific carrier protein is highly analogous to an enzyme–substrate interaction. The initial rate of such a reaction (before product accumulation) most accurately defines the enzyme kinetics. Similarly, the initial rate of entry into brain before test substance accumulation most accurately defines BBB carrier kinetics.

In the carotid injection methods, measurements of the amount of test substance removed by brain are made possible by expressing this brain uptake relative to a simultaneously injected reference substance.

In the method applied extensively by Crone, the reference substance does not measurably penetrate into brain, and measurement of test-substance uptake after carotid injection is based on measurements of both test and reference substances in the brain venous blood. By similarly measuring the injected mixture of substances, the amount of test substance taken up by brain can be calculated. The great advantage of the method is that the subject need not be killed, and it can even be performed on humans.[19] Because the uptake is based on the difference between the two substances in jugular blood, it does not lend itself to studies of substances less than about 10% cleared by brain. Below this, the venous curves become nearly equal, and it is difficult to accurately define their small differences.

An inverse of the Crone method[20] injects arterially the test substances mixed with a diffusible reference substance, which is taken up by brain to a known extent, and the test and reference substances are measured in brain tissue after decapitation a few seconds after injection rather than in jugular blood.[20]

The initial diffusible reference was tritiated water, which is 84% cleared in the barbiturate-anesthetized rat. This is suitable for test substances cleared more than about 10%. For substances cleared less than this, we have recently begun using [^{14}C]thiourea (4.5% cleared) as a reference. When studying low-uptake substances, we also include a nondiffusible reference ([^{113m}In]EDTA) in the injectate to correct for the incomplete washout of tracers from the brain blood compartment in the few seconds between injection and decapitation. The ^{113m}In is conveniently counted in the same liquid scintillation specimen as ^{3}H and ^{14}C.[21] Initially, we waited 15 s but later realized that this was unnecessarily long, allowing some washout, and so we reduced the delay to 5 s. With this method, it is possible to measure clearances of less than 1% with less than 10% S.D. It requires sacrificing the animal by decapitation and, accordingly, is best suited to small animals. As an experimental tool, however, it has been highly rewarding, particularly in the identification and characterization of carrier systems. By injecting a radiolabeled test substance and an unlabeled competitive substance of known concentration, changes in the uptake of the labeled species allow the calculation of carrier kinetics. This capability of allowing the creation of a wide range of fluid vehicle composition at the capillary exchange site (and thus allowing controlled perturbation of the exchange process) is the major advantage of the arterial bolus technique. Other advantages are its freedom from plasma protein binding effects, isolation of the BBB transport process from brain distribution space, and efficient use of isotopes.

The BBB permeability to water is substantially less than that to many other substances and approximates what one would expect from a high-resistance epithelium (a permeability–surface-area product of about 0.01 ml s^{-1} g^{-1}). During the course of examining the validity of the use of tritiated water as a reference substance in the carotid injection method,[22] it was noted that single-pass labeled water extraction is not constant and can increase by as much as one-third in response to noradrenergic influences[23] and to drugs such as tricyclic antidepressants.[24] The physiological significance of this interesting observation must await further examination, since single-pass extraction describes unidirectional flux of a tracer substance. It would be of great clinical interest if it could be shown that these apparent changes in permeability to unidirectional tracer flux into brain correlate in some way with net flux of water in the formation of brain edema.

Although BBB (and other biological membrane) permeability to water appears to be in a category all its own, to other substances it can usefully be considered either carrier mediated or lipid mediated. Lipid-mediated permeability is nonsaturable because there is an essentially unlimited lipid pathway on the surface of the brain capillary endothelial cells. Carrier-mediated permeability is saturable because of the finite number of carrier protein transport sites.

3.2. Lipid-Mediated BBB Permeability

The distribution of a plasma solute between the polar environment of plasma water and the relatively nonpolar environment of endothelial cell membrane lipid is governed by the polarity of the solute under the chemical conditions present at the plasma–BBB interface. This is the sum of the polarity caused by ionization and other effects, largely hydrogen bond formation, at various sites in the molecule.

When a plasma solute molecule is ionized, the charge site dominates the polar characteristics of the molecule. The polarity of nonionized molecules is dominated by their capacity to form hydrogen bonds. In the case of acids and bases, in which a fraction of the solute molecule is ionized and the remainder unionized, most of the membrane penetration probably is recruited from the less polar unionized species. This residual polarity can be greatly altered by minor changes in hydrogen-bond-forming groups with large effects on its membrane-penetrating characteristics. The striking differences in BBB permeabilities to morphine, codeine, and heroin are attributable to replacement of the two hydroxyls on morphine by less polar groups on its two analogues.[25] Similar structure–permeability correlations have been demonstrated for phenethylamine to which various polar groups have been attached.[26]

The acetylation of the two hydroxyl groups of morphine substantially reduces the polarity of the resultant heroin, thus allowing it to enter brain freely. It is then rapidly deacetylated back to morphine. This general process of making a molecule reversibly less polar has been called "latentiation"[27] and could usefully be tried with other drugs too polar to allow them to enter cells or penetrate the BBB.[13,28]

Using refined olive oil as the lipid phase, we have correlated the partition coefficients of a number of drugs with their percentage clearance by brain after carotid injection in the rat. Although the application of this model has allowed a degree of quantification of rates of drug penetration of the BBB, the results of lipid-mediated permeability could largely have been predicted from partition coefficient studies. The most interesting results, which could not be predicted, have come from studies of carrier-mediated permeability, since not only has it been possible to identify many independent carriers, but estimates can be made of their transport kinetics.[29]

The independent carrier systems identified to date have been for certain hexoses,[30] short-chain monocarboxylic acids,[31] neutral, basic,[32] and acidic[21] amino acids (one for each), two for nucleic acid precursors,[33] for choline,[34] and for T-3 and T-4.[35]

The hexose carrier exhibits significant affinities for D-glucose, mannose, galactose, 3-0-methyl-D-glucose, and 2-deoxyglucose. There is no measurable affinity for fructose or L-glucose.[30,36] The carrier is slightly less than half saturated at about normal blood glucose levels. It exhibits no affinity for brain metabolic substrates other than hexoses. The carrier appears to be passive, functioning in both directions (in and out of brain), and is sodium independent. Based on a wide range of injected glucose concentrations, there appears to be only a single species of carrier. Saturation of the carrier conformed to Michaelis–Menten kinetics. Linear transformation of the clearance data provided the following K_ms for various hexoses: 2-deoxy-D-glucose, 6 mM; D-glucose, 9 mM; 3-O-methyl-D-glucose, 10 mM; D-mannose, 21 mM; and D-galactose, 40 mM. Maximum transport velocity was 1.56 μmol/g for all five hexoses.[37] 2-Fluorodeoxyglucose has a K_m of 6.9.[38]

By variably prolonging the time of decapitation, it was possible to study the uptake and efflux of ^{3}HOH and [^{14}C]3-O-methyl-D-glucose.[37,39] The latter is not metabolized by brain, although it is transported into brain nearly identically with D-glucose. Its rate of efflux nearly equals its initial rate of influx, suggesting that the BBB hexose carrier functions bidirectionally with approximately equal efficiency and thus is equilibrative rather than active.

Hexose transport is strongly inhibited by phlorizin and more so by phloretin. The strong inhibition by phloretin is rapidly reversible, as indicated by the return to normal ($t_{1/2}$ approximately 4 s) when the uptake of glucose is studied with variable delays after a carotid bolus of phloretin.[37]

The kinetics of BBB glucose transport are not altered by several days of starvation. The amount of radiolabeled glucose taken up by various regions of rat brain is constant.

The brain is highly dependent on the hexose carrier to allow the considerable flux of glucose required by brain. The vulnerability of this carrier in various disease or toxic states is unknown, but if the carrier were to become inactive in any region of brain, within or adjacent to a lesion, that region would be subjected to what would, in effect, be a profound hypoglycemia, and irreversible damage would soon result. It is not known if this is a factor in any human disease. That various BBB carriers are vulnerable to poisoning has been demonstrated following carotid injection of mercury salts in the rabbit.[40,41] We have shown that intravenous phloretin or 3-O-methylglucose can cause an immediate drop to about one-half normal levels in brain free glucose (unpublished observation).

The monocarboxylic acid carrier accelerating BBB penetration of certain short-chain acids exhibits measurable affinities for acetic, propionic, butyric, lactic, pyruvic,[31] and β-hydroxybutyric acid.[31] For most of these short-chain acids, there are both lipid-mediated and carrier-mediated components contributing to their permeability. During starvation, this carrier appears to be induced to a considerable degree, and this probably results from the rise in blood levels of β-hydroxbutyric acid. In the rat, the transport capacity for ketones approximately doubled after 5 days of starvation.[42]

Although there is an appreciable carrier-mediated enhancement of short-chain monocarboxylic acids, there is no measurable permeability to the common di- or tricarboxylic acids. The penetration of short-chain acids such as lactate is increased at lower pH, suggesting that these acids are transported in those brief intervals when they are protonated and thus neutral.[43] A possible mechanism of the failure to transport acids with more than one carboxyl is that all of the groups are unlikely to be protonated simultaneously if the molecule is in a hydrogen ion concentration substantially above the p*K*s of its two or three carboxyl groups. The molecule would nearly always be ionized.

There are three carriers for amino acids, and affinities of the various amino acids for the three carriers appear to be based on their electrical charge at physiological pH. There are independent carriers for neutral, basic,[32] and acidic amino acids with no cross affinities between groups.[21] Within a group, all of the amino acids cross compete for their common carrier. There is a considerable range of stereospecificity, and the L-amino acids are always preferred.[44] The amino acid carriers appear to be substantially more than half saturated in the presence of normal blood free amino acid concentrations, and this may be a rate-limiting step in the brain metabolism of some amino acids. The high blood phenylalanine level in phenylketonuria substantially reduces the permeability of the BBB to other neutral amino acids but not to the basics, which move on another carrier.

We have studied the neutral amino acid carrier in humans by two methods. In the first method, we injected intravenously the methionine analogue selenomethionine (SEM) in which ^{75}Se (a γ emitter) is substituted for the sulfur in methionine. Selenomethionine is very nearly the biological equivalent of methionine and was developed because of its ability to display the human pancreas by external γ-ray scanning. Although not in common use for that purpose, it is commercially available. In a group of phenylketonuric children versus a matched mentally retarded control group, a smaller fraction of the injected SEM distributed to the brain in the phenylketonurics. This uptake was measured by an external γ-ray detector.[45,46]

Since such human studies are necessarily limited in scope, pooled blood sera from the two groups were produced and used as the vehicle for carotid injection in rat studies of the brain clearance of neutral and basic amino acids.[47] When we used rat serum as the vehicle for carotid injection, amino acid uptake was reduced by a factor of two to three relative to uptakes using Ringer's solution as a vehicle in which there were no free amino acids. The small amount of free amino acids in serum inhibits the uptake of the radioactive test amino acids. The pooled control retarded (nonphenylketonuric) serum similarly inhibited all amino acid uptakes. The pooled phenylketonuric serum reduced the uptake of only the neutral amino acids by a further factor of about three but had no effect on the basic amino acids. These effects of the phenylketonuric serum were shown to be caused by its increased phenylalanine content.

The BBB permeability to the essential amino acids is, in general, substantially greater than that to nonessentials,[48] indicating that, for amino acids, a small molecular size correlates with a reduced carrier affinity. There is no measurable uptake of glycine even though it is the smallest of the amino acids. The adult BBB lacks the alanine-prefering "A" system of Christensen.[49] The permeability of BBB closely resembles that of erythrocyte membrane. There appear to be two carriers for nucleic acid precursors, but the molecular criteria for their affinities remain unclear. The uptakes of 17 such precursors were studied in the rat model. Measurable and saturable uptakes were established for adenine, adenosine, guanosine, inosine, and uridine. Uptakes considered not significantly above the background level of the method (2%) were found for cytosine, thymine, uracil, guanine sulfate, hypoxanthine, orotic acid, folic acid, uric acid, cytidine, cytosine arabinoside, thymidine, and ATP. For those five precursors, our studies have identified two apparently independent carriers. One transported adenine (K_m = 0.03 mM) and could be inhibited by hypoxanthine. Adenosine, guanosine, inosine, and uridine all cross inhibited.[33] These studies were performed using tritiated water as a reference tracer and 15-s decapitation. Our current method using [^{14}C]thiourea as a reference with 5-s decapitation would probably reduce the variance seen in these studies considerably, and repeating these studies should better characterize these transport systems.

Because it is ionized and thus very polar, it would be expected that choline would exhibit very low lipid-mediated BBB penetration. There is a substantial uptake of labeled choline, however, and this can be saturable down to less than 1% during a single rat brain passage.[34] The uptake of labeled choline by brain after intravenous injection is saturable.[50]

3.3. Active Pumping by BBB

In addition to the passive diffusion of solutes between blood and plasma by lipid or carrier-mediated transport, it can be inferred that the BBB carries out some energy-consuming pumping because the composition of brain ECF, as represented by CSF composition, is, in several ways, quite different from plasma. The mitochondrial content of brain capillary endothelial cells is about four times that in most capillaries in other tissue,[51,52] suggesting that the work load of the BBB is substantially greater than that of other capillaries. The apparent "excess" work capability may be pumping ions in and out of brain to maintain the known ionic concentration gradients between brain and plasma. Iodide[53,54] and the acidic products of central monoamine metabolism[55,56] appear to be pumped out. It is reasonable to expect that K^+ is pumped into blood because of its low level in brain ECF and that magnesium and calcium are pumped into brain because they are in higher concentration in CSF.[57,58]

3.4. The Blood–CSF Barrier

This term is often used to designate the rate-limiting mechanism governing the equilibration of substances between blood and CSF. There is no single anatomic site for this "barrier." Accordingly, the term is not very useful and probably should be abandoned. Since the CSF is in continuity with the interstitial ECF of brain and spinal cord, the major route of exchange between blood plasma and CSF probably is the BBB, with lesser exchanges taking place through the epithelial layer of the choroid plexus and through the arachnoid membrane. For ventricular fluid, the choroid plexus may be the more important exchange site, and in the subarachnoid space, particularly around and caudal to the spinal cord, the arachnoid membrane may become significant.

4. SPECULATIONS ON FUNCTIONS OF THE BLOOD–BRAIN BARRIER

4.1. Exclusion of Blood-Borne Toxins

The first suggestion one might logically make concerning the BBB's function would be that it protects the brain from potentially toxic substances in blood. A good example of such a substance is bilirubin, which is strongly bound to plasma protein and thus is kept out of the brain even when present in the blood in high concentrations with clinical jaundice. Bilirubin is neurotoxic, and keeping it out of the brain clearly serves a useful function. The occasional patient with jaundice in later life is presumably made less symptomatic because of this, and the brains of infants with neonatal jaundice are usually protected against bilirubin.

4.2. Systemic Neurotransmitters

The BBB is highly impermeable to the known centrally active neurotransmitter agents such as norepinephrine. These agents may undergo strikingly rapid transient changes in concentration in the systemic blood, but the presence of the BBB prevents these wide fluctuations from being transmitted into the brain ECF, where they could bring about unwanted changes in CNS activity. For example, if an individual were to jump into cold water, the expected burst of systemic norepinephrine might result in a disabling cerebral excitation if there were no BBB.

Conversely, the impermeability of the BBB to central neurotransmitters serves to prevent their loss to blood after their synthesis in brain. This not only conserves these neurotransmitters but confines them near their site of origin. For example, norepinephrine released into a cortical synaptic cleft is removed by several mechanisms, such as diffusion from the synapse and reuptake into the nerve ending. If the BBB were freely permeable to norepinephrine, the blood (which would now be acting as a sink) would become an additional major source of removal; it would compete for nerve ending reuptake and require the formation of a considerably greater amount of norepinephrine. After passing into the blood, the norepinephrine might also reappear elsewhere in the brain. By this process, the ability of brain to restrict its neurotransmitter activity to small discrete regions would be impaired.

4.3. Modulation of Substrate Entry into Brain

The BBB contains a number of specific transport systems that exhibit saturation, thereby limiting the rate at which metabolic substrates in plasma can enter brain. These carriers facilitate the entry into brain of polar metabolic substrates. The K_ms of most of these (but not for amino acids) appear to be approximately their respective concentrations in blood plasma, as with many of the enzymes in the body, whose substrate concentrations approximate their K_ms. Thus, the BBB transport systems stabilize substrate influx into brain in the presence of changing blood substrate levels.

The entry of glucose into brain is almost entirely carrier mediated and is held relatively constant in the presence of hyper- or hypoglycemia. Lactate's penetration of the BBB is accelerated by a carrier shared by several other short-chain monocarboxylic acids. The level of blood lactate may fluctuate greatly in brief periods of time; a substantial rise can occur, for example, after bursts of extreme muscular activity. It would seem advisable that such rapid fluctuations not be transmitted into brain without some modulation. The short-chain monocarboxylic acid BBB carrier is about half saturated in the presence of usual blood lactate and pyruvate levels. In the presence of a sudden elevation of blood lactate, the brain is protected from the full lactate influx.

4.4. Brain Capillary Endothelial Cell Enzymes

The passage of substances through the brain capillary endothelial cells exposes them to the cytoplasmic enzymes of these cells before they enter brain. For some substances, this may enhance the effective BBB by making substances that might have passed through the luminal endothelial cell membrane more polar (by cytoplasmic enzyme actions) and thus preventing them from passing through the external endothelial cell membrane. Such a mechanism is probably effective with the monoamine transmitter agents, which are oxidized to an unknown extent within the endothelial cell cytoplasm and whose overall rate of penetration through the BBB is thereby impaired. In such cases, there is an enzymatic enhancement of the BBB. Since monoamine oxidase is largely confined to the surface of mitochondria, the large number of endothelial cell mitochondria in brain makes these cells several times as effective as sources of monoamine oxidase and thus should further enhance the enzymatic component of the BBB.

While these substances are in the endothelial cell cytoplasm, there may be metabolically useful enzymatic transformations, as in the case of dihydroxyphenylalanine (DOPA), which is probably converted to a substantial degree to dopamine. This DOPA is then free to diffuse either into the blood or into the brain ECF. That dopamine that diffuses into the brain results in a general level of extracellular dopamine that might not readily be obtained by intracellular CNS decarboxylation. Such an endothelial cell enzymatic mechanism may enable the BBB to provide brain ECF with potentially useful metabolic intermediates. Endothelial cytoplasmic dopamine resulting from decarboxylation no longer has an affinity for the BBB neutral amino acid carrier; thus, its rate of movement in either direction is greatly reduced relative to that of DOPA. Several enzymes occur either exclusively in brain capillaries (in distinction to capillaries in other tissues) or in much greater concentration. Among these are γ-glutamyl transpeptidase, DOPA decarboxylase, GABA transaminase, succinic semialdehyde dehydrogenase, and pseudocholinesterase.

4.5. The Large Apparent Metabolic Work Capacity of the BBB

Its mitochondrial content is a general indicator of the metabolic work capability of a cell. It has been shown that the mitochondrial content of the brain capillary endothelial cell is approximately four times that of nonneural capillary endothelial cells[51]; approximately 10–11% of the cytoplasm is mitochondria in brain and spinal cord capillary endothelial cells, and about 2.5–3% in other tissues.[52] The BBB's relatively large mitochondrial content suggests that it is capable of a correspondingly larger workload. Since the workload is approximately four times that of the usual general capillary, it suggests that about one-fourth of the workload of the brain capillary cell is shared by the general capillary cell, and three-quarters is not; this excess may well be concerned with pumping ions into and out of brain ECF.

It would be pointless to have a mechanism within the wall of the general capillary that would attempt to create a concentration gradient across this wall.

Any such gradient would immediately be abolished by diffusion through the nonspecific exchange routes in the capillary wall. Where such nonspecific diffusion does not exist, however, such as in the BBB, it would be possible to maintain a concentration gradient. The logical place for the pump required to create and maintain this gradient between blood plasma and brain ECF would be within the capillary wall. In the case of potassium ion, for example, the concentration in ECF in brain is approximately 40% lower than that in blood plasma. This undoubtedly serves to hyperpolarize and thus stabilize the neuron. To maintain this gradient, which is across the capillary wall, requires the expenditure of energy, and this could well be one of the functions being carried out by the BBB's apparent excess of mitochondria.

The BBB's apparent excess work capacity could also be related to the apparent capability of the brain parenchyma to generate some CSF. Although it is generally believed that most of the CSF is formed by choroid plexus under normal circumstances, several workers have shown that under experimental circumstances, the brain parenchyma is evidently capable of generating some measurable CSF. Some of the BBB's energy may be used to pump sodium into the brain ECF and thereby create the ionic imbalance necessary for CSF formation.

The carotid injection of hyperosmotic solutions results in a transient loss of the BBB, presumably by some effect of the osmotic shock on the capillary endothelial cell. This has been studied extensively[59–61] and has been suggested as a ploy for getting drugs through the BBB. This happens in a clinical setting when patients are subjected to the rapid carotid injection of approximately 1.6 M solutions during cerebral angiography.[62] There is no obvious change in cerebral function, suggesting that a brief (few hours) interruption of BBB function is not greatly deleterious to cerebral function.[63] It suggests further that the benefits derived from having a BBB are long range, such as protecting brain from growth hormone and other hormonal effects.

4.6. How Is the BBB Maintained?

How does a capillary cell know that it is in brain and should alter its fine structure in a way that will result in the BBB? It is attractive to propose that the CNS generates some humoral influence that causes nonspecific transendothelial transport to shut off. Such a humoral influence could be produced by the astrocytes in apposition to CNS capillaries.[64] That some cellular elements in brain induce capillaries to alter their structure has been supported by several experimental studies, especially that of Stewart and Wiley,[65] who implanted embryonic quail brain into chick embryos. When implanted into brain, capillaries in tissue of mesodermal origin developed BBB characteristics, whereas those implanted elsewhere did not.

The BBB appears to result from some ongoing function of healthy brain, since almost any significant abnormality of brain tissue results in a loss of BBB. In such lesions, the unique brain capillary cell characteristics, which cause the BBB, disappear, and the lesion's capillaries come to resemble the nonspecifically permeable capillaries in other tissues. This could be the result of malfunctioning astrocytes no longer able to produce the proposed humoral influence that normally maintains the BBB by inhibiting pinocytosis and fenestra formation and by causing tight interendothelial cell junction formation. These all represent specific characteristics of the endothelial cell plasma membrane and could be closely interrelated. A single substance created by the astrocyte could conceivably bring about all three of these effects. How the mitochondrial content of these cells is modified is unknown.

5. THE BLOOD–BRAIN BARRIER AS A SECOND-ORDER HOMEOSTATIC MECHANISM

Each of the many specialized cells in the total organism contributes something toward the optimization of the general ECF, and their concerted action results in a considerable homeostasis of the ECF in the presence of changing external environment. Thus, the internal temperature, pH, osmolarity, glucose concentration, oxygen tension, etc. are held within narrow optimum limits despite environmental influences that might change these parameters. The central neuronal environment appears to enjoy a substantially higher degree of homeostasis than does the general ECF, and, in several instances, optimization of the cellular fluid environment is at a different concentration from the general ECF. The BBB appears to offer a second-order homeostatic mechanism further optimizing the neural extracellular fluid environment as an ultrastable subcompartment of the general ECF.

REFERENCES

1. Wislocki, G. B., and King, L. S., 1936, *Am. J. Anat.* **58**:421–472.
2. Saunders, N. R., 1977, *Exp. Eye Res.* **25**:(Suppl.):523–550.
3. Braun, L. D., Cornford, E. M., and Oldendorf, W. H., 1980, *J. Neurochem.* **34**(1):147–152.
4. Landis, E. M., and Pappenheimer, J. R., 1963, *Handbook of Physiology*, Volume 11 (W. F. Hamilton, ed.), American Physiological Society, Washington, pp. 961–1074.
5. Brightman, M. W., and Reese, T. S., 1969, *J. Cell Biol.* **40**:648–677.
6. Crone, C., and Thompson, A. H., 1970, *Capillary Permeability* (C. Crone and N. Lassen, eds.), Academic Press, New York, pp. 447–453.
7. Krogh, A., 1946, *Proc. R. Soc. (Lond.)* [*Biol.*] **234**:171–177.
8. Siesjo, B. K., 1978, *Brain Energy Metabolism*, John Wiley & Sons, New York.
9. Bradbury, M. W. B., 1979, *The Concept of a Blood–Brain Barrier*, John Wiley & Sons, New York.
10. Bito, L. Z., Davson, H., and Fenstermacher, J. D. (eds.), 1977, *Experimental Eye Research: The Ocular and Cererospinal Fluids*, Volume 25, Supplement, Academic Press, New York.
11. Pardridge, W. M., and Mietus, L. J., 1979, *J. Clin. Invest.* **64**(1):145–154.
12. Oldendorf, W. H., 1974, *Proc. Soc. Exp. Biol. Med.* **147**(3):813–816.
13. Oldendorf, W. H., 1974, *Annu. Rev. Pharmacol.* **14**:239–248.
14. Collander, R., 1937, *Trans. Faraday Soc.* **33**:985–990.
15. Daniel, P. M., Donaldson, J., and Pratt, O. E., 1974, *J. Physiol.* (*Lond.*)**237**:8P–9P.
16. Chinard, F. P., Vosburgh, G. J., and Enns, T., 1955, *Am. J. Physiol.* **183**:221–234.
17. Crone, C., 1963, *Acta Physiol. Scand.* **58**:292–305.
18. Oldendorf, W. H., 1981, *Res. Methods Neurochem.* **5**:91–112.

19. Lassen, N. A., Trap-Jensen, J., Alexander, S. C., Olesen, J., and Paulson, O. B., 1971, *Am. J. Physiol.* **220**(6):1627–1633.
20. Oldendorf, W. H., 1970, *Brain Res.* **24**:372–376.
21. Oldendorf, W. H., and Szabo, J., 1976, *Am. J. Physiol.* **230**(1):94–98.
22. Raichle, M. E., Eichling, J. O., Straatmann, M. G., Welch, M. J., Larson, K. B., and Ter-Pogossian, M. M., 1976, *Am. J. Physiol.* **230**:543–552.
23. Raichle, M. E., Hartman, B. K., Eichling, J. O., and Sharpe, L. G., 1975, *Proc. Natl. Acad. Sci. U.S.A.* **72**(9):3726–3730.
24. Preskorn, S. H., Hartman, B. K., Raichle, M. E., and Clark, H. B., 1980, *J. Pharmacol. Exp. Ther.* **231**(2):313–320.
25. Oldendorf, W. H., Hyman, S., Braun, L., and Oldendorf, S. Z., 1972, *Science* **178**:984–986.
26. Oldendorf, W. H., 1971, *Eur. Neurol.* **6**:49–55.
27. Harper, N. J., 1959, *J. Med. Pharm. Chem.* **1**:467–500.
28. Creveling, C. R., Daly, J. W., Tokuyama, T., and Witkop, B., 1969, *Experientia* **25**:26–27.
29. Pardridge, W. M., Connor, J. D., and Crawford, I. L., 1975, *CRC Crit. Rev. Toxicol.* **3**(2):159–199.
30. Crone, C., 1965, *J. Physiol. (Lond.)* **181**:103–113.
31. Oldendorf, W. H., 1973, *Am. J. Physiol.* **224**:1450–1453.
32. Richter, J. J., and Wainer, A., 1971, *J. Neurochem.* **18**:613–620.
33. Cornford, E. M., and Oldendorf, W. H., 1975, *Biochim. Biophys. Acta* **394**(2):211–219.
34. Oldendorf, W. H., and Braun, L. D., 1976, *Brain Res.* **113**(1):219–224.
35. Pardridge W. M., 1979, *Endocrinology* **105**(3):605–612.
36. Oldendorf, W. H., 1971, *Am. J. Physiol.* **221**(6):1629–1639.
37. Pardridge, W. M., and Oldendorf, W. H., 1975, *Biochim. Biophys. Acta* **382**:377–392.
38. Crane, P. D., 1983, *J. Neurochem.* **40**:160–167.
39. Bradbury, M. W. B., Patlak, C. S., and Oldendorf, W. H., 1975, *Am. J. Physiol.* **229**(4):1110–1115.
40. Steinwall, O., and Klatzo, I., 1966, *J. Neuropathol. Exp. Neurol.* **25**:542–559.
41. Pardridge, W. M., 1976, *J. Neurochem.* **27**(1):333–335.
42. Gjedde, A., and Crone, C., 1975, *Am. J. Physiol.* **229**(5):1165–1169.
43. Oldendorf, W. H., Braun, L. D., and Cornford, E. M., 1979, *Stroke* **10**(5):577–581.
44. Oldendorf, W. H., 1973, *Am. J. Physiol.* **224**:967–969.
45. Oldendorf, W. H., Sisson, W. B., and Silverstein, A., 1971, *Arch. Neurol.* **24**:524–528.
46. Oldendorf, W. H., Sisson, W. B., Mehta, A. C., and Treciokas, L., 1971, *Arch. Neurol.* **24**:423–430.
47. Oldendorf, W. H., 1973, *Arch. Neurol.* **28**:45–48.
48. Oldendorf, W. H., 1971, *Proc. Soc. Exp. Biol. Med.* **136**:385–386.
49. Christensen, N. H., 1969, *Adv. Enzymol.* **32**:1–20.
50. Diamond, I., 1971, *Arch. Neurol.* **24**:333–339.
51. Oldendorf, W. H., and Brown, W. J., 1975, *Proc. Soc. Exp. Biol. Med.* **149**(3):736–738.
52. Oldendorf. W. H., Cornford, M. E., and Brown, W. J., 1977, *Ann. Neurol.* **1**(5):409–417.
53. Ahmed, N., and Van Harreveld, A., 1969, *J. Physiol. (Lond.)* **204**:31–50.
54. Davson, H., and Hollingsworth, J. R., 1973, *J. Physiol. (Lond.)* **233**:327–347.
55. Meek, J. L., and Neff, N. H., 1973, *Neuropharmacology* **12**:497–499.
56. Wolfson, L. I., Katzman, R., and Escriva, A., 1974, *Neurology (Minneap.)* **24**:772–779.
57. Bradbury, M. W. B., 1971, *Ion Homeostasis of the Brain* (B. K. Siesjo and S. C. Sorensen, eds.), Munksgaard, Copenhagen, pp. 138–153.
58. Bradbury, M. W. B., and Sorno, G. S., 1977, *Exp. Eye Res.* **25**(Suppl.):249–257.
59. Rapoport, S. I., 1970, *Am. J. Physiol.* **219**:270–274.
60. Rapoport, S. I., 1976, *Blood–Brain Barrier in Physiology and Medicine*, Raven Press, New York.
61. Rapoport, S. I., 1977, *Exp. Eye Res.* **25**(Suppl.):499–509.
62. Rapoport, S. I., Thompson, H. K., and Bidinger, J. M., 1974, *Acta Radiol.* **15**:21–32.
63. Rapoport, S. I., and Thompson, H. K., 1973, *Science* **180**:971.
64. Davson, H., and Oldendorf, W. H., 1967, *Proc. R. Soc. Med.* **60**:326–329.
65. Stewart, P. A., and Wiley, M. J., 1981, *Dev. Biol.* **84**(1):183–192.

18

Attenuated Blood–Brain Barrier

Maria Spatz

1. INTRODUCTION

The notion of restricted material exchange between the blood and the brain arose at the turn of the last century. It was derived from the observed absence of extravasation of synthetic dyes into the brain in contrast to their penetration of other tissues after intravascular administration of the dyes.[1–3] This event, which originally served to designate the existence of an interface between blood and brain as blood–brain barrier (BBB),[2,4] has been widely used as a marker for detecting the presence or absence of BBB disruption.[5,6] However, it later became evident that these dyes bind to albumin and form complexes that are actually responsible for their inability to cross the BBB.[7] Moreover, in the meantime, and particularly in the last two decades, great progress has been made in studying the properties and function of the BBB with the use of radiolabeled test molecules. Through these investigations, it became evident that the BBB represents a regulatory interface or highly selective barrier that controls the rate of in- and outflux of biological substances needed for the general and specific metabolic processes and neuronal activities in the brain.[5,6] Furthermore, the ultrastructural and biochemical studies conducted *in vivo* and *in vitro* in isolated cerebral microvessels as well as on cultured cerebral vascular endothelium demonstrated unequivocally that the selective and unique function of the barrier resides primarily within the endothelial cells, which might be under hormonal influence.[8]

Although on the whole the characterization of the properties of the altered BBB has not kept pace with the knowledge of the normal BBB function, substantial progress has been made in an understanding of the pathophysiology and the mechanisms involved in the attenuation of BBB permeability. These studies centered on the elucidation of the BBB changes and their role in the induction and/or progression of various disease processes as well as on clarification of the therapeutic usefulness of the transiently altered BBB permeability.

Maria Spatz • Laboratory of Neuropathology and Neuroanatomical Sciences, National Institute of Neurological and Communicative Disorders and Stroke, National Institutes of Health, Public Health Service, U.S. Department of Health and Human Services, Bethesda, Maryland 20205.

2. EFFECTS OF ENVIRONMENTAL POLLUTANTS

Both organic and inorganic mercury as well as lead are among the most extensively studied noxious environmental substances that are harmful to BBB function and in turn participate in the acceleration of toxic encephalopathy.

2.1. Mercury Ions

Mercuric compounds are known as potent metabolic inhibitors and have a high affinity for sulfhydryl (S-H) groups. They were shown to inhibit facilitated diffusion of monosaccharides in erythrocytes and affect the S-H groups involved in the Na^+ pump and K^+ transport.[9,10] In addition, $HgCl_2$ reacts with S-H groups responsible for the maintenance of membrane structure.[11,12] After their administration, both organic and inorganic mercurials were localized ultrastructurally in biological membranes. Thus, they may lead to alteration of membrane function, as is the case with the BBB. The possibility of mercurial dose-dependent selective impairment of BBB function was first suggested by Steinwall's studies.[13–17] He demonstrated an inhibition of either glucose or amino acid brain uptake (semiquantitative estimation of 20–40%) in the presence of a slightly increased BBB permeability to acid dyes with a low level of $HgCl_2$ but an obvious penetration of both glucose and a dye–albumin complex with a high dose of $HgCl_2$ perfusate.

Similar observations were made by us[18] by using the double-isotope technique of Oldendorf[19] (which measures the extraction of [^{14}C] test substance in the presence of H_2O as internal standard from brain during a single microcirculatory passage). A low concentration of $HgCl_2$ (10 μM) decreased brain uptake of the glucose analogue 2-deoxy-D-[1-^{14}C] glucose (2-DG) to the same minimal level as did the injection of unlabeled substrate (2-DG) (suggesting an inhibition of carrier-mediated facilitated transport), whereas a high dose of $HgCl_2$ (80 μM) enhanced 2-DG uptake above this level. The increase in brain uptake (BUI) of 2-DG above the value that could not be lowered by adding unlabeled substrate represents the passive passage of 2-DG into the brain. However, we could not demonstrate an increase in BUI above the control value except in animals with grossly hyperemic brains (Table I). The 2-DG inhibitory concentration of $HgCl_2$ observed in rabbits (10 μM) and that of glucose (50 μM) reported by Steinwall[15] were lower than that described in rats by Pardridge.[20] According to the latter studies, 100 μM $HgCl_2$ reduced the brain uptake of isoleucine and increased that of salicyclic acid but did not affect that of 3-0-methylglucose (3-0 MG), which decreased after the infusion of 1000 μM $HgCl_2$. This same concentration of $HgCl_2$ lowered the uptake of tryptophan and pyruvic acid but not the uptake of sucrose by the brain. Kinetic analysis of these results strongly suggested that alteration of BBB function by $HgCl_2$ is indeed dose dependent and selective. The lowest level of mercury ions affected the active efflux of acid metabolites (salicylic acid); a slightly higher level inhibited amino acid transport; high concentrations decreased glucose and pyruvate transport.

Table I
[^{14}C]Deoxy-D-Glucose and [^{14}C]3-O-Methyl-D-Glucose Brain Uptake

Solution	Addition: None	Addition: 60 mM cold 2-DG	Inhibition (%)
2-DG uptake[a]			
Ringer's	49.3 ± 2.1	—	—
Ringer's	—	14.4 ± 1.8	70.8
$HgCl_2$ 10 μM	16.3 ± 0.5	—	67.0
$HgCl_2$ 80 μM	29.2 ± 3.4	—	40.8
$HgCl_2$ 80 μM	—	26.7 ± 3.0	45.9
3-O MG uptake[b]			
None	30.0 ± 1.3	—	—
$HgCl_2$ 100 μM	27.8 ± 2.1	—	7.4
$HgCl_2$ 1000 μM	20.4 ± 1.5	—	32.0

[a] Mean ± S.E.M. based on data obtained from 3–9 rabbits Spatz *et al.*[18]
[b] Mean ± S.E.M. based on data obtained from 3–7 rats adapted from Pardridge.[20]

So far, the interrelationship between the biochemical events and ultrastructural changes occurring at the level of BBB was not investigated in the same experimental model of mercurial toxicity. Based on the studies available in the literature,[21,22] the disturbance of BBB transport might even occur prior to detectable changes in enzymes activity, since a small number of S-H group related to carrier transport are localized on the outer surface of the membrane.[11] The effect of mercury on the outer membrane of S-H groups could possibly also affect the glucose carrier transport in the presence of normal cerebral glycolysis,[23] which was observed in the preclinical state of mercurial toxicity, even though glucose transport may be regulated by brain glycolysis.

On the other hand, the decreased BBB transfer of glucose without increased permeability of protein-bound dyes could also be associated with mercuric alteration of some capillary enzyme activities, since such changes were observed in the cerebellum histochemically.[22] The appearance of passive diffusion of labeled 2-DG correlates well with the described extravasation of acid dye and [^{14}C]glucose reported by Steinwall and probably happens when the capillaries exhibit mitochondrial degeneration and an increased vesicular transport with widening of the endothelial leaflet spaces.

All of the observations described concerned with the degree of mercurial toxicity leave no doubt that its alteration of BBB transport, irrespective of the mercury dose, has an impact on the rest of the brain and in great part is responsible for the stage of injury.

2.2. *Lead*

Lead, similarly to other heavy metals (mercury, manganese), acts as a cytoplasmic poison that hinders the cellular activity of enzymes in the body, especially those of the kidney, liver, and brain.[24] It binds to erythrocytes and macrophages.[25]

The concept that the BBB is primarily involved in lead encephalopathy has been based on morphological studies and, in particular, those performed during postnatal development.[25,26] The observed cerebrovascular injury preceded neuronal and glial damage. Endothelial swellings, increased pinocytosis, and fragmentation of the basement membrane were seen in the cerebral capillaries. Moreover, in the developing brain, the vessels of the cerebellum were found to be the most severely affected in induced experimental lead encephalopathy. They displayed a variety of abnormalities manifested by malformation of endothelial buds and interendothelial and parajunctional regions of capillaries and venules. The tight junctions were partially separated along their length. Some of the capillaries were hypertrophic, showing an increase in rough endoplasmic reticulum, pinocytotic vesicles, and mitochondria. Ultrastructural studies associated with ^{210}Pb autoradiography[27] revealed the presence of the labeled substance in the endothelial cytoplasm earlier than in the astrocytic foot processes, indicating a role for primary endothelial exposure to this metal in its toxicity.

Further studies showed that the accumulated lead in the capillaries could be localized in intramitochondrial deposits in a noncrystalline matrix and appeared to be situated in the same areas as the accumulated calcium when examined by electron microscopy and x-ray analysis.[28] These observations were in agreement with previous studies[29] that showed that lead enhances the uptake of calcium in isolated brain microvessels, lowers the affinity for mitochondrial calcium uptake,[30] and inhibits calcium efflux from synaptosomal mitochondria.[28] Thus, it was suggested by Silbergeld *et al.*[28] that the endothelial changes in calcium metabolism caused by lead exposure may be associated with abnormal transport of fluid and electrolytes across the BBB and result in brain edema.

Lead was also shown to affect the transport of amino acids when [^{14}C] tryptophan passage was studied in neonatal rats.[31] Moreover, investigation of isolated cerebral microvessels in rats exposed to lead either *in vivo* or *in vitro* showed an inhibition of facilitated carrier-mediated uptake of 3-O MG with an increased uptake of diffusable mannitol.[32] The lead-induced blockage of 3-O MG capillary uptake was dose dependent and more marked in microvessels obtained from young than adult animals.

All of these studies are consistent with primary vascular lead toxicity and vascular sensitivity being greater in the developing than in the adult brain.

3. *EFFECT OF IRRADIATION*

Radiation such as x-rays, α-particles, and microwaves may cause BBB damage. Since this problem has been of great importance in clinical medicine, elucidation of their effects has been investigated in various experimental models. Most of the information available concerns x-ray and α-particle irradiation, which in large doses destroys all the cellular elements in the brain, opens the BBB, and produces a marked edema of the white matter.[33] The BBB is permeable to protein-bound dyes at this stage.

Experiments concerned with the clarification of the implied BBB involvement in the subacute transient manifestation of radiation-induced toxicity in central nervous system failed to demonstrate changes in BBB permeability with the use of [^{3}H]galactilol, [^{14}C]urea, ^{24}NaCl, [^{3}H]VM-26, and bleomycin[34,35] (the last two substances are used for the treatment of cancer). So far, subtle changes in the BBB transport system have not been observed even though such an event could initite or result in attenuation of the metabolism in the irradiated brain.[36]

Posttherapeutic radiation injury to the BBB may occur as late as 1–6 years. It is the result of capillary occlusion due to a progressive thickening of the basal membrane, ischemic tissue necrosis, and abnormal proliferation of new capillaries in the irradiated regions of the brain.[37]

All reported studies unequivocally indicate that radiation-induced x-ray and α-particle toxicity does not begin with BBB damage. Nevertheless, whenever BBB permeability is altered during the disease process, this might contribute to the exacerbation of brain injury.

There is general disagreement about the effects of microwave-induced radiation toxicity in the brain,[38,39] most probably as a result of a lack of standardization of exposure parameters. However, it appears that BBB alteration does not occur unless hyperthermia of brain tissue is induced by the microwave irradiation.

4. *EFFECT OF HYPERCAPNIA*

The inhalation of carbon dioxide may open the BBB to various substances such as phosphate, trypan blue, drugs, sulfate, iodinated serum albumin, sucrose, and Cr-EDTA.[40–47] The degree of BBB change depended on the dose and duration of CO_2 exposure irrespective of animal species,[44] although guinea pigs were found to be more sensitive than rabbits to CO_2. This effect is reversible as judged by the absence of increased BBB passage of trypan blue or [^{131}I]albumin 10 min or immediately after removal of CO_2, respectively. Originally, the hypercapnic alteration of the BBB was thought to be the result of either increased cerebral blood flow (CBF) or vasodilation.[43] However, Cutler and Barlow[44] found in some areas of the brain exposed to CO_2 an inverse relationship between vasodilation and the altered BBB by examining the plasma content of various brain regions. In the brainstem, the uptake of protein was high without vasodilation, whereas in the cortex a marked dilation was observed without protein leakage. These studies do not necessarily rule out the possibility that the BBB opening might be mediated by vasodilation[6] (as was shown to be the case in hypertension, see Section 5) but show that such a change may also occur without it.

In rabbits, severe hypercapnia (P_{CO_2} 98–150 mm Hg) reduced the brain uptake of glucose analogue (2-DG and 3-O MG)[48] as well as of various amino acids[49] (L-alanine, L-arginine, L-histidine, L-tyrosine, L-methionine, and D-leucine). However, it increased the uptake of L-leucine and had no effect on L-serine and glutamic acid uptake into the brain. Hypocapnia (P_{CO_2} 15–18 mm

Hg), on the other hand, had little effect on L-tyrosine and L-methionine but increased the uptake of the other amino acids into the brain, as determined with the double-tracer indicator technique of Oldendorf with H_2O as the reference standard. Although a variety of differently classified amino acids were screened in this experiment, the changes could not be attributed to any group except for glutamic acid, which was unaffected and belongs to the acidic amino acids. None of the experimental rabbits showed an elevation of systemic blood pressure.

At this point it should be mentioned that the Oldendorf technique using tritiated water (which was shown to be both diffusion and flow limited) as reference standard for the determination of the tested substances has not been considered to be the most suitable technique for a high-flow situation. However, the ratio of injected to recovered water in most experiments (either hypercapnia or hypoxia, which are associated with increased blood flow) was found similar to that in controls. These changes most likely did not entirely reflect the changes in blood flow since the various tested substances did not show a uniform brain uptake in response to the altered P_{CO_2}. As far as the uptake of glucose is concerned, a reduction of 2-DG brain uptake was observed in hypoxia (see Section 8) whether tritiated water or butanol was used as the standard reference substance.

Exposure to high CO_2 (15–30%) was also shown to increase BBB permeability to [^{51}Cr]EDTA without affecting the ^{24}Na brain uptake.[47] Moreover, in short-term, less severe hypercapnia (65–75 mm Hg), the transport of glucose, Na^+, Cl^-, and thiourea was similar to that observed in seizures, and the suggested increased transfer of these substances was related to increased CBF.[50]

This discussion of the effect of severe hypercapnia on the BBB should not be left without stressing the complexity of this problem, since high concentrations of CO_2 not only alter the BBB but increase the blood flow, abolish autoregulation by decreasing the periarterial pH, change the cerebral metabolism, and elevate the systemic and intracranial pressure.[5] Thus, future studies ought to include cerebral blood flow measurement in order to clarify subtle BBB changes in hypercapnia.

5. *EFFECT OF SEIZURES*

A variety of blood-borne substances have been shown to penetrate the brain in chemically or electrically induced convulsions. Most of the studies unequivocally showed that the BBB leakage occurs shortly after the onset and increases with the duration of seizure, regardless of the markers used (vital dyes, radiolabeled sulfate, phosphate, or albumin) and is reversible.[51–55] Moreover, studies on the blood–cerebrospinal barrier in prolonged convulsions induced with either pentylenetetrazole or electric shock in puppies also showed a progressive accumulation of CSF [^{131}I]albumin that correlated well with the duration of seizures.[52] Generally, there is agreement that the BBB of some brain regions is more susceptible than others to various convulsants, but there is no consensus as to the mechanism responsible for their production of abnormal BBB permeability (see also Section 4).

The thalamus and hypothalamus are among the brain areas most affected by convulsions. Lee and Olszewski[53] suggested that the effect may be related to the cerebral blood flow and neuronal activity, since the observed abnormal [^{131}I]albumin penetration of the brain in electrically induced seizure could be prevented by low blood pressure and pentobarbital. A similar regional distribution of [^{35}S]sulfate to that of labeled albumin was seen in the brain in seizures induced by pentylenetetrazole.[55] Lorenzo and Barlow,[54] who studied the BBB permeability of strychnine-induced seizures in paralyzed, artificially ventilated animals (to eliminate the possible effect of hypercapnia and/or hypoxia), correlated the observed increase of [^{35}S]sulfate brain penetration to sites known to undergo activation or rhythmic discharge to this stimulant. They also found that blinding decreased the penetration of [^{35}S]sulfate into the lateral geniculate bodies.[54]

The effect of cerebral blood flow on the BBB permeability in seizure is disputable. There have been a number of contradictory observations in the literature, even though the BBB disturbance in some epileptic seizures was shown to be mediated by increased systemic blood pressure (see Section 6). Moreover, there is still no evidence available as to whether the BBB changes in permeability might be directly or indirectly related to an alteration of vascular receptors by seizure activity, although such a possibility is feasible.

According to Lorenzo *et al.*, there was no clear connection between the regional distribution of increased BBB permeability and increased cerebral blood flow in pentylenetetrazole-induced seizures.[56] The greatest BBB leakage of protein occurred in the thalamus, which showed the lowest increase in CBF among brain areas. In addition, the greatest increase in plasma volume occurred in the cortex but not in areas of the thalamus. These results were similar to those of Cutler and Barlow[44] (see Section 4) showing dissociation between abnormal BBB protein permeability and vasodilatation. Therefore, they thought that neither increased CBF nor vasodilatation was the primary factor responsible for the increased BBB permeability. On the other hand, Lorenzo *et al.*[56] found that methantholine (an antagonist of acetylcholine) ameliorated the penetration of the radiolabeled protein, even though the EEG indicated the presence of ongoing seizures, suggesting a possible cholinergic mechanism for the BBB change.

Horseradish peroxidase (HRP) studies in seizure confirmed the regional distribution of BBB disruption.[57–59] The BBB passage of HRP was spotty and confined to large vessels and was diffusely spread through the neuropil from unidentified vessels. The capillaries showed the reaction product in the basement membrane of the endothelial cells as well as in intercellular spaces far beyond the vessel wall itself. The capillary endothelial cells displayed an increased number of vesicles without the reaction product. However, the larger vessels contained HRP in endothelial and muscle cell vesicles. Thus, the mode of the protein passage could not be ascertained with certainty, but it was thought to be vesicular in nature.

In contrast to these findings, short-lasting seizures in rats do not produce a disruption of BBB permeability to Evans blue.[50] Moreover, these studies, as well as those conducted in man[60] or in rats,[50] suggested that the observed BBB permeability change is related to the increased CBF rather than being a result of epileptic activity or hypercapnia. These conclusions were based on an investigation of BBB permeability utilizing a modification of the indicator dilution technique of Crane, involving fractional extraction of $^{24}Na^{+}$, $^{36}Cl^{-}$, [^{14}C]urea, and [^{14}C]thiourea with ^{131m}In-DTPA (a plasma tracer) as the main reference substance as well as measurements of cerebral blood flow with the ^{137}Xe injection method. A decreased E (extraction) for thiourea and an increased PS (permeability–surface product) for urea with increased CBF were found during electroconvulsive treatment of endogenous depression in man. In rats, glucose showed a decreased E with increased CBF during seizures and during short-lasting hypercapnia (see also Section 4). None of the other substances tested either in man or in rat showed increased penetration across the BBB during convulsions or hypercapnia. In addition, during neither single nor repeated seizures could increased extraction of ^{131m}In-DTPA tracer be detected in the animal studies, suggesting that if there is a disruption of BBB, then it probably is not large enough to permit the detection of a macromolecule during a single cerebrovascular passage.

6. *EFFECT OF HYPERTENSION*

The link between the cerebrovascular changes and the development of hypertensive encephalopathy was first recognized by Byrom[61] studying the pathogenesis of renal hypertension. Since that time, all experimental investigations of acute and chronic hypertension have focused on elucidation of the mechanism involved in the disturbances of the BBB. Acute hypertension induced either by various vasoactive substances or by compression of thoracic aorta increases the BBB permeability to protein tracers.[62] This BBB change was found to be reversible[63] as soon as the blood pressure returned to normal levels and has been considered to be the result of increased mechanical stress on the endothelium by the increased pressure and not a result of ischemia as thought previously.[62] The ischemic theory based on the possible presence of arterial vasospasm in acute hypertension has been abated in view of the observed arterial constriction and dilatation of the pial vessels[64–66] and the extensive studies of Johansson drawing attention to the significance of the vascular dilatation during hypertensive episodes.[66] Since the dilated arterioles are incapable of autoregulation, the capillaries and venules are exposed to higher pressure than normal and increase the permeability of BBB.[5,66]

In addition, a variety of factors have been shown to influence the BBB in hypertension. The severity of BBB damage was shown to depend on the suddenness, degree, and maximal level of blood pressure increase and a number of elements that affect the cerebrovascular tone of the endothelial cell membrane. For the elucidation of all the factors involved in the hypertensive disruption of the BBB, Johansson[62] applied the Laplace law (which states that the tension or stress on the wall of a sphere increases as the radius increases) to the blood vessels. Accordingly, hypercapnic and vasodilating substances were demonstrated to augment, whereas hyperventilation reduced, the hypertensive BBB leakage.[62,67,68] Hypercapnia without an increase in blood pressure did not alter the BBB permeability to protein even at $P_{a}CO_2$ levels as high as 300 mm Hg.[69] Similarly, the observed BBB leakage of macromolecules in epileptic seizures was found to be pressure related, since the BBB change was ameliorated by preventing the increase in blood pressure[66,69,71] (see also Sections 4 and 5). Moreover, the increased cerebral blood flow demonstrated in the same brain regions as the BBB changes induced by hypertension with amphetamine and norepinephrine implicated the presence of vascular dilatation in the affected areas.[68] Pretreatment with β-adrenergic antagonists was shown to reduce the hypertensive BBB leakage induced by these substances but not that by seizures. This effect had been thought to act either directly or indirectly (inhibition of the metabolic response) on the β receptors.[62] Furthermore, there has been ample evidence that the sympathetic stimulation of cervical ganglia protects the BBB against hypertensive injury.[72–75]

Recent quantitative studies evaluating the effect of sympathetic stimulation on the permeability of BBB to RISA and on the cerebral blood flow (using microspheres) showed that the greatest reduction of the increased BBB permeability to protein and of blood flow occurred in the brain regions (cortex) most susceptible to the effects of acute hypertension.[75] No detectable changes in BBB leakage and blood flow were observed in the brainstem and cerebellum, where the BBB and the blood flow are less affected by hypertension. Evidence is still lacking whether stimulation of the cervical ganglia affects the large and/or small arteries. However, Heistead *et al.*[75] suggested that the sympathetic stimulation in hypertension may constrict the large arteries and alter the rise of pressure in smaller distal vessels and in this way reduce the BBB injury. On the other hand, cervical sympathectomy led to an increase in the BBB permeability to Evans blue–albumin complex during drug-induced hypertension in anesthetized but not in conscious adult rats.[66]

In the latest studies by Mueller *et al.*,[76] the increased BBB permeability to Evans blue and radiolabeled albumin was observed to be greater in the chronically denervated (by unilateral cervical ganglionectomy) than in the intact cerebral hemisphere of spontaneously hypertensive (SHR) and Wistar–Kyoto (WKY) rats. The mechanism responsible for the observed neuronal effect on the BBB permeability is not as well understood as that on blood flow. The cervical sympathetic ganglia supply the innervation for the ipsilateral cerebrum, and at the present time, it is generally accepted that the blood flow is under the control of the sympathetic nervous system.[77] As far as the regulation of BBB permeability is concerned, it is still controversial whether it is under central and/or peripheral sympathetic control.[77] Grubb *et al.*[78] reported an augmented BBB permeability to water after stimulation of the locus coeruleus or cervical ganglia in monkey, whereas Edvinsson[73] observed an increase in brain insulin uptake following excision of a superior cervical ganglion.

X-ray irradiation was shown to increase the vulnerability of cerebral blood vessels to hypertensive BBB disruption.[79] A variety of drugs such as dexamethasone,[80,81] SITC (4-acetamido-4′-isothiocyanostilbene-2,2′-disulfonic acid), lidocaine, phenothiazine, and some anesthetics have been implicated in diminishing the hypertensive effect on BBB permeability by reacting in different ways with the endothelial cell membrane.[66,82]

Another approach to the study of hypertension in man has been to investigate various strains of hypertensive rats. The resistance vessels of the spontaneously hypertensive animals were observed to possess a structural adaptability or a primary abnormality manifested by a decreased lumen/vessel ratio similar to that reported in experimental renal hypertension and hypertensive individuals. These animals, exposed to sudden intraluminal pressure induced by vasoactive drugs, showed a lesser degree of BBB disruption than did normotensive rats treated in the same way. This protective effect is thought to be the result of diminished vascular wall tension because of its capacity to adjust better than the normal vessel during sudden elevation of blood pressure.[66,83]

The hypertensive BBB leakage of protein was considered to occur through widening of the tight junctions by dilatation and stretching of the vessel wall.[84] However, none of the investigations of cerebral or retinal microvessels in experimentally induced hypertension and in spontaneously hypertensive rats showed evidence of tight junction disruption.[85–91] On the contrary, ultrastructural studies using either exogenous HRP or the endogenous immunoglobulin G (IgG) as marker demonstrated a cytoplasmic vascular change manifested by an increased number of vesicles in the cerebral or retinal arterioles, venules, and capillaries. The number of endothelial vesicles was less pronounced with IgG than with exogenous HRP.[90] The retinal vascular vesicles not only contained HRP but were filled with lanthanum nitrate, ruthenium red, but not ferritin particles, which were seen in vacuoles when each of these markers was tested in spontaneously hypertensive rats.[91]

In the latest studies of chronic hypertensive encephalopathy, it was shown by immunocytochemical techniques that the extravasation of protein may occur in the absence of demonstrable leakage of Evans blue–albumin and that such an event may also occur in presymptomatic stroke-prone spontaneously hypertensive rats.[92] Similar diversity in the detection of extravasated protein in the brain was observed in the study of chronic renal hypertension by Nag *et al.*[93] These investigators demonstrated foci of endogenous plasma protein extravasation of varying ages without evidence of vascular occlusion and frequently in the absence of a visible BBB alteration to Evans blue or HRP. Based on these studies, it is evident that the immunochemical detection of extravascular plasma protein in the brain is the most sensitive method for the demonstration of changes in the BBB permeability that might not be recognized otherwise.

7. *EFFECT OF HYPEROSMOLAR PERFUSATES*

Almost four decades ago, Broman and Lindberg-Browman[94] demonstrated that the intravascular administration of hyperosmolar solutions may alter the BBB permeability to dyes that were thought to affect the vessels directly. This concept has been changed by the renewed interest in the study of hyperosmolar perfusates as a tool for transient opening of the BBB, which would be useful in chemotherapy of CNS diseases.

Thompson[95] brought forward the idea that BBB penetration by nonelectrolytes as a result of hyperosmolarity depends on their size and cellular permeability. This was based on the observed greater cerebral penetration by mannitol and sucrose than by urea and thiourea in rats perfused with hyperosmolar solutions. However, Rapoport and his co-workers[5] contributed the most extensive information concerning the characterization of hyperosmolar BBB alteration. The main properties of the BBB opening were originally established by the observed effects of topically applied hyperosmolar solutions on BBB (pia–arachnoid) permeability to Evans blue–albumin.[96] They were as follows: (1) hyperosmolar electrolytes and relatively lipid-insoluble nonelectrolytes opened the BBB reversibly; (2) the response was nonspecific; i.e., the molal threshold (the minimal osmolality producing Evans blue–albumin leakage of pia vessels after 10 min of topical application) was the same for all electrolytes and for the "reversible" nonelectrolytes; (3) in general, the extent of the barrier alteration increased with increasing osmolality; (4) the osmotic threshold above normal varied but corresponded to the projected cellular membrane permeability of the solutes. The opening of the barrier with substances of either ether–water or oil–water partition coefficients was found with high concentrations only. Similar responses were seen with hyperosmolar perfusates in the brain and eye.[5]

Both Thompson and Rapoport suggested that the effect of hyperosmolar BBB opening is the result of endothelial cell shrinkage. However, Thompson thought that the leakage might occur through normal tight junctions, whereas Rapoport's hypothesis visualized a widening of tight junctions as a result of endothelial cell shrinkage. These suggestions were to some extent substantiated by the earlier electron microscopic studies with peroxidase reported by Brightman *et al.*[97] and Sterrett *et al.*[98] Both groups demonstrated peroxidase spread across the basement membranes but without evidence of opened tight junctions. Moreover, Sterrett *et al.* described shrinkage of all layers of arterial wall, endothelium, smooth muscle, and pericyte with an increase in space, whereas Brightman *et al.* found the peroxidase tracer in the interjunctional pools between successive tight junctions of some endothelial cells. In view of these observations and the absence of a greatly increased number of pinocytotic vesicles, the authors concluded that the transfer of the peroxidase occurred through reversibly opened endothelial tight junctions.

However, recent studies by Houthoff *et al.*[99,100] strongly suggested that this process of hyperosmolar protein leakage across the BBB might not be the case. These workers investigated early and sequential hyperosmolar vascular changes to exogenous HRP as compared to endogenous antibodies to HRP tracer. For this evaluation, the level of IgG antibodies (anti-HRP) was enhanced by the immunization of the rats with HRP beforehand to facilitate their demonstration in the tissue. The patchy distribution of extravasated endogenous or exogenous HRP tracer was similar to the previously described focal vascular reponse to hyperosmolar perfusates seen with light microscopy. The extravasated tracer (HRP and anti-HRP) coincided with the visible edema of the brain.

Electron microscopically, a focal slight to moderate HRP reaction product was found in the basal lamina, extracellular spaces, and surrounding glial cells at 10-s and 5-min survival times but was virtually absent in these structures after survival times of 15 and 60 min. An abundance of vesicles was seen in the endothelium of capillaries and larger vessels at all times. Most of these vesicles were filled with the tracer. Moreover, channellike structures or continuous vesicular arrays were seen in the cytoplasm of tangentially cut endothelial cells. Although the HRP tracer was present in some interendothelial areas, a continuous filling of junctional spaces was rarely observed. Moreover, some intracytoplasmic distribution of the tracer was seen in shorter survival times. The patchy distribution of anti-HRP tracer was essentially similar to that of HRP at 10-s survival time, but the amount of tracer in the junctional areas and the numbers of endothelial vesicles with and without the marker were smaller than in the HRP group. On the other hand, the basal lamina of vessels, the extracellular space, and the parenchymal cells of the brain displayed an increased content of anti-HRP at 10 s and at 5 min, and a fair amount of tracer was still seen at 10 and 60 min of survival time, in contrast to the exogenous HRP.

These findings strongly indicate that hyperosmolar impairment of the BBB probably occurs early and appears to be complete 5 min after the insult. Thus, its reversibility appears to be as rapid as that reported for the amino acid γ-aminobutyrate.[101] In view of these observations, the authors suggested that the presence of the tracer at later times of survival in any of the observed endothelial areas cannot be the result of tracer outflow and therefore may not be indicative of the specific mechanism by which the passage of protein takes place across the BBB. They also suggested that the presence of estravasated tracer could have resulted from endothelial tracer overflow, since the analysis of their data showed only the extravasation of the protein markers during the period of their maximal presence in the endothelial cells.

The observed discrepancies between the localization of exogenous and endogenous HRP tracers at various sites do not support either the originally postulated mechanism of hyperosmotically induced passage of proteins across the BBB through widening of the endothelial tight junctions or the possibility of vesicular transport.

There is also some evidence that hyperosmotic perfusates can alter the BBB permeability to small molecular substances, which may or may not coincide with the disruption of the BBB to protein.[102,103]

Hyperosmotic urea treatment was found to augment the saturable brain uptake of glucose, as was shown by the double-indicator method of Oldendorf.[102] The data summarized in Table II show a reversible increase in the brain uptake (BUI) of [^{14}C]2-deoxy-O-glucose ([^{14}C]2-DG) in rabbits treated with intracarotid infusion of hypertonic urea. Hypertonic acetamide and NaCl also increased the brain uptake of [^{14}C]2-DG. However, lactamide inhibited the BUI of [^{14}C]2-DG irrespective of its tonicity. The increased BUI of [^{14}C]2-DG could be inhibited by unlabeled (cold) 2-DG.

A similar response to hypertonic urea was seen with [^{14}C]3-0-methyl-D-glucose ([^{14}C]3-O MG), with a BUI of 63.75 ± 1.85 S.E. compared to control

Table II
[^{14}C]Deoxy-D-Glucose Brain Uptake[18]

Perfusate			Uptake (Mean ± S.E.)		Inhibition
Tonicity	Solution	Molarity	No addition	60 mM cold 2-DG	(%)
Isotonic	Ringer's	0.3	48.83 ± 0.83 (10)[a]	14.35 ± 1.84 (5)[a]	70.6
Isotonic	Lactamide	0.06	35.10 ± 0.90 (5)	—	25.0[b]
Isotonic	Urea	0.3	44.62 ± 3.37 (3)	—	—
Hypertonic	Urea	2.3	79.92 ± 2.32 (14)	19.95 ± 2.02 (4)	73.7
Hypertonic	NaCl	0.5	67.21 ± 3.34 (3)	11.48 ± 0.79 (3)	82.9
Hypertonic	Acetamide	5.0	68.66 ± 0.99 (3)	18.67 ± 2.26 (4)	72.8
Hypertonic	Lactamide	2.0	34.4.0 ± 0.90 (5)	—	25.0[b]
Hypertonic	Urea	2.3	44.62 ± 3.37 (3)[c]		

[a] Number of animals in parentheses.
[b] Compared with Ringer's-perfused animals.
[c] Determined 40 min after hypertensive perfusion.

BUI of 30.39 ± 2.88 S.E. In both the experimental and control rabbits, the [^{14}C]3-O MG BUI could be reduced to the same degree as in the controls when [^{14}C]3-O MG was injected together with various concentrations of cold phlorizin (a known inhibitor of glucose carrier transport) or with 60 mM cold glucose.

The effect of hyperosmotic urea on glucose transport across the BBB was confirmed except for its reversibility by Pollay,[103] who studied not only [^{14}C]glucose but [^{14}C]fructose brain uptake by indicator diffusion techniques using [^{3}H]dextran (mol. wt. 100,000) as a reference substance as well as CBF with ^{133}Xe in dogs. In contrast to our findings, he found a slight inhibition of glucose transport with hyperosmolar saline solutions. The effect of this perfusate is difficult to evaluate, since we observed it sometimes to cause intravascular clotting of the blood. As far as the brain uptake of fructose was concerned, an abnormal passage of this substrate was observed in absence of Evans blue–albumin leakage, suggesting a graded alteration of the BBB permeability.

Recent studies by Pappius[104] demonstrating a hyperosmolar increase of local glucose utilization in the brain of rats suggested that the observed augmented glucose transport could be secondary, in response to the altered glucose metabolism, rather than a primary effect on the carrier or site that mediates glucose transport, as suggested by us[102] and by Bradbury.[6] Our postulate was based on the similarities between the hyperosmolar effect on glucose transport in the brain and those reported in adipose tissue and muscle. Perhaps future investigations will clarify whether the mechanism of the hyperosmolar stimulation of glucose transport across the BBB is the result of the altered cerebral metabolism only or of both metabolism and transport.

Increased penetration of norepinephrine and/or metabolites along with extravasation of protein tracers[105–107] (Evans blue, [^{125}I] serum albumin) without edema formation in the brain was demonstrated in rat and baboon after hyperosmotic arabinose or urea intercarotid injection (respectively). In both types of experiment, a two- to threefold increase of norepinephrine passage was seen across the BBB. The histofluorescence of norepinephrine in baboon was found at the same sites as that for protein. The osmotically enhanced penetration of norepinephrine was shown to increase CBF.

Consideration of the therapeutic usefulness of the application of hyperosmolar solutions draws attention to investigations concerned with possible central and/or systemic side effects of such treatment. Rapoport,[108] in his original work, demonstrated that hypertonic urea, like some of the hypertonic contrast media, produces two types of barrier disruption in the surviving animal. (1) The reversible type was thought to occur by uncomplicated opening of the tight junctions, whereas (2) the complicated BBB damage demonstrated by focal cerebral necrosis and contralateral motor deficit in the monkey was reported to be most likely a result of an additional cytotoxic effect of hypertonicity of urea or of hypertension and ischemia. These adverse long-term neurological sequelae associated with the hyperosmotic opening of the BBB were absent when permanent ligation of the carotid artery was avoided during the hyperosmolar perfusion. The findings of Pickard *et al.*[107] are in agreement with those of Rapoport. They showed only transient effects of hyperosmolar urea infusion on systemic and sagittal sinus pressure, heart rate, and end-tide P_{CO_2} within 5 min. Thereafter, they did not observe any ill effects on blood flow, metabolism, or physiological reactivity of cerebral circulation in baboons. On the other hand, Kozeniowska and Skolasinska[109] noticed a dissociation between the reversibility of BBB opening to Evans blue–protein and the CBF effect following the hyperosmotic urea perfusion. Their findings showed a lack of autoregulatory response of the CBF at the time of restored BBB permeability to protein tracer in the ipsilateral in contrast to the contralateral perfused hemisphere. Based on these results, the authors warned against the use of the dog model for the pharmacological investigation of hyperosmolar effects on CBF.

There have also been a number of diverse investigations in regard to the formation of edema following hyperosmolar infusion. The question of whether or not reversible (uncomplicated) hyperosmolar openings in the BBB may produce cerebral edema was evaluated by Rapoport and his colleagues[110] in detail using arabinose, which has a low cerebrovascular permeability, is freely soluble in water, and was shown not to produce long-term changes in the brain or in the animal's behavior. These studies clearly indicated that the cerebral hemisphere perfused with 1.4 to 1.6 M arabinose solution contained an increased amount of water (1.0–1.5%) relative to the unperfused side.

The edema formation (1.5% elevation of brain water is equivalent to brain swelling of 7%) was temporary, since the level of water declined at 24 h and disappeared entirely by 8 days. This transient increase in brain water content did not occur when the rats were perfused either with isotonic saline or with 1.4 M isotonic arabinose and when the cerebrovascular permeability to sucrose was not elevated, although some of the animals showed a slight brain staining with Evans blue. The correlation of brain edema with regional brain staining following carotid perfusion with 1.4–1.8 M arabinose showed the cerebral swelling to be proportional to the increase of cerebrovascular PA (the product of cerebral permeability, P, and capillary surface area, A), which is probably caused by penetration and retention of plasma salts and proteins of the brain as well as by elevated capillary hydraulic conductivity. Pappius *et al.*[104] had shown that this type of increased cerebral permeability caused by hyperosmolarity in conscious animals is accompanied by a transient increase in regional cerebral glucose utilization (mentioned above), a decrease in regional cerebral blood flow, and behavioral changes.

Fieschi *et al.*[111] also observed an immediate short-term depression of EEG and a tardive paroxysmal EEG activity with some tonic–clonic seizures in the same hyperosmolar model of rats as used by Pappius *et al.* They clearly showed that the paroxysmal activity depended on the effective opening of the BBB, since a normal EEG was found in animals without cerebral extravasation of Evans blue. Diazepam prevented the late paroxysmal EEG modification but not the early EEG depression and the opening of the BBB. From these[111] and Pappius' investigations,[104] it was concluded that focal seizure activity is responsible for the metabolic abnormalities associated with osmotic disruption of the BBB. Therefore, the authors cautioned against the use of this model for neuropharmacological studies. Nevertheless, the osmotic opening of the BBB has been used experimentally in animals and man to enhance the entry of various agents[112–114] (hexaminidase A, methotrexate, and neutralizing antibody to measles) for the evaluation of their effectiveness in the treatment of various disease processes.

One of the most important clinical experiments was the evaluation of the effects of short-duration intravascular hypertonicity on the blood–brain barrier in man in order to assess whether contrast media used for angiography could be harmful as a result of its transient intravascular hypertonicity.[115] These investigations showed that the BBB opening to sodium and chloride did not occur when the patients were injected with one of the following solutions: 5% saline, 25% mannitol, or contrast medium of the metrizoat group (isopaque-amine 280°). The short-term hypertonicity did not alter the BBB permeability to water, although the extraction of ^{3}HOH increased with the mannitol and isopaque-amine, corresponding to the lower concentration of water in these solutions. These findings did not rule out a possible minimal alteration of the BBB that could not be assessed by the markers used in these experiments but indicated that the transient hyperosmolarity that can be encountered in the clinical carotid angiography is inconsequential.

8. *EFFECT OF ISCHEMIA*

Numerous reports concerned with investigations of cerebral ischemia in various models concur that the cerebral vascular endothelium is more resistant than the other cellular elements of the brain to ischemia and/or hypoxia. However, it is still unclear whether and to what extent this phenomenon has a protective or damaging effect on the brain. Thus, the basic understanding of BBB reactivity to the reduction of blood flow and/or oxygen tension is of utmost importance in the elucidation of the pathophysiology of cerebrovascular disease and the possibilities of preventing and/or treating these processes.

Recent BBB studies, especially those in ischemia, have been focused on the multiplicity of this barrier system, and, therefore, investigations have been conducted in order to detect not only permeability changes to macromolecules (proteins) but to micromolecules, some of which are taken up by specific transport mechanisms into the brain. Moreover, the evaluation of the BBB response to ischemia had been greatly aided by the availability of techniques for the isolation of cerebral microvessels from the rest of the brain and the possibility of studying their reactivity *in vivo* and *in vitro*.

8.1. Blood–Brain Barrier Changes to Macromolecules

As mentioned above, the abnormal cerebral penetration of proteins derived from the blood has been extensively studied in various pathological conditions including ischemia. Again, the most commonly used marker has been trypan or Evans blue.

Until about the end of the 1970s, there was a great diversity of opinion, based on several investigations performed in a variety of ischemic models, as to whether or not, and if and when, ischemia induces an increased BBB permeability to protein.[116–120] These apparent contradictions were resolved by the careful studies of Ito *et al.*[121] in Mongolian gerbils subjected to unilateral common carotid artery occlusion. In this type of ischemia, in the symptom-positive animals, the degree of cerebral ischemia depends on the duration of the ischemic insult, whose manifestations are inversely proportional to the time of reestablished cerebral circulation. Thus, after a short period of brain ischemia, the extravasation of the Evans blue–albumin complex was seen at a later time than in cerebral ischemia of longer duration. For example, all gerbils with left common carotid artery clipping for 3 h showed blue discoloration of the cerebral hemisphere ipsilateral to occlusion only 3 h after clip release, whereas animals with a 6-h cerebral occlusion displayed similar changes 30 min after clip removal.

Ultrastructurally, symptom-positive (with histopathological picture of ischemic injury) animals demonstrated a somewhat enhanced transfer of HRP across the cerebral arterioles, venules, and capillaries, evidenced by a slightly increased number of endothelial vesicles after 3 h of ischemia and 1 h of reestablished cerebrovascular circulation. The gerbils subjected to 6 h of left common carotid occlusion and 1 h of release (symptom-positive) showed an increased number of vesicles containing HRP in the endothelium of all three types of vessels, with spread of the tracer from the basement membrane to the neuropil in the homolateral cerebral hemisphere. In addition, minimal pinocytotic uptake of the HRP was observed in the astroglial cells and neurons, as was focal swelling of the astrocytic end feet and other astrocytes. The most severe changes were found in the neuropil after 18 h of occlusion and 1 h of release. However, at none of the examined periods of time were the endothelial cells seen to be damaged, and the peroxidase was never found between the endothelial cells from the luminal to the first abluminal tight junction.[122]

In addition, Lossinsky *et al.* observed endothelial vesicles forming chains, autophagic vacuoles, and tubular structures and diffuse staining of some cells with HRP.[123] Occasional tubules were found to cross the entire wall of the endothelial cell when examined in thin sections (1500–2400 Å thickness). They were lined by a thin or thick membrane. The hemisphere contralateral to carotid artery occlusion showed also an extravasation of HRP but to a lesser degree than that seen in the ipsilateral brain.

With unilateral cerebral ischemia (1-h carotid artery clipping) in gerbils, it was clearly shown that the peak of the altered BBB permeability for substances of molecular size similar to albumin was the same (10 h after clip release) irrespective of the protein marker used ([^{14}C]dextran, [^{14}C]dextran + Evans blue, RISA).[124] The increased brain uptake of the radiolabeled dextran corresponded well with the presence of grossly visible Evans blue penetration of the ischemic hemisphere. On the other hand, the evaluation of the BBB permeability changes to proteins with Evans blue or HRP or IgG as markers revealed a diverse picture of ischemia induced by bilateral common carotid artery occlusion for 15 min and various periods of release.[125]

In this model, the extravasation of Evans blue–albumin complex could be seen neither macroscopically nor by electron microscopy. However, a faint perivascular Evans blue–albumin (EBA) fluorescence was observed in the region of amygdaloid nuclear complex at 72 h after clip release. Extravasation of HRP or its penetration of the endothelial basement membrane was not seen at any time after clip removal (5, 24, 48, 72 h), although an increased number of vesicles (some only filled with HRP) and cytoplasmic tracer were seen in the same endothelial cell at all surviving times. The perivascular presence of IgG was found at 48 and 72 h in 1-μm Epon® Vibratome sections, whereas with the EM, the IgG marker was seen in the basement membrane at 24 h after clip release. The IgG extravasation was always accompanied by the presence of this marker in the endothelial cytoplasm. Endothelial vesicles were not seen, but pinocytotic vesicles and multivesicular bodies were observed in pericytes, muscle cells, and occasionally in the astrocytic foot processes with this protein marker (Fig. 1).

A dependable demonstration of an altered BBB permeability by a protein marker was also found in the brain of cats with permanent occlusion of the middle cerebral artery.[126] The earliest appearance of focal perivascular protein extravasation was seen using the peroxidase–antiperoxidase method after 1 h, whereas Evans blue–albumin complex staining was seen several hours later at the periphery of ischemic lesion. Subsequently, the penetrated protein markers showed a striking affinity for ischemic brain parenchyma and a more marked preference for the gray over the white matter. However, at a later date, the extravasated serum protein [peroxidase–antiperoxidase (PAP)] was observed in the region of the white matter, while EBA remained in the gray matter. This demonstrable dissociation of the protein markers suggests a different pathway and process in the removal of extravasated protein markers and possibly various proteins itself. Ischemia may also cause a biphasic alteration of BBB to proteins, as was observed in gerbils with bilateral common carotid artery occlusion for 5 min[127] and in cats subjected to middle cerebral artery clipping for 1 h.[128] In these cases, the transient appearance of extravasated HRP and/or EBA complex seen shortly after release of the clip has been associated with hyperemia of the brain, whereas the delayed BBB leakage of proteins has been related to tissue injury. Moreover, in the cat, the first phase of the increased

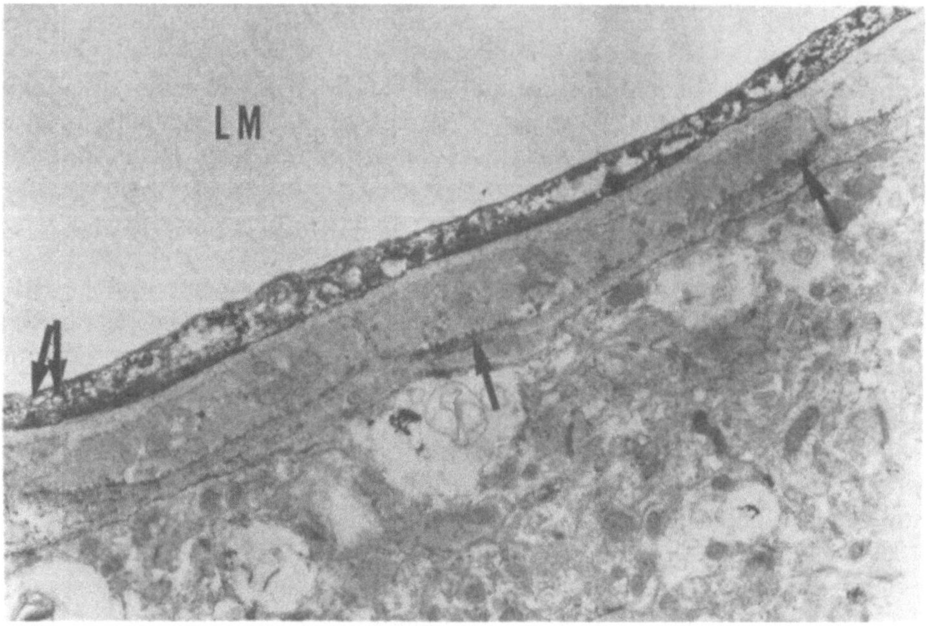

Fig. 1. Endogenous IgG (anti-HRP) tracer studies in Mongolian gerbils subjected to bilateral common carotid artery occlusion for 15 min and 72 h release. The reaction product is seen in the endothelial cytoplasm, basement membrane, and vesicles of the smooth muscle (arrow). Note the empty vesicles in the endothelium (two arrows). ×11,900. (Courtesy of H. Houthoff; based on K. G. Go *et al.*[125])

BBB permeability to proteins does not occur unless the reduction of CBF reached low levels (below 20 ml/min per 100 g).

8.2. Blood–Brain Barrier Changes to Micromolecules

The BBB permeability to small molecular substances such as sucrose increases prior to and prevails longer than that for proteins irrespective of whether cerebral ischemia was induced by occlusion of a major artery or by microembolization.[124,129] Ischemia not only causes BBB permeability changes to substances that do not easily cross the BBB normally but also involves substrates transported by carrier-mediated processes.

In the gerbil model of unilateral common carotid artery occlusion (6 and 18 h), an enhanced brain uptake of [^{14}C]2-DG or [^{14}C]3-O MG was seen in both symptom-positive and symptom-negative (without histological evidence of cerebral damage) animals.[130] The augmented [^{14}C]2-DG brain uptake could be self-inhibited (with unlabeled 2-DG) in all gerbils subjected to ischemia and a short time of release with the exception of 18-h symptom-positive animals. The same blocking effect was seen when the animals were injected simultaneously with the labeled 3-O MG and 2 mM phlorizin (cold), a known inhibitor of glucose carrier transport.

Those findings suggest that the increased glucose analogue transfer across the BBB took place by a facilitated carrier-mediated process with the exception of 18-h ischemic gerbils, in which passive diffusion contributed to the increased glucose uptake. At the same time, the animals also showed an increased uptake of [^{14}C]sucrose in the brain. In gerbils with a shorter period of ischemia (1 h), the increase of saturable glucose analogue uptake by the brain also occurred prior to the observed increased permeability to sucrose and protein tracers.[124] The augmented 2-DG brain uptake was already found 1 h after clip release, whereas increased brain uptake of sucrose was seen 5 h later. This enhancement of the carrier-mediated transport could be the result of an altered CBF and/or cerebral metabolism following the removal of the primary ischemic insult. However, there is also a likelihood that the functional disturbance may reside at the level of cerebral capillaries, which could be at least partly responsible for the observed increased glucose brain uptake. In support of this contention are the studies performed by us on isolated cerebral microvessels obtained from gerbils subjected to bilateral common carotid artery clipping and various periods of release. These microvessels showed an increased uptake of [^{14}C]2-DG that was inhibited by cold 3-O MG when they were separated from brain of gerbils recovering from the occlusion for 6 min and 15 min to 3 h of postischemia.[131]

On the other hand, severe deprivation of oxygen results in a reduction of undirectional glucose or glucose analogue transport into the brain, as reported by Betz and Gilboe[132] and Gilboe *et al.*[133] in dogs and by us in rabbits. In order to explore the mechanisms responsible for this phenomenon, Betz and Gilboe tried to relate the low rate of glucose transport to a decreased facilitated exchange diffusion of glucose, but they found a greater anoxic decrease in the unidirectional glucose flux (35%) than that foreseen (5%). Subsequently, Gilboe and associates investigated the kinetic characteristics of glucose transport, measuring the rate of unidirectional flux under control conditions in the presence of various glucose plasma concentrations in normal and hypoxic dogs and in animals under the influence of low blood pH (7.2). They found that the affinity constant between glucose and the transport mechanism was lower in hypoxic dogs and at low blood pH than in normal dogs, whereas the maximal transport rate was not changed significantly (hypoxia and low blood pH $K_1 = 16.0 \pm 4.0$ mM and 16.1 ± 4.1 mM, respectively; T_{max} 2.2 ± 0.4 and 1.7 ± 0.1 μmol/g per min, respectively; normal $K_1 = 8.0 \pm 0.7$ mM, and $T_{max} = 1.7 \pm 0.1$ μmol/g per min). Based on their data, they concluded that the cerebral hypoxic decrease in pH caused by the accumulation of lactate might change the carrier affinity for glucose.[133] Our studies conducted in isolated cerebral microvessels support the possibility of hypoxic and ischemic alteration of capillary function itself.[131,134]

In both instances, the cerebral microvessels separated from either ischemic gerbils or hypoxic rabbits showed a decreased uptake of [^{3}H]2-DG as compared to the capillaries obtained from sham-operated animals.[49,131] In contrast, the labeled neutral amino acid capillary uptake was normal at the time when an increased [^{3}H]2-DG uptake was seen in the capillaries (separated from brain of gerbils subjected to 6 min of bilateral ischemia).[135] However, the opposite effect was seen after a shorter period of ischemia (3 min), when the uptake of these amino acids was found to be increased and that of [^{3}H]2-DG uptake was found to be normal in microvessels. Similarly, a decreased uptake of [^{3}H]2-DG was observed with the cerebral microvessels when they were separated from brains of gerbils subjected to ischemically induced seizures in studies by Nell *et al.*[136] The reduction of the labeled [^{3}H]2-DG uptake was greater than that seen in ischemia only and persisted for 1.5–2.5 h after the cerebral circulation had been reestablished.

The marked reduction in capillary [^{3}H]2-DG uptake observed during exposure to nitrogen could be recovered either by substituting nitrogen for oxygen gas or by addition of free fatty acid (FFA)-free bovine serum albumin (BSA). Among various nucleotides tested, cyclicAMP had an effect similar to FFA-free BSA, but Na^+ and MgATP decreased the uptake of [^{3}H]2-DG in the microvessels irrespective of the type of gas exposure. Addition of EDTA similarly affected the [^{3}H]2-DG capillary uptake, which could be partly restored in the presence of $MgCl_2$ in the incubating medium. These results suggested that chelation of bivalent cations from the membrane could interfere with the uptake and/or metabolic process, since 2-DG is partially metabolized.[134]

Originally, it was thought that the capillaries do not possess the enzyme capable of phosphorylating 2-DG. However, later on, it was shown that hexokinase activity exists in the microvessels, but its kinetic characteristics differ from the parenchymatous enzyme, and therefore it was suggested that this enzyme could participate in the transcapillary passage of glucose, especially because glucose 6-phosphatase in high concentration is also present in these vessels.[137] However, this problem remains to be resolved because the glucose transport across the BBB has been demonstrated to occur by a facilitated carrier-mediated mechanism.

Nevertheless, these studies showed that the most striking influence on the capillary 2-DG uptake was the addition of a free fatty acid to the solution containing FFA-free BSA during incubation. Therefore, it appears that under anoxic, hypoxic, or ischemic conditions a release of free fatty acid may occur irrespective of its origin, whether it be from the endothelial cell proper or from the plasma membrane, since the protective effect of the FFA-free BSA on [^{3}H]2-DG uptake was decreased or abolished by the presence of free fatty acid in the incubating medium. In support of this concept are numerous reports of free fatty acid release in many models of cerebral ischemia. Even though the free fatty acids are a part of the normal cycle in CNS lipid metabolism, the membrane function may be impaired following their accumulation in ischemia.[138] The protective effect of FFA-free BSA could be the result of simple surface coating, which would prevent a physical and/or chemical structural change in plasma membrane. It could also have a more specific action, binding and/or serving as a transport vehicle for the release of fatty acids.

In conclusion, considering the effect of the individual factors on the capillary 2-DG uptake, it is very likely that a structural chemical alteration of plasma membrane could occur in hypoxia or ischemia in which the appearance of free fatty acid may be the result or the cause of altered capillary function.

Severe hypoxia may also affect amino acid transport into the brain. Among various substrates tested (L-leucine, alanine, arginine, histidine, tyrosine, methionine, serine, glutamic acid, and D-leucine), L-tyrosine and D-leucine showed the greatest decrease in the brain uptake (about 70%) in rabbits.[49] In contrast to our findings, Betz and Gilboe[132] found a normal L-leucine brain uptake in anoxic dogs. This difference could reflect blood flow and/or species specificity, since our studies *in vitro* are in agreement with the general concept of an oxygen- and Na-independent uptake of neutral amino acids in the cerebral microvessels. However, as mentioned above, ischemia may lead to a transient increase in neutral amino acids in the capillaries.[135] A similar increase of [^{3}H]isoleucine and [^{3}H]phenylalanine was seen in the microvessels incubated in the presence of Na^+ under nitrogen atmosphere. Since the brain effect of neutral amino acid was shown to be partially Na^+ dependent, it is possible that the ischemic cerebral microvessels may participate in removal of some of the accumulated amino acids.[135]

In summary, it is fair to say that BBB amino acid transport has not yet been fully explored in pathological conditions and especially in ischemia, and future investigations might prove helpful in the elucidation of the pathophysiology of ischemia and other disease processes.

The ischemic neuronal release of neurotransmitters has been implicated to adversely affect cerebral blood flow and and in this way exacerbates brain injury.[139] The possibility of abnormal BBB penetration of monoamines, which either act the same or add or even reverse their effects on blood flow, prompted above all the investigation of BBB permeability to monoamines in ischemia. Indeed, a BBB leakage to monoamine was detected in gerbils subjected to either unilateral[140] or bilateral ischemic insult.[141] The pargyline and pyrogallol (inhibitors of MAO and COMT, respectively)-pretreated animals showed an increased BBB passage of radiolabeled serotonin (5-HT), norepinephrine (NE) and dopamine (DA) as well as sucrose 72 h after the release of unilateral carotid artery occlusion for 1 h.[140] This increased permeability of the BBB to monoamines occurred at a later time than that for sucrose, 2-DG, dextran, and Evans blue (see above).

Fluorescence microscopic studies revealed bright fluorescent structures displaying the specific characteristics of NE fluorescence in the brain of ischemic animals injected with NE. They resembled bizarre, stretched, tortuous nerve fibers in the neostriatum at the periphery of the necrotic tissue. Some foci of red fluorescence (Evans blue) were also seen in these animals. A few days later, the animals sacrificed at 7 days revealed necrotic foci but without the presence of abnormal noradrenergic structures nor Evans blue extravasation in the affected brain. Hence, these results demonstrated a differential ischemic BBB change to monoamines. Moreover, ischemia can elicit a selective alteration of the BBB to monoamines as observed without the use of the inhibitors (for MAO and COMT) in bilateral common carotid artery clipped for 15 min (Fig. 2).[141] As may be seen from this figure, the enhanced passage of exogenous 5-HT occurred prior to that of NE and persisted longer than that of NE.

Since the abnormal leakage of the monoamines into the brain of unilateral ischemic animals took place at the time of a reduction in the cerebral activity of MAO,[142] it was thought that the bilateral model of ischemic BBB abnormality to monoamines[141] would be more useful for the examination of possible primary

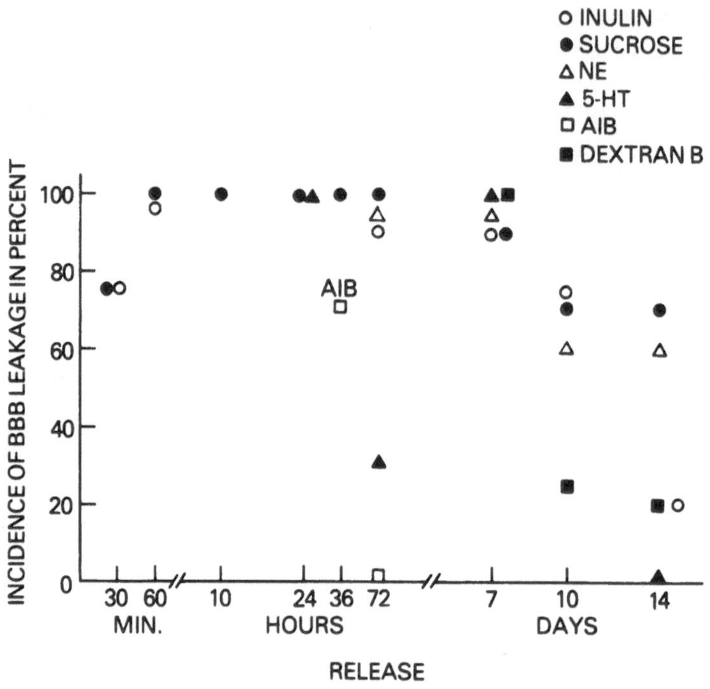

Fig. 2. The incidence of BBB permeability changes to various substances in Mongolian gerbils recovering from bilateral cerebral ischemia for 15 min. Each of the animals (in groups of 10–14) had radiolabeled sucrose together with one of the illustrated radioactive substances injected into the femoral vein 10 min prior to the termination of the designated experimental period. (Data based on Abe *et al.*[143] and M. Spatz, K. Abe, and T. Abe, unpublished observations.)

enzymatic changes in the microvascular barrier that normally restricts the passage of monoamines. Therefore, to elucidate the mechanism occurring during the selective BBB leakage of monoamines, experiments were designed to investigate the uptake and metabolism of the amines as well as the level of their degrading enzymes in cerebral microvessels obtained from ischemic gerbils at times coinciding with the absence and presence of increased BBB permeability to 5-HT and NE *in vivo*.[143] These experiments revealed that neither the capillary uptake of 5-HT, NE, and metaraminol [(M), the nonmetabolizable analogue of NE] nor the deamination of these amines was affected by the ischemic insult alone, although the activity of the MAO was reduced in these vessels. The increased specific capillary uptake of 5-HT (40% above normal) with a significantly decreased metabolism and the lowest activity of MAO (A type) was first observed 24 h after the release of occlusion. The augmented NE and M uptake into the cerebral microvessels was not seen until 72 h following the removal of the clips. However, the activity of capillary COMT and NE methylation was increased during ischemia and recovery as compared to the respective sham-operated controls.

When the effect of oxygen deprivation on uptake and metabolism in the cerebral microvessels was examined *in vitro*, it was shown that the exposure of normal microvessels to N_2 atmosphere (100%) for 5–60 min decreases both the deamination of exogenous 5-HT and the methylation of NE without significantly altering their capillary uptake. The presence of bivalent ions (essential for the determination of COMT activity) in the medium reduced the 5-HT deamination irrespective of the gases used for the incubation. However, these bivalent ions increased capillary methylation of NE in either air or oxygen but not in N_2 gas.[144]

Our latest studies concerned with the ischemic levels of endogenous catecholamines and their metabolites, although incomplete, showed that ischemia alone reduces the concentration of dopamine and 3,4-dihydroxyphenylacetic acid (DOPAC) but not that of NE.[145] However, within an hour after the clip's release, a decreased level of both catecholamines was observed in addition to a further reduction of DOPAC. The lowest concentration of NE was observed 24 h later, at the time of the greatest observed decrease in capillary MAO activity and of increased BBB permeability to exogenous 5-HT (see Table III).

Based on the results of the various experiments summarized above, the following conclusion can be made in regard to the nature of ischemic BBB change to monoamines. There is no doubt that the capillary MAO activity can be affected by either ischemia or O_2 deprivation. Thus, the reduction of the enzyme is responsible for the observed decreased deamination of exogenous 5-HT but not of NE, since the exogenous NE is methylated, whereas the endogenous amine is mainly deaminated. The presence of very low levels of MAO activity in the microvessels long after the release of occlusion could be the result of either inadequate supply or insufficient utilization of the oxygen or could be caused by inhibition of MAO activity by an ischemically induced alteration of the vascular or cerebral bivalent ion content.

The first occurrence of increased BBB permeability to monoamines (exogenous 5-HT) coincides with (1) already altered endothelial function demonstrable electron microscopically with HRP and IgG (see above), with little visible vascular extravasation of proteins, (2) increased capillary uptake of 5-HT but not NE, and (3) a marked decrease in the content of endogenous capillary NE and MAO activity but not DA content. The observed decrease in the endogenous capillary amines could reflect their release from the nerve endings and/or be the result of extraneuronal endothelial depletion of the monoamines, since we demonstrated that some of them can be synthesized in the cultured endothelium.[145] Therefore, it is possible that a certain level of microvascular amines controls the uptake, whereas the activity of MAO is responsible for the inactivation of the amines taken up by the capillaries, and in this way the microvessels as a constituent of the BBB regulate the cerebral in- and outflow of the monoamines. Whenever the equilibrium is disturbed, one might expect an increased BBB permeability, which would not require a frank disruption of the cerebral vessels.

The selectively increased BBB leakage and capillary uptake of monoamines seen in bilateral brain ischemia are probably valid only for exogenous 5-HT and NE because the latter is methylated rather than deaminated, in contrast to the endogenous NE. Therefore, the BBB leakage (outflow) of endogenous NE could take place at the same time as the observed exogenous 5-HT. The delayed ischemic enhancement of exogenous NE BBB permeability

Table III
Correlation of BBB Permeability to Monoamines with the Microvascular Levels of Monoamines and Their Degrading Enzymes as Well as with the Capillary Uptake of Exogenous Monoamines[143,145]

		Endogenous content of monoamines (% sham)			Enzyme activities (% sham)			Exogenous uptake of monoamines (% sham)		
	BBB[a] leakage	NE[a]	DA	DOPAC	MAO A	MAO B	COMT	5-HT	NE	M
Sham	None	100.0	100.0	100.0	100.0	100.0	100.0	100.0	100.0	100.0
Bilateral carotid artery occlusion 15 min	None	102.6	47.2	27.7	37.8	58.5	14.5-fold increase	102.9	106.0	100.8
+ Release 1 h	None	56.6	24.3	14.4	—	—	—	100.0	—	100.0
+ Release 24 h	5-HT	37.0	25.7	22.4	15.7	34.2	2.5-fold increase	140.5	104.4	103.5

[a] BBB, blood–brain barrier; MAO, monoamine oxidase; COMT, catechol-O-methyltransferase; 5-HT, 5-hydroxytryptamine; NE, norepinephrine; M, metaraminol; DA, dopamine; DOPAC, 3,4-dihydroxyphenylacetic acid.

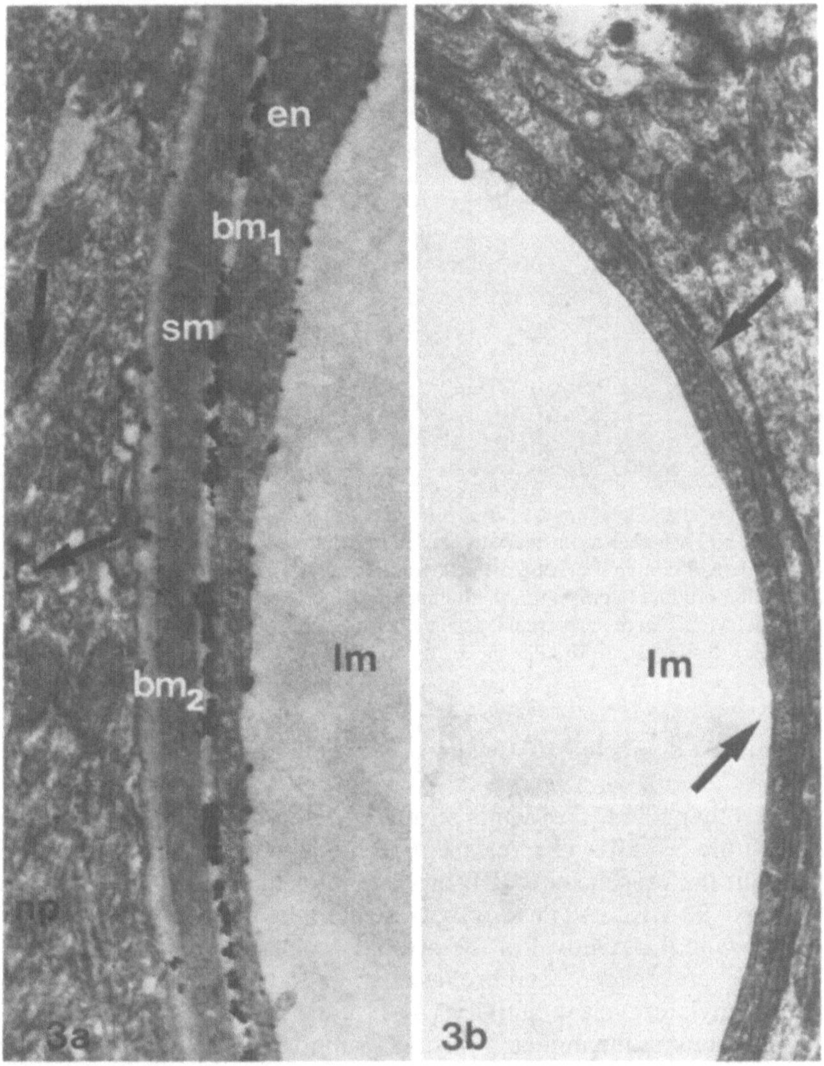

Fig. 3. Localization of the potassium *p*-nitrophenyl phosphatase (K-NPPase) activity in control Mongolian gerbil. a: Reaction end product is confined to the luminal (lm) and basement (bm) membranes. Basement membrane (bm_2) surrounding smooth muscle cell (sm) and perivascular neuropil (arrowheads) show moderate enzyme activity. ×8000. b: inclusion of 10 mM ouabain in the incubation medium. Reduction in membrane staining (arrows) is evident. ×8000. (Courtesy of B. J. Mršulja and B. B. Mršulja.)

and microvascular uptake most likely reflects a lower affinity for NE than 5-HT (K_m 14.7 and 2.3 μmol/mg protein[143] respectively) and the greater NE than 5-HT binding to albumin. In support of this concept is the marked increase in NA^+,K^+-ATPase activity (which is required for the uptake of these amines) observed in the vascular and perivascular regions (in particular, astrocytic cytoplasm[146]) (Figs. 3 and 4) and the demonstrated extravasation of IgG in

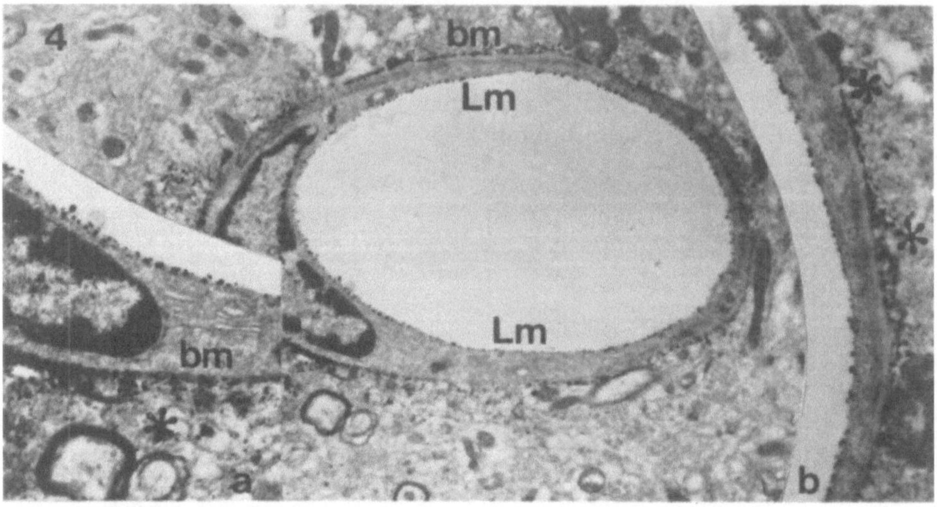

Fig. 4. The K-NPPase activity in the vascular endothelium and neuropil in the brain of gerbils subjected to bilateral common carotid artery occlusion for 15 min and 72 h of release. The conspicuous reaction product is present in the luminal (lm) and basement (bm) membranes. It is also abundant in the cytoplasm of perivascular astrocytic foot plates (asterisks). ×12,000. Inserts a and b ×19,250. (Courtesy of B. J. Mršulja and B. B. Mršulja, 1982.)

brain of animals subjected to 15 min of ischemia and 72 h of release[125] (see Fig. 1).

On the other hand, the demonstrable alteration of vascular endothelium observed within 5 h after clip release and the decrease of capillarly MAO activity without the presence of BBB monoamine leakage could be the result of the decreased Na^+,K^+-ATPase activity seen in the brains of these animals up to 20 h following the removal of the arterial occlusion.

Both the presence of cerebrovascular endothelial adrenergic receptors linked to adenylate cyclase activity (AC)[147] and the observed ischemic fluctuation of cerebral monoamines[148] strongly support the possible ischemic monoaminergic attenuation of endothelial AC regulatory system, which may lead to secondary cerebral blood flow and BBB change and thus adversely affect the function of the brain.

9. *EFFECT OF BRAIN INJURY*

Vasogenic edema observed in a variety of clinical conditions is an especially dreaded complication in patients with severe head injury. Therefore, it inspired many investigations concerned with the pathomechanisms of this disease process. Among many models, cold injury to the brain (introduced by Classen[149] and modified by Klatzo[150]) has proven to be the most useful for investigations of the development, spreading, and resolution as well as the treatment of brain edema.

This type of edema, as classified by Klatzo, was clearly shown to have its primary defect in the BBB, with an extracellular accumulation of fluid in white matter with some intracellular (astrocytic) swelling in the gray matter.[150–152] The vessels of the injured cortical region leak an exudate rich in protein, which was first identified as being serum albumin by Klatzo.[153] The radiolabeled protein studies also showed that large amounts of the protein enters the brain from the blood.[154] Moreover, the observed characteristically increased ratio of sodium to potassium in the edematous fluid has been attributed to the passage of sodium from blood to brain without the loss of potassium.[155] The actual sampling of edematous fluid obtained by placing inflected needles containing wicks into the cortical and underlying tissue had shown a similarity not only in ionic composition but also in colloid pressure of the edematous fluid to that of plasma. However, the colloid pressure of the edema fluid sometimes exceeded that of plasma, thus favoring the retention of the fluid in tissue.[156] The patterns of the extravasated proteins were the same whether tested with Evans blue or PAP tracer in cats sacrificed up to 4 days after cold injury only.[157]

Ultrastructurally, exogenous HRP was found to pass across the damaged endothelium in and near the region of cold injury, and endothelial pinocytosis as well as leaky interendothelial clefts were observed in adjacent areas.[158] A difference in protein passage across the cerebral microvessels was observed with the use of endogenous immunoglobulins (serum antibodies to HRP, raised by antigenic stimulation) and exogenous HRP (given intravenously as protein tracer in the marginal in contrast to the central zone of cold injury).[159] Time sequence studies (10s to 60 min) showed an absence of endothelial degeneration at the margin as compared to the central region, which revealed endothelial swelling, vacuolization, and disintegration of the cells. Some of the vessels contain platelets, red blood cells, and fibrin. Apart from the greater extravasation of both proteins in the marginal zone of freeze injury, the endothelial tracer-filled vesicles were scarcely present with the endogenous IgG, although a number of vesicles with and without exogenous HRP were seen frequently in the endothelial cells. Moreover, the anti-HRP marker rapidly appeared in the basement membrane and extracellular space as a marker of the spreading edema, whereas the exogenous HRP was not found in these locations outside the marginal zone. In contrast, both tracers were seen in the cytoplasm and basement membrane of many endothelial cells as well as to some degree in the extracellular spaces and intracellular in the center of the brain lesion. These changes were already present 10 s after injury and increased at larger survival times.

Pollay and Stevenes,[160] using a technique that simultaneously measures CBF (with [^{123}I]iodopyrine) and the brain uptake of radiolabeled tracers by single passage, demonstrated a decrease of CBF for the entire 58 days of observations, but with the greatest CBF depression (50%) at 48 h, when the largest increment in water content was also observed directly below the cold lesion. In the same area, the carrier-mediated glucose transport was decreased, but it was offset by an increased diffusion of glucose from blood to brain. The most significant leakage of glucose across the BBB was found at 14–21 days after cold injury, but [^{113m}In]diethylenetriaminepentaacetic acid (DPTA) brain uptake was enhanced in this area as compared to the control throughout the experimental period (2–58 days). A significant increase of ^{131m}In-DTPA but not of glucose brain uptake was also observed at the marginal zone for the first 14 days, whereas the tissue water content was significantly higher only at 48 h without CBF changes. Recently, Go *et al.* showed that the BBB permeability for [^{15}O]H_2O appears not to be affected by cold-induced injury.[161]

Experimentally marked extravasation of plasma protein into the brain was also found in mechanical type of brain and spinal cord injury.[162–167] However, the numerous techniques used for this type of injury do not result in a well-reproducible lesion except for the fluid-percussion model, which can be quantitated and thus has been studied extensively.[166] Animals subjected to this or to a similar trauma displayed penetration of the tested protein markers (either HRP or Evans blue or fluorescein isothiocyanate) in the brainstem, and the observed protein exudation correlated well with the onset of concussion.[162–166] Even though hypertension was observed to occur during the concussive injury, it is not implicated as the cause of the BBB protein leakage into the brainstem since the induction of hypertension *per se* resulted in an increase of BBB permeability to protein in the cortex and diencephalon but not in the stem. However, the prevention of arterial blood pressure rise eliminates the functional vasodilatations and (unresponsiveness to hypocapnia) morphological (endothelial) changes of the pial vessels.

Both light and electron microscopic studies of the brainstem have revealed arterial, venous, and capillary leakage of HRP.[166] The extent of the extravasation of HRP and vascular response depended on the severity of the lesion and the survival time (3 min to 2 h). The endothelial cells displayed an enhanced vesicular uptake of HRP throughout the cytoplasm. The HRP-filled vesicles and pits were open toward the abluminal side and appeared to be continuous with the tracer seen in the underlying basal lamina. In the arterioles, these changes were confluent with the basal membrane surrounding both the endothelial and muscle cells. Similar patterns were seen in the vesicles and capillaries. In addition, vacuoles and tubules filled with HRP and microvilli were observed in the endothelial cells. The number of microvilli and their apparent coalescence with endothelial surfaces free of the HRP tracer was found to be greater at 60 than at 15–30 min after injury. At 2 h, not very much HRP was found in the endothelial cells, but it was still present in the extracellular spaces, neurons, glia, and pericytes. Disruption of either tight junctions or cellular membranes was not seen at any time after the fluid-percussion injury. Therefore, the HRP penetration of brainstem was thought to occur by vesicular transport across the vessels.

In the same type of trauma, the dilated and unresponsive pial arterioles displayed luminal domelike projections and craterlike defects.[167] Analysis of serial electron microscopic sections suggested that they originated from collapsed or ruptured endothelial vesicles. These lesions were preventable either by inhibition of the hypertensive episode or by treatment with indomethacin or AHR-5850, which block the cyclooxygenase. Moreover, the angiotensin-induced hypertensive lesions could also be prevented by the inhibition of prostaglandin synthesis. Since the posttraumatic changes in the pial arterioles could be also avoided by pretreatment with free radical scavengers (astro blue, tetrazolium, and to a lesser degree by superoxide dismutose), free oxygen radicals (released during conversion of endoperoxide G_2 to H_2) were also implicated in the production of these the vascular responses. The involvement of these factors in the vascular damage was reproduced by topical application of either arachidonic acid or PGG_2 but not PGE_2, PGH_2, or 11,14,17-eicosatrienic acid (an unsaturated fatty acid that is not a substrate for cyclooxygenase).

In view of these observations, the authors concluded that increased production of prostaglandins mediates the arterial damage caused by brain injury and that their activity depends on the formation of free oxygen radicals. On the other hand, Awad *et al.*[168] reported recently that PGI intracarotidly administered prevented the BBB permeability to Evans blue and fluorescein–albumin complex but not the derangement of CBF or size of the infarct in cerebral ischemia of cat produced by middle cerebral artery occlusion.

Another model for studying the vasogenic type of edema has been the induction of altered BBB permeability to proteins using proteases. Intraventricularly administered collagenase has proven to be the most effective enzyme in producing, primarily, profound BBB changes and perivascular edema without affecting other cellular elements of the brain.[169–172] It enhances the BBB permeability to trypan or Evans blue–albumin complex, exogenous HRP, endogenous anti-HRP (IgG). Again, the vascular pattern of the tracer penetration depended on the marker used, although a disruption of endothelial cells was rarely seen. Both tracers were found in the endothelial cell, but a greater number of vesicles were observed with the exogenous HRP than the endogenous anti-HRP. On the other hand, the extent of anti-HRP penetration in vascular and perivascular space of the brain was greater than that of HRP. The tracer was seen not only in the thickened basement membrane, throughout the protenaceous material, and widespread in the extracellular spaces but also in the vesicles and multivesicular bodies of pericytes and in the vesicles of the muscle cells. The anti-HRP was also found in the endothelial cells and basement membrane of vessels that did not display a thickened basement membrane, which is characteristic of collagenase-induced injury. Thus, the extravasation of IgG was more widespread than the morphological signs of basement membrane injury and edema.

This model of BBB alteration is important for the study of the formation of vasogenic edema *per se*, since protease involvement may also occur in various cerebral cytotoxic processes, especially those associated with primary or metastatic brain tumors.

10. EFFECT OF HEPATIC FAILURE

Severe liver disease is often accompanied by an encephalopathy. During a variety of clinical investigations, high tryptophan and 5-hydroxyindoleacetic acid (5-HIAA) levels were observed in spinal fluid. This suggested that serotonin and other aromatic amino acids (phenylalanine and tyrosine, which could affect catecholamine metabolism) might play a role in the hepatic encephalopathy.[173,174]

Similar conditions can be reproduced in an animal model of acute hepatic failure, which is manifested by a constant change of plasma and brain amino acids and neurotransmitter levels in rats and dogs. The observed increased brain concentration of tryptophan could not be simply explained on the basis of the competition of existing amino acids at the BBB, since the branched amino acids (leucine, isoleucine, and valine) were found to be increased in the plasma in addition to other amino acids.[175] Thus, subsequent investigations of tryptophan transport across the BBB using the Oldendorf method showed that tryptophan, which usually has a low- and high-capacity transport system, is increased in hepatic failure. Since the high-capacity tryptophan transport was affected more than the low-capacity, the theoretical calculations suggested that a reduction in tryptophan passage by the low-capacity system occurs in the presence of high levels of competing amino acids. On the other hand, the increased rate of the high-capacity tryptophan BBB transport greatly contributes to the high content of brain tryptophan even though the level of free tryptophan in plasma has also some influence on the brain tryptophan influx.[176]

Other studies of the BBB transport in experimental hepatic failure showed selective changes in barrier transport; while glucose, inulin, and tyramine were unaffected, the basic amino acid, arginine, and the monocarboxylic acids, pyruvate and butyrate, were decreased in contrast to the increased passage of neutral amino acids across the BBB.[177–179]

A similar effect was seen in the uptake of amino acids in the microvessels separated from brain of rats subjected to portocaval anastomosis for 2 weeks.[180] A significant increase in the uptake of labeled neutral amino acids but not of lysine, glutamic acid, and α-methylaminoisobutyric acid was seen in microvessels of the experimental as compared to sham-operated animals. The kinetic studies of the uptake showed an increased V_{max} for phenylalanine and a decreased V_{max} for lysine without changes of K_m values.

As far as the microvascular BBB permeability to exogenous HRP is concerned, the studies of BBB in rats with portocaval anastomosis[181,182] demonstrated a uniformly patchy distribution of the HRP marker mainly in the cerebral cortex and basal ganglia. The brainstem and cerebellar cortex also showed the HRP reaction product, but to a lesser degree. The HRP tracer was confined to the vessel wall, but in some areas it extended into the adjacent neuropil, as examined by light microscopy.

Ultrastructurally, an increased number of endothelial vesicles containing HRP were found in arterioles, venules, and capillaries. The most marked reaction was seen in the arterioles, which revealed a great number of endothelial vesicles at the contraluminal side as well as in caveolae and multivesicular bodies but not outside membrane-bound structures of the cytoplasm. The basement membrane often contained the marker, which sometimes extended into the extracellular spaces. The reaction product was never observed between the endothelial cells, and the tight junctions were not disrupted. The electron microscopic examination of the vessels was essentially the same irrespective of the duration of portocaval anastomosis. However, the astrocytes in rats 10 days after surgery showed swollen end feet with sparsely appearing organelles. At 30 days following portocaval anastomosis, the swelling was absent, but an increased number of mitochondria and glial filaments were observed in these cells. Whenever the HRP tracer was seen in the astrocytic basement membrane, it could also be present in the cytoplasm but not in the membrane-bound structures. Morphometric evaluation of the endothelial mitochondrial content and the basement membrane thickness of cerebral capillaries in rats with portocaval anastomosis only showed significant differences in thickness of the basal membrane after 30 days.[183] Thus, the authors concluded that the portocaval anastomosis leads to increased vesicular transport of protein across the BBB and astrocytic damage, which is responsible for the intracellular uptake of substances from the extracellular space.

All of the experimental studies indicate that hepatic failure affects the specific BBB transport systems that are primarily responsible for the observed changes in the brain. On the other hand, the accumulated released tissue metabolites could in turn secondarily damage the BBB.

11. EFFECT OF INBORN ERRORS OF METABOLISM

In aminoacidurias, an abnormal plasma level of one of the competing amino acid groups may influence the BBB passage of others of the group and in this way modify the synthesis of proteins, myelin, and neurotransmitters, which in turn may alter cellular function.[184,185] Phenylketonuria, maple syrup urine disease, homocystinuria, and hyperlysinuria represent inborn errors of amino acid metabolism displaying cerebral ill effects of BBB amino acid competition.

Phenylketonuria is caused by a deficiency of phenylalanine hydroxylase activity. The tissues are unable to convert phenylalanine to tyrosine and accumulate phenylalanine and its derivatives in the blood even though large amounts are excreted in the urine.[184,186,187] Among all of these substances (phenylalanine, phenyl pyruvate, *o*-hydroxyphenylacetic acid, phenylethylamine, and phenylacetic acid), phenylalanine was shown to be directly neurotoxic.[188] It was also shown to inhibit the uptake of other essential amino acids into the brain. The γ-emitting synthetic amino acid [^{75}Se]selenomethionine was taken up to a lesser degree into the brain in phenylketonuric patients than in mental defectives without the amino acid metabolic abnormality.[189] Moreover, there was a decreased brain uptake index (50%) of labeled large neutral amino acids administered intracarotidly to rats in the presence of serum from phenylketonuric patients.[190] The phenylalanine/tyrosine ratio of 100:1 frequently found in phenylketonuria inhibits protein synthesis (70%) in brain slices.[191] A high concentration of phenylalanine interferes with the hydroxylation and transport of tryptophan, impairing the synthesis of serotonin in the brain. In phenylketonuria, the serum serotonin is low, and a decreased concentration of brain serotonin was found in phenylalanine-loaded animals.[192]

In addition, phenylacetate, the major metabolite of phenylalanine, accumulates in the brain in the animal model of phenylketonuria.[193] This substance was found to cross the BBB to a greater degree in early postnatal life than in adults. This increased passage of phenylacetate into the brain has been implicated as contributing to the mental derangement of phenylketonuria.

In *maple syrup urine disease*, a deficiency in the activity of decarboxylating enzymes responsible for decarboxylation of keto acids derived from leucine, isoleucine, and valine was found in hepatic mitochondria and leukocytes.[194] The accumulation of branched-chain keto acids as well as their precursor amino acid in the plasma, brain, and urine is characteristic of this disease. The excessive levels of the plasma amino acids completely inhibit the passage of other amino acids through the BBB,[191] and the keto acids interfere with the decarboxylation of pyruvic acid and with brain diglycinase and therefore impair myelination.[194–196] This biochemical dysfunction is manifested by neurological abnormalities during the first week of life.

In *hyperlysinuria*, an inhibition of arginine transport into the brain has been implicated in interference with cerebral growth.[197] *Galactosemia* is another inborn error of metabolism in which high plasma concentrations inhibit the BBB glucose transport competitively.[198,199] The observed cerebral dysfunction displayed by an abnormal cerebral metabolism and convulsions is the result not only of the blocked glucose brain uptake but also of the accumulation of galactose metabolites. It is characterized by mental retardation, hepatomegaly, and cataracts as a result of the absence of galactose-l-phosphate methyltransferase activity. The convulsions were reproduced in experimental galactosemia of chicken.[200]

12. EFFECT OF ANESTHESIA AND DIMETHYL SULFOXIDE

12.1. Anesthesia

Since barbiturates reduce CBF to the same degree as they decrease the metabolic rate, it has been expected that barbiturates might increase the glucose content of the brain if the transport is constant. However, the sparsely available data have been conflicting. Nemoto *et al.*[201] observed an increase of T_{max} for glucose transport, whereas Betz *et al.*[202] and Bachelard *et al.*[203] did not find any effect of barbiturate on glucose transport. Moreover, Crane *et al.*[204] did not observe an increase in the levels of brain glucose with pentobarbital anesthesia. Recently, the effect of barbiturate on glucose transport was reexamined by Gjedde and Rasmussen[205] by simultaneously determining the CBF and the blood–brain glucose flux. They found a decrease of unidirectional glucose transport (50% reduction in T_{max} and K_m) that was almost proportional to the change in CBF and the reported decrease of cortical glucose utilization. Moreover, this study showed little change in brain glucose content in pentobarbital anesthesia.

These results are in agreement with reported effects of other anesthetics (such as halothane, nitrous oxide) on the BBB passage of glucose. However, the reduced rate of transport (K_m) was seen only with nitrous oxide and pentobarbital but not with halothane. The mechanism involved in the observed anesthetic change of glucose BBB transport is not entirely clear and cannot be completely explained on the basis of reduced blood flow, especially since halothane anesthesia induced an increased blood flow but a decreased BBB transport of glucose. In view of the recent description of two independent BBB transport systems for glucose, a high-affinity, low-capacity system and a low-affinity, high-capacity system,[206] it will be feasible to reevaluate these phenomena.

Nitrous oxide anesthesia may not only influence the BBB permeability of glucose but also change the passage of protein across the BBB when it is accompanied by an increased systemic arterial blood pressure above 160 mm Hg. This leakage may be even more pronounced in hypercapnic than normocapnic animals.[207] The greater penetration of barbital across the BBB with probenecid pretreatment as compared to with saline has been explained by their inhibitory interaction on the acid transport system out of the brain rather than by metabolism, since barbital is only slightly metabolized in the rat.[208]

12.2. Dimethyl Sulfoxide

The dimethyl sulfoxide (DMSO) alteration of BBB permeability has been somewhat controversial until recently. In the late 1960s, Brink and Stein[209] observed that when DMSO was used as a vehicle for the intraperitoneal injection of [^{14}C]pemoline, an increased level of the radioactive substance was found in the brain of rats. However, the DMSO disruption of the BBB was disputed, since a similar effect was not seen with [^{14}C]*p*-aminohippuric acid dissolved in DMSO but administered intravenously.[210] On the other hand, Thompson and Hart[211] clearly demonstrated a DMSO-induced change in the transport of [^{14}C]urea but not of [^{14}C]sucrose, thus suggesting that it may selectively facilitate the brain uptake of substances. Subsequently, it was also shown that DMSO increases the uptake of L-DOPA, 5-HTP, norepinephrine, and epinephrine in rats and in neonatal chicks.[212,213] Lately, DMSO (10–15%) injected together with HRP (Sigma type VI) stimulated endothelial pinocytosis and filled the extracellular spaces within 2 h.[214] However, DMSO given intraperitoneally, HRP administered intravenously, or HRP injected with 0.25 ml of DMSO did not show the same effect. The mechanisms of the DMSO opening of the BBB remain unknown but require further evaluation, since this compound is widely used as a solvent in the laboratory.

13. EFFECT OF CHEMICAL AGENTS

The integrity of BBB can be also altered by individual administration of several vasoactive substances, cyclic nucleotides, and the previously mentioned prostaglandins. Most of these compounds, such as serotonin, norepinephrine, histamine, cyclicAMP, or the precursors of cyclicAMP induced endothelial vesicular activity in the capillaries and/or arterioles, as was demonstrated with the use of exogenous HRP.[215–218] In addition, an increased BBB permeability to ferritin was also observed with intravenous injection of dibutyryl cyclic 3′,5-adenosine monophosphate,[217] whereas an increased brain penetration of radiolabeled sucrose and γ-aminobutyric acid was detected after intraarterial administration of histamine too.[218] The histamine effect was dose dependent and could be prevented by pretreatment with metramide but not by mepyramine (inhibitors of H_2 and H_1 receptors, respectively). A selective change in the BBB permeability was also achieved by topical application of bradykinin, which caused a BBB leakage of small- but not large-molecular-weight fluorescein-labeled substances.[219] Furthermore, in the last few years, it was clearly shown that the BBB permeability to water can be changed by either intraventricular injection of angiotensin II[220] or the administration of antidepressant drugs such as amitriptyline hydrochloride (AMI).[221–223] Both of these substances raised the brain uptake of $[^{15}O]H_2O$ when the extraction of the tracer bolus and its rate of washout (CBF) were monitored by external annihilation radiation produced by $[^{15}O]H_2O$.

In studies on primates,[223] AMI induced a dose-dependent and reversible increase in BBB permeability to the labeled water. No side effects were observed with the first injection, but the second one produced a slight elevation of mean arterial pressure, and a third injection caused some reduction in $[^{15}O]H_2O$ uptake (but without a return to normal levels) and a decrease in CBF. The reduction in CBF suggested a loss of autoregulation that was responsive to metabolic demands when the reactivity was tested with CO_2. The BBB permeability response to AMI could be blocked by the adrenergic antagonist phenoxybenzamine and by 6-hydroxydopamine ablation of central adrenergic system. Similar diffusibility of radiolabeled water across the BBB was observed by electrical or chemical stimulation of the locus coeruleus. The opposite effect was demonstrated after intraventricular administration of the α-adrenergic blocker phentolamine.

Amitriptyline at low concentration is known to act as an indirect agonist by augmenting the release of norepinephrine from the adrenergic fibers and inhibiting its uptake; at higher concentrations, it blocks the α-adrenergic receptors. Thus, the results of both experimental models strongly suggest that the permeability of BBB to water is mediated by the central adrenergic system.[73,76,78,221] Moreover, the demonstration of α_2- and β_2-adrenergic receptors as well as the presence of an adenylate cyclase system responsive to vasoactive substances in cerebral endothelial cultures[147] strengthen the possibility that the BBB could be modulated by the listed substances, although the exact mechanism of its action will have to be elucidated in the future.

14. COMMENTS

This summarized review of transport phenomena in pathological conditions studied by a variety of methods illustrates that BBB permeability may be indeed altered selectively in many disease processes. Although the scope of investigation in some cases was limited by technical difficulties existing at the time of these studies, it seems that not all of the observed BBB changes reflect and/or are related directly to an altered CBF. Thus, it appears that a decoupling between BBB permeability and CBF might take place under certain circumstances. Certainly, the time of kinetic studies is forthcoming for the use of techniques allowing simultaneous measurement of CBF and BBB permeability, which will help in elucidation of these phenomena. However, there is no doubt that the alteration of BBB may be primary in nature or may occur secondary to previous brain damage. There is also clearly more evidence available that the primary BBB changes might be subtle and that they are localized in the cerebral capillaries, which represent the main barrier between the blood and brain. Moreover, the studies of cerebral microvascular fractions have proven that isolated preparations of the microvessels are very useful in investigating some of the factors involved in the alteration of BBB permeability.

As far as the choice of suitable tracers is concerned for the detection of changes occurring in the BBB level, it depends to some extent on the type of BBB injury and the parameters selected for such a study. Nevertheless, the radiolabeled substances have been preferable, especially for the quantitative assessment of the BBB abnormalities.

In view of the frequent use of protein tracers for the evaluation and assessment of BBB disruption, it is useful to recapitulate and elaborate on the available information about these markers, although some were already described in various sections. The most commonly used tracer has been the Evans blue dye, which has a high affinity for serum albumin and remains in the serum longer than NaF1 (NaF1 has sometimes also been administered for the assessment of BBB protein leakage, but a firm tracer binding to protein occurs only at low concentration of NaF1). The administration of Evans blue and the detection of the extravasated complex are rather simple, and it can be seen either grossly or microscopically since it has fluorescent properties. However, the dissociation of the dye–albumin complex and the binding of free dye to tissues are more likely in pathological states whenever a longer interval of time is allowed to pass between the injection of the dye and sacrifice of the animal. Therefore, the presence of tracer staining reflects the presence of an increased amount of albumin either at the time of observation or at some time past, in other words, a present or past alteration of BBB.[224] Moreover, this concept was reinforced by evaluating the fate of the Evans blue–albumin complex and serum proteins (using the PAP method) during the more recent pathogenetic studies of vasogenic edema induced by cold injury.

These studies clearly indicate that the intensity of Evans blue staining was longer lasting than that of PAP. Furthermore, in this model of edema, the pattern of the PAP reaction progressively changed from the third day after injury, when the PAP first appeared in astrocytes and started to disappear from the extracellular space. Eventually, after a few weeks, the PAP staining was only found in small astrocyte inclusions most likely related to lysozymes, thus suggesting an astrocytic involvement in the resolution of edema by metabolism of the extravasated albumin.[225]

At the present time, it can be unequivocally concluded that so far endogenous antisera to serum protein as well as to the IgG have been the most sensitive markers used for the detection of BBB protein leakage in a variety of pathological models. The observed greater sensitivity of endogenous than exogenous protein tracers (HRP) in establishing opening of the BBB most probably rests with the continuous and probably constant plasma level of the former in contrast to the latter tracer, which remains in the circulation for a short period only.

A comparative evaluation of both endogenous protein markers in the same model is still unavailable, and therefore their different advantages and disadvantages for the detection of BBB alteration can be only speculative.

Without going into a lengthy discussion of this problem, it should be noted that the immunohistochemical visualization of each endogenous marker will also depend to some extent on its preexisting concentration in the blood. It appears that the possibility of raising the level of circulating anti-HRP provides a favorable and specific situation (overcoming the main disadvantage of detecting high-molecular-weight antibody) for an easier determination of a slow and/or slight BBB leakage of protein, which so far has been difficult to detect with the use of antisera to serum protein ultrastructurally. Certainly, there are many other factors that may play a role in the visualization of each tracer, and investigations should be addressed to all of them in the future.

On the other hand, the observed difference resulting from the use of either exogenous or endogenous protein markers is not only limited to their detection but also to their pattern of vascular uptake and passage.

Generally, in almost all pathological conditions, the increased number of endothelial vesicles observed with exogenous HRP was lower than with endogenous IgG (anti-HRP). Although the endothelial presence of protein tracer was found with both tracers depending on the model of BBB disruption being studied, as a rule this type of reaction was prevalent with IgG. The cytoplasmic localization of the protein tracer has been interpreted either as an artifact of fixation[226] or as a sign of possible irreversible cell damage.[227] However, Houthof *et al.*[99,100] noticed that the cytoplasmic pooling, especially of the IgG tracer, was best related to tracer extravasation in the brain and could not be influenced by various fixatives. Even though other procedural factors related to an artifactual presence of the tracers in the endothelial cytoplasm could not be excluded, the mere absence of tracer in the cellular organelles and its disappearance with the time of the animal's survival were suggested by the authors to reflect a reactive rather than irreversible change in the endothelial cells.

On the basis of all of the studies of abnormal BBB opening, it is evident that the pattern of the vascular protein uptake depends more on the type of tracer used than on the injury. Moreover, the localization of protein markers was recently shown to depend on their molecular charge, whereas the molecular radius and weight have been also implicated in BBB penetration.[229]

In any case, the variable pathological responses of cerebral microvessels described have been important in discussing their function and the mechanism responsible for the increased protein passage across the BBB. Westergaard was the first to draw attention to the functional meaning of the endothelial vesicles.[228] According to his and other observations, a limited number of endothelial vesicles are present in some cerebral arterioles, but their number in microvessels increases in a variety of pathological conditions.[57–59,70,71,86,88–90,99,100,122,125,164–166,171,215,217,218,226,227] Moreover, their appearance can be induced by the previously mentioned vasoactive substances and cyclic nucleotides. Thus, it has been thought that the augmented passage of protein across the altered BBB (from either the blood or the brain) takes place by vesicular transport. Whether the implicated vesicular transfer of protein through the BBB is an active transport has lately been disputed, since the HRP tracer has been found in essentially the same locations irrespective of the animal's survival time (10 s to 60 min), whereas the extravasation of the protein was only seen at 10 s and 5 min, and IgG was seen at all times after hyperosmolar perfusion.

The observed presence of the tracer in tubular structures has been explained as the result of vesicular fusion, but Lossinsky *et al.*[123] suggested that the tubular network observed in gerbils might exist in gerbils and in other species normally and might serve as a rapid route for BBB penetration of proteins as observed during ischemia and other diseases. Thus, they suggested that such a pathway for proteins would be more efficient than the energy-dependent fusion of endothelial vesicles. Last but not least, the presence of the tracers in the endothelial cytoplasm has to be considered, especially since this type of response was seen with the use of either exogenous or endogenous protein markers (although it was mainly seen with the latter), as suggesting endothelial cytoplasmic pooling of the tracer. Whether a luminal surface charge[230,231] may be responsible for the cytoplasmic appearance of protein tracer remains to be resolved in the future.

In conclusion, it is still not certain whether the abnormal penetration of protein across the BBB occurs by active or passive process. Moreover, there is still insufficient evidence that the chemical induction of endothelial vesicles can be equated with transport, except as the mechanism directly responsible for the abnormal BBB permeability, since these changes were produced, by and large, with high doses of these substances.

Nonetheless, the presently available observations strongly suggest that some neurotransmitters might be involved in regulating the BBB permeability.[73,76,78,145,221] Moreover, the observed changes in BBB permeability induced by either stimuli or ablation of central or peripheral nervous sytem are highly suggestive of neurogenic control of the BBB.[8,145,147] The demonstration of adrenergic receptors in the brain microvessels and in the cerebrovascular endothelial culture strengthens this supposition. Thus, it is expected that the direction of future investigations of BBB changes will focus on these and other, still unknown factors to shed some more light on the mechanisms involved in regulation of both normal and altered BBB permeability.

REFERENCES

1. Erlich, P., 1897, *Ther. Monatshefte* **1**:88–90.
2. Lewandowsky, M., 1900, *Z. Klin. Med.* **40**:480–494.
3. Goldmann, E. E., 1909, *Beitr. Klin. Chir.* **64**:192–265.
4. Stern, W. E., and Gautier, R., 1921, *Arch. Int. Physiol.* **17**:138–192.
5. Rapoport, S. I., 1976, *Blood–Brain Barrier in Physiology and Medicine*, Raven Press, New York.
6. Bradbury, M., 1979, *The Concept of a Blood–Brain Barrier*, John Wiley & Sons, New York.
7. Tschirgi, R. D., 1950, *Am. J. Physiol.* **163**:756.

8. Spatz, M., and Mrsulja, B. M., 1982, *Advances in Cellular Neurobiology*, Volume 3 (S. Federoff, and L. Hertz, eds.), Academic Press, New York, pp. 311–337.
9. Passow, H., Rothstein, A., and Clarkson, T. W., 1961, *Pharmacol. Rev.* **13**:185–224.
10. Sutherland, R. M., Rothstein, A., and Weed, R. J., 1967, *J. Cell Biol.* **69**:185–198.
11. Benesch, R. E., and Benesch, R., 1954, *Arch. Biochem. Biophys.* **48**:38–42.
12. Rothstein, A., 1966, *AEC Symp.* **6**:383–397.
13. Steinwall, O., 1968, *Prog. Brain Res.* **29**:357–365.
14. Flodmark, S., and Steinwall, O., 1963, *Acta Physiol. Scand.* **57**:446–453.
15. Steinwall, O., and Klatzo, I., 1966, *J. Neuropathol. Exp. Neurol.* **25**:542–549.
16. Steinwall, O., 1968, *Prog. Brain Res.* **29**:357–364.
17. Steinwall, O., and Snyder, S. H., 1969, *Acta Neurol. Scand.* **45**:369–375.
18. Spatz, M., Rap, Z. M., Rapoport, S. I., and Klatzo, I., 1976, *Neuropathol. Appl. Neurobiol.* **2**:53–61.
19. Oldendorf, W. H., 1971, *Am. J. Physiol.* **221**:1629–1639.
20. Pardridge, W. M., 1976, *J. Neurochem.* **27**:333–335.
21. Chang, L. W., Ware, R. A., and Desnoyers, P. A., 1973, *Food Cosmet. Toxicol.* **11**:283–286.
22. Ware, R. A., Chang, L. W., and Burkholder, P. M., 1974, *Acta Neuropathol.* (*Berl.*) **30**:211–224.
23. Yoshino, Y., Mozai, T., and Nakao, K. 1966, *J. Neurochem.* **13**:1223–1230.
24. Osetowska, E., 1971, *Pathology of the Nervous System*, Volume 2 (J. Minckler, ed.), McGraw-Hill, New York, pp. 1644–1651.
25. Press, M. E., 1977, *J. Neuropathol. Exp. Neurol.* **36**:169–193.
26. Lampert, P., Garro, F., and Pentschew, A., 1967, *Adv. Neurol.* **28**:207–222.
27. Thomas, J. A., Dallenbach, F. D., and Thomas, M., 1973, *J. Pathol.* **19**:45–50.
28. Silbergeld, E. K., Wolinsky, J. S., and Goldstein, G. W., 1980, *Brain Res.* **189**:369–376.
29. Goldstein, G. W., Wolinsky, J. C., and Csejtey, J., 1977, *Ann. Neurol.* **1**:235–239.
30. Goldstein, G. W., 1977, *Brain Res.* **136**:185–188.
31. Lorenzo, A. V., and Gewirtz, M., 1977, *Brain Res.* **132**:386–392.
32. Kolber, A. R., Krigman, M. R., and Morrell, P., 1980, *Brain Res.* **192**:513–521.
33. Miquel, J., and Haymaker, W., 1965, *Prog. Brain Res.* **15**:89–114.
34. Levin, V. A., Edwards, M. S., and Byrd, A., 1979, *Int. J. Radiat. Oncol. Biol. Phys.* **5**:1627–1631.
35. Edwards, M. S., Levin, V. A., and Byrd, A., 1979, *Int. J. Radiat. Oncol. Biol. Phys.* **5**:1633–1635.
36. Schettler, T., and Sheely, C. N., 1970, *J. Neurosurg.* **32**:89–94.
37. Eyster, E. F., Nielson, S. L., Sheline, G. E., and Wilson, L. B., 1974, *J. Neurosurg.* **40**:267–271.
38. Oscar, K. J., and Hawkins, D. T., 1977, *Brain Res.* **126**:281–293.
39. Merritt, H. J. H., Chamners, A. F., and Allen, S. J., 1978, *Radiat. Environ. Biophys.* **15**:367–377.
40. Brierley, J. B., 1951, *J. Physiol.* (*Lond.*) **116**:24P.
41. Clemedson, C. J., Hartelius, H., and Holmberg, G., 1958, *Acta Pathol. Microbiol. Scand.* **42**:137–149.
42. Goldberg, M. A., Barlow, C. F., and Roth, L. J., 1961, *J. Pharmacol.* **131**:308–318.
43. Goldberg, M. A., Barlow, C. F., and Roth, L. J., 1963, *Arch. Neurol.* **9**:496–507.
44. Cutler, W. P., and Barlow, C. F., 1966, *Arch Neurol.* **14**:54–63.
45. Cameron, I. R., Davson, H., and Segal, M. B., 1970, *Yale J. Biol. Med.* **42**:241–247.
46. Evans, C. A. N., Reynolds, J. M., Reynolds, M. L., and Saunders, N. R., 1976, *J. Physiol.* (*Lond.*) **255**:701–714.
47. Sorensen, S. C., 1974, *Brain Res.* **70**:174–178.
48. Berson, F. G., Spatz, M., and Klatzo, I., 1975, *Stroke* **6**:691–696.
49. Spatz, M., and Klatzo, I., 1976, *Transport Phenomena in the Nervous System* (G. Levi, L. Battistin, and A. Lajtha, eds.), Plenum Press, New York, pp. 479–495.
50. Bolwig, T. G., Hertz, M. M., and Holm-Jensen, J., 1977, *Eur. J. Clin. Invest.* **7**:95–100.
51. Bjerner, B., Broman, T., and Swenson, A., 1944, *Acta Psychiatr. Neurol.* **19**:431–452.
52. Lending, M., Slobody, L. B., and Mestern, J., 1958, *Am. J. Physiol.* **197**:465–468.

53. Lee, J., and Olszewski, J., 1961, *Neurology (Minneap.)* **11**:515–519.
54. Lorenzo, A. V., and Barlow, C. F., 1967, *J. Pharmacol.* **157**:555–564.
55. Lorenzo, A. V., Barlow, C. F., and Roth, A. J., 1967, *Am. J. Physiol.* **212**:1277–1287.
56. Lorenzo, A. V., Shirahige, I., Liang, M., and Barlow, C. F., 1972, *Am. J. Physiol.* **223**:268–277.
57. Lorenzo, A. V., Hedley-Whyte, E. W., Eisenberg, H. M., and Hsu, D. W., 1975, *Brain Res.* **88**:136–140.
58. Petito, C. K., Schaefer, J. A., and Plum, F., 1977, *Brain Res.* **127**:251–267.
59. Westergaard, E., Hertz, M. M., and Bolwig, T. G., 1978, *Acta Neuropathol. (Berl.)* **41**:73–80.
60. Bolwig, T. G., Hertz, M. M., Paulson, O. B., Spotoft, H., and Rafaelsen, O. J., 1977, *Eur. J. Clin. Invest.* **7**:87–93.
61. Byrom, F. B., 1954, *Lancet* **2**:201–211.
62. Johansson, B. B., 1976, *Transport Phenomena in the Nervous System* (G. Levi, L. Battistin, and A. Lajtha, eds.), Plenum Press, New York, pp. 517–527.
63. Johansson, B. B., and Linder, L. E., 1978, *Acta Neurol. Scand.* **57**:345–348.
64. McKenzie, E. T., Strandgaard, S. Graham, D. I., Jouer, J. V., Harper, A. M., and Farrar, J. K., 1976, *Circ. Res.* **39**:33–41.
65. Auer, L., 1978, *Eur. Neurol.* **17**:166–178.
66. Johansson, B. B., 1980, *Adv. Exp. Med. Biol.* **131**:211–226.
67. Johansson, B. B., 1976, *Clin. Sci.* **51**:41S–43S.
68. Carlsson, C., and Johansson, B. B., 1978, *Acta Neuropathol. (Berl.)* **41**:125–129.
69. Johansson, B., and Nilsson, B., 1977, *Acta Neuropathol. (Berl.)* **38**:153–158.
70. Petito, C. K., Schaefer, J. A., and Plum, F., 1976, *Dynamics of Brain Edema* (H. M. Pappius and W. Feindel, eds.), Springer-Verlag, New York, pp. 38–42.
71. Bolwig, T. G., Hertz, M. M., and Westergaard, E., 1977, *Acta Neurol. Scand.* **56**:226–234.
72. Bill, A., and Linder, J., 1976, *Acta Physiol. Scand.* **96**:114–121.
73. Edvinsson, I., Hardebo, J. E., and Owman, C., 1977, *Acta Neurol. Scand. [Suppl.]* **64**:50–51.
74. Beausang-Linder, M., and Bill, A., 1981, *Acta Physiol. Scand.* **111**:193–199.
75. Heistad, D. D., and Marcus, M. L., 1979, *Circ. Res.* **45**:331–338.
76. Mueller, S. M., Ertel, P. J., Felten, D. L., and Overhage, J. M., 1982, *Stroke* **13**:83–88.
77. Heistad, D. D., 1981, *J. Cereb. Blood Flow Metab.* **1**:447–450.
78. Grubb, R. L., Jr., Raichle, M. E., and Eichling, J. O., 1978, *Brain Res.* **144**:204–207.
79. Blomstrand, C., Johansson, B., and Rosengren, B., 1975, *Acta Neuropathol. (Berl.)* **31**:97–102.
80. Eisenberg, H. M., Barlow, C. F., and Lorenzo, A. V., 1970, *Arch. Neurol.* **23**:18–22.
81. Johansson, B. B., and Lund, S., 1978, *Acta Physiol. Scand.* **104**:281–286.
82. Johansson, B. B., Auer, L. M., and Linder, L. E., 1982, *Stroke* **13**:220–225.
83. Johansson, B. B., and Nordberg, C., 1978, *Pathology of Cerebrospinal Microcirculation* (J. Cervós-Navarro and E. Fritschka, eds.), Raven Press, New York, pp. 349–357.
84. Giacomelli, F., Wiener, J., and Spiro, D., 1970, *Am. J. Pathol.* **59**:133–159.
85. Hazama, F., Amano, S., Haebara, H., Yamori, Y., and Okamoto, K., 1976, *The Cerebral Vessel Wall* (J. Cervós-Navarro, E. Betz, F. Matakas, and R. Wullenweber, eds.), Raven Press, New York, pp. 245–252.
86. Westergaard, E., van Deurs, B., and Bronsted, H. E., 1977, *Acta Neuropathol. (Berl.)* **37**:141–152.
87. Nag, S., Robertson, D. M., and Dinsdale, H. B., 1977, *Lab. Invest.* **36**:150–161.
88. Nag, S., Robertson, D. M., and Dinsdale, H. B., 1979, *Acta Neuropathol. (Berl.)* **46**:107–116.
89. Nagy, Z., Mathieson, G., and Huttner, I., 1979, *Acta Neuropathol. (Berl.)* **48**:45–53.
90. Houthoff, H. J., Go, K. G., and Molenaar, I., 1981, *Acta Neuropathol. (Berl.) [Suppl.]* **7**:13–16.
91. Yoshimoto, H., and Irinoda, K., 1976, *Jpn. Heart J.* **17**:365–366.
92. Klatzo, I., 1981, *Advances in Physiological Sciences*, Volume 7, *Cardiovascular Physiology, Microcirculation and Capillary Exchange* (A. G. B. Kovách, J. Hamer, and L. Szabó, eds.), Akadémiai Kiadó, Budapest, pp. 343–348.

93. Nag, S., Robertson, D. M., and Dinsdale, H. B., 1984, *Recent Progress in the Study and Therapy of Brain Edema* (K. G. Go and A. Baethmann, eds.), Plenum Press, New York, pp. 93–105.
94. Broman, T., and Lindberg-Broman, A. M., 1945, *Acta Physiol. Scand.* **10**:102–125.
95. Thompson, H. M., 1970, *Capillary Permeability* (C. Crone and N. A. Lassen, eds.), Munksgaard, Copenhagen, pp. 459–467.
96. Rapoport, S. I., Hori, M., and Klatzo, I., 1972, *Am. J. Physiol.* **223**:323–331.
97. Brightman, M. W., Hori, M., Rapoport, S. I., Reese, T. S., and Westergaard, E., 1973, *J. Comp. Neurol.* **152**:317–326.
98. Sterrett, P. R., Thompson, A. M., Chapman, A. L., and Matzke, H. A., 1974, *Brain Res.* **77**:281–295.
99. Houthoff, H. J., and Go, K. G., 1980, *Adv. Neurol.* **28**:75–81.
100. Houthoff, J. H., Go, K. G., and Gerrits, P. O., 1982, *Acta Neuropathol.* (*Berl.*) **56**:99–112.
101. Gazendam, J., Blasberg, R. G., Patlak, C. S., Fenstermacher, J. D., and Rapoport, S. I., 1980, *Intracranial Pressure IV* (K. Shulman, A. Marmaroli, J. D. Miller, D. P. Becker, A. M. Hochwald and M. Brock, eds.), Springer-Verlag, New York, pp. 312–313.
102. Spatz, M., Rap, Z., Rapoport, S. I., and Klatzo, I., 1976, *Neuropathol. Appl. Neurobiol.* **2**:53–61.
103. Pollay, M., 1975, *Neurology* (*Minneap.*) **25**:852–856.
104. Pappius, H. M., Savaki, H. E., Fieschi, C., Rapoport, S. I., and Sokoloff, L., 1979, *Ann. Neuro.* **5**:211–219.
105. Chiueh, C. C., Sun, C. L., Kopin, I. J., Fredericks, W. R., and Rapoport, S. I., 1978, *Brain Res.* **145**:291–301.
106. Hardebo, J. E., Edvinsson, L., MacKenzie, E. T., and Owman, C., 1979, *Acta Neuropathol.* (*Berl.*) **47**:145–150.
107. Pickard, J. C., Durity, F., Welsh, F. A., Langfitt, T. W., Harper, A. M., and MacKenzie, E. T., 1977, *Brain Res.* **122**:170–176.
108. Rapoport, S. I., 1973, *Small Vessel Angiography*, C. V. Mosby, St. Louis, pp. 137–151.
109. Kozeniowska, E., and Skolasinska, K., 1980, *Acta Physiol. Pol.* **31**:385–389.
110. Rapoport, S. I., Fredericks, K., Ohno, K., and Pettigrew, K. D., 1980, *Am. J. Physiol.* **238**:R421–R431.
111. Fieschi, C., Lenzi, G. L., Zanette, E., Orti, F., and Passero, S., 1980, *Life Sci.* **27**:239–243.
112. Hicks, J. T., Albrecht, P., and Rapoport, S. I., 1976, *Exp. Neurol.* **53**:768–779.
113. Neuwelt, E. A., Frenkel, E. P., Rapoport, S., and Barnett, P., 1980, *Neurosurgery* **7**:36–43.
114. Neuwelt, E. A., Barranger, J. A., Brady, R. O., Pagel, M., Furbish, F. S., Quirt, J. M., Mook, G. E., and Frenkel, E., 1981, *Proc. Natl. Acad. Sci. U.S.A.* **78**:5838–5841.
115. Paulson, O. B., and Hertz, M. M., 1978, *Eur. J. Clin. Invest.* **8**:391–396.
116. Broman, T., 1949, *The Permeability of Cerebrospinal Vessels in Normal and Pathological Conditions*, Munksgaard, Copenhagen.
117. Eich, J., and Wiemers, K., 1950, *Dtsch. Z. Nervenheilkd.* **164**:537–539.
118. Plum, F., Posner, J. B., and Alvord, E. C., 1963, *Arch. Neurol.* **9**:563–570.
119. Hossmann, K. A., and Olsson, Y., 1971, *Acta Neuropathol.* (*Berl.*) **18**:113–122.
120. Olsson, Y., Cromwell, R. M., and Klatzo, I., 1971, *Acta Neuropathol.* (*Berl.*) **18**:89–102.
121. Ito, W., Go, K. G., Walker, J. T., Jr., Spatz, M., and Klatzo, I., 1976, *Acta Neuropathol.* (*Berl.*) **34**:1–6.
122. Westergaard, E., Go, K. G., Klatzo, I., and Spatz, M., 1976, *Acta Neuropathol.* (*Berl.*) **35**:307–325.
123. Lossinsky, A. S., Garcia, J. H., Iwanowski, L., and Lightfoote, W. E., 1979, *Acta Neuropathol.* (*Berl.*) **47**:105–110.
124. Spatz, M., Fujimoto, T., and Go, K. G., 1976, *Dynamics of Brain Edema* (H. M. Pappius and W. Feindel, eds.), Springer-Verlag, Berlin, Heidelberg, pp. 181–186.
125. Go, K. G., Houthoff, H. J., Huitema, S., and Spatz, M., 1984, *Recent Progress in the Study and Therapy of Brain Edema* (K. G. Go and A. Baethmann, eds.), Plenum Press, New York, pp. 539–550.
126. Klatzo, I., Laursen, H., Orzi, F., Chui, E., Wilmes, F., Suzuki, R., and Horie, R., 1981, *Cerebrovascular Diseases: New Trends in Surgical and Medical Aspects* (H. Barnett, P.

Paoletti, E. Flamm, and G. Brambilla, eds.), Elsevier/North Holland Biomedical Press, Amsterdam, pp. 5–18.
127. Kirino, T., Suzuki, R., Laursen, H., Yamaguchi, I., and Klatzo, I., 1982, *Stroke* **13**:117.
128. Kuroiwa, T., Ting, P., Suzuki, R., Fenton, I., and Klatzo, I., 1982, *J. Neuropathol. Exp. Neurol.* **41**:352.
129. Nishimoto, K., Wolman, M., Spatz, M., and Klatzo, I., 1978, *Adv. Neurol.* **20**:237–244.
130. Spatz, M., Go, K. G., and Klatzo, I., 1974, *Pathology of Cerebral Microcirculation* (J. Cervós-Navarro, ed.), Walter de Gruyter, Berlin, pp. 361–366.
131. Spatz, M., Mršulja, B. B., Mićić, D., Mršulja, B. J., and Klatzo, I., 1977, *Brain Res.* **120**:141–145.
132. Betz, A. L., and Gilboe, D. D., 1974, *Brain Res.* **65**:368–372.
133. Gilboe, D. D., Costello, D., and Fitzpatrick, J. H., Jr., 1980, *Adv. Exp. Med. Biol.* **131**:279–293.
134. Mićić, D., Mićić, J., Swink, M. E., and Spatz, M., 1978, *Proc. Soc. Exp. Biol. Med.* **158**:318–332.
135. Spatz, M., Mićić, D., Fujimoto, T., Mršulja, B. B., and Klatzo, I., 1979, *Pathophysiology of Cerebral Energy Metabolism* (B. B. Mršulja, L. M. Rakić, I. Klatzo, and M. Spatz, eds.), Plenum Press, New York, pp. 143–153.
136. Nell, J. H., Chabi, E., and Welch, K. M. A., 1978, *Adv. Neurol.* **20**:183–187.
137. Djuričić, B. M., Rogač, L., Spatz, M., Rakić, L., and Mršulja, B. B., 1978, *Adv. Neurol.* **20**:197–205.
138. Bazan, N. G., and Rodriguez de Turco, E. B., 1980, *Adv. Neurol.* **28**:197–205.
139. Moskowitz, M. A., and Wurtman, R. J., 1976, *Cerebrovascular Diseases* (P. Scheinberg, ed.), Raven Press, New York, pp. 153–166.
140. Hervonen, H., Steinwall, O., Spatz, M., and Klatzo, I., 1980, *Adv. Exp. Med. Biol.* **131**:295–305.
141. Abe, K., Abe, T., Klatzo, I., and Spatz, M., 1980, *Adv. Neurol.* **28**:429–441.
142. Mićić, D., Abe, K., Rausch, W. D., Abe, T., and Spatz, M., 1980, *Circulatory and Developmental Aspects of Brain Metabolism* (M. Spatz, B. B. Mršulja, L. M. Rakić, and W. D. Lust, eds.), Plenum Press, New York, pp. 81–95.
143. Abe, T., Abe, K., Mićić, D., Djuričić, B., Mršulja, B. B., and Spatz, M., 1980, *Circulatory and Developmental Aspects of Brain Metabolism* (M. Spatz, B. B. Mršulja, L. M. Rakić, and W. D. Lust, eds.), Plenum Press, New York, pp. 215–223.
144. Spatz, M., Abe, T., Rausch, W. D., Abe, K., Merkel, N., and Maruki, C., 1981, *Cerebral Microcirculation and Metabolism* (J. Cervós-Navarro and E. Fritschka, eds.), Raven Press, New York, pp. 23–28.
145. Spatz, M., Maruki, C., Karnushina, I., Nagatsu, I., Bembry, J., and Merkel, N., 1982, *Stroke. Animal Models. Proceedings of the International Symposium, Wiesbaden, 1981* (V. Stefanovick, ed.) pp. 27–40.
146. Mršulja, B. J., and Mršulja, B. B., 1984, *Recent Progress in the Study and Therapy of Brain Edema* (K. G. Go and A. Baethmann, eds.), Plenum Press, New York, pp. 683–689.
147. Karnushina, I., Spatz, M., and Bembry, J., 1982, *Life Sci.* **30**:849–858.
148. Cvejić, V., Mićić, D. V., Djuričić, B. M., Mršulja, B. J., and Mršulja, B. B., 1980, *Acta Neuropathol. (Berl.)* **51**:71–77.
149. Clasen, R. A., Brown, D. V. L., Leavitt, S., and Hass, G. M., 1953, *Surg. Gynecol. Obstet.* **96**:605–616.
150. Klatzo, I., Piraux, A., and Laskowski, E. J., 1958, *Exp. Neurol.* **17**:548–564.
151. Klatzo, I., 1967, *J. Neuropathol. Exp. Neurol.* **26**:1–14.
152. Klatzo, I., 1972, *Steroids and Brain Edema* (H. J. Reulen and K. Schürmann, eds.), Springer-Verlag, Berlin, Heidelberg, New York, pp. 1–8.
153. Klatzo, I., Wisniewski, H., and Smith, D. E., 1965, *Prog. Brain Res.* **15**:73–88.
154. Pappius, H. M., and McCann, W. P., 1969, *Arch. Neurol.* **20**:207–216.
155. Pappius, H. M., and Gulati, D. R., 1963, *Acta Neuropath. (Berl.)* **2**:451–460.
156. Go, K. G., Patberg, W. R., Teelken, A. W., and Gazendam, J., 1976, *Dynamics of Brain Edema* (H. M. Pappius and W. Feindel, eds.), Springer-Verlag, Berlin, Heidelberg, New York, pp. 63–67.

157. Chui, E., Wilmes, F., Sotelo, J. E., Horie, R., Fujiwara, K., Suzuki, R., and Klatzo, I., 1981, *Cerebral Microcirculation and Metabolism* (J. Cervós-Navarro and E. Fritschka, eds.), Raven Press, New York, pp. 121–127.
158. Baker, R. N., Cancilla, P. A., Pollock, P. S., and Frommes, S. P., 1971, *J. Neuropathol. Exp. Neurol.* **30**:668–679.
159. Houthoff, H. J., Go, K. G., and Huitema, S., 1981, *Cerebral Microcirculation and Metabolism* (J. Cervós-Navarro and E. Fritschka, eds.), Raven Press, New York, pp. 331–336.
160. Pollay, M., and Stevenes, F. A., 1980, *Neurol. Res.* **1**:239–245.
161. Go, K. W., Lammertsma, A. A., Paans, A. M. J., Vaalburg, W., and Woldring, M. G., 1981, *Arch. Neurol.* **38**:581–584.
162. Ommaya, A. K., Rockoff, S. D., and Baldwin, M., 1964, *Neurosurgery* **21**:249–265.
163. Rinder, L., and Olsson, Y., 1968, *Acta Neuropathol. (Berl.)* **11**:183–200.
164. Persson, L., and Hansson, H. A., 1976, *Acta Neuropathol. (Berl.)* **35**:333–342.
165. Beggs, J. L., and Waggener, J. D., 1976, *Lab. Invest.* **34**:428–439.
166. Povlishock, J. T., Becker, D. P., Sullivan, H. G., and Miller, J. D., 1978, *Brain Res.* **153**:223–239.
167. Kontos, H. A., Wei, E. P., Ellis, E. F., Dietrich, W. D., and Povlishock, J. T., 1981, *Fed. Proc.* **40**:2326–2330.
168. Awad, I., Little, J. R., Skrinska, V., Lucas, F., and Lesser, R., 1984, *Recent Progress in the Study and Therapy of Brain Edema* (K. G. Go and A. Baethmann, eds.), Plenum Press, New York, pp. 413–427.
169. Robert, A. M., and Godeau, G., 1974, *Biomedicine* **21**:36–39.
170. Robert, A. M., Godeau, G., Misculin, M., and Moati, F., 1977, *Neurochem. Res.* **2**:449–455.
171. Godeau, G., and Robert, A. M., 1979, *Cell Biol. Int. Rep.* **3**:747–751.
172. Gazendam, J., Houthoff, H. J., Huitema, S., and Go, K. G., 1984, *Recent Progress in the Study and Therapy of Brain Edema* (K. G. Go and A. Baethmann, eds.), Plenum Press, New York, pp. 159–173.
173. Young, S. N., Lai, S., Sourke, T. L., Feldmuller, F., Aronoff, A., and Martin, J. B., 1975, *J. Neurol. Neurosurg. Psychiatry* **38**:322–330.
174. Fischer, J. E., and Baldessarini, R. J., 1976, *Prog. Liver Dis.* **5**:363–397.
175. Mans, A. M., Saunders, S. J., Kirsch, R. E., and Biebuyck, J. F., 1979, *J. Neurochem.* **32**:285–292.
176. Mans, A. M., Biebuyck, J. F., Saunders, S. J., Kirsch, R. E., and Hawkins, R. A., 1979, *J. Neurochem.* **33**:409–418.
177. Cremer, J. E., Lai, J. C. K., and Sarna, G. S., 1977, *J. Physiol. (Lond.)* **266**:70P–71P.
178. James, J. H., Escourrou, J., and Fischer, J. E., 1978, *Science* **200**:1395–1397.
179. Sarna, G. S., Bradbury, M. W. B., Cremer, J. E., Lai, J. C. K., and Teal, H. M., 1979, *Brain Res.* **160**:69–83.
180. Cardelli-Cangiano, P., Cangiano, C., James, J. H., Jeppsson, B., Brenner, W., and Fischer, J. E., 1981, *J. Neurochem.* **36**:627–632.
181. Laursen, H., Schroder, H., and Westergaard, E., 1975, *Acta Pathol. Microbiol. Scand. [A]* **83**:266–268.
182. Laursen, H., and Westergaard, E., 1977, *Neuropathol. Appl. Neurobiol.* **3**:29–43.
183. Laursen, H., 1980, *Neuropathol. Appl. Neurobiol.* **6**:375–386.
184. Wiltse, H. E., and Menke, J. H., 1972, *Handbook of Neurochemistry*, Volume 7 (A. Lajtha, ed.), Plenum Press, New York, pp. 143–167.
185. Rosenberg, L. E., 1980, *Harrison's Principles of Internal Medicine*, 9th ed., McGraw-Hill, New York, pp. 461–468.
186. Jervis, G. A., 1947, *J. Biol. Chem.* **169**:651–656.
187. Kaufman, S., and Fischer, D. B., 1974, *Molecular Mechanism of Oxygen Activation* (O. Hayashi, ed.), Academic Press, New York, pp. 285–369.
188. Oates, J. A., Nirenberg, P. Z., Jepson, J. B., Sjoerdsma, A., and Udenfriend, S., 1963, *Proc. Soc. Exp. Biol. Med.* **112**:1078–1081.
189. Oldendorf, W. H., Sisson, W. B., and Silverstein, A., 1971, *Arch. Neurol.* **24**:524–528.
190. Oldendorf, W. H., 1973, *Arch. Neurol.* **28**:45–48.
191. Appel, S. H., 1966, *Trans. N.Y. Acad. Sci.* **29**:63–70.

192. Pare, C. M. B., Sandler, M., and Stacey, R. S., 1957, *Lancet* **272:**551–553.
193. Loo, Y. H., Fulton, T., and Wisniewski, H. M., 1979, *J. Neurochem.* **32:**1697–1698.
194. Bowden, J. A., and Connelly, J. L., 1968, *J. Biol. Chem.* **243:**3526–3531.
195. Morris, M. D., Lewis, B. D., Doolan, P. D., and Harper, H. A., 1961, *Pediatrics* **28:**918–923.
196. Silberberg, D. H., 1969, *J. Neurochem.* **16:**1141–1146.
197. Banos, G., Daniel, P. M., and Pratt, O. E., 1974, *J. Physiol.* (*Lond.*) **236:**29–41.
198. Nadler, H. L., Inouye, T., and Hsia, D. Y. Y., 1969, *Galactosemia* (D. Y. Y. Hsia, ed.). Charles C. Thomas, Springfield, Illinois, pp. 127–139.
199. Kalcklar, H. M., Kinoshita, J. H., and Donnell, G. H., 1973, *Biology of Brain Dysfunction* (G. E. Gaull, ed.), Plenum Press, New York, pp. 31–88.
200. Knull, H. R., and Wells, W. W., 1973, *J. Neurochem.* **20:**415–422.
201. Nemoto, E. M., Stezoski, S. W., and MacMurdo, D., 1978, *Anesthesiology* **49:**170–176.
202. Betz, A. L., Gilboe, D. D., Yudilevich, D. L., and Drewes, L. R., 1973, *Am. J. Physiol.* **225:**586–592.
203. Bechelard, H. S., Daniel, P. M., Love, E., and Pratt, O. E., 1973, *Proc. R. Soc. Lond.* [*Biol.*] **183:**71–82.
204. Crane, P. D., Braun, L. D., Cornford, E. M., Cramer, J. E., Glass, J. M., and Oldendorf, W. H., 1978, *Stroke* **9:**12–18.
205. Gjedde, A., and Rasmussen, M., 1980, *J. Neurochem.* **35:**1382–1387.
206. Gjedde, A., 1981, *J. Neurochem.* **36:**1463–1471.
207. Johansson, B. B., and Linder, L. E., 1978, *Acta Anaesthesiol. Scand.* **22:**463–466.
208. Wahlstrom, G., 1978, *Acta Pharmacol. Toxicol.* (*Kbh.*) **43:**260–265.
209. Brink, J. J., and Stein, D. G., 1967, *Science* **158:**1479–1480.
210. Kocsis, J. J., Harkaway, S., and Vogel, W. H., 1968, *Science* **160:**1472–1473.
211. Thompson, A. M., and Hart, E. J., 1968, *Fed. Proc.* **27:**333.
212. De La Torre, J. C., 1970, *Experientia* **26:**1117–1118.
213. Hanig, J. P., Morrison, J. M., and Krop, S., 1971, *J. Pharm. Pharmacol.* **23:**386–387.
214. Broadwell, R. D., Salcman, M., and Kaplan, R. S., 1982, *Science* **217:**164–166.
215. Westergaard, E., 1975, *J. Ultrastruct. Res.* **50:**383.
216. Joó, F., 1972, *Experientia* **28:**1470–1471.
217. Joó, F., 1979, *Pathophysiology of Cerebral Energy Metabolism* (B. B. Mrsulja, L. M. Rakić, I., Klatzo, and M. Spatz, eds.), Plenum Press, New York, pp. 211–237.
218. Gross, P. M., Teasdale, G. M., Graham, D. I., Angerson, W. J., and Harper, A. M., 1982, *Am. J. Physiol.* **243:**307–317.
219. Unterberg, A., Maier-Hauff, K., Wahl, L., Schürer, M., Lange, M., and Baethmann, A., 1984, *Recent Progress in the Study and Therapy of Brain Edema* (K. G. Go and A. Baethmann, eds.), Plenum Press, New York, pp. 175–182.
220. Grubb, R. L., and Raichle, M. E., 1981, *Brain Res.* **210:**426–430.
221. Preskorn, S. H., Hartman, M. E., Raichle, M. E., Swanson, L. W., and Clark, H. B., 1980, *Adv. Exp. Med. Biol.* **131:**127–138.
222. Preskorn, S. H., Hartman, B. K., and Clark, H. B., 1980, *Psychopharmacology* **70:**1–4.
223. Preskorn, S. H., Raichle, M. E., and Hartman, B. K., 1982, *Science* **217:**250–252.
224. Clasen, R. A., Pandolfi, S., and Hass, G. M., 1970, *J. Neuropathol. Exp. Neurol.* **29:**266–284.
225. Klatzo, I., Chui, E., Fujiwara, K., and Spatz, M., 1980, *Adv. Neurol.* **28:**359–373.
226. Hirano, A., Backer, N. H., and Zimmerman, H. M., 1969, *Arch. Neurol.* **20:**300–308.
227. Petito, C. K., 1979, *J. Neuropathol. Exp. Neurol.* **38:**222–234.
228. Westergaard, E., 1974, *Pathology of Cerebral Microcirculation* (J. Cervós-Navarro, ed.), Walter de Gruyter, Berlin, New York, pp. 218–227.
229. Houthoff, H. J., Rennke, H. G., and Wisniewski, H. M., 1984, *Recent Progress in the Study and Therapy of Brain Edema* (K. G. Go and A. Baethmann, eds.), Plenum Press, New York, pp. 67–79.
230. Zweifach, B. W., 1980, *Advances in Microcirculation* Volume 9, (B. M. Altura, ed.), Karger, Basel, pp. 206–225.
231. Nagy, Z., Peters, H., and Huettner, I., 1981, *Acta Neuropathol. Appl. Neurobiol.* **1:**59–68.

19

Immunohistochemistry of Cell Markers in the Central Nervous System

O. Keith Langley, M. Saïd Ghandour, and Giorgio Gombos

1. INTRODUCTION

In a tissue such as the central nervous system (CNS), where cellular heterogeneity is at the very basis of the functioning of the system, techniques that permit the dissection of the tissue at the cellular and possibly ultrastructural level are fundamental. For years, the chasm between neurochemistry on the one hand and neurophysiology and neuromorphology on the other has always been that existing between homogenates of whole tissue and the function of a particular cell type. Regional distribution studies performed sometimes after lesions, cellular and subcellular fractionations, and "punch techniques" were for years the only methods available to the neurochemist for localization studies. A sole exception is that for molecules with activity that could be revealed by histochemical methods.

One of the major methodological advances of recent years has been the coupling of a visible probe to an antibody for revealing *in situ* the corresponding antigen. This new technique was called immunohistochemistry or immunohistology or immunocytology (ICC), depending on the level of the microscopic observation and the taste of the authors. The limitation of the ICC methods and the possible errors of interpretation, the search for appropriate controls, and the elimination of artifacts are discussed below. However, in careful hands and after a critical examination of the data, ICC has now become a reliable and essential tool in neurobiology.

This chapter attempts to demonstrate how ICC has been used both to distinguish between different types of cells in the nervous system (e.g., neurons from astroglial or oligodendroglial cells) and to subdivide populations of a given

O. Keith Langley, M. Saïd Ghandour, and Giorgio Gombos • Centre de Neurochimie du CNRS, 67084 Strasbourg, France.

general cell type (e.g., different types of neurons). It also outlines some of the implications in the dissection of brain pathways in terms of immunohistochemically identifiable enzymes, neuroactive peptides, proteins, or lipids of the brain during its development and under pathological conditions and in mutants concerning brain anatomy. It should be emphasized, however, that one of the unsolved problem is that of identification of cells at a precocious stage of development. Early antigens specific for cells of a given type and that persist in the same cell in the adult await discovery. Those already known are usually present in very low amounts and are difficult to demonstrate. The discovery of such antigens will be of great value, mainly for identifying cell types in embryo and in cultures *in vitro*.

Immunohistochemistry of cells of the central nervous system has advanced essentially along four main lines of approach. One early approach concerned the search for molecules (either lipids or proteins) that were remarkable by their exclusive presence in the nervous system: the proteins S-100[1] and 14-3[2] were among such molecules studied. If in the time that has elapsed since their discovery such molecules have been shown to be present in cells outside the brain, this does not detract from the inherent interest in their cellular localization and their possible use as markers for particular cell types within the brain. It could be debated at length whether there is much point in searching for molecules "specific" to CNS except when the hypothesis is that the tissue or cellular specificity of the molecules is at the basis of the specificity of a function (i.e., glutamate decarboxylase or other enzymes specific for a given neurotransmitter synthesis). It should not be forgotten, however, that in some cases molecules also present in tissues other than CNS have in the CNS a specific cellular localization (e.g., isoenzyme II of carbonic anhydrase[3]) and possibly specific functions. Thus, the search should be for molecules that in the CNS have a specific function. Such molecules may not necessarily be specific to CNS.

Enzymes concerned with particular neurotransmitters have provided a functionally related basis for the immunohistochemical labeling of different neurons.[4] The catecholaminergic system represents an example that has been extensively studied in this way. In this case, correlation with function may be easy [i.e., neurons with glutamate decarboxylase (GAD) immunoreactivity, hence presumably GABAergic neurons, are inhibitory, whereas neurons with choline acetyltransferase immunoreactivity, hence possibly cholinergic neurons, are excitatory, etc.].

In the last 20 years, more and more interest has been attracted towards an increasing list of frequently very small peptides, certain of which have subsequently achieved the status of neurotransmitter.[5] These compounds, the neuropeptides, which in many cases were originally considered peripheral hormones, represent perhaps one of the most significant advances discussed here. By use of antibodies raised against these natural or synthetic peptides, it has not only been possible to distinguish different peptide-containing nerve cells and nerve pathways, but it has also been possible to subsequently discuss their functional implication in different neurons.

A more recent approach, and one that is still in its infancy, which will be oulined here, concerns the application of the techniques of monoclonal antibodies[6] to discover new brain-specific proteins or antigens that in the brain are limited either to one cell type or to a particular cell organelle (e.g., surface membrane, mitochondria, etc.). This technique lends a new twist to the search for specific molecules, since whereas with classical antibodies the antigens are first purified, and then antibodies are raised against them, with the monoclonal antibody technique, in many cases the antigens are not purified: they frequently consist of crude membrane fractions or whole cells. Thus, the antibodies are raised, and the hybridoma producing them is first cloned; the antigen is localized as a second step, and only the final step involves the identification and the purification of the antigen.

To deal exhaustively with even one aspect of brain immunohistochemistry (e.g., that of the neuropeptides) would demand an entire volume. It has thus been decided to collate here the principal information in the form of tables and to refer the specialized interested reader to the vast bibliography for detailed information concerning specific neuronal pathways related to given antigens. A few examples are underlined to illustrate the potential of the technique and the great contribution it has brought to the fields of both neuroanatomy and brain function. Although a great deal of effort has been devoted to the immunohistochemistry of neural cell cultures, this chapter limits its remarks largely to the intact CNS.

2. *SHORT REVIEW OF IMMUNOHISTOLOGICAL METHODS*

All of the techniques currently used for immunohistology owe their origin to the immunofluorescence method introduced in 1941 by Coons,[7] later improved by Coons,[8] and subsequently modified by Riggs *et al.*[9] The method depends on attaching a probe that may be visualized by fluorescence to the antibody used to reveal tissue antigenic sites. Various probes have been employed. Fluorescein isocyanate was the choice of Coons[10], later to be replaced by the more stable fluorochrome fluorescein isothiocyanate (FITC),[9] which remains the predominant marker currently used for immunofluorescence. A second fluorescent probe useful in double-labeling studies is rhodamine B isothiocyanate (RITC). The preparation and purification of fluorochrome-labeled antibodies are described by Nairn[11] and by Sternberger.[12] Although fluorescent probes have been used extensively, they present certain limitations, and enzyme-conjugated antibody methods were developed to overcome such difficulties. For these, the most commonly employed probe is horseradish peroxidase.[13,14] Diaminobenzidine with hydrogen peroxide is normally used to reveal tissue-bound peroxidase-conjugated antibody, producing a reddish brown insoluble product,[15] though a different color (useful for multiple-staining methods[16]) may be obtained using 4-Cl-1-naphthol as substrate. Alkaline phosphatase and glucose oxidase have been employed successfully as an alternative to peroxidase.[17,18] Problems of conjugation of peroxidase to antibodies have been discussed by Vandesande.[19]

In addition to these probes, heavy metals have been found useful in labeling antibodies, particularly for electron microscope studies. Ferritin has been commonly employed but suffers from a relatively high level of nonspecific staining and the relatively large size of the conjugate, which may limit penetration into cells.[20–22] A more recently developed method using colloidal gold particles coated with protein A appears to display certain advantages for ultrastructural studies.[23,24] Protein A–peroxidase conjugates have also been employed.[25]

Three principal immunocytochemical methods are in current use. These are designated the direct method, the indirect method, and the unlabeled antibody (PAP) method. In the first, the primary antibody specific for tissue antigenic sites is directly coupled to a visible probe and used in a single incubation step. In the indirect (or sandwich) method first described by Weller and Coons,[26] the primary antibody (prepared, for example, in the rabbit) is unlabeled and, after incubation with the tissue, antigen–antibody sites are revealed by a second antibody (in this case against rabbit immunoglobulins, IgG) to which the chosen probe is attached. The third method developed independently by Sternberger[12,27] and Mason *et al.*[28] consists of essentially three steps in which both the primary antibody (prepared, for example, in rabbit) and the second antibody (prepared in goat against rabbit IgG) are both unlabeled. A final incubation with a soluble complex of peroxidase–rabbit antiperoxidase (PAP) permits the localization of tissue antigenic sites. The relative advantages and limitations of the three methods have been extensively discussed.[19] The direct method suffers from its relative insensitivity, whereas the indirect method frequently suffers from nonspecific binding of the second antibody to unwanted tissue sites. Many of these problems are overcome in Sternberger's PAP method, which is reported to be, in addition, 20–125 times more sensitive than other methods. This amplification is achieved both by a reduction in the background staining and by enabling several immunoglobulin molecules of the secondary antibody to interact with each molecule of the primary antibody bound to the tissue. In addition, the PAP complex contains three peroxidase molecules. In attempts to increase sensitivity, other minor variations in the indirect method have been employed that either involve double bridge sequences or take advantage of the known affinity of avidin for biotin, in which, after incubation with a primary rabbit antibody, biotin-conjugated sheep antirabbit serum is used as secondary antibody followed by avidin-conjugated peroxidase.[29] Since several molecules of avidin may bind to each biotin molecule, a net amplification of the signal is obtained. However, the lowest backgrounds are most frequently obtained with the PAP method.

A crucial consideration in the interpretation of immunohistochemical data concerns how much the staining pattern obtained by a given method reflects the tissue distribution of a particular antigen and how far the possibility of artifacts can be ruled out. The validity of a particular method may be discussed on various levels. The specificity of the primary antiserum for the antigen under consideration is of utmost importance[30] and may in general be tested by adsorption with the purified antigen, if this is available (which is not always the case for monoclonal antibodies). Although polyclonal antibodies may contain several species of immunoglobulin directed against different antigenic determinants on any one antigen, such is not the case for monoclonal antibodies. Here, the antibody is a single species directed against a single antigenic determinant of the antigen. If this determinant appears in several tissue constituents, the term "specificity" has to be modified accordingly. Thus, a monoclonal antibody prepared against the GFA protein was found to recognize a determinant common to several intermediate filamentous proteins.[31] The problem is acute in the study of neuropeptides, and antibodies are frequently described as having "antigenlike" specificity (e.g., somatostatinlike specificity). Some problems of specificity may arise from the use of excessively high concentrations of primary antiserum.[30] The possible presence of Fc binding sites in some cells should also not be neglected. Preincubation with normal serum of an unrelated animal may be used to reduce such nonspecific staining.

A second criterion of specificity is that reagents (including antibodies) other than the primary antibody should not produce additional tissue staining. The causes of such "method nonspecificity" are different for fluorescent and enzyme-conjugated probes, and the reader is referred to Vandesande[19] for details of the problem. Although the PAP method enjoys much greater sensitivity over other methods, it sometimes yields false negatives because of a bad choice of antibody dilutions.[32,33] A problem common to all methods, concerning the amount of primary antibody that will bind to the tissue, is related to antigen concentration itself. This may be dramatically modified by fixation and embedding procedures. This factor is particularly important where the immunohistochemical techniques are employed after dehydration and tissue embedding.[20,34] Although fixation is of prime importance in preserving tissue morphology and also in preventing the antigen from being displaced during tissue handling, excessive fixation, particularly with glutaraldehyde, results in dramatic loss of antigenicity and negative immunohistochemical results. A compromise thus has to be sought with regard to morphological preservation and retention of the immunoreactivity for individual antigens, which is, for immunoelectron microscopy, one of the remaining obstacles to be overcome.

3. CELL-SPECIFIC MARKERS IN THE CNS AS REVEALED BY IMMUNOCYTOLOGY IN TISSUE SECTIONS

3.1. Macroglia

3.1.1. Astrocytic Antigens

3.1.1a. S100 Protein. The S-100 protein was among the first so-called brain-specific proteins to receive detailed attention,[1,2] deriving its name from the fact that it is precipitated from soluble brain extracts with 100% ammonium sulfate. Its cellular localization in the brain has been surrounded by much confusion and debate, but in recent years the situation has been clarified. Early immunohistochemical data appeared to support a nuclear localization in both oligodendrocytes and neurons, and, furthermore, a nerve terminal membrane localization was reported in addition to a localization in oligodendrocyte membranes and the cytoplasm of astrocytes.[35–42] The first clear-cut ICC evidence of an exclusively astrocytic localization of this protein was provided by Matus and Mughal,[43] a view supported by other more recent studies.[44,45] It is now evident that the putative neuronal and oligodendrocyte localization of S-100 was a result of artifacts resulting in part from fixation problems of a small highly soluble protein and in part from the use of excessively high concentrations of antisera.[46] In the cerebellum, the astrocyte is the only cell that contains S-100; however, it is present in other satellite glial cells in dorsal root ganglia and in Müller cells in the retina and also in Schwann cells.[47–49] Despite the fact that the presence of this protein has been known for more than 20 years, its function in the brain still awaits discovery.

3.1.1b. Glial Fibrillary Acidic Protein. Originally isolated by Eng and coworkers from multiple sclerosis plaques (a tissue rich in fibrous astrocytes) as a water-soluble protein,[50,51] by ammonium sulfate precipitation, isoelectric focusing, and preparative PAGE, it was later isolated from normal white matter, though in an impure state and in low yield.[52] Its preparation by hydroxylapatite chromatography has been much criticized because of possible contamination with tubulin.[53] However, antisera obtained with such preparations have been widely used with no apparent problems related to such contamination. The correlation of this soluble protein with the glial intermediate filaments is now well established.[53] It should be underlined that this protein is distinct from other filamentous proteins, although recently a monoclonal antibody prepared against GFA was found to recognize a sequence in common with other 10-nm filaments including neurofilaments.[31] The GFA proteins isolated from different pathological tissues appear to be identical. Thus, antiserum against human fibrous astrocytoma cross reacts with antisera to MS plaque GFA protein.[54] NSA-1 (neurospecific antigen) also elicits an antiserum that cross reacts with GFA.[55] Another protein first described before the discovery of GFA protein as a water-soluble protein from a pathological brain biopsy[56] was termed α-albumin and was subsequently shown to be identical to GFA protein.[57]

Immunochemical studies using anti-GFA serum are abundant (see ref. 53); GFA is currently the most widely used of cell markers. This is in part because of its undisputed exclusive localization in astrocytes or in related satellite cells such as the Müller cells of the retina.[48] Ultrastructural studies show a free cytoplasmic localization in addition to an association with glial filaments.[53,58] Differences in intensity of astrocyte labeling have been reported, and these are related essentially to variations in the content of glial filaments in astrocytes in different situations. The antigenicity of the protein is well conserved after fixation and tissue processing and can be demonstrated in paraffin-embedded material many years after initial preparation. For this reason, it has been adopted as the classical astrocyte marker in pathological tissue.

3.3.1c. Vimentin. Although adult rat brain cytoskeletal proteins in the 50-kilodalton range comigrate with the GFA protein, a 57-kilodalton polypeptide comigrating with purified rat vimentin (a protein associated with a well-defined class of 10-nm filaments found in fibroblasts) was prominent in the newborn rat brain.[59] Antisera against vimentin showed the protein to be present mainly in immature glia.[59] It was not apparent in the neuroblastic hippocampal and cerebellar germinal layers, but it was found associated with meninges and blood vessels.

3.3.1d. α_2-Glycoprotein. This brain-specific water-soluble acidic glycoprotein extracted from human brain and first described in 1967[60] has recently been studied by immunocytochemical methods. Although circumstantial evidence had earlier suggested a glial localization, immunoelectron histochemistry has provided[61] clear-cut proof of the essentially astrocyte localization of this glycoprotein. Oligodendrocytes do not appear to contain it.

3.1.1e. $\alpha\alpha$-Enolase (Nonneuronal Enolase). One of the isoenzymes of the glycolytic enzyme enolase (E.C. 4.2.1.11), the $\alpha\alpha$ form, which is abundant in liver, is also present in the brain.[62,63] A report indicating a glial localization[64] was confirmed by an ultrastructural study[29] showing that the enzyme was, in adult cerebellum, only to be found in astrocytes. A switch from the $\alpha\alpha$ isoenzyme to the $\gamma\gamma$ form in certain neurons has been suggested to occur,[66] but further confirmation has not been obtained.

Oligodendrocytes do not contain immunohistochemically detectable levels of the $\alpha\alpha$ enolase,[29] nor has the $\gamma\gamma$ form been detected outside neurons or paraneurons, nor indeed has the putative hybrid $\alpha\gamma$.[29] In addition, the form particularly rich in muscle ($\beta\beta$) has not been detected in oligodendrocytes.[67] This would seem to indicate an unusual metabolism of the oligodendrocytes compared with other cells in the CNS.

3.1.1f. Glutamine Synthetase (E.C. 6.3.1.2). Glutamine synthetase, a key enzyme involved in the detoxification of ammonia in the brain and one that is also important in the metabolism of the putative neurotransmitters γ-aminobutyric and glutamic acids, has been shown by both light and electron microscopic ICC[68,69] to be confined to astrocytes. These results confirm that the astrocytes, and, in particular, the astroglial processes surrounding synaptic terminals, represent the site at which the neurotransmitter glutamate is taken up, converted to glutamine, and released to be taken up by the neurons as glutamine.

3.1.2. Oligodendrocytes and Myelin Antigens

3.1.2a. Carbonic Anhydrase (Carbonate Hydrolyase E.C. 4.2.1.1.). The two isoenzymes of carbonic anhydrase, CAI and CAII, are widely distributed in different mammalian tissues. Although both forms exist in blood cells, CAII is the sole endogenous form in the brain. Early histochemical data suggested a glial localization,[70] but ICC has shown the enzyme to be confined to one cell type, the oligodendrocyte.[71] Electron immunohistochemistry showed both a cytoplasmic and membrane localization, but myelin membranes were never found to be labeled.[72]

3.1.2b. Glycerol-3-Phosphate Dehydrogenase. This soluble NAD-linked enzyme has been found by ICC in cerebellum exclusively in oligodendrocytes.[73]

The ultrastructural localization of the enzyme was not unlike that found for CAII. No association with myelin membranes was observed.

3.1.2c. Myelin Constituents. Myelin Basic Protein (MBP). Myelin is characterized by a very particular and relatively simple protein composition in comparison with plasma membranes. The myelin basic proteins constitute an ensemble of up to four related proteins depending on species and together represent major myelin constituents. Antisera prepared against MBP cross react with all four related proteins. Early in brain development, the MBPs appear in the perikarya of oligodendrocytes[74,75]; later, its cytoplasmic levels are reduced when it is found in myelin sheaths.[74]

Wolfgram Proteins. Two immunologically related proteins, W1 and W2, isolated as more minor constituents from myelin preparations, have been demonstrated by ICC in both oligodendrocytes and myelin sheaths.[76] In the young animal, these are essentially membrane-associated antigens.[77]

Proteolipid Protein. The major protein of CNS myelin is an extremely hydrophobic molecule, proteolipid protein. Antisera against this molecule have shown PLP to be present in both myelin and the oligodendrocyte cell bodies and processes in the young rat.[78] In the adult, no immunoreactivity can be detected in oligodendrocyte cytoplasm.

CNPase [2',3'-Cyclic Nucleotidase 3'-Phosphodiesterase (E.C. 3.1.4.16)]. CNPase is an enzyme found widely in many tissues that has a high specific activity and is apparently enriched in myelin fractions. It has also been found in high levels in bulk preparations of oligodendrocytes.[79] The enzyme has been shown immunohistologically in oligodendrocytes and myelin.[80] A suggestion, which awaits confirmation, has been made that CNPase is identical to the Wolfgram proteins.[81,82]

Myelin-Associated Glycoproteins (MAG). Isolated myelin contains relatively little glycoprotein compared with plasma membrane. However, some glycoprotein bands with high apparent molecular weights have been demonstrated by PAGE of myelin preparations.[83] It has been shown by optical ICC using antisera raised against MAG that these are indeed associated with myelin sheaths.[84]

Galactocerebroside. Galactocerebroside is the major glycolipid in myelin. It has been shown by immunofluorescence to be present on the surface of oligodendrocytes both in neural cell cultures and in bulk cellular preparations.[85,86] It has proved more difficult to demonstrate its presence in tissue slices, however. Some reports of its presence on frozen sections treated cautiously with organic solvents to remove a certain amount of membrane lipids and possibly also to expose antigenic sites have appeared.[87] In the same study, the presence of sulfatide in myelin was demonstrated immunohistochemically.

3.1.2d. Nervous System Antigen-1 (NS-1). This antigen, first detected in neural cultures by a polyclonal antibody produced by immunizing with cerebellar cultures, has been localized in oligodendrocytes.[88] This antigen awaits biochemical characterization.

3.1.2e. Ran-1. Rat neural antigen-1 (Ran-1), a surface antigen defined by mouse antiserum against a rat neural cell line, is essentially absent from the CNS but is expressed strongly by Schwann cells.[85]

3.2. Endothelial Cell Markers

γ-Glutamyltranspeptidase (E.C. 2.3.2.2, γ-GTP), an enzyme that catalyses the transfer of the γ-glutamyl residue of glutathione to amino acids, is widely distributed in animal tissues. In the brain it appears by immunohistochemistry to be limited to the endothelial cells lining blood vessels.[89] Immunoelectron microscopy showed the highest concentrations to be on the luminal surface.[89] An additional ependymal and possibly glial localization was also reported.[90]

Another immunohistological marker for vascular endothelial cells is the *serum of patients with Chagas' disease* (a South American trypanosomiasis). Such serum[91] contains immunoglobulins that bind to antigens in the interstitium and blood vessels of the heart and skeletal muscle, endocardium, and vascular structures in the liver, kidney, stomach, and placenta. In the brain, this antigen seems to be confined to capillary endothelial cells.[91] Fibroblasts in culture are not labeled, and thus the antigen differs from *fibronectin*[92] (cold insoluble globulin, LETS protein), which is found by ICC to be present in both endothelial and leptomeningeal cells.

More recently, a monoclonal antibody (termed *MESA-1*) has been isolated that recognizes a surface antigen confined to endothelial cells in mouse brain.[93] This antigen is expressed by a subpopulation of fibronectin-positive cells in neural cultures. The antigen is not brain specific: blood vessels in lung, liver, kidney, and heart were also labeled, but so far it is species specific. Indirect evidence suggests that it may be a glycolipid.

3.3. Neuronal Antigens

3.3.1. General Neuronal Markers

3.3.1a. Tetanus Toxin. Brief mention should be made of the widespread value of this neuronal marker in cultures. Tetanus toxin binding sites revealed by antibody against the toxin appear to be abundant on the surface of cultured neurons (both perikarya or dendrites), though subpopulations of GFA-positive cells (astrocytes) that are weakly labeled with the toxin have also been detected.[86] It has not been used to a great extent on sections of nervous tissue.

3.3.1b. 14-3-2 Protein. Among the first so-called brain-specific proteins separated by ion-exchange chromatography, a soluble protein termed 14-3-2[2] was later shown to be an isoenzyme (the form γγ) of the glycolytic enzyme enolase E.C. 4.2.1.11.[94] Although other forms of the enzymes occur in both the CNS and other tissues, this protein was, at least for several years, considered to be confined to the nervous system.[62] Antibodies raised against this form of the enzyme do not cross react with the other isoenzymes (αα and ββ) forms. With such antisera, ICC has shown its exclusive localization in neurons,[62] though considerable variation in intensity among different neuronal classes has been noted.[65] Such findings have led to the adoption of the name "neuron-specific enolase" (NSE). However, more recently, it has been detected in paraneurons[95] outside the CNS.

3.3.1c. 14-3-3 Protein. A protein isolated at the same time as 14-3-2 and which was eluted from the DEAE column immediately after the 14-3-2 protein has only recently received detailed attention. This protein, called 14-3-3,[96] has no associated enolase activity but, like 14-3-2, appears at least in the cerebral cortex to be localized in neurons.

3.3.1d. Neurofilament Proteins and Tubulin. Much effort has been devoted to the biochemical study of the various polypeptides that constitute the 10-nm filaments in neurons—the neurofilaments. The 210 K and 155 K polypeptides have recently been identified immunohistochemically as specifically neuronal components.[97] In the cerebellum, evident differences among different neuronal types in the cellular content of neurofilaments have been demonstrated by immunolabeling techniques.[98,99] Monoclonal antibodies against neurofilaments have recently been obtained.[100] In contrast, tubulin (a protein present in microtubules) is found by ICC in both neurons and glial cells.[100] In addition, the availability of antisera against high-molecular-weight microtubule-associated proteins has proved useful in the immunohistochemical dissection of events occurring in dendrite growth during brain development.[101,102]

3.3.1e. D-2 Protein. Originally isolated by Jørgensen and Bock[103] as a 140-kilodalton glycoprotein from rat brain synaptosomal fractions, this antigen was later shown to be composed of at least two polypeptides and was subsequently localized by immunoperoxidase techniques on the surface of neurons.[104,105] It has recently been shown by immunochemical methods to be identical to a glycoprotein triplet BSP-2[106] (brain surface protein-2) recognized by a monoclonal antibody.[107] This antigen was shown in the cerebellum to be present on both neurons and astroglial cells (but not oligodendrocytes.)[108]

3.3.2. Markers Not Related to Neurotransmission, Specific for Certain Neurons

3.3.2d. Olfactory Bulb Protein. A protein isolated from the olfactory bulb[109] was later shown by immunofluorescence[110] to be localized in olfactory epithelium, olfactory nerves, and synaptic glomeruli. This protein appears to be limited to mature primary olfactory chemoreceptor neurons in normal mice, but it is also expressed in mitral and tufted cells after lesions of the fila olfactoria.[111]

3.3.2b. Purkinje Cell Antigens. A high-molecular-weight *glycoprotein P400* (of 400 kilodaltons), the levels of which were found to be considerably reduced in mutant mice with Purkinje cell deficiencies,[112] was later shown by immunofluorescence to be located exclusively in the Purkinje cells.[113]

More recently, a small glycoprotein subunit has been isolated by preparative polyacrylamide electrophoresis from the Triton X-100-insoluble Con-A-positive glycoprotein fraction of rat brain.[114] It was found to represent only 0.03% of total brain protein and to possess a relatively high content of carbohydrate. It was found by immunoelectron microscopy[115] to be present exclusively in the Purkinje cell in both cell bodies and prolongations, both in the cerebellar molecular layer and in the central deep nuclei. Immunoprecipitate was found associated both with cytoplasmic and plasma membranes and polyribosomes. It is present precociously in cerebellar development and persists in the adult. Its function, like that of P400, is as yet unknown.

Another protein, a *cyclic-GMP-dependent protein kinase*, has also recently been shown by immunohistological techniques to be localized solely in the Purkinje cell neurons in the cerebellum.[116] This protein, however, is not brain specific, being found extensively in other tissues.

3.3.2c. Vitamin-D-Dependent Calcium-Binding Protein (*CaBp*). An antiserum raised against this calcium-binding protein (different from calmodulin, S-100 protein, and other calcium-binding proteins) was used in ICC. It would appear that specific neurons in chick nervous system contain this protein[117]: in particular, Purkinje cell perikarya and fibers in chicken cerebellum.[117,118] The CaBp is also present in intestine, kidney, and pancreas.

3.3.3. Markers Related to Neurotransmission Specific for Certain Neurons

3.3.3a. Enzymes of Neurotransmitter Synthesis. To date, enzymes of synthesis of all of the known neurotransmitters have been found only in neurons, although in some cases they are more abundant in nerve terminals than in perikarya, where they can be shown only after axonal transport is blocked by colchicine (see Table I for references). Thus, neurons utilizing one or another neurotransmitter can be recognized by their immunoreactivity for one or another enzyme of neurotransmitter synthesis: cholinergic neurons can be recognized by the presence of choline acetyltransferase, GABAaergic by that of glutamate decarboxylase (GAD), and histaminergic neurons, if they exist in the CNS, should contain histidine decarboxylase (HDC). In the case of monoaminergic neurons, all of these neurons can be recognized at the optical microscopy level by their content of monoamine by using the fluorescence obtained by the Falk-Hillarp method[119,120] or by the glyoxylic acid method,[121–123] but a finer subdivision can be obtained according to their immunoreactivity for one or another enzyme in the chain of monoamine synthesis.

Immunocytology with these enzymes is also necessary for recognition of the different types of monoaminergic neurons by electron microscopy: all monoaminergic neurons contain aromatic amino acid decarboxylase (AADC or DOPA decarboxylase, DDC) (see Fig. 1), but serotonergic neurons also contain tryptophan hydroxylase (TRYH), whereas catecholaminergic neurons contain tyrosine hydroxylase (TH). Catecholaminergic neurons may then be subdivided into dopaminergic (only TC and DDC present), noradrenergic [dopamine β-

Table I
Cellular and Ultrastructural Localization of Enzymes of Neurotransmitter Synthesis Revealed by Immunocytology

Antigen purified from	Antiserum raised in	Immunohistochemical technique used		CNS region studied (optical microscopy)	Ultrastructural localization
		Technique	Procedure		
Tyrosine hydroxylase [L-*tyrosine, tetrahydropteridine : oxygen oxidoreductase (3-hydroxylating E.C. 1.14.16.2*] (TH)[162]					
Human pheochromocytoma[129]	Rabbit	Indirect immunofluoresc.	Paraformaldehyde. Cryostat sections	Maps of catecholamine neurons in mes-, di-,[130] and telencephalon[131] also in relation to the distribution of substance P[132] in the rat. Ontogenesis (TH, DDC, DBH, PNMT immunohistochemistry) of rat adrenal medulla[134]	
		Indirect immunofluorescence and PAP	Paraformaldehyde and glutaraldehyde; Vibratome sections		The use of Triton X-100 or digitonin allowed a penetration of antibody sufficient for ultrastructural studies of TH[133]
				Dopamine neurons in rat olfactory bulb[136]	Ref. 136
		PAP	Formalin fix. paraffin-embedded sections	TH-immunoreactive neurons in different areas of human brain in the adult[137] and in the embryo[138]	

Rat pheochromocytoma[139]	Rabbit	Indirect immunofluorescence	Formalin fix. cryostat sections	Dopaminergic mesencephalic neurones projecting to limbic areas contain a cholecystokininlike peptide[140]
Rat pheochromocytoma[141]	Rabbit	Immunoperoxidase (Fab)	Periodate–lysine–paraformal-dehyde fix. cryostat sections	TH activation by stress and electroconvulsive shock and TH subcellular distribution in rat adrenal glands[141]
Bovine adrenal medulla[142]	Rabbit	Indirect immunofluorescence and PAP	Different formalin fixations; sliding microtome sections	Catecholinergic neuronal perikarya and fibers (TH, DBH, and PNMT immunocytology) in the paraventricular and supraoptic nuclei of the hypothalamus of normal and Brattleboro rats, also in relation to oxytocin- and vasopressin-containing cells[143]
		Immunoperoxidase	Paraformaldehyde glutaraldehyde. Vibratome section	Dopaminergic dendrites in the pars reticulata of rat substantia nigra and their striatal input[144]
Bovine adrenal medulla[145]	Rabbit	Indirect immunofluorescence	Frozen sections then Hartman tech. for DBH[178]	TH- and DBH-containing neurons mainly in rat hypothalamic nuclei[145]
				Axoplasmic transport of TH and DBH studied in ligated rat and guinea pig sciatic nerves[146]

(Continued)

Table I. (*Continued*)

Antigen purified from	Antiserum raised in	Immunohistochemical technique used: Technique	Immunohistochemical technique used: Procedure	CNS region studied (optical microscopy)	Ultrastructural localization
Bovine adrenal medulla[147,148]	Rabbit	PAP	Paraformaldehyde fixation. picric acid–formalin postfix. paraffin or cryostat sections (optic microscopy) glutaraldehyde–paraformal-dehyde fix. for EM; Vibratome sect.	Chromaffin cells of rat adrenal medulla and TH immunoreactive cells in cervical ganglia and brain[149]	
				Dopaminergic (TH, DBH immunocytology) and serotonergic (TRYH immunocytology) neurons in rat brain; different ultrastructural distribution of the enzymes[150]	
				TH and TRYH in locus coeruleus neurons in optical and electron microscopy (serotonergic innervation of noradrenergic neurons in locus coeruleus)[151]	
				Changes in TH in dopamine neurons of substantia nigra in response to axonal injury[152]	
			Paraformaldehyde fixation and postfixation, cryostat section	TH in neuronal processes (distal processes of primordial catecholaminergic neurons) within the ventricular zone of prenatal rat brain[153]	
		PAP modified (Triton-X100)	Vibratome sections, Paraformaldehyde	TH, DBH, PNMT, Substance P, and Leu^3-enkephalin in	

		PAP modified (Triton X100)	fixation. Cryostat sections.	neuronal perikarya and terminals in area postrema and medial nucleus tractus solitarii of rat brain[154] and in human fetal sympathetic ganglia[155]
A new antiserum against trypsinized TH	Rabbit	PAP		TH in neurons of human fetal nervous system[156]
				TH in neurons of fetal rat brain during early[157] and late[158] ontogeny
		PAP	Paraformaldehyde–glutaraldehyde fixation, Vibratome sections	Ultrastructural localization of TH in dopaminergic axons and terminals in adult rat brain neostriatum[159]
				Immunocytochemistry with TH was used to study the developmental morphology and synaptic associations in the nigrostriatal anlage in fetal rat brain[160]
				TH in noradrenergic neurons in locus coeruleus: ultrastructural association of TH with microtubules[161]
Rat striatum	Four monoclonal antibodies produced by hybridomas of mouse immunocytes fused with mouse plasma cytoma cells	PAP	Paraformaldehyde fixed and postfixed, Vibratome sections	Neurons in rat brain tissue sections[163]

(Continued)

Table I. (Continued)

Antigen purified from	Antiserum raised in	Immunohistochemical technique used: Technique	Immunohistochemical technique used: Procedure	CNS region studied (optical microscopy)	Ultrastructural localization
L-Aromatic-amino-acid decarboxylase (*L-3,4-dihydroxyphenylalanine carboxy-lyase, E.C. 4.1.1.26*) (AADC), also called dopamine decarboxylase (DDC), which is immunologically identical to 5-hydroxytryptophan decarboxylase					
Bovine adrenal medulla[164,165]	Rabbit	Indirect immunofluorescence	In early experiments cryostat sections fixed with acetone or methanol chloroform; then in later experiments formalin fixation, Vibratome sections	AADC-immunoreactive cells were dopamine- and 5-HT-containing neuronal perikarya in rat mesencephalic raphe nuclei, substantia nigra, and paranigral areas.[166–168] Also, immunofluorescent terminals were detected where DA and 5-HT systems are known to terminate.[169] Possible 5-HT terminals were detected in rat olfactory bulb[135] AADC was studied during development of rat adrenal medulla[134]	
		PAP (Triton X-100 or digitonin)			Ultrastructural distribution of AADC[134]
Dopamine-β-hydroxylase [3,4-dihydroxyphenylethylamine, ascorbate : oxygen oxidoreductase (β-hydroxylating), E.C. 1.14.17.1] (DBH)					
Sheep adrenal medulla[170]	Rabbit	Indirect immunofluorescence or peroxidase	Cryostat sections; ethanol fixation or (for EM) paraformaldehyde fixation	DBH in chromaffin cells of adrenal medulla, sympathetic neurons, and nerves fibers[170–172] Axonal transport of DBH in sympathetic nerves[173] DBH immunofluorescence in central nervous system[174]	

Bovine adrenal medulla[175,176]	Rabbit	Indirect immunofluorescence	Cryostat sections fixed with chloroform–methanol (Hartman method)[178]	Cellular localization of DBH in different tissues, PNS and CNS[177–180] and locus coeruleus[181]	
Bovine adrenal medulla[182]	Rabbit	Indirect immunofluorescence	Hartman method; later formalin fixation was used, Vibratome sections	Localization of DBH (together with localization of AADC and PNMT) in different areas and tissues,[167,168,182,183] DBH immunoreactivity in neuronal perikarya and fibers of locus coeruleus, nucleus anterior ventralis thalami, and Forel's field H1[169]	
		Indirect immunofluorescence and PAP	Formaldehyde fixation and postfixation; cryostat sections	Noradrenergic nerve terminals in the different layers of rat olfactory bulb[135]	
		Direct immunoperoxidase with HRP conjugated to anti-DBH antibody	Paraformaldehyde fixation, Vibratome sections		DBH in nerve terminals (vesicles) of rat brain[184]
		Indirect immunofluorescence	Formaldehyde fixation; cryostat sections	Appearance of DBH (TH, DDC, PNMT) during ontogenesis of rat adrenal medulla[134]; DBH-(PNMT-, enkephalin-, and somatostatin-)-like immunoreactivity in human adrenal medulla and pheochromocytoma[185]	

(*Continued*)

Table I. (Continued)

Antigen purified from	Antiserum raised in	Immunohistochemical technique used: Technique	Immunohistochemical technique used: Procedure	CNS region studied (optical microscopy)	Ultrastructural localization
Bovine adrenal medulla[147]	Rabbit	PAP (Triton X100)	Paraformaldehyde fixation, Vibratome	Dopaminergic neurons in rat brain and ultrastructural distribution of DBH, TH, and TRYH[150] DBH in neurons of rat area postrema and of medial nucleus tractus solitarii,[154] human fetal sympathetic ganglia,[155] catecholamine and enkephalin neurons of locus coeruleus, basal ganglia, area postrema[186]	
Bovine adrenal medulla[187]	Rabbit	Indirect immunofluorescence and immunoperoxidase	Glutaraldehyde–paraformal-dehyde fixation	DBH in rat and bovine chromaffin cells[187]	See ref. 187
Bovine adrenal medulla[188]	Rabbit	Indirect immunofluorescence	Hartman method[178]	Axonal transport of DBH from locus coeruleus to hypothalamus[188]	
Bovine adrenal medulla[189]	Rabbit	PAP	Periodate–lysine)parafor-maldehyde fixation, Vibratome sections	Noradrenergic innervation to the neurons of different preganglionic nuclei of rat thoracic cord studied with immunocytochemistry with DBH[190]	

Bovine adrenal medulla[191,192]	Rabbit	PAP (Fab) (Triton X100)	Paraformaldehyde–glutaraldehyde fixation Paraformaldehyde fixation, cryostat sections	DBH in neurons and nerve terminals in rat locus coeruleus and hypothalamus[191,192]
Bovine adrenal medulla[193]	Goat	Indirect immunofluorescence	Hartman method[178]	Specific uptake and retrograde transport in rat sympathetic elements and CNS of anti-DBH antibodies[193,194]
		PAP (Triton X100)	Paraformaldehyde fixation, freezing microtome section	Projection of locus coeruleus and subcoeruleus/medial parabrachial nuclei in monkey identified by DBH immunocytochemistry and axonal transport studies[195]
Rat adrenal medulla[196]	Rabbit	Indirect immunofluorescence or PAP (Triton X100)	Formalin fixation	DBH (TH and PNMT, oxytocin and vasopressin immunocytochemistry) in neurons and nerve terminals in paraventricular and supraoptic nuclei of normal and Brattleboro rats[143]
Rat adrenal medulla[197]	Guinea pig	Indirect immunofluorescence (Triton X100)[200,202]; PAP (Triton X100)[201,202]	Paraformaldehyde fixation or Hartman[178] method[198,200,203], Paraformaldehyde–glutaraldehyde, Vibratome sections[201,202]	DBH-immunoreactive neuronal perikarya and/or fibres and/or fiber patterns and orientation in rat locus coeruleus,[198,201,202] cerebellum,[198,202] neocortex,[198,199,202] pontine and mesencephalic tegmenta, periaqueductal gray in caudal midbrain, roof of IV ventricle,[201] DBH-like immunoreactivity in submandibular ganglion cells, which lack norepinephrine,[203] all studied with a homologous immune serum[197]

(*Continued*)

Antigen purified from	Antiserum raised in	Immunohistochemical technique used: Technique	Immunohistochemical technique used: Procedure	CNS region studied (optical microscopy)	Ultrastructural localization
Human pheochromocytoma[146,147]	Rabbit	Indirect immunofluorescence	Hartman method[178]	DBH- and TH-containing neurons mainly in rat hypothalamic nuclei[145] Axoplasmic transport of TH and DBH in ligated rat and guinea pig sciatic nerves[146]	
Norepinephrine N-methyltransferase (*S-adenosyl-L-methionine:phenylethanolamine N-methyltransferase E.C. 2.1.1.28*) (PNMT)					
Bovine adrenal medulla[164–166,204]	Rabbit	Indirect immunofluorescence	formalin fixation, Vibratome or cryostat sections	In rat weakly immunofluorescent terminals in dorsomedial hypothalamic nucleus and perifornical area; strongly fluorescent perikarya in medulla oblongata[169,183]; PNMT-containing cell bodies in reticular formation of medulla oblongata (nerve cell groups called C1 and C2) with long descending (to spinal cord) and ascending fibers (PNMT-positive axonal bundles in pons and medulla oblongata) with terminals around affe- and efferent visceral nuclei of lower brainstem, locus coeruleus, hypothalamus, and periventricular gray[205]; these neurons resist 6-	

hydroxydopamine[206]; in rats, this distribution parallels that obtained by biochemistry,[207] but in primate brain (monkey and men), PNMT is also present in amaygdala, septum, habenula, and basal ganglia[207]; PNMT-positive neurons are absent in rat olfactory bulb[135]; PNMT appearance during development[134] and its ultrastructural localization in adult[207] were studied in rat adrenal medulla: only a subpopulation of chromaffin cells contains PNMT[169]; no PNMT-containing cells are found in sympathetic ganglia neurons and SIF cells, but few are found in adrenal medulla of human fetuses.[155] PNMT-immunoreactive perikarya were found in the medial nucleus tractus solitarii of rat brain; PNMT-positive terminals were found throughout the area postrema and in the medial nucleus tractus solitarii[154]

(Continued)

Table I. (Continued)

Antigen purified from	Antiserum raised in	Immunohistochemical technique used: Technique	Immunohistochemical technique used: Procedure	CNS region studied (optical microscopy)	Ultrastructural localization
Bovine adrenal medulla[208]	Rabbit	Indirect immunofluorescence	Formalin fixation, cryostat sections	Noradrenergic (PNMT-containing) fibers abundantly innervate the parvocellular portion of the paraventricular nucleus; only few fibers are seen in the magnocellular portion and in the supraoptic nucleus in normal and Brattleboro rats in relation to DBH-, TH, oxytocin-, and vasporessin-containing neurons[143]	
Bovine adrenal medulla[209]	Rabbit	Indirect immunofluorescence	Formaldehyde fixation and postfixation, Vibratome sections	In rat medulla oblongata, three groups of PNMT-containing cell bodies were found, two correspond to C1 and C2 of Hökfelt *et al.*; another, called C3, made of cells scattered in the fasciculus longitudinal medial[107]	

Tryptophan hydroxylase [*L-tryptophan tetrahydropteridine : oxygen oxidoreductase (5-hydroxylating)*] E.C. 1.14.16.4) (TRYH)					
Rat midbrain (raphe nuclei)[210]	Rabbit	PAP	Paraformaldehyde fixation; picric acid–formalin postfixation, paraffin sect. (O.M.), glutaraldehyde–formaldehyde fixation, Vibratome section (E.M.)	TRYH-immunoreactive neurons were detected in rat midbrain, pons, medulla, particularly in raphe nuclei[210]; the serotonergic innervation of adrenergic neurons in nucleus locus coeruleus in rat[151] and other CNS regions[150] was also studied with the anti-TRYH immune sera in immunocytology; serotonergic neurons were found in rat gut[211]	The ultrastructural distribution of TRYH within serotonergic neurons was studied[150,210] and compared to that of TH in dopaminergic and noradrenergic neurons[151]
Histidine decarboxylase (*L-histidine carboxy-lyase, E.C. 4.1.1.22*) (HDC)					
Rat fetus (whole body)[212] Fetal rat liver[213]	Rabbit	Indirect immunofluorescence or PAP	Paraformaldehyde fixation, cryomicrotome sections	Localization of HDC in rat stomach (apparently parietal cells) and brain[212]: all areas biochemically rich in HDC, such as hypothalamus, mammillary body, amygdala, and dentate girus of hippocampus, are immunoreactive; the stria terminalis is particularly rich in HDC[213]	

(Continued)

Table I. (*Continued*)

Antigen purified from	Antiserum raised in	Immunohistochemical technique used: Technique	Immunohistochemical technique used: Procedure	CNS region studied (optical microscopy)	Ultrastructural localization
	Choline-0-acetyltransferase (*acetyl CoA : choline-0-acetyltransferase, E.C. 2.3.1.6*) (ChAT)				
Bovine brain[214]	Guinea pig	Indirect immunofluorescence	Cryomicrotome sections	ChAT immunoreactivity was shown in bovine spinal cord, anterior horn[215]	
		PAP	Formalin-fixed paraffin embedded	Rabbit spinal cord and cerebellum,[216] rabbit forebrain,[217] human cerebellum from normal and Huntington disease[218]	
Human neostriatum[219]	Rabbit	Indirect immunofluorescence	Cryomicrotome sections	Bovine spinal cord anterior horn, dorsal roots; rat and beef cerebral cortex[220]	
		PAP	Paraformaldehyde fixation, Vibratome section	Guinea pig neostriatum[221]	In addition to cytosol, DAB reaction products are seen bound to microtubules, ribosomes, membranes (mitochondrial, plasma membranes, RER, vesicles), present, at low levels, in cell nuclei;

					endings containing DAB reaction products were those damaged[221]
					Localization and identification in the interpeduncular nucleus of the cholinergic boutons of fibers derived from the medial habenula[222]
Human neostriatum[233,224]	Rabbit	PAP	Paraformaldehyde/ glutaraldehyde fixation, cryomicrotome sections	Rat spinal cord ventral horn, septal nuclei, neostriatum, rostral forebrain; guinea pig neostriatum and neuromuscular junction (diaphragm)[225]; atlas of ChAT localization in cat CNS in serial section from olfactory bulb to upper cervical spinal cord (except cerebellum)[226]; different brain areas of various mammals[227]	
Bovine caudate[228]	Rabbit	Indirect immunofluorescence	Paraformaldehyde fixation, cryomicrotome sections	Magnocellular neurons in pontine reticular formation and neurons in medullary reticular formation[228]; two forms (A and B) of ChAT were found	

(Continued)

Table I. (Continued)

Antigen purified from	Antiserum raised in	Immunohistochemical technique used: Technique	Immunohistochemical technique used: Procedure	CNS region studied (optical microscopy)	Ultrastructural localization
Pig brain[229]	Rabbit				
Bovine caudate nucleus[230,231]	Monoclonal antibodies produced by hybridomas obtained by the fusion of immunocytes of Lewis rats and mouse myeloma cells (Sp2/0-Ag 14).			Two hybridoma lines producing anti-ChAT were obtained[230]	
Glutamate decarboxylase (*L-glutamate-1-carboxy-lyase, E.C. 4.1.1.15*) (GAD)					
Mouse brain[232,233]	Rabbit	Indirect immunoperoxidase	Paraformaldehyde fixation, tissue sectioner or cryomicrotome sections	GAD-positive nerve endings but not perikarya are detected in different areas of rodents CNS[234,235] The nerve terminal of inhibitory cerebellar neurons in adult rats and the corresponding growing neurites during development but not the perikarya contain GAD[236–238]; after colchicine, the perikarya also become GAD positive[239] The same in Ammon's horn[239]	

				GAD-containing nerve terminals are present in spinal cord (dorsal horn lamellae); some of these terminals are presynaptic to other axonal terminals[240]
				GAD-positive nerve terminals decrease at the site of experimental epileptic foci in monkey cerebral cortex[241]
				A large number of terminals in the substantia nigra contain GAD,[242] mostly of pallidostriatal origin, and some from intrinsic neurons[243]
		PAP		In cell cultures some neurons derived from rat substantia nigra contain GAD[244]
		Indirect immunoperoxidase	Paraformaldehyde, tissue sectioner or cryomicrotome sections	GAD-positive neurons of rat olfactory bulb[245]
		PAP		GAD-positive neurons of rat cerebellum and benzodiazepine receptors[246]
		PAP (Fab)		In rat GAD-immunoreactive striatopallidal, striatoentopeduncular, pallidonigral pathways and neostriatal local circuits were shown[247]
				Gabaergic neurons and circuits were shown in substantia gelatinosa of rat spinal cord[248]

(*Continued*)

Table I. *(Continued)*

Antigen purified from	Antiserum raised in	Immunohistochemical technique used: Technique	Immunohistochemical technique used: Procedure	CNS region studied (optical microscopy)	Ultrastructural localization
		PAP (Prot. A)	Periodate–lysine paraformal-dehyde	Gabaergic neurons were detected in rabbit retina[249]	
				Gabaergic projection of stria medullaris fibers to the lateral habenula were shown in rat[250]	
		Indirect PAP or immunofluorescence	Formaldehyde or Bouin fixation; cryomicrotome sections	In rat[251–253] or rat, mouse, and monkey cerebellum,[254] only some Purkinje cells contain GAD; others contain motilin[254]	
		Indirect immunofluorescence	Formaldehyde fixation, cryomicrotome sections	In rat (colchicine treated), three clusters of magnocellular GAD-containing neurons were found in posterior hypothalamus; dispersed GAD-immunoreactive cell bodies were present in several other hypothalamic nuclei, and GAD-immunoreactive fibers were evenly distributed through hypothalamus[255]	
Rat brain[256]	Sheep	PAP	Paraformaldehyde with or without glutaraldehyde, Vibratome sections	No need for colchicine to show GAD in perikarya of rat cerebellum[256]; GAD-immunoreactive terminals in rat substantia nigra are reduced after kainic-acid-induced lesion of striatum[258]	

Mouse brain[259]	Rabbit	Indirect immunofluorescence	Paraformaldehyde fixation, cryomicrotome sections	Systematic study of GAD immunoreactivity in the whole rat CNS[259]	
Catfish brain[260]	Rabbit	Indirect immunoperoxidase (Prot. A)	Formalin fixation, Vibratome sections	GAD immunoreactivity in goldfish retina[260]	
Rat brain[261–263]	Rabbit	PAP	Paraformaldehyde–glutaraldehyde, tissue sectioner	GAD immunoreactivity in rat cerebellum, substantia nigra, nucleus raphe dorsalis,[264] and habenular complex[265]	In nucleus raphe dorsalis, only nerve terminals are GAD positive, but after colchicine, perikarya are also GAD positive[264]

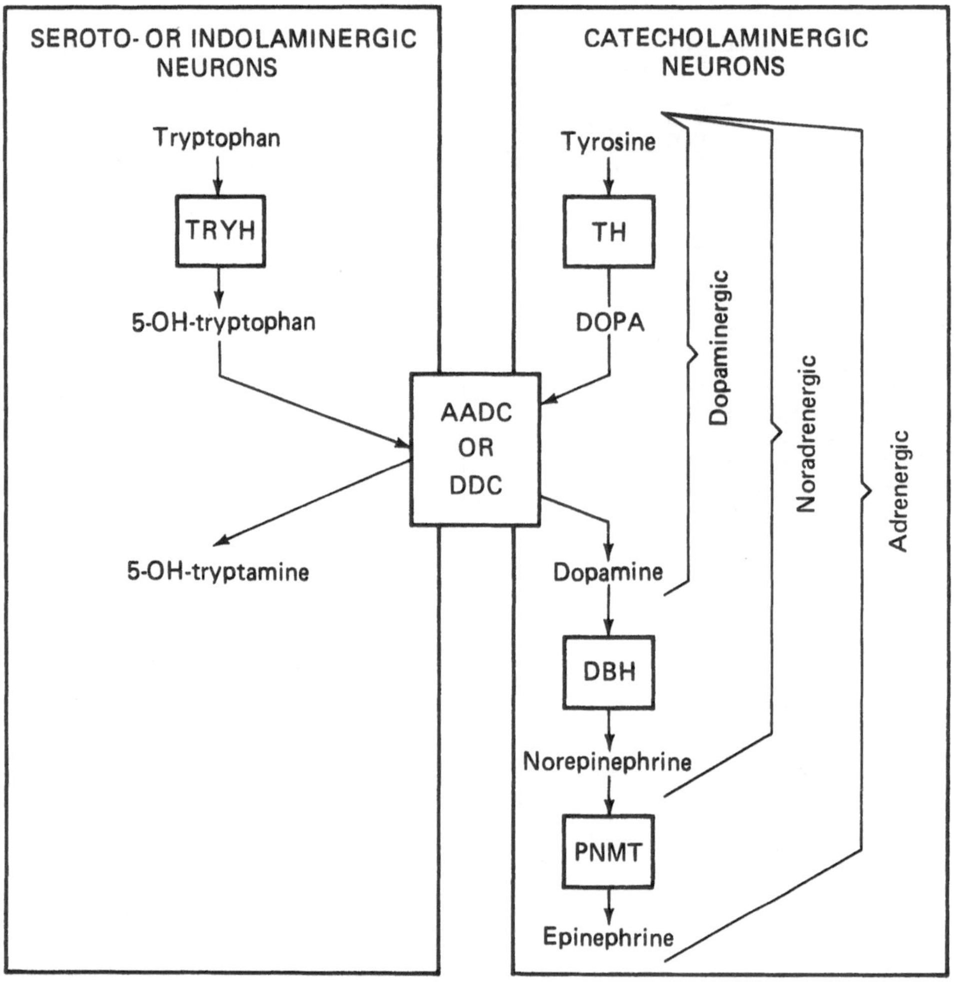

Fig. 1 Enzymes of monoamine synthesis present in different types of monoaminergic neurons. TRYH, tryptophan hydroxylase; AADC, aromatic amino acid decarboxylase, immunologically identical to dopamine decarboxylase, DDC; TH, tyrosine hydroxylase; DBH, dopamine-β-hydroxylase; PNMT, norepinephrine methyltransferase.

hydroxylase (DBH) also present], and adrenergic [TH, dDC, DBH, and norepinephrine methyltransferase (PNMT) all present].

The lack of specificity of enzymes synthesizing the amino acid putative neurotransmitters (aspartate, glutamate, glycine) other than GABA does not permit the use of the same approach, but in some cases attempts have been carried out by developing antibodies directed against the neurotransmitter itself. Recent attempts to obtain antiglutamate antiserum have been reported, and an antiserotonin antibody has been obtained (see Section 3.3.3c). But the approach of mapping neurons according to their immunoreactivity for the neurotransmitter and/or neuromodulator itself has been greatly developed in the case of neuropeptides. Such experiments, particularly those in which a double immunoreactivity has been shown in the same neurons now cast certain doubts on the validity of Dale's principle (one neurotransmitter per neuron), since cells containing both a peptide and a "classical" neurotransmitter have now been described (see Table I for references). More detailed comments on this point may be found in ref. 5.

Another classical assumption that has virtually crumbled is that morphologically identical neurons use the same neurotransmitters. Some neurons such as Purkinje cells appear to be separable into three classes according to their glutamate decarboxylase and motilin content (see ref. 254 in Table I and Section 3.3.3c). But in any case, immunocytology of neurotransmitters and neurotransmitter-related enzymes is helping to draw at the morphological level a functional map of nerve cells and their connecting pathways, particularly when immunocytology is used in conjunction with lesions (produced by either surgical or electrochemical methods or, more recently, by neurotoxins or excitotoxins). Autoradiography of neurotransmitters and neurotransmitter precursors taken up by high-affinity mechanisms, autoradiography of receptor ligands, and autoradiography or other histological methods revealing material taken up by endocytosis and transported by retrograde or anterograde axonal transport have also been used in conjunction with immunohistochemical methods to provide functional data on characterized neurons (for examples, see references in Table I).

3.3.3b. Enzymes of Neurotransmitter Catabolism and/or Inactivation. Enzymes involved in neurotransmitter catabolism when different from those involved in neurotransmitter synthesis, do not have a specific neuronal localization.

With regard to the enzymes of monoamine catabolism, monoamine oxidase (MAO) is a mitochondrial enzyme present in all cells, and catechol-O-methyltransferase (COMT) is not detectable by immunohistochemistry in neurons (see Table II).

With regard to GABAergic neurotransmission, the immunoreactivity of GABA transaminase (GABA-T), the enzyme that catabolyzes GABA, appears not to have any particular cellular specificity (see Table II), but the immunocytochemical results do not agree with former histochemical data.[124–128]

With regard to the cholinergic system, to our knowledge acetylcholinesterase (AChE) has not yet been localized by immunocytology, but histochemical methods show a neuronal localization that, however, is not limited to cholinergic neurons.

Glutamine synthetase, an enzyme possibly involved in the glutamate–glutamine cycle of neurotransmitter glutamate, appears to have an astrocytic localization (see Section 3.1.1.). Antisera against the other enzyme of the cycle, glutaminase, have not been used in immunocytology.

3.3.3c. Neurotransmitters or Neuromodulators. "Classical" Neurotransmitters or Neuromodulators. An approach that only recently appeared to be successful is to use a neurotransmitter as hapten and immunize a rabbit with a protein–neurotransmitter complex. The immunoglobulin specific for the neurotransmitter hapten is then purified. In this way, pure monospecific antibodies

Table II
Cellular and Ultrastructural Localization of Enzymes Involved in Neurotransmitter Catabolism or Inactivation as Revealed by Immunocytology

Antigen purified from	Antiserum raised in	Immunohistochemical technique used		CNS region studied (optical microscopy)	Ultrastructural localization
		Technique	Procedure		
Catechol-0-methyltransferase (E.C. 2.1.1.6) (COMT)					
Rat liver[266–268]	Rabbit	PAP (Triton X100)	Formaldehyde fixation. Cryomicrotome sections.	COMT immunoreactivity was shown in ductus deferens, seminal vesicles, oviduct, pregnant uterus, placenta, pancreatic islets, parotid and sweat glands, lymph nodes, thymus, kidney, and liver[267]	
		Indirect immunofluorescence (Triton X100)		COMT immunoreactivity was shown in rat liver, kidney, and brain and in chinchilla and bovine brain[269]; in the brain of the three species COMT was detected in cells of the ependyma lining lateral, third, and fourth ventricles, in choroid plexus, in myelinated tracts (not in myelin itself but in interfascicular oligodendrocytes and astrocytes); oligodendrocytes, astrocytes, and cerebellar Bergmann glia,[269–272] leptomeninges and ciliary epithelium contain COMT immunoreactivity[270]	

				COMT immunofluorescence was studied in detail in several circumventricular organs of rat brain; in all of these organs COMT immunofluorescence was found in glial elements, but ependymal cells did not show COMT immunoreactivity in the subformical organ and in the area postrema, which contrasts with the immunofluorescence of ependyma in organum vasculosum of the lamina terminalis and in the subcomissural organ[271]; the relationship between the cellular localization of COMT and its physiological function is discussed in ref. 272	
GABA transaminase (*4-aminobutyrate:2-oxoglutarate aminotransferase, E.C. 2.6.1.19*) (GABA-T)					
Mouse brain[234,274]	Rabbit	Indirect immunoperoxidase,[234] PAP,[275] indirect immunofluorescence[275]	Paraformaldehyde fixation,[234,235] tissue sectioner or cryomicrotome,[234] Vibratome[275]	In cerebellum, Purkinje neurons (not all, though), perikarya, and dendrites, Golgi neurons, stellate and basket cells, few large neurons of deep nuclei, Bergmann glia (perikarya and radial fibers), astrocytes in the granular layer are GABA-T positive; granule cells[234,275] are possibly not positive; also, some neurones in the nucleus interpositus seem to be GABA-T positive[234]	Cytoplasm, plasma membranes, outer membrane of cellular organelles, microtubules, filaments, but not mitochondrial matrix (as expected according to subcellular fractionations) of pre- and postsynaptic cells of a GABAergic synapse show peroxidase reaction products[234]

were obtained against serotonin[276,277] and glutamate[278] and used in immunohistochemistry. Similarly, antiguanosine-3′,5′-cyclic monophosphate (cyclic GMP) and antiadenosine-3′,5′-cyclic monophosphate (cyclic AMP) immune sera were obtained.[279,280]

The anti-cyclic-GMP immune serum used in immunocytology in adult rat cerebellum labeled mainly neuroglial cells, some stellate and basket cells, and some large neurons in deep nuclei.[279] In striatum,[280] cyclic GMP immunoreactivity was also found in both neurons and glial cells, but not in all of them. At the EM level, with a PAP method, the cyclic GMP immunoreactivity was found in the cytosol matrix and on the surface of intracellular and surface membranes.[279] The anti-cyclic-AMP immune serum also labeled Purkinje cells, granule cells, and Golgi cells in addition to neuroglial cells.[279] Drugs that increase cerebellar activity enhance cyclic GMP levels.[279] Electron microscopy[280] shows that in the striatum both cyclic GMP and cyclic AMP immunoreactivities were stronger in the postsynaptic area of terminal boutons and glial cells processes; cyclic GMP immunoreactivity, however, predominated postsynaptically, whereas cyclic AMP immunoreactivity predominated in the glial cell processes.

Neuropeptides. In addition to classical chemical neurotransmitters, including acetylcholine, GABA, and the monoamines, interest has been aroused in recent years in the likely implication of small peptides in the process of neurotransmission.[5,281] Today, approximately 30 peptides have been shown to be present in the brain. Many of these neuropeptides were known as neuro- or adenohypophyseal hormones or as hypophysiotropic hormones (for reviews see refs. 281–285). They are produced by the hypothalamus and the hypophysis and are secreted in the bloodstream of the hypothalamic–hypophyseal portal system.[286] The areas of the brain that release neurohormones are characterized by a high vascularization, fenestrated capillaries, absence of glial sheaths around neurons, and absence of cilia and tight junctions in the ependyma; hence, the blood–brain barrier appears to be absent there.[12] Seven such "windows" in the CNS have been described,[12] all around the third ventricle and in the neurohypophysis. It is thus not surprising to find that frequently the highest concentrations of peptides are found in the hypothalamus.

But many of these peptides were also found in areas other than those with neuroendocrine activity. This raised the possibility that although those peptides localized in brain sites from which they can be released into the bloodstream have hormone function, the same peptides (or at least some of them) localized in "nonsecretory" regions of the CNS may have an additional role distinct from their endocrine function.[5,12,281,282] Such a hypothesis was supported by experiments showing the presence of these peptides in synaptosomes from which they were released in a Ca^{2+}-dependent fashion and by the effects of these peptides on cell physiology and animal behavior, which are unrelated to their known hormone action.[282] Thus, the presence of neuropeptides in neurons in CNS sites not normally associated with endocrine activity has been taken as an indication that these neuropeptides have a role either as neurotransmitters and/or as modulators of the action of "classical" neurotransmitters or a role as "factors" involved in long-term events.

Adrenal medulla was found to be an abundant source of some neuropeptides, and this was not surprising since chromaffin cells share many characteristics of neurons, and for this they were classified as "paraneurons." More surprising was the fact that many peptide hormones of the gut, stomach, and pancreas were also found in the CNS. This apparent random distribution of neuropeptide-containing cells is explained by Pearse[287] with the APUD (amine precursor uptake and decarboxylation) system theory, according to which cells secreting neuropeptides all derive from a common neuroectodermal precursor cell line. Migration of these cells during embryogenesis is responsible for their distribution in such different areas as the CNS, the adrenal glands, the gut, and associated glands. Whether the APUD cell theory also can explain the cellular origin of peptides first found in the bloodstream (see below) we cannot say.

Immunohistochemistry with specific antisera raised against the different neuropeptides has played an essential part in the identification of neurons containing them, in distinguishing the different peptide systems, and in mapping the CNS pathways of peptide-containing neurons.[5,12,288] The relative ease with which antibodies against the various peptides can now be made available has been possible thanks to the utilization of a series of recent technological developments.

The limiting factor in studies of peptide hormones was the very limited amount of purified peptides that could be obtained, although the analysis of their amino acid sequences required at least milligram quantities of them. This limitation was in part overcome by the introduction of mass spectrometry and of more refined chromatographic techniques (HPLC), permitting analysis of minute quantities of peptides. This development in analytical techniques and progress in techniques of peptide chemical synthesis enabled copies of peptides or slightly modified copies to be chemically synthesized in large amounts once the amino acid sequence of the peptide had been deduced. Since these synthetic peptides are usually too small to provoke antibody production, they were coupled to large proteins to use as immunogens. Such antibodies could be radiolabeled, and peptides containing radioactive amino acids were synthesized. This has allowed the direct measure of peptide amounts by radioimmunoassay (RIA) in tissue homogenates and/or made available "probes" to be used for molecular cloning.

The most recent and highly rewarding technical development in peptide research is the use of molecular cloning, mainly for studies of the precursor proteins of the neuropeptides. In fact, all neuropeptides except carnosine (see below) are formed by sequential proteolysis first of a large-molecular-weight protein precursor and then of the large peptides derived from it. This proteolysis has been termed prohormone "processing." Nucleic acid hybridization techniques and molecular cloning have allowed the isolation of the mRNA of the preprohormones when only the neuropeptides or antibodies against the neuropeptide were available. In addition to this, base sequence analysis of the cDNA copy of the mRNA of the preprohormones has been used for deducing the amino acid sequence more rapidly than with the classical amino acid sequencing technique of peptides.

The very nature of the process of formation of neuropeptides can be a source of problems in ICC and/or RIA, since antibodies could react not only with the corresponding antigenic determinants on the neuropeptide but also with those on its precursor peptides and protein; however, this is not always the case, since examples are known (e.g., LHRH) where the antiserum reacts with the peptide but not with the high-molecular-weight precursor. Evidently, spatial rearrangement of the peptide or of its precursor(s) is important in exposing or hiding the antigenic determinant amino acid sequence. In many cases, the demonstration by ICC of the presence in nerve fibers and in terminals of immunoreactivity associated with a given peptide is easy, but this immunoreactivity cannot be detected in the corresponding cell bodies unless the animal is treated with colchicine to block the axonal transport of the peptides and their precursor from the cell body to the terminals.

Another point to be considered for the interpretation of ICC results is that some prohormones are multihormone precursors (see below, i.e., peptides derived from proopiomelanocortin), and, thus, one cell might be immunoreactive to different antisera raised against different peptides, all derived from the same precursor. Furthermore, short amino acid sequences are sometimes common to different neuropeptides possibly derived from different precursor proteins; thus, the possibility exists that different cells containing different peptides have similar immunoreactivity. However, the presence of a given amino acid sequence does not necessarily imply an immunoreactivity with a specific antiserum (e.g., β-endorphin contains the sequence of Met-enkephalin, but neurons reacting with antisera directed against β-endorphin were not the same neurons as those reacting with antienkephalin[289,290]) These problems explain the caution exercised by some authors when speaking of hormonelike immunoreactivity when an antiserum against a peptide hormone reacts in ICC or RIA with antigens within a given cell.

Neuropeptides can be classified into separate groups according to the hormone activity by which they were first known and then be grouped into subclasses according to the following different criteria: either according to the precursor prohormone from which they derived or according to chemical similarity or according to the tissue in which they were first found. In view of the extensive bibliography on the subject, it has not been possible in this brief review to be exhaustive.

Neurohypophyseal Neuropeptide Hormones. The first group of neuropeptides[286,291–322] includes the hypothalamic neurohypophyseal hormones such as oxytocin (OXT) (nine amino acids) and vasopressin (VP) or antidiuretic hormone (ADH) (also nine amino acids) (Lys-VP in pigs and Arg-VP in the other mammals). These hormones are related to vasotocin, valitocin, aspargtocin, glumitocin, isotocin, and mesotocin (other nine-amino-acid peptide hormones present in various nonmammalian animal species), which share with oxytocin and vasopressin a common sequence of at least six amino acids. Neurophysin (Np) I and II (for review see ref. 321), the carrier proteins (mol. wt. 10,000) for OXT and VP, respectively, appear to have a similar distribution to that of OXT and VP, respectively.

The amino acid sequence of the 166-amino-acid precursor protein from which VP and Np II derive by proteolysis has been deduced from the nucleotide sequence of the cloned cDNA for the bovine precursor proteins.[324] Oxytocin- and VP-containing neurons and fibers have identical distribution, but OXT neurons and fibers are distinct from VP neurons and fibers. The largest concentration of neurons containing OXT and VP is found in the hypothalamus, particularly in the supraoptic and paraventricular nuclei. Their fibers terminate in the median eminence with recurrents to the magnocellular hypothalamic neurons. However, VP-, OXT-, and Np-positive neuronal processes also project to the brainstem (locus coeruleus, n. parabranchialis, Edinger–Westphal nucleus, various vagal nuclei) and the spinal cord (substantia gelatinosa), whereas other nerve terminals are found in the thalamus, the hippocampus, the amygdala (medial nucleus), the septum, the subfornical organ, and the interstitial (bed) nucleus of the stria terminalis. By electron microscopy, VP in nerve terminals was found to be confined to granules 85–300 nm in diameter.[296] In addition to the VP hormone action on water retention and to that of OXT on uterus contraction, it would appear that the CNS action of VP and OXT is on behavior: OXT injected intraventricularly appears to induce maternal behavior in virgin female rats,[325] and VP, which appears to affect learning behavior, may be simply raising attention level (for review see ref. 282).

Adenohypophyseal Neuropeptide Hormones. The second group of peptides consists of hormones produced by the adenohypophysis. The first subgroup consists of all peptides derived from a large protein called proopiomelanocortin. The amino acid sequence of proopiomelanocortin has been deduced from the nucleotide sequence of cloned cDNA for the bovine protein[326] (263 amino acids; molecular weight 31,000). This protein is mainly synthesized by cells in the arcuate nucleus of the hypothalamus and in the adenohypophysis. Those opiatelike peptides, called endorphins (endogenous morphins), that derive from this protein precursor are also included in this group. Proopiomelanocortin appears to be cleaved into equimolar amounts of corticotropin (or ACTH) (39 amino acids), β-lipotropin (β-LPH) (91 amino acids), a "61-K fragment," and a "signal peptide" sequence. The ACTH peptide is split into α-MSH (melanocyte-stimulating hormone) and CLIP (corticotropinlike intermediate-lobe peptide), whereas β-LPH is split into γ-LPH (58 amino acids) and β-endorphin (30 amino acids). These two peptides can be further cleaved: β-endorphin gives α-endorphin (16 amino acids), and γ-LPH yields β-MSH (a peptide with MSH action weaker than that of α-MSH). The amino acid sequence of γ-endorphin is the same as that of α-endorphin with one additional amino acid, Leu.

An interesting point is that the first five amino acids of α-endorphin (starting from the NH_2-terminal side) share a common sequence with Met-enkephalin (opiatelike pentapeptide). However, enkephalins (see second subgroup of adenohypophyseal peptides) are derived from another protein precursor (proenkephalin). Nucleotide sequencing of the cDNA copy of preproenkephalin mRNA from human pheochromocytoma[327] and from bovine adrenal medulla[328–329] show that the 263–267 amino acid precursor is similar in the two species: it contains four copies of Met-enkephalin, one copy of Leu-enkephalin (also a pentapeptide), one heptapeptide (Met-enkephalin-Arg-Phe), and one octapeptide (Met-enkephalin-Arg-Gly-Leu). As in the case of the peptides derived from proopiomelanocortin, enkephalins are also sandwiched between Lys-Arg, Lys-Lys, or Arg-Arg sequences in the proenkephalin molecule. The complete sequences of any endorphin or of dynorphin or of α-neoendorphin do not appear in the proenkephalin molecule.

The cellular localization of the immunoreactivity of neuropeptides derived from proopiomelanocortin[289,290,330,340,347] is similar for all of these peptides: neurons in the arcuate nucleus projecting to the median eminence, preoptic area, supraoptic nucleus, locus coeruleus, and, in man, to the infundibular nucleus.

In contrast with endorphin-containing neuronal structures, enkephalin-containing neuronal perikarya have a widespread distribution, at least in the rat CNS,[289,300,309,332,341–354,424] with far-ranging projections. The nucleus medialis habenulae, the paraventricular nucleus of the thalamus, the amygdala (central nucleus), the n. accumbens, and the bed nucleus striae terminalis also contain enkephalin-positive neurons and fibers. The hippocampus receives enkephalin-positive terminals, and the median forebrain bundle contain enkephalin-positive fibers. Other areas such as the periaqueductal gray and the mesoencephalic nuclei all contain enkephalin-positive structures. Unambiguous localization of enkephalin inside neuronal perikarya, which is clearly distinct from the outline of neuronal perikarya revealed by a great number of enkephalin-containing nerve terminals, has been obtained in the brainstem of colchicine-pretreated rats[345]: enkephalin-containing cell bodies were found in the following brainstem nuclei or areas: *n. tractus solitarii, substantia gelatinosa and interpolaris zone of trigeminal nuclear complex*, near the n. raphe magnus, n. cochlearis dorsalis, n. vestibularis medialis, n. paraolivaris, *n. parabranchialis*, n. tegmenti dorsalis of Gudden, n. interpedencularis, neuronal perikarya in the periaqueductal gray matter, at the border of the lemniscus lateralis, and in the *medial geniculate*. Enkephalin-containing fibers are abundant around those nuclei here shown in *italics*. Enkephalin-containing nerve fibers and terminals, but no enkephalin-containing neuronal perikarya, were observed in the locus coeruleus and the nucleus of facial nerve. Enkephalin-containing terminals and neurons were observed in the dorsal horn of the spinal cord. In view of the common precursor of Met-ENK and of Leu-ENK, the data of Larsson *et al.*[355] showing neurons containing one but not the other neuropeptide are difficult to explain.

Dynorphin is a very potent opioid polypeptide extracted from porcine hypophysis[356] and duodenum.[357] Dynorphin and α- and β-neoendorphin all contain at their NH_2-terminal the five-amino-acid sequence of Leu-enkephalin, but their precursor differs from that of enkephalin. In fact, the amino acid sequence of this precursor, named preproenkephalin B (to distinguish it from the adrenal Leu- and Met-enkephalin precursor cited above, which was called proenkephalin A), was deduced from the nucleotide sequence of the cloned cDNA for the porcine peptide precursor.[358] From this sequence, it would appear that three opiate peptides derive from this precursor: one molecule of dynorphin (17 amino acids), one of β-neoendorphin (nine amino acids), and one molecule of Leu-enkephalin. No α-enkephalin sequence (14 amino acids) was detected.

A synthetic dynorphin consisting of 13 amino acids (dynorphin 1–13) has been synthesized, and antibodies against this sequence were raised.[359] With this antiserum, the regional distribution of dynorphin was determined by RIA on dissected CNS areas.[360,361] Dynorphin distribution was as widespread as that of enkephalin, but the distribution of the two opiates is not identical. Very recently,[362] ICC mapping of the CNS with an antidynorphin immune serum, which did not cross react either with enkephalin or with α-neoendorphin, was carried out in parallel with a study with an anti-α-neoendorphin immune serum that did not cross react either with enkephalin or dynorphin or β-neoendorphin. Since β-endorphin is the 1–9 segment of the 14-amino-acid sequence of α-neoendorphin, it would appear that these antisera react with amino acid sequences not common to the two neoendorphins (i.e., the sequence 10–14). The ICC mapping shows that α-neoendorphin and dynorphinan have identical cellular distribution. This does not agree either with the protein precursor structure or with the antisera cross reactivity. Structures such as the hypothalamic magnocellular nuclei are rich in both dynorphin/α-neoendorphin-containing neurons and enkephalin-containing neurons. The nucleus accumbens, the median forebrain bundle, the hippocampus, and the medulla oblongata are rich in both dynorphin/α-neoendorphin-containing and enkephalin-containing fibers, whereas other areas rich in enkephalin such as the interpeduncular nucleus or the central nucleus of the amygdala do not react with anti-α-neoendorphin and antidynorphin immune sera. The substantia nigra and the internal capsule, where enkephalinlike immunoreactivity is absent, contain α-neoendorphin/dynorphin-positive fibers.

Other endogenous opiate peptides are known but less well studied. As far as we know, the possible localization of other adenohypophyseal hormones in CNS neurons has not been studied yet. Such hormones include the somatomammotropins [growth hormones (GH) or somatotropin and lactotropin (LTH), also called either prolactin or mammotropic hormone] and the glycoprotein hormones, which include the thyroid-stimulating hormone (TSH) or thyrotropin and the gonadotropins [follicle-stimulating hormone (FSH) and lutenizing hormone (LH) or interstitial cell stimulating hormone (ICSH)].

Hypophysiotropic Neuropeptide Hormones. The third group of neuropeptides consists of hypophysiotropic hormones.

Luteinizing hormone-releasing hormone (LHRH or LRH) or factor (LRF), also called gonadoliberin or gonadotropin-releasing hormone (GnRH) or luliberin, is a decapeptide whose main hormone action is to stimulate the release of both FSH and LH from the adenohypophysis.[5,281–284] LHRH-containing structures[363–386] appear to be mainly confined to areas on the surface of the third ventricle. LHRH-containing neuronal perikarya were seen in the arcuate nucleus and the preoptic area; their fibers are probably those that reach the median eminence via the tuberoinfundibular tract. The LHRH-containing terminals in the median eminence, medial septal area, diagonal band, olfactory tubercle, and organum vasculosum of the lamina terminalis terminate on capillaries, and this must be related to the endocrine action of the LHRH. But many LHRH-containing fibers rich in varicosities also run over the surface of the third ventricle. Close to this surface, small bipolar LHRH-containing neurons are also found, suggesting the possible secretion of LHRH into the CSF.[380] These are the fibers that form the bed nucleus of the stria terminalis and send processes (via the anterior commissure) to the subfornical organ. Structures containing LHRH immunoreactivity were also observed in the amygdala, paraolfactory cortex, mammillary bodies, and central tegmental area. There is evidence that LHRH also directly controls behavior, and the intraventricular or subcutaneous injection of LHRH evokes the copulatory posture.[282]

Thyrotropin-releasing factor (TRF) or hormone (TRH), also called thyroliberin, is a tripeptide whose hormone action is to induce release of TSH from the hypophysis.[5,281–285] At the behavioral level, TRH seems to elicit excitation and anorexia and to antagonize barbiturate- and alcohol-induced sedation.[282] Only 20% of TRH in CNS is in the hypothalamus, where the hormone is the most concentrated. TRH-containing fibers[387–391] were detected by ICC in the medial part of the external layer of the median eminence, the dorsomedial hypothalamic nucleus, the perifornical region, and the parvocellular part of the hypothalamic paraventricular nucleus. TRH-containing fibers at lower density were observed in the other hypothalamic areas and nuclei. In the tele- and diencephalon, TRH-positive fibers were observed in nucleus accumbens, in the organum vasculosum, and in basketlike structures around cell bodies in the bed nucleus, stria terminalis and in the nucleus septilateralis. In the brainstem, TRH-positive fibers were observed in and around several motor nuclei (of the oculomotor, trigeminal, facial, and hypoglossal nerves) and the nucleus intercalatus. Also, the anterior horn of spinal cord contains TRH-positive fibers. In addition, single fibers or low-density networks were observed in different CNS regions. In colchicine-treated rats,[389] TRH-containing perikarya were detected in several cell groups in the hypothalamus and medulla oblongata. Some of these cell groups in the brainstem also contain serotonin and substance P.[391] At the ultrastructural level,[390] TRH immunoreactivity was found in the cytoplasm of some hypothalamic neurons and in terminal boutons in large (~1000 Å) granular vesicles with a highly electron dense core. Terminal boutons of the median eminence seem to occupy a "secretory position," whereas in other hypothalamic nuclei, they were frequently in contact with dendrites and neuronal perikarya without clearly demonstrable synapses. Such synapses were found in the TRH-positive boutons in spinal cords. The small size and easy diffusibility of the TRH molecule account for the difficulties encountered in its detection in ICC.

Somatostatin (SS) or growth hormone release-inhibiting hormone (GH-RIH) or somatotropin release-inhibiting factor (SRIF) is a quadridecapeptide first isolated from the hypothalamus that inhibits the secretion of somatotropic hormone and of TSH but not of the other adenohypophyseal hormones.[5,281–285,405] This peptide hormone is a classical example of gut–brain peptides (see below) and is abundant both in the gut and in the pancreas. These localizations explain the inhibitory action on insulin, glucagon, gastrin, secretin, and motilin release. Somatostatin also inhibits acetylcholine release, and this might explain, at least in part, the inhibitory effect on gastric secretion and intestinal contraction. Somatostatin also appears to be present in sympathetic ganglia and adrenal medulla.[5] This was the first example of the presence of peptides in the autonomic ganglia, which had previously been considered to be simply either noradrenergic or cholinergic. The behavioral effects of somatostatin injections are generally of a "sedative" nature.[282]

The pituitary–hypothalamic localizations of SRIF[350,392–411] (preoptic, supraoptic, suprachiasmatic, and paraventricular nuclei) account for 25% of the total SRIF in the CNS, and it is possibly related to the SRIF hormone action. The remaining 75% is widely distributed in the CNS.[349,392–411,425] Neuronal perikarya and fibers that contain SRIF-like immunoreactivity were found even in the cerebral and hippocampal cortex with subcortical projections. This immunoreactivity was also present in the amygdala and other limbic structures and in mes-, met, and myel-encephalon. No SRIF-like immunoreactive cell bodies were observed in spinal cord, but SRIF-like immunoreactive fibers were detected in the substantia gelatinosa.[354,392–411] A cell population in the dorsal root ganglia and one among the primary sensory neurons contain somatostatinlike immunoreactivity.[354] Many SRIF-like immunoreactive neurons and fibers in the diencephalon of newborn rats are no longer immunoreactive in the adult.[412]

A hypothalamic factor inducing the release of corticotropinlike and β-endorphinlike activities,[413] which probably corresponds to the long sought after CRF (corticotropin-releasing factor) or corticoliberin, was recently isolated as a 41-amino-acid peptide that is very similar (half of the amino acids are identical, and another 12 represent single base changes) to sauvagine, a 40-amino acid peptide extracted from the skin of a South American frog (*Phylomedusa sauvagei*), which also stimulates ACTH and β-endorphin release and inhibits TSH, GH, and prolactin secretion. Also, similarities exist with angiotensinogen (four amino acids) and angiotensin I (three amino acids). An anti-CRF immune serum that did not cross react with other hypothalamic peptides has been obtained and used in ICC[414] for localizing in the central nervous system neuronal fibers and perikarya that contained the CRF-like immunoreactivity. In parallel, antisera to various other peptides were also used. Most CRF-like immunoreactive fibers originate from neurons in the paraventricular nucleus of the hypothalamus and run to the fibrillar layer of the zona extrema, where they give radiating collaterals that terminate on capillaries. The pattern of distribution of these fibers is different from the pattern of VP-, OXT-, SRIF-, LHRH-, and endorphin-containing fibers. Also the CRF-containing perikarya are distributed differently from the neuronal perikarya containing OXT and VP immunoreactivity. Other perikarya containing CRF-like immunoreactivity have been found in the lateral and mediobasal hypothalamus but not in the supraoptic and suprachiasmatic nuclei.

Whether CRF or other hypophysiotropic hormones[282,284] [growth hormone-releasing hormone (GHRH) or factor (GRF), prolactin-releasing factor, and, possibly, MSH release-inhibiting factor (MIF1) or PLG (Pro-Leu-Gly-NH_2)][284] also have other functions in the CNS remains to be clarified.

Brain–Gut Peptide Hormones. The fourth group of neuropeptides consists of the so-called gut–brain peptides, which are present and manifest their action in the gut and associated glands as well as in the CNS. As already mentioned, somatostatin and enkephalin may also be considered as gut peptides. Insulin, although of larger molecular weight (86 amino acids) than the peptides so far described, might turn out to be another gut–brain peptide if its synthesis in the CNS can be clearly demonstrated.[281] The other brain–gut peptides can be subdivided into subgroups. The first subgroup is made of peptides that could be related to physaelemin, the undecapeptide extracted from frog skin. All of these peptides have part of their animo acid sequence in common, and all promote the contraction of smooth muscle fiber preparations. They include substance P, neurotensin, and bombesin in mammals and eleidosin in the octopus.

Substance P (SP) (for reviews see refs. 5,281,282,284,285,405,415) is a undecapeptide that was at first described as a potent hypotensive substance extracted from equine brain and intestine. Substance-P-containing neurons and fibers are present in many areas of the nervous system.[344,349,354,391,402,405,416–437] At present, SP is the most probable candidate for the role of "pain neurotransmitter." This role is suggested by several observations: (1) nerve fibers in the tooth pulp (where only "pain" nerve fibers are present) contain SP; (2) nerve terminals in the most superficial laminae of the posterior horn of the medulla and the substantia gelatinosa of the trigeminal nuclear complex contain SP; (3) about 20% of dorsal root spinal ganglion cells contain SP (other cells contain VIP or CCK or somatostatin); (4) SP is present in a class of primary sensory neurons. Experiments with capsaicin (a substance from hungarian paprika that affects SP-containing neurons) show that SP-containing fibers are also necessary for the vasodilatation around injured areas that is mediated by sensory nerve fibers.

Evidence of the relationship of SP to pain is given by experiments showing that enkephalins block the release of SP and induce analgesia. Thus, it has been suggested that enkephalin released by "enkephalinergic" terminals in the spinal cord (these enkephalin-containing fibers could derive from either propriospinal neurons or from supraspinal descending neurons) acts presynaptically on the SP terminals and blocks the release of SP. Possibly, neurotensin-containing local propriospinal neurons also produce a similar effect on SP release. Descending serotonergic fibers originating from cells in the raphe nuclei, which apparently also contain substance P, also participate in these circuits involved in "pain transmission." Recent data show also the presence of TRH in some of these cells.[391] (Thus, two peptides and one classical neurotransmitter may be present in a single neuron.)

Neurons containing 5-HT and SP were also observed in the nucleus interfascicularis hopoglossi. Other CNS areas in which 5-HT and SP are both abudant are the amygdala, the periaqueductal gray, and the substantia nigra. Substance-P-containing neuronal pathways are very often paralleled by juxtaposed pathways of neuronal perikarya, fibers, and terminals containing another neurotransmitter, e.g., an "SPergic pathway" doubles the GABAergic striatonigral pathway (N.B., GABA inhibits SP release in substantia nigra), and another SPergic pathway doubles the cholinergic pathway between the habenular and the interpeduncular nuclei. Abundant SP-containing and enkephalin-containing cells coexist in close vicinity not only in the dorsal horn of spinal cord and substantia gelatinosa of the spinal trigeminal nuclei and in the central gray but also in many other areas such as raphe nuclei, amygdala, ventral segmental area, and septum (most of these cells are interneurons). In fact, SP, VIP, and CKK are frequently observed in local neurons making short-range connections, but never has a neuron been found to contain both SP and enkephalinlike immunoreactivity; e.g., in the amygdala, SP-containing perikarya are in the medial nucleus; these neurons send their terminals to the central nucleus, where the NT- and enkephalin-containing neuronal perikarya of amygdala are located.

Another SP-containing pathway is that of neurons in the bed nucleus of stria terminalis, which send fibers to the medial preoptic area. Substance P fibers were not observed in adult mammalian cerebellum. A recent report shows, however, that exogenous SP fibers that disappear by the end of the second postnatal week are present in neonatal rat cerebellum together with a few SP-containing neurons in the nucleus interpositus.[437]

Neurotensin (NT) (for review see refs. 5,281,282,284,285,405,415), a potent vasoactive (lowers blood pressure by dilating blood vessels) tridecapeptide, was discovered as a byproduct of substance P purification from bovine hypothalamus. Neurotensin is also present in CNS, gut, and associated glands. Besides its multiple hormonal actions, neurotensin injected intracisternally in mice induces an analgesic effect at doses $1/1000$ of those of enkephalin. Neurotensin is also a putative neurotransmitter, since it is found in whole-brain synaptosomes and is released from them in a Ca^{2+}-dependent fashion. Neurons with NT-like immunoreactivity are adundant in the hypothalamus (medial eminence, preoptic, and basal hypothalamic areas) and are also widespread in the CNS but are found neither in cerebellum nor in cerebral cortex.

The regional distribution of NT-positive neurons[402,405,438,441] is similar to that of enkephalin-positive neurons, but NT and enkephalin immunoreactivity are not to be found in the same neurons [e.g., the NT-containing neurons in the central nucleus of the amygdala, whose fibers pass through the ventrolateral stria terminalis and terminate at least in part in the interstitial (bed) nucleus of the stria terminalis, duplicate a similar pathway made of neurons and fibers that contain enkephalin]. The localization of NP in substantia gelatinosa and in the dorsal gray of spinal cord and in the substantia gelatinosa of the caudal portion of the trigeminal nuclei complex suggests a possible role of NT in pain perception and in the "pain-gating" phenomenon.

Other areas in which NT-immunoreactive neuronal perikarya are dense and NT-immunoreactive fibers and terminals are concentrated are the nucleus parabrachialis, the n. raphe dorsalis, the locus coeruleus (high density of NT neurons), the periaqueductal gray, and the ventral tegmental area of Tsai. Sparse NT-positive perikarya and rare fibers are observed in the dorsal cochlear nucleus; NT-positive fibers and terminals, but no NT-positive neuronal cell bodies, are detectable on the floor of the fourth ventricle and in some areas of the ventrolateral reticular formation.

Bombesin is a 14-amino acid peptide (isolated from the skin of certain European frogs) that stimulates gastrin release and has other effects on the gastrointestinal tract.[415] A bombesinlike immunoreactivity has been described in a peptide of the mucosa of mammalian gut that stimulates gastrin release. Gastrin-releasing peptides (GRP) with bombesinlike immunoreactivity have been synthesized and antisera raised against them; ICC with these antisera show bombesinlike immunoreactivity in fine varicose nerve terminal systems with a distribution similar to that of SP.[442] Since the immune serum used in these experiments does not react with SP, it would appear that a subpopulation of SP-containing fibers also contains GRP–bombesinlike immunoreactivity.

The second subgroup of gut–brain peptides consists of the peptides of the so-called secretin group since these share part of their amino acid sequence with secretin.[415] These peptides are secretin, glucagon, GIP, and VIP.[415]

Secretin[415] is a 17-amino-acid peptide (with similarities to glucagon) secreted by the mucosa of the small intestine. Secretin inhibits gastric acid secretion and smooth muscle contraction and stimulates pancreatic bicarbonate and water secretion. Enteroglucagon[281,415] is identical to pancreatic glucagon and is a 29-amino-acid peptide (although smaller and larger peptides with less glucagonlike activity but of unknown composition that only sometimes cross react immunologically with glucagon have also been described).

Gastric inhibitory peptide (GIP)[415] is an intestinal 43-amino-acid peptide (15 of which are the same and in the same position as in glucagon) that stimulates insulin release and, like glucagon, inhibits gastric and stimulates intestinal secretion. To our knowledge these three intestinal peptides have not yet been found in the CNS. In contrast, VIP,[443] the fourth peptide of the secretin group of peptides, is present not only in the intestinal and gastric mucosa but also in esophagoeal and rectal mucosa, pancreas, adrenal glands, neurons of the myenteric plexus, and in a wide range of CNS areas and numerous malignant tumors.[281,284,285,415] As its name indicates, vasoactive intestinal peptide (VIP) causes dilatation and in general relaxes smooth muscles, inhibits gastric secretions, stimulates intestinal and gastric secretions, insulin release, glycogenolysis, and lipolysis, and has other minor actions.[281,284,285,415]

As reviewed by Snyder and Innis,[285] in the peripheral nervous system, VIP has been found in sympathetic ganglia and in the myenteric plexus, where it appears that some VIP-containing fibers are intrinsic to the intestine. VIP immunoreactivity is also present in nerve fibers and terminals of pancreas and salivary glands. In peripheral nerve terminals, VIP is clearly localized in vesicles. In the CNS, VIP release from synaptosomes is K^+ stimulated and Ca^{2+} dependent.

If VIP is a neurotransmitter, it is an excitatory one. Biological effects of VIP appear to be mediated by cyclic AMP. The localization of VIP in the CNS by ICC[5,281,405,444–453] shows two aspects that differentiate VIP from the previously described neuropeptides: the first is that VIP-containing neuronal perikarya are present in the cerebral cortex (layers II–IV in the neocortex, in all layers in the paleocortex), where they account for 1–5% of the nerve cell bodies; the other is that most of these neurons are local, associative bipolar neurons oriented perpendicularly to the cortical surface. This suggests a function in activating or synchronizing neuronal activity within vertical columns of cerebral cortical cells. Although VIP is most concentrated in the cerebral cortex in general and in the hippocampus, this localization is not exclusive. A localization in hypothalamus and pituitary pars nervosa has been reported. Three long pathways of VIP-immunoreactive neurons are so far known: a major pathway from the amygdala to the bed nucleus and the hypothalamus via the stria terminalis,[450] another from the neurons of dorsal root ganglia to the dorsal horn, and, finally, a pathway from as yet unidentified neurons in the brainstem to the ipsilateral hypothalamus, n. accumbens, amygdala (central nucleus), and bed nucleus of stria terminalis via the medial forebrain bundle.[454] VIP is also found in nerve fibers around cerebral blood vessels, where it possibly exerts a vasodilatory function.

The third subgroup of brain–gut peptides consists of the gastrin–cholecystokinin peptides. Gastrin[415] is an intestinal peptide that exists in multiple forms: gastrin 17 (17 amino acids) is predominant in the gastric antral mucosa, whereas "big gastrin" or gastrin 34 predominates in blood. An even longer gastrin peptide is called component I-gastrin. The probable physiological effect of gastrin is to increase antral mobility; the other effects on stomach, pancreas and intestine are possibly of a pharmacological nature.[415] Gastrinlike immunoreactivity was reported in the CNS,[455] but subsequent studies have shown that gastrinlike peptides are limited to the magnocellular hypothalamic neurons that project to the neurohypophysis,[456] whereas when antigastrin immune sera were used, most of the CNS immunoreactivity was attributible to the presence of cholecystokinin, a peptide that shares with gastrin the first five amino acids at the C-terminal end.

Cholecystokinin (CCK)[415] is a peptide originally extracted from duodenum that owes its name to its action on the contractions of gallbladder (contractions provoked by the presence of lipids in the small intestine). Because of its stimulating action on the secretion of pancreatic enzymes, CCK was also called pancreozymin. A variety of other effects of CCK on the digestive tracts and associated glands are known.[415] Cholecystokinin is a peptide present in different sizes: counting from the C-terminus, the different CCK are CCK-33, CCK-8, and CCK-4; a form of CCK longer than CCK-33 is also known.[281,282,284,285,415]

In the brain, CCK-8 is the most abundant (80%), followed by CCK-4 (15%) and CCK-33 (5%). Recent studies indicate that, at least at the peripheral level, the receptors for CCK-4 are quite distinct from those for CCK-8. The amount of CCK in the whole human brain has been calculated as being of the order of 1–2 mg as compared to the microgram quantities of other peptides. The CCK levels in the cerebral cortex are also very high, and CCK is a rapid and powerful excitant of cerebral cortical cells.[281] Cholecystokinin-containing neuronal cell bodies and fibers are present, as expected, in the cerebral cortex but are also widespread in the CNS.[456–466] In the rabbit brain, a great number of cortical neurons react with anti-CCK immune sera,[458] but this is not the case for rat and guinea pig brain, where the distribution of CCK-containing neurons is more discrete: the highest CCK-positive cortical neuron concentration is in the pyriform cortex and mostly in layers II and III.

Another collection of CCK-positive cells and fibers was found in the hippocampus. Cholecystokinin, like VIP, is present in cerebral cortical neuronal perikarya, and both CCK and VIP might be major cortical neurotransmitters.[285,405] The CCK-containing neurons of cerebral cortex have the same morphological characteristics as VIP-containing cortical neurons,[405,459] i.e., bipolar neurons with the major axis perpendicular to the cortical surface. A subpopulation of amacrine cells in the retina contains CCK.[464] Amygdala contains CCK-positive fibers (very abundant in the central nucleus) but no positive neuronal perikarya. In the diencephalon, CCK neurons were found in the dorsomedial nucleus of the hypothalamus and in the lateral thalamus. In the brainstem, the most dense collection of CCK-positive cells and fibers is in the periaqueductal gray; neurons in the interstitial nucleus of the ventral tegmental decussation were less immunofluorescent. Cholecystokinin appears to coexist with dopamine in those brainstem neurons that project to the caudate.[462,466] Another group of CCK-positive perikarya was found in the caudal portion of the dorsal raphe nucleus, where many 5-HT-containing neurons are present, but 5-HT-containing and CCK-like immunoreactive neurons are different.[465] There are also CCK-positive cell bodies in the dorsal root ganglia and in terminals in the dorsal gray matter of the spinal cord.[285,402] Some authors suggest that CCK is also involved in the pain process.[285] Injected peripherally, CCK induces satiety in previously hungry rats, presumably by a peripheral action.[281,282,285]

The last group of gut peptides is a miscellaneous group in which we have included motilin, pancreatic peptides, and chymodenin. So far only the last two have also been found in brain, and, thus, they can be considered as gut–brain peptides; to our knowledge, no antiserum raised against chymodenin is yet available for ICC studies. Chymodenin[415] is a large (mol. wt. 5000) polypeptide of, so far, unknown amino acid composition extracted from porcine duodenum. It appears to stimulate pancreatic secretion and particularly that of chymotrypsinogen.

Motilin is a 22-amino-acid peptide contained in a cell population of enterochromaffin cells of duodenum, jejunum, and ileum.[415] It has no structural similarity to any known gut peptide. It appears to be released on alkalinization of the duodenum and to stimulate gastric secretion and contraction. An antiserum raised against porcine motilin directed against the midportion and the carboxyl end of the peptide was used in ICC.[256,467] In rat cerebellum, about 50% of Purkinje cells in the vermis contain motilin,[256] and in the hemisphere a much higher percentage was found.[467] It would appear that short or long ranks of motilin-containing Purkinje cells (motilinergic?) running rostrocordally in the sagittal plane alternate with GAD-containing Purkinje cells (GABAergic).[256,467] In coronal sections of cerebellum, typical symmetrical palisades of motilin-containing Purkinje cells are seen. It would appear that many Purkinje cells contain only motilin, whereas many others contain only GAD.[254] Motilin-containing terminals are not only seen, as expected, around the neurons of deep cerebellar and vestibular nuclei but also, unexpectedly, around neurons of other brainstem nuclei, suggesting the possibility of hitherto unknown direct projections of Purkinje cells to several brainstem nuclei.[467] Motilin-reactive cells were also observed in a few hypothalamic nuclei and pituitary gland. From a physiological point of view, the discovery of motilin in Purkinje cells is important, since motilin appears to be excitatory for cerebral cortex neurons, whereas GABA is inhibitory, and Purkinje cells have been considered, so far, to be inhibitory cells.

Pancreatic polypeptide is a 36-amino-acid polypeptide isolated from chicken pancreas as a byproduct of insulin purification. It was found to display its own hormone action alone or in synergism or antagonism with gastrin and secretin.[414] The amino acid sequence of this peptide, named avian pancreatic peptide (APP), was determined, and antisera raised against it.[468] A similar 36-amino-acid peptide was also extracted from bovine pancreas and named bovine pancreatic peptide (BPP). There is no great similarity in amino acid sequence between APP and BPP,[470] and there is no cross reactivity between the two different pancreatic peptides.[411] A peptide with a somewhat similar amino acid sequence to APP, called PYY, has been isolated from porcine intestine and pancreas,[470] but it too does not react with APP immune serum.[412]

Several CNS neurons were found to react with APP immune serum, but, as reported by Loren,[469] they do not react with anti-BPP immune serum. APP-like immunoreactivity[411,469–471,472] (APP-LI) was observed in neuronal cell bodies in the hippocampus, in the neocortex, in the striatum, in the lateral septum, in the nucleus accumbens, and in olfactory tubercles. Often, some of these neurons also reacted with antisomatostatin immune sera. In hypothalamus, where somatostatin-containing perikarya are present, no APP-LI neuronal cell perikarya but some APP-LI fibers were observed. This is similar to the pancreas, where somatostatin and APP-LI are in different cells. The APP-LI was found to be present within TH-positive neurons of the locus coeruleus, of the groups A1/A3 of a reticular formation, and of the vagal–nucleus solitarius complex (A2). Also, APP-LI terminals were observed in fibers in the periaqueductal gray, caudal raphe, arcuate nucleus, and locus coeruleus. A high density of axons and terminals with APP-LI that in part originate from intrinsic cell bodies was found within the dorsal horn of spinal cord and also around motor neurons in the sacral and lower lumbar spinal cord; APP-LI neurons were found in sympathetic lateral thoracid ganglia; APP and Met-enkephalinlike immunoreactivity coexist in sacral parasympathetic neurons.

The fifth group of neuropeptides consists of two peptides derived from a larger precursor present in the blood: one is angiotensin II; the other is bradykinin. Angiotensin II (Ang-II), an eight-amino-acid peptide, is a potent vasoconstrictor, and stimulates aldosterone release from the adrenal cortex.[281,282,285] Centrally administered Ang-II stimulates drinking behavior and raises blood pressure. Angiotensin II is formed by the enzymatic cleavage (angiotensin-converting enzyme) of a dipeptide from the decapeptide angiotensin I. Antiotensin I is produced by the activity of renin (or hypertensin), a protease first found in kidney, on angiotensinogen (a protein of the α_2-globulin fraction of blood plasma). Angiotensins, angiotensin receptors, renin, angiotensin-converting enzyme, and angiotensinase (the enzyme of angiotensin catabolism) have been detected in several CNS areas, particularly in hippocampus but also in cerebellum and midbrain.[281–283] The major problem for the ICC detection of Ang-like activity in the CNS is the fact that angiotensinase activity is quite high in the brain, and, thus, Ang might be rapidly proteolyzed during tissue processing. It must be emphasized that some data indicate that the antiotensinlike immunoreactivity does not correspond to the octapeptide but to a much larger molecule.[473] The ICC data[474–477] show Ang-II-like immunoreactivity in neuronal perikarya in paraventricular nuclei and perifornical areas of the hypothalamus and in terminals and fibers in periventricular gray, locus coeruleus, amygdala (central nucleus), spinal nuclei of trigeminal nerve, and substantia gelatinosa in the spinal cord. These appear to be terminals possibly originating from sensory neurons. Changaris *et al.*[478] also find Ang-II-like immunoreactivity in fiber projections in the hippocampus and in cerebellar white matter. Cerebellar deep nuclei appear to be studded with angiotensin-II-immunoreactive terminals.

Another way to study the presence of angiotensin I or II in neurons is to examine by ICC the localization of renin, the angiotensin-I-producing enzyme. However, the results obtained with two different antisera raised against mouse maxillary gland renin are contradictory.[479,480] Fuxe *et al.*[479] found reninlike immunoreactivity in some of the neurons (probably some of the VP- and/or OXT-containing neurons) in supraoptic and paraventricular nuclei of the hypothalamus as well as in fibers in the posterior pituitary gland. This corresponds to the localizations described for Ang-II. Also, some neurons in the medulla oblongata and Purkinje cells in cerebellum were labeled even in rats or mice not pretreated with colchicine. In contrast with these data, Inagami *et al.*[480] report a localization of reninlike immunoreactivity in subpopulations of oligodendrocytes and possibly of astrocytes in different brain areas.

Bradykinin (*BK*) is a nine-amino-acid peptide discovered[281,282,285] as a biologically active (causes slow contractions of guinea pig ileum) factor obtained by proteolysis (trypsin or *Bothrops jaraca* venom) of the α_2-globulin blood fraction. Other kinins such as BK-10 (H_2N-Lys-BK) and BK-11 (H_2N-Met-Lys-BK), all possibly derived from kininogen(s), are known. Bradykinin is centrally hypertensive, antinociceptive, and has several other central, peripheral, and behavioral effects, many of which must be mediated through other neurotransmitter substances.[281,284,285] (Circumstantial evidence indicates that BK might be the transmitter in kininergic neurons. In colchicine-injected rats, bradykininlike activity was found by ICC[481] to be localized only in hypothalamic neuronal cell bodies gathered in dense cell clusters overlying the periventricular and dorsomedial nuclei. From these neurons, fibers with large varicosities spread within the hypothalamus or reach the perirhinal and cingulate cortices, the ventral portion of caudate–putamen, the lateral septal area (where they could mediate the hypertensive action of centrally administered BK, and the periaqueductal gray (where they could mediate the analgesic action of centrally administered BK).

The sixth group of neuropeptides includes a miscellaneous group of peptides against which, to our knowledge, no antibodies have been raised or at least not used in ICC. Among these are the several learning-related peptides (scotophobin, amelitin, and catabathmophobin discovered by Ungar *et al.*) and the sleep-inducing peptides (for review see ref. 282). Another peptide, which is the only neuropeptide not produced by proteolysis of a larger protein precursor, is the dipeptide carnosine (β-alanylhistidine).[281,282,285] Carnosine is synthesized directly by carnosine synthetase (an ATP-requiring enzyme which condenses β-alanine and L-histidine) and is destroyed by carnosinase (a peptidase). Carnosine, anserine, (β-alanyl-3-methylhistidine), and homocarnosine (γ-aminobutyrylhistidine) are present in skeletal muscles and brain of many vertebrates.[281–283] Carnosine, carnosine-binding sites (stereospecific), carnosine synthetase, and carnosinase are highly concentrated in the olfactory bulb;

$ZnSO_4$-induced olfactory deafferentiation reduces the level of the two enzymes in the olfactory bulb in parallel with the decrease of the specific olfactory bulb protein. This and other data point to the possible role of carnosine as neurotransmitter of the olfactory receptor neurons.[482–484]

Recently,[485] studies on the processing of RNA transcripts of the cloned gene encoding for calcitonin (a 32-amino-acid peptide hormone secreted by cells, presumably APUD cells, located in the thyroid) showed that the RNA processing is different depending on the tissue: in thyroid tissue, the processing of the RNA transcribed from the calcitonin gene produces predominantly a mRNA coding for calcitonin; in hypothalamus, the processing of the RNA transcribed from the same gene produces a partially different mRNA, which codes for a protein with a 76-amino-acid peptide segment identical to preprocalcitonin, whereas the segment corresponding to calcitonin is replaced by another peptide sequence in which 37 amino acids account for the so-called calcitonin gene-releated peptide (CGRP). This is a good example of the possibilities of RNA splicing in neuropeptide preprohormone synthesis. If CGRP peptide turns out to be a neurohormone or a neuromodulator, it will represent the first example of a hormone whose genetic coding was known even before the hormone was known.

To conclude this section on neurotransmitters as neuronal markers, we recall other examples of studies in which, in the same (or in serial) tissue section(s), neuropeptides revealed by ICC and neurotransmitters revealed by histochemistry were simultaneously observed. These examples can be found in references.[486–491] We also refer the reader to ref. 492 for an example of methods that combine ICC with retrograde transport of fluorescent dyes for tracing biochemically defined neuronal pathways.

3.4. Antigens Present in Several Cell Types

The glycoprotein Thy-1, deriving its name from its presence on thymocytes, has been widely used in tissue cultures as a marker for fibroblasts.[493] Some neurons[494–497] also appear to express the antigen, but this is not found on the majority of leptomeningeal cells or on oligodendrocytes or astrocytes in short-term cultures. In long-term cultures, astrocytes appear to express the molecule.[86] In contrast, the anti-Thy-1 immune sera used for ICC in cerebellar sections label the surface of cerebellar neurons and their processes in the adult and during postnatal development; the label in the pia and arachnoid disappears after day 16 postnatally.[497]

Cathepsin D is an acid peptidase that probably also plays a role in neuropeptide function. Cathepsin D immunoreactivity has been localized by ICC mainly in neurons of all CNS regions, but also in oligodendrocytes and choroid plexus epithelial cells. In the cytoplasm of all of these cells, cathepsin D is distributed in a granular (putative lysosomal) pattern.[498]

Guanylate cyclase [GTP pyrophosphate-lyase (cyclizing), E.C. 4.6.1.2] has been localized, by ICC, in rat cerebella and found to be present in all cells, both neuronal and glial.[499] In contrast, a recent ICC study carried out with four monoclonal antibodies against different antigenic determinants of guanylate cyclase together indicate that in cerebella, the guanylate cyclase immunoreactivity is present primarily in Purkinje cells and their dendrites and also in other cells including glial cell processes.[500] The immunoreactivity was detected in caudate–putamen mainly within medium spiny neurons and in neocortex within pyramidal cells. Many other structures, however, except neuronal fibers, were also immunoreactive.

The use of ICC with immune sera raised against whole rat blood serum proteins showed accumulation of immunoperoxidase reaction products not only, as expected, in perivascular pericytes and in the neuropil of brain areas outside the blood–brain barriers (see "the windows" in Section 3.3.3c) but also inside the neurons. Apparently, neurons projecting to the above-described area pick up blood proteins by endocytosis that, by retrograde axonal transport, reach the neuronal perikarya.[501]

Biochemical data support the concept that gangliosides are constituents of nerve endings,[502] astroglia,[503,504] and myelin.[505] ICC with antibodies against gangliosides has shown surface labeling of Purkinje cells and granule cells in the cerebellum, though some glial cells were also immunoreactive.[506]

4. *MOLECULES OF CELL ORGANELLES REVEALED BY IMMUNOCYTOLOGY WITH POLYCLONAL ANTIBODIES*

4.1. *Protein I*

Protein I is a principal natural substrate for cyclic-AMP- and Ca^{2+}-dependent protein kinases of mammalian brain. With antisera against this protein, immunoreactivity has been detected as punctate deposits in certain regions rich in nerve terminals.[507] Variations in staining intensity among different zones were observed. At the electron microscope level,[508] reaction product was found to be associated with the perimeter of synaptic vesicles and with submembranous material in the postsynaptic neuron.

4.2. *Calmodulin*

Antisera to both calmodulin, a Ca^{2+}-dependent modulator protein abundant in brain,[509] and a heat-labile calmodulin-binding protein have been used to localize these proteins in mouse caudate–putamen. Whereas calmodulin is widely distributed outside the CNS, in the brain, it appears to be confined to neuronal elements, primarily at postsynaptic sites within neuronal somata and dendrites.[510]

4.3. *Synaptic Vesicle Protein*

A polyclonal antibody produced against synaptic vesicles purified from the electric organ of *Narcine brasiliensis* has been found to cross react with nerve terminals in mammalian CNS and the PNS,[511] lending support to the concept that synaptic vesicle antigenic determinants are well conserved in evolution.

4.4. Postsynaptic Density Antigens

When large-scale preparations of bovine postsynaptic densities (PSD) were used to elicit antibodies in rabbits, few of the polypeptide components proved to be good immunogens.[512] Protein I, which is present in PSD, appeared to have been lost during subcellular fractionation. The major antigen (a 95-kilodalton protein), representing less than 0.3% of total PSD protein, was found in mouse and rat brain to be predominantly or exclusively localized in the PSD.[512]

5. MISCELLANEOUS POLYCLONAL ANTIBODIES

Several polyclonal antisera to neural tumoral cell lines have been used to differentiate between cultured cells.[513] One antiserum recognizing cell lines with both neuronal and glial properties called anti-NG2 recognized a cell surface glycoprotein of high molecular weight.[514] Subsequently, a monoclonal antibody recognizing the same molecule was isolated. A battery of antisera prepared against either whole neural tumor cells or brain particulate fractions have been studied by ICC in the cerebellum.[515] Three of these, anti-NS-2, anti-NS-3, and anti-NS-4, showed certain similarities. Anti-NS-2 and anti-NS-3 labeled both neurons and glia.

6. MONOCLONAL ANTIBODIES

The inherent limitations of conventional antibodies have forced neurobiologists to turn to new techniques for obtaining monospecific antibodies. Although a pure antigen preparation is the general requisite for a monospecific polyclonal antibodies, immunogens consisting of a multitude of antigenic components may be used to elicit monoclonal antibodies by the techniques developed by Kohler and Milstein.[6] Although the use of such techniques in the field of neurobiology is still in its infancy, a large catalogue of different antibodies has already been obtained in different laboratories,[516–529] which define antigens present in the nervous system. An enormous effort will be required to characterize biochemically the antigens discovered by such applications; however, this chapter would be incomplete without a passing mention of some of the more interesting antibodies so far studied.

Barnstable,[516] using crude membrane fractions from adult rat retina, obtained a battery of monoclonal antibodies. Some had specificity against photoreceptor cells, others against Müller cells, the radial glia of retinal tissue, and one was directed against adult rat and retinal neurons. The latter also labeled some neurons in cerebral cortex and cerebellum.

Dissociated cerebellar cultures and crude cerebellar membranes have frequently been chosen as immunogens for the production of monoclonal antibodies. Thus, an antibody called A4 (elicited by cultured cerebellar cells as immunogens) has been described[517]; A4 recognized a surface molecule common to all CNS neurons examined but absent in PNS neurons. In contrast, another antibody (elicited by dorsal root ganglia cells as immunogen) called 38/D7 specifically labeled PNS neurons but not fibroblasts, Schwann cells, or CNS neurons.[518]

Two antibodies called BSP-2 and BSP-3 (elicited either by cultures or by cerebellar glycoproteins as immunogens) recognized antigens that were expressed by both astrocytes and neurons in the cerebellum.[519–521] One of these, BSP-2, appeared to be identical to the D-2 antigen studied by polyclonal antibodies.[521]

Another antibody called C 4/12 recognized an antigen present both in astrocytes and granule cell perikarya in cerebellar cell cultures, but in cerebellar slices the antigen was detected only in granule cell bodies.[522]

A monoclonal antibody capable of distinguishing between subpopulations of astrocytes was obtained by Schachner *et al.*[523] using crude cerebellar membranes, and others (called O_1, O_2,O_3, and O_4) against oligodendrocytes and against astrocytes (called C_1) were elicited by immunization with bovine corpus callosum. Many of these antibodies demonstrated the presence of common antigens on different cell types.[523]

An antibody (A2B5) produced by immunization with fixed chick retinal cells appears to recognize a tetrasialoganglioside in all neuronal perikarya of chicken retina but not in neuronal processes or in other nonneuronal cells of the retina.[524] Also, other CNS area bound the antibody.[524] A2B5 also labeled peptide-producing endocrine cells in different glands. This is in agreement with the suggested common embryological origin of peptide-producing endocrine cells and neurons (see Section 3.3.c).[525]

Using synaptosomal plasma membrane and synaptic junction subcellular fractions as immunogens, Matthew *et al.*[526] produced and isolated several monoclonal antibodies. One recognizing a minor 65 K component of SPM was found by immunoelectron microscopy to label both synaptic vesicles and SPM in the cerebellum and also chromaffin granules in the adrenal medulla. It was proposed that this protein is present on all vertebrate neurosecretory vesicles. But sometimes the results obtained even with relatively pure subcellular fractions have been unexpected. For example, synaptosomal plasma membrane fractions from bovine hippocampus were used as antigens, and one of the monoclonal antibodies obtained reacted with galactocerebroside, a good oligodendrocyte marker. The development of oligodendrocyte and Schwann cells in culture *in vitro* has been studied by using this antigen in ICC.[527]

It is not surprising that the use of relatively crude fractions as immunogens for monoclonal antibody production occasionally produces antibodies with unexpected specificity. Thus, among several antibodies obtained by Matus, one recognized an antigen present on mitochondria[528] that appears in cerebellar neurons only after their final division and migration. In another series, a single monoclonal antibody recognized a sequence common to more than 30 proteins.[100]

A monoclonal antibody (UCHT1) generated against human T-lymphocytes not only is a good T-cell marker but also specifically stains, in ICC, Purkinje cells of different animal species.[529]

Mesa-I, the monoclonal antibody recognizing an antigen specific to mouse endothelial cells, has been already mentioned.[93]

7. CONCLUSION

The immense contribution made by immunohistochemistry to neurobiology extends across many domains. It initially proved to be and continues as a powerful tool for cellular identification and for the recognition of subtypes of cells, problems that are most acute in the developing brain and in cultures in which adult morphology is not expressed. The purity of neural cultures may now be routinely verified with the aid of cell-type-specific markers. In the field of neuropathology, cells may be categorized by their immunohistochemistry rather than their morphology. Whereas the functional roles of some of the first discovered "brain-specific" proteins continue to remain an enigma, those of others have been, by contrast, well defined. This is particularly true for the neurotransmitter-associated enzymes. Considerable advances have also been made on functional aspects of the neuropeptides. The potential of immunohistochemistry in mapping nerve pathways related to a particular neurotransmitter or peptide has been exploited to provide more information on the functioning of the brain in health and disease. Recent applications of monoclonal antibodies provide further exciting possibilities for the discovery of molecules that play fundamental roles in both the function and development of the nervous system. Immunohistochemistry continues to provide essential data on the tissue localization of such molecules. If it is agreed that in recent years immunohistochemical study of the nervous system has been rewarding, there is every reason to believe that its future will reap a bumper harvest.

REFERENCES

1. Moore, B. W., 1965, *Biochem. Biophys. Res. Commun.* **19**:739–744.
2. Moore, B. W., and McGregor, S., 1965, *J. Biol. Chem.* **240**:1647–1653.
3. Ghandour, M. S., Langley, O. K., Vincendon, G., and Gombos, G., 1979, *J. Histochem. Cytochem.* **27**:1634–1637.
4. Pickel, V. M., 1979, *Fed. Proc.* **38**:2374–2380.
5. Hökfelt, T., Johansson, O., Ljungdahl, A., Lundberg, J. M., and Schultzberg, M., 1980, *Nature* **284**:515–521.
6. Köhler, G., and Milstein, C., 1975, *Nature* **356**:495–497.
7. Coons, A. H., Creech, H. J., and Jones, R. N., 1941, *Proc. Soc. Exp. Biol.* **47**:200–202.
8. Coons, A. H., and Kaplan, M. H., 1950, *J. Exp. Med.* **91**:1–13.
9. Riggs, J. L., Loh, P. C., and Eveland, W. C., 1960, *Proc. Soc. Exp. Biol.* **105**:655–658.
10. Coons, A. H., 1961, *J. Immunol.* **87**:499–503.
11. Nairn, R. C., 1969, *Fluorescent Protein Tracing*, Churchill Livingstone, Edinburgh, p. 503.
12. Sternberger, L. A., 1979, *Immunocytochemistry*, John Wiley & Sons, New York.
13. Nakane, P. K., and Pierce, B. G., Jr., 1967, *J. Cell Biol.* **33**:307–318.
14. Avrameas, S., 1969, *Immunochemistry* **6**:43–52.
15. Graham, R. C., Jr., and Karnovsky, M. J., 1966, *J. Histochem. Cytochem.* **14**:291–302.
16. Nakane, P. K., 1968, *J. Histochem. Cytochem.* **16**:557–560.
17. Massayeff, R., and Maiolini, R., 1975, *J. Immunol. Methods* **8**:223–234.

18. Mason, D. Y., and Sammons, R. E., 1978, *J. Clin. Pathol.* **31**:454–460.
19. Vandesande, F., 1979, *J. Neurosci. Methods* **1**:3–23.
20. Striker, G. E., Donati, E. J., Petrali, J. P., and Sternberger, L. A., 1966, *Exp. Mol. Pathol.* [*Suppl.*] **3**:52–58.
21. Singer, S. J., 1975, *Immunochemistry* **12**(6–7):615–616.
22. Tokoyasu, T. K., and Singer, S. J., 1976, *J. Cell Biol.* **71**:894–906.
23. Roth, J., Bendayan, M., and Orci, L., 1978, *J. Histochem. Cytochem.* **26**:1074–1081.
24. Roth, J., Bendayan, M., and Orci, L., 1980, *J. Histochem. Cytochem.* **28**:55–57.
25. Dubois-Dalcq, M., McFarland, H., and McFarlin, D., 1977, *J. Histochem. Cytochem.* **25**:1201–1206.
26. Weller, T. H., and Coons, A. H., 1954, *Proc. Soc. Exp. Biol.* **86**:789–794.
27. Sternberger, L. A., 1969, *Mikroskopie* **25**:346–361.
28. Mason, T. E., Phifer, R. F., Spicer, S. S., Swallow, R. A., and Dreskin, R. B., 1969, *J. Histochem. Cytochem.* **17**:563–569.
29. Langley, O. K., and Ghandour, M. S., 1981, *Histochem. J.* **13**:137–148.
30. Swaab, D. F., Pool, C. W., and van Leeuwen, F. W., 1977, *J. Histochem. Cytochem.* **25**:388–391.
31. Pruss, R. M., Mirsky, R., Raff, M. C., Thorpe, R., Dowding, A. J., and Anderson, B. H., 1981, *Cell* **27**:419–428.
32. Bigbee, J. W., Kosek, J. C., and Eng, L. F., 1977, *J. Histochem. Cytochem.* **25**:443–447.
33. Vandesande, F., Dierickx, K., and DeMey, J., 1977, *Cell Tissue Res.* **180**:443–452.
34. Heyderman, E., and Monaghan, P., 1979, *Invest. Cell. Pathol.* **2**:119–122.
35. Rusca, G., Calissano, P., and Alema, S., 1972, *Brain Res.* **49**:223–227.
36. Haglid, K. G., Hamberger, A., Hansson, H. A., Hydén, H., Persson, L., and Rönnbäck, L., 1974, *Nature* **251**:523–534.
37. Haglid, K. G., Hamberger, A., Hansson, H. A., Hydén, H., Persson, L., and Rönnbäck, L., 1975, *Nature* **258**:748–749.
38. Haglid, K. G., Hamberger, A., Hansson, H. A., Hydén, H., Persson, L., and Rönnbäck, L., 1976, *J. Neurosci. Res.* **2**:175–191.
39. Donato, R., and Michetti, F., 1974, *Experientia* **30**:511–512.
40. Tabuchi, K., and Kirsch, W. N., 1975, *Brain Res.* **92**:175–180.
41. Hydén, H., and Rönnbäck, L., 1975, *Brain Res.* **100**:615–628.
42. Vacca, L., and Hutchings, D., 1977, *Dev. Psychobiol.* **10**:171–176.
43. Matus, A., and Mughal, S., 1975, *Nature* **258**:746–748.
44. Ghandour, M. S., Labourdette, G., Vincendon, G., and Gombos, G., 1981, *Dev. Neurosci.* **4**:98–109.
45. Legrand, C., Clos, J., Legrand, J., Langley, O. K., Ghandour, M. S., Labourdette, G., Gombos, G., and Vincendon, G., 1981, *Neuropathol. Appl. Neurobiol.* **7**:299–306.
46. Ghandour, M. S., Langley, O. K., Labourdette, G., Vincendon, G., and Gombos, G., 1981, *Dev. Neurosci.* **4**:66–78.
47. Cocchia, D., and Michetti, F., 1980, *Synaptic Constituents in Health and Disease* (M. Brzin, D. Sket, and H. Bachelard, eds.), Pergamon Press, New York, p. 258.
48. Linser, P., and Moscona, A. A., 1981, *Dev. Neurosci.* **4**:433–441.
49. Eng, L. F., Kosek, J. C., Fornd, L., Deck, J., and Bigbee, J., 1976, *Trans. Am. Soc. Neurochem.* **7**:211.
50. Eng, L. F., Gerstt, B., and Vanderhaegen, J. J., 1970, *Trans. Am. Soc. Neurochem.* **1**:42.
51. Eng, L. F., Vanderhaegen, J. J., Bignami, A., and Gerstt, B., 1971, *Brain Res.* **28**:351–354.
52. Dahl, D., and Bignami, A., 1973, *Brain Res.* **57**:343–360.
53. Eng, L. F., and Bigbee, J. W., 1978, *Advances in Neurochemistry*, Volume 3 (B. W. Agranoff and M. H. Aprison, eds.), Plenum Press, New York, pp. 43–97.
54. Palfreman, J. W., Thomas, D. G. T., Ratcliffe, J. G., and Graham, D. I., 1979, *J. Neurol. Sci.* **41**:101–113.
55. Delpech, B., and Delpech, A., 1975, *Immunochemistry* **12**:691–697.
56. Adriaenssens, K., Lowenthal, A., Karcher, D., Mardens, Y., van Sande, M., and van Heule, R., 1968, *Pathol. Eur.* **3**:194–199.
57. Knoppe, M., 1980, *Etude Biochemique et Radioimmunochimique d'une Protéine Spécifique du Système Nerveux: l'α-albumine (GFA)*, Thesis, Université Louis Pasteur, Strasbourg.

58. Schachner, M., Hedley-Whyte, E. T., Hsu, D. W., Schoonmaker, G., and Bignami, A., 1977, *J. Cell Biol.* **75**:67–73.
59. Dahl, D., Rueger, D. C., Bignami, A., Weber, K., and Osborn, M., 1981, *Eur. J. Cell Biol.* **24**:191–196.
60. Warecka, K., and Bauer, H. J., 1967, *J. Neurochem.* **14**:783–787.
61. Langley, O. K., Ghandour, M. S., Vincendon, G., Gombos, G., and Warecka, K., 1982, *J. Neuroimmunol.* **2**:131–143.
62. Marangos, P. J., Zis, A. P., Clark, R. L., and Goodwin, F. K., 1978, *Brain Res.* **150**:117–133.
63. Zomzely-Neurath, C., and Keller, A., 1977, *Mechanisms, Regulation and Special Function of Protein Synthesis in the Brain* (S. Roberts, A. Lajtha, and W. H. Gispen, eds.), Elsevier, Amsterdam, pp. 279–298.
64. Schmechel, D. E., Marangos, P. J., Zis, A. P., Brightman, M. W., and Goodwin, F. K., 1978, *Science* **199**:313–314.
65. Langley, O. K., Ghandour, M. S., Vincendon, G., and Gombos, G., 1980, *J. Neurocytol.* **9**:783–798.
66. Schmechel, D. E., Brightman, M. W., and Marangos, P. J., 1980, *Brain Res.* **190**:195–214.
67. Ghandour, M. S., Langley, O. K., and Keller, A., 1981, *Exp. Brain Res.* **41**:271–279.
68. Norenberg, M. D., 1979, *J. Histochem. Cytochem.* **27**:756–762.
69. Norenberg, M. D., and Martinez-Hernandez, A., 1979, *Brain Res.* **161**:303–310.
70. Giacobini, E., 1961, *Science* **134**:1524–1525.
71. Ghandour, M. S., Langley, O. K., Vincendon, G., Gombos, G., Filippi, D., Limozin, N., Dalmasso, C., and Laurent, G., 1980, *Neuroscience* **5**:559–571.
72. Langley, O. K., Ghandour, M. S., Vincendon, G., and Gombos, G., 1980, *Histochem. J.* **12**:473–483.
73. Leveille, P. J., McGinnis, J. F., Maxwell, D. S., and de Vellis, J., 1980, *Brain Res.* **196**:287–305.
74. Sternberger, N. H., Itoyama, Y., Kies, M. W., and Webster, H. de F., 1978, *J. Neurocytol.* **7**:251–263.
75. Sternberger, N. H., Itoyama, Y., Kies, M. W., and Webster H. de F., 1978, *Proc. Natl. Acad. Sci. U.S.A.* **75**:2521–2524.
76. Roussel, G., Delaunoy, J. P., Nussbaum, J. L., and Mandel, P., 1977, *Neuroscience* **2**:307–313.
77. Roussel, G., and Nussbaum, J. L., 1981, *Histochem. J.* **13**:1029–1047.
78. Agrawal, H. C., Hartman, B. K., Shearer, W. T., Kalmbach, S., and Margolis, F. L., 1977, *J. Neurochem.* **28**:495–508.
79. Poduslo, S. E., and Norton, W. T., 1972, *J. Neurochem.* **19**:727–736.
80. Nishizawa, Y., Kurihara, T., and Takahashi, Y., 1981, *Brain Res.* **212**:219–222.
81. Drummond, R. J., and Dean, G., 1980, *J. Neurochem.* **35**:1155–1165.
82. Sprinkle, T. J., Wells, M. R., Garver, F. A., and Smith, D. B., 1980, *J. Neurochem.* **35**:1200–1208.
83. Zanetta, J.-P., Sarliève, L. L., Mandel, P., Vincendon, G., and Gombos, G., 1977, *J. Neurochem.* **29**:827–838.
84. Sternberger, N. H., Quarles, R. H., Itoyama, Y., and Webster H. de F., 1979, *Proc. Natl. Acad. Sci. U.S.A.* **76**:1510–1514.
85. Raff, M. C., Mirsky, R., Fields, K. L., Lisak, R. P., Dorfman, S. H., Silberberg, D. H., Gregson, N. A., Leibowitzs, S. L., and Kennedy, M. G., 1978, *Nature* **274**:813–816.
86. Raff, M. C., Fields, K. L., Hakomori, S. I., Mirsky, R., Pruss, R. M., and Winter, J. 1979, *Brain Res.* **174**:283–308.
87. Zalc, B., Monge, M., Dupouey, P., Hauw, J. J., and Baumann, N. A., 1981, *Brain Res.* **211**:341–354.
88. Schachner, M., 1974, *Proc. Natl. Acad. Sci. U.S.A.* **71**:1795–1799.
89. Ghandour, M. S., Langley, O. K., and Varga, V., 1980, *Neurosci. Lett.* **20**:125–129.
90. Shine, H. D., and Haber, B., 1981, *Brain Res.* **217**:339–349.
91. Wilkin, G. P., Woodhams, P. L., and dos Santos, R. R., 1981, *Dev. Neurosci.* **4**:296–306.
92. Schachner, M., Schoonmaker, G., and Hynes, R. O., 1978, *Brain Res.* **158**:149–158.

93. Ghandour, M. S., Langley, K., Gombos, G., Hirn, M., Hirsch, M.-R., and Goridis, G., 1982, *J. Histochem. Cytochem.* **30**:165–170.
94. Bock, E., and Dissing, J., 1975, *Scand. J. Immunol.* [*Suppl.*] **24**:31–36.
95. Schmechel, D., Marangos, P. J., and Brightman, M., 1978, *Nature* **276**:834–836.
96. Jackson, P., and Thompson, R. J., 1980, *Biochem. Soc. Trans.* **8**:616–617.
97. Anderton, B. H., Thorpe, R., Cohen, J., Selvendram, S., and Woodhams, P., 1980, *J. Neurocytol.* **9**:835–844.
98. Matus, A. I., Meelian, N. G., and Hugh Jones, D., 1979, *J. Neurocytol.* **8**:513–525.
99. Yen, S.-H., and Fields, K. L., 1981, *J. Cell Biol.* **88**:115–126.
100. Hawkes, R., Niday, E., and Matus, A., 1982, *Proc. Natl. Acad. Sci* **79**:2410–2414.
101. Matus, A., Bernhardt, R., and Hugh-Jones, T., 1981, *Proc. Natl. Acad. Sci. U.S.A.* **78**:3010–3014.
102. Bernhardt, R., and Matus, A., 1982, *J. Cell Biol.* **92**:589–593.
103. Jørgensen, O. S., and Bock, E., 1974, *J. Neurochem.* **23**:879–880.
104. Jørgensen, O. S., and Møller, M., 1980, *Brain Res.* **194**:419–429.
105. Jørgensen, O. S., and Møller, M., 1981, *Brain Res.* **221**:15–26.
106. Hirn, M., Ghandour, M. S., Deagostini-Bazin, H., and Goridis, C. 1983, *Brain Res.* **265**:87–100.
107. Hirn, M., Pierres, M., Deagostini-Bazin, H., Hirsch, M. R., and Goridis, C., 1981, *Brain Res.* **214**:433–439.
108. Langley, O. K., Ghandour, M. S., Gombos, G., Hirn, M., and Goridis, C., 1982, *Neurochem. Res.* **7**:349–362.
109. Margolis, F. L., 1972, *Proc. Natl. Acad. Sci. U.S.A.* **69**:1221–1224.
110. Hartman, B. K., and Margolis, F. L., 1975, *Brain Res.* **96**:176–180.
111. Monti Graziadei, G. A., and Graziadei, P. P. C., 1981, *Brain Res.* **209**:405–410.
112. Mallet, J., Huchet, M., Pougeois, R., and Changeux, J.-P., 1975, *FEBS Lett.* **52**:216–220.
113. Mikoshiba, K., Huchet, M., and Changeux, J.-P., 1979, *Dev. Neurosci.* **2**:254–275.
114. Reeber, A., Vincendon, G., and Zanetta, J.-P., 1981, *Brain Res.* **229**:53–65.
115. Langley, O. K., Reeber, A., Vincendon, G., and Zanetta, J.-P., 1982, *J. Comp. Neurol.* **208**:335–344.
116. Lohmann, S. M., Walter, U., Miller, P. E., Greengard, P., and de Camilli, P., 1981, *Proc. Natl. Acad. Sci. U.S.A.* **78**:653–657.
117. Roth, J., Baetens, D., Norman, A. W., and Garcia-Segura, L.-M., 1981, *Brain Res.* **222**:452–457.
118. Jande, S. S., Tolnai, S., and Lawson, D. E. M., 1981, *Histochemistry* **71**:99–116.
119. Falck, B., Hillarp, N. Å., Thieme, G., and Torp, A., 1962, *J. Histochem. Cytochem.* **10**:348–354.
120. Falck, B., and Owman, C., 1965, *Acta Univ. Lund II* **7**:1–23.
121. Lindvall, O., Björklund, A., Höckfelt, T., and Ljungdahl, Å., 1973, *Histochemie* **35**:31.
122. Lindvall, O., and Björklund, A., 1974, *Histochemistry* **39**:97–127.
123. de la Torre, J. C., and Surgeon, J. W., 1976, *Neuroscience* **1**:451–453.
124. van Gelder, N. M., 1965, *J. Neurochem.* **12**:231–237.
125. van Gelder, N. M., 1965, *J. Neurochem.* **12**:239–244.
126. Kuriyama, K., Sisken, B., Ito, J., Simonsen, D. G., Haber, B., and Roberts, E., 1968, *Brain Res.* **11**:412–430.
127. Hyde, J. C., and Robinson, N., 1974, *Brain Res.* **82**:109–116.
128. Davies, W. E., 1975, *Brain Res.* **83**:27–33.
129. Park, D. H., and Goldstein, M., 1976, *Life Sci.* **18**:55–60.
130. Hökfelt, T., Johansson, O., Fuxe, K., Goldstein, M., and Park, D., 1976, *Med. Biol.* **54**:427–453.
131. Hökfelt, T., Johansson, O., Fuxe, K., Goldstein, M., and Park, D., 1977, *Med. Biol.* **55**:21–40.
132. Ljungdahl, A., Hökfelt, T., Nilsson, G., and Goldstein, M., 1978, *Neuroscience* **3**:945–976.
133. Johansson, O., Hökfelt, T., Biberfeld, P., and Goldstein, M., 1976, *J. Ultrastruct. Res.* **54**:478–479.
134. Verhofstad, A. A. J., Hökfelt, T., Goldstein, M., Steinbusch, H. W. M., and Joosten, H. W. J., 1979, *Cell Tissue Res.* **200**:1–14.

135. Halász, N., Ljungdahl, A., Hökfelt, T., Johansson, O., Goldstein, M., Park, D., and Biberfeld, P., 1977, *Brain Res.* **126**:455–474.
136. Halász, N., Johansson, O., Hökfelt, T., Ljungdahl, Å., and Goldstein, M., 1981, *J. Neurocytol.* **10**:251–259.
137. Pearson, J., Goldstein, M., and Brandeis, L., 1979, *Brain Res.* **165**:333–337.
138. Pearson, J., Brandeis, L., and Goldstein, M., 1980, *Dev. Neurosci.* **3**:140–150.
139. Markey, K. A., Kondo, S., Shenkman, L., and Goldstein, M., 1980, *Mol. Pharmacol.* **17**:79–85.
140. Hökfelt, T., Skirboll, L., Rehfeld, J. F., Goldstein, M., Markey, K., and Dann, O., 1980, *Neuroscience* **5**:2093–2124.
141. Stephens, J. K., Masserano, J. M., Vulliet, P. R., Weiner, N., and Nakane, P. K., 1981, *Brain Res.* **209**:339–354.
142. Berod, A., 1978, *Purification de la Tyrosine Hydroxylase: Applications Immunochimiques au Niveau des Neurones Catécholaminergiques Centraux*, Thèse de 3ème cycle à l'Université Claude Bernard, Lyon.
143. Swanson, L. W., Sawchenko, P. E., Bérod, A., Hartman, B. K., Helle, K. B., and van Orden, D. E., 1981, *J. Comp. Neurol.* **196**:271–285.
144. Wassef, M., Berod, A., and Sotelo, C., 1981, *Neuroscience* **6**:2125–2139.
145. Nagatsu, I., Inagaki, S., Kondo, Y., Karasawa, N., and Nagatsu, T., 1979, *Acta Histochem. Cytochem.* **12**:20–37.
146. Nagatsu, I., Kondo, Y., Inagaki, S., Karasawa, N., Kato, T., and Nagatsu, T., 1977, *Acta Histochem. Cytochem.* **10**:494–499.
147. Joh, T. H., Geghman, C., and Reis, D., 1973, *Proc. Natl. Acad. Sci. U.S.A.* **70**:2767–2771.
148. Joh, T. H., and Reis, D. J., 1975, *Brain Res.* **85**:146–151.
149. Pickel, V. M., Joh, T. H., Field, P. M., Becker, C. G., and Reis, D. J., 1975, *J. Histochem. Cytochem.* **23**:1–12.
150. Pickel, V. M., Joh, T. H., and Reis, D. J., 1976, *J. Histochem. Cytochem.* **24**:792–806.
151. Pickel, V. M., Joh, T. H., and Reis, D. J., 1977, *Brain Res.* **131**:197–214.
152. Reis, D. J., Gilad, G., Pickel, V. M., and Joh, T. H., 1978, *Brain Res.* **144**:325–342.
153. Specht, L. A., Pickel, V. M., Joh, T. H., and Reis, D. J., 1978, *Brain Res.* **156**:315–321.
154. Armstrong, D. M., Pickel, V. M., Joh, T. H., Reis, D. J., and Miller, R. J., 1981, *J. Comp. Neurol.* **196**:505–517.
155. Hervonen, A., Pickel, V. M., Joh, T. H., Reis, D. J., Linnoila, I., and Miller, R. J., 1981, *Cell Tissue Res.* **214**:33–42.
156. Pickel, V. M., Specht, L. A., Sumal, K. K., Joh, T. H., Reis, D. J., and Hervonen, A., 1980, *J. Comp. Neurol.* **194**:465–474.
157. Specht, L. A., Pickel, V. M., Joh, T. H., and Reis, D. J., 1981, *J. Comp. Neurol.* **199**:233–254.
158. Specht, L. A., Pickel, V. M., Joh, T. H., and Reis, D. J., 1981, *J. Comp. Neurol.* **199**:255–276.
159. Pickel, V. M., Beckley, S. C., Joh, T. H., and Reis, D. J., 1981, *Brain Res.* **225**:373–385.
160. Specht, L. A., Pickel, V. M., Joh, T. H., and Reis, D. J., 1981, *Brain Res.* **218**:49–66.
161. Pickel, V. M., Joh, T. H., and Reis, D. J., 1975, *Proc. Natl. Acad. Sci. U.S.A.* **72**:659–663.
162. Pickel, V. M., 1978, *Electron Microscopy of Enzymes*, Volume 5, *Principles and Methods* (M. A. Hayat ed.), van Nostrand Reinhold, New York, pp. 147–161.
163. Ross, M. E., Reis, D. J., and Joh, T. H., 1981, *Brain Res,* **208**:493–498.
164. Goldstein, M., 1972, *Research Methods in Neurochemistry*, Volume 1 (N. Marks and R. Rodnight, eds.), Plenum Press, New York, pp. 317–340.
165. Hökfelt, T., Fuxe, K., Goldstein, M., and Joh, T. H., 1973, *Histochemie* **33**:231–254.
166. Hökfelt, T., Fuxe, K., and Goldstein, M., 1973, *Brain Res.* **53**:175–180.
167. Goldstein, M., Fuxe, K., Hökfelt, T., and Joh, T. H., 1971, *Experientia* **27**:951–952.
168. Goldstein, M., Fuxe, K., and Hökfelt, T., 1972, *Pharmacol. Rev.* **24**:293–309.
169. Hökfelt, T., Fuxe, K., and Goldstein, M., 1973, *Brain Res.* **62**:461–469.
170. Geffen, L. B., Livett, B. G., and Rush, R. A., 1969, *J. Physiol.* (*Lond.*), **204**:593–605.
171. Frydman, R., and Geffen, L. B., 1973, *J. Histochem. Cytochem.* **21**:166–174.
172. Livett, B. G., Geffen, L. B., and Rush, R. A., 1971, *Phil. Trans. R. Soc.* (*Lond.*) [*Biol.*] **261**:359–361.

173. Livett, B. G., Geffen, L. B., and Rush, R. A., 1969, *Biochem. Pharmacol.* **18**:923–924.
174. Cheah, T. B., and Geffen, L. B., 1970, *Proc. Aust. Phys. Pharm. Soc.* **1**:22–23.
175. Hartman, B. K., and Udenfriend, S., 1969, *Anal. Biochem.* **30**:391–394.
176. Hartman, B. K., and Udenfriend, S., 1970, *Mol. Pharmacol.* **6**:85–92.
177. Hartman, B. K., Zide, D., and Udenfriend, S., 1972, *Proc. Natl. Acad. Sci. U.S.A.* **69**:2722–2726.
178. Hartman, B. K., 1973, *J. Histochem. Cytochem.* **21**:312–322.
179. Hartman, B. K., 1974, *J. Psychiatr. Res.* **11**:283–288.
180. Swanson, L. W., and Hartman, B. K., 1975, *J. Comp. Neurol.* **163**:467–506.
181. Swanson, L. W., 1976, *Brain Res.* **110**:39–56.
182. Fuxe, K., Goldstein, M., Hökfelt, T., and Joh, T. H., 1970, *Res. Commun. Chem. Pathol. Pharmacol.* **1**:627–636.
183. Fuxe, K., Goldstein, M., Hökfelt, T., and Joh, T. H., 1971, *Prog. Brain Res.* **34**:127–138.
184. Lundberg, J., Bylock, A., Goldstein, M., Hansson, H.-A., and Dahlstrom, A., 1977, *Brain Res.* **120**:549–552.
185. Lundberg, J. M., Hamberger, B., Schultzberg, M., Hökfelt, T., Granberg, P.-O., Efendic, S., Terenius, L., Goldstein, M., and Luft, R., 1979, *Proc. Natl. Acad. Sci. U.S.A.* **76**:4079–4083.
186. Miller, R. J., and Pickel, V. M., 1980, *Adv. Biochem. Psychopharmacol.* **25**:349–360.
187. Aunis, D., Hesketh, J. E., and Devilliers, G., 1980, *J. Neurocytol.* **9**:255–274.
188. Levin, B. E., 1978, *Brain Res.* **150**:55–68.
189. Ross, R. A., Joh, T. H., and Reis, D. J., 1978, *J. Neurochem.* **31**:1491–1502.
190. Glazer, E. J., and Ross, L. L., 1980, *Brain Res.* **185**:39–49.
191. Saito, K., Cimarusti, D. L., and Roberts, E., 1977, *Neurochem. Res.* **2**:338.
192. Cimarusti, D. L., Saito, K., Vaughn, J. E., Barber, R., Roberts, E., and Thomas, P. E., 1979, *Brain Res.* **162**:55–67.
193. Jacobowitz, D. M., Ziegler, M. G., and Thomas, J. A., 1975, *Brain Res.* **91**:165–170.
194. Silver, M. A., and Jacobowitz, D. M., 1979, *Brain Res.* **167**:65–76.
195. Westlund, K. N., and Coulter, J. D., 1980, *Brain Res. Rev.* **2**:235–264.
196. Helle, K. B., Fillenz, M., Stanford, C., Pihl, K. E., and Srebro, B., 1979, *J. Neurochem.* **32**:1351–1355.
197. Grzanna, R., and Coyle, J. T., 1976, *J. Neurochem.* **27**:1091–1096.
198. Grzanna, R., Morrison, J. H., Coyle, J. T., and Molliver, M. E., 1977, *Neurosci. Lett.* **4**:127–134.
199. Morrison, J. H., Grzanna, R., Molliver, M. E., and Coyle, J. T., 1978, *J. Comp. Neurol.* **181**:17–40.
200. Grzanna, R., and Molliver, M. E., 1980, *The Reticular Formation Revisited: Specifying Function for a Nonspecific System*, Volume 6 (J. A. Hobson and M. A. B. Brazier, eds.), Raven Press, New York, pp. 83–97.
201. Grzanna, R., and Molliver, M. E., 1980, *Neuroscience* **5**:21–40.
202. Grzanna, R., Molliver, M. E., and Coyle, J. T., 1978, *Proc. Natl. Acad. Sci. U.S.A.* **75**:2502–2506.
203. Grzanna, R., and Coyle, J. T., 1978, *Brain Res.* **151**:206–214.
204. Joh, T. H., and Goldstein, M., 1973, *Mol. Pharmacol.* **9**:117–129.
205. Hökfelt, T., Fuxe, K., Goldstein, M., and Johansson, O., 1974, *Brain Res.* **66**:235–251.
206. Jonsson, G., Fuxe, K., Hökfelt, T., and Goldstein, M., 1977, *Med. Biol. (Helsinki)* **54**:421–426.
207. Lew, J. Y., Matsumoto, Y., Pearson, J., Goldstein, M., Hökfelt, T., and Fuxe, K., 1977, *Brain Res.* **119**:199–210.
208. van Orden, L. S., Burke, J. P., Redick, J. A., Rybarczyk, K. E., van Order, D. E., Baker, H. A., and Hartman, B. K., 1977, *Neuropharmacol.* **16**:129–133.
209. Howe, P. R. C., Costa, M., Furness, J. B., and Chalmers, J. P., 1980, *Neuroscience* **5**:2229–2238.
210. Joh, T. H., Shikimi, T., Pickel, V. M., and Reis, D. J., 1975, *Proc. Natl. Acad. Sci. U.S.A.* **72**:3575–3579.
211. Gershon, M. D., Dreyfus, C. F., Pickel, V. M. Joh, T. H., and Reis, D. J., 1977, *Proc. Natl. Acad. Sci. U.S.A.* **74**:3086–3089.

212. Fukui, H., Watanabe, T., and Wada, H., 1980, *Biochem. Biophys. Res. Commun.* **93**:333–339.
213. Tran, V. T., and Snyder, S. H., 1981, *J. Biol. Chem.* **256**:680–686.
214. Chao, L.-P., and Wolfgram, F., 1973, *J. Neurochem.* **20**:1075–1081.
215. Eng, L. F., Uyeda, C. T., Chao, L.-P., and Wolfgram, F., 1974, *Nature* **250**:243–245.
216. Kan, K.-S. K., Chao, L.-P., and Eng, L. F., 1978, *Brain Res.* **146**:221–230.
217. Chao, L.-P., Kan, K.-S. K., and Hung, F.-M., 1982, *Brain Res.* **235**:65–82.
218. Kan, K.-S. K., Chao, L.-P., and Forno, L. S., 1980, *Brain Res.* **193**:165–171.
219. Singh, V. K., and McGeer, P. L., 1974, *Life Sci.* **15**:901–913.
220. McGeer, P. L., McGeer, E. G., Singh, V. K., and Chase, W. H., 1974, *Brain Res.* **81**:373–379.
221. Hattori, T., Singh, V. K., McGeer, E. G., and McGeer, P. L., 1976, *Brain Res.* **102**:164–173.
222. Hattori, T., McGeer, E. G., Singh, V. K., and McGeer, P. L., 1977, *Exp. Neurol.* **55**:666–679.
223. Peng, J. H., McGeer, P. L., Kimura, H., Sung, S. C., and McGeer, E. G., 1980, *Neurochem. Res.* **5**:943–962.
224. Peng, J. H., McGeer, E. G., and McGeer, P. L., 1981, *Neurosci. Lett.* **21**:281–285.
225. Kimura, H., McGeer, P. L., Peng, J. H., and McGeer, E. G., 1980, *Science* **208**:1057–1059.
226. Kimura, H., McGeer, P. L., Peng, J. H., and McGeer, E. G., 1981, *J. Comp. Neurol.* **200**:151–201.
227. Kimura, H., McGeer, P. L., Peng, J. H., and McGeer, E. G., 1981, *Acta Histochem. Cytochem.* **14**:77.
228. Cozzari, C., and Hartman, B. K., 1980, *Proc. Natl. Acad. Sci. U.S.A.* **77**:7453–7457.
229. Eckenstein, F., Barde, Y.-A., and Thoenen, H., 1981, *Neuroscience* **6**:993–1000.
230. Levey, A. I., Aoki, M., Fitch, F. W., and Wainer, B. H., 1981, *Brain Res.* **218**:383–387.
231. Levey, A. I., and Wainer, B. P., 1982, *Brain Res.* **234**:469–473.
232. Wu, J.-Y., Matsuda, T., and Roberts, E., 1973, *J. Biol. Chem.* **248**:3029–3034.
233. Saito, K., 1976, *GABA in Nervous System Function* (E. Roberts, T. N. Chase, and D. B. Tower, eds.), Raven Press, New York, pp. 103–111.
234. Barber, R., and Saito, K., 1976, *GABA in Nervous System Function* (E. Roberts, T. N. Chase, and D. B. Tower, eds.), Raven Press, New York, pp. 113–132.
235. Wood, J. G., McLaughlin, B. J., and Vaughn, J. E., 1976, *GABA in Nervous System Function* (E. Roberts, T. N. Chase, and D. B. Tower, eds.), Raven Press, New York, pp. 133–148.
236. Saito, K., Barber, R., Wu, J.-Y., Matsuda, T., Roberts, E., and Vaughn, J. E., 1974, *Proc. Natl. Acad. Sci. U.S.A.* **71**:269–273.
237. McLaughlin, B. J., Wood, J. A., Saito, K., Barber, R., Vaughn, J. E., Roberts, E., and Wu, J.-Y., 1974, *Brain Res.* **76**:377–391.
238. McLaughlin, B. J., Wood, J. A., Saito, K., Roberts, E., and Wu, J.-Y., 1975, *Brain Res.* **85**:355–371.
239. Ribak, C. E., Vaughn, J. E., and Saito, K., 1978, *Brain Res.* **140**:315–332.
240. McLaughlin, B. J., Barber, R., Saito, K., Roberts, E., and Wu, J. Y., 1975, *J. Comp. Neurol.* **164**:305–322.
241. Ribak, C. E., Harris, A. B. Vaughn, J. E., and Roberts, E., 1979, *Science* **205**:211–214.
242. Ribak, C. E., Vaughn, J. E., Saito, K., Barber, R., and Roberts, E., 1976, *Brain Res.* **116**:287–298.
243. Ribak, C. E., Vaughn, J. E., and Roberts, E., 1980, *Brain Res.* **192**:413–420.
244. Panula, P., Wu, J.-Y., Emson, P., Liesi, P., and Rechardt, L., 1981, *Neurosci. Lett.* **22**:303–307.
245. Ribak, C. E., Vaughn, J. E., Saito, K., Barber, R., and Roberts, E., 1977, *Brain Res.* **126**:1–18.
246. Möhler, H., Richards, J. G., and Wu, J.-Y., 1981, *Proc. Natl. Acad. Sci. U.S.A.* **78**:1935–1938.
247. Ribak, C. E., Vaughn, J. E., and Roberts, E., 1979, *J. Comp. Neurol.* **187**:261–284.
248. Barber, R., Vaughn, J. E., Saito, K., McLaughlin, B. J., and Roberts, E., 1978, *Brain Res.* **141**:35–55.

249. Brandon, C., Lam, D. M. K., and Wu, J.-Y., 1979, *Proc. Natl. Acad. Sci. U.S.A.* **76**:3557–3561.
250. Cottesfeld, Z., Brandon, C., and Wu, J.-Y., 1981, *Brain Res.* **208**:181–186.
251. Chan-Palay, V., Krogsgaard-Larsen, P., Bornstein, M., Wu, J.-Y., Koller, C., and Palay, S. L., 1979, *Brain Res. Bull.* **4**:688.
252. Chan-Palay, V., Krogsgaard-Larsen, P., Cocalis, M., Wu, J.-Y., and Palay, S. L., 1979, *Brain Res. Bull.* **4**:687.
253. Chan-Palay, V., Palay, J. L., and Wu, J.-Y., 1979, *Anat. Embryol.* **157**:1–4.
254. Chan-Palay, V., Nilaver, G., Palay, S. L., Beinfeld, M. C., Zimmerman, E. A., Wu, J.-Y., and O'Donohue, T. L., 1981, *Proc. Natl. Acad. Sci. U.S.A.* **78**:7787–7791.
255. Vincent, S. R., Hökfelt, T., and Wu, J.-Y., 1982, *Neuroendocrinology* **34**:117–125.
256. Oertel, W. H., Schmechel, D. E., Tappaz, M. L., and Kopin, I. J., 1981, *Neuroscience* **6**:2689–2700.
257. Oertel, W. H., Schmechel, D. E., Mugnaini, E., Tappaz, M. L., and Kopin, I. J., 1981, *Neuroscience* **6**:2715–2735.
258. Oertel, W. H., Schmechel, D. E., Brownstein, M. J., Tappaz, M. L., Ransom, D. H., and Kopin, I. J., 1981, *J. Histochem. Cytochem.* **29**:977–980.
259. Pérez de la Mora, M., Possani, L. D., Tapia, R., Teran, L., Palacios, R., Fuxe, K., Hökfelt, T., and Ljungdahl, A., 1981, *Neuroscience* **6**:875–895.
260. Lam, D. M. K., Su, Y. Y. T., Swain, L., Marc, R. E., Brandon, C., and Wu, J. Y., 1979, *Nature* **278**:565–567.
261. Blindermann, J.-M., Maître, M., and Mandel, P., 1979, *J. Neurochem.* **32**:245–246.
262. Blindermann, J.-M., Maître, M., Ossola, L., and Mandel, P., 1978, *Eur. J. Biochem.* **86**:143–152.
263. Maître, M., Blinderman, J. M., Ossola, L., and Mandel, P., 1978, *Biochem. Biophys. Res. Commun.* **85**:885–890.
264. Nanopoulos, D., Belin, M. F., Maître, M., Vincendon, G., and Pujol, J. F., 1982, *Brain Res.* **232**:375–389.
265. Belin, M. F., Aguera, M., Nanopoulos, D., Gamrani, H., Maître, M., Calas, A., and Pujol, J. F., 1982, *Neurochem. Int.* **4**:303–312.
266. Creveling, C. R., Borchardt, R. T., and Isersky, C., 1973, *Frontiers in Catecholamine Research* (E. Usdin and S. H. Snyder, eds.), Pergamon Press, New York, pp. 117–119.
267. Inoue, K., Tice, L. W., and Creveling, C. R., 1977, *Structure and Function of Monoamine Enzymes* (E. Usdin, N. Weiner, and M. B. H. Youdim, eds.), Marcel Dekker, New York, pp. 835–859.
268. Kaplan, G. P., Hartman, B. K., and Creveling, C. R., 1980, *Neurochem. Res.* **5**:869–877.
269. Kaplan, G. P., Hartman, B. K., and Creveling, C. R., 1979, *Brain Res.* **167**:241–250.
270. Kaplan, G. P., Hartman, B. K., and Creveling, C. R., 1981, *Brain Res.* **204**:353–360.
271. Kaplan, G. P., Hartman, B. K., and Creveling, C. R., 1981, *Brain Res.* **229**:323–335.
272. Kaplan, G. P., Hartman, B. K., and Creveling, C. R., 1979, *Catecholamines: Basic and Clinical Frontiers* (I. Kopin, E. Usdin, and J. Barchas, eds.), Pergamon Press, New York, pp. 1354–1356.
273. Creveling, C. R., and Hartman, B. K., 1982, *Biochemistry of S-Adenosylmethionine and Related Compounds* (E. Usdin, R. T. Borchardt, and C. R. Creveling, eds.), Macmillan, London, New York, pp. 479–486.
274. Saito, K., Schousboe, A., Wu, J.-Y., and Roberts, E., 1974, *Brain Res.* **65**:287–296.
275. Chan-Palay, V., Wu, J.-Y., and Palay, S. L., 1979, *Proc. Natl. Acad. Sci. U.S.A.* **76**:2067–2071.
276. Steinbusch, H. W. M., Verhofstad, A. A. J., and Joosten, H. W. J., 1978, *Neuroscience* **3**:811–819.
277. Lidov, H. G. W., Grzanna, R., and Molliver, M. E., 1980, *Neuroscience* **5**:207–227.
278. Storm-Mathisen, J., Leknes, A., and Bore, A. B., 1982, *Neuroscience* (*Suppl.*) **7**:S203.
279. Chan-Palay, V., and Palay, S. L., 1979, *Proc. Natl. Acad. Sci. U.S.A.* **76**:1485–1488.
280. Ariano, M. A., and Matus, A. I., 1981, *J. Cell Biol.* **91**:287–292.
281. Snyder, S. H., 1980, *Science* **209**:976–983.

282. Prange, A. J., Jr., Nemeroff, C. B., Lipton, M. A., Breese, G. R., and Wilson, I. C., 1978, *Handbook of Psychopharmacology*, Volume 13 (L. L. Iversen, S. D. Iversen, and S. H. Snyder, eds.), Plenum Press, New York, pp. 1–107.
283. Guillemin, R., 1978, *Science* **202**:390–402.
284. Vale, W., Rivier, C., and Brown, M., 1977, *Annu. Rev. Physiol.* **39**:473–527.
285. Snyder, S. H., and Innis, R. B., 1979, *Annu. Rev. Biochem.* **48**:755–782.
286. Zimmerman, E. A., Carmel, P. W., Husain, M. K., Ferin, M., Tannebaum, M., Frantz, A. G., and Robinson, A. G., 1973, *Science* **182**:925–927.
287. Pearse, A. G. E., and Takor, T., 1976, *Clin. Endocrinol.* **5**(Suppl.):229.
288. Pickel, V. M., 1981, *Neuroanatomical Tract-Tracing Methods* (L. Heimer and M. J. Robards), Plenum Press, New York, pp. 483–509.
289. Bloom, F., Battenberg, E., Rossier, J., Ling, N., and Guillemin, R., 1978, *Proc. Natl. Acad. Sci. U.S.A.* **75**:1591–1595.
290. Watson, S. J., Akil, H., Richard, C. W. III, and Barchas, J. D., 1978, *Nature* **275**:226–228.
291. Silverman, A. J., 1975, *Am. J. Anat.* **144**:433–443.
292. Silverman, A. J., and Zimmerman, E. A., 1975, *Cell Tissue Res.* **159**:291–301.
293. Krisch, B., 1976, *Cell Tissue Res.* **174**:109–127.
294. Watkins, W. B., and Choy, V. J., 1976, *Neurosci. Lett.* **3**:293–297.
295. Dierickx, K., Vandesande, F., and DeMey, J., 1976, *Cell Tissue Res.* **168**:141–151.
296. Castel, M., and Hochman, J., 1976, *Cell Tissue Res.* **174**:69–81.
297. Dube, D., Leclerc, R., and Pelletier, G., 1976, *Am. J. Anat.* **147**:103–108.
298. Vandesande, F., Dierickx, K., and DeMey, J., 1977, *Cell Tissue Res.* **180**:443–452.
299. van Leeuwen, F. W., and Swaab, D. F., 1977, *Cell Tissue Res.* **177**:493–501.
300. Watkins, W. B., and Choy, V. J., 1977, *Cell Tissue Res.* **180**:491–503.
301. Buijs, R. M., Swaab, D. F., Dogterom, J., van Leeuwen, F. W., 1978, *Cell Tissue Res.* **186**:423–433.
302. van Leeuwen, F. W., Swaab, D. F., and De Raay, C., 1978, *Cell Tissue Res.* **193**:1–10.
303. Vandesande, F., Dierickx, K., and DeMey, J., 1975, *Cell Tissue Res.* **156**:377–380.
304. Buijs, R. M., 1978, *Cell Tissue Res.* **192**:423–435.
305. Buijs, R. M., 1980, *J. Histochem. Cytochem.* **28**:357–360.
306. Antunes, J. L., Carmel, P. W., and Zimmerman, E. A., 1977, *Brain Res.* **137**:1–10.
307. Hawthorn, J., Ang, V. T. Y., and Jenkins, J. S., 1980, *Brain Res.* **197**:75–81.
308. Glasmann., W., and Sofroniew, M. V., 1980, *Neurosci. Lett.* [*Suppl.*] **5**:S252.
309. Micevych, P., and Elde, R., 1980, *J. Comp. Neurol.* **190**:135–146.
310. Swanson, L. W., 1978, *Neurosci. Abstr.* **4**:415.
311. Dierickx, K., and Vandesande, F., 1978, *Cell Tissue Res.* **174**:25–33.
312. Pelletier, G., Leclerc, R., Labrie, F., and Puviani, R., 1974, *Mol. Cell. Endocrinol.* **1**:157.
313. Vandesande, F., DeMey, J., and Dierickx, K., 1974, *Cell Tissue Res.* **151**:187–200.
314. Watkins, W. B., 1976, *Cell Tissue Res.* **175**:165–181.
315. Wolf, G., 1976, *Endokrinologie* **68**:288–299.
316. Swanson, W., 1977, *Brain Res.* **128**:346–353.
317. Kozlwoski, G. P., Frenk, S., and Brownfield, M. S., 1977, *Cell Tissue Res.* **179**:467–473.
318. McNeill, T. H., Hoffman, D. L., and Kozlowski, G. P., 1977, *Am. J. Anat.* **149**:613–618.
319. Sofroniew, M. V., and Weindl, A., 1978, *Endocrinology* **102**:334–337.
320. Sar, M., and Stumpf, W. E., 1980, *Neurosci. Lett.* **17**:179–184.
321. Loren, I., Schwandt, P., Alumets, J., Hakanson, R., Neureuther, G., Richter, W., and Sundler, F., 1980, *Cell Tissue Res.* **205**:349–359.
322. McNeill, T. H., and Sladek, J. R., Jr., 1980, *J. Comp. Neurol.* **193**:1023–1033.
323. Breslow, E., 1979, *Annu. Rev. Biochem.* **48**:251–274.
324. Land, H., Schütz, G., Schmale, H., and Richter, D., 1982, *Nature* **295**:299–303.
325. Pedersen, C. A., and Prange, A. J., Jr., 1979, *Proc. Natl. Acad. Sci. U.S.A.* **76**:6661–6665.
326. Nakanishi, S., Inoue, A., Kita, T., Nakamura, M., Chang, A. C. Y., Cohen, S. N., and Numa, S., 1979, *Nature* **278**:423–427.
327. Comb, M., Seeburg, P. H., Adelman, J., Eiden, L., and Herbert, E., 1982, *Nature* **295**:663–666.
328. Noda, M., Furutani, Y., Takahashi, H., Toyosato, M., Hirose, T., Inayama, S., Nakanishi, S., and Numa, S., 1982, *Nature* **295**:202–206.

329. Gubler, U., Seeburg, P., Hoffman, B. J., Gage, L. P., and Udenfriend, S., 1982, *Nature* **295**:206–208.
330. Watson, S. J., Barchas, J. D., and Li, C. H., 1977, *Proc. Natl. Acad. Sci. U.S.A.* **74**:5155–5158.
331. Zimmerman, E. A., Liotta, A., and Krieger, D. T., 1978, *Cell Tissue Res.* **186**:393–398.
332. Chretien, M., Benjannet, S., Seidat, N. G., and Lis, M., 1979, *Clinical Neuroendocrinology, A Patholophysiological Approach* (G. Tolis, ed.), Raven Press, New York, pp. 147–152.
333. Bugnon, C., Bloch, B., and Lenys, D., 1981, *Neuroscience* **6**:1299–1313.
334. Swaab, D. F., and Fisser, B., 1977, *Neurosci. Lett.* **7**:313–317.
335. Nilaver, G., Zimmerman, E. A., Defendini, R., Liotta, A. S., Krieger, D. T., and Brownstein, M. J., 1979, *J. Cell Biol.* **81**:50–58.
336. Leranth, C., Williams, T. H., Chretien, M., and Palkovits, M., 1980, *Cell Tissue Res.* **210**:11–19.
337. Larsson, L.-I., 1980, *J. Histochem. Cytochem.* **28**:133–141.
338. Bloch, B., Bugnon, C., Fellman, D., and Lenys, D., 1978, *Neurosci. Lett.* **10**:147–152.
339. Ibata, Y., Watanabe, K., Kinoshita, H., Kubo, S., Sano, Y., Sakura, N., Yanaihara, C., and Yanaihara, N., 1980, *Neurosci. Lett.* **17**:185–189.
340. Haynes, L. W., Smyth, D. G., and Zakarian, S., 1982, *Brain Res.* **232**:115–128.
341. Elde, R., Hökfelt, T., Johansson, O., and Terenius, L., 1976, *Neuroscience* **1**:349–351.
342. Simantov, R., Kuhar, M. J., Uhl, G. R., and Snyder, S. H., 1977, *Proc. Natl. Acad. Sci. U.S.A.* **74**:2167–2171.
343. Lewis, R. V., Stein, S., Gerber, L. D., Rubinstein, M., and Udenfriend, S., 1978, *Proc. Natl. Acad. Sci. U.S.A.* **75**:4021–4023.
344. Pickel, V. M., Joh, T. H., Leemna, S. E., and Miller, R. J., 1978, *Anat. Rec.* **190**:511–512.
345. Uhl, G. R., Goodman, R. R., Kuhar, M. J., Childers, S. R., and Snyder, S. H., 1979, *Brain Res.* **166**:75–94.
346. Schwarcz, R., Fuxe, K., Hökfelt, T., Terenius, L., and Goldstein, M., 1980, *J. Neurochem.* **34**:772–778.
347. Bayon, A., Koda, L., Battenberg, E., Azad, R., and Bloom, F. E., 1980, *Neurosci. Lett.* **16**:75–80.
348. Johnson, R. P., Sar, M., and Stumpf, W. E., 1980, *Brain Res.* **194**:566–571.
349. Graybiel, A. M., Ragsdale, C. W., Yoneoka, E. S., and Elde, R. P., 1981, *Neuroscience* **6**:377–397.
350. Tramu, G., Beauvillain, J. C., Croix, D., and Leonardelli, J., 1981, *Brain Res.* **215**:235–255.
351. Naik, D. R., Sar, M., and Stumpf, W. E., 1981, *J. Comp. Neurol.* **198**:583–601.
352. Schulman, J. A., Finger, T. E., Brecha, N. C. and Karten, H. J., 1981, *Neuroscience* **6**:2407–2416.
353. Glazer, E. J., and Basbaum, A. I., 1981, *J. Comp. Neurol.* **196**:377–390.
354. Jancso, G., Hökfelt, T., Lundberg, J. M., Kiraly, E., Halasz, N., Nilsson, G., Terenius, L., Rehfeld, J., Steinbusch, H., Verhofstad, A., Elde, R., Said, S., and Brown, M., 1981, *J. Neurocytol.* **10**:963–980.
355. Larsson, L.-I., Childers, S., and Snyder, S. H., 1979, *Nature* **282**:407–410.
356. Goldstein, A., Tachibana, S., Lowney, L. I., Hunkapiller, M., and Hood, L., 1979, *Proc. Natl. Acad. Sci. U.S.A.* **76**:6666–6670.
357. Tachibana, S., Araki, K., Ohya, S., and Yoshida, S., 1982, *Nature* **295**:339–340.
358. Kakidani, H., Furutani, Y., Takahashi, H., Noda, M., Morimoto, Y., Hirose, T., Asai, M., Inayama, S., Nakanishi, S., and Numa, S., 1982, *Nature* **298**:245–249.
359. Ghazarossian, V. E., Chavkin, C., and Goldstein, A., 1980, *Life Sci.* **27**:75–86.
360. Goldstein, A., and Ghazarossian, V. E., 1980, *Proc. Natl. Acad. Sci. U.S.A.* **77**:6207–6210.
361. Botticelli, L. J., Cox, B. M., and Goldstein, A., 1981, *Proc. Natl. Acad. Sci. U.S.A.* **78**:7783–7786.
362. Weber, E., Roth, K. A., and Barchas, J. D., 1982, *Proc. Natl. Acad. Sci. U.S.A.* **79**:3062–3066.
363. Barry, J., Dubois, M. P., and Poulain, P., 1973, *Z. Zellforsch.* Mikrosk. Anat. **146**:351–366.
364. Pelletier, G., Labrie, F., Puviani, R., Arimura, A., and Shally, A. V., 1974, *Endocrinology* **95**:314–315.

365. King, J. C., Parsons, J. A., Erlandsen, S. L., and Williams, T. H., 1974, *Cell Tissue Res.* **153**:211–217.
366. Barry, J., Dubois, M. P., and Canetto, B., 1974, *Endocrinology* **95**:1416–1423.
367. Baker, B. L., Dermody, W. C., and Reel, J. R., *Endocrinology* **97**:125–135.
368. Gross, D. S., 1976, *Endocrinology* **98**:1408–1417.
369. Doerr-Schott, J., and Dubois, M. P., 1976, *Cell Tissue Res.* **172**:477–486.
370. Silverman, A. J., and Desnoyers, P., 1976, *Cell Tissue Res.* **169**:157–166.
371. Pelletier, G., Leclerc, R., and Dube, D., 1976, *J. Histochem. Cytochem.* **24**:864–871.
372. Mazzuca, M., 1977, *Neurosci. Lett.* **5**:123–127.
373. Silverman, A. J., Autunes, J. L., Ferin, M., and Zimmerman, E. A., 1977, *Endocrinology* **101**:134–142.
374. Paulin, C., Dubois, M. P., Barry, J., and Dubois, P. M., 1977, *Cell Tissue Res.* **182**:341–345.
375. Kami, K., Makino, T., Shiina, M., and Wada, M., 1977, *Okajinas Folia Anat. Jpn.* **54**:259–281.
376. Barry, J., 1977, *Cell Tissue Res.* **181**:1–14.
377. McNeill, T. H., and Sladek, J. R., Jr., 1978, *Science* **200**:72–74.
378. Hoffman, G. E., Knigg, K. M., Moynihan-McCourt, J. A., Melnyk, V., and Arimura, A., 1978, *Neuroscience* **3**:219–231.
379. Nozaki, M., and Kobayashi, H., 1979, *Arch. Histol. (Jpn.)* **42**:201–220.
380. Burchanowski, B. J., Knigge, K. M., and Sternberger, L. A., 1979, *Proc. Natl. Acad. Sci. U.S.A.* **76**:6671–6674.
381. Barry, J., 1979, *Int. Rev. Cytol.* **60**:179–221.
382. Jennes, L., and Stumpf, W. E., 1980, *Cell Tissue Res.* **209**:239–256.
383. Kawano, H., and Daikoku, S., 1981, *Neuroendocrinology* **32**:179–186.
384. Nozaki, M., 1981, *Fol. Endocrinol. Jpn.* **57**:364–371.
385. Joseph, S. A., Piekut, D. T., and Knigge, K. M., 1981, *J. Histochem. Cytochem.* **29**:247–254.
386. Dees, W. L., Sorensen, A. M., Kemp, W. M., and McArthur, N. H., 1981, *Brain Res.* **208**:123–134.
387. Hökfelt, T., Fuxe, K., Johansson, O., Jeffcoate, S., and White, N., 1975, *Eur. J. Pharmacol.* **34**:389–392.
388. Hökfelt, T., Fuxe, K., Johansson, O., Jeffcoate, S., and White, N., 1975, *Neurosci. Lett.* **1**:133–139.
389. Johansson, O., Hökfelt, T., Elde, R. P., Jeffcoate, S. L., and White, N., 1978, *Neurosci. Lett. (Suppl.)* **1**:220.
390. Johansson, O., Hökfelt, T., Jeffcoate, S. L., White, N., and Sternberger, L. A., 1980, *Exp. Brain Res.* **38**:1–10.
391. Johansson, O. Hökfelt, T., Pernow, B., Jeffcoate, S. L., White, N., Steinbusch, H. W. M., Verhofstad, A. A. J., Emson, P. C., and Spindel, E., 1981, *Neuroscience* **6**:1857–1881.
392. Pelletier, G., Labrie, F., Arimura, A., and Schally, A. V., 1974, *Am. J. Anat.* **140**:445–450.
393. Hökfelt, T., Efendic, S., Johansson, O., Luft, R., and Arimura, A., 1974, *Brain Res.* **80**:165–169.
394. Sétáló, G., Vigh, S., Schally, A. V., Arimura, A. V., and Flerkó, B., 1975, *Brain Res.* **90**:352–356.
395. Pelletier, G., Leclerc, R., Dube, D., Labrie, F., Puviani, F., Arimura, A., and Schally, A. V., 1975, *Am. J. Anat.* **142**:397–401.
396. King, J. C., Gerall, A. A., Fishback, J. B., Elkind, K. E., and Arimura, A., 1975, *Cell Tissue Res.* **160**:423–430.
397. Elde, R. P., and Parsons, J. A., 1975, *Am. J. Anat.* **144**:541–548.
398. Dubois, M. P., and Kolodziejczyk, E., 1975, *C.R. Acad. Sci. [D] (Paris)* **281**:1737–1740.
399. Hökfelt, T., Efendic, S., Hellerstrom, C., Johansson, O., Luft, R., and Arimura, A., 1975, *Acta Endocrinol. (Kbh.) [Suppl.]* **200**:1–41.
400. Alpert, L. C., Brawer, J. R., Patel, Y. C., and Reichlin, S., 1976, *Endocrinology* **98**:255–258.
401. Baker, B. L., and Yu, Y.-Y., 1976, *Anat. Rec.* **186**:343–356.
402. Hökfelt, T., Elde, R., Johansson, O., Luft, R., Nilsson, G., and Arimura, A., 1976, *Neuroscience* **1**:131–136.

403. Krisch, B., 1977, *Cell Tissue Res.* **179**:211–224.
404. Dierickx, K., and Vandesande, F., 1979, *Cell Tissue Res.* **201**:349–359.
405. Emson, P. C., and Lindvall, O., 1979, *Neuroscience* **4**:1–30.
406. Krisch, B., and Leonhardt, H., 1980, *Cell Tissue Res.* **205**:327–331.
407. Shiosaka, S., Takatsuki, K., Sakanaka, M., Inagaki, S., Takagi, H., Senba, E., Kawai, Y., Minagawa, H., and Tohyama, M., 1981, *Neurosci. Lett.* **25**:69–73.
408. Finley, J. C. W., Maderdrut, J. L., Roger, J. L., and Petrusz, P., 1981, *Neuroscience* **6**:2173–2192.
409. Dalsgaard, C.-J., Hökfelt, T., Johansson, O., and Elde, R., 1981, *Neurosci. Lett.* **27**:335–339.
410. Romagnano, M. A., Pilcher, W. H., Bennet-Clarke, C., Chafel, T. L., and Joseph, S. A., 1982, *Brain Res.* **234**:387–398.
411. Vincent, S. R., Skirboll, L., Hökfelt, T., Johansson, O., Lundberg, J. M., Elde, R. P., Terenius, L., and Kimmel, J., 1982, *Neuroscience* **7**:439–446.
412. Shiosaka, S., Takatsuki, K., Sakanaka, M., Inagaki, S., Takagi, H., Senba, E., Kawai, Y., Minagawa, H., and Tohyama, M., 1981, *Neurosci. Lett.* **25**:69–73.
413. Vale, W., Spiess, J., Rivier, C., and Rivier, J., 1981, *Science* **213**:1394–1397.
414. Bugnon, C., Fellmann, D., Gouget, A., and Cardot, J., 1982, *Nature* **298**:159–161.
415. McGuigan, J. E., 1978, *Annu. Rev. Med.* **29**:307–318.
416. Hökfelt, T., Kellerth, J. O., Nilsson, G., and Pernow, B., 1975, *Science* **190**:889–890.
417. Ljungdahl, A., Hökfelt, T., Nilsson, G., and Goldstein, M., 1975, *Neuroscience* **3**:945–976.
418. Kanazawa, I., Emson, P., and Cuello, A. C., 1977, *Brain Res.* **119**:447–453.
419. Pickel, V. M., Reis, D. J., and Leeman, S. E., 1977, *Brain Res.* **122**:534–540.
420. Chan-Palay, V., and Palay, S. L., 1977, *Proc. Natl. Acad. Sci. U.S.A.* **74**:4050–4054.
421. Hökfelt, T., Ljungdahl, A., Terenius, L., Elde, R., and Nilsson, G., 1977, *Proc. Natl. Acad. Sci. U.S.A.* **74**:3081–3085.
422. Chan-Palay, V., Jonsson, G., and Palay, S. L., 1978, *Proc. Natl. Acad. Sci. U.S.A.* **75**:1582–1586.
423. Hökfelt, T., Ljungdahl, A., Steinbusch, H., Verhofstad, A., Nilsson, G., Brodin, E., Pernow, B., and Goldstein, M., 1978, *Neuroscience* **3**:517–538.
424. Ljungdhal, A., Hökfelt, T., and Nilsson, G., 1978, *Neuroscience* **3**:861–943.
425. Cuello, A. C., and Kanazawa, I., 1978, *J. Comp. Neurol.* **178**:129–156.
426. Pickel, V. M., Joh, T. H., Reis, D. J., Leeman, S. E., and Miller, R. J., 1979, *Brain Res.* **160**:387–400.
427. Itoga, E., Kito, S., Kishida, T., Yanaihara, N., and Wakabayashi, I., 1979, *Neurosci. Lett.* [*Suppl.*] **2**:S10.
428. Cuello, A. C., Galfre, G., and Milstein, C., 1980, *Proc. Natl. Acad. Sci. U.S.A.* **76**:3532–3536.
429. Paulin, C., Charnay, Y., Dubois, P. M., and Chayvialle, J.-A., 1980, *C.R. Acad. Sci.* [*D*] (Paris) **291**:257–260.
430. Vacca, L. L., Abrahams, S. J., and Naftchi, N. E., 1980, *J. Histochem. Cytochem.* **28**:297–307.
431. Vincent, S. R., Kimura, H., and McGeer, E. G., 1981, *J. Comp. Neurol.* **199**:113–123.
432. Iversen, L. L., Nagy, J., Emson, P. C., Lee, C. M., Hanley, M., Sandberg, B., Ninkovic, H., and Hunt, S., 1981, *Chemical Transmission, 75 Years* (L. Stjarne, P. Hedquist, H. Lagercrantz, and A. Wenmaln, eds.), Academic Press, New York, pp. 501–512.
433. Edvinsson, L., McCulloch, J., and Uddman, R., 1981, *J. Physiol.* (*Lond.*) **318**:251–258.
434. Sakanaka, M., Shiosaka, S., Takatsudi, K., Inagaki, S., Takagi, H., Senba, E., Kawai, Y., Matsuzaki, T., and Tohyama, H., 1981, *Brain Res.* **221**:231–242.
435. Kondo, H., and Yui, R., 1981, *Brain Res.* **222**:134–137.
436. DiFiglia, M., Aronin, N., and Leeman, S. E., 1981, *Brain Res.* **233**:381–388.
437. Inagaki, S., Sakanaka, M., Shiosaka, S., Senba, E., Takagi, H., Takatsuki, K., Kawai, Y., Matsuzaki, T., Iida, H., Hara, Y., and Tohyama, M., 1982, *Neuroscience* **7**:639–645.
438. Uhl, G., and Snyder, S., 1976, *Life Sci.* **19**:1827–1832.
439. Uhl, G. R., Kuhar, M. J., and Snyder, S. H., 1977, *Proc. Natl. Acad. Sci. U.S.A.* **74**:4059–4063.

440. Uhl, G. R., Goodman, R. R., and Snyder, S. H., 1979, *Brain Res.* **167**:77–91.
441. Uhl, G. R., and Snyder, S. H., 1979, *Brain Res.* **161**:522–526.
442. Fuxe, K., Locatelli, V., McDonald, T., Hökfelt, T., Agrati, L. F., Yanaihara, N., and Mutt, V., 1982, *Neuroscience* **7**(Suppl.):S74.
443. Said, S. I., and Faloona, G. R., 1975, *N. Engl. J. Med.* **293**:155–160.
444. Larsson, L.-I., Fahrenkrug, J., Schaffalitzky de Muckadell, O., Sundler, F., Håkanson, R., and Rehfeld, J. F., 1976, *Proc. Natl. Acad. Sci. U.S.A.* **73**:3197–3200.
445. Fuxe, K., Hökfelt, T., Said, S. I., and Mu, T. T. V., 1977, *Neurosci. Lett.* **5**:241–246.
446. Hanko, J., Edvinsson, L., Fahrenkrug, J., Håkanson, R., Larsson, L.-I., Owman, C., Schaffalitzky de Muckadell, O., and Sundler, F., 1977, *Acta Neurol. Scand.* [*Suppl.*] **56**:216–217.
447. Emson, P. C., Gilbert, R. F. T., Loren, I., Fahrenkrug, J., Sundler, F., and Schaffalitzky de Muckadell, O. B., 1979, *Brain Res.* **177**:437–444.
448. van Noorden, S., Polak, J. M., Bloom, S. R., and Bryant, M. G., 1979, *Neuropathol. Appl. Neurobiol.* **5**:149–154.
449. Lorén, I., Emson, P. C., Fahrenkrug, J., Björklund, A., Alumets, J., Hakanson, R., and Sundler, F., 1980, *Neuroscience* **4**:1953–1976.
450. Roberts, G. W., Woodhams, P. L., Crow, T. J., and Polak, J. M., 1980, *Brain Res.* **195**:471–475.
451. Roberts, G. W., Woodhams, P. L., Bryant, M. G., Crow, T. J., Bloom, S. R., and Polak, J. M., 1980, *Histochemistry* **65**:103–119.
452. Sims, K. B., Hoffman, D. L., Said, S. I., and Zimmerman, E. A., 1980, *Brain Res.* **186**:165–183.
453. Pelletier, G., Leclerc, R., Puviani, R., and Polak, J. M., 1981, *Brain Res.* **210**:356–360.
454. Marley, P. D., Emson, P. C., Hunt, S. P., and Fahrenkrug, J., 1981, *Neurosci. Lett.* **27**:261–266.
455. Vanderhaeghen, J. J., Signeau, J. C., and Gepts, W., 1975, *Nature* **257**:604–605.
456. Vanderhaeghen, J. J., Lotstra, F., DeMey, J., and Gilles, C., 1980, *Proc. Natl. Acad. Sci. U.S.A.* **77**:1190–1194.
457. Rehfeld, J. F., Larsson, L.-I., Goltermann, N. R., Schwartz, T. W., Holst, J. J., Jensen, S. L., and Morley, J. S., 1980, *Nature* **284**:33–38.
458. Straus, E., Muller, J. E., Choi, H.-S., Paronetto, F., and Yalow, R. S., 1977, *Proc. Natl. Acad. Sci. U.S.A.* **74**:3033–3034.
459. Dockray, G. J., 1976, *Nature* **264**:568–570.
460. Innis, R. B., Correa, F. M. A., Uhl, G. R., Schneider, B., and Snyder, S. H., 1979, *Proc. Natl. Acad. Sci. U.S.A.* **76**:521–525.
461. Larsson, L. I., and Rehfeld, J. F., 1979, *Brain Res.* **165**:201–218.
462. Hökfelt, T., Rehfeld, J. F., Skirboll, L., Ivemark, B., Goldstein, M., and Markey, K., 1980, *Nature* **285**:476–478.
463. Gilles, C., Lotstra, F., and Vanderhaeghen, J. J., 1981, *Neurosci. Lett.* [*Suppl.*] **7**:S320.
464. Osborne, N. N., Nicholas, D. A., Cuello, A. C., and Dockray, G. J., 1981, *Neurosci. Lett.* **26**:31–35.
465. van der Kooy, D., Hunt, S. P., Steinbusch, H. W. M., and Verhofstad, A. A. J., *Neurosci. Lett.* **26**:25–30.
466. Skirboll, L. R., Grace, A. A., Hommer, D. W., Rehfeld, J., Goldstein, M., Hökfelt, T., and Bunney, B. S., 1981, *Neuroscience* **6**:2111–2124.
467. Nilaver, G., Defendini, R., Zimmerman, E. A., Beinfeld, M. C., and O'Donohue, T. L., 1982, *Nature* **295**:597–598.
468. Kimmel, J. R., Hayden, L. J., and Pollock, H. G., 1975, *J. Biol. Chem.* **250**:9369–9376.
469. Lorén, I., Alumets, J., Håkanson, R., and Sundler, F., 1979, *Cell Tissue Res.* **200**:179–186.
470. Tatemoto, K., and Mutt, V., 1980, *Nature* **285**:417–418.
471. Hökfelt, T., Lundberg, J. M., Terenius, L., Jancsó, G., and Kimmel, J., 1981, *Peptides* **2**:81–87.
472. Hunt, S. P., Emson, P. C., Gilbert, R., Goldstein, M., and Kimmel, J., 1981, *Neurosci. Lett.* **21**:125–130.
473. Meyer, D. K., Phillips, M. I., and Eiden, L., 1982, *J. Neurochem.* **38**:816–820.
474. Fuxe, K., Ganten, S., Hökfelt, T., and Bolme, P., 1976, *Neurosci. Lett.* **2**:229–234.

475. Changaris, D. G., Demers, L. M., Keil, L. C., and Severs, W. B., 1976, *Exp. Neurol.* **51**:699–704.
476. Phillips, M. I., Quinlan, J. T., and Weyhenmeyer, J., 1980, *Life Sci.* **27**:2589–2594.
477. Zimmerman, E. A., Krupp, L., Hoffman, D. L., Matthew, E., and Nilaver, G., 1980, *Peptides* **1**:3–10.
478. Changaris, D. G., Severs, W. B., and Keil, L. C., 1978, *J. Histochem. Cytochem.* **26**:593–607.
479. Fuxe, K., Ganten, D., Hökfelt, T., Locatelli, V., Poulsen, K., Stock, G., Rix, E., and Taugner, R., 1980, *Neurosci. Lett.* **18**:245–250.
480. Inagami, T., Clemens, D. L., Celio, M. R., Brown, A., Sandru, L., Herschkowitz, N., Hoffman, L. H., and Kasselberg, A. G., 1980, *Neurosci. Lett.* **18**:91–98.
481. Corrêa, F. M. A., Innis, R. B., Uhl, G. R., and Snyder, S. H., 1979, *Proc. Natl. Acad. Sci. U.S.A.* **76**:1489–1493.
482. Margolis, F. L., Roberts, N., Ferriero, D., and Feldman, J., 1974, *Brain Res.* **81**:469–483.
483. Harding, J., and Margolis, F. I., 1976, *Brain Res.* **110**:351–360.
484. Hirsch, J. D., Grillo, M., and Margolis, F. L., 1978, *Brain Res.* **158**:407–422.
485. Amara, S. G., Jonas, V., Rosenfeld, M. G., Ong, E. S., and Evans, R. M., 1982, *Nature* **298**:240–244.
486. Pickel, V. M., Joh, T. H., Leeman, S. E., and Miller, R. J., 1978, *Anat. Rec.* **190**:511–512.
487. Tramu, G., Beauvillain, J. C., Croix, D., and Leonardelli, J., 1981, *Brain Res.* **215**:235–255.
488. McNeill, T. H., Sladek, J. R., Jr., 1980, *J. Comp. Neurol.* **193**:1023–1033.
489. Johnson, R. P., Sar, M., and Stumpf, W. E., 1980, *Brain Res.* **194**:566–571.
490. Graybiel, A. M., Ragsdale, C. W., Yoneoka, E. S., and Elde R. P., 1981, *Neuroscience* **6**:377–397.
491. Lundberg, J. M., Hökfelt, T., Schultzberg, M., Uvnäs-Wallensten, K., Köhler, C., and Said, S. I., 1979, *Neuroscience* **4**:1539–1559.
492. Sawchenko, P. E., and Swanson, L. W., 1981, *Brain Res.* **210**:31–51.
493. Mirsky, R., and Thompson, E. J., 1975, *Cell* **4**:95–101.
494. Barclay, A. N., and Hydén, H., 1978, *J. Neurochem.* **31**:1375–1391.
495. Morris, R. J., Mancini, P. E., and Pfeiffer, S. E., 1980, *Brain Res.* **182**:119–135.
496. Schachner, M., 1982, *J. Neurochem.* **39**:1–8.
497. Barclay, A. N., 1979, *J. Neurochem.* **32**:1249–1257.
498. Whitaker, J. N., Cass Terry, L., and Whetsell, W. O., Jr., 1981, *Brain Res.* **216**:109–124.
499. Zwiller, J., Ghandour, M. S., Revel, M. O., and Basset, P., 1981, *Neurosci. Lett.* **23**:31–36.
500. Ariano, M. A., Lewicki, J. A., Brandwein, H. J., and Murad, F., 1982, *Proc. Natl. Acad. Sci. U.S.A.* **79**:1316–1320.
501. Sparrow, J. R., 1981, *Brain Res.* **212**:159–163.
502. Wiegandt, H., 1967, *J. Neurochem.* **14**:671–674.
503. Hamberger, A., and Svennerholm, L., 1971, *J. Neurochem.* **18**:1821–1829.
504. Norton, W. T., and Poduslo, S. E., 1971, *J. Lipid Res.* **12**:84–90.
505. Suzuki, K., Poduslo, J. I., and Poduslo, S. E., 1968, *Biochim. Biophys. Acta* **152**:576–586.
506. Laev, H., Rapport, M. M., Mahadik, S. P., and Silverman, A. J., 1978, *Brain Res.* **157**:136–141.
507. De Camilli, P., Ueda, T., Bloom, F. E., Battenberg, E., and Greengard, P., 1979, *Proc. Natl. Acad. Sci. U.S.A.* **76**:5977–5981.
508. Bloom, F. E., Ueda, T., Battenberg, E., and Greengard, P., 1979, *Proc. Natl. Acad. Sci. U.S.A.* **76**:5982–5986.
509. Grand, R. J. A., and Perry, S. V., 1980, *Biochem. Soc. Trans.* **8**:487–489.
510. Wood, J. G., Wallace, R. W., Whitaker, J. N., and Cheung, W. Y., 1980, *J. Cell Biol.* **84**:66–76.
511. Kelly, R. B., Carlson, S. S., von Wedel, R. J., Hooper, J. E., Miljanich, G. P., and Brasier A. R., 1981, *Cold Spring Harbor Reports in the Neurosciences*, Volume 2, *Monoclonal Antibodies to Neural Antigens* (R. McKay, M. C. Raff, and L. F. Reichardt, eds.), Cold Spring Harbor Laboratory, New York, pp. 153–161.
512. Sampedro, M. N., Bussineau, C. M., and Cotman, C. W., 1981, *J. Cell Biol.* **90**:675–686.
513. Stallcup, W. B., and Cohn, M., 1976, *Exp. Cell Res.* **98**:285–297.

514. Stallcup, W. B., Levin, E. J., and Raschke, W., 1981, *Cold Spring Harbor Reports in the Neurosciences*, Volume 2, *Monoclonal Antibodies to the Neural Antigens* (R. McKay, M. C. Raff, and L. F. Reichardt, eds.), Cold Spring Harbor Laboratory, New York, pp. 39–49.
515. Schachner, M., Ruberg, M. Z., and Carhow, T. B., 1976, *Brain Res. Bull.* **1**:367–377.
516. Barnstable, C. J., 1980, *Nature* **286**:231–235.
517. Cohen, J., and Selvendran, S. Y., 1981, *Nature* **291**:421–423.
518. Vulliamy, R., Rattray, S., and Mirsky, R., 1981, *Nature* **291**:418–420.
519. Hirn, M., Pierres, M., Deagostini-Bazin, H., Hirsch, M., and Goridis, C., 1981, *Brain Res.* **214**:433–439.
520. Hirn, M., Pierres, M., Deagostini-Bazin, H., Demierre, M., Goridis, C., Ghandour, S., Langley, K., and Gombos, G., 1982, *Neuroscience* **7**:239–250.
521. Langley, O. K., Ghandour, M. S., Gombos, G., Hirn, M., and Goridis, C., 1982, *Neurochem. Res.* **7**:343–356.
522. Giotta, G. J., Heitzmann, J., and Cohn, M., 1982, *Dev. Brain Res.* **4**:209–221.
523. Schachner, M., Sommer, I., Lagenaur, C., and Schnitzer, J., 1981, *Cold Spring Harbor Reports in the Neurosciences*, Volume 2, *Monoclonal Antibodies to the Neural Antigens* (R. McKay, M. C. Raff, and L. F. Reichardt, eds.), Cold Spring Harbor Laboratory, New York, pp. 15–20.
524. Eisenbarth, G. S., Walsh, F. S., and Nirenberg, M., 1979, *Proc. Natl. Acad. Sci. U.S.A.* **76**:4913–4917.
525. Eisenbarth, G. S., Shimizu, K., Conn, M., Mittler, R., and Wells, S., 1981, *Cold Spring Harbor Reports in the Neurosciences*, Volume 2, *Monoclonal Antibodies to the Neural Antigens* (R. MacKay, M. C. Raff, and L. F. Reichardt, eds.), Cold Spring Harbor Laboratory, New York, pp. 209–218.
526. Matthew, W. D., Reichardt, L. F., and Tsavaler, L., 1981, *Cold Spring Harbor Reports in the Neurosciences*, Volume 2, *Monoclonal Antibodies to the Neural Antigens* (R. MacKay, M. C. Raff, and L. F. Reichardt, eds.), Cold Spring Harbor Laboratory, New York, pp. 163–180.
527. Ranscht, B., Clapshaw, P. A., Price, J., Noble, M., and Seifert, W., 1982, *Proc. Natl. Acad. Sci. U.S.A.* **79**:2709–2713.
528. Hawkes, R., Niday, E., and Matus, A., 1982, *Cell* **28**:253–258.
529. Garson, J. A., Beverley, P. C. L., Coakham, H. B., and Harper, E. I., 1982, *Nature* **298**:375–377.

20

Neurochemistry of Invertebrates

Marilyn W. McCaman

1. INTRODUCTION

This chapter attempts to cover some of the recent studies in invertebrates concerned with the biochemical properties of the "classical" neurotransmitters, ACh, 5-HT, DA, and GABA,* as well as some "putative" neurotransmitters, including glutamate and a few peptides. Such properties as distribution, synthesis and breakdown, uptake and transport, and isolation of receptors are discussed when information warrants. Detailed discussion of analytical procedures has been avoided, although it goes without saying that many of these studies could not have been carried out without sophisticated techniques for separation and detection. Information regarding specific methods that concerns sensitivity, specificity, reproducibility, etc. will have to be obtained from the original articles.

The information presented here is intended to update and extend the topics covered in chapters appearing in the earlier edition of this *Handbook* (Volume 2, Chapter 23, Neurochemistry of Invertebrates, and Volume 5, Chapter 10, Chemistry of Isolated Invertebrate Neurons). Since only a single chapter has been allotted to invertebrate neurochemistry, some of the observations are discussed only briefly. Many significant and classic studies published before 1970 are not referenced individually since they were discussed in these earlier chapters. Even so, the number of references is astounding and reflects the upsurge of interest in these preparations.

Earlier chemical studies of invertebrate nervous systems focused largely on the problems of energy generation, of nitrogen and carbohydrate metabolism, and of the chemical composition of various parts of the nervous system, e.g., individual ganglia, nerve cords. More recently, the increased sensitivity

* The following abbreviations are used throughout this chapter: ACh, acetylcholine; 5-HT, 5-hydroxytryptamine or serotonin; DA, dopamine or 3,4-dihydroxyphenylethylamine; GABA, γ-aminobutyric acid; HPLC, high-pressure liquid chromatography; GCMS, gas chromatography–mass spectrometry; RIA, radioimmunoassay.

Marilyn W. McCaman • Division of Neurosciences, Beckman Research Institute of the City of Hope, Duarte, California 91010.

and selectivity of various analytical procedures, such as HPLC, GCMS, and RIA, have allowed accurate measurements of smaller and smaller amounts of cellular constituents. Thus, endogenous levels of various amino acids, amines, and peptides have even been quantitated in some isolated, individual neurons. Such measurements have provided critical information for establishing specific biogenic amines, e.g., ACh, as neurotransmitters. Some of the most direct evidence that these compounds satisfy the criteria set forth has been obtained from invertebrate preparations such as the inhibitory, GABA-containing neurons in lobster ganglia (see Section 3.2) or the 5-HT-containing neurons in cerebral ganglia of gastropod molluscs (see Section 2.3). In addition, the wide range of preparations available has often allowed specific neurochemical problems to be studied more advantageously here than could be done using higher animals.

Sections of this chapter are subdivided into the major phyla and arranged according to the organization of the nervous system as described in several texts dealing with invertebrate nervous systems.[1–3] For a chapter on neurochemistry, these subdivisions are somewhat artificial, since many of the biochemical reactions are common throughout these phyla, even throughout the whole animal kingdom. Nevertheless, such categories may be helpful when attempting to correlate chemical information with the wealth of other information from physiology, morphology, and behavior. A brief description of the nervous system within these groups is given for purposes of orientation.

The "lower invertebrates," which include the Acoelomata and the Aschelminths, are the most primitive animals to possess a real central nervous system. The distinguishing characteristics of this group are a brain and a set of longitudinal medullary cords connected by commissures. These commissures send out processes that form a synaptic nerve net of unipolar, bipolar, and multipolar neurons, but there is no morphological distinction between dendritic and axonic neuron endings. Among the preparations used in this group are the platyhelminths, *Schistosoma mansonii* and *Notoplana actiola*, and the nematode, *Caenorhabditis elegans*.

Molluscs include chitons, clams, snails, octopi, and squid. These animals possess well-defined ganglia, usually paired, joined by connectives and commissures that form a ring around the esophagus and a loop into the viscera. The marine gastropods *Aplysia californica* and *Tritonia diomedia*, as well as the garden snail *Helix*, have been used extensively for chemical, physiological, and behavioral studies.[4–6] Most of the somata in marine animals are pigmented, and, after the connective tissue sheath is cut away, many can be identified and isolated. The cephalopods have the most highly developed nervous systems among the invertebrates and, correspondingly, have a complex behavior pattern. Areas corresponding to the individual ganglia of lower molluscs can be recognized here, but they are now centralized into a complex brain.

The annelids and arthropods comprise much of the Articulata and possess nervous systems of similar basic design. This consists of a ventral "ladderlike" system containing pairs of ganglia connected by a commissure and a brain connected to the rostral end of the nerve cord. In oligochaetes and in polychaetes, the nerve cells may be scattered along the length of the ventral cord, whereas in leeches the ganglia are sharply demarcated. A system of giant fibers occurs in many of these animals as well as in crustacea and insects. Among the annelids, the earthworm *Lumbricus terrestris* and the leech *Hirudo medicinalis* have been intensively studied. General characteristics of arthropods include fusion of ganglia, increased complexity of the brain, and development of a distinct stomatogastric nervous system supplying the digestive system. The horseshoe crab *Limulus polyphemus*, the lobster *Homarus americanus*, the spiny lobster *Panulirus interruptus* among the marine arthropods, *Procambarus clarkii*, a freshwater crustacean, and the cockroach *Periplaneta americana* and the locust *Schistocerca gregaria* among the insects have been used for many neurochemical experiments.

Most invertebrate ganglia are made up of a connective tissue sheath, a rind or cortex of neuronal cell bodies, and a central neuropil free of cell bodies. This arrangement is seen throughout the various invertebrate phyla and is never seen in vertebrates.[7] However, both echinoderms and protochordates are distinctly different from the animals discussed above. The echinoderm nervous system, like those of most radially symmetrical animals, is notable in its lack of centralization. There is a diffuse nerve plexus and a circumoral nerve ring from which nerve cords (radial nerves) arise. There are no true ganglionic complexes; instead, cell bodies are located along the nerve cords or in the epithelium. The protochordates are a diverse group of primitive chordates that lack a backbone but possess, at some time in their life cycle, the three distinguishing chordate characteristics—a notochord, a dorsal hollow nerve cord, and pharyngeal gill slits.

Wherever possible, information regarding biochemical properties of individual neurons rather than heterogeneous populations is noted. Neurochemical studies have been carried out in representatives of all of these phyla, but because of size, accessibility, etc., some animals are used repeatedly. In this way, a wealth of information has accumulated for physiological and chemical maps of neurons and neuronal circuits in *Aplysia, Tritonia, Helix, Panulirus*, and *Homarus*, particularly. It is here in the area of single cell chemistry that invertebrates have made, and will continue to make, a unique contribution to our understanding of neurochemistry in general and of chemical transmission in particular.

2. *BIOGENIC AMINES*

2.1. *Acetylcholine*

Acetylcholine (ACh) is a biogenic amine involved in chemical transmission in both the central and the peripheral nervous systems of almost all invertebrates. Studies have shown that ACh can mediate either excitatory or inhibitory transmission depending on the receptor profile of the postsynaptic cell. In some preparations, it has been identified as the transmitter used by specific neurons, including motor and sensory neurons as well as interneurons. It is synthesized from choline and acetyl-CoA by the enzyme choline acetyltransferase (CAT), and its distribution correlates well with the distribution of this enzyme but not with that of its degradative enzyme, acetylcholinesterase (AChE).

The typical cholinergic neuron is characterized chemically by an ACh concentration of 0.3–3 mM and by the presence of CAT activity in its soma and axon.[8] In addition, the cholinergic cell can transport ACh from cell body to terminals, accumulate choline by a high-affinity uptake system, and release ACh from terminals in a stimulation-dependent, Ca^{2+}-dependent manner.[9] In some preparations, ACh receptors corresponding to the classical nicotinic and muscarinic receptors have been described. However, there appear to be additional classes of receptors (in insects, for example), where some have characteristics of both and others have characteristics distinct from either of the more classical types. Synaptic transmission and receptor pharmacology in invertebrate ganglia have been reviewed in depth elsewhere.[4,8–13]

2.1.1. Lower Invertebrates

In coelenterates, no ACh has been detected and there is conflicting physiological evidence regarding the action of ACh in *Hydra*.[14] Studies of CAT activity, which might resolve the issue of whether ACh is a transmitter in these animals, have not yet been reported. Most of the recent information regarding ACh as a transmitter in the platyhelminths and nematodes is indirect. Few quantitative assays of ACh in nervous tissues have been carried out, although earlier studies indicated that ACh could be detected in planaria, trematodes, and cestodes (Table I; see ref. 4 for references). Choline acetyltransferase activity was demonstrated only in trematodes[6,13] and in the nematode *Nippostrongylus brasiliensis*. In a neuromuscular preparation consisting of the proboscis of the nemertine *Paranemertes peregrina*, ACh had an excitatory effect on the longitudinal muscle but inhibited peristalsis. These effects appeared to be mediated by different receptors, since curare blocked only the latter effect.[15] In another preparation, application of ACh to the muscle of the nematode *Ascaris lumbricoides* also produced contraction. In this case, the effect was enhanced by cholinesterase blocking agents but was not blocked by atropine. Histochemical studies of AChE in this preparation showed positive areas only in the longitudinal muscle. In different studies, AChE was purified from extracts of the nematode *Parascaris squorem* by conventional methods. Analysis of the kinetics of the reaction showed that the preferred substrate was ACh, but the enzyme was only partially inhibited by eserine.[16]

2.1.2. Molluscs

Among the molluscs, the marine gastropod *Aplysia californica* has been studied the most extensively, and several reviews dealing with the biochemical, physiological, and pharmacological properties of cholinergic neurons in these ganglia have appeared.[4,8,10–12,17] Because the somata of these neurons are accessible for physiological studies and can be isolated for chemical studies, the content, metabolism, and transport of transmitters in specific neurons can be studied directly.

2.1.2a. Distribution. Acetylcholine is found in the ganglia of all representatives of this large phyla that have been assayed (Table I), and, in a general way, the concentration of ACh in ganglia tends to increase as the complexity of the nervous system increases. For example, in the nervous systems of various molluscs, including *Patella, Helix*, and *Aplysia*, the ACh levels range from 50 to 200 nmol/g wet tissue, whereas in bivalves and in cephalopods, even higher levels (as much as tenfold) were found.[18]

Surveys of individuals, identified neurons in *Aplysia californica* and in *Tritonia diomedia* showed that high concentrations of ACh were found in only a few neurons, and it was virtually undetectable in others[8,11,19,20] (Table II). The levels of ACh present in different cholinergic neurons varied from approximately 6 to 30 pmol per soma; however, when expressed on the basis of either protein or volume, these values agreed more closely. Thus, among these somata, an average ACh level of 4.2 ± 0.2 μmol/g protein or 0.35 ± 0.1 mM was calculated.[8,11]

In contrast to ACh, choline levels were not significantly different between cholinergic and noncholinergic cells, and the average was approximately 10 pmol/soma. Even though choline is the immediate precursor of ACh, its concentration was equal to or less than that of ACh even in the cholinergic neurons.[19] This relationship is quite different from that in the mammalian CNS, where choline levels are considerably higher than ACh levels.[21]

Both cytoplasmic and particle-bound ACh can be found in these ganglia, and some of the more recent data indicate that the cytoplasmic compartment is more intimately involved with neurotransmission than had been appreciated previously. Subcellular fractionation of *Aplysia* ganglia showed ACh divided between free and bound forms,[22,23] although at the time of these experiments it was supposed that the free ACh was an artifact of homogenization and that ACh was completely bound in vesicles. In later studies, newly synthesized ACh, formed after injection of [^{3}H]choline into a cholinergic neuron, was found predominantly in the "free" form.[24] A novel way of demonstrating the effects of free ACh was reported in experiments in which *Aplysia* cells were injected with AChE.[25] Under these circumstances, newly synthesized ACh could no longer be recovered from tissue extracts, and transmission was blocked after a period of time sufficient for the enzyme to reach the terminals. In noncholinergic control cells, transmission was unimpaired[25] (also see Section 4.9). These observations support the idea that much of the ACh important for initiating synaptic events is, indeed, unbound. Other studies regarding cytoplasmic ACh and its relationship to the vesicular hypothesis have been reviewed in detail[26–30] (also see Section 4.9).

Cephalopods are one of the richest sources of ACh, particularly the head ganglion of the octopus and the optic lobes of the squid. In these animals, however, no chemical substance has as yet been identified as the transmitter at any synapse in the CNS, although ACh has been identified as the transmitter in specific parts of the peripheral nervous system, e.g., in the control of chromatophores.[4,8]

2.1.2b. Metabolism, Uptake, and Transport. Activity of the enzyme CAT has been measured in ganglia of many molluscs (Table III) and in individual

Table I
The Quantitative Distribution of Amines in Some Representative Invertebrate Species[a]

Phylum/class/species	Tissue	DA	NE	OA	5-HT	ACh	HM
Coelenterates							
Actinia equina[109]	Whole body	—	—	—	N.D.	—	2000–5800
Bunodactis regnaudi[109]	Whole body	—	—	—	4300	—	N.D.
Pseudoactinia vayia[109]	Whole body	—	—	—	N.D.	—	N.D.
Anthotoe stimpsoni[109]	Whole body	—	—	—	N.D.	—	N.D.
Nematode							
Caenorhabditis elegans[198]	Whole body	9.5	N.D.	8.5	—	—	—
Molluscs							
Gastropods							
Prosobranch							
Patella vulgata[18]	Pooled g.	620	—	—	39	<10	—
Bivalve							
Mytilus edulis[211]	CNS	2300	220	—	423	—	—
Tapes watlinga[248]	Visceral g.	372	—	215	—	—	—
Anodonta cygnea[212]	CNS	1600	110	—	2300	—	—
Opistobranch							
Aplysia californica	Pleural g.	62	N.D.	.3[258]	108	564[19]	53[297]
	Abdominal g.	388	N.D.	2.0[247]		321[19]	132[297]
				<3[258]			
				2.3[247]	490		
	Buccal g.	286	N.D.	300[258]	191	298[19]	206[297]
				8.5[247]			
	Pedal g.	1738	N.D.	130[258]	520	616[19]	546[297]
				10[247]			
	Cerebral g.	223	N.D.	80[258]	294	483[19]	108[297]
				18[247]			
Tritonia diomedia[159]	Gastroesophageal g.	N.D.	N.D.	—	N.D.	—	—
	Cerebral pleural g.	1260	N.D.		93	—	—
	Buccal g.	348	—	206	—	—	—
	Pedal g.	1877	—	210	—	—	—

Pulmonata							
Lymnaea stagnalis	Pleural g.	340[202]	N.D.	120[b]	710[b]	—	150[b]
	Buccal g.	1320[202]	—	720[b]	1070[b]	—	150[b]
	Left parietal g.	870[202]	—	100[b]	350[b]	—	210[b]
	Right parietal g.	1150[202]	—	3[b]	860[b]	—	100[b]
	Visceral g.	1270[202]	—	30[b]	830[b]	—	90[b]
	Cerebral g.	1080[202]	—	140[b]	690[b]	—	330[b]
	Pedal g.	2570[202]	—	410[b]	2500[b]	—	230[b]
Helix pomatia	Subesophageal g.	530[204]	10[204]	10[204]	303[120]	—	—
	Supraesophageal g.[204]	610	23	6.5	—	—	—
	Buccal g.[204]	1112	12	7.8	—	—	—
	Optic g.[204]	450	11	12.4	—	—	—
Planorbis corneus	Circumesophageal g.	—	—	—	340[226]	—	—
Cephalopods							
Octopus vulgaris	Optic lobe	1140[120]	320[120]	74[248]	303[120]	7800[76]	—
	Vertical lobes[200]	110	334	—	—	—	—
	Circumesophageal g.[264]	777	78	219	—	—	—
	Superior buccal lobe[248]	6530	1010	301	440	—	—
	Stellate g.[248]	55	<4	14	—	—	—
	Gastric g.	366	240	22	—	—	—
Sepia officinalis	Optic lobes	453[264]	189[120]	140[248]	148[120]	—	—
	Vertical lobes[200]	38	68	—	—	—	—
Loligo vulgaris	Optic lobes	329[120]	172[120]	28[264]	50[120]	—	—
	Vertical lobes	33[200]	36[200]	12[264]	—	—	—
Annelids							
Oligochaeta							
Lumbricus terrestris[218]	Cerebral g.	—	—	483	—	—	—
	Subpharyngeal g.			530	—	—	—
	Ventral nerve cord	129	82	349	Trace	—	—

(*Continued*)

Table 1. (Continued)

Phylum/class/species	Tissue	DA	NE	OA	5-HT	ACh	HM
Hirudinea							
Hirudo medicinalis	Whole CNS[161]	111	47	230	—	—	—
Macrobdella decora	Segmental g.	57[158]	Trace[158]	160–650[273]	650[158]	—	—
	Brain	—	—	163[273]	—	—	—
	Ventral nerve cord	—	—	169[273]	1700[160]	—	—
Erpobdella octoculata[273]	Connective	—	—	65–170[273]	—	—	
	Brain	—	—	—	176	—	—
	Ventral Nerve Cord Connective	—	—	—	222	—	—
Arthropods							
Chilepeds							
Limulus polyphemus	Brain	25.8[225]	42[225]	—	118[168]	—	—
	Circumesophageal g.[225]	18.9	9.5	—	—	—	—
	Abdominal nerve cord	38[225]	26[225]	—	565[168]	—	—
	Cardiac g.[225]	54	15	—	—	—	—
Crustaceans							
Homarus americanus	Brain	—	N.D.[385]	14[385]	5.5[171]	12[171]	—
	1st Thoracic nerve cord[385]	—	N.D.	54	—	—	—
	Abdominal nerve cord[385]	—	N.D.	2.7	—	—	—
	Circumesophageal g.[171]	—	—	0.75	N.D.	—	—
	Subesophageal g.[171]	—	—	2.0	3.2	—	—
	Second thoracic roots[171]	—	—	3000	16.6	—	—
	Pericardial organs[171]	—	—	1090	10	—	—
Pacifastacus Leniusculus[174]	Whole CNS	131	8.2	39	8.8	—	—

(Continued)

Insects							
Periplaneta americana	Brain	119[178]	—	—	121[178]	—	—
	Cerebral g.[281]	155	32	85	—	—	—
	Optic lobes[281]	116	10	190	—	—	—
	Subesophageal g.[281]	62	6	105	—	—	—
	Pro-thoracic g.[281]	35	4	71	—	—	—
	6th abdominal g.[281]	29	1	99	—	—	—
	Supraesophageal g.[248]	57	7.2	159	—	—	—
Locusta migratoria	CNS[210]	86	14	—	—	—	—
Manduca sexta, larva	Brain[231]	1410	586	—	93	—	—
	Nerve cord[231]	4840	99	—	114	—	—
Manduca sexta, adult	Brain[109]	—	—	—	—	—	—
Echinoderms							
Asteroids							
Pycnopodia helianthoides	Arm nerve[296]	392	127	16	Trace	—	—
	Tube foot[296]	23	5	5	Trace	—	—

[a] Where necessary original values have been recalculated from ng to nmol using the molecular weight of the free amine or converted to a protein basis by assuming 10% of the tissue wet weight to be protein. All data are expressed as nmol/g protein. N.D., not detected.

[b] J. Ono, unpublished data.

Table II
Neurotransmitters in Individual Identified Neurons

Organism	Cell	Amine	Content (pmol/cell)	Comments
Molluscs				
Aplysia				
californica	C-1	5-HT	10[127]	Also contains 0.2 pmol ACh and 0.18 pmol HM[297,a]
	C-2	HM	1.4[297]	
	C-3	HM	0.64[300]	
	R-2	ACh	26[19]	Also contains 0.15 pmol HM[a]
	L-10	ACh	3.8[19]	
	L-11	ACh	6.2[19]	
	R-14	ACh	0.2[19]	Also contains 0.14 pmol HM and 200 pmol glycine[297,347]
Tritonia diomedia	C1	5-HT	7.1[116]	No DA or ACh detected in these neurons
	Pd1	5-HT	3.5[116]	
	Pd4	5-HT	3.3[116]	
	Pd20	5-HT	0.5[116]	
	Pd21	5-HT	2.2[116]	
Lymnaea	C1	5-HT	0.38[b]	
stagnalis	R.Pd GDC	DA	0.52[202]	
Helisoma trivialis	C1	5-HT	0.09[b]	
	L.Pd GDC		0.80[202]	
Planorbis corneus	GDC	DA	5.4[201]	
Helix pomatia	GSC	5-HT	6.3[48]	
	GSC	ACh	0.8[47]	
Annelids				
Hirudo	Retzius cell	5-HT	2.5[160]	
medicinalis	Retzius cell	5-HT	4.3[158]	Also contained 0.05 pmol DA and 0.02 pmol OA
Macrobdella	Retzius cell	5-HT	0.4[161]	
decora	VL cell	5-HT	0.2[161]	
Arthropods				
Periplaneta	DUM cell	OA	0.08[281]	From abdominal g.
americana	DUM cell	OA	0.14[281]	From metathoracic g.
Panulirus interruptus	Stomatogastric cell	HM	0.12[c]	

[a] R. E. McCaman, unpublished data.
[b] J. Ono, unpublished data.
[c] B. Claiborne, unpublished data.

neurons of a few species.[22,31,32] In *Aplysia*, only those neurons with significant CAT activity also contain measurable amounts of ACh.[19,22] The multiaction interneuron L10 in the abdominal ganglion is one such example, and this neuron has been used for many other studies of cholinergic synaptic transmission.[4,5] In another marine mollusc, *Tritonia*, ACh could be measured in various ganglia and in a few individual, identified cells, but no CAT activity could be detected. These tissues did not contain an inhibitor of CAT activity, so this lack of CAT activity has been difficult to explain.[11] In *Helix*, CAT activity was found in

Table III
Enzyme Activities in Representative Invertebrates[a]

Organism	Tissue	CAT	AChE	AAD	MAO	NAT	HD
Aplysia californica	Buccal g.	0.243[32]	11.5[35]	0.134[139]	N.D.	N.D.	0.032[297]
	Pleural g.	0.196[32]	7.9[35]	0.054[139]	N.D.	N.D.	
	Cerebral g.	0.163[32]	10.5[35]	0.150[139]	N.D.	N.D.	0.156[297]
	Abdominal g.	0.122[32]	7.5[35]	0.105[139]	N.D.	N.D.	0.014[297]
	Pedal g.	0.234[32]	12.5[35]	0.265[139]	N.D.	N.D.	0.058[297]
Helix aspersa	Parietal–visceral g.	0.17[31]	0.74[31]	0.003[31]	—	N.D.[267]	—
	Cerebral	0.25[31]	0.73[31]	0.004[31]	0.9[207]	N.D.[207]	—
	Optic	0.30[31]	1.25[31]	0.004[31]	—	N.D.[267]	—
	Pedal	0.29[31]	0.99[31]	0.007[31]	—	N.D.[267]	—
	Buccal	0.16[31]	1.04[31]	0.001[31]	1.8[207]	N.D.[267]	—
Arthropods							
Homarus americanus	Brain[72]	1.45	144	—	—	—	—
	Subesophageal g.[72]	1.05	194	—	—	—	—
	Thoracic g.[72]	0.92	211	—	—	—	—
	Abdominal g.[72]	0.53	165	—	—	—	—
Astacus leptodactylus	Brain	—	—	0.11[94]	—	0.13[267]	—
Drosophila melanogaster	Brain[181]	5.6	34.9	0.19	0.004	0.18	
	Thoracic g.	0.86	6.4	0.05	—	0.16	—
	Whole flies	0.66	3.9	0.11	0.002	0.25	
Locusta migratoria	Brain	—	—	0.21[94]	N.D.[267]	0.88[267]	
	Thoracic g.[267]	—	—	—	N.D.	0.66	—

[a] Where necessary, original values have been recalculated from ng to nmol using the molecular weight of the free amine, and converted to a protein basis by assuming 10% of the tissue wet weight to be protein. Activities expressed as nmol/μg protein per h.

both the cell body and neuropil layers of the various ganglia. Only about 20% of the individual neurons assayed contained measurable activity, indicating that there was some cell specificity in this species also.[31]

Partially purified CAT has been prepared from ganglia of *Helix*, *Aplysia*, and squid.[22,31,33] Although all preparations required Na^+ or K^+ for activation, there were considerable differences in other kinetic parameters, e.g., K_ms for choline and acetyl-CoA, solubilities, molecular weight. The *Helix* enzyme could not be purified extensively because it was unstable. Much of the information regarding purification and characterization of this enzyme in both invertebrate and vertebrates has been compared and summarized in a recent article.[34]

In most molluscan tissues, activity of the degradative enzyme AChE is very high and can be found in all parts of the nervous system (Table III). Furthermore, this enzyme activity was comparable in all individual neurons, indicating no selective association with cholinergic functions.[23,32] From studies of the effects of different inhibitors or substrates, the enzyme in *Aplysia* has been characterized as "true" AChE, and histochemical evidence indicates that this AChE is associated with the glial sheaths.[35]

Two additional enzymes involved with choline metabolism were measured in ganglia and neurons of *Aplysia*. Choline phosphokinase catalyzes the formation of choline phosphate from choline and ATP, and enzyme activity was found in all ganglia and in all neurons tested.[36] Choline phosphate (sometimes called phospho- or phosphorylcholine) was also one of the products formed by all neurons after injection of [^{3}H]choline (*vide infra*). This enzyme appears to be more directly involved with the synthesis of choline-containing phospholipids than with neurotransmission. Similarly, the activity of phosphorylcholine phosphatase, the hydrolytic enzyme that regenerates choline from phosphorylcholine, was also evenly distributed in all ganglia and in all neurons. Again, this enzyme does not appear to be involved directly in the metabolism of ACh.[11] At present, there is no evidence that choline itself is synthesized by *Aplysia* ganglia.[37]

Isolated *Aplysia* ganglia rapidly take up [^{3}H]choline and convert it to [^{3}H]ACh. Label was recovered almost exclusively from the neuropil and from the axons of cholinergic neurons, and little from the cell bodies.[37–39] This apparent lack of synthesis in cell bodies must be related to the absence of uptake mechanisms for choline, since CAT activity in these structures is high. Choline was taken up by both high-affinity and low-affinity systems. The high-affinity process had a Michaelis constant of 2–8 μM, whereas the low-affinity process was not saturated at more than 400 μM. It appeared that only the high-affinity process was operating under physiological conditions.[38]

When choline was injected into the soma of a cholinergic cell, e.g., L10 of *Aplysia*, it was quickly incorporated into ACh and transported from the cell body to the terminals.[39–41] Within a few hours, radioactivity could be released into the medium, and this amount increased during periods of stimulation in the presence of Ca^{2+}.[40] However, the released label could only be recovered in the form of choline. This observation has raised some doubt that the radioactivity did, indeed, represent release of ACh transported from the cell body to terminals, since many substances taken up by nonneural elements can also be released by electrical stimulation.[30,43,50] On the other hand, it is virtually impossible to inhibit AChE totally, and the small amounts of ACh released could have been hydrolyzed before the sample was removed. Cholinergic function demonstrated in other cells is consistent with the presence of CAT activity and/or the ability to synthesize ACh from [^{3}H]choline.[5,8] For example, the LD_{HI} (cardioinhibitory neurons in the abdominal ganglion of Aplysia) and the LD_G (a motor neuron innervating the gill muscle) have been shown to synthesize [^{3}H]ACh.[9,22,44]

In the *Aplysia* neuron R2, newly synthesized ACh formed after the intrasomatic injection of [^{3}H]choline into the soma was rapidly transported out of the cell body along its axons and into the right connective. This wave of radioactivity advanced along the nerve at a rate of about 17 mm/day.[39,41] A transport system for ACh unique to cholinergic neurons was postulated, supported by the fact that ACh was transported more rapidly than other labeled materials and that this radioactivity in the nerve was associated with a sedimentable tissue fraction.[39] Both orthograde and retrograde transport of ACh could be demonstrated in these neurons.[24]

A different choline transport system has been described in the 5-HT-containing cell (C1) of *Aplysia*.[45,46] In this case, the labeled choline was almost completely metabolized in the cell body into betaine and phosphorylcholine, whereas nerves formed only betaine. Neither ACh nor CAT activity was detected in isolated C1 somata,[19,32] although small amounts of ACh and low levels of CAT activity have been reported in the homologous giant serotonin-containing (GSC) neuron in *Helix*.[31,47] The amount reported was 1% or less of the values found in cholinergic neurons. These observations have generated considerable argument that this is an example of the presence in a single neuron of multiple transmitters. However, whether or not this ACh or CAT activity has functional significance in these neurons has yet to be determined[11,48–50,55–57] (also see Section 6).

In other molluscs, there is much less information available regarding uptake and transport of ACh. Squid optic lobes also contain both a high- and low-affinity uptake system, and here, too, only the high-affinity system is believed to be present in the cholinergic terminals.[51] These tissues and other parts of the cephalopod nervous system, including the giant squid axon and the stellate ganglion, have been used extensively for exploring the properties of vesicles, ACh receptors, etc., and reviews of such studies have been published elsewhere.[27,52]

2.1.2c. Receptors. The receptor profile of gastropod neurons is quite complex, and ACh applied to individual somata hyperpolarizes some and depolarizes others. This is attributed to the presence on the somal membrane of several pharmacologically distinct ACh receptors, each of which mediates a change in a specific ionic conductance. Conductance changes associated with Na^+, K^+, or Cl^-, either singly or in various combinations, have been observed. These different receptor types have been shown to coexist on the same cell membrane and, presumably, produce the multicomponent responses.[4,5,10,12,53–55] α-Bungarotoxin, as well as other snake venoms, has been used to distinguish ACh receptors in *Aplysia*.[56–59] Kinetic parameters that described the binding of α-bungarotoxin in homogenates from *Aplysia* ganglia were remarkably similar to those obtained in vertebrate CNS preparations and are characteristic of nicotinic cholinergic receptors. However, in parallel physiological experiments, the toxin effect was not specific to ACh-induced responses, and snake toxins blocked chloride-mediated responses to both cholinergic and noncholinergic agonists.[59] These data indicated that the specificity of the toxin for a cholinergic receptor, as demonstrated in the binding studies, does not correlate well with the physiological effects of the toxins. Similar discrepancies between binding characteristics and physiological effects of toxins have also been demonstrated in several vertebrate CNS preparations.[58] This is in contrast to the data obtained from neuromuscular preparations, where the effects do correlate well.[60,61]

2.1.3. Annelids

Acetylcholine is probably the excitatory transmitter at the neuromuscular junction of most annelids. Application of ACh to the body wall of various representatives, including the earthworm *Lumbricus terrestris* and the leech *Hirudo medicinalis*, produced a contraction of the muscle that was blocked by curare and was augmented by eserine.[62,63] Many of the earlier studies also proposed that ACh was the excitatory transmitter at these neuromuscular junctions and have been summarized elsewhere.[4,6]

Among this group of animals, only a few studies have been made in which ACh or CAT activity have been assayed (See Table I and ref. 193). In leech ganglia, individual neurons (inhibitory or excitatory motoneurons, mechanosensory neurons, and Retzius cells) were isolated and analyzed for CAT activity.[19] Only the excitatory motoneurons contained measurable levels of CAT activity, although low levels had been reported previously in isolated Retzius cells. When this group of neurons was isolated after incubation with [^{3}H]choline, the excitatory motoneurons had synthesized seven times more ACh than other neurons.[63] The enzyme AChE in leech ganglia had many properties similar to "true" AChE, and activity in detergent-solubilized extracts sedimented in sucrose gradients as two peaks.[66] Recent studies localized AChE activity in individual, identified leech neurons both histochemically and by direct assay. Pretreating ganglia with a nonpenetrating esterase inhibitor, echothiophate, virtually eliminated extracellular AChE, and staining was revealed primarily in motoneurons. The biochemical and histochemical data strongly supported the assumption that this residual activity represented intracellular AChE. Many of the stained cells corresponded to the CAT-containing cells described earlier; however, there were a few cells, such as "anterior pagoda" or the "nut" cells, that contained CAT but did not stain for AChE, or the Leydig cell, which stained for AChE but did not contain CAT.[64,66] Thus, it appeared that under these conditions AChE activity was also a marker for cholinergic neurons but was not as specific as CAT activity.

During embryological development, ACh and the ability to synthesize ACh cannot be detected in the leech *Haementeria ghilianii* until the later stages. Acetylcholine synthesis was quite low even when the ganglia were morphologically complete and their neurons had begun to grow axons. However, once the neurons sent axons into the connectives, and segmental nerves began to display action potentials, the ability to synthesize ACh increased rapidly, reaching a value 25 times higher than the initial value.[67,68] In contrast to the detection of ACh itself, the degradative enzyme, AChE, could be detected early in the development of the leech *Helobdella triserialis* and was traced through embryonic development into adult stages. It could even be detected histochemically in various parts of the germinal plate before formation of nervous tissue. After the nervous system was formed, AChE was found in the neuropil of the segmental ganglia and in the musculature of the body wall, a pattern of staining similar to that seen in adult preparations.[67]

2.1.4. Arthropods

Within this phylum there is good evidence that ACh is the neurotransmitter of many chemosensory and mechanosensory neurons. This evidence comes from pharmacological studies, from experiments measuring the synthesis of ACh from radioactive precursors, as well as from direct assays of ACh or of CAT activity. More recently, some evidence has been presented that some motoneurons are also cholinergic, but such examples are not common. Much of this work has been summarized earlier.[4,8,11,69–72]

Acetylcholine is found in many parts of these nervous systems, but the highest concentrations and the highest synthetic capabilities are associated with sensory, rather than motor, structures (Table I). Data for CAT activity (Table III) and for the synthesis of ACh from [^{3}H]choline[71,72] also gave the highest values in sensory structures. For example, in lobster *Homarus americanus*, ACh synthesis could be detected only in axons of afferent sensory neurons and not in axons of efferent excitatory or inhibitory neurons.[72] Similarly, levels of CAT activity were also approximately 500 times higher in sensory than in efferent axons, whereas levels of choline and AChE activity were comparable in all three axon types.[71] Further, antennular hair cells and nerve, an exclusively sensory preparation, showed the highest level of ACh, in contrast to the dactyl opener and other muscles, which, along with their motor nerves, contained no significant ACh.[72] Only during periods of sensory stimulation could ACh be detected in perfusates, and this amounted to approximately 13 pmol of ACh per minute from thoracic ganglia of the lobster.[69] Acetylcholine was also synthesized from [^{3}H]choline in abdominal ganglia isolated from the ventral nerve cord of the horseshoe crab *Limulus polyphemus* but not in the photoreceptors associated with the nerve cord.[73] The ganglia readily took up choline from the medium, and the newly synthesized ACh was released by high K^+ pulses in a Ca^{2+}-dependent manner.[74]

In the moth, *Manduca sexta*, ACh is also produced and stored in large quantities in the segmental ganglia, in various sensory centers of the brain, in the optic lobes, and in the antennal lobes.[75] These structures synthesized and stored large amounts of ACh, approximately 100 times more than that found for catecholamines or for 5-HT. Direct measurements of CAT activity also indicated that the highest levels were associated with the sensory system[76]; however, localization of CAT activity to individual neurons has not yet been reported. Individual neurons have been isolated from locusts, but none of those tested contained measurable CAT activity.[77]

The activities of CAT and AChE have been measured through the life cycle of the fruitfly *Drosophila melanogaster*[78] and during the metamorphosis of the moth *Manduca sexta*.[76,79] In the former, both enzymes showed little activity in the egg, increasing slowly throughout the various larval stages. Acetylcholinesterase increased rapidly during pupation, reaching adult levels prior to eclosion, whereas CAT remained low until the time of eclosion, when it increased dramatically. In *Manduca*, changes in the level of ACh and in activities of CAT and AChE in the antennal lobes were measured during pupation. These lobes are the target of the axons that arise from sensory neurons in the antennae. The activities of both enzymes (as well as the level of ACh) increased markedly at the sixth day of pupation and maintained this level through eclosion.

Though it has not been demonstrated often, ACh may also play a role as a transmitter at some arthropod neuromuscular junctions. For example, in the stomatogastric ganglion of the crustacean *Panulirus interruptus*, specific motoneurons were assayed and found to contain measurable amounts of ACh and significant CAT activity.[80] Application of ACh to the muscles innervated by these cells produced excitation, and curare blocked the response in the muscle fibers. Although excitatory effects produced by the application of ACh had been reported at the crayfish neuromuscular junction,[81] in more recent experiments with this preparation, ACh did not depolarize the muscle.[82] Neither ACh nor CAT activity was found in the lobster abdominal slow flexor muscle and its innervating nerve,[72] indicating that ACh is not concentrated at these neuromuscular junctions.

Synaptosomal fractions prepared from various arthropods have been used to study the distribution of ACh as well as the uptake of choline. In preparations from the cockroach brain, synaptosomes were enriched in ACh compared with the original homogenates, but only 30% of the total ACh could be sedimented. This finding was attributed to the fragility of the nerve ending particles rather than to the presence of substantial amounts of ACh in the cytoplasm.[83] Synaptosomes were also prepared from the locust brain using a microscale "flotation" technique that yielded well-preserved organelles presumed to be derived from predominantly cholinergic nerve endings. These synaptosomes showed both a low- and high-affinity component, but only the high-affinity uptake was Na^+ dependent and inhibited by hemicholinium-3.[84] Synaptosomes prepared from ventral cords of *Homarus* and *Limulus* also took up [^{3}H]choline into an osmotically labile compartment, and these fractions also showed enrichment of the CAT activity.[72,85,86]

The mechanism of ACh synthesis and the kinetics of the reaction were studied in partially purified CAT preparations from *Drosophila* and from the locust *Schistocerca migratoria*.[41,87,88] More recently, by a series of affinity chromatography procedures, the enzyme from *Drosophila* has been purified more than 10,000-fold, and the final preparation showed three major protein bands (67,000, 54,000, and 13,000 daltons) after polyacrylamide gel electrophoresis.[89] Structural and immunologic studies showed that these proteins were closely related. It was proposed that the two lower-molecular-weight forms resulted from a specific posttranslational modification of the largest form and that these two polypeptides were tightly associated in the native state. Antibodies selective for the *Drosophila* CAT were produced using monoclonal antibody techniques, and two stable cell lines were isolated and maintained in culture.[90] Neither antibody reacted with CAT-containing extracts of nervous tissue from vertebrates, molluscs, or even other insects. Further, by Scatchard analysis of their reactions with *Drosophila* CAT, these antibodies appeared to be nonidentical, monoclonal, and highly selective for *Drosophila*.

Highly purified AChE has been prepared from many different sources. In *Homarus*, Triton X-100-solubilized extracts yielded a fraction with a single peak of activity by velocity sedimentation analysis that had the characteristics of a "true" AChE.[72,85] In insects, AChE is also a particulate enzyme tightly bound to membranes, but here it appears to be more susceptible to disaggregation by solubilizing agents. Multiple forms of the enzyme were always found in preparations from the brains of houseflies, fruit flies, house crickets, and the tobacco hornworm.[91–93] Many studies have also been directed toward determining whether the enzyme structure has been modified in insecticide-resistant strains,[92–94] but to date, the results are inconclusive.

Mutants of *Drosophila* that have no AChE activity usually die early in development, but some mutants displaying abnormalities in CAT and/or AChE activities do survive. Often these animals also show abnormalities in neuroanatomy, electrophysiological properties, and/or behavior patterns. In a few instances, behavior deficits in genetic mosaics could be correlated with morphological defects in specific regions of the nervous system.[95] For example, mosaics lacking AChE on one side of the optic lobe or in the posterior inferior protocerebrum showed deficits in oculomotor behavior, although they appeared normal in other movements and in posture. These genetic mosaics present some interesting opportunities to correlate biochemical, morphological, and behavioral studies.

Three different ACh receptors have been identified in the insect CNS. These receptors are defined as "nicotinic" binding sites characterized by their ability to bind α-bungarotoxin, as "muscarinic" binding sites characterized by the ability to bind quinuclidinylbenzilate, and as mixed "nicotinic–muscarinic" sites defined by the binding of decamethonium.[96,97] In addition to these pharmacological specificities, some receptor preparations display differences in solubility characteristics, and this property has been used in their purification.[98–101]

Fractions with high-affinity binding for α-bungarotoxin have been studied in nervous tissue extracts of *Drosophila, Musca, Manduca*, and *Periplaneta*.[100–104] These preparations possess many of the characteristics expected of an ACh receptor: heat lability, saturable binding properties, and inhibition of toxin binding by various nicotinic blocking agents as well as by ACh itself.

Autoradiographs of frozen sections showed that toxin binding in *Drosophila* was restricted to synaptic areas of the CNS and in *Manduca* to the neuropile of the antennal lobes.[103]

The nicotinic ACh receptor in *Drosophila* and in *Periplaneta* has been suggested as a site of insecticide action rather than the enzyme AChE. Two compounds, nicotine and a new synthetic compound, 2-isothiocyanatoethyltrimethylammonium iodide, are both insecticides as well as agonists for the receptor. Further, they inhibited α-bungarotoxin binding at physiologically relevant concentrations.[105] Parallel electrophysiological experiments on the cercal nerve giant fiber synapse in the abdominal ganglion of *Periplaneta americana* showed that both of these compounds also depressed the evoked EPSP at these concentrations. If these conclusions remain valid when applied to a wider variety of insects and insecticides, they will have a profound effect on the design of compounds to be synthetized for this purpose.

2.1.5. Remarks

The ACh system was the first neurotransmitter system to be defined, and its effects in peripheral nervous systems has been studied very intensively. Within the last 10 years tremendous progress has been made in describing the complex actions of this biogenic amine in central nervous systems, particularly on individual nerve cells, and in describing the capability of such cells to synthesize, inactivate, store, and transport ACh. In some cases the amount released by specific neurons could be measured.

Invertebrate preparations, such as the brains of cephalopod molluscs or fruitflies, have proved to be excellent sources of the purification of CAT and for purification of ACh binding proteins, although more recent binding studies used the electric organs from elasmobranchs such as *Torpedo*.[106] In addition, the association of ACh and cholinergic function with CAT rather than AChE activity has been demonstrated convincingly in the isolated, identified neurons of the gastropod molluscs. The studies described in this section represent one of the best examples in which the multidisciplinary approach to neurobiology has been employed productively. The combination of biochemistry, physiology, pharmacology, and morphology has provided a synergistic effect that has increased our understanding of cholinergic actions in addition to cataloging the results. These studies serve as a model for the study of other neurotransmitters, for which the evidence, as well as the understanding of their effects, is less complete.

2.2. 5-Hydroxytryptamine

5-Hydroxytryptamine (5-HT, serotonin) is another biologically active amine widely distributed among all invertebrates and has been demonstrated to be a neurotransmitter in many systems. In most nervous tissues 5-HT can be visualized using formaldehyde-induced or glyoxylic acid-induced fluorescence, and fluorescing material has been seen in neuropil, in axons, and occasionally in specific neurons. Bioassays were used previously to detect 5-HT but have now been replaced to a large extent by more direct chemical procedures, e.g., radioenzymatic, HPLC, and GCMS assays.

The synthesis of 5-HT from the amino acid precursor, tryptophan, is a pathway common to both invertebrates and vertebrates. In this pathway, tryptophan is hydroxylated to 5-hydroxytryptophan (5-HTP), which is subsequently decarboxylated to 5-HT. In all systems studied, the hydroxylation step, catalyzed by tryptophan hydroxylase, appears to be rate limiting. The second enzyme in this sequence, aromatic acid decarboxylase (AAD), is a relatively nonspecific enzyme and catalyzes the decarboxylation of 5-HTP to 5-HT as well as the decarboxylation of DOPA to DA (see Section 2.3).

Degradation of 5-HT by special enzymes is not a general feature of invertebrate nervous tissue. Whereas in vertebrates, monoamine oxidase (MAO) is the most common metabolic route, this pathway is found only occasionally among invertebrates. Brains of cephalopods are exceptions and contain an active MAO that deaminates 5-HT in addition to other biogenic amines. Instead of metabolism, uptake of transmitter from the synaptic regions is postulated most often as the mechanism for terminating the actions of 5-HT.

The physiological role of 5-HT in neurotransmission in various invertebrates is discussed here only briefly. Details of the pharmacology of 5-HT and various antagonists, and the characterization of 5-HT receptors in terms of ionic conductance changes, have been discussed elsewhere.[4,6,10,107,108]

2.2.1. Lower Invertebrates

5-Hydroxytryptamine was detected in extracts of whole animals of various coelenterates and platyhelminths using a bioassay procedure.[13,14,109] Formaldehyde-induced fluorescence showed the yellow, light-sensitive fluorescence typical of 5-HT in neurons in the cerebral ganglia and in the ventral nerves of the planaria *Phagocata oregonensis*, *Dugesia tigrina*, and *Procotyla fluviatilis*[110] and of the parasitic trematode *Schistosoma mansoni*.[111] Studies using glyoxylic acid-induced fluorescence showed that two compounds were present in the nemertine *Paranemertes perigrina*. One, whose spectrum closely matched that of 5-HT, was present in the proboscideal plexus; a different compound, whose spectrum matched that of NE, was found in the proboscideal nerve cords.[15] These findings were consistent with the physiological finding that application of 5-HT (or NE) stimulated electrical and contractile activity. Earlier studies had also reported 5-HT in extracts of various platyhelminths (see ref. 6 for references); however, since it has been pointed out that the amino acid lysine interferes with 5-HT assays using the ninhydrin fluorescence procedure,[112] these results need to be repeated.

A combination of techniques localized 5-HT in the rectal ganglion and in the body wall muscles of the nematode *Aspiculuris tetraptera*.[113] Nerve processes containing 5-HT were also found in the pharynx of another nematode, *Caenorhabditis elegans*,[114] where 5-HT appears to play a role in pharyngeal pumping. These cell bodies, located only after tissue was incubated with 5-HT, were identified as neurosecretory motoneurons. It had been suggested earlier that 5-HT may also act as a cilioregulatory agent, since application of 5-HT-blocking agents inhibited locomotion.[115,116]

Only a few studies have been concerned with synthesis and metabolism of 5-HT in these animals. Aromatic amino acid decarboxylase (AAD) activity was measured in *Caenorhabditis*,[114] and high-affinity uptake systems in the trematode *Schistosoma mansoni* and in the nematode *Phocanema decipiens* have been described.[117,118] Histochemical studies in *Aspiculuris* showed no MAO activity.[113] In another nematode, *Faciola hepatica*, a 5-HT-sensitive adenylate cyclase was found. Enzyme activity was increased fivefold by low concentrations of 5-HT and augmented further by GTP.[119] These observations on cyclase activities were similar to those seen in higher animals.

2.2.2. Molluscs

2.2.2a. Distribution. 5-Hydroxytryptamine was first isolated from octopus salivary glands and from bivalve hypobranchial glands some 30 years ago, and many of the earlier studies of its actions, e.g., the effects on heart and other muscles, were carried out in molluscs. Some values for 5-HT in the nervous systems of representatives of this phylum are given in Table I. It has been found in various classes of molluscs, although within a given nervous system, it is not uniformly distributed. For example, in *Octopus vulgaris*, it was concentrated in the optic lobes and in the retina (which contains projections from the optic lobe) but could not be detected in other regions.[120] The concentrations in some of these molluscan ganglia are among the highest found in all invertebrates (Table I).

Formaldehyde-induced fluorescence in ganglia of gastropods revealed several large, yellow-fluorescing somata in the cerebral ganglia of various gastropod molluscs.[121,122] In the bivalve *Mytilus*, such cells were also seen in the visceral ganglion.[123] The fluorescence characteristics of various identified cells in the brains of *Helix aspersa* and *Planorbis corneus*, along with their distinguishing electrophysiological and pharmacological properties, have been summarized.[124–126] The neurons studied in most detail are the giant serotonin-containing cells (GSC) found in cerebral ganglia. In *Aplysia*, they are called C-1 cells and occur as a pair of large (250 μm diameter), pigmented cell bodies in the medial rostral portion of the ganglion. They contain amounts of 5-HT ranging from 0.5 to 7 pmol per cell[127,128] (Table II). When calculated on the basis of volume, these values range from 1 to 10 mM, typical of the concentration of other neurotransmitters in specific cells.[8] In addition, GSCs have also been described in *Tritonia, Lymnaea, Helix*, and *Helisoma*. These neurons appear to be homologous among the various gastropods on the basis of physiological, morphological, and embryological evidence (see Section 6.1).[121,122,126–131] Detailed descriptions of the ion conductances involved, the effects of various blocking agents, the synaptic connections, and the role of these neurons in various aspects of feeding behavior have been reviewed in recent publications.[5,8,131–133]

A few other identified neurons contained measurable amounts of 5-HT. In the abdominal ganglion of *Aplysia*, it was found in some neurons isolated from the RB cluster; however, individual neurons can only be identified physiologically (R. McCaman, unpublished observations). One of the neurons in this cluster, RB_{HE}, produced a long-lasting accelerating action on the heart, a stimulation that could be mimicked by 5-HT.[9] In *Tritonia*, several identified neurons in the pedal ganglia were also among those that contained 5-HT. One pair (L and R pedal 21) induced pedal ciliary beating on direct stimulation, and this effect was mimicked by adding 5-HT to the bathing fluid or by injecting it into the animal.[116]

2.2.2b. Metabolism, Uptake, Release, and Transport. Both direct and indirect procedures have been used to demonstrate the synthesis of 5-HT from its precursors. After ganglia were incubated with radiolabeled tryptophan, only 5-HT-containing neurons contained labeled 5-HTP and 5-HT, showing that tryptophan hydroxylase activity was localized to these cells.[129,134] Hydroxylase activity measured in bivalves was found to be inhibited by DA in a dose-dependent manner. Since DA lowered 5-HT levels in ganglia of *Mytilus edulis*, it was suggested that through this inhibition of tryptophan hydroxylase activity, DA regulates the intracellular 5-HT concentrations.[135,136] Another enzyme inhibitor, *p*-chlorphenylalanine, reduced 5-HT in *Planorbis* brain when administered to the whole animal.[137]

Further evidence of 5-HT synthesis has been obtained by injecting radiolabeled precursor amino acids into individual *Aplysia* neurons. Injections of tryptophan into various cells in the abdominal ganglion produced 5-HT only in some cells in the RB cluster, again indicating specificity of hydroxylase activity, whereas injection of 5-HTP produced 5-HT in all cells.[138] This latter observation is consistent with direct measurements of AAD in *Aplysia*, where decarboxylase activity was found in all cells, although the activity was approximately 20-fold higher in 5-HT-containing cells.[139] In contrast, AAD activity in individual cells in *Tritonia* was restricted exclusively to the 5-HT-containing neurons. In these somata, activity ranged from 50 to 80 pmol/μg protein per h, whereas no activity could be detected in other identified neurons.[116,127] Thus, only in some animals does the presence of AAD parallel precisely the distribution of 5-HT, whereas tryptophan hydroxylase is a more reliable marker for serotoninergic cells.

In Table III, some values for the activities of various synthetic and degradative enzymes in ganglion preparations of representative molluscs are given. Of the enzymes listed, only AAD is widely distributed. Most gastropod molluscs show very little MAO activity,[134,139] although the enzyme was present in the bivalve *Mytilus*.[140] In cephalopods, MAO is a major degradative pathway.[52] and administration of enzyme inhibitors such as pargyline to *Octopus vulgaris* produced a significant decrease in the 5-HT oxidation products detected in various lobes of the brain.[120] N-Acetyltransferase (NAT) activity was also very low in molluscs and is not considered to be a major metabolic pathway for 5-HT (see Section 5.2).

High-affinity uptake systems have been described in several studies. In *Helix*, this system was Na^+ dependent and operated at 0°C.[142] Uptake was blocked by imipramine and other antidepressant drugs, substances that also potentiated and prolonged the synaptic potentials mediated by the 5-HT-containing neurons.[126,129,134] In *Mytilus*, both high- and low-affinity systems were described.[143] This high-affinity system (K_m approximately 3×10^{-7} M) also required Na^+ and was sensitive to chlorimipramine.

5-Hydroxytryptamine taken up by ganglia of various molluscs both *in situ* and *in vitro* has been localized at the level of both the light and the electron microscope. Autoradiographs of *Helix* ganglia after exposure to [^{3}H]5-HT showed silver grains not over cell bodies but rather only over fine axon branches and processes that contained small, dense-cored vesicles.[129] In the marine mussel *Mactra stultorum*, labeled 5-HT also accumulated in both glial cells and their processes in the cortical areas and the neuropil axons but not in cell bodies.[144] Earlier results in *Aplysia* had also shown that most of the radioactivity taken up from the medium was associated with elements of the connective tissue sheath (see ref. 4 for earlier references). Thus, it appears that only nerve endings can take up 5-HT and that somata lack this ability. In contrast, after incubation with 5-HTP, perikarya of the GSC were filled with silver grains, and few were seen at nerve endings.[129] These neurons can take up 5-HT precursors, convert them into 5-HT, and then transport this 5-HT along their axons to nerve terminals.

Metabolism of 5-HT was studied after injection of the labeled compound directly into cell bodies or after it was taken up by ganglia and nonnervous tissues. Injected neurons, C-1 as well as others, and various tissues, such as kidney, formed a metabolite described as a glucuronide of 5-HT. In addition to this metabolite, incubated ganglia formed a second metabolite, which appeared to be a more complicated sugar conjugate.[145] These metabolites were separated by electrophoresis and chromatography and partially characterized by the recovery of [^{3}H]5-HT after acid hydrolysis and after β-glucuronidase digestion, but no further structural studies have been carried out. This second product may be formed in glia, and it has been proposed that these elements play a dual role; i.e., they inactivate monoamines released at synapses in the neuropil via uptake and protect somata from monoamines diffusing into the ganglia via metabolism.[144]

5-Hydroxytryptamine is rapidly transported down the C-1 neurons of *Aplysia*. After injection of [^{3}H]5-HT into the soma, radioactivity appeared in the axons within 2 h, and export continued at a rate ranging from 60 to 120 mm/day thereafter. The label, recovered largely as free 5-HT, was located primarily in the axons of the injected cell and was sequestered by unique lysosomal organelles that appear to function as storage depots for the transmitter.[146] When 5-HTP, GABA, or choline was injected into the soma, they moved more slowly along the axon and at rates consistent with diffusion rather than with fast transport. Fast transport of 5-HT was not observed in cholinergic neurons, also indicating that the transport system described above was selective for the 5-HT-containing neurons.[145–149]

Transport systems for 5-HT have also been described in other molluscs. Morphological studies in *Helix* indicated that [^{3}H]5-HT was transported to nerve endings after being injected into GSCs.[150] Further, the particulate fractions isolated by high-speed centrifugation contained more than 80% of the 5-

HT injected into the cell body, whereas non-5-HT cells contained only 6% of the label in this fraction.[126] These observations are consistent with a vesicle-bound distribution of 5-HT within the GSC. In the octopus, labeled 5-HT taken up by brain was also transported. In this case, it moved from the superior buccal lobes to the posterior salivary glands, where it accumulated in large quantities.[151]

Release of 5-HT from nerve endings in response to stimulation has been demonstrated in only a few preparations. It could be detected in the bathing fluid after repetitious stimulation of the anterior byssus retractor muscle in *Mytilis*[152] and after direct stimulation of the C-1 neuron in *Aplysia*.[153] In the latter case, 5-HT could only be detected when an uptake-blocking agent was also present.

Adenylate cyclases in *Aplysia* ganglia are sensitive to low concentrations of both 5-HT and DA. The total amount of cyclic AMP as well as the synthesis of cyclic AMP from [^{3}H]adenosine were stimulated by 5-HT, and these effects were observed in whole ganglia, in connectives, and in identified cell bodies.[154] However, pharmacological evidence suggests that these effects are associated with structures other than postsynaptic receptors, since antagonists that block 5-HT receptors of these neurons do not block the cyclic AMP effects.[10]

2.2.3. *Annelids*

By use of fluorescence microscopy, numerous 5-HT-containing neurons can be seen in the nervous systems of *Nereis* and *Lumbricus*.[141,155] These cells were present along the entire length of the cord but were more numerous within ganglia and were often grouped at the site of entry of segmental nerves. A similar arrangement of 5-HT-containing neurons is seen in arthropods (*vide infra*). The fluorescence intensity was increased by administration of MAO inhibitor, iproniazid, and decreased by reserpine.[155] Chemical measurements also confirmed these histochemical results.[156] In all leeches, the Retzius cells contained a high concentration of 5-HT measured both by direct chemical assay and by histofluorescence.[17,107] In *Hirudo medicinalis*, they are large, easily identified cells and exist as pairs in each of 21 segmental ganglia. There is reasonable agreement that each cell contains approximately 2–4 pmol 5-HT[157–160] (Table II). In another leech, *Macrobdella decora*, where the Retzius cells are smaller, they contained only 0.4 pmol/cell. When calculated on the basis of volume, the concentration in these latter Retzius cells was approximately 10–20 mM. Another group of neurons isolated from *Macrobdella* ganglia, the VL cells, also contained 5-HT, and the concentration in these cells was also very high (>100 mM).[161] Synthesis of 5-HT occurred only in Retzius cells after ganglia were incubated with radiolabeled precursors, and assays of AAD activity in isolated somata showed that these cells have about 75 times more activity than control somata.[63,65] Even in the very early stages of development of the leech, AAD activity was found in Retzius cells, and this activity increased in proportion to cell size during the later stages.[66]

Localization at the EM level of radioactivity in leech ganglia incubated with labeled 5-HT showed silver grains concentrated in Retzius cells, primarily in granules located within the pigment layers.[162] When labeled 5-HT was applied to the body wall, grains were also found in chromaffin cells of the vasofibrous tissue as well as in axons and terminals of Retzius cells.[163]

The function of 5-HT in these annelids is not well understood. Application of 5-HT *in vitro* to muscle strips of *Lumbricus* caused contractions, whereas application to intact polychaete larvae elicited mucus secretion, a function also attributed to the Retzius cell of leeches.[17,107] Structure–activity studies carried out on Retzius cells from both *Hirudo* and *Hemopsis* showed that the 5-HT receptor was of only a single type and different from 5-HT receptors characterized in other preparations.[107,164] In *Hirudo*, application of 5-HT to other identified neurons produced a depolarization of certain sensory neruons but a hyperpolarization of a motoneuron.[161] Recently, 5-HT-containing cells in embryos of the giant leech *Haementeria ghilianii* have been selectively destroyed by treatment with a cytotoxic agent, 5,7-dihydroxytryptamine. These embryos develop into juvenile leeches that have difficulties in swimming, yet display normal swimming movements after injection with 5-HT.[165] Further studies correlating 5-HT levels with function at various stages of development and in adults are in progress, and these may be helpful in determining the physiological role of these neurons.

2.2.4. Arthropods

Quantitative measurements of 5-HT in the nervous systems of chilepeds crustaceans, and insects have been reported (Table I). Significant quantities were found in the brain and in the abdominal ganglion of *Limulus*, and in the brains of the crustaceans *Pacifastacus, Uca*, and *Carcinus*,[168] and in adult blowflies.[176]

Quantitative measurements of 5-HT in these nervous systems show that it is found in most ganglia but is present in low concentrations relative to other biogenic amines (Table I). Among the various arthropods listed, 5-HT levels were considerably higher in insects than in crustaceans or in chilepeds.

Formaldehyde-induced fluorescence has revealed 5-HT in limited parts of these nervous systems. In the horseshoe crab *Limulus*, yellow fluorescence was seen in specific cells in the eye. These cells, the eccentric cells, took up 5-HT from the medium and were depolarized by the application of 5-HT,[166] and it was postulated that 5-HT is associated with lateral inhibition between ommatidia in these animals.[167]

In the crab *Carcinus maenus*, a pair of yellow-fluorescing neurons could be seen in the last thoracic segment of the ventral ganglion mass.[168] The pericardial organ of this and other crabs are also rich in yellow as well as green (DA-containing) fibers.[169,170] In the lobster *Homarus americanus*, 5-HT was concentrated in the pericardial organ in addition to another neurosecretory area associated with the second thoracic roots.[171,179] In the spiny lobster *Panulirus interruptus*, the ligamental nerve plexuses (structures homologous with the pericardial organs of other crustacea) also showed intense fluorescence.[172] Chemical measurements confirmed that these structures contained 5-HT and DA as well as large amounts of another biogenic amine, octopamine (see Section 2.5). Other areas rich in 5-HT are the brain and eyestalks of the fiddler crab *Uca pugilator*[173] and of the crayfish *Pacifastacus leniusculus*.[174] In the eyestalks of *Uca* and of the crayfish *Cambarellus shufeldi*, 5-HT is believed to control the release of the red-pigment-dispersing hormone rather than to act directly on the chromatophores.[175]

Formaldehyde-induced 5-HT fluorescence was also described in parts of the CNS of the cockroaches *Blaberus giganteus* and *Periplaneta americana*[71] and could be localized to specific nerve cells in the locusts *Schistocerca gregaria* and *Locusta migratoria* and in the noctuid moth *Spodoptera littoralus*.[177] These specific cells were found in areas associated with the visual system, however, a number of insects do not appear to contain 5-HT in their nervous tissues.[178]

Synthesis of 5-HT from radiolabeled precursors was demonstrated in isolated thoracic roots and in the pericardial organ of *Homarus*[171] and in the ligamental nerve plexuses of *Panulirus*.[180] In *Homarus*, both 5-HT synthesis and high-affinity uptake were localized to one of the four morphologically distinct nerve endings found in the roots where high concentrations of 5-HT were found. Both high- and low-affinity uptake mechanisms for 5-HT, as well as for tryptophan, were described. The 5-HT taken up into these tissues did not appear to be metabolized further, and it could be released by a pulse of K^+ in a Ca^{2+}-dependent manner.[171,179]

In insects, high rates of 5-HT synthesis were found both in the optic lobes and in the abdominal ganglion of the sphinx moth, *Manduca sexta*, and populations of 5-HT-containing neurons are also more concentrated in these areas.[75] Direct measurement of AAD activity in *Drosophila melanogaster* showed higher activity in brain than in thoracic ganglia (Table III). Another enzyme, N-acetyltransferase (NAT), has also been measured in *Drosophila*[181] and in the honeybee, *Apis mellifera*,[182] and activity was found to be evenly divided among the ganglia. Although N-acetylation appears to be a major degradative pathway for other amines in insects, little N-acetyl-5-HT accumulated in these tissues[178] (see Section 5.2). The only uptake system described was in the antennal lobe of the locust, where formaldehyde-induced fluorescence was observed in specific fibers after incubation with various indolylethylamines.[183]

Receptors for 5-HT have been studied in the salivary gland of the blowfly *Calliphora erthyrocephala*.[184] Low concentrations of 5-HT enhanced the rate of secretion, and this system has been used to define the structural requirements for the 5-HT receptor. The model formulated from these data resembled that proposed for this 5-HT receptor in leech[164] but differed considerably from 5-HT receptors in molluscs.[4,10]

Adenylate cyclases stimulated specifically by low concentration of 5-HT have also been described in the nervous systems of some insects.[185] This enzyme was found in homogenates of the thoracic ganglia of the cockroach *Periplaneta americana*[186] and in the larval nerve cord of the moth *Manduca sexta*[187] but not in the brain of the berthan armyworm *Maemestra configurata*[188] or in preparations from *Drosophila* heads.[189] The significance of the distribution of this enzyme and/or its relationship to the effects of 5-HT in these systems is still unclear.

2.2.5. Remarks

A tremendous amount of evidence has accumulated indicating that 5-HT is a transmitter in almost every invertebrate nervous system. In each phylum, there are examples of its presence, of synthetic pathways and uptake systems, and, in a few cases, axoplasmic transport, vesicular localization, and release have also been noted. The Retzius cell of the leech was the first 5-HT-containing neuron to be studied intensively.[18,107,158] More recently, the thoracic roots of the lobster have provided an example where synthesis, storage, and high-affinity uptake systems can be demonstrated within a single nerve ending.[171] Other preparations, notably the 5-HT-containing neurons of the gastropod molluscs,[126–128] have also provided some of the most definitive evidence currently available in any nervous system correlating distribution and metabolism of 5-HT with physiology, morphology, and behavior.

Yet in spite of all these data, it seems curious that the total number of 5-HT neurons found, for example, in leech ganglia is very small, fewer than ten out of several hundred cells. Similar limited numbers of 5-HT-containing cells, detected either by direct chemical measurement or by histofluorescence, are found in molluscs and crustacea. On the other hand, the numbers of fluorescent fibers coursing through many different nerve tracts is quite large. Thus, either the few recognizable neurons branch extensively to produce magnificent and far-flung axonal trees or many potential 5-HT-containing neurons have been overlooked by the present detection methods. It will be interesting in future studies to see whether new histochemical procedures, i.e., by using antibodies to 5-HT, will reveal significantly more positive-reacting neurons than have been revealed to date.

One suggestion for the high concentration of 5-HT seen in neuropil and nerve tracts versus somata concerns selective processing of the transmitter as it travels from the soma through the axon toward the target nerve terminals. In this case, 5-HT might be sequestered within somata in a chemically altered state and, therefore, would not be accessible to the reagents ordinarily used for fluorescent or radioenzymatic assay. Peptidoamines containing 5-HT could represent such an altered state,[190] although none have yet been found in invertebrates (also see Section 5.1). On the other hand, some processes capable of transforming peptides during transport have been described in axons of specific *Aplysia* neurons (see Section 4.6). There is little doubt that 5-HT acts as a transmitter in many invertebrate systems, but more precisely how it functions remains obscure.

2.3. Dopamine

Dopamine (DA; 3,4-dihydroxyphenylethylamine) is the most abundant of the catecholamines in invertebrate nervous systems. Histochemical studies using formaldehyde- or glyoxylic acid-induced fluorescence show that DA is associated with a large number of neurons and their processes, and recent improvements in chemical assays have allowed DA to be measured in a few somata in gastropod molluscs.[8,191–194] Like many of the other biogenic amines,

DA is a potent agonist, and application of low concentrations to various neuronal preparations can produce both excitatory and inhibitory effects. Discussions of the different DA receptors, classified according to ion conductance changes, can be found in other reviews.[4,6,10,107,108,192,193]

Synthetic enzymatic pathways for DA are similar throughout the invertebrates. The amino acid tyrosine is the precursor, and it is hydroxylated to form DOPA, which, in turn, is decarboxylated to form DA. The enzymes catalyzing these reactions are tyrosine hydroxylase (TH) and AAD. The latter is a relatively nonspecific enzyme that decarboxylates both DOPA and 5-HTP to their respective amines, and some of its properties and its distribution have been discussed in the previous section. Tyrosine hydroxylase is the rate-limiting step in this synthesis, and its activity has been measured in some DA-containing neurons. Oxidation of DA via MAO is a pathway present in only a few species, and inactivation of DA is usually attributed to uptake mechanisms at nerve endings.

2.3.1. Lower Invertebrates

Animals as primitive as protozoa have systems that synthesize DA from tyrosine and DOPA[192]; however, the earliest example of catecholaminelike material in an organized nervous system has been seen in sea anemones. Green formaldehyde-induced fluorescence was detected in the sensory nerves of both *Metridium senile* and *Taelia felina*, and analysis of extracts prepared from the oral zone of these animals showed that both DA and DOPA were present.[1,14,195] In a separate study, histochemical studies localized MAO activity to a discrete layer in the neural zone.[196]

In planaria and in the trematodes *Schistosoma mansoni* and *Fasciola hepatica*, green-fluorescing cells and tracts were found throughout the central ganglia and nerve cords,[197] whereas in the nematode *Caenorhabditis elegans*, fluorescence was associated with specific mechanosensory cells.[198] In the latter study, reserpine depleted fluorescence from the axons but not from the cell bodies of these neurons. More detailed summaries of physiological and pharmacological studies have been published recently.[193]

2.3.2. Molluscs

Dopamine is found in large amounts in most molluscan nervous systems, and its distribution in various representatives is given in Table I. In these examples, DA concentrations were usually higher than those of any of the other biogenic amines. In the marine gastropod *Aplysia californica*, all the major ganglia except the pleural were conspicuously high. The largest proportion of this DA was found in the neuropil, and the connective tissue sheath and the outer core of somata and glia contained very little.[160,199,208] The richest source of DA reported was the superior buccal lobe of the cephalopod *Octopus vulgaris*.[200]

Direct chemical measurements of DA in isolated neurons of some freshwater gastropods have confirmed earlier observations based on formaldehyde-induced fluorescence (see Table II).[48,201,202] In these neurons from *Planorbis*, *Lymnaea*, and *Helisoma*, the concentration of DA was estimated to be 1–5 mM. The left pedal ganglion of the freshwater pulmonate snail *Planorbis corneus* and the right pedal ganglion of another freshwater pulmonate snail *Lymnaea stagnalis* each contain a large neuron that fluoresces green after formaldehyde treatment, typical of DA. Direct chemical assay of isolated cells confirms the presence of DA.[201,202] Mapping of the synaptic connections of these and other cells has shown that these nervous systems are mirror images of each other and that the DA-containing cells are homologous.[48,233] No DA-containing neurons have been found in the marine gastropods *Aplysia* and *Tritonia*, although there was an earlier report of a green-fluorescing neuron in the latter.[203]

Assays of TH activity in *Helix* showed that this enzyme was associated with nervous rather than nonnervous tissues.[191] whereas in *Planorbis* it could be localized to specific DA-containing cells.[204,205] As noted above, the distribution of AAD was cell specific in some molluscs and not in others. In *Helix*, *Planorbis*, and *Tritonia*, the activity of this enzyme was limited to amine-containing neurons; however, in *Aplysia* AAD activity was detected in all neurons sampled (see also Section 2.2.2).[31,124,139,191]

Inactivation of DA in molluscan tissues is postulated to be primarily via reuptake mechanisms at nerve endings, although enzymes capable of inactivating DA are also present. The MAO pathway appears to be significant in *Patella, Mytilus, Helix*, and in several cephalopods[120,207]; however, in the freshwater bivalves and in *Aplysia*, no MAO activity could be detected[139,205,206] (Table III). Studies of this enzyme in cephalopods show that, like the vertebrate enzyme, it acts on a wide range of monoamines.[52] Another enzyme that metabolizes DA, catechol-O-methyltransferase, has been found in a variety of tissues in addition to nervous tissue, and its activity was evenly distributed among all the ganglia and all individual neurons.[35]

Uptake of DA into various ganglia has been characterized in the bivalves *Quadrula pustulosa*[206] and *Mytilus edulis*,[217] in the gastropods *Aplysia*[199] and *Helix*,[191,210] and in the cephalopod *Loligo pealis*.[212] Both high- and low-affinity uptake processes were present, but only the high-affinity system was Na^+ dependent. In *Mytilus* ganglia, the uptake of DA was influenced by the presence of other amines such as 5-HT, suggesting a feedback mechanism for regulating tissue levels of each amine.[211] In both bivalves, the majority of the labeled DA taken up was recovered unchanged as DA,[143,209] whereas in *Aplysia* substantial amounts of DA were converted to a new, unidentified compound (M. McCaman, unpublished observations). Ultrastructural studies of [^{3}H]DA taken up by *Quadrula* ganglia showed that the majority of this radioactivity was localized over synaptic vesicles.[209] Detailed morphological studies of the giant DA-containing neuron in *Planorbis* were carried out after injection of [^{3}H]DA. Autoradiographs showed extensive branching of the axon extending through several other ganglia, and synaptic connections appeared to be of the *en passage* type.[213] Similarly, autoradiographs of another mussel, *Mactra stultorum*, showed that silver grains were concentrated over dense-core granules in many axons. However, in this case, grains were also associated with glial processes, and cell bodies were not labeled.[144]

Pharmacological studies of the effects of DA in various molluscan neurons characterized the complex population of receptors, and these results have been summarized in other articles.[4,6,10,214] Dopamine binding in *Mytilis* tissues showed high specificity and high affinity properties, and analysis of the kinetic parameters suggested that more than a single type of DA receptor was present.[135] Opiate binding sites have also been described in molluscs, and it has been proposed that these drugs modulate dopaminergic systems in invertebrates via a presynaptic release mechanism analogous to that proposed in vertebrates. Inhibition of DA release by both morphine and the enkephalins, and reversal of this effect with naloxone, has been noted in *Mytilis* and *Octopus* ganglia,[217] and an increase in DA concentration in the presence of opiates was reported in *Anodonta* and in *Helix*.[215,216] In contrast, no binding of [^{3}H]spiroperidol, a marker for DA receptors in vertebrate tissues, was detected in the cephalopods *Sepia* and *Octopus*.[218] At present, it is not clear whether this reflects structural differences in the receptors or methodological differences in the assays or both.

2.3.3. Annelids

Dopamine has been measured quantitatively in only a few of these nervous systems (Table I; see also ref. 193). In general, values for DA were higher than those for NE but were considerably lower than those for OA in these tissues. Incubation of reserpine with isolated nerve cords reduced DA values by 30% but reduced NE even more.[155]

In the ventral nerve cord of the leeches *Hirudo medicinalis* and *Macrobdella decora*, the first three segmental ganglia have two catecholamine-containing neurons whose somata lie within the anterior roots rather than in the ganglion itself.[107] In embryos of another leech, *Haementeria ghilianii*, these cells appeared late in the course of development, about the time of body closure.[68,219] Although the formaldehyde-induced fluorescence of DA is green in most tissues, areas in these nervous systems that fluorescence blue are also considered to be catecholamine containing. However, until DA is assayed quantitatively in a system that resolves it from other catecholamine compounds, it may be in error to attribute all of this fluorescence to DA *per se* (*vide infra*). Such blue fluorescence was also seen in the neuropil of these leeches,[107] in the longitudinal fibers in the nerve cord, and in segmental nerves of the earthworm *Lumbricus terrestris*[193] and the polychaete *Nereis*.[141]

2.3.4. Arthropods

Histochemical fluorescence (using both formaldehyde and glyoxylic acid) and chemical assays were used to study the distribution of DA in arthropods. In spiders (*Arachnida: Araneida*), fluorescence was seen in many areas of the nervous system, but it was not associated with specific cell bodies,[220] whereas in crustaceans many green intensely fluorescent cell bodies could be seen. Some examples include the cerebral ganglia and eyestalks of the crab *Carcinus maenus*,[169] the stomatogastric system and the pericardial organs (or ligaments) of the lobsters *Homarus* and *Panulirus*,[221,222] and the optic lobes and eye stalks of the crayfish *Pacifastacus leniusculus*.[195,223,224] In insects, the most intense fluorescence was seen in the protocerebral neuropile, although the distribution varied considerably from species to species.[71,177,178,230]

Quantitative measurements of DA are consistent with the histofluorescent studies, and values for various parts of the nervous system of representatives of this phylum are given in Table I.[168,174,225–228] The values for insects are comparable to those found in molluscs and are much higher than values for other arthropods.

Although the amount of DOPA in most nervous tissues is relatively small relative to the amount of DA, in various parts of the crayfish brain it was equal to or greater than the amount of DA, and in connective tissue it was considerably higher than DA.[174] In these studies, DOPA was separated from DA by HPLC, and the results showed that the content of DA in crude extracts was overestimated without an adequate separation scheme. The presence of both fluorescent compounds may also explain the lack of effect of reserpine in depleting the "DA" stores in connective tissue.[174,227] Extracts of the ventral nerve cord of the boll weevil analyzed after HPLC separation revealed DOPA, NE, and a new catecholic compound tentatively identified as deoxyepinephrine.[228]

Synthesis of DA from [^{3}H]tyrosine was studied in the stomatogastric nervous system of *Panulirus interruptus*. The commissural ganglia and the stomatogastric ganglia together with their connectives contained the highest amounts of [^{3}H]DA.[229] Synthesis of DA, as well as of NE, was also measured in the thoracic nerves of *Homarus*.[179] However, no synthesis occurred in the lateral eyes and optic nerve of the horseshoe crab *Limulus*.[345] In the moth *Manduca sexta*, synthesis was detected primarily in the segmented ganglia, with only small amounts of labeled DA found in the optic lobes and the protocerebrum.[231]

More direct studies of the enzymes involved in DA synthesis were made in the locust *Locusta migratoria* and in the fruit fly *Drosophila melanogaster*. The locust AAD was more specific than most, and the enzyme decarboxylated DOPA but not 5-HTP, although in other respects (pH optimum, etc.) it was similar to preparations described in other invertebrates.[233] The activity of AAD was measured during the various developmental stages in *Drosophila* and was found to increase dramatically during the latter periods of pupation, just prior to eclosion.[234] There was no evidence of MAO activity in either insect,[181,194] although it could be detected histochemically in the brain of cockroaches and of spiders.[220,226] Another enzyme, DA-sensitive adenylate cyclase, was studied in the brain of the moth *Mamestra configurata*, and this activity could be distinguished from the octopamine-sensitive enzyme.[232] N-Acetyltransferase activity was also surveyed in various arthropods, but significant activity was found only in insects. These studies are discussed in more detail in a later section of this chapter (Section 5.2).

2.3.6. Remarks

Although there is little doubt that DA is an important neurotransmitter in many invertebrate systems, it has been difficult to establish this role in a completely satisfactory manner. In any one preparation, only a few of the criteria have been satisfied, and no specific example stands out as a good model for dopaminergic neurotransmission. On the other hand, fragments of information have been collected from a variety of sources. Specific DA-containing neurons have been visualized, and both synthetic reactions and uptake mechanisms have been studied in many different animals, whereas DA receptors have been characterized in only a few. The most complete evidence in a single preparation is that obtained in the GDC of the snail *Planorbis*, where both physiological and biochemical results support the candidacy of DA as a neurotransmitter.[48]

As in the case of 5-HT, the large amounts of DA measured in nerves and in neuropil of various ganglia, e.g., *Aplysia*, contrast markedly with the absence of cell bodies that contain DA. Even in those species in which DA-containing cell bodies have been found, their number is very few; yet the neuropil and nerve trunks are intensely fluorescent. Similarly, although DA uptake mechanisms seem to be absent from *Aplysia* soma (M. McCaman, unpublished observations), DA-synthesizing enzymes are present in individual neurons,[127] so that it has been surprising that no free DA accumulates in cell bodies.

2.4. *Norepinephrine and Epinephrine*

Norepinephrine (NE) is not usually considered to be an important neurotransmitter in the nervous system of most invertebrates, although recent studies have indicated that platyhelminths may be an exception. High concentrations are found in octopus and in some crayfish, but significant amounts were found in gastropod molluscs only rarely.[191–195] Quantitative estimations of these catecholamines rely on adequate separation by chromatography or by solvent extraction from the (usually) much larger amounts of DA. Most of the assay procedures are quite tedious, although recently described HPLC methods using electrochemical detectors hold promise of increased resolution and sensitivity.[174]

Synthesis of NE from DA is catalyzed by the enzyme dopamine β-hydroxylase (DBH), and this activity usually parallels the presence of NE. The degradative pathways (MAO, COMT, reuptake into the tissue) are similar to those described previously for DA and will not be discussed in detail (see Section 2.3).

Norepinephrine has been reported to be the catecholamine present in the nervous system of various platyhelminths and is postulated to be the predominant monoamine in the nervous system of polyclads.[235,236] This identification is based primarily on fluorescence techniques, although it has recently been shown that this identification may be in error. Not only do the fluorescence spectra of glyoxylic acid derivatives of various catecholamines show a great deal of overlap, but some amino acids such as tyrosine, glutamate, and aspartate also produce fluorescent compounds in the presence of this reagent.[236] The most recent findings indicate that the fluorescent compound in the nerve cord of *Gyrocotyle fimbriata* is not a catecholamine, and, although it was not identified, it had many characteristics of an amino acid.[237] A high-affinity uptake system for NE has recently been described in the human blood fluke, *Schis-*

tosoma mansoni,[238] but there have been no reports on synthesis and metabolism.

Norepinephrine has also been found in the nervous systems of the bivalves, *Mytilus* and *Anodonta*, and of cephalopods, but it is absent from most gastropods (Table I). In cephalopods NE concentrations are exceptionally high in the buccal and the optic lobes, and the distribution parallels, in general, that of DA in these animals.[239] Among the gastropods, NE has only been reported in the terrestrial snail *Helix pomatia*.[204,240]

In arthropods, modest concentrations were found in the crab supraesophageal ganglion, in *Limulus* ganglia and nerve cord, in crayfish brain and eyestalks, and in all portions of the cockroach and blowfly nervous systems (Table I and ref. 176). Levels of NE comparable to those of DA and 5-HT were reported in the crayfish *Pacifastacus leniusculus*.[174] The only example of a NE-containing cell was found in the syn-ganglion of the tick, *Bophilus*. This posterodorsal neuron contained NE at concentrations greater than those of other catecholamines.[241]

Administration of reserpine depleted, whereas MAO inhibitors increased, levels of NE in those tissues where it was found in relatively high concentrations, e.g., in nerve cord of *Lumbricus*,[157] in octopus ganglia,[239] and in crayfish brain.[174] In slices of octopus brain, dihydroxyphenylacetic acid was the only MAO metabolite formed, indicating that DA rather than NE was oxidized via this pathway.[242] Synthesis of NE from radioactive precursors was reported in *Mytilus*[143] and in octopus brain[242] but could not be demonstrated in the stomatogastric ganglion of the spiny lobster, *Panulirus*,[229] or in the adult brain of the moth, *Manduca sexta*.[243] High-affinity uptake systems were described in the optic lobes of the squid, *Loligo pealis*,[244] in *Mytilus*,[143] and in the abdominal nerve cords of the cockroach.[245]

Application of NE to the eyes of the fiddler crab *Uca pugilator* releases the melanin-dispersing hormone. This peptide hormone (also see Section 5.1) triggers the rapid movement of pigments in the melanophore, a mechanism by which the function of the eye changes.[3] By use of a series of adrenoreceptor blocking agents, this hormone-releasing effect of NE was shown to be mediated predominantly by postsynaptic α_1 receptors.[246]

Epinephrine (E, N-methyl-NE) appears to be absent from most molluscan and arthropod nervous tissues,[177,178] even from the optic lobes of octopi that contain high concentrations of NE.[239] However, E was recently detected in *Limulus* but could not be quantitated because it was incompletely resolved from other catecholic compounds.[225]

2.5. Octopamine

For many years neurobiologists considered octopamine (OA) more as a pharmacological curiosity than as a bona fide transmitter. In spite of the fact that it was found in much greater concentrations in invertebrate than in vertebrate systems, it had previously been considered only as an intermediate in the synthesis of NE or as a metabolic product of the trace amine, tyramine. Only recently has its own unique role in nervous tissues been more clearly defined, and, based on these data, OA has been proposed as a neurotransmitter in invertebrates.[193,247–251] In the crustacean nervous system, many of the criteria for establishing it as a neurotransmitter have been satisfied, i.e., it is present in individual neurons, it has an action identical with nerve stimulation, and synthesizing systems have been demonstrated. In addition, high-affinity uptake systems in annelids and in insects have also been described. Many physiological effects evoked by OA are, indeed, neurotransmitterlike, although its actions at the peripheral nervous system are best interpreted at the moment as "modulatory."[250,252–254]

For many of these studies, a radioenzymatic procedure was used for the determination of OA, although more recently, methods involving HPLC and GCMS have also been introduced.[174,255–260,266] The identification of "natural" OA as the D-isomer has now been firmly established. By a HPLC separation of diasteroisomers, it was shown that D-octopamine was formed from radiolabeled L-tyrosine in brains and in homogenates of snails, crayfish, and lady beetles.[260] Thus, all results for which a racemic DL mixture of OA had been used for quantitation need to be reevaluated.

2.5.1. Lower Invertebrates

Very little information regarding OA has been reported in the lower invertebrates. Octopamine has recently been measured in extracts from the adult nematode *Caenorhabitis elegans* using the radioenzymatic procedure (Table I). During development, OA concentrations increase fivefold from the level found in eggs and larva to that found in the adults.[114] Mutant strains of nematodes with well-characterized neuronal defects have been developed, and some of these lack detectable levels of OA. In addition, animals are being studied in which individual neurons have been ablated by a laser microbeam.[261] From correlations of presence or absence of neurons with physiological, behavioral, and chemical studies, a role for OA in these animals may be uncovered.

2.5.2. Molluscs

Salivary glands of the octopus were the source from which OA was isolated originally. Later, OA was also found in the nervous system of cephalopods, and these values, e.g., in the buccal lobes, are still among the highest reported in invertebrates (Table I). Subcellular fractions prepared from octopus nerves show that the greatest portion of the OA was in the synaptosomal fraction.[262]

Another rich source of OA in molluscs is the visceral ganglion of the clam *Tapes watlinga*.[257] This value was much higher (10- to 30-fold) than those reported in gastropods[264] (Table I). In *Aplysia*, the buccal ganglion contained the highest concentration, whereas in the abdominal and pleural ganglia, no OA could be detected.[258,265,266] The neuropil region, an area rich in nerve endings and synaptic connections, contained most of the amine. No OA could be detected among the more than 50 individual neurons sampled,[265] although an earlier study had reported OA in several neurons in the abdominal ganglion.[256]

In general, the levels of OA found in gastropod ganglia were approximately 20–100 times less than those for the other biogenic amines, e.g., DA and 5-HT (Table I).

In metabolic pathways in which tyrosine is the precursor, hydroxylation rather than decarboxylation appears to be the rate-limiting step,[248] as illustrated in studies of β-hydroxylation using deuterium-labeled tyramine and isolated ganglia from *Tapes*.[257] These studies are noteworthy because the isotopically labeled compound ([^{2}H]tyramine) is not radioactive, and the product had to be assayed using a GCMS procedure. In *Octopus*, where levels of catecholamines, OA, and tyramine are all high, incubation of brain tissues with [^{3}H]tyrosine produced primarily [^{3}H]DA and NE rather than [^{3}H]OA.[242]

No MAO activity (using OA as a substrate) could be detected in the ganglia of the gastropods *Helix* and *Aplysia*, although considerable activity was observed in the anterior byssus retractor muscle of *Mytilus*.[267] In *Octopus*, where activity is very high, MAO has many characteristics similar to the mammalian enzyme (see also Section 2.3.3). Administration of pargyline sufficient to cause inhibition of enzyme activity resulted in an increase in OA levels in various parts of these brains.[248]

Radiolabeled OA was actively taken up by isolated ganglia of *Helix* or of *Aplysia*. In *Helix*, this uptake was shown to be Na^+ dependent, but it was not a high-affinity process.[210] In *Aplysia*, preliminary results indicated that label was taken up by the connective tissue elements rather than by neurons and that OA was partially metabolized to a new, uncharacterized product.[265]

Increased cyclic AMP synthesis, as well as increased incorporation of ^{32}P into phosphoproteins, was observed when OA was incubated with isolated *Aplysia* ganglia.[268] Such phosphorylated proteins may be involved in mediating the action of cyclic AMP in these tissues (see Section 5.3).[269,270]

When applied iontophoretically to spontaneously firing individual neurons of *Helix* or of *Aplysia*, OA inhibited this activity.[249] Responses in *Aplysia* could be elicited only when OA was applied in the neuropil region, and the induced hyperpolarization probably resulted from increased K^+ conductance.[271] In *Helix*, five cells could be identified that were inhibited by the application of OA, and one cell was found that was excited. Usually, cells that responded to application of OA also responded to DA and NE; however, OA receptors could be distinguished pharmacologically from these other receptors.[272]

2.5.3. Annelids

All portions of the annelid nervous system, i.e., the brain, the ventral nerve cord, as well as the connectives, were rich sources of OA. Values for the earthworm *Lumbricus terrestris* and the leeches *Erpobdella octoculata, Macrobdella decora*, and *Hirudo medicinalis* are given in Table I. The CNS of *Lumbricus* contained the highest concentration, yet outside the nervous system, very little OA was found.[273]

In leech, both tyrosine and tyramine have been shown to be precursors for OA in the ventral nerve cords. When the nerve cord of *Erpobdella* was incubated with [^{3}H]tyrosine, substantial amounts of tyramine and lesser amounts of OA and DA were formed. In the same system, OA was the principal product formed from [^{3}H]tyramine, and only traces of DA were detected.[63,274] Oxidation deamination is an important degradative pathway, since *p*-hydroxymandelic acid, the end product in the MAO pathway for the metabolism of OA, has been identified in leech ganglia, and, further, incubation in the presence of a MAO inhibitor, iproniazid, increased the OA concentration in these tissues.[273]

Both high- and low-affinity uptake systems were described in leech ganglia. The high-affinity system ($K_m = 4 \times 10^{-7}$ M) is saturable and is sensitive to Na^+, and it is this system, presumably, that removes neurally released OA.[274] In these studies, nerve cords preloaded with [^{14}C]OA released label after direct stimulation or after depolarization with high concentrations of K^+, and this release was dependent on Ca^{2+}.

Reports of the effects of OA applied to the annelid nervous system are limited, and receptors specific for OA have not yet been described.[248,275] However, an adenylate cyclase has been described in *Lumbricus* that is stimulated about threefold in the presence of 10–100 μM OA, although here, as in other systems, the precise relationship between cyclase and receptor is still unclear. Of the various biogenic amines tested, only OA and 5-HT produced the stimulation, and these effects were additive, suggesting that the stimulatory actions were separable.[275] In contrast, the effect of NE and histamine were slightly inhibitory.

2.5.4. *Arthropods*

Octopamine has been measured in the brains and nerve cords of various members of this phylum (Table I). Low concentrations of OA are found in most crustacean tissues, although a few, such as the second thoracic roots of lobsters or pericardial organs of various crustaceans, are dramatic exceptions. In *Homarus*, there are specific cells located within the second roots of each thoracic ganglion that stain with neutral red and contain high concentrations of OA. Morphological studies of these roots revealed that four distinct types of nerve endings were present, one of which was associated with the OA-containing cells.[171,252,276] Terminals of these cells contained large dense-cored vesicles but made no obvious contacts with target cells, which is typical of neurosecretory cells. Some root cells also send processes to the pericardial organs, structures that release biogenic amines including OA into the hemolymph.

There is some indication that OA is also associated with the photoreceptor system in the horseshoe crab *Limulus polyphemus*, where it was found in the optic ganglion. Photoreceptor-enriched fractions prepared from the ventral nerve readily synthesized OA from labeled precursors, although no free OA was detected in these fractions.[73] In the locust, OA was also high in optic lobes,[251] and in the firefly *Photuris*, OA has been reported in the terminal section of the nervous system that contains both nerve endings and the lantern. The luminescence of the lantern elicited by nerve stimulation can be mimicked by the application of both OA and its N-methylated derivative, synephrine[277];

however, there are no reports that synephrine is present in significant amounts in this or any other invertebrate.

In insects, relatively high concentrations of OA were found in various areas of the brain and ventral nerve cords[247–251] in addition to the stomatogastric system and neural–hemal structures such as the corpora cardiaca.[278] Octopamine has been measured in individual neurons in the metathoracic ganglion of the locust[254,279,280] and the cockroach[281] by both direct and indirect chemical methods. These neurons, the dorsal unpaired median (DUM) neurons, form a distinct cluster on the dorsal midline of each segmental ganglion and modulate both neuromuscular transmission and the responsiveness of the muscle, although they are not motoneurons.[253,254] The OA concentration in one of the DUM neurons was at least 800 times that of an identified motorneuron. Octopamine was detected in all individual DUM neurons assayed, whereas another biogenic amine, 5-HT, could not be detected in single cells.[282]

Octopamine levels in DUM cells have been correlated with embryological differentiation in the grasshopper, *Schistocerca niger*.[282] No OA was detected in ganglia before day 13 even when pooled samples were assayed. From day 13 to day 20, there was a continuous increase in OA that correlated well with the morphological differentiation of the DUM neuron and with the onset of neutral red staining of these neurons.[283] Similarly, during the metamorphosis of the moth *Mamestra configurata*, the levels of OA is also rose rapidly during the 10 days prior to eclosion.[284]

Synthesis of OA from radiolabeled precursors has been studied in many different arthropods. Initial studies showed that injection of the radioactive precursor [^{14}C]tyrosine into lobster nerves resulted in the formation of both OA and DA, but not NE.[179] Further, these authors postulated that this synthesis took place in two different kinds of nerve cells. The OA-synthesizing cells would contain AAD and dopamine-β-hydroxylase activities but not tyrosine hydroxylase, whereas the DA-synthesizing cells would contain tyrosine hydroxylase and AAD but not dopamine-β-hydroxylase. Later, it was demonstrated that, indeed, the OA-containing root cells incubated with [^{3}H]tyrosine took up the amino acid and converted it to OA only. These root cells also released newly synthesized OA when exposed to a pulse of K^+, and the process was Ca^{2+} dependent.[276]

In the ligamental nerve plexuses and in the stomatogastric and commissural ganglia of the spiny lobster *Panulirus interruptus*, the ability to synthesize OA from labeled tyrosine or tyramine was also demonstrated.[180,229] The ligamental nerve plexuses in *Panulirus* appear to be homologous with the pericardial organs of other crustacea and also to release OA by a Ca^{2+}-dependent mechanism following electrical stimulation.[180] Synthesis of OA was also demonstrated in nervous tissue from the moth *Manduca sexta*.[75] In this preparation, the segmental ganglia formed more OA from [^{3}H]tyrosine than other ganglia, but no synthesis occurred in the antennal lobe.

The activity of N-acetyltransferase (NAT) is very high, particularly in insects, and could be the preferred pathway for inactivating amines in these animals. In *Drosophila*, NAT is a relatively nonspecific enzyme and acetylates all biogenic amines,[181] whereas NAT preparations from the crayfish *Astacus leptodactylus* appear to be more specific[267] (see Section 5.2). Monoamine oxidase activity is very low or completely absent in nervous tissues of most arthropods (Table III).

Uptake systems in the cockroach can be divided into three components: a high- and a low-affinity system, both of which are sensitive to Na^+, and a Na^+-insensitive component that is not saturated up to 100 μM.[245] The high-affinity system was postulated to be the inactivation mechanism for neurally released OA, thereby satisfying one of the criteria for neurotransmitter candidacy. In insects, OA is readily taken up by neural tissue but does not appear to be metabolized,[250] whereas in lobster, two novel metabolites of OA were detected.[285,286] When ganglia from *Homarus americanus* were incubated with radiolabeled OA, one compound formed was identified as the O-sulfate conjugate of OA, and the other as the peptidoamine, β-alanyl-OA (for further discussion, see Section 5.1).

Application of OA to certain individual neurons in the *Limulus* CNS elicited a hyperpolarization,[287] whereas other actions of OA seem to be more indirect. Octopamine increased the rate and strength of the heart beat and modulated synaptic transmission at the crustacean neuromuscular junction.[250,252,288] In the crayfish, OA has a strong facilitatory action in the nerve–muscle system rather than a direct action on the muscle, and the effects are long lasting.[289] These observations have been interpreted as indicating that OA has a modulatory, rather than a transmitterlike, effect in these preparations.

Structural requirements for OA receptors, defined as the OA-sensitive adenylate cyclase in the cockroach brain and thoracic ganglia of the cockroach and by the development of luminescence in the firefly lantern system, have been carried out.[249,277,290] Except for the fact that, in the firefly system, synephrine was considerably more potent than OA itself, there was a definite similarity in the effects of various structural analogues in these systems. These OA receptors were characterized as the α-adrenergic type, clearly differentiated from those of dopamine and/or adrenoreceptors.[185,186] In lantern homogenates, adenylcyclase activity was stimulated approximately 20-fold by 10^{-5} M OA. Octopamine-stimulated cyclases have also been studied in the brains of the moth *Mamestra configurata* at various stages of metamorphosis,[232] in the heads of *Drosophila melanogaster*,[189,263] and in *Limulus*.[291] In the latter, adenylate cyclase was found in both the protocerebrum and the circumesophageal ganglionic ring; however, only in the latter tissue was the cyclase responsive to OA. The cyclic AMP content of intact ganglia of the cockroach was increased after incubation with OA, and these effects were potentiated by theophylline[186] (see Section 5.3).

Cyclases have also been suggested as possible mediators of insecticide action in addition to their known effects on glycogenolysis.[188,290] The insecticides, chlordimeform and N-demethylchlordimeform, mimic the effect of OA as an activator of adenylate cyclase and bind specifically and reversibly to OA receptors. In addition, they also mimic the actions of OA on the intact firefly light organ and on the locust neuromuscular junction.[292,293] Whether these effects are directly related to the "cidal" action of these compounds has yet to be firmly established, although the correlations observed so far are very good.

2.5.5. Remarks

This biogenic amine is now established as a neurotransmitter in invertebrates, although the number of systems using it is probably more limited than those using DA and 5-HT. There is no sensitive histochemical method for OA, so that distribution studies have relied heavily on "hot zap" techniques demonstrating synthesis or on chemical surveys in order to locate OA-containing neurons. Neutral red staining has been used to locate individual neurons in some species, although this stain is also taken up by DA- and 5-HT-containing neurons in many species.[294] Unfortunately, this stain is not taken up by such cells in all invertebrates. Although it has been used effectively in annelid and arthropod preparations, it has been of limited use in molluscs, particularly marine gastropods such as *Aplysia*, where selective uptake mechanisms are absent.

In most invertebrates, OA is found in much higher concentrations than NE, although the ratio is reversed among vertebrates.[248] Representatives from two of the higher invertebrate phyla, Echinoderms and Chordata, are exceptions in that the OA/NE ratio is low. In fact, in a recent report, no OA could be detected in the tunicate *Ciona intestinalis* (Phylum: Chordate).[293,295,296] These phyla together with the vertebrate phyla comprise the deuterosomes, a classification based on embryological development,[3] whereas the remaining invertebrate phyla above the level of coelenterates are protostomes. Although the reasons for this divergence are unknown, the shift from OA- to NE-mediated transmission appears to be recapitulated in mammals, where OA-synthesizing regions were replaced by NE-synthesizing regions during the early phases of development in the rat.[296] Although OA appears to be more plentiful in invertebrate nervous systems, there are still few preparations in which OA appears to act as a classical neurotransmitter. Investigations into its more indirect actions may contribute to a greater understanding of this amine in both vertebrates and invertebrates.[272,289,320]

2.6. Histamine

Histamine (HM) has been added to the group of biogenic amine neurotransmitters only recently, now that evidence has been presented demonstrating this role in a convincing manner. Among the gastropod molluscs, two preparations have been described in which HM-containing neurons are accessible for direct study. These are the cerebral ganglion of *Aplysia* and the visceral ganglion of *Lymnaea*, where HM has been localized to specific neurons and their processes.[297–300] In gastropods, physiological, pharmacological, and morphological studies all support the candidacy of HM as a neurotransmitter.[300–305] However, in other preparations, very little information regarding the distribution and chemistry of HM is presently available. Values for HM in the nervous systems of some invertebrates are given in Table I.

Handling of the common sponge, *Suberites inconstans*, causes itching and swelling of the fingers, and this phenomenon may be be caused by the presence of large amounts of HM in these tissues (approximately 300 nmol/g protein).[307]

The tentacles and body structures of some sea anemones also contained high concentrations of HM (220–1000 nmol/g protein), although others contained little to none.[109] In the group of anemones assayed, it was curious that those with high levels of HM contained no 5-HT, and, conversely, those with high levels of 5-HT contained no HM. There were also several species that contained neither HM nor 5-HT.

Among the molluscan ganglia studied, the highest concentrations of HM were found in the cerebral and pedal ganglia in *Aplysia californica*.[297] It could also be detected in those nerve trunks emerging from the cerebral ganglion that contained processes from histaminergic cells (*vide infra*). No HM could be detected in the connective tissue sheath overlying the ganglia. Individual identified neurons containing high concentrations of HM have been discovered thus far only in the cerebral ganglion.[298,300] These cells (C-2, C-3) were located symmetrically in each lateral rostral quadrant and contained HM at a concentration of approximately 0.4–4 mM (0.4–4 pmol per cell; Table II). Although HM was undetectable in most other isolated somata assayed, a few cells (R-2, R-14, and C-1) contained HM at a level approximately 1% of that found in the C-2 soma.[298,299] Whether this HM is functionally significant in these neurons has yet to be established.

In the freshwater snail, *Lymnaea stagnalis*, HM was detected in a neuron from the visceral ganglion,[301] although none could be detected in other neurons assayed. This "giant visceral neuron" also took up [^{3}H]HM when the isolated ganglion was incubated *in vitro*, and electron micrographs prepared from these neurons showed aggregations of large dense-cored vesicles.[301,308]

Histamine is synthesized by decarboxylation of the amino acid histidine. In *Aplysia*, the enzyme catalyzing this reaction, histidine decarboxylase (HD), has biochemical properties that distinguish it from other amino acid decarboxylases such as AAD (Table III). Surveys of enzyme activity in isolated cells showed that HD activity was located almost exclusively in the HM-containing neurons.[297,298,306] Similarly, after incubation of intact ganglia with the precursor [^{3}H]histidine, labeled HM was only found in these cells, whereas all cells contained the amino acid. Inactivation of HM appears to occur via reuptake mechanisms or by conjugation to glutamate or β-alanine to form peptidoamines[310] (see Section 5.1). Only in a few animals, such as cephalopods, does oxidative deamination of HM occur.[52] Inactivation by N-acetylation has been described in arthropods (see Section 3.2), but inactivation by N-methylation, the major degradative pathway in mammals, does not appear to be important in invertebrates.

Transport of [^{3}H]HM and other labeled substances after injection in these cell bodies showed that some label moved rapidly out of the cell body and into the axon at a rate of approximately 50 mm per day, although some moved more slowly. Exposure to high concentrations of colchicine or to low temperatures only partially blocked transport, suggesting that both fast transport and diffusion processes were operating.[309] The portion of radioactivity found in the axon was recovered as unchanged HM, whereas that in the cell body consisted of both HM and its metabolite, γ-glutamyl-HM. Subcellular fractionation of these injected somata showed label incorporated into particulate material, and this sedimentable radioactivity was not affected by reserpine.

Uptake mechanisms for HM have also been studied. In *Helix*, the kinetic properties of HM uptake were typical of a carrier-mediated process with a K_m of 10^{-5} M, and this process could be divided into Na^+-sensitive and Na^+-insensitive components. Uptake could be inhibited by various structural analogues as well as by the α-adrenoreceptor blocking agent, phenoxybenzamine and by chlorimipramine. Histamine accumulated by soaking ganglia in a solution of [^{14}C]HM was released by pulses of K^+ in a Ca^{2+}-dependent manner.[301] Incubations longer than 5 min were complicated by the rapid appearance of a HM metabolite. In both *Aplysia* and *Helix*, radiolabeled HM was taken up by the isolated ganglia, but only part of this radioactivity could be recovered as HM.[302,310] In *Aplysia*, HM was rapidly converted to a new product characterized as γ-glutamyl-HM (see Section 5.1); however, the labeled product in *Helix* has not yet been identified.

In *Aplysia*, HM-containing neurons make monosynaptic connections with a series of other neurons within the cerebral ganglion, and application of HM to the postsynaptic cells mimicked the synaptically-evoked response even where these responses were multicomponent in nature. Pharmacological studies of the effects of various blocking agents, e.g., cimetidine, were consistent with the postulated histaminergic function of these cells.[300,303–305] Histamine receptors have also been studied in *Helix*, where both H_1 and H_2 receptors have been defined using various pharmacological agents.[311]

Ultrastructural studies of the *Aplysia* neuron C-2 showed that within the axons were varicosities filled with a heterogeneous population of vesicles. Most were large, electron-dense structures; however, there were also numbers of smaller, electron-lucent vesicles. "Morphologically-conventional" synapses between C-2 and postsynaptic processes were also observed.[312]

It has been known for a long time that the cerebral ganglion of the cephalopod *Octopus vulgaris* contains substantial amounts of HM and that intravascular injection of HM causes chromatophore expansion and general agitation of the animal. High concentrations of HM have also been found in optic lobes of other octopi and in squid.[313] "Histamine oxidase" activity, probably a broad-range MAO rather than a true histaminase, has been extracted from optic lobes of the octopus, and this pathway is important for the metabolism of HM in these molluscs.[52]

Histamine is found in the venom of various insects, e.g., wasps, and probably plays an important role in the cardiovascular reactions produced by stings, but its effects in the nervous systems of arthropods have not been studied in great detail. In early studies, HM was detected in the stellate ganglion and in the heart of the crab *Carcinus maenus*, and a unique metabolic product, β-alanylhistamine, has been identified in cardiac tissue (see Section 5.1). Synthesis of HM from radiolabeled histidine was also demonstrated in the CNS of the moth, *Manduca sexta*.[75]

Histamine has been measured in the stomatogastric nervous system of the spiny lobster, *Panulirus interruptus*. Among the nerves and ganglia assayed, the stomatogastric ganglion contained the highest concentration of HM, although this HM was found primarily in the neuropil region. In the supraesophageal ganglion (brain), on the other hand, two identifiable neurons were located that contained HM at a concentration greater than 1 mM (B. Claiborne, personal communication). This report is the first in which HM has been localized to a single neuron in any arthropod.

Information regarding HM as a transmitter in invertebrate systems is relatively recent and has been collected using a variety of disciplines in addition to neurochemistry. These studies have shown that there is strong evidence that HM plays a role as a neurotransmitter in the molluscan nervous system, and data for a similar role in crustacea are accumulating. Although synthesis from histidine and inactivation via MAO are pathways common to many vertebrates and invertebrates, other metabolic systems, such as formation of peptidoamines or N-acetyl-HM, appear to be more specialized reactions. Neither of these latter pathways has been described in vertebrates, whereas N-methylation, an important degradative pathway in mammals, has yet to be described in invertebrate systems. Thus, at present, only a few of the biochemical reactions are helpful in understanding the actions of histamine and its role in neurotransmission. Further, there appear to be few similarities between actions of HM in nervous systems of invertebrates and in peripheral nervous systems of vertebrates, where much is known of HM effects, or in central nervous systems, where its actions are poorly understood.

2.7. *Trace Amines*

Within this group are the endogenous amines, tyramine and tryptamine, which are usually considered to be intermediates in the synthesis of the known transmitters or end products of minor metabolic pathways. Often these products are found in concentrations several orders of magnitude less than those of the more common biogenic amines, DA and 5-HT (compare Tables I and IV). There are a few notable exceptions, and these are discussed below. In addition, these compounds are usually less potent pharmacologically than the neurotransmitter to which they are related. Some of the earlier observations on these amines in invertebrates have been reviewed.[193,248] Recently, there has been increased interest in the role of these trace amines in various clinical disorders. Phenylethylamine, for example, has been studied intensely in mammalian systems, and a link between amine levels and pathological, neurological, and psychiatric states including schizophrenia has been sought.[314–316]

The polyamines, a series of aliphatic primary amines, are also included here. Neither their function nor, more specifically, their role in neurochemistry is well understood; nevertheless, they represent a group of compounds that can undergo a wide range of reactions, some of which are related to specialized cell functions such as membrane integrity.

2.7.1. *Tyramine, Phenylethanolamine, and Phenylethylamine*

Tyramine (*p*-hydroxyphenylethylamine), the decarboxylation product of tyrosine, was first detected in the posterior salivary glands of the octopus and has been found in high concentrations in the nervous systems of only a few animals (Table IV). Of all the animals listed in Table IV, the visceral ganglion

Table IV
Concentrations of β-Phenylethylamine, p-Tyramine, m-Tyramine, and Tryptamine in Nervous Tissues of Representative Invertebrates[a]

Animal	Tissue	β-Phenylethylamine	*p*-Tyramine	*m*-Tyramine	Tryptamine
Clam					
Tapes watlinga	Visceral g.[257]	—	70	—	—
Snail					
Helix aspersa	Circumesophageal g.[248]	0.54	1.8	—	0.2
Octopus					
Octopus dofeini martini	Optic lobes	0.24[318]	13[248]	0.04[248]	0.03[248]
	Sup. buccal lobe	—	13[248]	N.D.[318]	—
	Inf. buccal lobe	—	0.3[248]	N.D.[318]	—
Leech					
Hirudo medicinalis	Segmental g.	N.D.[273]			N.D.[158]
Lobster					
Homarus americanus	Supraesophageal g.[248]	—	15	—	—
Locust					
Schistocerca gregaria	Whole nervous system[248]	<0.2	5.4	<0.1	<0.1
Starfish					
Pycnopodia helianthoides	Arm nerve[296]	0.36	0.16	<0.05	78
	Tube feet[296]	0.18	0.08	Trace	7

[a] Where necessary, original values have been recalculated from ng to nmol using the molecular weight of the free amine, and converted to a protein basis by assuming 10% of the tissue wet weight to be protein. Values expressed as nmol/g protein.

of the mollusc *Tapes watlinga* was the richest source of tyramine.[257,317] In *Octopus dofleini*, it was unevenly distributed among the various lobes, and these levels were increased dramatically by pargyline (a MAO inhibitor) and decreased by reserpine, similar to their effects on catecholamine levels.[318] In *Helix*, ganglia metabolized a small amount of [^{14}C]tyramine into *p*-hydroxyphenylacetic acid via the MAO pathway and into OA via the β-hydroxylase pathway. However, a larger amount (more than 50%) was converted into a new unidentified substance. This metabolite had an R_f value similar to, but not identical with, that of tyrosine and was not DA or any known phenolic acid.[191]

Synthesis of tyramine from [^{3}H]tyrosine was greater in the protocerebrum, optic lobes, and segmental ganglia of the adult tobacco hornworm *Manduca* than in the antennal lobes and subesophageal ganglion.[75] In the locust, the amount of label incorporated into tyramine from [^{3}H]tyrosine was tenfold greater than that incorporated into DA.[248] Several metabolities of tyramine were found after incubation with the brain of pharate adults of *Manduca*. One metabolite was identified as N-acetyltyramine (see Section 5.2), and another was identified tentatively as β-alanyltyramine, a new peptidoamine[243] (see Section 5.1). N-Acetyltyramine was also produced in cerebral and thoracic ganglia from the locusts.[319]

In lobster, [^{3}H]tyramine was taken up by thoracic roots of the nerve cord, and specific nerve endings were labeled. Electron microscopic studies showed silver grains over the "type 3" nerve endings, identified as OA-containing cell endings, but not over the cell bodies. Chemical analysis of these roots showed radioactivity associated with OA as well as with other metabolites rather than with tyramine.[171] Active uptake systems for tyramine have also been demonstrated in locusts and in the cockroach[245]; however, uptake and metabolism were not distinguished in these tissues.

In invertebrates, as in mammals, both the *meta* and *para* isomers of tyramine have been found, although the latter is present in much greater quantities (Table IV). It has been shown that *p*-tyramine is readily formed by the decarboxylation of tyrosine, but whether or not the *meta* isomer is formed by the dehydroxylation of a catechol intermediate, as has been postulated in mammals, or by another mechanism remains to be established.[314] No significant physiological role for *m*-tyramine has yet been defined.

Significant amounts of phenylethanolamine have only been found in nervous tissues of a very few invertebrates. The clam *Tapes watlinga* contains this amine in addition to large amounts of tyramine. Unlike tyramine, however, its concentration in the gut was almost ten times higher than that in the nervous system.[317] Phenylethanolamine had also been reported to be present in ganglia and in nerve trunks of *Aplysia californica*,[321] although subsequent studies have not confirmed this finding.[265,266] No physiological role has been defined for this compound either, although specific receptors for this amine mediating both excitatory and inhibitory responses were described in *Aplysia* ganglia.[321] Relative to other amines, only small amounts of phenylethylamine were found in molluscs and in echinoderms (Table IV), and it was not detectable in insects and other arthropods.

2.7.2. Tryptamine

Tryptamine (β-indolylethylamine) is not widely distributed in invertebrates (Table IV) and in most systems is much less potent than the biogenic amine, 5-HT. It is synthesized from tryptophan by decarboxylation, but it is usually not readily hydroxylated to form the neurotransmitter 5-HT. This pathway is usually considered to be subordinate to the primary pathway for 5-HT synthesis in which hydroxylation to 5-HTP occurs before decarboxylation (see Section 2.3). By use of dansyl derivatives and microchromatography, tryptamine was identified in the circumesophageal and buccal ganglia of the snail *Helix aspersa*.[322] Quantitative measurements showed that it also accumulated to a remarkable degree in the arm nerves of a starfish, whereas in all other species studied it was found in very small amounts (Table IV). These starfish, on the other hand, do not contain 5-HT, nor could 5-HT be detected in tunicates,[295,296,322] suggesting that tryptophan hydroxylase activity may be low in these higher invertebrates. It is not known whether tryptamine serves a function similar to or separate from that of 5-HT in these nervous systems.

Tryptamine was the preferred substrate for the enzyme MAO in the tissues of the snail *Biomphalaria glalerata*[238] and for the enzyme NAT in bees.[182] The latter observation suggests that acetyl-transferring reactions are important in the metabolism of indolealkylamines in many insects, as has been shown for catecholamines, although little is known regarding the effects of acetyltryptamine in these nervous systems (see Section 5.2).

2.7.3. Piperidine

This cyclic amine mimics the action of ACh when applied to some individual neurons of the terrestrial snail, *Helix pomatia*, and inhibits its action when applied to others. Small quantities of piperidine have been detected in the brain of *Helix* and of another snail, *Otala lactea*, and these amounts decreased during periods of inactivity.[324] Radiolabeled piperidine injected into snails was taken up by the nervous tissue only slowly (over a period of days), demonstrating the presence of an effective blood–brain barrier.[324,325]

2.7.4. Putrescine, Cadaverine, and Other Polyamines

The polyamines, putrescine (2,4-diaminobutane), cadaverine (2,5-diaminopentane), spermine, and spermidine, along with their acetyl derivatives, have been implicated in various cellular functions, particularly those involved with fertilization and development. For many of these studies, the sea urchin has been particularly useful[326]; however, the role of polyamines in the nervous system is still poorly understood. Some suggestions have included maintenance of the integrity of membranes or transport of proteins, and putrescine has been suggested as a precursor of GABA, and cadaverine as a precursor of piperidine.[324,328] Putrescine is presumed to be synthesized from ornithine via ornithine decarboxylase, whereas cadaverine is formed by the decarboxylation of lysine.

The distribution of putrescine in the nervous system of molluscs is quite variable. High levels were found in both ganglia and individual neurons of the freshwater clam *Anodonta cygneus* and of the garden snail *Helix pomatia*, but it was almost undetectable in the marine clams *Spisula solida* and *Mytilus edulis*. In the marine snail *Aplysia californica*, ganglion levels of putrescine and cadaverine ranged from 30 to 60 nmol/mg protein, whereas the levels of the polyamines spermidine and spermine were considerably higher (200–1200 nmol/g protein).[327,328] Cadaverine was found to be about four times higher in the brain of active versus dormant *Helix*.[329] Reports regarding polyamines in other phyla are very meager and are not related to the nervous system.

2.7.5. *Remarks*

Interest in the trace amines tyramine, phenylethylamine, etc., once considered only as metabolic abnormalies, has increased in recent years. In mammalian systems, these compounds now appear to be involved in certain neurological conditions, e.g., phenylketonuria. Further, it has been shown that chronic abuse of amphetamine produces a condition similar to paranoid schizophrenia, and administration of tyramine mimics this condition.[330] Tyramine can also release catecholamines from adrenergic terminals, and it has been suggested that it and other "noncatecholic phenylethylamines" function as neuromodulators in various systems.[314,331] However, in invertebrates there is not yet sufficient information to assess their role in the nervous system.

The variable distribution of the various polyamines is intriguing, but the significance of their high concentrations in some nervous systems is not known. Perhaps further investigations will reveal whether such compounds have a unique function, particularly in those nervous systems in which they accumulate in large quantities.

3. *AMINO ACIDS*

This section deals to a large extent with two amino acids, glutamate and GABA, which are prime candidates for the excitatory and the inhibitory transmitter, respectively, at invertebrate neuromuscular junctions. From studies in crustaceans, it had been assumed for some time that glutamate and other acidic amino acids were exclusively excitatory agents and that GABA and other neutral amino acids were exclusively inhibitory agents.[4,6,332,333] However, other studies, e.g., those using snail neurons,[10] showed that these designations were oversimplified and that the responses to amino acids were determined by the character of the receptors present in the particular muscle or neuronal membrane. In some preparations, it has even been possible to demonstrate excitatory, inhibitory, or multicomponent responses to the application of a single amino acid. This literature has been reviewed recently.[334,345]

In invertebrate systems, two other acidic amino acids besides glutamate are of importance. In many marine animals, aspartate and taurine are present in remarkably high concentrations and may control the intracellular ionic environment. In addition to these osmoregulating effects, however, aspartate can be a potent agonist and has been proposed as a neurotransmitter at some crustacean neuromuscular junctions and for *Aplysia* neurons.[346,351,363] There is little evidence at present that taurine serves such a function in invertebrates (see Section 3.4).

Among the neutral amino acids, GABA has been studied most intensively, and its role as an inhibitory transmitter is well established.[4,344] Recently, interest in glycine as a possible neurotransmitter was sparked when it was found to be concentrated in specific neurons of *Aplysia*.[347] Studies on other neutral amino acids are quite limited and are discussed only briefly.

Methods for measuring amino acids in tissues include both GC and LC procedures as well as enzymatic assays for specific compounds. Fluorescence (dansyl or *o*-phthalaldehyde) or spectrophotometric (ninhydrin) assays have been used most commonly. Continued improvements in methodology have resulted in procedures in which 15–25 amino acids can now be quantitated routinely at the level of 1–10 pmol in less than an hour.

3.1. Glutamate and Aspartate

These two amino acids are among the highest of those found in extracts of various invertebrate nervous systems (Table V). A high concentration of glutamate has been reported in the platyhelminths *Schistosoma mansoni* and *Spirometia mansonoides*.[349] In the longitudinal nerve cord of the parasitic cestodarian *Gyrocotyle fimbriata*, substances with fluorescence emission spectra similar to those of glutamate and aspartate have also been reported. These emission spectra obtained *in situ* from tissue treated with glyoxylic acid could be distinguished from the spectra obtained from 5-HT and from the catecholamines.[237] Assays of amino acids in annelids showed glutamate, aspartate, glycine, and taurine to be the predominant ones,[379] and values of 4–6 μmol/g were found in the nervous systems of the earthworm *Lumbricus* and the sea mouse *Aphrodite aculeata*.

Among the gastropod molluscs *Helix, Limax, Buccinium*, and *Aplysia*, the concentrations of glutamate and aspartate were very similar in the various ganglia.[347,350,351] Likewise, among the individual *Aplysia* neurons assayed, there were only minor differences. Both glutamate and glutamine were detected in all somata, and they were found in a ratio of about 2:1, respectively.[352] The "white" cells of the abdominal ganglion (R-3–R-14) of *Aplysia californica*, which also contain high concentrations of glycine (*vide infra*), contained somewhat lower concentrations of both glutamate and aspartate, presumably to maintain osmotic balance.

The concentration of glutamate in cephalopod brain ranges from 10 to 40 μmol/g for octopus, squid, and cuttlefish, whereas that of aspartate ranges from 8 to 111 μmol/g.[350,353] The common isomer of all amino acids is the L configuration, yet the D-isomer of aspartate has also been found in high concentration in the optic lobes of various cephalopods. In octopus, for example, this isomer comprises about 40% of the total aspartic acid present, and these values increased to 60–85% in squid and in cuttlefish. None of this isomer was found

Table V
Free Amino Acids in the Nervous Systems of Various Invertebrates[a]

Amino acid	*Aplysia californica* Cerebral g.	*Limulus polyphemus*[c] Abdominal g.	*Oronectes immunis*[d] Ventral nerve	*Carcinus maenus*[e] Peripheral n.	*Carcinus maenus*[e] Cerebral g.	*Manduca sexta* larvae[f] Nerve cords	*Manduca sexta* larvae[f] Brain
Taurine	34[b]	0.48	2.6	75.5	77.2	tr	tr
Aspartate	37[g]	8.4	7.2	198.6	33.5	0.66	1.95
Glutamine	—	—	5.2	23.7	—	—	—
Serine	6[b]	0.28	1.5	3.0	2.8	10.4	12.7
Glutamate	20[g]	7.2	5.0	36.1	18.2	7.11	24.8
Proline	N.D.[b]	5.4	1.8	24.4	53.2	—	—
Glycine	2.5[g]	1.4	2.7	7.3	35.5	10.5	13.0
Alanine	15[b]	2.7	6.9	21.6	20.5	8.13	17.7
Valine	0.4[b]	—	0.09	0.68	—	3.2	5.6
Methionine	—	—	0.08	0.65	0.93	tr	tr
Isoleucine	1.5[b]	0.2	0.11	0.27	0.73	1.0	1.1
Leucine	1.2[b]	0.52	0.21	0.46	1.08	1.1	1.6
Tyrosine	0.8[b]	0.38	0.21	0.52	1.74	1.16	1.71
Phenylalanine	0.5[b]	0.12	0.08	0.17	0.58	0.23	0.34
Lysine	2[b]	—	0.25	0.59	1.98	1.2	3.55
Histidine	12[b]	0.43	0.46	0.48	0.97	3.9	8.7
Arginine	8[b]	0.43	7.0	13.99	10.73	2.97	5.83

[a] Where necessary, original values have been recalculated from ng to nmol using the molecular weight of the free amine, and converted to a protein basis by assuming 10% of the tissue wet weight to be protein.
[b] M. McCaman, unpublished observations.
[c] James *et al.*[377]
[d] Lin and Cohen.[219]
[e] Evans.[336]
[f] Taylor and Newburgh.[231]
[g] Iliffe *et al.*[347]

in nonnervous tissues, nor was any found in ganglia of the gastropod *Aplysia* or in vertebrate brains.[354]

The acidic amino acids were among the most prominent in arthropods. In Table V, it is also apparent that all amino acids are higher in marine than in freshwater crustaceans (compare *Carcinus* and *Oronectis*, respectively). These nervous systems contained more aspartate than glutamate,[336,345] and, similarly, in lobster, both excitatory and inhibitory axons contained three to five times more aspartate than glutamate.[342] In larval and adult forms of *Manduca sexta*, glutamate was the most abundant amino acid, and aspartate values were about tenfold lower. Glutamate was very abundant in the nervous systems of various spiders, but aspartate values were quite variable.[364] In the deuterostome groups (sea urchins, starfish, and tunicates), the concentrations of both glutamate and aspartate were markedly lower.[295,350]

Studies of the uptake of [^{3}H]glutamate by ganglia of the snail *Helix pomatia* showed that the radioactivity was concentrated primarily in the glial cells and that neurons themselves appeared to take up only small amounts of the label.[337,365] In cockroach nerve–muscle preparations, glutamate was taken up by a high-affinity system, and this mechanism has been proposed for the removal of glutamate from the cleft of the "glutamate" synapse.[366,367] Morphological studies showed that radiolabeled glutamate was taken up by different tissue elements depending on the rate of stimulation of the nerves. In the locust extensor tibia nerve–muscle preparation, label accumulated in the glia surrounding axon terminals after low-frequency stimulation, whereas after high-frequency stimulation, it was found in both glial and axonal terminals.[368] These results support the hypothesis that glutamate exists in axons in more than a single pool and that each pool may use a different mechanism for uptake and release of its glutamate.[337,369]

Glutamate can be released from neurons by nerve stimulation or by a pulse of K^+; however, usually many other amino acids are also released by this treatment.[369] In most of these experiments, glutamate released in the absence of nerve stimulation was high and variable, obscuring effects of the excitatory nerve stimulation. Earlier experiments demonstrating glutamate release in the lobster, crab, and locust have already been reviewed.[332–337,369] However, the abdominal slow flexor muscle of *Cambarus* could be isolated cleanly from the surrounding tissues and provided an ideal preparation for release studies.[82] In these experiments, glutamate release was proportional to frequency of stimulation and partially dependent on Ca^{2+}. This information provides strong support for the suggestion that glutamate is a transmitter in this system. Further, release of radioactivity after uptake of [^{14}C]glutamate has also been demonstrated. This release occurred only after excitatory nerve stimulation and not after inhibitory nerve stimulation or after application of GABA[361]; however, this kind of evidence is not as compelling as the release of the endogenous amino acid.

Studies of the metabolism of these amino acids has not proved to be very helpful in understanding their role in nervous systems, since only a few enzymatic reactions, e.g., synthesis of GABA from glutamate, appear to be associated specifically with nervous tissue function. Other enzymatic reactions studied include amidation to form glutamine, transamination or deamination to form α-ketoglutarate, and the analogous reactions involving aspartate. However, these enzymatic reactions appear to be indicators of general metabolic systems such as protein synthesis and energy production rather than indicators of special neuronal functions such as neurotransmission. Nor do any of these enzymes appear to inactivate glutamate released at nerve endings. Instead, reuptake is the mechanism that has been postulated for inactivating glutamate as well as other amino acids at nerve terminals.[337]

In coelenterates, the simplest multicellular animals with an organized nervous system, glutamate usually elicits an inhibitory response. Recently, it has been proposed that these effects of glutamate are produced not by a direct effect on muscle cells but by inhibiting the release of ATP, which has been proposed as the actual neurotransmitter in this system.[337] In molluscs, receptors for glutamate are found on most *Aplysia* neurons, and the application of glutamate most often produces a Na^+-dependent depolarization.[4,8] Receptors for aspartate are much more limited.[363] *Helix* neurons displayed both single- and multiple-component response to glutamate application.[356,357] In annelids, the effects of glutamate are almost always excitatory and include contraction of the body wall of the earthworm *Lumbricus terrestris*[355] and of the leech *Hirudo medicinalis*[62] and stimulation of the Retzius cells. Glutamate or a glutamatelike compound appears to be the transmitter at excitatory synapses on muscles in insects. Recent evidence demonstrated the similarity of the reversal potential for the EPSP and the glutamate potential in the neuromuscular junction of *Drosophila* larvae[343] and the blocking effects of various glutamyl esters in the mealworm *Tenebrio*.[358]

The most complete evidence of glutamate as a neurotransmitter has been found in the neuromuscular junction preparations of crustaceans. Here, it has been shown that motoneurons are rich in glutamate, that nerve stimulation significantly increases its release, that the ionic mechanisms and elementary current events evoked by applied glutamate are identical to those observed in the response evoked by nerve stimulation, and that receptors to glutamate and to natural transmitter have the same pharmacological profile[332,340,359–362] (also see refs. 4, 337, and 343, where some of the earlier results are summarized).

The isolation of "glutamate synaptic receptors" has been reported in several studies. Fractions prepared from the brine shrimp *Artemia longinaris* specifically bound L-glutamate but not aspartate. Binding was inhibited by diethylglutamate and other glutamate analogues, but at relatively high concentrations.[370] Fractions prepared from the intrathoracic muscle of the locust *Schistocerca gregaria* also bound glutamate preferentially. This fraction contained two hydrophobic proteins (or proteolipids) that showed high-affinity binding ($K_d = 8 \times 10^{-6}$ M).[371] The study of glutamate receptors, both chemically defined and physiologically defined, has been hampered by the lack of specific pharmacological agents that are antagonists to glutamate. Compounds presently available, e.g., glutamate esters, block responses only at high concentrations, whereas others, e.g., kainic acid, can produce irrevsible changes. Recent reports have indicated that the vasodilatory drug diltiazem is a selective antagonists for glutamate, and this compound may prove useful in future pharmacological studies in invertebrates.[334]

In lobster axon preparations, aspartate had little excitatory action of its own but appeared to potentiate the effects of glutamate. Such interactions suggested that extrajunctional receptor sites occupied by one amino acid may represent a source of long-term modulation on the effects of a second amino acid at its receptor sites.[342] However, since aspartate is so highly concentrated in all peripheral nerves, there has been some speculation as to whether or not this effect has any physiological significance.[338]

3.2. GABA

Early studies on GABA were concerned with its chemical characterization and with its identification as the inhibitory transmitter in various invertebrate preparations. As a result of many interdisciplinary studies, most of the criteria necessary to establish GABA as a neurotransmitter have now been satisfied.[4,6,8,372–378] During the course of these investigations, preparations such as the crayfish stretch receptor and neuromuscular junction systems of the crayfish, lobster, and cockroach as well as preparations involving individual neurons of *Aplysia* and *Limulus* have been used.

Although GABA has been most widely studied in arthropods, there are a few reports of its presence and of its effects in other phyla.[193] For example, among the annelids, GABA has been detected in the nerve cords of the polychaetes *Aphrodita aculeata* and *Nereis virens* and in the oligochaete *Lumbricus terrestris*,[355,379] although these concentrations are quite low relative to the values found in arthropods. In molluscan ganglia, quantitative measurements showed that it was present in all ganglia of *Aplysia*[380]; however, with a sensitive GCMS procedure, none could be detected in individual neurons.[381] Earlier, a GABA-like compound in extracts of individual *Aplysia* neurons was reported on the basis of studies using a dansyl-TLC procedure, but the exact chemical nature of this compound was not positively established.[379,382]

Inhibitory neurons isolated from the thoracic ganglia of the lobster *Homarus americanus* contained high concentrations of GABA.[383] More recent studies confirmed earlier results showing approximately 100 pmol GABA/μg protein (30 pmol/soma) in neurons I-2 and I-3. In addition, two other populations of GABA-containing neurons in these ganglia were described.[381] These other cells contained considerably less GABA than the inhibitory cells, and nothing is known of their physiological role. GABA was also concentrated in peripheral inhibitory nerves of the lobster, where it was found at concentrations as high as 0.1 M, whereas much lower concentrations were found in sensory and in motor neurons.[384]

Synthesis of GABA occurs predominantly by decarboxylation of glutamic acid via the enzyme glutamic decarboxylase (GAD). These pathways have been detected in most arthropods, and GAD has been studied in detail in *Limulus*,[73] lobster,[385,386] honeybee,[387] and cockroach brain.[388] No GAD activity was found in homogenates of molluscan ganglia, even though GABA was detected in some species, e.g., *Aplysia* and *Helix*.[380,381] This suggested that it may be synthesized via an alternative pathway in these animals.[389]

Although metabolizing enzymes such as GABA–α-ketoglutarate transaminase have been described in a few invertebrates (for example, lobster and crab), GABA is assumed to be inactivated primarily by specific uptake systems.[6,386,390] Kinetic analysis showed that these were high-affinity systems, but of these, only the systems in lobster and *Aplysia* were sodium dependent.[351,367,391] In crustacean nerve–muscle preparations, [^{3}H]GABA was also taken up by this tissue; however, most of this label was localized in the Schwann cells and not in neuronal endings.[378] In *Aplysia*, it was also taken up predominantly into extraneuronal rather than into neuronal sites.[351] On the other hand, in the cockroach brain, the label appeared as dense accumulations of silver grains over distinct nerve cell groups and nerve fiber bundles, and little accumulated over glia.[392] It still is not clear whether this uptake into nonneuronal tissues has any relationship to the removal of GABA from the synaptic cleft by nerve endings or by glial processes, the mechanism postulated for terminating its action on postsynaptic receptors.[335]

Calcium-dependent release of transmitter from nerve endings is one criterion that is often difficult to demonstrate because of the limited quantities involved. However, at crustacean neuromuscular junctions, GABA could be measured in perfusates after stimulation of inhibitory but not excitatory axons, and this release was reduced in the present of low Ca^{2+}.[394] (Also see ref. 394 for earlier citations.)

GABA receptors have been described in muscle cells in the body wall of the nematode *Ascaris lumbricoides*[395] and in various molluscan neurons.[4,8] In *Aplysia*, membrane responses involving distinct ionic mechanisms have been described, and these results indicated the presence of at least six different GABA receptors and/or GABA receptors coupled to different ionophores.[4,335,376,396] In various arthropod neuromuscular preparations, applied GABA duplicated the effects of the inhibitory transmitter released by nerve stimulation. It had the same reversal potential and was blocked by low concentrations of the GABA antagonists, picrotoxin and bicuculline.[4,337,377,393]

Studies on isolated GABA receptors prepared from invertebrate tissues are limited. A hydrophobic protein fraction isolated from muscle of the brine shrimp *Artemisia longinaris* showed high-affinity binding for GABA at a single site.[397] Another preparation, membrane fractions from crayfish muscle, also displayed high-affinity binding characteristics for GABA; however, these binding sites were not inhibited by the classical GABA inhibitors, bicuculline, picrotoxin, and muscimol.[373,398] Although most of the effects of these drugs have been attributed to blocking GABA receptor sites, these studies suggested that the drug binding site might be the chloride ionophore and that such sites were quite separate from the GABA binding sites.[399]

3.3. Glycine

In recent years evidence has accumulated that glycine may be a transmitter at specific neurons in the mammalian spinal cord; however, few glycine-sensitive systems have been described in invertebrates.[4,6] There are a few examples where the concentration of glycine is unusually high, but the other biochemical and physiological requirements for neurotransmitter candidacy have yet to be satisfied.

Glycine is one of the more abundant amino acids found in extracts of most nervous tissues (Table V). In the nerve cord of the polychaete *Nereis virens*, glycine concentrations were higher than those of any of the other amino acids, and application of glycine to the body wall produced muscular contractions.[355] Excitatory effects were also described in the "N" sensory cell in the leech and in a specific cell (V3) of the visceral ganglion of the snail *Onchidum verruculatum*, but application of glycine to *Helix* neurons had no effect.[335,396]

In the abdominal ganglion of the marine snail *Aplysia californica*, glycine is concentrated in a unique group of identifiable neurons. These neurons (R-3–R-14) contain glycine at levels about 20-fold higher than those found in most other *Aplysia* neurons.[347] They appear white when viewed through the microscope under incident light, a characteristic that also differentiates them from other neurons in this ganglion. Cells R-3–R-14 have been categorized as neurosecretory cells because of the large, "neurosecretory"-type granules they contain and because of the absence of typical synapses at nerve endings in the sheath and at the surface of various lacunae.[400] Some of their axons have been traced through the branchial nerve to the area of the heart, where they appear to terminate on or near specific muscle fibers.[401]

When radiolabeled glycine was incubated with whole abdominal ganglia, most of the label recovered from cells R-3–R-14 was unchanged glycine, although smaller amounts had been incorporated into serine and into macromolecules.[347] Other studies showed that part of the label in the macromolecular fraction was associated with a specific polypeptide of approximately 6000 daltons.[402–404] Autoradiographs of these neurons showed that silver grains were distributed randomly throughout the cytoplasm and nucleus and were not concentrated over the dense-core neurosecretory granules. Adjacent glial cells were only lightly labeled. The peptide-containing bag cells (see Section 4.2) also incorporated large amounts of label, but these cells showed silver grains over areas believed to be the Golgi apparatus, i.e., in the high protein-synthesizing areas.[401,405] Only a small fraction of the total label taken up by the intact ganglion could be found in the R-3–R-14 neurons, and, in fact, most of it was recovered from the connective tissue sheath. Here, too, radioactivity was in the form of unchanged glycine, although a significant portion had been incorporated into the peptide glutathione and into a novel peptide, β-aspartylglycine. Very little radioactivity found in the sheath was high-molecular-weight (acid-precipitable) peptidic material (M. McCaman, unpublished results).

The uptake mechanism for glycine in *Aplysia* cells R-3–R-14 has been characterized as a low-affinity (K_m about 2 mM), Na^+-dependent system.[406] It was taken up by these cells at about twice the rate found in other neurons, whereas other amino acids such as leucine were taken up equally well by all cells. Glycine label was also transported away from the ganglion and localized primarily in the axons of the white cells. The rate of transport was estimated to be 70 mm per day, and transport was inhibited by Hg^{2+}, vinblastine, nocodazole, and low Ca^{2+}.[406]

In arthropods, glycine has also been measured in various nervous system preparations, including *Limulus, Carcinus, Oronectus*, and *Manduca* (Table V). In all of these examples, glycine was the most abundant (or one of the most abundant) amino acid. In various species of arachnids, glycine comprised about 7–10% of the total free amino acids based on dansyl-TLC.[364] The fact that glycine was present in higher amounts in the central than the peripheral nervous system of *Carcinus* (Table V) suggested that it may play a special role in the former, but other supporting evidence is lacking.[336] Likewise, little is known of the role of this amino acid in insects, although extrajunctional glycine receptors have been discovered on somata, axons, and axon terminals of certain motoneurons.[407] In echinoderms and tunicates, glycine comprises an even higher percentage of the free amino acid composition of the nervous systems (as much as 40%). In these marine animals, glycine together with taurine may regulate osmolarity rather than being involved with neurotransmission.[350]

3.4. Taurine

This sulfur-containing amino acid (β-aminoethanesulfonic acid) is found in all invertebrate nervous tissues but is remarkably high in the nervous systems of many of the marine species (Table V). For example, in annelids, taurine values in the nerve cord of the sea mouse *Aphrodite aculeata* were almost 100 times higher than those in the nerve cord of the earthworm *Lumbricus*.[379] Similarly, values for taurine in ganglia of marine molluscs such as the whelk *Buccinium undulatum*, the sea slug *Aplysia californica*, and the octopus *Eledone cirrhosa* are more than 35 μmol/g wet tissue, whereas in the garden snail *Helix aspersa*, the values were only 1 μmol/g.[408,409] Individual neurons isolated from *Aplysia* ganglia also contained taurine at high concentration, and, except for cells R-3–R-14, it was the most abundant amino acid found. In the latter, high glycine content appeared to be balanced by lower taurine content.[409]

Marine crustaceans also contain high levels of taurine. In the cerebral ganglion of the crab *Carcinus maenus*, taurine was almost 30 times higher than in the ventral nerve cord of *Oronectes*, a freshwater crayfish.[336,379] In insects, taurine concentrations are comparable to those in marine crustaceans, e.g., about 30 μmol/g, although these levels change dramatically during development. In the larva and prepupal stages of the moth *Mamestra configurata*, only small amounts were found, but during early pupation, brain taurine levels increased 20-fold. The highest levels were detected during the time of intensive brain development and appeared to be related to the development of the massive optic lobes.[410,411] In other arthropods, especially high levels of taurine were detected in the salticid species of spiders, e.g., *Marpissa muscosa*, where it comprised 34% of the total free amino acids.[364] It may be only an interesting coincidence that these animals also have well-developed optic lobes and use sight for orientation. The only exception to the presence of high taurine levels in marine animals was found in two echinoderms, the sea urchin *Echinus esculentes* and the starfish *Asterias rubeas*, both of which contained relatively small amounts of this amino acid (3 μmol/g).[379]

In vertebrates, taurine is synthesized by two major pathways. One proceeds via hypotaurine, and the other via cysteinesulfinic acid, but few studies have been made in invertebrates to determine the relative importance of these pathways.[412] In insects, synthesis of taurine via the cysteine–cysteamine–hypotaurine–taurine pathway was demonstrated after injection of [^{35}S]cysteine into pupae of the moth *Mamestra*.[410] Under these conditions, no synthesis via the cysteinesulfinic acid–cysteic acid–taurine pathway could be detected. Hypotaurine in unusually large amounts was detected in various tissues including the nervous tissue of the clam *Noetia ponderosa*; however, neither the origin nor the disposition of this compound was investigated.[413]

The actions of taurine on membrane permeability, as well as its uptake and release from synaptosomes, have been put forth as arguments that it plays a role as a neurotransmitter or as a neuromodulator.[412,414] The presence of receptors specific for taurine (and for β-alanine) in the antennular system of the spiny lobster and the high levels associated selectively with the optic lobes in insects are observations consistent with a transmitter role for this amino acid.[412] Yet, the fact that it is ubiquitously distributed throughout the nervous systems of marine molluscs argues more for osmoregulation in these animals. Until more definitive evidence is available, taurine must be considered a candidate for both roles.

3.5. β-*Alanine*

β-Alanine is the lower homologue of GABA and is found in highest concentration in those crustaceans in which GABA also is highly concentrated. It is widely distributed in the nervous system of the lobster *Homarus americanus* and may be a cotransmitter of GABA at peripheral inhibitory synapses.[415] Although GABA was limited to nerve and muscle tissues, β-alanine was found in these tissues as well as in hepatopancreas and in hemolymph. When applied at the crayfish neuromuscular junction or to *Helix* neurons, it had an inhibitory action similar to that of GABA but usually was much less potent.[375,396] Uptake mechanisms for this amino acid in the lobster abdominal ganglion were independent of those previously characterized for GABA.[418] In these tissues, synthesis was from [^{3}H]uracil rather than from [^{3}H]aspartate, indicating that β-alanine was synthesized by a more indirect route than simply by decarboxylation of the acidic amino acid.[418]

Conjugates of β-alanine and histamine have been isolated from the heart of the crab *Carcinus maenus*,[417] and conjugates of β-alanine and octopamine have been isolated from the ventral nerve cord of the lobster *Homarus*.[286] These peptidoamine compounds appear to be inactive physiologically and are discussed in more detail in a later section (see Section 5.1).

3.6. *Proline*

Proline has been measured in the nervous systems of various invertebrates (Table V). In the polychaete *Nereis*, in various spiders, and in crab tissues, it was one of the most abundant amino acids, although it was very low in the crayfish.[355,364] In the clam *Anodonta*, proline comprised as much as 30% of the total free amino acids of the ganglia, in contrast to the marine gastropod *Aplysia*, where it was virtually absent from nervous tissue.[419]

Although the metabolism, uptake, and transport of proline have not been studied extensively in invertebrates, there is some evidence that it may be important in neurotransmission. Recent studies have shown that it has inhibitory actions at the neuromuscular junction of the crayfish *Procambarus clarkii*, where it inhibited endogenous and glutamate-induced muscle contractions.[420] In the body wall preparation of the echiuroid worm *Urechis unicinctus*, proline induced muscle contraction when applied at low concentrations, however, here this effect was shown to be indirect, i.e., proline stimulated the chemoreceptor systems located in the epidermis, which, in turn, released ACh which caused muscle contraction.[421]

3.7. Remarks

It has been difficult to establish any amino acids other than glutamate and GABA as neurotransmitters in invertebrate systems. Both aspartate and glycine have some properties similar to their more effective, higher homologues, but the data accumulated thus far are much less compelling. Unusually high concentrations of glycine in some *Aplysia* neurons[347] or of aspartate in some lobster nerves[346] along with specific aspartate receptors in *Aplysia*[351] constitute much of the available evidence.

One of the problems regarding amino acids as transmitters centers around the fact that they are already present in the fluids that bathe the postsynaptic nerve terminals. One argument put forth proposed that the glutamate is released into such a small volume within the synaptic cleft that its concentration is raised momentarily above the threshold necessary to fire the postsynaptic cell. Further, this released glutamate appears to come from a special pool. The concept of compartmentalization of glutamate within neurons has been suggested as a mechanism by which specific neurons could segregate "neuroactive" glutamate from "metabolic" glutamate. Evidence has accumulated that such a mechanism may be operating in mammalian systems, but it has not been demonstrated convincingly in invertebrates.

With technological improvements it has been possible to identify and to quantitate all of the protein amino acids as well as many additional amino acids in the extracts of various nervous systems and even of some individual neurons. As more and more preparations are surveyed, they may reveal new amino acids or unusual amounts of known amino acids that could be of significance in nervous system function. Although the recent trend in neurochemical investigations has been toward the assay and identification of peptides and polypeptides, investigations into the physiological effects of proline and taurine point up the fact that some of these common amino acids, though overlooked for many years, still could be important.

4. PEPTIDES

4.1. Introduction

Secretion of peptides has also been shown to be a major activity of some neurons in both vertebrate and invertebrate systems. In recent years, this concept of neurons as specialized secretory cells has been widely accepted, and the common features between exocrine gland cells and neurons have been explored.[422–425] In invertebrates, several systems have been studied in detail. These include the bag cells of *Aplysia*, the pericardial organs and the eyestalks of some crustaceans, as well as the corpora cardiaca of insects. Many of the peptides secreted are relatively low-molecular-weight compounds (<2000 daltons), and their effects include initiating or modulating reproduction, water balance, heart rate, color changes, and growth.[422,426,427]

Although more than 50 biologically active peptides associated with invertebrate systems have been characterized, only a few of these compounds are discussed here. This section focuses primarily on FMRFamide isolated from bivalve ganglia, on egg-laying hormone (ELH) isolated from *Aplysia* ganglia, and on proctolin isolated from cockroach hindgut. Some recent studies of other peptides are included where their presence or actions are relevant to neurochemistry.

The biochemical mechanisms for ribosomal synthesis, processing, and transport of peptides appear to be similar in vertebrate and invertebrate systems.[428–430] One hypothesis proposes that mRNA codes for the synthesis of signal sequences in peptides in addition to coding for the biologically active amino acid sequences. The "signal" sequence of the polypeptide chain would not only determine its final destination but also provide the means of insertion into, or release from, neuronal membranes. Studies have shown that newly synthesized proteins can be modified in at least two ways: by uncoupling the linear polypeptide chain (via peptidase or protease action) or by chemical alterations (via hydroxylation, crossbridging, or addition of carbohydrate side chains).[431–433] Many of these processes have been explored in greater depth in vertebrate systems; however, some comparable examples of peptide synthesis with posttranslational modifications have also been described in invertebrates.[428]

The increasing use of immunocytochemical procedures using antibodies directed against specific compounds has revealed the presence of many peptides hitherto unsuspected in invertebrates.[435] This has become apparent in the case of the hypothalamic peptides, TRH, enkephalin, and substance P, and it is of particular interest that these peptides are not uniformly distributed in invertebrate nervous systems.[437–442] Some notable structural similarities between such peptides as the hyperglycemic hormone in the crab *Cancer* and mammalian insulin[443] or between eledoisin in cephalopods and mammalian substance P have also been found.[444] However, some of the pitfalls associated with the use of fluorescent-conjugated antibodies have been reviewed recently, and it has been pointed out that some positive reactions may reflect overlapping immunosequences in peptides that are unrelated physiologically.[436] Nevertheless, these studies have provided many examples in which peptides are neuron specific, although it still remains to be established how similar the functions really are and whether "enkephalinlike" or "insulinlike" are meaningful terms to use in describing their effects in leech, crabs, etc.[422,441,442]

Various terms have been used to describe the actions of peptides in nervous systems. Proctolin, the gut-stimulating hormone of insects, has many characteristics associated with the term "classical neurotransmitter", i.e., it is fast-acting, localized in specific nerve terminals, and has actions identical with those evoked by nerve stimulation.[479] On the other hand, the terms "neuromodulator" or "neuroregulator" have been used to describe peptides that appear to act in a nonsynaptic manner, i.e., they act by modifying the postsynaptic membrane or by affecting the mechanism that releases transmitter from nerve endings.[422,424,433,445,446] The vagueness associated with these latter descriptions reflects the fact that we understand these processes only poorly, and we expect that our definitions will become more precise as our knowledge increases.[422]

4.2. FMRFamide and Other Cardioactive Peptides

Since the isolated snail heart is a standard bioassay preparation for the determination of ACh and 5-HT, it is not surprising that a number of nervous tissue extracts containing cardioactive factors have been described. In addition to containing biogenic amines, many of these extracts also contain active peptidic material.[447] One such cardioactive factor isolated from ganglia of the bivalve *Macrocallista nimbosa* is FMRFamide, a tetrapeptide named according to the sequence of the amino acids, Phe-Met-Arg-Phe-amide.[448] Subcellular fractionation of ganglion homogenates showed FMRFamide activity to be enriched in certain fractions containing neurosecretory granules, but these fractions were different from those containing cholinergic and serotoninergic granules.[448]

When tested on various molluscan hearts, FMRFamide stimulated some and inhibited others and sometimes produced contractions of noncardiac muscles.[449] When applied to *Helix* neurons, the synthetic peptide evoked a variety of responses, including hyperpolarization with increased potassium conductance, sodium-dependent depolarization, and a voltage-sensitive inward current.[451] Using the radula protractor muscle of the whelk *Busycon contrarium* as a bioassay, FMRFamide-like activity in extracts of individual neurons of *Helix aspersa* was surveyed. The only neuron to show activity consistently was F1, a large neuron in the right parietal ganglion[452]; however, purification of extracts from pooled ganglia from *Helix* or from the pond snail *Lymnaea stagnalis* yielded material that was similar to, but not identical with, FMRFamide.[452,453] This *Helix* peptide is like another small cardioactive peptide (SCP) also isolated from these ganglia.[454] The SCP had a similar molecular weight but was not characterized further. No FMRFamidelike activity was detected in ganglion extracts of *Aplysia californica.*

FMRFamide *per se* cannot be a general molluscan transmitter or neurohormone since the factors present in bivalves and in gastropods are different.[450] Instead, these peptides may belong to a family of peptides similar to those found in higher animals.[455] FMRFamide is related to the C-terminal tetrapeptide of cholecystokinin (CCK) and gastrin and is identical to the primary sequences of the C-terminal tetrapeptide of an enkephalin-related heptapeptide isolated from mammalian brain. Radioimmune and immunocytochemical studies demonstrated that FMRFamide-like material was present in the CNS and in the gut endocrine cells of vertebrates, although its distribution pattern was different from that of gastrin and CCK.[456,467] There is no direct chemical or physiological evidence at present to suggest that FMRFamide plays a role in neurotransmission or in modulation in mammalian brain, even though it has been detected in various neurons.

In addition to the smaller FMRFamide-like peptide found in extracts of *Helix* ganglia, a "large cardioactive peptide" (LCP) with a molecular weight of approximately 7000 was isolated by molecular sieve chromatography.[454] This peptide was found in especially high concentration in the auricles but was also found in subesophageal ganglia and in nerve trunks. Subcellular fractionation of nervous tissue showed this peptide associated with microsomal elements. Stimulation of the visceral nerves released LCP from auricle preparations, whereas pulses of high K^+ released LCP from both auricle and ganglia in a Ca^{2+}-dependent manner. The receptors for LCP could be differentiated from those for 5-HT, another potent cardioactive agent, by various pharmacological agents, and it was suggested that the peptide may act as a neuroregulator in the heart and in the pharyngeal retractor muscle of this snail.[457]

4.3. Egg-Laying Hormone (ELH) and Other Peptides Involved with Reproduction

The bag cells of *Aplysia* synthesize and release ELH and are one of the most intensively studied neurosecretory systems in invertebrates. This system consists of bilateral clusters of small, electrically connected neurons located in the rostral portion of the abdominal ganglion. The ELH released by these cell clusters generates a series of behaviors, one of which results in egg laying.[458] Stimulation of the atrial glands during mating may be the endogenous trigger responsible for discharging ELH from the bag cells and for initiating the egg-laying process. Immunocytochemical studies confirmed earlier ones indicating that these clusters were a homogeneous population of neurons and that their processes extend into the connective tissue sheath overlying the ganglion as well as into the connective and other nerves emerging from the ganglion.[470]

Extracts from bag cells contain a mixture of peptides, but only ELH has been thoroughly characterized and sequenced.[459,460] It is a basic polypeptide with an isoelectric point at about 9.0 and contains 36 amino acids (calculated mol. wt. 4385). During incubation with [^{3}H]leucine, these cells first synthesized a high-molecular-weight precursor peptide (of about 29,000 daltons), which was subsequently cleaved to form several lower-molecular-weight peptides. One of these was ELH, and another was called "acid peptide."[461–463] Both peptides were transported from cell body toward secretory terminals and released when the bag cells were activated.[464]

Recently, the genes encoding for ELH have been cloned, and their organization and expression have been studied. At least five distinct genes for ELH have been described, and, in the sequence analysis of one recombinant clone, a contiguous pattern of nucleotides coding for the 36 amino acids of ELH was found.[465] Additional RNA transcripts for ELH may be expressed in other tissues, e.g., the atrial gland.[466] These mRNAs are believed to synthesize polypeptide precursors containing the ELH sequence that are processed by a series of hydrolytic reactions, one of which produces the biologically active peptide.[428,433] This is consistent with the signal hypothesis, which proposes that newly synthesized peptides must attain a critical length (approximately 70 amino acids) before they can be properly sequestered and secreted, and then later the biological activity is unmasked by a sequential hydrolytic cleavage.[432]

In addition to causing egg laying, ELH induces cessation of feeding and initiates a series of head-waving movements prior to actual egg laying. Many of these inhibitory effects on feeding can be correlated with activation of several pairs of neurons in the buccal ganglia that send processes toward the buccal musculature.[467] Egg-laying hormone is also involved with the regulation of activity of the neuron R-15 in the *Aplysia* abdominal ganglion.[468,469] This action is thought to be both nonsynaptic and selective with regard to which cells will respond. It has been proposed that a small set of genes coding for a series of peptides, each controlling a particular behavior, can coordinate these behaviors because they all contain a common peptide, in this case ELH.[465]

Purified ELH prepared from *Aplysia californica* induced egg laying in three other species of *Aplysia* (*A. vaccaria, A. braziliana*, and *A. dactylomela*), and antibody directed against ELH prepared from *A. californica* also cross reacted with ELH from those species. In each animal, antibody reacted only with bag cells and not with other neurons.[472] These observations illustrate that both structure and function of this peptide were conserved within this limited group.

In the atrial gland of *Aplysia*, three additional peptides have been characterized: the egg-laying-releasing hormone (ERH) and two smaller peptides, A and B. The ERH hormone had N-terminal sequences that overlapped with bag cell ELH, and the C-terminal residues overlapped with both the A and B peptides. The ERH not only mediated release of ELH from bag cells but also had direct actions on the egg-laying system, whereas the smaller atrial peptides acted only on the bag cells.[466,471,472] Egg-laying peptides have also been reported in other molluscan nervous systems, e.g., in *Pleurobranchaea californica* and in the cerebral caudodorsal cells of *Lymnaea stagnalis*.[473,474] A somewhat different process, the laying of egg capsules, has been described in the snail *Busycon canaliculatum*, and peptide factors isolated from the parietal ganglion initiated this process.[475]

Under the influence of a gonad-stimulating hormonal peptide released from its nervous system, spawning can be induced in the starfish *Asterias amurensis*. The peptide is believed to stimulate production of 1-methyladenine, which is the factor involved in maturation of oocytes and shedding of the gametes. The peptide purified from *Patiria miniata* had a mol. wt. of approximately 4800, whereas the peptide isolated from *Asterias forensis* had a mol. wt. of 2100.[476]

4.4. Proctolin

The pentapeptide proctolin has been postulated to be an excitatory transmitter at the insect hindgut. This compound produced a graded contraction when applied to the longitudinal muscle of the gut of the cockroach *Periplaneta americana* similar to that evoked by repetitive nerve stimulation. It also increased the rate and amplitude when applied to the heart.[477–479] A convenient and sensitive bioassay was developed using the locust leg extensor muscle preparations, where proctolin increases the myogenic rhythm and amplitude of the contraction.[480]

Proctolin has been isolated from the hindgut of cockroaches and shown to have the structure Arg-Tyr-Leu-Pro-Thr.[477,479] These studies also demonstrated that the peptide could be synthesized in the cockroach brain and transported through the nerves to the corpus cardiacum, from which it was released. Repeated stimulation of the proctodeal nerve released the peptide into the perfusates of an isolated nerve–hindgut preparation.[478,483] In addition to these sources, proctolin was also detected in extracts of the sixth abdominal ganglion and in thoracic and supraesophageal ganglia of cockroaches, locusts and crickets. Subcellular fractionation of these extracts located activity in the fractions enriched with neurosecretory granules.[483,484]

Immunocytochemical studies localized proctolinlike substances in many cells in ganglia from *Periplaneta*.[481] The highest number of cell bodies was found in the terminal ganglion, and the lowest number in the cerebral ganglion. Among the lateral white cells of the abdominal ganglion, radioimmunoassay and bioassay demonstrated that each neuron contained the equivalent of 0.05–0.1 pmol of proctolin, and individual ganglia contained the equivalent of 0.3–0.8 pmol. Treatment with leucine aminopeptidase but not trypsin abolished activity, and the active material was cochromatographic with the synthetic peptide.[480]

In a different cockroach, *Leucophaea maderae*, all of the myogenic activity found in the fore- and hindgut extracts could be accounted for by the presence of proctolin; however, assays for proctolin in the head extracts were negative. Instead, a separate peptide, the "hindgut-stimulating neurohormone," was found, and this compound was distinct chromatographically from proctolin.[485]

Another compound closely resembling or identical to proctolin was isolated from the pericardial organs of the crab, *Cardisoma carnifex*. Both the chemical and physiological characteristics of the isolated peptide were indistinguishable from those of proctolin. When applied to the isolated lobster cardiac ganglion, this peptide produced excitatory effects, and it has been proposed that a proctolinlike peptide may serve a neurohumoral role in this tissue.[486] In the lobster *Homarus*, proctolin at concentrations as low as 10^{-10} M acted directly on the muscle, causing a sustained contraction of the opener muscle of the dactyl of the walking leg.[488]

4.5. Peptides in Crustacean Eyestalks

The sinus gland, a neurosecretory organ of decapod crustacea usually situated in the eyestalks, stores and releases several peptide hormones that control movement of pigment in the eye. The crustacean eye contains a system of pigments used for screening and reflecting light. One of these peptides, the light-adapting hormone, causes pigment in the distal cells of the retina to move into a more proximal position. This compound has recently been isolated from the eyestalks of the shrimp *Pandalus borealis*, and studies of its structure indicate that it is a polypeptide with 18 amino acids.[489] Eyestalks also contain peptides that act on chromatophores, specific pigment cells in the cuticle that control the color. One of these, "red pigment-concentrating hormone," has been isolated from *Pandalus* and has the structure, pGlu-Leu-Asn-Phe-Ser-Pro-Gly-Trp-NH_2. The synthetic compound acted on erythrophores in picogram amounts but was completely inactive on leukophores or melanophores.[490] Other eyestalk hormones include a hyperglycemic factor from *Cancer magister* that decreases the glycogen content of various organs,[443] a "white pigment-concentrating hormone" from the shrimp *Crangon crangon*,[491] and a "black pigment-dispersing activity" from the fiddler crab *Uca pugilator*. Application of NE triggered the release of the latter hormone, and α-adrenergic blocking agents inhibited this effect.[246]

4.6. Mammalian Neuropeptides

Following the characterization and synthesis of the enkephalins, substance P, and other low-molecular-weight peptides found in mammalian brain, immunocytochemical procedures for visualizing these peptides were developed.[492,493] The antibodies directed against these peptides have also been used in invertebrate preparations, and many positive-reacting neurons have been revealed. For example, endorphin- or enkephalinlike immunoreactivity has been demonstrated in specific neurons in the earthworm, in the leech, in the bivalve *Mytilis*, and in other molluscs, and radioimmunoassay detected this peptide in the locust CNS.[193,441,494–498] The primary photoreceptors, the retinular cells, as well as some fiber tracts in the spiny lobster *Panulirus interruptus* also reacted very strongly to enkephalin.[442]

The nervous system of the cnidarian *Hydra* contains substance P-like immunoreactivity, and addition of this peptide to cultures of these animals stimulated the regeneration of the head regions.[499] In lobster, substance P reactivity could be traced to discrete cell bodies in the proximal fringes of the lamina of the retina.[442] In *Helix*, this peptide was localized in distinct groups of neurons in various ganglia, while cholescystokinin-like (CCK) reactivity was found primarily in 5-HT-containing cells (C-1) of the cerebral ganglion.[500] CCK/gastrin-like immunoreactivity was also demonstrated in cell bodies and axon terminals from a few unidentified neurons in the sub-esophageal ganglion.[594]

Positive immunocytochemical responses were also seen to other neural as well as nonneural mammalian peptides. Insulinlike material was seen in the median neuroscretory cells of the brain of the blowfly *Calliphora vomitoria*. Of the more than 20 visible cells, only six to eight were stained with this antibody.[501] Peptides isolated from the hemolymph of the tobacco hornworm *Manduca sexta* were similar to vertebrate insulin with regard to solubility and to chromatographic, immunologic, and biological properties.[502] Gastrinlike activity was revealed in neuroendocrine systems of both molluscs and insects by immunocytochemical procedures.[438,503] Vasopressinlike activity was seen in the neurosecretory system of the octopus[496] and in the subesophageal ganglia of locust and in the silkworm, *Bombyx mori*,[504,505] and vasotocinlike activity was detected in *Aplysia* ganglia.[506]

By use of antibodies directed against portions of the ACTH molecule, two giant neurons of the pond snail *Lymnaea stagnalis* were identified, one in the visceral ganglion and the other in the right parietal ganglion.[439] Positive sites for immunoreactive thyrotropic releasing factor were also demonstrated in the brain of another gastropod, *Physa*,[507] whereas calcitonin-reacting sites were demonstrated in specific neurons of several invertebrates, including the snail *Aplysia*.[500] Among the protochordates *Ciona intestinalis, Styela clava, Styela plicata*, and *Ascidiella aspersa* immunoreactivity of specific neurons to many of the mammalian neuropeptides has been demonstrated. Examples include ACTH, LH-RH, somatostatin, Substance P, calcitonin, and CCK/gastrin.[508,596–599] It is not yet clear what physiological role, if any, such peptides play in these animals, but the fact that specific amino acid sequences have been conserved is significant and may be related to the fact that these peptides appear very early in the course of their embryological development (see Section 4.10).

4.7. *Unidentified Peptides Isolated from Invertebrates*

In addition to the peptides discussed above that have been isolated, sequenced, and synthesized, a large number of uncharacterized factors that are probably peptides have been found in various nervous systems of invertebrates. Within this heterogeneous group are compounds that induce or inhibit electrical activity when applied to nerve cells or when injected into the animal, that regulate water balance, or that initiate developmental changes.[426–428,433,509]

A "pacemaker" peptide isolated from molluscan ganglia that acts directly on nerve cells is similar to but not identical with either of the mammalian hypothalamic peptides, vasopressin or oxytocin.[510,511] The compound isolated from *Aplysia* or *Otala* ganglia, like the mammalian peptides, induced bursting pacemaker potential activity when tested on a specific neuron of *Otala lactea*[511] and on neurons of dormant *Helix*.[512] It has a molecular weight of about 1000 and is inactivated by pronase but not by trypsin or chymotrypsin.

The R-15 neuron in the abdominal ganglion of *Aplysia* is involved in the regulation of water balance, and a "neuroendocrine reflex" consisting of R-15 and the osphradium (an organ specifically responsive to the osmolarity of the sea water) has been proposed as part of the mechanism controlling water uptake. The rate of firing of R-15 is influenced by the osphradium, which, in turn, influences the release of its water-regulating hormone. Extracts of this cell contain one or more heat-stable, low-molecular-weight polypeptides and cause a rapid weight gain when injected into intact animals.[513] The neuron synthesizes a 12,000-dalton polypeptide, processing it into two smaller fragments of about 1500 and 9000 daltons, which are then transported out of the soma.[428,463,514,515] Whereas the 12,000-dalton polypeptide was synthesized by many neurons in *Aplysia*, only R-15 appeared to process it further, and the 1500-dalton fragment is the presumed water-regulating hormone.

Water balance in other animals also is regulated by peptide factors. In the freshwater snail *Lymnaea stagnalis*, several neurosecretory cells in the pleural ganglion (the "dark green cells") appear to control water balance,[516] whereas in the earthworm *Lumbricus*, a heat-stable factor isolated from brain is also involved in this function, although it has yet to be established that the active principle is a peptide.[517]

In contrast, water balance in insects appears to be controlled by two peptides, a diuretic hormone and an antidiuretic hormone. Extracts from the abdominal or mesothoracic ganglia of the cockroach *Periplaneta americana* contained both peptidelike factors. The diuretic factor has a molecular weight of approximately 30,000, whereas that of the antidiuretic factor is approximately 8000.[518] The diuretic hormone from *Rhodnius prolixus* is a polypeptide of about 60,000 daltons and is inactivated by heating.[519] It is synthesized by neurosecretory cells in the brain and then transported to and stored in the corpora cardiaca.[520] Increased levels of cyclic AMP during those periods when the Malphigian tubules were stimulated by the hormone have also been reported.[521]

Peptides that initiate changes in development are very numerous. In the cnidarian *Hydra attenuata*, "head-activator," "foot-activator," and "neck-inducing" factors have been described. The head activator has been characterized as the undecapeptide, pGlu-Pro-Pro-Gly-Gly-Ser-Lys-Val-Ile-Leu-Phe. Peptides with identical amino acid sequence have also been isolated from human hypothalamus, bovine hypothalamus, and rat intestine.[527,528] This was one of the first demonstrations that a neuropeptide found in one of the most elementary nervous systems was conserved without modification in more sophisticated nervous systems. The "foot-activator" factor also appears to be a peptide of about 1000 daltons, but details of its composition and distribution have not yet been reported.

A few other peptides are included here because of their unusual effects on postsynaptic cells or because they are examples of unique peptides associated with identified neurons. In the marine mollusc *Tritonia diomedia*, a peptidelike material of approximately 1400 daltons was isolated from somata of the C2 neurons in the cerebral ganglion. When applied to postsynaptic neurons, this compound mimicked four different postsynaptic responses evoked by stimulation of the C2 neurons. Ultrastructural studies of the C2 somata showed that they had many of the characteristics of neurosecretory cells; i.e.., they contained rough endoplasmic reticulum and numerous large, dense secretory vesicles.[526]

Synthesis of peptides has also been studied in some *Aplysia* neurons other than R-15 (*vida supra*). High-resolution gels showed that each of the cells, R-2, L-10, L-11, and the white cells, R-3–R-13, incorporated [^{3}H]leucine into distinctly different peptides. Only R-2 synthesized significant amounts of high-molecular-weight protein (>55,000 daltons), whereas the other cells synthesized unique smaller peptides of 12,000 daltons or less.[403,522,523] Polypeptides synthesized in the circumesophageal ganglia of *Aplysia* were transported through connectives to the abdominal ganglion and released from there. Considerable processing of these peptides occurred during the transport process, since the peptides released by electrical stimulation or by a pulse of K^+ were low molecular weight (<2000 daltons).[524,525] Their physiological role is still under investigation. Many additional peptidelike factors have been extracted from insect brains and other sources, but they have not yet been characterized in detail.[427,428,509]

4.8. Proteins

This short section deals only with some recent studies on calmodulin and glycoproteins and with the injection of enzymes into individual neurons. Other aspects of protein neurochemistry, e.g., the isolation of ACh receptors and the properties of specific enzymes concerned with synthesis and degradation of neurotransmitters, were discussed briefly in the appropriate sections. Topics such as venoms and other neurotoxins found in invertebrates are not covered here, and recent reviews should be consulted.[428,529–531]

Calmodulin is an acidic protein (approximately 17,000 daltons) that changes configuration depending on the Ca^{2+} concentration and is closely associated with axonal transport.[532] Although originally obtained from bovine brain, this protein has also been isolated from the sea pansy *Renilla reniformis*[533] and from the octopus optic lobes.[534] The structural composition of all calmodulins is quite similar, although the invertebrate proteins differ by at least one tyrosine residue from the bovine protein. A protein isolated from octopus and described originally as a protein specific for the molluscan nervous system has now been identified as calmodulin.[535] The unusual amino acid trimethyllysine is characteristic of this protein, and in preparations from optic lobes this amino acid was found at a concentration slightly less than 1 mol/mol calmodulin. A preparation from octopus brain also contained significant amounts of the monomethyl- and dimethyllysines, which may be breakdown products of the trimethyl derivative.[536]

Glycoproteins are complex, high-molecular-weight proteins found in all tissues, primarily in cell membranes. In these membranes, they are oriented such that their carbohydrate chains face the extracellular environment and, because of this orientation, are considered to play key roles in a variety of surface interactions including cell recognition and the binding of hormones and antibodies. Glycoproteins have also been shown to be involved in intracellular membrane phenomena such as axonal transport and with the interaction of synaptic vesicles with release sites.

Synthesis and transport of glycoproteins was studied by injecting [^{3}H]fucose or [^{3}H]N-acetylgalactosamine into the soma of an identified neuron and analyzing the cell and its processes at various times after injection. In the leech *Hirudo medicinalis*, glycoproteins were transported along Retzius cell axons at a rate of 5–10 mm per day.[537] In the soma of the R-2 neuron of *Aplysia californica*, SDS gel electrophoresis showed that the label was associated with only five major proteins among all of the components separated by this technique.[538] Newly synthesized glycoprotein was also detected in the R-2 axon, and the amount of this radioactivity increased with time, whereas fucose itself remained in the cell body. Only three of the glycoproteins that were synthesized were also transported, however.[539] Injection of isotopic sugar directly into the axon also labeled the glycoproteins, demonstrating that mechanisms exist in the axon as well as in the soma for the modification of these proteins. Autoradiographic studies of injected cells confirmed these findings and showed intense labeling over the Golgi apparatus in the cell body, over vesicles in both the cell body and the axon, and over external membranes.[540,541]

Radioactivity not incorporated into these proteins remained as fucose or as one of two metabolites tentatively identified as fucose phosphate or as GDP-fucose.[540] Label injected into soma remained primarily within the confines of that cell and its processes, whereas [^{3}H]fucose taken up from solution was incorporated into glial cells and into the connective tissue as well as into the R-2 soma.[539] Recent evidence indicates that specific membrane glycoproteins are even routed to particular sites in that cell.[543] More detailed discussion of the mechanisms of axonal transport and the synthesis and processing of transported proteins have been presented elsewhere.[542–545]

Enzyme isolation and purification are necessary in order to be able to study the molecular mechanism of their catalytic actions. Purified enzymes have also been used as analytical tools by chemists for many years, and, recently, they have even been used as antigens to generate specific antibodies for use in certain immunocytochemical procedures.[546] Clear-cut results require "clean" enzyme preparations, and much effort has gone into achieving adequate purification.[89,90] Enzymes have also been used to study certain properties of neurotransmission after being injected into specific neurons. Highly purified preparations of AChE, various phospholipases, and neuraminidase were injected into the somata of *Aplysia* neurons, and their effects on synaptic events were recorded. For example, a short time after injection of the enzyme AChE into cholinergic neurons in the buccal ganglion, transmission between the injected cell and its follower progressively decreased and eventually failed. These results implied that ACh from the "free" cytoplasmic compartment rather than "bound" ACh was being released into the synaptic cleft. This proposed explanation has stimulated much discussion regarding cholinergic mechanisms, since it casts some doubt on the role of vesicles and on the release of ACh by exocytosis.[546,547]

The effects of injected phospholipases on synaptic transmission were not as dramatic as those of AChE, but various lipases could be classified according to their actions on active or passive membrane properties.[546] Neuraminidase, an enzyme that hydrolyzes only sialic-acid-containing compounds, was selected to determine the role of glycoproteins in these events. Here, too, transmission failed a short time after somal injection. The model that was suggested to explain these results proposed that Ca^{2+} ions were bound to the inner-facing sialic acid moieties of the membrane glycoproteins and that the reversibility of this intracellular binding of Ca^{2+} controlled the release of transmitter.[548]

4.9. Remarks

Within the last few years, a wide range of peptides (or at least of specific amino acid sequences in peptides) has been revealed by immunocytochemical techniques. The conservation of hormones through evolutionary development is an area that is of interest both to neurobiologists and to molecular geneticists. It has been proposed that genes specifying some mammalian peptide hormones appeared very early in evolution and have been conserved unchanged. For example, immunocytochemical reactions have not only detected *Hydra* peptides in mice, but have also detected hypothalamic peptides in molluscs.[455,456,496] Other organisms may have acquired these genes by transfer of genetic material from other organisms, perhaps viruses,[425] a mechanism only recently considered in a serious manner as a means of inducing species specificity. The detection of sites throughout the invertebrate system that react with mammalian peptides and of sites in vertebrate systems that react with invertebrate peptides emphasizes their universality, but we do not yet appreciate what functions these compounds serve in such diverse animals.[425,552,553]

Some of the low-molecular-weight peptides that influence nervous activity have been synthesized and are commercially available. Proctolin is at present the only neuroactive peptide that is considered to be a transmitter in at least one system, whereas all of the others are labeled as "neuromodulators" or "neuroregulators," Proctolin has only been positively identified in insects. Its presence is suspected in crustaceans, but it has not yet been detected in vertebrates, although the distinction between "invertebrate" and "vertebrate" peptides is rapidly becoming meaningless.

In addition to studies on their actions as neurotransmitters and neuromodulators, many investigations are in progress that involve the action of invertebrate peptides as neurotoxins. Specific invertebrate toxins such as black widow spider venom,[549] cephalotoxins from *Octopus vulgaris*,[550] and various toxins from sea anemones[551] have been used as tools for the investigation of membrane properties of neurons in ways similar to those using toxins from vertebrates, such as tetrodotoxin or bungarotoxin.

A specific amine and a specific peptide occurring together in the same neuron have been reported in a few studies. One recent example was in *Helix*, where both 5-HT and CCK were found in the GSC.[500] Several examples have been described in mammalian systems, where reactivity to CCK was also associated with 5-HT- (or DA-) containing structures or where reactivity to vasointestinal peptide was associated with ACh.[492,554] A group of neuroendocrine cells, the amine precursor uptake and decarboxylation (APUD) cells originally described by Pearse,[555] appear to have as their primary function the synthesis of polypeptide hormones. These cells also contain marker enzymes that are associated with the synthesis or secretion of amines. It is proposed that these cells, only some of which arise from the neural crest during development, are part of a specialized, loosely defined neuroendocrine control system. At present, there is little information regarding APUD cells in invertebrates; however, some of the molecular markers for these cells, such as tryptophan hydroxylase and dopamine-β-hydroxylase, are those very enzymes found specifically in neurotransmitter-containing neurons. Whether or not such cells are related embryologically to APUD cells and the definition of the role of both amine and peptide in these cells will be the subject of future investigation.

Recently, criteria for the physiological function of peptides were formalized,[556] and these criteria appear remarkably similar to those proposed earlier for neurotransmitters.[557,558] To define the function of a peptide, one must demonstrate that: (1) it is present in the presumed "releasing" cell; (2) it is synthesized or accumulated in this cell; (3) there is a system for release on activation; (4) released peptide has access to appropriate receptor sites; (5) receptors are present on these "receiving" cells; (6) exogenously applied peptide mimics physiological activation; (7) an inactivation system exists; and (8) pharmacological agents antagonize both exogenous and endogenous systems to the same degree.

To measure peptide constituents in various nervous tissues by direct chemical assay presents a real challenge to the neurochemist. Existing methods for quantitative analysis of many endogenous peptides are already pushed to their limit by the amounts of tissue available at the ganglion level, much less at the cell level. The past decade has seen vast improvements in methodology so that as little as 0.05–0.1 pmol of many of the biogenic amines could be measured quantitatively, amounts found in a few isolated molluscan neurons (Table III). However, it is unlikely that many peptides are present in individual neurons even in these amounts. More probably, even large neurons will contain amounts that are at least an order of magnitude less. This means either that many samples need to be pooled for a single analysis or that new, more sensitive methods must be devised. Previously, the scaling down of older procedures or the use of new techniques, e.g., fluorometric or radiometric assays, was adequate to lower detection limits to the level required to carry out determinations of most neurotransmitters. However, few of these modifications are relevant to peptide analyses in crude tissue extracts.

Recent improvements in various phases of adsorption chromatography have enabled investigators to achieve separations quickly and efficiently that were considered impossible earlier. Further, the ability to sequence and assay smaller and smaller amounts has increased the number of neuroactive peptides that have been characterized and, thus, the number of peptides for which assays are needed. Certainly, the progress in immunologic techniques such as radioimmunoassays is impressive and, with the availability of more synthetic compounds, it will permit more of these peptides to be assayed. Monoclonal antibodies, cloning of genes, and analysis of specific mRNAs can also be exploited. However, radioimmunoassays and immunocytochemical techniques each have their own unique set of technical and biological problems and often are not the most desirable alternatives. Thus, we must also find ways to increase the sensitivity of direct chemical assays in order to measure these compounds in nervous tissues and, more specifically, in neurons.

5. *MISCELLANEOUS*

5.1. *Peptidoamines*

This new class of compounds is composed of specific biogenic amines linked covalently to a small peptide or to a single amino acid. The amino acid or peptide is N-terminal, so that these compounds are the equivalent of a peptide with a blocked C-terminus, i.e., the carboxyl group of the amino acid has condensed with the primary amine. The first peptidoamines described were a series of oligopeptides synthesized in mouse brain homogenates from radioabeled histamine and a mixture of amino acids, including N-acetylaspartic acid.[559] Later, peptidoamines incorporating DA and 5-HT were also reported.[560] In mouse brain, these peptides were of variable length and composition, but all contained N-acetylaspartic acid in the N-terminal position and the amine in the C-terminal position. In invertebrates, on the other hand, the only peptidoamines reported thus far have consisted of a single amino acid (either β-alanine or glutamate rather than a peptide) linked to the biogenic amines, OA, tyramine, or HM.

β-Alanyl derivatives have been reported in several preparations. In the lobster *Homarus interruptus*, incubation of isolated ganglia with OA led to the production of several new products, one of which was the peptidoamine β-alanyl-OA.[285,286] These metabolites were formed primarily in the connective tissue and in glia surrounding the neurons rather than in neurons themselves. The brains of pharate adults from *Manduca sexta* metabolized tyramine in a similar manner. In this case, the peptidoamine, β-alanyltyramine,was identified, and the two additional metabolites were shown to be N-acetyltyramine and a compound tentatively identified as the O-β-glucoside of N-acetyltyramine.[243] An endogenous ninhydrin-positive constituent, "carcicine," was isolated from cardiac muscle of the crab *Carcinus maenus* and identified as β-alanylhistamine.[417] Of all the tissues assayed, this compound was found only in heart muscle. However, when carcicine was added to an isolated heart preparation, it had no effect on either frequency or amplitude. Since no carcicine was formed from carnosine (β-alanylhistidine), a peptide often found in large quantities in muscle tissues, this indicated that the peptidoamine was produced via a novel synthetic pathway rather than via the decarboxylation of a preformed peptide.

The only glutamate-containing peptidoamines reported are in *Aplysia*, where γ-glutamylhistamine was the product formed by the ganglia during incubation with histamine.[310] As much as 90% of the label from [^{14}C]HM taken up by the ganglia was recovered in the form of the peptidoamine. Although γ-glutamylhistamine accumulated rapidly in isolated tissue, only trace amounts were found endogenously (D. Weinreich, personal communication).

The enzyme catalyzing the formation of this peptidoamine has been characterized as γ-glutamylhistamine synthetase. It is a soluble enzyme requiring ATP as an energy source, and the apparent K_ms for HM and for glutamate are 0.6 and 10.5 mM, respectively. Of the amino acids tested, only glutamate was effectively incorporated into the peptidoamine. Activity could be detected in most *Aplysia* tissues but was highest in nervous tissue. All ganglia and all individual cells contained the enzyme, but capsular tissue, e.g., the connective tissue sheath, was a richer source of activity than neurons.[561] The kinetic properties of this enzyme and its distribution differentiate it clearly from other γ-glutamyl-transferring enzymes such as glutathione synthetase, γ-glutamyltranspeptidase, etc.

In most preparations, these peptidoamines have been detected only as labeled products after incubation with a radiolabeled amine, although in the crab, endogenous carcicine could be measured in heart extracts. They do not appear to be attacked by tissue peptidases, since large amounts were recovered from the tissue even after long periods of incubation.[310] In those systems in which they have been tested, the peptidoamines, like the N-acetylamines (*vide infra*), are inactive biologically. However, there is no evidence at present to suggest that synthesis of these compounds is a mechanism for terminating the physiological activation of these amines at nerve endings.

5.2. N-Acetyldopamine and Other N-Acetyl Compounds

N-Acetyl derivatives of biogenic amines do not appear to be important constituents in the nervous system of lower invertebrates, although N-acetylation has been suggested for inactivating amines in tissues that contain no MAO. Neither N-acetyl-DA (NADA), N-acetyl-HM, nor N-acetyl-5-HT could be detected in molluscan ganglia, and only low levels of N-acetyltransferase activity could be measured.[208,310]

In insects, catecholamines are involved in cuticle formation, and NADA is a precursor of the melaninlike material formed in the sclerotization process. There is also increasing evidence that it may have a specific function in these nervous systems. In the cerebral ganglion of the cockroach, NADA values were very high (approximately 100 nmol/g protein), whereas in abdominal ganglia values were only 10% of this number. In general, the concentrations of NADA and DA paralleled each other.[562] Values for NADA in cerebral ganglia from insects representing other orders, including Hymenoptera, Lepidoptera, Diptera, Coleoptera, and Odonata, were also reported.

N-Acetyl-transferring enzymes have been described in *Drosophila*, in the honeybee, in the locust, and in the cockroach, but these activities are not limited to nervous tissues.[181,182,319] The *Drosophila* enzyme efficiently acetylates 5-HT, DA, and tyramine, and partially purified preparations have been used in radioenzymatic assays for these amines.[208] Some characteristics of these enzymes have already been discussed in sections dealing with individual biogenic amines (see Sections 2.2, 2.3, and 2.5). Only in the cornborer *Ostrinia* has kinetic analysis of NAT indicated that there are differences between the enzyme in brain and the enzyme in other tissues.[564] A novel enzyme that hydroxylates N-acetyltyramine to form NADA has been described in nervous tissues of the locust. Unlike tyrosine hydroxylase or tyrosinase, this enzyme hydroxylates tyramine specifically.[565] These studies further suggested that, in this nervous system, N-acetyltyramine represented a major intermediate in the pathway from tyrosine to NADA, whereas in cuticle formation this was only a minor pathway.

Few N-acetyl compounds other than N-acetylamines have been reported in invertebrates. One compound, N-acetylaspartic acid, which is present in high concentrations in mammalian brain, has also been found in the stellate ganglion of the squid, *Loligo vulgaris*,[566] but this compound is absent from most other molluscs and from arthropods.

5.3. Cyclic AMP

The precise mechanism by which the cyclic nucleotides, cyclic AMP and cyclic GMP, influence either the metabolism of the neuron or the generation of the impulse at the neuronal membrane is not yet understood. Nevertheless, the rapid and dramatic changes in cellular concentrations of these nucleotides induced by electrical activity and/or by various neurotransmitters indicates that these events are closely related.[269,567]

Activity in the 5-HT-containing neuron C-1 of *Aplysia* potentiates contraction and increases the concentration of cyclic AMP in its target, the buccal musculature. These effects are enhanced in the presence of phosphodiesterase inhibitors and certain analogues of cyclic AMP as well as in the presence of 5-HT itself. Even in homogenates of these muscles, increases in cyclic AMP concentration could be demonstrated after exposure to 5-HT. All of the results indicate that these effects are mediated primarily by a 5-HT-sensitive adenylate cyclase.[569,570] Earlier studies had shown that in the abdominal ganglion, electrical activity as well as 5-HT and DA stimulated the synthesis of cyclic AMP.[154] These ganglia readily took up [^{3}H]adenine and converted some into ATP and a smaller amount into cyclic AMP.[570] At the crayfish neuromuscular junction, several agents that elevated cyclic AMP levels also increased dramatically the release of transmitters in response to nerve stimulation, suggesting that presynaptically located adenylate cyclase was activated by 5-HT and that the release of transmitter at the nerve endings was modulated by this increased cyclic AMP.[571]

In the planarian *Dugesia gonocephala*, dopaminergic agonists also increased cyclic AMP levels, whereas antagonists, including morphine, decreased the levels.[572] A DA-sensitive adenylate cyclase in the pedal ganglion of the mussel *Mytilus edulis* has been characterized. In this system, the catecholamines, as well as apomorphine, stimulated cyclic AMP synthesis, whereas amine-blocking agents, e.g., fluphenazine, opiods, endorphin, and enkepalin, inhibited these effects.[573] In arthropods, specific 5-HT-, DA-, and OA-sensitive adenylate cyclases were discussed earlier under each amine. The distribution and specificity of these cyclases further support the suggestion that they are closely related to the molecular events of neurotransmission.

Increased levels of cyclic AMP and cyclic GMP in *Aplysia* ganglia were also produced by the application of the peptides oxytocin and vasopressin.[567] These peptides produced long-term responses in the identified neuron R-15 that could be mimicked by extracellular or intracellular application of phosphodiesterase inhibitors. The presence of both cyclic nucleotides was necessary to produce the pattern of enhanced bursts and interburst hyperpolarizations that was recorded as increased activity. Similar effects were obtained with an uncharacterized peptide fraction isolated from *Helix* ganglia. This factor not only regulated burst cell activity but also increased intracellular cyclic AMP and the adenylate cyclase activity of membrane preparations.[574] Other peptides that increase levels of cyclic AMP include the adipokinetic hormone from crustacean eyestalk glands,[510] the purified crustacean hyperglycemic hormone,[422] and the molluscan neuropeptide FMRFamide.[450]

The accumulation of intracellular cyclic AMP may explain the slow onset of certain peptide responses but is unlikely to explain directly the long duration of action of many of these hormones. Often, the hormonal effect is considerably longer than the time of maximal cyclic AMP (or cyclic GMP) levels. An alternate explanation centers around the phosphorylation of key proteins. A variety of neurotransmitters, hormones, and other regulatory agents affect the phosphorylation of specific proteins in their target tissue, not all of which act through the mediation of cyclic AMP.[576] Examples where the incorporation of label from [^{32}P]ATP into endogenous phosphoproteins was stimulated by cyclic AMP were found in *Helix* and in *Aplysia* ganglia.[268,575] In the mollusc *Hermissenda crassicornis*, increased incorporation of ^{32}P into a phosphoprotein of 20,000 daltons was correlated with a learning task involving light. This is a simple system involving only five photoreceptors, a lens, and a few pigment and epithelial cells and appears to be well suited for the correlation of chemical changes with behavior.[577]

5.4. Nucleic Acids

Most of the studies concerned with nucleic acids are concerned with RNA and the synthesis of polypeptides and proteins. It has long been known that ribosomes are absent from axons of adult animals and that most of the protein synthesis occurs in the cell bodies. This newly synthesized protein is then transported down the axons to supply the needs of the nerve terminals. Some early studies had indicated that isolated nerve fibers incorporated labeled amino acids into proteins; however, more detailed studies demonstrated that the protein appearing in the cytoplasm was synthesized in glial cells and then transferred to axons.[578–582]

Nucleic acid synthesis in the cell body of the giant neuron R-2 in *Aplysia* was studied using labeled uridine or cytidine.[580] Label was incorporated readily into ribosomal RNA, but very little was incorporated into the DNA. Stimulation of the connective increased the amount of label incorporated into RNA but did not increase the uptake of uridine or the amount of label appearing in UTP. After injection of labeled uridine or orotic acid into Retzius cells of the leech (*Hirudo medicinalis* or *Haemopsis marmorata*), label was incorporated into macromolecules in both the nucleus and cytoplasm. The data also suggested that label was transported from the soma into cell processes.[537] More direct evidence for RNA in axoplasm was obtained using axons of the giant squid *Loligo pealii* and the giant fiber of the polychaete *Myxicola*. These preparations contained a large portion of low-molecular-material (4 S) that had all the appropriate characteristics of RNA, i.e., it was charged with amino acids by a synthetase prepared from squid brain, amino acylation was prevented by prior incubation with RNase, and it was labile at pH 9.[583]

5.5. Lipids, Phospholipids, and Carotenoids

Lipids and phospholipids are important constituents of membranes and influence the permeabilities of ions and molecules, the integrity of cell surfaces, and various conduction properties of nerve cells. Detailed analyses of lipids, particularly of free fatty acids, diglycerides, and cholesterol, in molluscs and some other invertebrates have been summarized in an earlier publication.[584] As in vertebrates, the nervous system of molluscs contains higher levels of phospholipids than other tissues. In ganglia from *Helix pomatia*, phosphoinositides and phosphatidylserine comprised almost half of the total phospholipids, and phosphatidylcholine, phosphatidylethanolanine, and sphingomyelin were represented about equally. Concentrations of phospholipids in cephalopod brain were even higher than those in gastropod ganglia.[585]

Phospholipid-to-cholesterol ratios varied between 0.5 and 1.0 for *Helix, Lymnaea, Murex, Sepia*, and *Homarus*.[586,587] Neither gangliosides nor sialic acid derivatives could be detected in the molluscs *Helix, Lymnaea, Murex*, and octopus, whereas sulfolipids were found at about 1/20 of the phospholipid value in these animals.

Carotenoids are hydrophobic pigment compounds whose presence in molluscan cells has been known for many years, but their function is unclear. Since they are oxidizable, it has been proposed that they play a role in oxidative metabolism; however, more recent studies indicate that they may be involved in movement of Ca^{2+} instead. Some neurons are activated in the presence of visible light, and the absorption spectrum of these pigment molecules is similar to the action spectrum of light in these cells.[588–590] Morphological studies have shown that carotenoids in *Lymnaea* are associated with Ca^{2+}-sequestering organelles. The ability of these organelles to bind and to transport Ca^{2+} across membranes has been attributed to their presence. It is theorized that visible light suppresses the Ca^{2+}-sequestering function of these carotenoids, resulting in increased cytoplasmic Ca^{2+}, and that this increased Ca^{2+} accounts for the neuronal discharge. However, isolated lipochondria prepared from whole *Aplysia* ganglia produced many of the same morphological changes on exposure to light that were seen in intact R-2 cells, including loss of Ca^{2+}, so that these characteristics appear to be general ones and not necessarily related to the electrophysiological response. Two pigments, β-carotene and a compound tentatively identified as retinol, have been isolated from lipochondria, but neither has been shown to act as a chromophore in the light-sensitive neurons.[590]

5.6. *Remarks*

This group of miscellaneous compounds is a collection of chemical loose ends that do not fit into any of the neat categories discussed above, yet they all have something special to do with nervous activity either as a special metabolite or as a component of a special system (transport, light transduction, etc.). Only further study will show which of them are only interesting curiosities and which will become focal points for future research.

6. *SUMMARY*

Neurochemists are continually reaffirming the universality of fundamental cellular mechanisms, and favorable invertebrate preparations have contributed much to our understanding of these mechanisms. In the last 10 years, remarkable advances have been made in our knowledge of the unique biochemistry of individual neurotransmitters. Various biogenic amines and amino acids have been measured in individual neurons of *Helix, Aplysia*, and lobster, and many of the synthetic and degradative enzymatic reactions have been characterized. Extensive use of radiolabeled compounds has made it possible to follow the transport of substances from cell body toward target organs and, in selected cases, to measure the release of the transmitter after stimulation. When these data were consistent with physiological and pharmacological studies, many of the criteria set forth for establishing a compound as a neurotransmitter were fulfilled.

Among the few examples where this has been accomplished are the 5-HT-containing metacerebral neurons of *Aplysia* (see Section 3.1.3). These neurons appear to be serotoninergic in function and have been shown to synthesize, transport, and release 5-HT. An impressive array of evidence also attests to the cholinergic nature of L-10 and L-11 in the abdominal ganglion, to the histaminergic nature of the C-2 and C-3 neurons of the cerebral ganglion of *Aplysia*, and to the GABAergic nature of the "I" cells of the lobster thoracic ganglion. However, these "textbook" examples represent only a few percent of the total number of neurons present in any one ganglion.

The vast majority of individual neurons surveyed contain none of these biogenic amines. Even when ultrasensitive analytical procedures were used, a remarkably few number of cell bodies contained measurable amounts, even though substantial amounts may have been present in whole ganglion extracts, e.g., DA or GABA in molluscan ganglia. Since it has been demonstrated that few of these neurons are electrically coupled, they must use some chemical compound to transmit information across synapses. Therefore, our inability to detect a chemical transmitter could mean that our methods are not sensitive enough to detect the amounts that are present. Perhaps many neurons contain somal concentrations less than 100 μM (a lower limit of most identified cells) or are of such small size that the total they contain falls below our detection limits. On the other hand, some transmitters may not accumulate in cell bodies and may be rapidly transported into axons or may even be synthesized in axons and nerve terminals. Another possibility is that we are looking at the wrong compounds. Perhaps we should be looking for a small peptide like proctolin, whose presence is obscured by the overwhelming amounts of free amino acid in tissue extracts.[591] Although contrary to precedent, these new transmitters could even be "non-amine" compounds. In some vertebrate systems, strong arguments have been made that adenosine or ATP are, indeed, transmitter substances[593,595]; however, little information is available in invertebrates that suggests such a role here. To characterize chemically some of these remaining populations of neurons is one area where investigations must continue.

The presence of more than one transmitter in a single neuron has been reported as chemical methods for the analysis of biogenic amines have become more sensitive. Up to this point in this review, I have avoided discussion of multiple transmitters, since, at the moment, the concept remains an interesting curiosity rather than a unifying principle.[323,592,593] To date, there are at least ten examples in invertebrate preparations where small amounts of a second (or third) "transmitter" substance have been detected in certain serotoninergic or cholinergic neurons.[8,11,31,46,48,49,247] Usually, these "non-ergic" compounds were found in quantities that amounted to 1–5% of the "ergic" compounds; nevertheless, investigators have focused on the fact that they were found at all rather than on the fact that they were found at very low concentrations. Much energy has been expended speculating as to why more than one neurotransmitter may be needed to convey all the information inherent in neuronal communication, and many of these arguments are highly speculative. It is true that some observations are not explained satisfactorily in terms of the known actions of a single transmitter, but then, how many of them could be better explained by postulating a second transmitter as a "neuromodulator"?

The demonstration of homology of neurons, on the other hand, is a unifying concept that helps to bring a small degree of order out of the perplexing mass of data that has accumulated. Loosely defined, it is the identification of neurons with certain common characteristics among different animals. Identifiable neurons are a prominent feature of leech, crustaceans, insects, and gastropods, and some of these are so consistent as to be recognized in individual animals. Sometimes neurons can be recognized as homologous by their size and location or color; sometimes by input and output connections or by pharmacological responses; most recently, by formaldehyde-induced fluorescence or direct chemical measurements of amines.[1,121,122,127] Some examples include the 5-HT-containing cells in the cerebral ganglia of various gastropod molluscs, *Aplysia, Tritonia, Helix*, and *Helisoma*; the DA-containing cells of *Planorbis, Lymnaea*, and *Helisoma*[191,201,202,204]; and the ligamental plexus in *Panulirus* and the pericardial organs of *Homarus* and other crustaceans.[171,172] This once rare kind of consistency among neurons actually extends to many types and groups. If this concept can be extended even further to include more diverse animals, it may provide additional clues regarding the conservation of structure and function (or lack of it) throughout a portion of the animal kingdom.

The capability of direct chemical analysis has improved to the point where the measurements of biogenic amines in individual somata, albeit large molluscan cells, are no longer a novelty. Even the problem of assaying cells for specific peptides, still considered a formidable task, will probably be solved by technological improvements of HPLC. However, although individual cell bodies can be isolated, and individual axons in nerve trunks can be analyzed, the information obtained from these measurements may or may not be relevant to events or conditions occurring at the synapse, the *raison d'être* of the neuron.

The search for "simple" nervous systems among the invertebrates has stimulated much research, yet the more we study them, the more we find that even these systems are complicated and interdependent. However, it is anticipated that advances in analytical capabilities will increase our knowledge and understanding of these simpler systems, which, in turn, may be applied to higher animals.

ACKNOWLEDGMENTS. The author wishes to thank Drs. R. E. McCaman, J. Ono, R. Hammerschlag, G. Stone, D. Pearson, and G. Crawford for their advice and encouragement in this project and Ms. Cheryl Denison and Ms. Vivian Thacher for their patience and skill in the preparation of the manuscript.

REFERENCES

1. Bullock, T. H., Orkand, R., and Grinnell, A. M., 1977, *Introduction to the Nervous System*, W. H. Freeman, San Francisco.
2. Bullock, T. H., and Horridge, G. A., 1965, *Structure and Function in the Nervous System of Invertebrates*, W. H. Freeman, San Francisco.
3. Meglitsch, P. A., 1972, *Invertebrate Zoology*, Oxford University Press, New York.
4. Gerschenfeld, H. M., 1973, *Physiol. Rev.* **53**:1–119.
5. Kandel, E. R., 1976, *Cellular Basis of Behavior*, W. H. Freeman, San Francisco.
6. Leake, L. D., and Walker, R. J., 1980, *Invertebrate Neuropharmacology*, Blackie, Glasgow.
7. Cobb, J. L. S., and Pentreath, V. W., 1978, *Prog. Neurobiol.* **10**:231–252.
8. Kehoe, J. S., and Marder, E., 1976, *Annu. Rev. Pharmacol.* **16**:245–268.
9. Liebeswar, G., Goldman, J. E., Koester, J., and Mayeri, E., 1975, *J. Neurophysiol.* **38**:767–779.
10. Ascher, P., and Kehoe, J. S., 1975, *Handbook of Psychopharmacology*, Volume 4 (L. L. Iversen, S. D. Iversen, and S. Snyder, eds.), pp. 265–310.
11. McCaman, R. E., and McCaman, M. W., 1976, *Biology of Cholinergic Function* (A. M. Goldberg and I. Hanin, eds.), Raven Press, New York, pp. 485–513.
12. McCaman, R. E., and Ono, J. K., 1982, *Progress in Cholinergic Biology: Model Cholinergic Synapse* (I. Hanin and A. M. Goldberg, eds.), Raven Press, New York, pp. 22–43.
13. Mansour, T. E., 1979, *Science* **205**:462–469.
14. Martin, S. M., and Spencer, A. N., 1983, *Comp. Biochem. Physiol.* **74C**:1–14.
15. Solon, M. H., and Koopowitz, H., 1981, *Marine Behav. Physiol.* **7**:331–343.
16. Aisa, E., Principato, G. B., Biagioni, M., and Giovannini, E., 1982, *Comp. Biochem. Physiol.* **71C**:119–122.
17. Schwartz, J. H., 1974, *Synaptic Transmission and Neuronal Interactions*, Volume 28 (M. V. L. Bennett, ed.), Raven Press, New York, pp. 239–257.
18. Leake, L. D., Evans, T. G., and Walker, R. J., 1975, *Comp. Biochem. Physiol.* **51C**:205–213.
19. McCaman, R. E., Weinreich, D., and Borys, H., 1973, *J. Neurochem.* **21**:473–476.
20. McCaman, R. E., and Stetzler, J., 1977, *J. Neurochem.* **28**:669–671.
21. Goldberg, A. M., and McCaman, R. E., 1974, *Choline and Acetylcholine: Handbook of Chemical Assay Methods* (I. Hanin, ed.), Raven Press, New York, pp. 385–407.
22. Giller, E., and Schwartz, J. H., 1971, *J. Neurophysiol.* **34**:93–107.
23. Giller, E., and Schwartz, J. H., 1971, *J. Neurophysiol.* **34**:108–115.
24. Triestman, S. N., and Schwartz, J. H., 1977, *J. Gen. Physiol.* **69**:725–741.
25. Tauc, L., Hoffmann, A., Tsuji, S., Hinzin, D. H., and Faille, L., 1974, *Nature* **250**:496–498.
26. Tauc, L., 1977, *Physiol. Rev.* **46**:521–593.
27. Israel, M., Dunant, Y., and Manaranche, R., 1979, *Prog. Neurobiol.* **13**:237–275.
28. Boyne, A. F., 1978, *Life Sci.* **22**:2057–2066.
29. Tauc, L., 1979, *Biochem. Pharmacol.* **27**:3493–3498.
30. Kelly, R. B., Deutsch, J. W., Carlson, S. S., and Wagner, J. A., 1979, *Annu. Rev. Neurosci.* **10**:399–446.
31. Emson, P. C., and Fonnum, F., 1974, *J. Neurochem.* **22**:1079–1088.
32. McCaman, R. E., and Dewhurst, S. A., 1970, *J. Neurochem.* **17**:1421–1426.
33. Prempeh, A. B. A., Prince, A. K., and Hide, E. G. J., 1972, *Biochem. J.* **129**:991–994.
34. Chao, L.-P., 1980, *J. Neurosci. Res.* **5**:85–115.
35. McCaman, R. E., and Dewhurst, S. A., 1971, *J. Neurochem.* **18**:1329–1335.
36. Dewhurst, S. A., 1972, *J. Neurochem.* **19**:2217–2219.
37. Schwartz, J. H., Eisenstadt, M. L., and Cedar, H., 1975, *J. Gen. Physiol.* **65**:255–273.
38. Eisenstadt, M. L., Triestman, S. N., and Schwartz, J. H., 1975, *J. Gen. Physiol.* **65**:275–291.

39. Eisenstadt, M. L., and Schwartz, J. H., 1975, *J. Gen. Physiol.* **65**:293–313.
40. Koike, H., Kandel, E. R., and Schwartz, J. H., 1974, *J. Neurophysiol.* **37**:815–827.
41. Koike, H., Eisenstadt, M., and Schwartz, J. H., 1972, *Brain Res.* **37**:152–159.
42. Triestman, S. N., and Schwartz, J. H., 1982, *J. Gen. Physiol.* **69**:725–741.
43. Dennis, M. J., and Miledi, R., 1974, *J. Physiol. (Lond.)* **237**:431–452.
44. Carew, J. J., Pinsker, H., Rubinson, K., and Kandel, E. R., 1974, *J. Neurophysiol.* **37**:1020–1040.
45. Cohen, J. L., Weiss, K. R., and Kupfermann, I., 1978, *J. Neurophysiol.* **41**:157–180.
46. Goldberg, D. J., and Schwartz, J. H., 1980, *J. Physiol. (Lond.)* **307**:259–272.
47. Hanley, M. R., Cottrell, G. A., Emson, D. C., and Fonnum, F., 1974, *Nature (New Biol.)* **251**:631–633.
48. Cottrell, G. A., 1977, *Neuroscience* **2**:1–18.
49. Osborne, N. N., 1977, *Nature* **270**:622–623.
50. Burnstock, G., 1976, *Neuroscience* **1**:239–248.
51. Dowdall, M., and Simon, E. J., 1973, *J. Neurochem.* **21**:969–982.
52. Tansy, E. M., 1979, *Comp. Biochem. Physiol.* **64C**:173–182.
53. Kehoe, J. S., 1972, *J. Physiol. (Lond.)* **225**:85–114.
54. Kehoe, J. S., 1972, *J. Physiol. (Lond.)* **225**:115–146.
55. Kehoe, J. S., 1972, *J. Physiol. (Lond.)* **225**:147–172.
56. Kehoe, J. S., Sealock, R., and Bon, C., 1976, *Brain Res.* **107**:527–540.
57. Shain, W., Greene, L. A., Carpenter, D. O., Sytkowski, J., and Vogel, Z., 1974, *Brain Res.* **72**:225–240.
58. Parmentier, J., and Carpenter, D., 1976, *Animal, Plant and Microbial Toxins* (A. Ohsaka, K. Hayaishi, and Y. Sawai, eds.), Plenum Press, New York, pp. 179–191.
59. Ono, J. K., and Salvaterra, P. M., 1981, *J. Neurosci.* **1**:259–270.
60. Morley, B. J., Kemp, G. E., and Salvaterra, P., 1979, *Life Sci.* **24**:859–872.
61. Brown, D. A, 1979, *Advances in Cytopharmacology*, Volume 3 (B. Ceccarelli and F. Clementi, eds.), Raven Press, New York, pp. 225–230.
62. Gardner, C. R., 1981, *Comp. Biochem. Physiol.* **68C**:85–90.
63. Sargent, P. B., 1977, *J. Neurophysiol.* **40**:453–460.
64. Wallace, B. G., and Gillon, J. W., 1982, *J. Neurosci.* **2**:1108–1118.
65. Coggeshall, R. E., Dewhurst, S. A., Weinreich, D., and McCaman, R. E., 1972, *J. Neurobiol.* **3**:259–265.
66. Wallace, B., 1981, *Neurobiology of the Leech* (K. J. Muller, J. G. Nicholls, and G. S. Stent, eds.), Cold Spring Harbor Laboratory, New York, pp. 147–167.
67. Fitzpatrick-McElligott, S., and Stent, G., 1981, *J. Neurosci.* **1**:901–907.
68. Stent, G., and Weisblat, D. A., 1982, *Sci. Am.* **246**:136–147.
69. Florey, E., 1973, *J. Comp. Physiol.* **83**:1–16.
70. Pittman, R. M., 1971, *Comp. Gen. Pharmacol.* **2**:347–371.
71. Barker, D. L., Herbert, E., Hildebrand, J. G., and Kravitz, E. A., 1972, *J. Physiol. (Lond.)* **226**:205–229.
72. Hildebrand, J. G., Townsel, J. G., and Kravitz, E. A., 1974, *J. Neurochem.* **23**:951–963.
73. Battelle, B. A., Kravitz, E. A., and Stieve, H., 1979, *Experientia* **35**:778–780.
74. Newkirk, R. F., Maleque, M. A., and Townsel, J. G., 1980, *Neuroscience* **5**:303–311.
75. Maxwell, G. D., Tait, J. F., and Hildebrand, J. G., 1978, *Comp. Biochem. Physiol.* **61C**:109–119.
76. Prescott, D. J., Hildebrand, J. G., Sanes, J. R., and Jewett, S., 1977, *Comp. Biochem. Physiol.* **56C**:77–84.
77. Emson, P. C., Burrows, M., and Fonnum, F., 1974, *J. Neurobiol.* **5**:33–42.
78. Dewhurst, S. A., McCaman, R. E., and Kaplan, W. D., 1970, *Biochem. Gen.* **4**:499–508.
79. Sanes, J. R., Prescott, D. J., and Hildebrand, J. G., 1977, *Brain Res.* **119**:389–402.
80. Marder, E., 1976, *J. Physiol. (Lond.)* **257**:63–86.
81. Futamachi, K. J., 1972, *Science* **175**:1373–1374.
82. Kawagoe, R., Onodera, K., and Takeuchi, A., 1981, *J. Physiol. (Lond.)* **312**:225–236.
83. Takeno, K., Hiromori, T., and Yanigiya, I., 1981, *Insect Biochem.* **11**:527–535.
84. Breer, H., 1982, *J. Neurobiol.* **13**:103–117.

85. Denburg, J. L., 1973, *Biochim. Biophys. Acta* **298**:967–972.
86. Newkirk, R. F., Sukumar, R., Thomas, W. E., and Townsel, J. G., 1981, *Comp. Biochem. Physiol.* **70C**:177–184.
87. Emson, P., Malthe-Sorenssen, O., and Fonnum, F., 1974, *J. Neurochem.* **22**:1089–1098.
88. Driskell, W. J., Weber, B. H., and Roberts, E., 1978, *J. Neurochem.* **30**:1135–1141.
89. Slemmon, J. R., Salvaterra, P. M., Crawford, G. D., and Roberts, E., 1982, *J. Biol. Chem.* **257**:3847–3852.
90. Crawford, G., Slemmon, J. R., and Salvaterra, P. M., 1982, *J. Biol. Chem.* **257**:3853–3856.
91. Zingde, S., and Krishnan, K. S., 1980, *Development and Neurobiology of Drosophila* (O. Siddqi, P. Batu, L. M. Hall, and J. C. Hall, eds.), Plenum Press, New York, pp. 87–103.
92. Tripathi, R. L., Telford, J. N., and O'Brien, R. D., 1978, *Biochim. Biophys. Acta* **525**:103–111.
93. Silver, L. H., and Prescott, D. J., 1982, *J. Neurochem.* **38**:1709–1718.
94. Devonshire, A. L., 1975, *Biochem. J.* **149**:463–469.
95. Greenspan, J. R., Finn, J. A., Jr., and Hall, H. C., 1980, *J. Comp. Neurol.* **189**:741–774.
96. Dudai, Y., 1979, *Trends Biochem. Sci.* **4**:40–44.
97. Salvaterra, P. M., and Foders, R. M., 1979, *J. Neurochem.* **32**:1509–1517.
98. Mansour, N. A., Eldefrawi, M. E., and Eldefrawi, A. T., 1977, *Biochemistry* **16**:4126–4132.
99. Harris, R., Cattell, K. J., and Donnellan, J. F., 1981, *Insect Biochem.* **11**:371–385.
100. Satelle, D. B., 1980, *Receptors for Neurotransmitters, Hormones and Pheromones in Insects* (L. M. Hall, J. G. Hildebrand, and D. B. Satelle, eds.), Elsevier, Amsterdam, pp. 71–124.
101. Jimenez, F., and Rudloff, E., 1978, *Biochim. Biophys. Acta* **535**:505–517.
102. Schmidt-Nielsen, B. K., Gepner, J. I., Teng, N. H., and Hall, L. N., 1977, *J. Neurochem.* **29**:1013–1029.
103. Hildebrand, J. G., Hall, L. M., and Osmond, B. C., 1979, *Proc. Natl. Acad. Sci. U.S.A.* **76**:499–503.
104. Dudai, Y., 1978, *Biochim. Biophys. Acta* **539**:505–517.
105. Gepner, J. I., Hall, L. M., and Satelle, D. B., 1978, *Nature* **276**:188–190.
106. Conti-Tronconi, B. M., and Raftery, M., 1982, *Annu. Rev. Biochem.* **51**:491–530.
107. Lent, C. M., 1977, *Prog. Neurobiol.* **8**:81–117.
108. Evans, P. D., 1980, *Advances in Insect Physiology*, Volume 15 (M. J. Berridge, J. E. Treherne, and V. B. Wigglesworth, eds.), Academic Press, New York, pp. 317–473.
109. Mazzanti, G., and Piccinelli, D., 1979, *Comp. Biochem. Physiol.* **63C**:215–219.
110. Welsh, J. H., and Williams, L. D., 1970, *J. Comp. Neurol.* **138**:103–116.
111. Bennett, J., and Beuding, E., 1971, *Comp. Biochem. Physiol.* **39A**:859–868.
112. Tomosky-Sykes, T., Jardine, I., Mueller, J. F., and Bueding, E., 1977, *Anal. Biochem.* **83**:99–108.
113. Anya, A. O., 1973, *Comp. Gen. Pharmacol.* **4**:149–156.
114. Horvitz, H. R., Chalfie, M., Trent, C., Sulston, J. E., and Evans, P. D., 1982, *Science* **216**:1012–1014.
115. Tomosky, T. K., Bennett, J. L., and Bueding, E., 1974, *J. Pharmacol. Exp. Ther.* **190**:260–271.
116. Audesirk, G., McCaman, R. E., and Willows, A. O. D., 1979, *Comp. Biochem. Physiol.* **62C**:87–91.
117. Bennett, J., and Bueding, E., 1973, *Mol. Pharmacol.* **9**:311–319.
118. Goh, S. L., and Davey, K. G., 1976, *Tissue Cell* **8**:421–435.
119. Northrup, J. K., and Mansour, T. E., 1978, *Mol. Pharmacol.* **14**:820–833.
120. Juorio, A. V., and Killick, S. W., 1972, *Comp. Gen. Pharmacol.* **13**:283–295.
121. Sakharov, D. A., 1970, *Annu. Rev. Pharmacol.* **10**:335–352.
122. Sakharov, D. A., 1976, *Neurobiology of Invertebrates: Gastropod Brain* (J. Salanki, ed.), Akademiai Kiado, Budapest, pp. 27–40.
123. Stefano, G. B., and Aiello, E., 1975, *Biol. Bull.* **148**:141–156.
124. Loker, J. E., Kerkut, G. A., and Walker, R. J., 1975, *Comp. Biochem. Physiol.* **50A**:443–452.
125. Loker, J. E., Kerkut, G. A., and Walker, R. J., 1975, *Comp. Biochem. Physiol.* **51C**:83–90.
126. Osborne, N. N., 1978, *Biochemistry of Characterized Neurones* (N. N. Osborne, ed.), Pergamon Press, Oxford, pp. 47–80.

127. Weinreich, D., McCaman, M. W., McCaman, R. E., and Vaughn, J. E., 1973, *J. Neurochem.* **20**:969–976.
128. Cottrell, G. A., 1974, *J. Neurochem.* **22**:557–559.
129. Pentreath, V. W., and Cottrell, G. A., 1973, *Z. Zellforsch.* **143**:21–35.
130. Gerschenfeld, H. M., and Paupardin-Tritsch, D., 1974, *J. Physiol.* (*Lond.*) **243**:457–481.
131. Weiss, K. R., Cohen, J., and Kupfermann, I., 1975, *Brain Res.* **99**:381–386.
132. Granzow, B., and Rowell, C. H. F., 1981, *J. Exp. Biol.* **90**:283–305.
133. Kupfermann, I., Cohen, J. L., Mandelbaum, D. E., Schonberg, M., Susswein, A. J., and Weiss, K. R., 1979, *Fed. Proc.* **38**:2095–2102.
134. Osborne, N. N., and Neuhoff, V., 1974, *J. Neurochem.* **22**:363–371.
135. Smith, J. R., 1982, *Comp. Biochem. Physiol.* **71C**:57–61.
136. Hiripi, L., and Stefano, G. B., 1980, *Life Sci.* **27**:1205–1209.
137. Marsden, C. A., 1976, *Neurobiology of Invertebrates. Gastropod Brain* (J. Salanki, ed.), Akademiai Kiado, Budapest, pp. 177–189.
138. Eisenstadt, M., Goldman, J. E., Kandel, E. R., Koike, H., Koester, J., and Schwartz, H., 1973, *Proc. Natl. Acad. Sci. U.S.A.* **70**:3371–3375.
139. Weinreich, D., Dewhurst, S. A., and McCaman, R. E., 1972, *J. Neurochem.* **19**:1125–1130.
140. Stefano, G. B., and Aiello, E., 1978, *Experientia* **34**:749–750.
141. White, D., and Marsden, J. R., 1978, *Biol. Bull.* **155**:395–409.
142. Stahl, W. L., Neuhoff, V., and Osborne, N. N., 1977, *Comp. Biochem. Physiol.* **56C**:13–18.
143. Burrell, D. E., and Stefano, G. B., 1981, *Comp. Biochem. Physiol.* **70C**:71–76.
144. Elekes, K., 1978, *Neuroscience* **3**:48–58.
145. Goldman, J. E., and Schwartz, J. H., 1977, *Brain Res.* **136**:77–88.
146. Schwartz, J. H., Shkolnik, L. J., and Goldberg, D. J., 1979, *Proc. Natl. Acad. Sci. U.S.A.* **76**:5967–5971.
147. Goldman, J. E., Kim, K. S., and Schwartz, J. H., 1976, *J. Cell Biol.* **70**:304–318.
148. Schwartz, J. H., and Shkolnik, L. J., 1981, *J. Neurosci.* **1**:606–619.
149. Goldman, J. E., and Schwartz, J. H., 1974, *J. Physiol.* (*Lond.*) **242**:61–76.
150. Pentreath, V., 1976, *J. Neurocytol.* **5**:43–61.
151. Barlow, J. J., Juorio, A. V., and Martin, R., 1974, *J. Comp. Physiol.* **89**:105–122.
152. York, B., and Twarog, B. M., 1973, *Comp. Biochem. Physiol.* **44A**:423–430.
153. Gerschenfeld, H. M., Hamon, M., and Paupardin-Tritsch, D. B., 1978, *J. Physiol.* (*Lond.*) **274**:265–278.
154. Cedar, H., and Schwartz, J. H., 1971, *J. Gen. Physiol.* **60**:570–587.
155. Gardner, C. R., and Cashin, C. H., 1975, *Neuropharmacology* **14**:493–500.
156. Osborne, N. N., Briel, G., and Neuhoff, V., 1972, *Experientia* **28**:1015–1018.
157. Robertson, H. A., and Osborne, N. N., 1979, *Comp. Biochem. Physiol.* **64C**:7–14.
158. McAdoo, D. J., and Coggeshall, R. E., 1976, *J. Neurochem.* **26**:163–167.
159. McCaman, M. W., Weinreich, D., and McCaman, R. E., 1973, *Brain Res.* **53**:129–137.
160. McAdoo, D. J., 1978, *Biochemistry of Identified Neurons* (N. N. Osborne, ed.), Pergamon Press, Oxford, pp. 19–45.
161. Lent, C. M., Ono, J. K., Keyser, K. T., and Karten, H. J., 1979, *J. Neurochem.* **32**:1559–1563.
162. Coggeshall, R. E., 1972, *Anat. Rec.* **172**:489–498.
163. Coggeshall, R. E., and Yakstra-Sauerland, B. A., 1974, *J. Comp. Neurol.* **156**:459–470.
164. Smith, P. A., and Walker, R. J., 1975, *Comp. Biochem. Physiol.* **51C**:195–203.
165. Glover, J. C., and Kramer, A. P., 1982, *Science* **216**:317–319.
166. Adolph, A. R., and Ehinger, B., 1975, *Cell Tissue Res.* **163**:1–14.
167. Adolph, A. R., 1976, *J. Gen. Physiol.* **67**:417–431.
168. Roberts, C. J., Radley, T., Poat, J. A., and Walker, R. F., 1983, *Comp. Biochem. Physiol.* **74C**:437–440.
169. Goldstone, M. W., and Cooke, I. M., 1971, *Z. Zellforsch. Mikrosk. Anat.* **116**:7–19.
170. Cooke, I. M., and Goldstone, M. W., 1970, *J. Exp. Biol.* **53**:651–668.
171. Livingstone, M. S., Schaeffer, S. F., and Kravitz, E. A., 1981, *J. Neurobiol.* **12**:27–54.
172. Sullivan, R. E., 1978, *Life Sci.* **22**:1429–1437.
173. Fingerman, M., Julian, W. E., Spirtes, M. A., and Kostrzewa, R. M., 1974, *Comp. Gen. Pharmacol.* **5**:299–303.

174. Elofsson, R., Laxmyr, L., Rosengren, E., and Hansson, C., 1982, *Comp. Biochem. Physiol.* **71C**:195–201.
175. Rao, K. R., and Fingerman, M., 1975, *Comp. Biochem. Physiol.* **51C**:53–58.
176. Nassel, D. R., and Laxmyr, L., 1983, *Comp. Biochem. Physiol.* **75C**:259–265.
177. Elofsson, R., and Klemm, N., 1972, *Z. Zellforsch. Mikrosk. Anat.* **133**:475–499.
178. Klemm, N., 1976, *Prog. Neurobiol.* **7**:99–169.
179. Evans, P. D., Kravitz, E. A., Talamo, B. R., and Wallace, B. G., 1976, *J. Physiol. (Lond.)* **262**:51–70.
180. Sullivan, R. E., Friend, B. J., and Barker, D. L., 1977, *J. Neurobiol.* **8**:581–605.
181. Dewhurst, S. A., Crocker, S. G., Ikeda, K., and McCaman, R. E., 1972, *Comp. Biochem. Physiol.* **43B**:975–981.
182. Evans, P. H., and Fox, P. M., 1975, *J. Insect Physiol.* **21**:343–355.
183. Klemm, N., and Schneider, L., 1975, *Comp. Biochem. Physiol.* **50C**:177–182.
184. Berridge, M. J., 1972, *J. Exp. Biol.* **56**:311–321.
185. Nathanson, J. A., 1977, *Physiol. Rev.* **57**:157–256.
186. Harmar, A. J., and Horn, A. S., 1977, *Mol. Pharmacol.* **13**:512–520.
187. Taylor, D. P., and Newburgh, R. W., 1976, *Comp. Biochem. Physiol.* **61C**:73–79.
188. Bodnaryk, R. P., 1982, *Insect Biochem.* **12**:1–6.
189. Uzzan, A., and Dudai, Y., 1982, *J. Neurochem.* **38**:1542–1550.
190. Reichelt, K. L., and Edminson, P. D., 1977, *Peptides in Neurobiology* (H. Gainer, ed.), Plenum Press, New York, pp. 171–181.
191. Osborne, N. N., 1976, *Neurobiology of Invertebrates Gastropod Brain* (J. Salanki, ed.), Akademia Kiado, Budapest, pp. 141–161.
192. Kerkut, G. A., 1973, *Br. Med. Bull.* **29**:100–104.
193. Gardner, C. R., and Walker, R. J., 1982, *Prog. Neurobiol.* **18**:81–120.
194. Murdock, L. L., 1971, *Comp. Gen. Pharmacol.* **2**:254–274.
195. Elofsson, R., Falck, B., Lindvall, O., and Myhrberg, H., 1977, *Cell Tissue Res.* **182**:525–536.
196. Lenicque, P. M., Toneby, M. I., and Doumenc, D., 1977, *Comp. Biochem. Physiol.* **56C**:31–34.
197. Gianutsos, G., and Bennett, J. L., 1977, *Comp. Biochem. Physiol.* **58C**:157–159.
198. Sulston, J., Dew, M., and Brenner, H., 1975, *J. Comp. Neurol.* **163**:215–226.
199. Carpenter, D. O., Breese, G., Schanberg, S., and Kopin, I., 1971, *Intern. J. Neurosci.* **2**:49–56.
200. Juorio, A. V., and Barlow, J. J., 1974, *Comp. Gen. Pharmacol.* **5**:281–289.
201. Powell, B., and Cottrell, G. A., 1974, *J. Neurochem.* **22**:605–606.
202. McCaman, M. W., Ono, J. K., and McCaman, R. E., 1979, *J. Neurochem.* **32**:1111–1113.
203. Manokhina, M. S., and Kuzmina, L. V., 1971, *J. Evol. Biochem. Physiol.* **7**:296–301.
204. Guthrie, P. B., Neuhoff, V., and Osborne, N. N., 1975, *Comp. Biochem. Physiol.* **52C**:109–111.
205. Osborne, N. N., Priggemeier, E., and Neuhoff, V., 1975, *Brain Res.* **90**:261–271.
206. Myers, P. R., and Sweeney, D. C., 1972, *Comp. Gen. Pharmacol.* **3**:277–282.
207. Guthrie, P. B., Neuhoff, V., and Osborne, N. N., 1975, *Experientia*, **31**:775–776.
208. McCaman, M. W., McCaman, R. E., and Stetzler, J., 1979, *Anal. Biochem.* **95**:175–180.
209. Myers, P. R., 1974, *Tissue Cell* **6**:49–64.
210. Hiripi, L. and S.-Rozsa, K., 1973, *J. Insect Physiol.* **19**:1482–1485.
211. Stefano, G. B., Catapane, E. J., and Aiello, E., 1976, *Science* **194**:539–541.
212. Brown, M., Burrell, D. E., Stefano, G. B., 1981, *Comp. Biochem. Physiol.* **70C**:215–221.
213. Pentreath, V. W., and Berry, M. S., 1975, *J. Neurocytol.* **4**:249–260.
214. Ascher, P., 1972, *J. Physiol. (Lond.)* **225**:173–209.
215. Osborne, N. N., and Neuhoff, V., 1979, *J. Pharm. Pharmacol.* **31**:481–492.
216. Kream, R. M., Zukin, R. S., and Stefano, G. B., 1980, *J. Biol. Chem.* **255**:9218–9224.
217. Stefano, G. B., Hall, B., Makman, M. H., and Duorkin, B., 1981, *Science* **213**:928–930.
218. Corelli, V., Memo, M., Spano, P. F., and Trabucchi, M., 1981, *Neuroscience* **6**:2077–2079.
219. Lin, S., and Cohen, H. P., 1973, *Comp. Biochem. Physiol.* **45B**:249–263.
220. Meyer, W., and Jehnen, R., 1980, *J. Morphol.* **64**:69–81.

221. Osborne, N. N., and Dando, D., 1970, *Comp. Biochem. Physiol.* **32:**327–331.
222. Kushner, P. D., and Maynard, E. A., 1977, *Brain Res.* **129:**13–28.
223. Aramant, R., and Elofsson, R., 1976, *Cell Tissue Res.* **166:**1–24.
224. Aramant, R., 1980, *Comp. Biochem. Physiol.* **66C:**29–36.
225. O'Connor, E. F., Watson, W. H. III, and Wyse, G. A., 1982, *J. Neurobiol.* **13:**49–60.
226. Richter, V. D., and Reutsche, E., 1977, *Acta Histochem.* **60:**304–311.
227. Myhrberg, H. E., Elofsson, R., Aramant, R., Klemm, N., and Laxmyr, L., 1979, *Comp. Biochem. Physiol.* **62C:**141–150.
228. Nestler, C., Brown, A., and Wheeler, A. P., 1981, *Comp. Biochem. Physiol.* **69C:**53–60.
229. Barker, D. L., Kushner, P. D., and Hooper, N. K., 1979, *Brain Res.* **161:**99–113.
230. Bjorklund, A., Falck, B., and Klemm, N., 1970, *J. Insect. Physiol.* **16:**1147–1154.
231. Taylor, D. P., and Newburgh, R. W., 1979, *Insect Biochem.* **9:**265–272.
232. Bodnaryk, R. P., 1979, *Insect Biochem.* **9:**155–162.
233. Cottrell, G. A., Abernathy, K. B., and Barrand, M. A., 1979, *Neuroscience* **4:**685–689.
234. McCaman, M. W., McCaman, R. E., and Lees, G. L., 1972, *Anal. Biochem.* **45:**242–252.
235. Chou, T. C., Bennett, J. L., and Beuding, E., 1972, *Int. J. Parasitol.* **58:**1098–1102.
236. Keenan, L., and Koopowitz, H., 1981, *Science* **214:**1151–1152.
237. Keenan, L., and Koopowitz, H., 1982, *J. Neurobiol.* **13:**9–21.
238. Mermel, L., Guchwait, R. B., Bourgeois, J. G., and Bueding, E., 1981, *Comp. Biochem. Physiol.* **69C:**227–234.
239. Juorio, A. V., 1971, *J. Physiol.* (*Lond.*) **216:**213–226.
240. Osborne, N. N., and Cottrell, G. A., 1970, *Comp. Gen. Pharmacol.* **1:**1–9.
241. Stone, B. F., Binnington, K. C., and Neish, A. L., 1978, *Experientia* **34:**1173–1174.
242. Juorio, A. V., and Barlow, J. J., 1973, *Experientia* **29:**943–945.
243. Maxwell, G. D., Moore, M. M., and Hildebrand, J. G., 1980, *Insect Biochem.* **10:**657–665.
244. Pollard, H. B., Backer, J. L., Bohr, W. A., and Dowdale, M. J., 1975, *Brain Res.* **85:**23–31.
245. Evans, P. D., 1978, *J. Neurochem.* **30:**1015–1022.
246. Hanumante, H., and Fingerman, M., 1981, *Comp. Biochem. Physiol.* **70C:**27–34.
247. Axelrod, J., and Saavedra, J. M., 1977, *Nature* **265:**501–504.
248. Robertson, H. A., and Juorio, A. V., 1976, *Int. Rev. Neurobiol.* **19:**173–224.
249. Harmar, A. J., 1980, *Noncatecholic Phenylethylamines* (A. Mosnaim and M. Wolf, eds.), Marcel Dekker, New York, pp. 97–150.
250. Talamo, B. R., 1980, *Noncatecholic Phenylethylamines* (A. Mosnaim and M. Wolf, eds.), Marcel Dekker, New York, pp. 261–292.
251. Evans, P. D., 1978, *J. Neurochem.* **30:**1009–1013.
252. Evans, P. D., Talamo, B. R., and Kravitz, E. A., 1975, *Brain Res.* **90:**340–347.
253. Evans, P. D., and O'Shea, M., 1978, *J. Exp. Biol.* **73:**235–260.
254. Batelle, B. A., Evans, J. A., and Chamberlain, S. C., 1982, *Science* **216:**1250–1252.
255. Molinoff, P. B., Landsberg, L., and Axelrod, J. P., 1969, *J. Pharmacol. Exp. Ther.* **179:**253–261.
256. Saavedra, J. M., Brownstein, M. J., Carpenter, D. O., and Axelrod, J., 1974, *Science* **185:**364–365.
257. Dougan, D. F. H., Duffield, P. H., Wade, D. N., and Duffield, A. M., 1977, *Comp. Biochem. Physiol.* **70C:**277–280.
258. McCaman, M. W., and McCaman, R. E., 1978, *Brain Res.* **141:**347–352.
259. Mell, L. D., Jr., and Carpenter, D. O., 1980, *Neurochem. Res.* **5:**1089–1096.
260. Starratt, A. N., and Bodnaryk, R. P., 1981, *Insect Biochem.* **11:**645–648.
261. Chalfie, M., and Sulston, J., 1981, *Dev. Biol.* **82:**358–363.
262. Juorio, A. V., and Molinoff, P. B., 1974, *J. Neurochem.* **22:**271–280.
263. Dudai, Y., and Svi, S., 1982, *J. Neurochem.* **38:**1551–1558.
264. Walker, R. J., Ramage, A. G., and Woodruff, G. N., 1974, *Experientia* **30:**11–13.
265. McCaman, M. W., 1980, *Noncatecholic Phenylethylamines* (A. Mosnaim and M. Wolf, eds.), Marcel Dekker, New York, pp. 193–201.
266. Farnham, P. J., Novak, R. A., and McAdoo, D. J., 1978, *J. Neurochem.* **30:**1173–1175.
267. Hayashi, S., Murdock, L. L., and Florey, E., 1977, *Comp. Biochem. Physiol.* **58C:**183–191.
268. Levitan, I. B., and Barondes, S. H., 1974, *Proc. Natl. Acad. Sci. U.S.A.* **71:**1145–1148.

269. Levitan, I. B., 1979, *The Neurosciences. Fourth Study Program*, (F. O. Schmitt and F. G., Worden, ed.), MIT Press, Cambridge, pp. 1043–1055.
270. Nathanson, J. A., 1976, *Trace Amines and the Brain* (E. Usdin and S. Snyder, eds.), Marcel Dekker, New York, pp. 161–190.
271. Carpenter, D. O., and Gaubatz, G. L., 1974, *Nature* **252**:483–485.
272. Batta, S., Walker, R. J., and Woodruff, G. N., 1979, *Comp. Biochem. Physiol.* **64C**:43–51.
273. Webb, R. A., and Orchard, I., 1980, *Comp. Biochem. Physiol.* **67C**:135–140.
274. Webb, R. A., and Orchard, I., 1981, *Comp. Biochem. Physiol.* **70C**:201–207.
275. Robertson, H. A., and Osborne, N. N., 1979, *Comp. Biochem. Physiol.* **64C**:7–14.
276. Wallace, B. G., Talamo, B. R., Evans, P. D., and Kravitz, E. A., 1974, *Brain Res.* **74**:349–355.
277. Robertson, H. A., and Carlson, A. D., 1976, *J. Exp. Zool.* **195**:159–164.
278. David, J. C., and Lafon-Cazol, M., 1979, *Comp. Biochem. Physiol.* **64C**:161–164.
279. Hoyle, G., 1975, *J. Exp. Zool.* **193**:425–431.
280. Evans, P. D., and O'Shea, M., 1977, *Nature* **270**:257–259.
281. Dymond, G. R., and Evans, P. D., 1979, *Insect Biochem.* **9**:535–545.
282. Goodman, C. S., O'Shea, M., McCaman, R. E., and Spitzer, N. C., 1979, *Science* **204**:1219–1222.
283. Goodman, C., Bate, M., and Spitzer, N. C., 1981, *J. Neurosci.* **1**:94–102.
284. Bodnaryk, R. P., 1980, *Insect Biochem.* **10**:169–173.
285. Kennedy, M. B., 1978, *J. Neurochem.* **30**:315–320.
286. Kennedy, M. B., 1980, *Soc. Neurosci. Abstr.* **3**:252.
287. Roberts, C. J., and Walker, R. J., 1978, *Comp. Biochem. Physiol.* **69C**:301–306.
288. Battelle, B. A., and Kravitz, E. A., 1978, *J. Pharmacol. Exp. Ther.* **205**:438–448.
289. Florey, E., and Rathmeyer, M., 1978, *Comp. Biochem. Physiol.* **61C**:229–237.
290. Nathanson, J. A., 1980, *Science* **203**:65–68.
291. Atkinson, P. W., Herman, W. S., and Sheppard, J. R., 1977, *Comp. Biochem. Physiol.* **58C**:107–110.
292. Hollingworth, R. M., and Murdock, L. L., 1980, *Science* **208**:74–76.
293. Nathanson, J. A., and Hunnicutt, E. J., 1981, *Mol. Pharmacol.* **20**:68–75.
294. Stuart, A. E., Hudspeth, A. S., and Hall, Z. W., 1974, *Cell Tissue Res.* **153**:55–69.
295. Osborne, N., Neuhoff, V., Ewers, E., and Robertson, H. A., 1979, *Comp. Biochem. Physiol.* **63C**:209–213.
296. Juorio, A. V., and Robertson, H., 1977, *J. Neurochem.* **28**:573–579.
297. Weinreich, D., 1976, *Neurobiology of Invertebrates. Gastropoda Brain* (J. Salanki, ed.), Akademiai Kiado, Budapest, pp. 191–206.
298. Weinreich, D., 1978, *Biochemistry of Characterized Neurons* (N. N. Osborne, ed.), Pergamon Press, Oxford, pp. 153–175.
299. Weinreich, D., Weiner, C., and McCaman, R. E., 1975, *Brain Res.* **84**:341–345.
300. Ono, J. K., and McCaman, R. E., 1980, *Neuroscience* **5**:835–840.
301. Turner, J. D., and Cottrell, G. A., 1977, *Nature* **267**:447–448.
302. Osborne, N. N., Wolter, K. D., and Neuhoff, V., 1979, *Biochem. Pharmacol.* **28**:2799–2805.
303. Weinreich, D., 1977, *Nature* **267**:854–856.
304. McCaman, R. E., and McKenna, D., 1978, *Brain Res.* **141**:165–171.
305. Gruol, D. L., and Weinreich, D., 1979, *Brain Res.* **162**:281–301.
306. Weinreich, D., and Yu, Y. T., 1977, *J. Neurochem.* **28**:361–370.
307. Das, N. P., Lim, H. S., and Teh, Y. F., 1971, *Comp. Gen. Pharmacol.* **2**:473–475.
308. Turner, J. D., Powell, B., and Cottrell, G. A., 1980, *J. Neurocytol.* **9**:1–14.
309. Gotoh, H., and Schwartz, J. H., 1982, *Brain Res.* **242**:87–98.
310. Weinreich, D., 1979, *J. Neurochem.* **32**:363–369.
311. Carpenter, D. O., and Gaubatz, G. L., 1975, *Nature* **254**:343–344.
312. Bailey, C. H., Chen, M. C., Weiss, K. R., and Kupfermann, I., 1982, *Brain Res.* **238**:205–210.
313. Roseghini, M., and Ramorino, L. M., 1970, *J. Neurochem.* **17**:489–492.
314. Boulton, A. A., 1976, *Adv. Biochem. Psychopharmacol.* **15**:57–67.
315. Edward, D. J., and Antelman, S. M., 1980, *Non-Catecholic Phenylethylamines*, Part I. *Phenylethylamines*, Marcel Dekker, New York, pp. 21–46.

316. Axelrod, J., Saavedra, J. M., and Usdin, E., *Trace Amines and the Brain* (E. Usdin and M. Sandler, eds.), Marcel Dekker, New York, pp. 1–20.
317. Duffield, P. H., Dougan, D. F. H., and Duffield, D. M., 1981, *Biomed. Mass Spectrom.* **8**:170–174.
318. Juorio, A. V., and Philips, S. R., 1975, *Brain Res.* **83**:180–184.
319. Vaughn, P. F. T., and Neuhoff, V., 1976, *Brain Res.* **117**:175–180.
320. Evans, P. D., 1978, *Trends Neurosci.* **1**:154–157.
321. Saavedra, J. M., Ribas, J., Swann, J., and Carpenter, D. O., 1977, *Science* **195**:1004–1006.
322. Osborne, N. N., 1973, *Br. J. Pharmacol.* **48**:546–549.
323. Osborne, N. N., 1979, *Trends Neurosci.* **2**:73–75.
324. Giacobini, E., 1976, *Adv. Biochem. Psychopharm.* **15**:17–56.
325. Dolezalova, H., Giacobini, E., and Stepita-Klauco, M., 1973, *Intern. J. Neurosci.* **5**:53–59.
326. Manen, C., and Russell, D. H., 1973. In: *Polyamines in Normal and Neoplastic Growth* (D. H. Russell, ed.), Raven Press, New York, pp. 23–48.
327. Gould, R. M., and Cottrell, B. A., 1974, *Comp. Biochem. Physiol.***48B**:591–597.
328. Kremzner, L. R., and Ambron, R. T., 1982, *J. Neurochem.* **38**:1719–1727.
329. Dolezalova, H., Stepita-Klauco, M., and Seiler, N., 1972, *Brain Res.* **67**:349–351.
330. Wu, P. H., Baker, G. B., and Henwood, R. W., 1980, *Non-Catecholic Phenylethylamines*, Part I (A. Mosnaim and M. Wolf, eds.), Marcel Dekker, New York, pp. 307–340.
331. Sabelli, H. C., Borison, R. L., Diamond, B. I., May, J., and Havdala, H. S., 1980, *Non-Catecholic Phenylethylamines*, Part I (A. Mosnaim and M. Wolf, eds.), Marcel Dekker, New York, pp. 345–376.
332. Kravitz, E. A., Slater, C. R., Takahashi, K., Bownds, M. D., and Grossfeld, R. M., 1970, *Excitatory Spnaptic Mechanisms* (P. Andersen, and J. K. Jansen, eds.), Universitet Forlaget, Oslo, pp. 85–93.
333. Kerkut, G. A., Ralph, K., Walton, R. J., Woodruff, G., and Woods, R., 1970, *Excitatory Synaptic Mechanisms* (P. Andersen and J. K. S. Jansen, eds.), Universitet Forlaget, Oslo, pp. 94–108.
334. Shinozaki, H., 1980, *Prog. Neurobiol.* **14**:121–155.
335. Nistri, A., and Constanti, A., 1979, *Prog. Neurobiol.* **13**:117–235.
336. Evans, P. D., 1973, *J. Neurochem.* **21**:11–17.
337. Usherwood, P., 1978, *Adv. Comp. Physiol. Biochem.* **7**:225–309.
338. Shank, R. P., and Graham, L. T., Jr., 1978, *Adv. Neurochem.* **3**:165–201.
339. Johnson, L., 1972, *Brain Res.* **37**:1–19.
340. Takeuchi, A., Onodera, K., and Kawagoe, R., 1981, *Adv. Biochem. Psychopharmacol.* **29**:365–368.
341. Florey, E., and Rathmayer, M., 1981, *Adv. Biochem. Psychopharmacol.* **29**:351–380.
342. Constanti, A., and Nistri, A., 1979, *Br. J. Pharmacol.* **65**:287–301.
343. Jan, L. Y., and Jan, Y. N., 1976, *J. Physiol. (Lond).* **262**:215–236.
344. Hertz, L., 1979, *Prog. Neurobiol.* **13**:277–323.
345. Lin, S., and Cohen, H. P., 1973, *Comp. Biochem. Physiol.* **45B**:249–263.
346. McBride, W. J., Shank, R. P., Freeman, A. R., and Aprison, M. H., 1974, *Life Sci.* **14**:1109–1120.
347. Iliffe, T. M., McAdoo, D. J., Beyer, C. B., and Haber, B., 1977, *J. Neurochem.* **28**:1037–1042.
348. VandenBerg, C. J., Matheson, D. F., Ronda, G., Reijnierse, G., Blokhius, G., Kroon, M., Clarke, D. D., and Garfinkel, D., 1978, *Metabolic Compartmentation and Neurotransmission* (S. Berl, J. Davison, and D. Schneider, eds.), Plenum Press, New York pp. 516–543.
349. DeFeudis, F. V., 1979, *Int. Rev. Neurobiol.* **21**:129–216.
350. Osborne, N. N., 1972, *Comp. Biochem. Physiol.* **43B**:579–585.
351. Zeman, G. H., and Carpenter, D. O., 1975, *Comp. Biochem. Physiol.* **52C**:23–26.
352. Borys, H., Weinreich, D., and McCaman, R. E., 1973, *J. Neurochem.* **21**:1345–1351.
353. Shank, R. P., Freeman, A. R., McBride, W. J., and Aprison, M., *Comp. Biochem. Physiol.* **50C**:127–131.
354. D'Aniello, A., and Guiditta, A., 1977, *J. Neurochem.* **29**:1053–1057.
355. Jost, J., Cain, H., and Marsden, J. R., 1981, *Comp. Biochem. Physiol.* **68C**:43–47.

356. Sczepaniak, A. C., and Cottrell, G. A., 1973, *Nature* **241**:62–64.
357. Lowagie, C., and Gerschenfeld, H., 1973, *Nature* **248**:533–535.
358. Yamamoto, N., and Washio, H., 1979, *Comp. Biochem. Physiol.* **63C**:75–80.
359. Atwood, H. L., 1976, *Prog. Neurobiol.* **7**:291–391.
360. Phillis, J. W., and Wu, K. S., 1981, *Prog. Neurobiol.* **16**:187–239.
361. Wang, L. D. L., and Boyarsky, C. C., 1979, *Life Sci.* **24**:1011–1014.
362. Levitan, I. B., and Benson, J. A., 1981, *Trends Neurosci.* **4**:574–578.
363. Yarowsky, P. J., and Carpenter, D. O., 1976, *Nature* **192**:807–809.
364. Meyer, W., Poehling, H. M., and Neuhoff, V., 1980, *Comp. Biochem. Physiol.* **67C**:83–86.
365. Reinecke, M., 1977, *Cell Tissue Res.* **169**:361–382.
366. Faeder, I. R., Matthews, J. A., and Salpeter, M. M., 1974, *Brain Res.* **80**:53–70.
367. Crawford, A. C., and McBurney, R. N., 1977, *J. Physiol.* (*Lond.*) **268**:711–729.
368. Botham, R. P., Beadle, D. J., Hart, R. J., Potter, C., and Wilson, R. G., 1979, *Cell Tissue Res.* **203**:379–386.
369. Daoud, A., and Miller, R., 1976, *J. Neurochem.* **26**:119–123.
370. Fiszer de Plazas, S., and deRobertis, E., 1974, *J. Neurochem.* **23**:1115–1120.
371. Cull-Candy, S. G., James, R. W., and Lunt, G. G., 1976, *Nature* (*New Biol.*) **246**:62–64.
372. Takeuchi, A., and Takeuchi, N., 1972, *Adv. Biophys.* **3**:45–95.
373. Olsen, R. W., 1976, *GABA in Nervous System Function* (E. Roberts, T. N. Chase, and D. B. Tower, eds.), Raven Press, New York, pp. 287–304.
374. Johnston, G. A. R., 1978, *Annu. Rev. Pharmacol. Toxicol.* **18**:269–289.
375. Takeuchi, A., and Takeuchi, N., 1975, *Neuropharmacology* **14**:627–634.
376. Yarowsky, P., and Carpenter, D. O., 1978, *J. Neurophysiol.* **41**:531–541.
377. James, V. A., Roberts, C. J., and Walker, R. J., 1982, *Comp. Biochem. Physiol.* **71C**:229–237.
378. Usherwood, P., Atwood, H., Orkand, P. M., and Kravitz, E. A., 1971, *J. Cell Biol.* **49**:73–89.
379. Osborne, N. N., 1971, *Comp. Gen. Pharmacol.* **2**:433–438.
380. Kuroda, Y., and Okada, Y., 1977, *Experientia* **33**:1623–1625.
381. McCaman, M. W., Colby, B. N., and McCaman, R. E., 1979, *J. Neurochem.* **33**:967–971.
382. Cottrell, G. A., 1974, *J. Neurochem.* **22**:557–559.
383. Otsuka, M., Kravitz, E. A., and Potter, D. D., 1967, *J. Neurophysiol.* **30**:725–752.
384. McBride, W., Freeman, A. R., Graham, L. T., and Aprison, M. H., 1975. *J. Neurobiol.* **6**:321–328.
385. Molinoff, P. B., and Kravitz, E. A., 1968, *J. Neurochem.* **15**:391–409.
386. Hall, Z. W., Bownds, M. D., and Kravitz, E. A., 1970, *J. Cell Biol.* **46**:290–299.
387. Fox, P. M., and Larson, J. R., 1972, *J. Insect Physiol.* **18**:439–457.
388. Baxter, C. F., and Torralba, G. F., 1975, *Brain Res.* **84**:383–397.
389. Koidl, B., 1974, *J. Comp. Physiol.* **94**:49–55.
390. Martin, D. L., 1976, *GABA in Nervous System Function* (E. Roberts, T. N. Chase, and D. Tower, eds.), Raven Press, New York, pp. 347–386.
391. Horvitz, I. S., and Orkand, R. K., 1980, *J. Neurobiol.* **11**:447–458.
392. Frontali, N., and Pierantini, R., 1973, *Comp. Biochem. Physiol.* **44A**:1369–1372.
393. Olsen, R. W., 1982, *Annu. Rev. Pharmacol. Toxicol.* **22**:245–277.
394. Craelius, W., and Fricke, R. A., 1981, *J. Neurobiol.* **12**:249–258.
395. Brading, A. F., and Caldwell, P. C., 1971, *J. Physiol.* (*Lond.*) **217**:605–624.
396. Walker, R. J., Azanza, M. J., Kerkut, G. A., and Woodruff, G. N., 1975, *Comp. Biochem. Physiol.* **50C**:147–154.
397. DeRobertis, E., and Fiszer de Plazas, S., 1974, *J. Neurochem.* **23**:1121–1125.
398. Olsen, R. W., Lee, J. M., and Ban, M., 1975, *Mol. Pharmacol.* **11**:566–577.
399. Ticku, M. K., and Olsen, R. W., 1977, *Biochem. Biophys. Acta* **464**:519–529.
400. Price, C. H., Coggeshall, R. E., and McAdoo, D. J., 1978, *Amino Acids as Chemical Transmitters* (F. Fonnum, ed.), Plenum Press, New York, pp. 213–219.
401. Price, C. H., and McAdoo, D. J., 1979, *J. Comp. Neurol.* **188**:647–677.
402. Loh Y.-P., and Gainer, H., 1975, *Brain Res.* **92**:193–205.
403. Aswad, D. W., 1979, *J. Neurobiol.* **9**:267–284.

404. Wilson, D., 1976, *J. Neurobiol.* **7**:407–416.
405. Price, C. H., Coggeshall, R. E., and McAdoo, D. J., 1978, *Brain Res.* **154**:25–40.
406. McAdoo, D. J., Iliffe, T. M., Price, C. H., and Novak, R. A., 1978, *Brain Res.* **154**:41–51.
407. Acher, R., 1981, *Trends Neurosci.* **4**:425–428.
408. Hokfelt, T., Johansson, O., Ljungdahl, A., Lundberg, I. M., and Schultzberg, M., 1980, *Nature* **284**:515–521.
409. McCaman, R., and Stetzler, J., 1977, *J. Neurochem.* **29**:739–741.
410. Bodnaryk, R., 1981, *Insect Biochem.* **11**:199–205.
411. Bodnaryk, R., 1981, *Insect Biochem.* **11**:9–16.
412. Rassin, D. K., 1981, *Adv. Biochem. Psychopharmacol.* **29**:127–133.
413. Amende, L. M., and Pierce, S. K., Jr., 1978, *Comp. Biochem. Physiol.* **59B**:257–261.
414. Mandel, P., Pasantes-Morales, H., and Urban, P. F., 1976, *Transmitters in the Visual System* (S. L. Bonting, ed.), Pergamon Press, Oxford, pp. 89–102.
415. Grossfeld, R., 1975, *Comp. Biochem. Physiol.* **51C**:1–4.
416. Grossfeld, R., 1976, *Comp. Biochem. Physiol.* **53C**:41–49.
417. Arnould, J. M., and Frentz, R., 1975, *Comp. Biochem. Physiol.* **50C**:59–66.
418. Felix, D., and Kuntzle, H., 1976, *Adv. Biochem. Psychopharmacol.* **15**:165–173.
419. Hiripi, L., and Osborne, N. N., 1976, *Comp. Biochem. Physiol.* **53B**:549–553.
420. Van Harreveld, A., 1980, *J. Neurobiol.* **11**:519–529.
421. Muneoka, Y., Ichimua, Y., Shiba, Y., and Kanno, Y., 1981, *Comp. Biochem. Physiol.* **69C**:171–177.
422. Haynes, L. W., 1980, *Prog. Neurobiol.* **15**:205–245.
423. Scharrer, B., 1976, *Prog. Brain Res.* **45**:125–135.
424. Fingerman, M., 1973, *Fed. Proc.* **32**:2195–2203.
425. Acher, R., 1981, *Neuroscience* **4**:225–229.
426. Golding, D. W., 1974, *Biol. Rev.* **49**:161–224.
427. Frontali, N., and Gainer, H., 1977, *Peptides in Neurobiology* (H. Gainer, ed.), Plenum Press, New York, pp. 259–294.
428. Gainer, H., Loh, Y.-P., and Sarne, Y., 1977, *Peptides in Neurobiology* (H. Gainer, ed.), Plenum Press, New York, pp. 183–219.
429. Pickering, O. V., 1978, *Essays in Biochemistry* (P. N. Campbell and W. N. Aldridge, eds.), Academic Press, New York, **14**:45–81.
430. Livett, B., 1981, *Neurosci. Res. Prog. Bull.* **20**:39–45.
431. Mains, R. E., and Eipper, B. D., 1979, *Proc. Soc. Exp. Biol. Med.* **33**:37–55.
432. Blobel, G., 1980, *Proc. Natl. Acad. Sci. U.S.A.* **77**:1496–1500.
433. Hales, C. N., 1978, *FEBS Lett.* **94**:10–16.
434. Marks, N., 1977, *Peptides in Neurobiology* (H. Gainer, ed.), Plenum Press, New York, pp. 221–258.
435. Sternberger, L., 1979, *Immunocytochemistry*, 2nd ed., John Wiley & Sons, New York.
436. Van de Sande, F., 1979, *J. Neurosci. Methods* **1**:3–17.
437. Van Noorden, S., Fritsch, H. A. R., Grillo, T. A. E., Polak, J. M., and Pearse, A. G. E., 1979, *Gen. Comp. Endocrinol.* **37**:54–72.
438. Dockray, G. J., 1979, *Fed. Proc.* **38**:2295–2301.
439. Boer, H. H., Schot, L. P. C., Roubos, E. W., ter Maat, A., Lodder, J. C., Reichelt, D., and Swaab, D. F., 1979, *Cell Tissue Res.* **202**:231–240.
440. Duve, H., and Thorpe, A., 1979, *Cell Tissue Res.* **200**:187–191.
441. Zipser, B., 1980, *Nature* **283**:857–858.
442. Mancillas, J. R., McGinty, J. F., Selverston, A. I., Karten, H., and Bloom, F. E., 1981, *Nature* **293**:576–578.
443. Kleinholz, L. H., 1975, *Nature* **258**:265–267.
444. Leeman, S. E., Mroz, E. A., and Carraway, R. E., *Peptides in Neurobiology* (H. Gainer, ed.), Plenum Press, New York, pp. 99–144.
445. Goldworthy, G. J., and Mordue, W., 1974, *J. Endocrinol.* **60**:529–558.
446. Kupfermann, I., Lynch, G., Mayeri, C., and Weight, F. F., 1979, *Fed. Proc.* **38**:2115–2126.
447. Agarwal, R. A., Leon, P. J. B., and Greenberg, M. J., 1972, *Comp. Gen. Pharmacol.* **3**:249–260.

448. Price, D. A., and Greenberg, M. J., 1977, *Science* **197**:670–671.
449. Nagle, G., 1981, *J. Neurobiol.* **12**:599–611.
450. Greenberg, M. J., and Price, D. A., 1983, *Ann. Rev. Physiol.* **45**:271–288.
451. Nagle, G. T., and Greenberg, M. J., 1982, *Comp. Biochem. Physiol.* **71C**:101–105.
452. Cottrell, G. A., Price, D. A., and Greenberg, M. J., 1981, *Comp. Biochem. Physiol.* **70C**:103–107.
453. Geraerts, W. P. M., deWith, N. D., Roubos, E. W., and Joosse, J., 1981, *Neurosecretion: Molecules, Cells, Systems* (D. Farner and K. Lederis, eds.), Plenum Press, New York, pp. 337–347.
454. Morris, H., Panico, M., Karplus, A. and Lloyd, P., *Nature* **300**:643–645.
455. Boer, H. H., Schot, L. P. C., Veetra, J. H., and Reichelt, D., 1980, *Cell Tissue Res.* **213**:21–30.
456. Dockray, C. J., Vaillant, C., and Williams, R. G., 1981, *Nature* **293**:656–657.
457. Lloyd, P. E., 1980, *J. Comp. Physiol.* **138**:265–270.
458. Strumwasser, F., Kaczmarek, L. K., Jennings, K. R., and Chiu, A. Y., 1981, *Neurosecretion: Molecules, Cells, Systems* (D. Farner and K. Lederis, eds.), Plenum Press, New York, pp. 249–268.
459. Arch, S., Early, P., and Smock, T., 1976, *J. Gen. Physiol.* **68**:197–210.
460. Chiu, A. Y., Hunkapiller, M. W., Heller, E., Stuart, D. K., Hood, L. E., and Strumwasser, F., 1979, *Proc. Natl. Acad. Sci. U.S.A.* **76**:6656–6660.
461. Berry, R. W., 1981, *Biochemistry* **21**:6200–6205.
462. Berry, R. W., Trump, M. J., and Baylen, J. T., 1981, *Biochemistry* **21**:6206–6211.
463. Loh, Y.-P., Sarne, Y., and Gainer, H., 1975, *J. Comp. Physiol.* **100**:283–295.
464. Stuart, D. K., Chiu, A. Y., and Strumwasser, F., 1980, *J. Neurophysiol.* **43**:488–498.
465. Scheller, R. H., Jackson, J. F., McAllister, L. B., Schwartz, J. H., Kandel, E. R., and Axel, R., 1982, *Cell* **28**:707–719.
466. Arch, S., Smock, T., Gurvis, R., and McCarthy, C., 1978, *J. Comp. Physiol.* **128**:67–70.
467. Weber, E., Evans, C. J., Samuelsson, S. J., and Barchas, J. D., 1981, *Science* **214**:1248–1250.
468. Branton, W. D., Arch, S., Smock, T., and Mayeri, E., 1978, *Proc. Natl. Acad. Sci. U.S.A.* **75**:5732–5736.
469. Mayeri, E., and Rothman, B. S., 1981, *Neurosecretion: Molecules, Cell, Systems* (D. Farner and K. Lederis, eds.), Plenum Press, New York, pp. 305–316.
470. Chiu, A. Y., and Strumwasser, F., 1981, *J. Neurosci.* **1**:812–826.
471. Heller, E., Kaczmarek, L. K., Hunkapiller, M. W., Hood, L. E., and Strumwasser, F., 1980, *Proc. Natl. Acad. Sci. U.S.A.* **77**:2328–2332.
472. Schlesinger, D. H., Babarik, S. P., and Blankenship, J. E., 1981, *Symposium on Neurohypophyseal Peptide Hormones and other Biologically Active Peptides* (D. H. Schlesinger, ed.), Elsevier, New York, pp. 135–150.
473. Davis, W. J., Mpitsos, G. J., and Pinneo, J. M., 1974, *J. Comp. Physiol.* **90**:225–243.
474. Ram, J., 1975, *Biol. Bull.* **149**:443–451.
475. Dogterom, G. E., and Geraerts, W. P. M., 1981, *Neurosecretion: Molecules, Cells, Systems* (D. Farner and K. Lederis, eds.), Plenum Press, New York, pp. 507.
476. Kubota, J., and Kanatani, H., 1974, *Science* **187**:654–655.
477. Starratt, A. N., and Brown, B. E., 1975, *Life Sci.* **17**:1253–1256.
478. Brown, B. E., 1975, *Life Sci.* **17**:1241–1252.
479. Starratt, A. N., and Steele, R. W., 1980, *Neurohormonal Techniques in Insects* (T. A. Miller, ed.), Springer-Verlag, New York, pp. 1–30.
480. O'Shea, M., and Adams, M. E., 1981, *Science* **213**:567–569.
481. Bishop, C. A., and O'Shea, M., 1982, *J. Comp. Neurol.* **207**:223–238.
482. Marks, E. P., and Holman, G. M., 1974, *J. Insect Physiol.* **20**:2087–2093.
483. Holman, G. M., and Cook, B. J., 1972, *Biol. Bull.* **142**:446–460.
484. Sowa, B. A., and Berg, T. K., 1975, *J. Insect Physiol.* **21**:511–516.
485. Holman, G. M., and Cook, B. J., 1979, *Insect Biochem.* **9**:149–154.
486. Sullivan, R. E., 1979, *J. Exp. Zool.* **210**:543–552.
487. Kingan, T., and Titmus, M., 1983, *Comp. Biochem. Physiol.* **74C**:75–78.

488. Schwartz, T. L., Harris-Warrick, R. M., Glusman, S., and Kravitz, E. A., 1980, *J. Neurobiol.* **1**:623–628.
489. Fernlund, P., 1976, *Biochim. Biophys. Acta* **439**:17–25.
490. Fernlund, P., 1974, *Biochim. Biophys. Acta* **371**:312–322.
491. Skorkowski, E. F., and Kleinholz, L. H., 1973, *Gen. Comp. Endocrinol.* **20**:595–597.
492. Hokfelt, T., Lundberg, J. M., Schultzberg, M., Johansson, O., Ljungdahl, A., and Rehfeld, J., 1980, *Adv. Biochem. Psychopharmacol.* **22**:1–24.
493. Bloom, F. E., 1980, *Peptides: Integrators of Cell and Tissue Function*, Raven Press, New York.
494. Alumets, J., Hakanson, R., Sundler, F., and Threll, J., 1979, *Nature* **279**:805–806.
495. Martin, R., Frosch, D., Weber, E., and Voight, K. H., 1979, *Neurosci. Lett.* **15**:253–257.
496. Van Noorden, S., Fritsch, H. A. R., Grillo, T. A. E., Polak, J. M., and Pearse, A. G. E., 1980, *Gen. Comp. Endocrinol.* **40**:47–62.
497. Stefano, G. B., Kream, R. M., and Zukin, R. E., 1980, *Brain Res.* **181**:440–445.
498. Grimmelikhuizen, C. J. P., Balfe, A., Emson, P. C., Powell, D., and Sundler, F., 1981, *Histochemistry* **71**:325–333.
499. Taban, C. H., and Cathieni, M., 1979, *Experientia* **35**:811–812.
500. Osborne, N. N., Cuello, A. C., and Dockray, G. J., 1982, *Science* **216**:409–411.
501. Duve, H., and Thorpe, A., 1979, *Cell Tissue Res.* **200**:187–191.
502. Kramer, K. J., Childs, C. N., Speirs, R. D., and Jacobs, R. M., 1982, *Insect Biochem.* **12**:91–98.
503. Kramer, K. R., Speirs, R. D., and Childs, C. N., 1977, *Gen. Comp. Endocrinol.* **32**:423–426.
504. Remy, C., and Girardie, J., 1980, *Gen. Comp. Endocrinol.* **40**:27–45.
505. Remy, C., Girardie, J., and Dubois, M. P., 1979, *Gen. Comp. Endocrinol.* **37**:93–100.
506. Moore, G., Thornhill, J. A., Gill, V., Lederis, K., and Lukowiak, K., 1980, *Brain Res.* **206**:213–218.
507. Grimm-Jorgensen, M., 1978, *Gen. Comp. Endocrinol.* **35**:387–390.
508. Fritsch, H. A. R., Van Noorden, S., and Pearse, A. G. E., 1979, *Cell Tissue Res.* **202**:263–274.
509. Stone, J., and Mordue, W., 1981, *Insect Biochem.* **11**:353–361.
510. Mordue, W., and Stone, J. V., 1980, *Insect Biochem.* **10**:219–239.
511. Ifsin, M. S., Gainer, H., and Barker, J. L., 1975, *Nature* **254**:72–74.
512. Konenko, N. I., 1979, *Neuroscience* **4**:2055–2069.
513. Kupfermann, I., and Weiss, R., 1976, *J. Gen. Physiol.* **67**:113–123.
514. Aswad, D., 1977, *J. Neurochem.* **28**:1137–1140.
515. Strumwasser, F., and Wilson, D. L., 1976, *J. Gen. Physiol.* **67**:691–702.
516. Roubos, E. W., and Moorer Van Delft, C., 1976, *Cell Tissue Res.* **174**:221–231.
517. Carley, W. W., 1981, *Neurosecretion: Molecules, Cells, Systems* (D. Farner, and K. Lederis, eds.), Plenum Press, New York, pp. 506.
518. Rowell, H. G., 1976, *Adv. Insect Physiol.* **12**:63–123.
519. Aston, R. J., and White, A. F., 1974, *J. Insect Physiol.* **20**:1673–1682.
520. Aston, R. J., and Hughes, R., 1980, *Neurohumoral Techniques in Insects* (T. A. Miller, ed.), Springer-Verlag, Berlin, Heidelberg, New York, pp. 15–34.
521. Aston, R. J., 1975, *J. Insect Physiol.* **21**:1873–1877.
522. Gainer, H., and Wollberg, Z., 1974, *J. Neurobiol.* **5**:243–261.
523. Koike, H., and Nagata, Y., 1979, *J. Physiol.* (*Lond.*) **295**:397–417.
524. Berry, R. W., and Geinisman, T., 1979, *J. Neurobiol.* **10**:489–498.
525. Berry, R. W., 1979, *J. Neurobiol.* **10**:499–508.
526. Snow, R. W., 1982, *J. Neurobiol.* **13**:267–277.
527. Schaller, H. C., 1979, *Trends Neurosci.* **2**:120–122.
528. Bodenmuller, H., and Schaller, H. C., 1981, *Nature* **293**:579–580.
529. Fenical, W., 1982, *Science* **215**:923–928.
530. Premuzic, E., 1971, *Prog. Chem. Org. Nat. Prod.* **29**:417–485.
531. Norton, T. R., 1981, *Fed. Proc.* **40**:21–25.
532. Shecket, G., and Lasek, R. J., 1982, *J. Neurochem.* **38**:827–832.

533. Vanaman, T. C., Shareif, F., and Watterson, D. M., 1979, *Calcium Binding Proteins and Calcium Function* (R. H. Wasserman, R. A. Carradino, E. Carafoli, R. H. Kretsinger, D. H. MacLennan, and F. L. Siegel, eds.), Elsevier, New York, pp. 107–148.
534. Seamon, K. R., and Moore, B. W., 1980, *J. Biochem.* **255**:11644–11647.
535. Ciuditta, A., Moore, B. W., and Prozzo, N., 1977, *J. Neurochem.* **29**:235–244.
536. Molla, A., Kilhoffer, M. C., Ferraz, C., Audemard, E., Walsh, M. P., and Demaille, J. G., 1981, *J. Biol. Chem.* **256**:982–987.
537. Rieske, E., Schubert, P., and Kreutzberg, G. W., 1975, *Brain Res.* **84**:365–382.
538. Ambron, R. T., Goldman, J. E., and Schwartz, J. H., 1974, *J. Cell Biol.* **61**:665–675.
539. Ambron, R. T., and Treistman, S. N., 1977, *Brain Res.* **121**:287–309.
540. Ambron, R. T., Goldman, J. E., Thompson, E. B., and Schwartz, J. H., 1974, *J. Cell Biol.* **61**:649–664.
541. Thompson, E. B., Schwartz, J. H., and Kandel, E. R., 1976, *Brain Res.* **112**:251–281.
542. Ambron, R. T., Sherbany, A. A., and Schwartz, J. H., 1981, *Brain Res.* **207**:33–48.
543. Ambron, R. T., 1982, *Brain Res.* **239**:489–505.
544. Schwartz, J. H., 1979, *Annu. Rev. Neurosci.* **2**:467–504.
545. Hammerschlag, R., and Stone, G., 1982, *Trends Neurosci.* **5**:12–15.
546. Tauc, L., 1980, *Trends Neurosci.* **3**:241–245.
547. Tauc, L., 1979, *Biochem. Pharmacol.* **27**:3493–3498.
548. Tauc, L., and Hinzen, L., 1974, *Brain Res.* **80**:340–344.
549. Ornberg, R. L., Smyth, T., and Benton, A. W., 1976, *Toxicon* **14**:329–333.
550. Cariello, L., and Zanetti, L., 1977, *Comp. Biochem. Physiol.* **57C**:169–173.
551. Beress, L., and Beress, R., 1975, *FEBS Lett.* **50**:311–314.
552. Boyd, C. A. R., 1979, *J. Theor. Biol.* **76**:415–417.
553. Niall, H. D., 1982, *Annu. Rev. Physiol.* **44**:615–624.
554. Skirkoll, L. R., Grace, A. A., Hommer, D. W., Rehfeld, J., Goldstein, M., Hokfelt, T., and Bunney, B. S., 1981, *Neuroscience* **6**:2111–2124.
555. Pearse, A. G. E., 1978, *Centrally-Acting Peptides* (E. J. Hughes, ed.), Macmillan, New York, pp. 49–59.
556. Dunn, A. J., 1978, *Neurosci. Res. Prog. Bull.* **16**:600–614.
557. Werman, R., 1966, *Comp. Biochem. Physiol.* **18**:745–766.
558. Orrego, F., 1979, *Neuroscience* **4**:1035–1059.
559. Reichelt, K. L., and Kwamme, E., 1973, *J. Neurochem.* **21**:849–859.
560. Reichelt, K. L., and Edminson, P. D., 1977, *Peptides in Neurobiology* (H. Gainer, ed.), Plenum Press, New York, pp. 171–181.
561. Stein, C., and Weinreich, D., 1982, *J. Neurochem.* **38**:204–214.
562. Murdock, L., and Omar, D., 1981, *Insect Biochem.* **11**:161–166.
563. Mir, A. K., and Vaughn, P. F. T., 1981, *J. Neurochem.* **36**:441–446.
564. Evans, P. H., Soderlund, D., and Aldrich, J. R., 1980, *Insect Biochem.* **10**:375–380.
565. Mir, A. K., and Vaughn, P. F. T., 1981, *Insect Biochem.* **11**:571–577.
566. Curatolo, A., Cecchi, L., Eusebi, F., and de Santis, A., 1979, *J. Neurochem.* **32**:1349–1350.
567. Berridge, M. J., 1979, *The Neurosciences, Fourth Study Program* (F. O. Schmitt and F. G. Worden, ed.), MIT Press, Cambridge, pp. 873–892.
568. Kaczmarek, L. K., Jennings, K., and Strumwasser, F., 1978, *Proc. Natl. Acad. Sci. U.S.A.* **75**:5200–5204.
569. Kupfermann, I., Cohen, J. L., Mandelbaum, D. E., Schonberg, M., Susswein, A. J., and Weiss, K. R., 1979, *Fed. Proc.* **38**:2095–2102.
570. Cedar, H., Kandel, E. R., and Schwartz, J. H., 1972, *J. Gen. Physiol.* **60**:558–569.
571. Enyeart, H., 1982, *J. Neurobiol.* **12**:505–513.
572. Venturini, G., Carolei, A., Palladini, G., Margotta, V., and Cerbo, R., 1981, *Comp. Biochem. Physiol.* **69C**:105–108.
573. Stefano, G., Catapane, E. J., and Kream, R. M., 1981, *Cell Mol. Neurobiol.* **1**:57–68.
574. Levitan, I. B., and Treistman, S. N., 1977, *Brain Res.* **136**:307–317.
575. Bandle, E. F., and Levitan, I. B., 1977, *Brain Res.* **125**:325–331.
576. Greengard, P., 1978, *Cyclic Nucleotides, Phosphorylated Proteins and Neuronal Function*, Raven Press, New York.

577. Neary, J. R., Crow, T., and Alkon, D. I., 1981, *Nature* **293**:658–660.
578. Lasek, R., Gainer, H., and Przybylskii, R., 1974, *Proc. Natl. Acad. Sci. U.S.A.* **71**:1188–1192.
579. Peterson, R. P., 1970, *J. Neurochem.* **17**:325–338.
580. Lasek, R. J., Dabrowski, C., and Nordlander, R., 1973, *Nature* (*New Biol.*) **244**:162–165.
581. Black, M. M., and Lasek, R. J., 1977, *J. Neurobiol.* **8**:229–237.
582. Voogt, P., 1973, *Chemical Zoology* (M. Florkin and B. T. Scheer, eds.), Academic Press, New York, pp. 245–300.
583. Osborne, N. N., Althaus, H. A., and Neuhoff, V., 1972, *Comp. Biochem. Physiol.* **43B**:671–679.
584. Bolognani, L., Masserini, M., Bodini, P. A., Bolognani, F. A. M., and Ottaviani, E., 1981, *J. Neurochem.* **36**:821–825.
585. Noren, R., and Svennenholm, L., 1973, *Z. Evol. Biokhim. Fiziol.* **9**:225–234.
586. Segler, K., Rahmann, H., and Rosner, H., 1978, *Biochem. Syst. Ecol.* **6**:87–93.
587. Henkart, M., 1975, *Science* **188**:154–157.
588. Brown, A. M., Baur, P. S., Jr., and Tuley, F. H., Jr., 1975, *Science* **188**:157–160.
589. Petrunyaka, V. V., 1982, *Cell Mol. Neurobiol.* **1**:11–20.
590. Krauhs, J. M., Sordahl, L. A., and Brown, A. M., 1977, *Biochim. Biophys. Acta* **471**:25–31.
591. O'Shea, M., 1982, *Trends Neurosci.* **5**:69–72.
592. Brownstein, M., Axelrod, J., and Saavedra, J., Zeman, G. H., and Carpenter, D. O., 1974, *Proc. Natl. Acad. Sci. U.S.A.* **71**:4662–4665.
593. Burnstock, G., 1980, *Nerve Cells, Transmitters, and Behavior* (R. Levi-Montalcini, ed.), Elsevier, Amsterdam, pp. 45–81.
594. Aletta, J. M., and Goldberg, D. J., 1982, *Science* **218**:913–916.
595. Burnstock, G., 1981, *J. Physiol.* (*Lond.*) **313**:1–35.
596. Bokisch, A. J., Osborne, N. N., and Walker, R. J., 1983, *Comp. Biochem. Physiol.* **75C**:171–177.
597. Georges, D., and Dubois, M. P., 1980, *C.r. Acad. Sci. Paris, Ser. D.* **290**:29–32.
598. Thorndyke, M. C., 1982, *Regul. Peptides* **3**:281–293.
599. Pestarino, M., 1983, *Experientia* **39**:1156–1158.

Index